ESSENTIAL CLINICAL ANATOMY

To Marion, my wife and best friend, who was presented with the Williams & Wilkins Golden Crab Award for her contribution to the editing and preparation of this book

Keith L. Moore

To my husband, Enno, and for my children, Erik and Kristina

Anne M. R. Agur

ESSENTIAL CLINICAL ANATOMY

Keith L. Moore, PhD, FIAC, FRSM

Professor Emeritus of Anatomy and Cell Biology
Faculty of Medicine, University of Toronto
Toronto, Ontario, Canada
Visiting Professor of Clinical Anatomy
Department of Anatomy, Faculty of Medicine
University of Manitoba, Winnipeg, Manitoba, Canada
Former Head of Anatomy, University of Manitoba
Former Chair of Anatomy, University of Toronto

Anne M. R. Agur, BSc (OT), MSc

Associate Professor of Anatomy and Cell Biology, Occupational Therapy,
Physical Therapy, and Biomedical Communications
Faculty of Medicine, University of Toronto
Toronto, Ontario, Canada

With the assistance of

Marion Moore, BA
Kam Yu, BSc, AAM

Williams & Wilkins
A WAVERLY COMPANY
BALTIMORE • PHILADELPHIA • LONDON • PARIS • BANGKOK
BUENOS AIRES • HONG KONG • MUNICH • SYDNEY • TOKYO • WROCLAW

Editor: Patricia Coryell
Development Editors: Laura M. Bonazzoli, Lisa Donohoe
Project Editor: Therese Grundl
Designer: Norman W. Och
Illustration Planners: Lorraine C. Wrzosek, Image Foundry Ltd., MK Designs
Cover Designer: Tom Scheuerman
Production Coordinators: Anne Stewart Seitz, Carol Eckhart

Accurate indications, adverse reactions, and dosage schedules for drugs are provided in this book, but it is possible that they may change. The reader is urged to review the package information data of the manufacturers of the medicines mentioned.

Printed in the United States of America

Library of Congress Cataloging in Publication Data

Moore, Keith L.
Essential clinical anatomy \ Keith L. Moore, Anne M.R. Agur.
p. cm.
Compilation of material extracted from Clinically oriented anatomy
/ Keith L. Moore. 3rd ed. and Grant's atlas of anatomy / Anne M.R. Agur. 9th ed.
Includes bibliographical references and index. ISBN 0-683-06128-3
1. Human anatomy. 2. Human anatomy—Atlases. I. Agur, A. M. R.
II. Moore, Keith L. Clinically oriented anatomy. III. Agur, A. M.
R. Grant's atlas of anatomy. IV. Title.
[DNLM: 1. Anatomy—handbooks. QS 39 M822e 1995]
OM23.2.M673 1995
61 1—dc2O
DNLM/DLC
for Library of Congress

94-36235
CIP
97 98 99
3 4 5 6 7 8 9 10

Reprints of chapters may be purchased from Williams & Wilkins in quantities of 100 or more. Call our Special Sales Department, (800) 358-3583.

PREFACE

It is clear that many students and practitioners in the health care professions and related disciplines require a compact yet thorough textbook of clinical anatomy. The parent of this book, *Clinically Oriented Anatomy* (COA), by the senior author, is recommended as a resource for more complete and detailed descriptions of human anatomy and its relationship and importance to medicine and surgery.

Essential Clinical Anatomy is an overview of the important aspects of anatomy described in COA. The number of structures described is limited to those deemed likely to be important to the practitioner. Furthermore, the structures receive an amount of attention that is roughly proportionate to their importance. Presentations are brief and

- Provide a basic text of human anatomy for use in current health sciences curricula
- Present an appropriate amount of anatomical material in a readable and interesting form
- Provide a concise clinically oriented anatomical reference for clinical courses in subsequent years
- Serve as a rapid review when preparing for examinations, particularly the national boards
- Offer enough information for those wishing to refresh their knowledge of anatomy

Essential Clinical Anatomy is a concise text with a strong clinical orientation and many descriptive figures and tables. Most illustrations are in full color and are designed to highlight important facts and show their relationship to clinical medicine and surgery. Some illustrations are from *Grant's Atlas*, by the junior author; others are from COA. Current diagnostic imaging techniques (radiographs, CTs, and MRIs) are also included to demonstrate anatomy as it is often viewed clinically. Interspersed in blue boxes and white boxes with blue borders are clinical comments that relate anatomy to clinical practice. They are introduced with the intention of illustrating the importance of correlating preclinical and clinical subjects.

Surface anatomy is emphasized because the examination of every patient involves applied knowledge of this approach to the study of anatomy. Bony and other anatomical landmarks are used as points of reference during physical examinations and for surgical approaches to internal organs. The fundamental aim of surface anatomy is visualization of the structures that lie beneath the skin. Surface anatomy information is presented in white boxes headed with a pink bar.

The terminology conforms with the sixth edition of *Nomina Anatomica* (1989). To facilitate communication, unofficial widely used alternative terms appear in parentheses [e.g., uterine tube (Fallopian tube), omental bursa (lesser sac), and rectouterine pouch (pouch of Douglas)]. Many terms are anglicized for those who prefer not to use Latin terms [e.g., deep brachial artery (profunda brachii artery) and hip bone (os coxae)].

We welcome your comments and suggestions for improvements in the next edition.

Keith L. Moore
Anne M. R. Agur

CONTENTS

ACKNOWLEDGMENTS

We thank Marion Moore for her indefatigable support and encouragement during the preparation of the manuscript and for providing constructive comments during its development. We also thank the medical artist, Kam Yu, who prepared the superb color illustrations. Other drawings from *Clinically Oriented Anatomy* and *Grant's Atlas* were prepared previously by Angela Cluer, Nina Kilpatrick, Stephen Mader, David Mazierski, Sari O'Sullivan, and Bert Vallecoccia. The photographs were taken by Paul Schwartz of IMS Creative Communications, University of Toronto.

We are grateful to the following for their constructive comments during the preparation of this book: Dr. Joseph Bast, The University of Kansas Medical Center; Dr. Richard Drake, University of Cincinnati Medical Center; Marilyn Li, student at Brown University School of Medicine; Dr. Margaret H. Hines, The Ohio State University College of Medicine; Dr. Todd R. Olson, Albert Einstein College of Medicine; Dr. Charles Pincus, University of Toronto, Faculty of Medicine; Dr. E. George Salter, The University of Alabama at Birmingham; Dr. William J. Swartz, Louisiana State University Medical Center; and Dr. Linda L. Wright, Boston University School of Medicine.

We also acknowledge the assistance of many physicians who have helped us. We are greatly indebted to Dr. W. Kucharczyk, Professor and Chair of Radiology, and Dr. E. Becker, Associate Professor of Radiology, University of Toronto, Faculty of Medicine, Toronto, Ontario, Canada. They and other colleagues, acknowledged in the figure and table credits, provided most of the radiographs, CTs, and MRIs. Dr. Tom White, Department of Radiology, College of Medicine, The Health Science Center, University of Tennessee at Memphis, also provided some excellent CTs.

We are indebted to many people at Williams & Wilkins for their patience and invaluable cooperation: Timothy S. Satterfield, Nancy Evans, Norman W. Och, Lorraine C. Wrzosek, Therese Grundl, Carol Eckhart, and Anne Stewart Seitz. We should also recognize the friendly support of the former President, Professional and Reference Group, Sara Finnegan, who invited us several years ago to join the Williams & Wilkins team of authors.

Keith L. Moore
Anne M. R. Agur
University of Toronto
Toronto, Ontario, Canada

1 / INTRODUCTION TO CLINICAL ANATOMY

Approaches to Studying Anatomy

Anatomy is the science of the structure of the body. There are three main approaches to studying anatomy:

- Systemic anatomy
- Regional anatomy
- Clinical anatomy

Surface anatomy is the study of the living body at rest and in action and is used in all three approaches. The main aim of surface anatomy is the visualization of structures that lie beneath the skin. Regardless of which approach is used, one must visualize the three-dimensional structure of the body.

SYSTEMIC ANATOMY

Systemic anatomy is the study of the body as a series of organ systems.

- The integumentary system (dermatology) consists of the skin (integument) and its appendages (e.g., hair and nails); the skin forms a protective covering for the body
- The skeletal system (osteology) consists of bones and cartilage; it provides support for the body and protects vital organs (e.g., ribs and sternum protect the heart and lungs)
- The articular system (arthrology) consists of joints and their associated ligaments
- The muscular system (myology) is composed of muscles that move parts of the body (e.g., bones at joints)
- The nervous system (neurology) consists of the central nervous system (brain and spinal cord) and the peripheral nervous system (cranial and spinal nerves, together with their motor and sensory endings); the nervous system controls and coordinates functions of organs (e.g., heart) and other structures (e.g., muscles) and relates the body to the environment
- The circulatory system (angiology) consists of the cardiovascular and lymphatic systems, which function in parallel: the cardiovascular system consists of the heart and blood vessels that conduct blood through the body; the lymphatic system is a network of lymphatic vessels that withdraws excess tissue fluid (lymph) from the body's interstitial (intercellular) fluid compartment, filters it through the lymph nodes, and returns it to the bloodstream
- The digestive system (gastroenterology) is composed of the organs associated with ingestion, digestion, and absorption of food
- The respiratory system (respirology) is composed of the air passages and lungs that supply oxygen and eliminate carbon dioxide
- The urinary system (urology) consists of the kidneys, ureters, urinary bladder, and urethra, which produce, transport, store, and intermittently excrete urine, respectively
- The reproductive system consists of the genital organs that are concerned with reproduction
- The endocrine system (endocrinology) consists of the ductless glands (e.g., thyroid gland) which produce hormones that are carried by the circulatory system to all parts of the body; they influence metabolism and other body processes (e.g., menstrual cycle)

REGIONAL ANATOMY

Regional anatomy (topographic anatomy) is the study of the regions of the body (e.g., thorax and abdomen) (Fig. 1.1). This approach deals with structural relationships of the parts of the body in the region that is being studied. Most laboratory courses in human anatomy are based on regional dissection.

CLINICAL ANATOMY

Clinical anatomy emphasizes aspects of structure and function of the body that are important in the practice of medicine, dentistry, and the allied health sciences. It incorporates the regional and systemic approaches and stresses clinical applications. In addition, surgical and imaging techniques (e.g., radiography) are used to demonstrate living anatomy. Case studies are an integral part of clinical anatomy.

The structure of people varies considerably; hence, anatomical variations are common. For example, bones of the skeleton vary among themselves, not only in their basic shape, but in lesser details of surface structure. There is also a wide variation in the size, shape, and form of the attachment of muscles. Similarly, there is variation in the method of division of nerves and arteries. A marked deviation from normal is called an *anomaly* (malformation) (e.g., a limb defect caused by the prenatal administration of thalidomide).

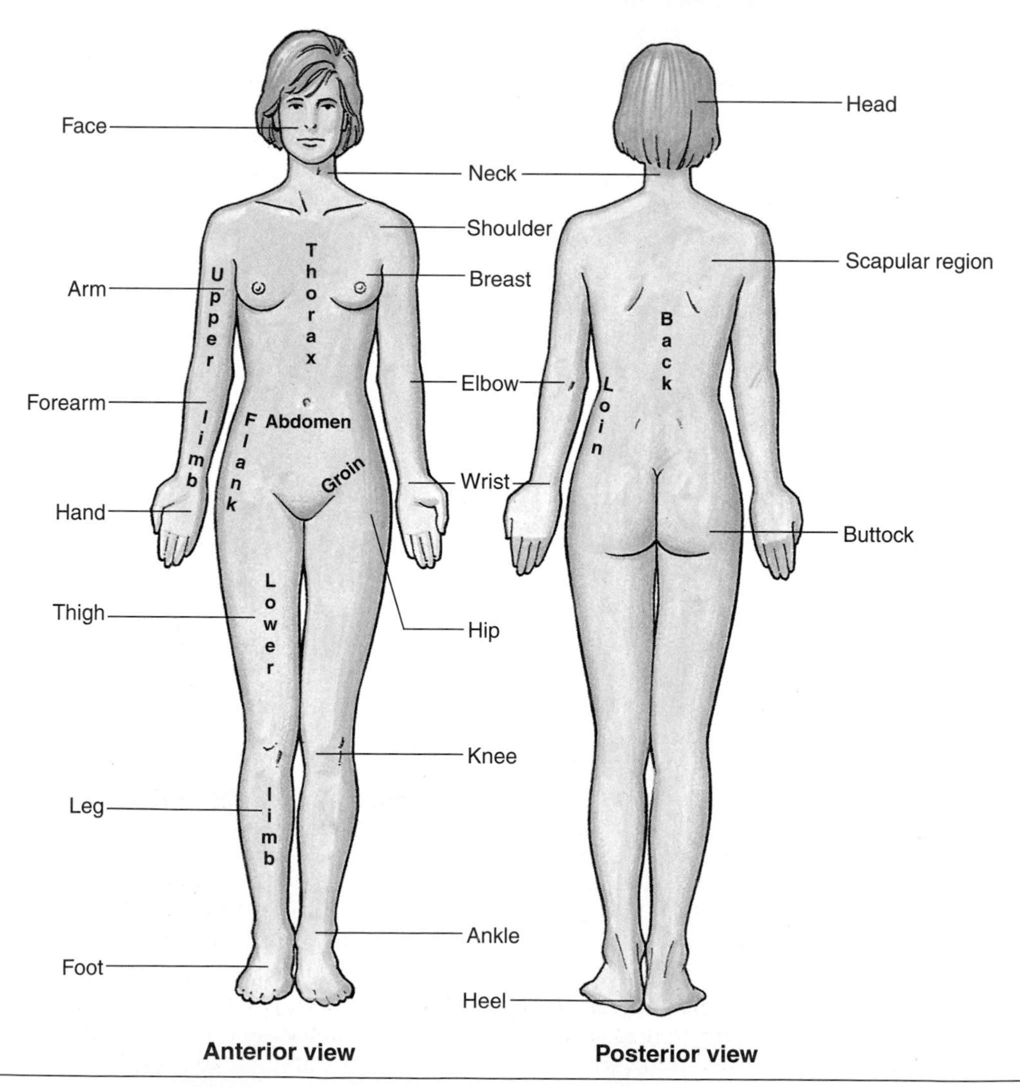

Figure 1.1. Anatomical position and regions of the body. All descriptions are based on the assumption that the person is standing in the anatomical position.

Anatomical and Medical Terminology

Anatomy has an international vocabulary, so accurate use of the words is important. Anatomical terms have precise meanings and form a major part of medical terminology. Although eponyms are not used in official anatomical terminology, those commonly used appear in parentheses throughout this book to avoid ambiguity and misunderstanding because some clinicians still use them [e.g., cerebral arterial circle (circle of Willis)]. Similarly, unofficial alternative terms appear in parentheses for clarity [e.g., internal thoracic artery (internal mammary artery)].

ANATOMICAL POSITION

All anatomical descriptions are expressed in relation to the anatomical position to ensure that the descriptions are unambiguous (Fig. 1.1). A person in the anatomical position

- Is standing erect or lying supine (on one's back) as if erect, with head, eyes, and toes directed anteriorly (forward)
- Has upper limbs by the sides with palms facing anteriorly
- Has lower limbs together with the feet directed anteriorly

ANATOMICAL PLANES

Anatomical descriptions are based on four anatomical planes that pass through the body in the anatomical position:

- Median plane (midsagittal plane) is the vertical plane passing longitudinally through the body, dividing it into right and left halves
- Sagittal planes are vertical planes passing through the body parallel to the median plane (it is helpful to give a point of reference such as a sagittal plane through the midpoint of the clavicle)
- Coronal planes are vertical planes passing through the body at right angles to the median plane, dividing it into anterior (front) and posterior (back) portions
- Horizontal planes are transverse planes passing through the body at right angles to the median and coronal planes; a horizontal plane divides the body into superior (upper) and inferior (lower) parts (it is helpful to give a reference point such as a horizontal plane through the umbilicus).

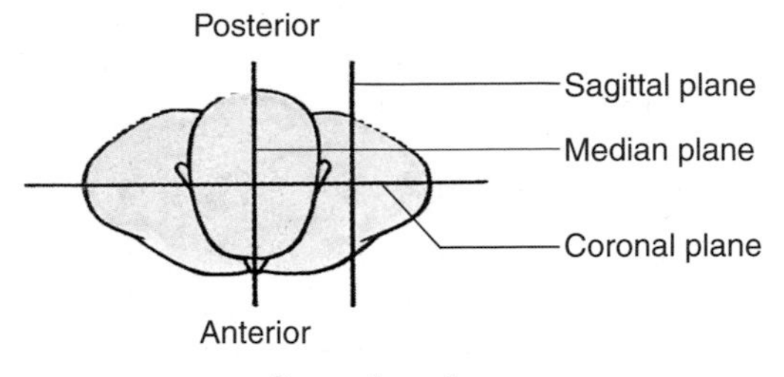

Superior view

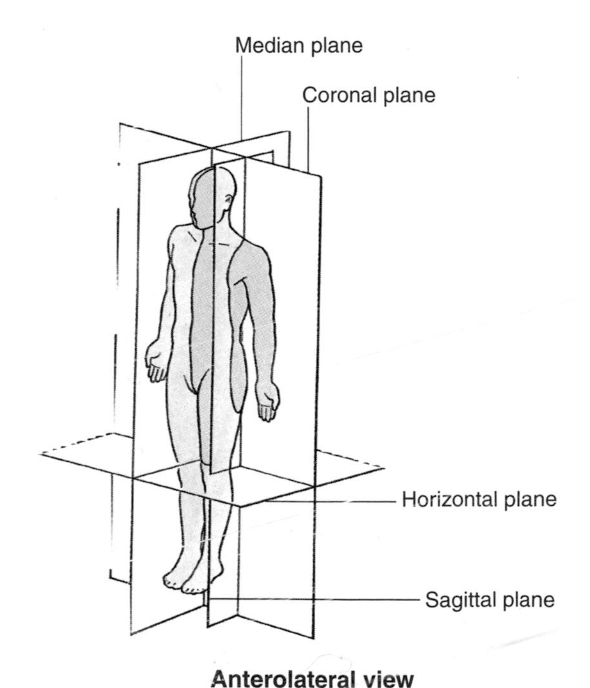

Anterolateral view

TERMS OF RELATIONSHIP AND COMPARISON

Various adjectives, explained and illustrated in Table 1.1, are arranged as pairs of opposites (e.g., superior and inferior). They are used to describe the relationship of parts of the body in the anatomical position and to compare the relative position of two structures with each other. For example, the great (big) toe is on the medial side of the foot, whereas the thumb is on the lateral side of the hand.

Table 1.1
Commonly Used Terms of Relationship and Comparison

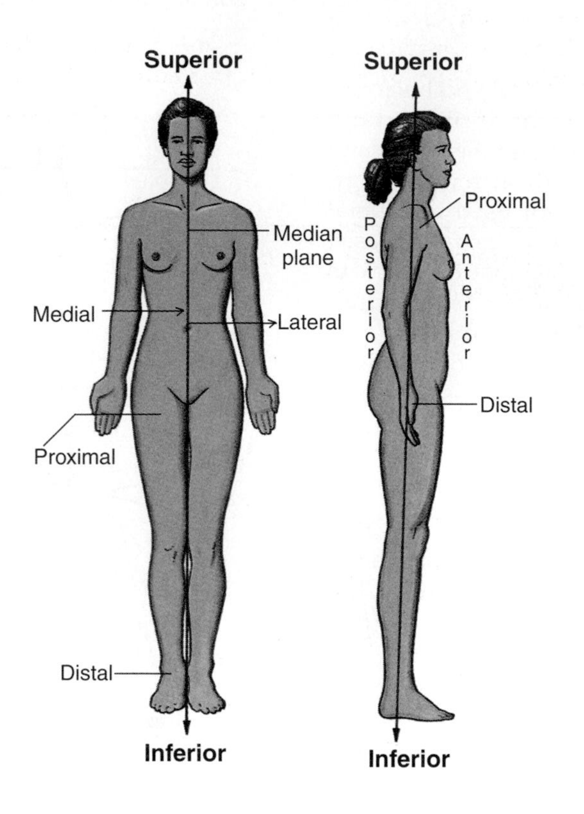

Term	Meaning	Usage
Superior (cranial)	Nearer to head	The heart is superior to the stomach
Inferior (caudal)	Nearer to feet	The stomach is inferior to heart
Anterior (ventral)	Nearer to front	The sternum is anterior to the heart
Posterior (dorsal)	Nearer to back	The kidneys are posterior to the intestine
Medial	Nearer to median plane	The fifth digit (little finger) is on the medial side of the hand
Lateral	Farther from median plane	The first digit (thumb) is on the lateral side of the hand
Proximal	Nearer to trunk or point of origin (e.g., of a limb)	The elbow is proximal to the wrist, and the proximal part of an artery is its beginning
Distal	Farther from trunk or point of origin (e.g., of a limb)	The wrist is distal to the elbow and the distal part of the lower limb is the foot
Superficial	Nearer to or on surface	The muscles of the arm are superficial to its bone (humerus)
Deep	Farther from surface	The humerus is deep to the arm muscles

TERMS OF MOVEMENT

Various terms are used to describe movements of the body [e.g., flexion of the limbs (Table 1.2)]. Movements take place at joints where two or more bones or cartilages articulate with one another. They are described as pairs of opposites (e.g., flexion and extension).

Diagnostic Imaging Techniques

Familiarity with diagnostic imaging techniques commonly used in clinical settings

Table 1.2
Terms of Movement

Flexion means bendlng of a part or decreasing the angle between body parts. *Extension* means straightening a part or increasing the angle between body parts.

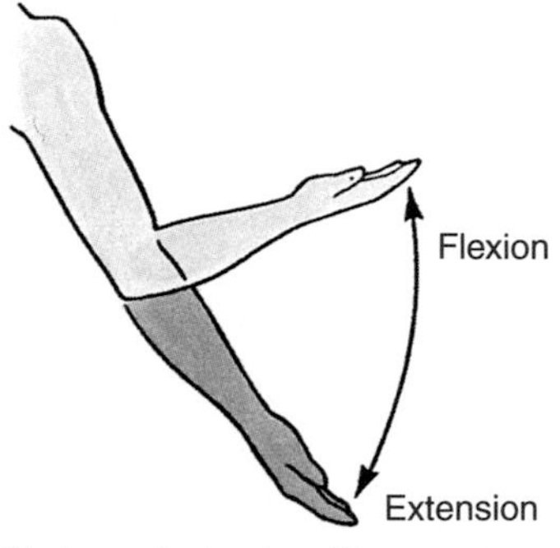

Flexion and extension of forearm at elbow joint

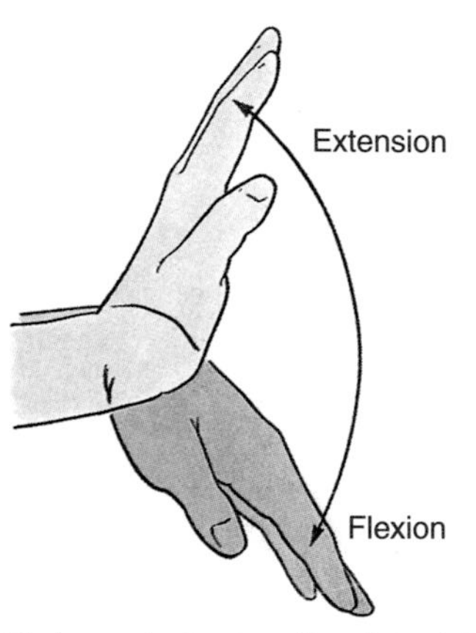

Flexion and extension of hand at wrist joint

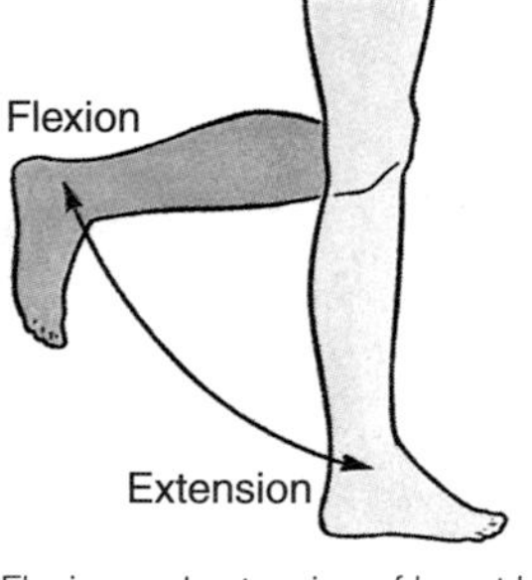

Flexion and extension of leg at knee joint

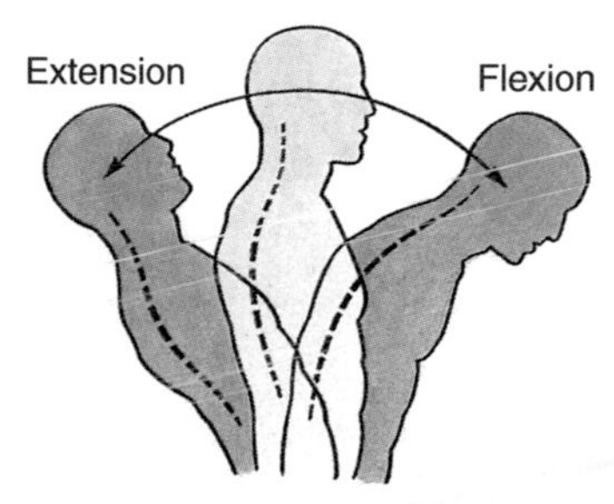

Flexion and extension of vertebral column at intervertebral joints

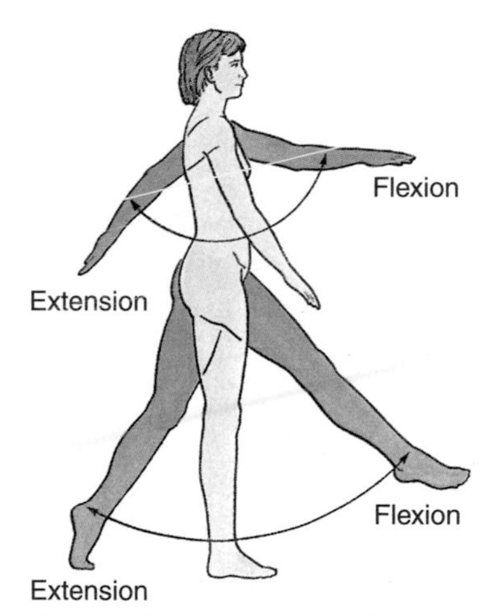

Flexion and extension of upper limb at shoulder joint and lower limb at hip joint

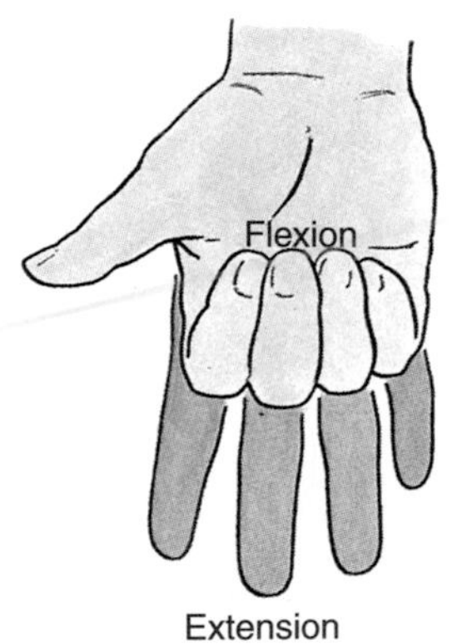

Flexion and extension of digits (fingers) at interphalangeal joints

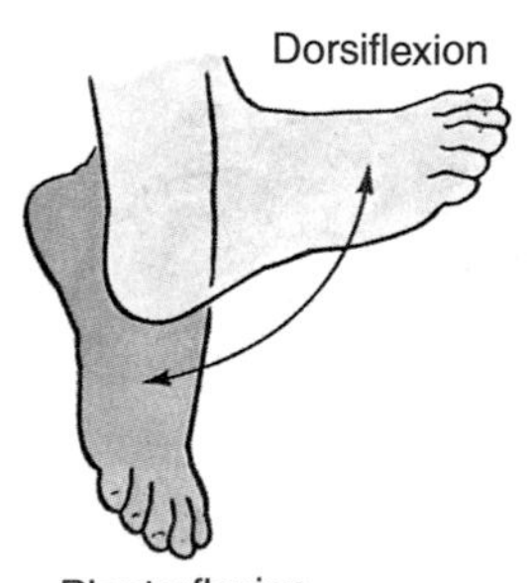

Dorsiflexion and plantarflexion of foot at ankle joint

Table 1.2. *Continued*

Abduction means moving away from the median plane of the body in the coronal plane. *Adduction* means moving toward the median plane of the body in the coronal plane. In the digits (fingers and toes), abduction means spreading them, and adduction refers to drawing them together. *Rotation* means moving a part of the body around its long axis. Medial rotation turns the anterior surface medially and lateral rotation turns this surface laterally. *Circumduction* is the circular movement of the limbs, or parts of them, combining in sequence the movements of flexion, extension, abduction, and adduction. *Pronation* is a medial rotation of the forearm and hand so that the palm faces posteriorly. *Supination* is a lateral rotation of the forearm and hand so that the palm faces anteriorly, as in the anatomical position. *Eversion* means turning sole of foot outward. *Inversion* means turning sole of foot inward. *Protrusion* (protraction) means to move the jaw anteriorly. *Retrusion* (retraction) means to move the jaw posteriorly.

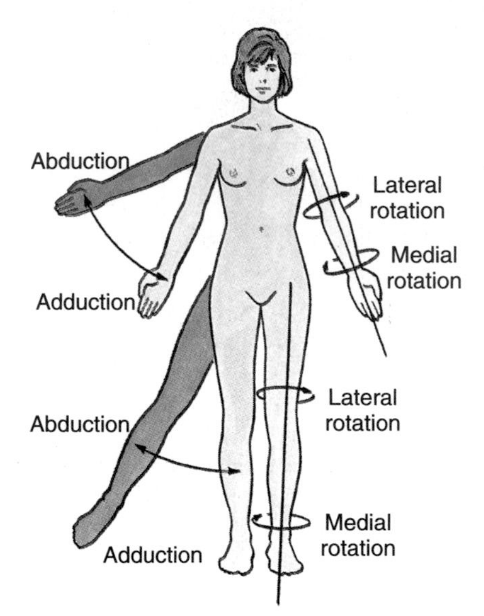

Abduction and adduction of right limbs and rotation of left limbs at the shoulder and hip joints, respectively

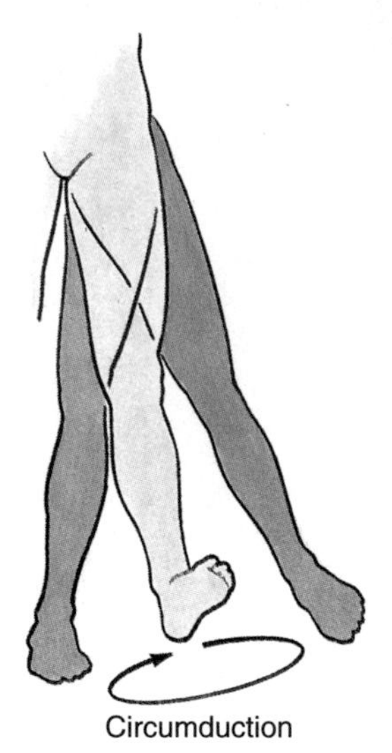

Circumduction (circular movement) of lower limb at hip joint

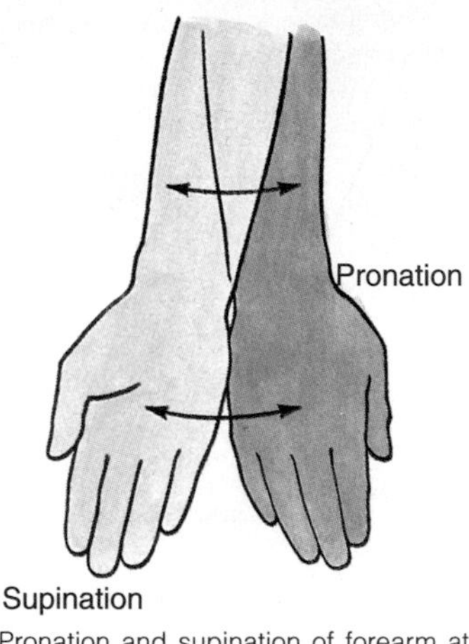

Pronation and supination of forearm at radioulnar joints

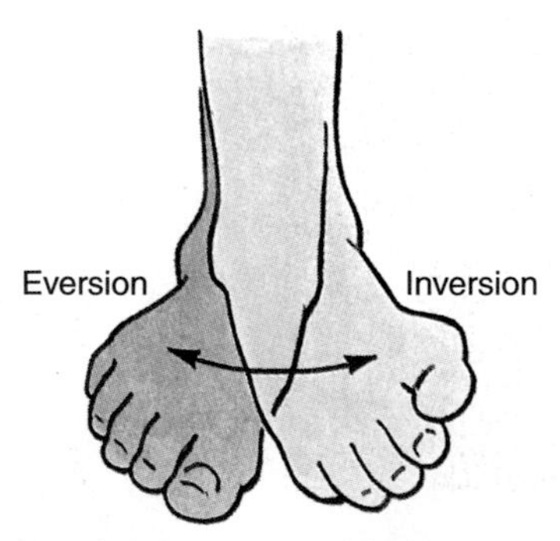

Inversion and eversion of foot at subtalar and transverse tarsal joints

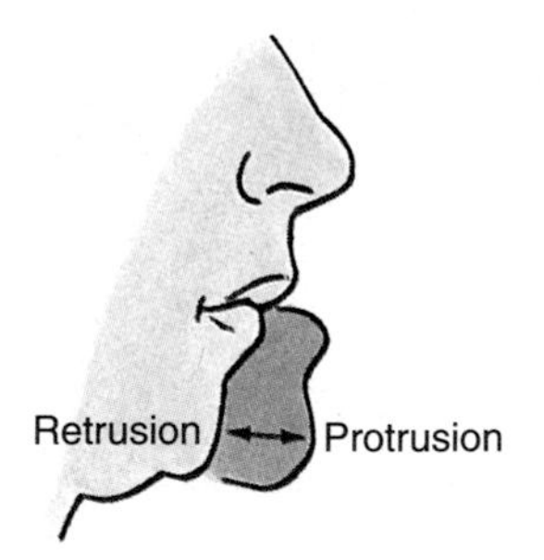

Protrusion and retrusion of jaw at temporomandibular joints

enables one to recognize abnormalities such as congenital anomalies, tumors, and fractures. The most commonly used diagnostic imaging techniques follow (Table 1.3):

- Radiography
- Computerized tomography
- Magnetic resonance imaging
- Ultrasonography

Table 1.3
Diagnostic Imaging Techniques

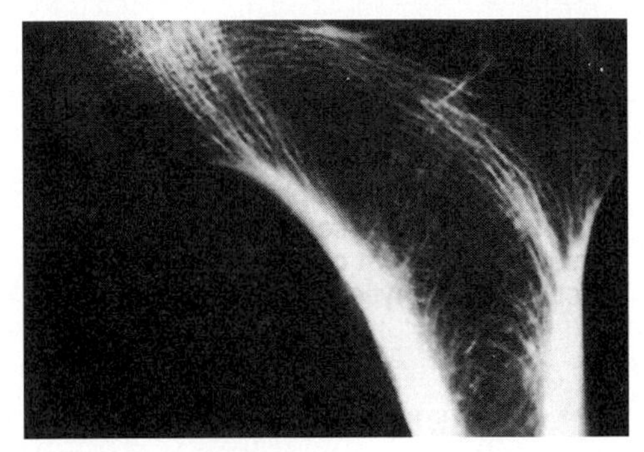

Radiograph of a coronal section of the proximal end of the femur, showing compact and spongy bone, which appears white and dark, respectively.

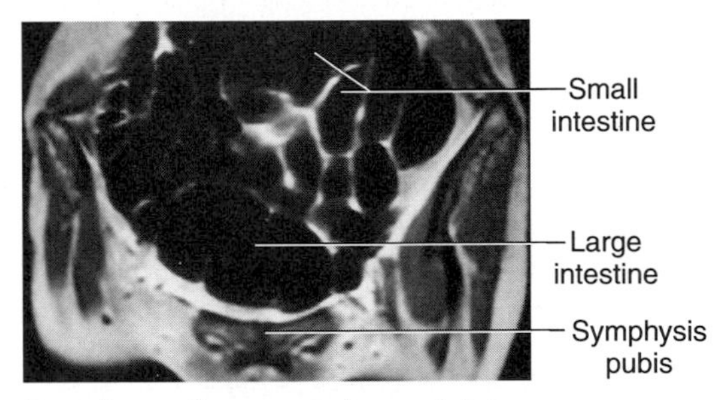

Coronal magnetic resonance image of abdomen.

Computerized tomography shows images of sections of the body. A beam of x-rays is passed through the body as the x-ray tube moves in a circle around the body. The amount of radiation absorbed by each different type of tissue of the chosen body plane varies with the amount of fat, cancellous and compact bone, and water density of the tissue in each element. A multitude of linear energy absorptions is measured and stored in a computer that compiles and generates images.

Ultrasonography (sonography) gives images of deep structures in body by recording reflections of pulses of ultrasonic waves directed into the tissues. A common use of diagnostic ultrasound imaging in pregnancy is to assess fetal age and well-being.

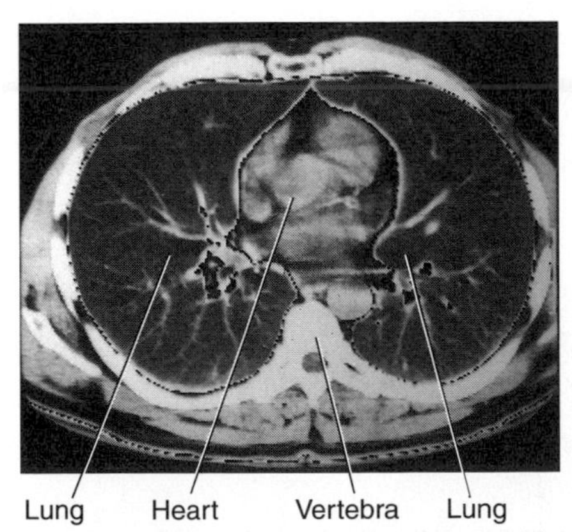

Computerized tomography scan of the thorax showing the heart, lungs, and vertebra. It is customary to view such sections so that the body is supine and the inferior aspect of the section is seen from below. Hence, the right side of the body is on your left.

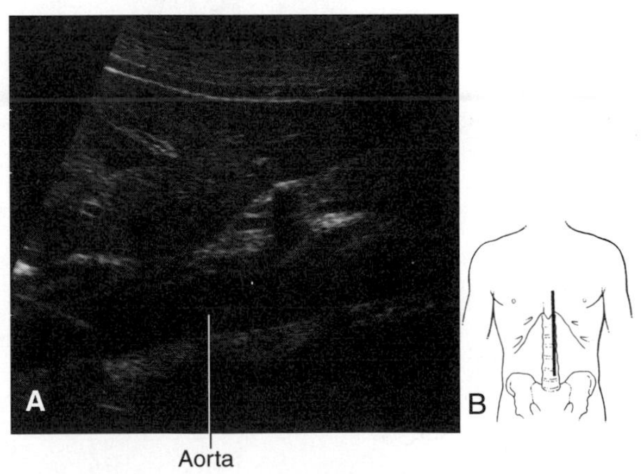

A. Sagittal ultrasound scan of abdomen through aorta. **B.** Orientation sketch. *Black line*, plane of scan.

Skin and Fascia

The skin, the largest organ of the body, consists of a superficial cellular layer called the *epidermis* and a deep connective tissue layer called the *dermis* (Fig. 1.2). The skin provides

- Protection from injury and fluid loss (e.g., in minor burns)
- Heat regulation, through sweat glands and blood vessels
- Sensation, by way of superficial nerves and their sensory endings (e.g., for pain)

The *superficial fascia* is composed of loose connective tissue and fat. Located between the dermis and underlying deep (investing) fascia, it contains sweat glands, blood and lymphatic vessels, and cutaneous nerves. The *deep fascia* is a dense, more organized connective tissue layer that invests deep structures (e.g., muscles).

Skeletal System

The skeleton is composed of bones and cartilages (Fig. 1.3). *Bone* is a rigid form of connective tissue that forms most of the skeleton and is the chief supporting tissue of the body. *Cartilage* is a resilient form of connective tissue that forms some parts of the skeleton [e.g., costal (rib) cartilages]. The proportion of bone and cartilage in the skeleton changes as the body grows; the younger a person is, the greater the contribution of cartilage.

The *axial skeleton* consists of the bones of the head (skull), neck (hyoid bone and cervical vertebrae), and trunk (ribs, sternum, vertebrae, and sacrum). The *appendicular skeleton* consists of the bones of the limbs (extremities, appendages), including those forming the pectoral (shoulder) and pelvic girdles.

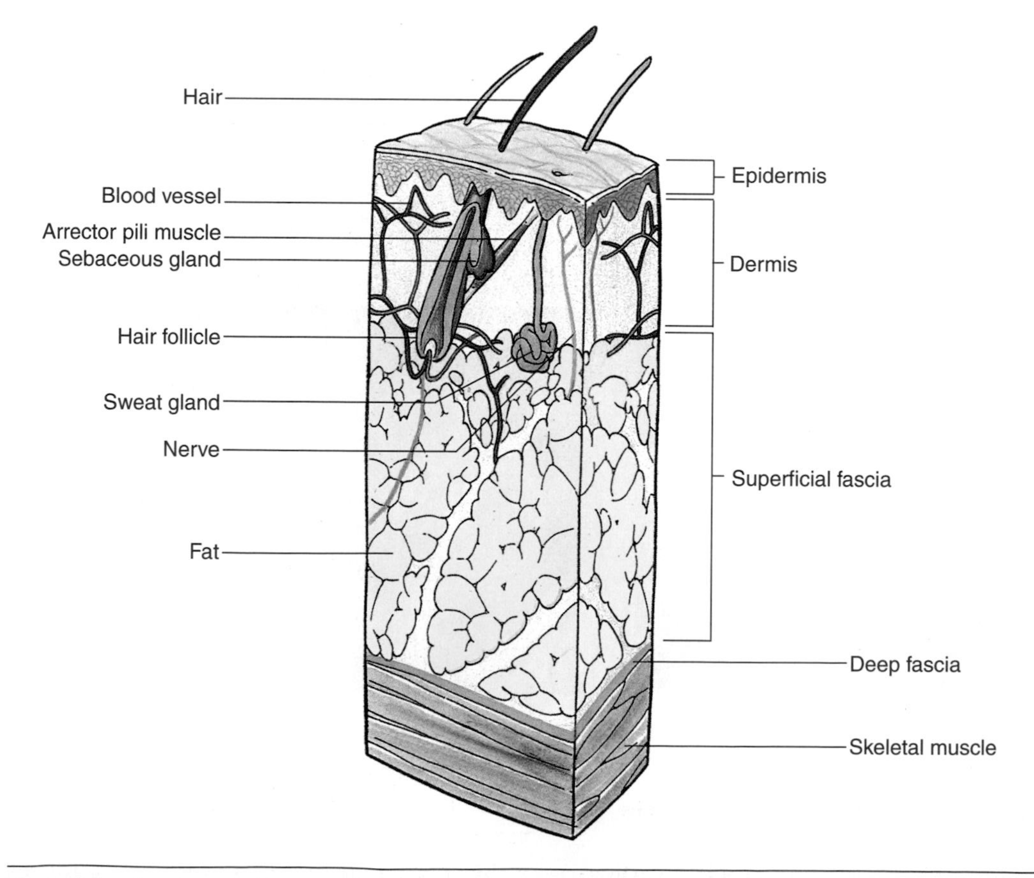

Figure 1.2. Structure of skin and subcutaneous tissue.

Bones provide

- Protection for vital structures
- Support for the body
- A mechanical basis for movement
- Blood cells (*red bone marrow* is the site for development of red blood cells, some lymphocytes, granulocytic white cells, and platelets)
- Storage for salts (calcium, phosphorus, and magnesium)

There are two types of bone, *spongy* (cancellous) and *compact* (Fig. 1.4). The differences between these types depend on the relative amount of solid matter and the number and size of the spaces they contain. All bones have an outer shell of compact bone around a central mass of spongy bone, except where the latter is replaced by a *marrow (medullary) cavity*.

CLASSIFICATION OF BONES

Bones are classified according to their shape (Fig. 1.3):

- Long bones are tubular (e.g., humerus)
- Short bones are cuboidal and are found only in the ankle (tarsus) and wrist (carpus)
- Flat bones usually serve protective functions (e.g., flat bones of the skull protect the brain)
- Irregular bones have various shapes (e.g., those in the face)
- Sesamoid bones develop in certain tendons (e.g., patella) and are found where tendons cross the ends of long bones in the limbs; they protect tendons from excessive wear and often change the angle of the tendons as they pass to their attachments

BONE MARKINGS

Bone markings appear wherever tendons, ligaments, and fascia are attached. The various markings and features of bones are given names (Fig. 1.3):

- Condyle (rounded articular area) (e.g., lateral condyle of femur)
- Crest (ridge of bone) (e.g., iliac crest)
- Epicondyle (eminence superior to a condyle) (e.g., lateral epicondyle of humerus)
- Facet (smooth flat area where a bone articulates with another bone)
- Foramen (passage through a bone) (e.g., obturator foramen)
- Fossa (hollow or depressed area) (e.g., infraspinous fossa of scapula)
- Line (linear elevation) (e.g., soleal line of tibia)
- Malleolus (rounded process) (e.g., lateral malleolus of fibula)
- Notch (indentation at edge of a bone) (e.g., greater sciatic notch)
- Protuberance (projection) (e.g., external occipital protuberance)
- Spine (thornlike process) (e.g., spine of scapula)
- Spinous process (projecting spinelike part) (e.g., spinous process of a vertebra)
- Trochanter (large blunt elevation) (e.g., greater trochanter of femur)
- Tubercle (small raised eminence) (e.g., greater tubercle of humerus)
- Tuberosity (large rounded elevation) (e.g., ischial tuberosity)

Bones are living organs that hurt when injured, bleed when fractured, and change with age. Like other organs, bones have blood vessels, lymphatic vessels, and nerves, and they may become diseased. Unused bones atrophy (decrease in size) (e.g., in a paralyzed limb). Bone may be absorbed, which occurs in the mandible when teeth are extracted. Bones hypertrophy (enlarge) when they have increased weight to support for a long period.

Fractures are more common in children than in adults because of the combination of their slender growing bones and carefree activities. Fortunately many of these breaks are *greenstick fractures* (incomplete breaks caused by bending of the bone). Fractures in growing bones heal faster than those in adult bones. During old age both the organic and inorganic components of bone decrease, producing a condition called *osteoporosis*. There is a reduction in the quantity of bone (atrophy of skeletal tissue); hence the bones lose their elasticity and fracture easily.

BONE DEVELOPMENT

All bones are derived from mesenchyme (embryonic cellular connective tissue) but by two different processes (Cormack, 1993; Moore and Persaud, 1993, listed under "Suggested Readings"). Bones develop in two ways by replacing either mesenchyme or cartilage; the histology of a bone is the same whether it is of membranous or cartilaginous origin.

- Mesenchymal models of bone form during the embryonic period and begin to undergo

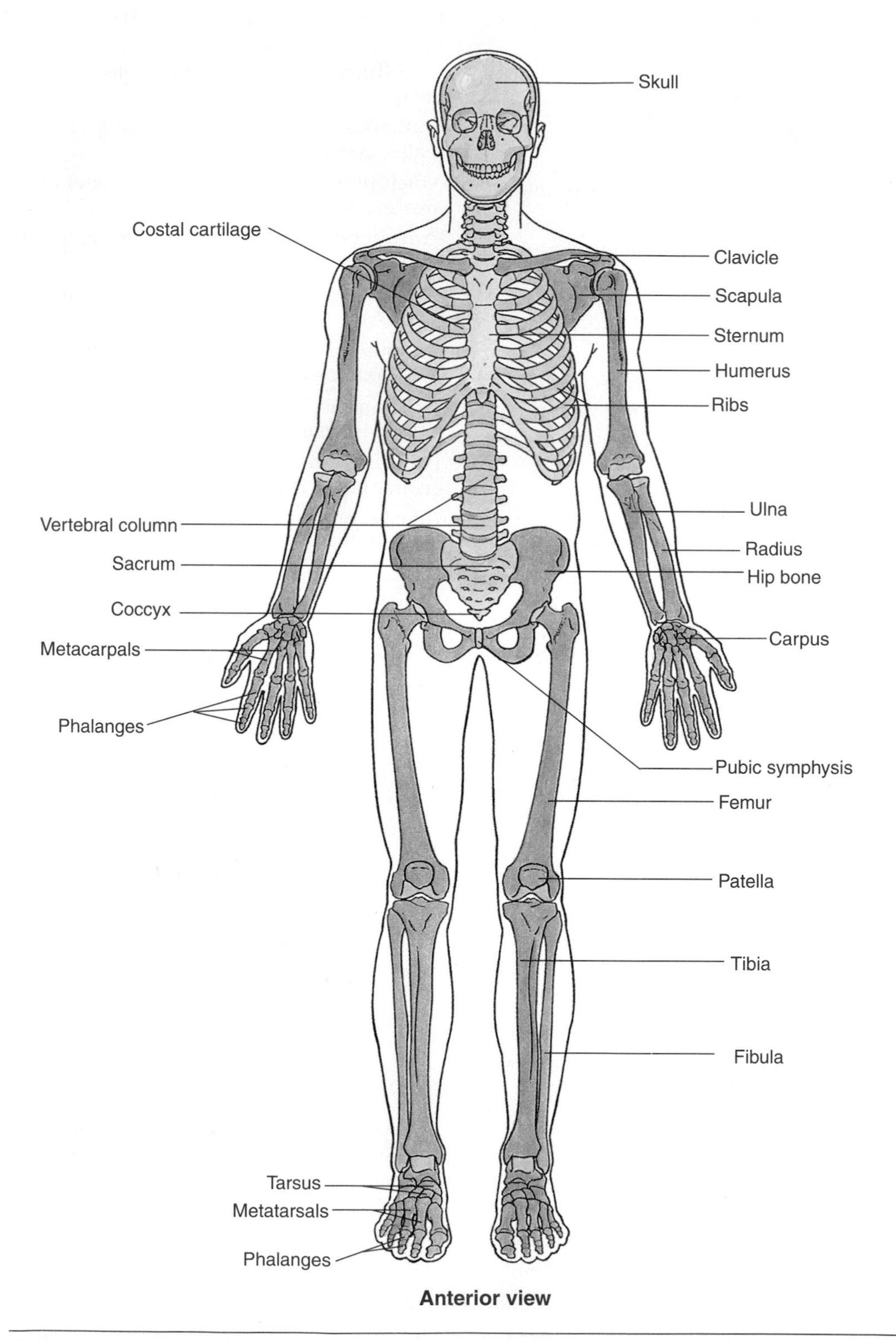

Figure 1.3. Bones (p. 10) and bone markings (p. 11). Appendicular skeleton is *purple* to distinguish it from axial skeleton. *Blue*, cartilages.

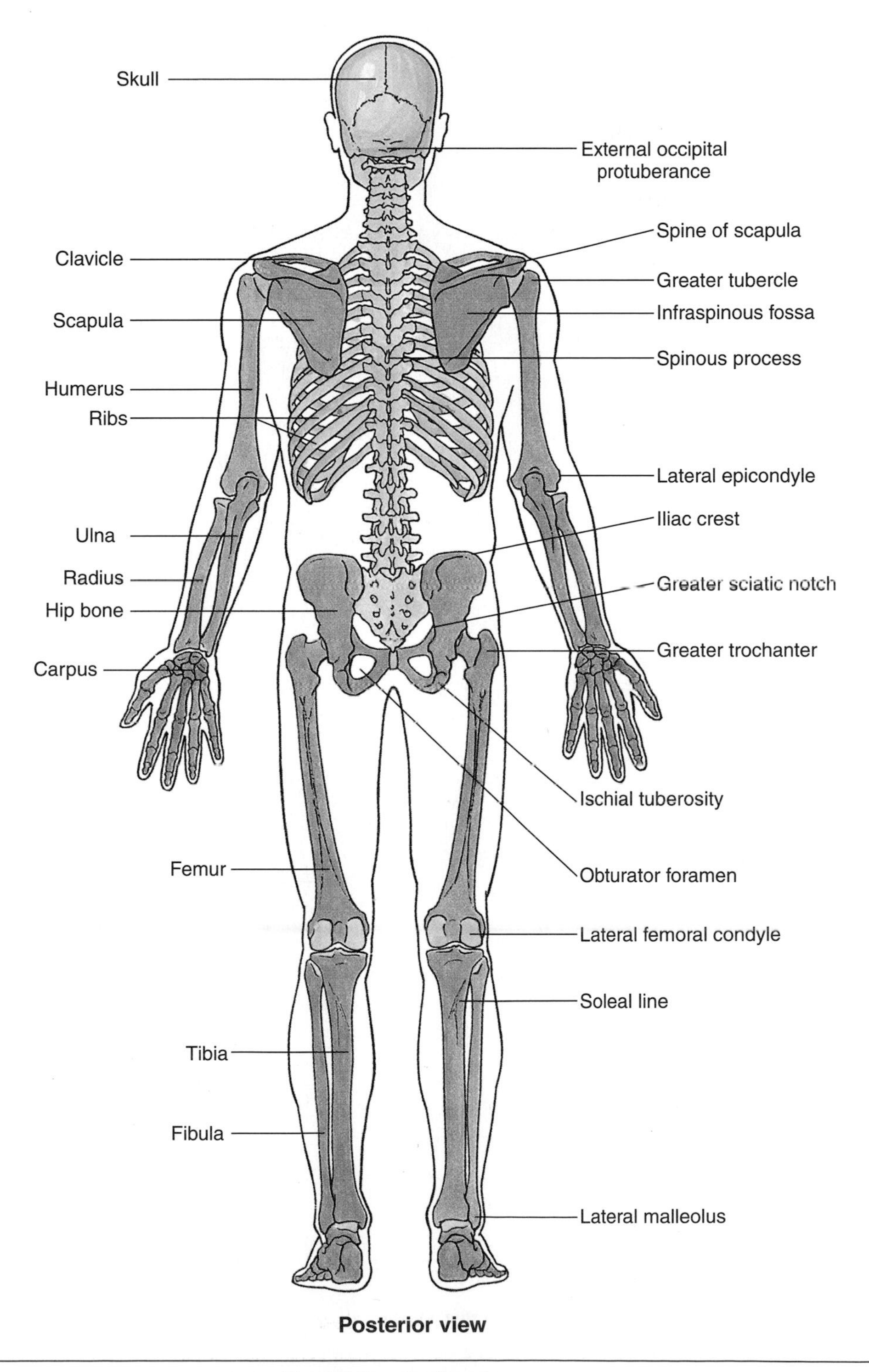

Posterior view

Figure 1.3. *Continued.*

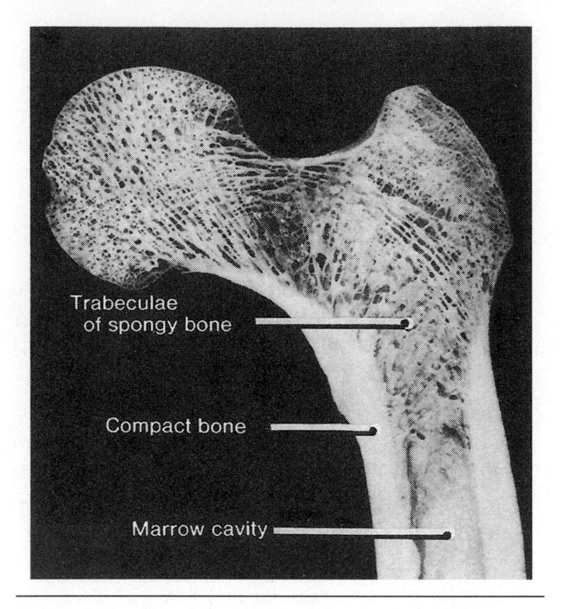

Figure 1.4. Coronal section of proximal end of femur.

direct ossification in the fetal period; this type of bone development is called *intramembranous ossification*

- Cartilage models of bones form from mesenchyme during the fetal period, and bone then replaces most of the cartilage in the model; this type of bone development is called *endochondral ossification*

A brief description of *endochondral ossification* is necessary to understand how long bones grow (see Cormack, 1993, listed under "Suggested Readings" for details). The body (shaft) of a bone that is ossified from a primary ossification center is called the *diaphysis* (Fig. 1.5). Most secondary ossification centers appear after birth; the parts of a bone ossified from these secondary centers are called *epiphyses*. The flared part of the diaphysis nearest the epiphysis is referred to as the *metaphysis*. To enable growth in length to continue, the bone formed

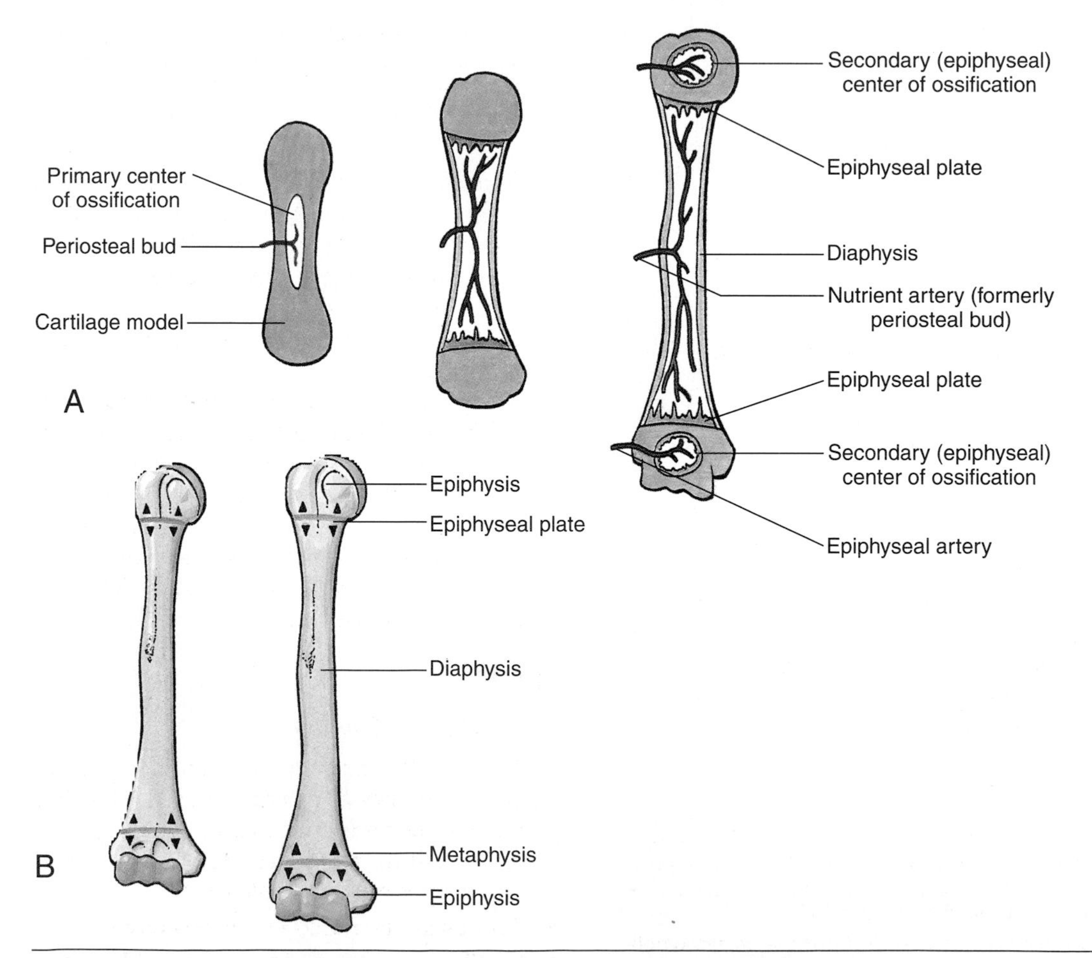

Figure 1.5. Growth of a long bone. **A.** Formation of primary and secondary centers of ossification. **B.** Growth in length occurs on both sides of epiphyseal plates (*arrowheads*).

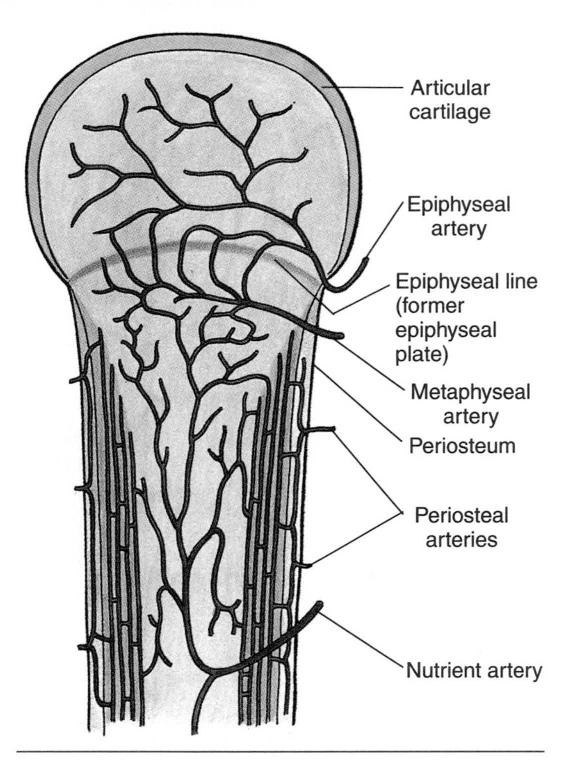

Figure 1.6. Proximal part of a long bone illustrating its arterial supply. Note that epiphysis and metaphysis have their own arteries.

from the primary center in the diaphysis does not fuse with that formed from the secondary centers in the epiphyses until the adult size of the bone is reached. During growth of a bone, plates of cartilage known as the *epiphyseal (cartilaginous) plates* intervene between the diaphysis and epiphyses. These growth plates are eventually replaced by bone at each of its two sides, diaphyseal and epiphyseal. When this occurs, bone growth ceases, and the diaphysis fuses with the epiphyses. The bone formed during this fusion process (*synostosis*) is particularly dense and is recognizable as an *epiphyseal line* (Fig. 1.6). The epiphyseal fusion of bones occurs progressively from puberty to maturity.

A general index of growth during infancy, childhood, and adolescence is indicated by *bone age*, as determined from radiographs. The age of a person can be determined by studying ossification centers. The two main criteria are

- Appearance of calcified material in the diaphysis and/or epiphyses
- Disappearance of the dark line representing the epiphyseal plate (absence of this line indicates that epiphyseal fusion has occurred); fusion occurs at specific times for each epiphysis

VASCULATURE AND INNERVATION OF BONES

Arteries enter bones from the *periosteum,* the fibrous connective tissue membrane investing bones (Fig. 1.6). *Periosteal arteries* enter at numerous points and supply the compact bone; they are responsible for its nourishment. Consequently, a bone from which the periosteum has been removed will die. Near the center of the body, a *nutrient artery* passes obliquely through the compact bone and supplies the spongy bone and bone marrow. Metaphyseal and epiphyseal arteries supply the ends of bones.

Loss of the arterial supply to an epiphysis or other parts of a bone results in death of bone tissue (*avascular necrosis*). After every fracture, small areas of adjacent bone undergo necrosis.

Veins accompany the arteries, and many of the large veins leave through foramina near the articular ends of the bones. Bones containing red bone marrow have numerous large veins.

Lymphatic vessels are abundant in the periosteum.

Nerves accompany the blood vessels supplying bones. The periosteum is rich in sensory nerves called *periosteal nerves;* some of these carry pain fibers. These nerves are especially sensitive to tearing or tension, which explains why pain from broken bones is severe. Within bones, *vasomotor nerves* cause constriction or dilation of blood vessels.

JOINTS

Classification of Joints

A joint is the connection between any of the rigid parts (bones or cartilages) of the skeleton. Joints are classified according to the type of material uniting the articulating bones. There are three types of joint (Table 1.4):

- Fibrous joints united by fibrous tissue
- Cartilaginous joints united by cartilage or a combination of cartilage and fibrous tissue

Table 1.4
Types of Joints

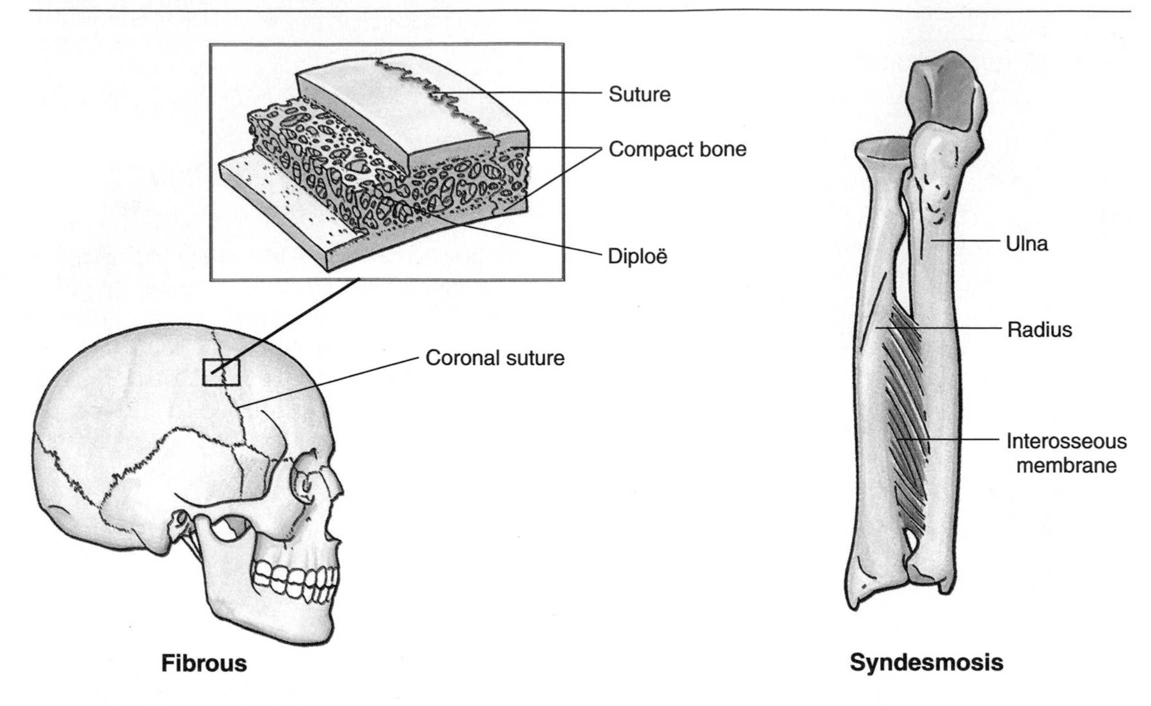

In fibrous joints, articulating bones are joined by fibrous tissue. Sutures of the skull are examples of type of fibrous joint where bones are close together and united by fibrous tissue, often interlocking along a wavy line. Flat bones consist of two plates of compact bone separated by spongy bone and marrow (diploë). In a syndesmosis, the bones are joined by a sheet of fibrous tissue (e.g., interosseous membrane joining forearm bones).

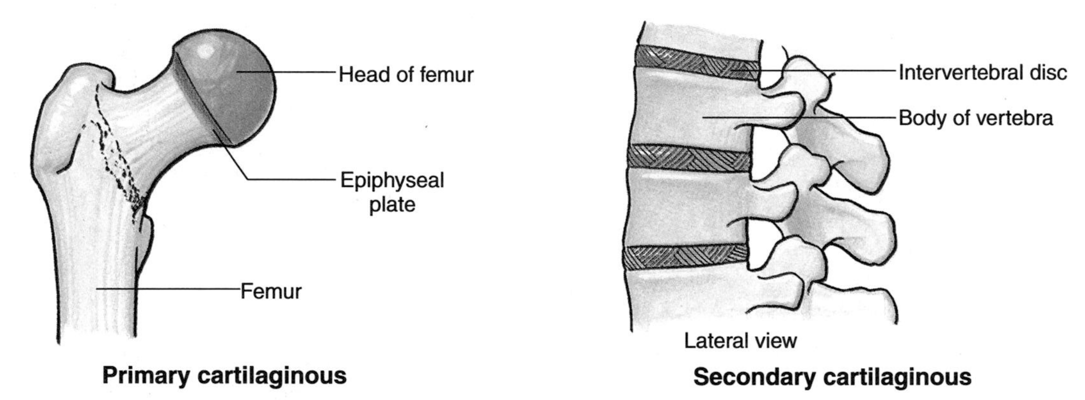

In cartilaginous joints, articulating bones are united by fibrocartilage or hyaline cartilage. Common type of synchondrosis is that in a developing long bone where bony epiphysis and body are joined by an epiphyseal plate. In a symphysis the binding tissue is a fibrocartilaginous disc (e.g., between two vertebrae).

A synovial joint (articulation) is characterized by a joint cavity; the two bones are separated by a joint cavity containing synovial fluid but are joined by an articular capsule (fibrous capsule lined with synovial membrane). The bearing surfaces of the bones are covered with articular cartilage. Synovial joints are the most common and important type of joint functionally. They provide free movement between the bones they join and are typical of nearly all joints of the limbs.

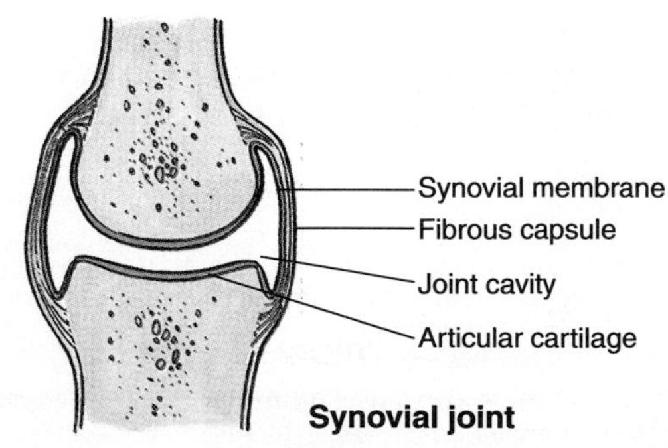

Table 1.5
Types of Synovial Joint

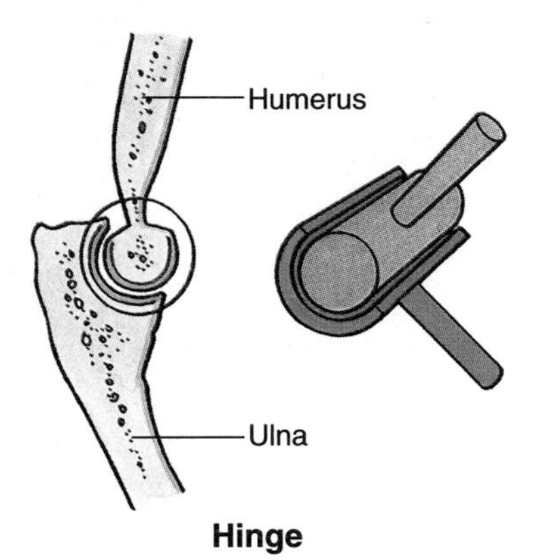

Hinge

Hinge joints (uniaxial) permit flexion and extension only (e.g., elbow joint).

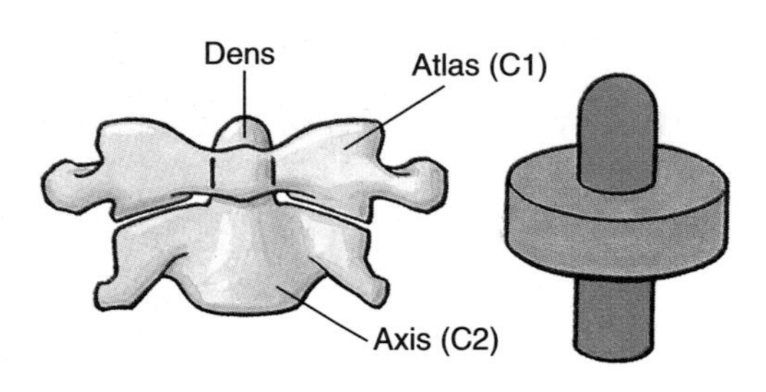

Pivot

Pivot joints (uniaxial) allow rotation. A round process of bone fits into a bony ligamentous socket [e.g., atlantoaxial joint between the atlas (C1) and axis (C2)].

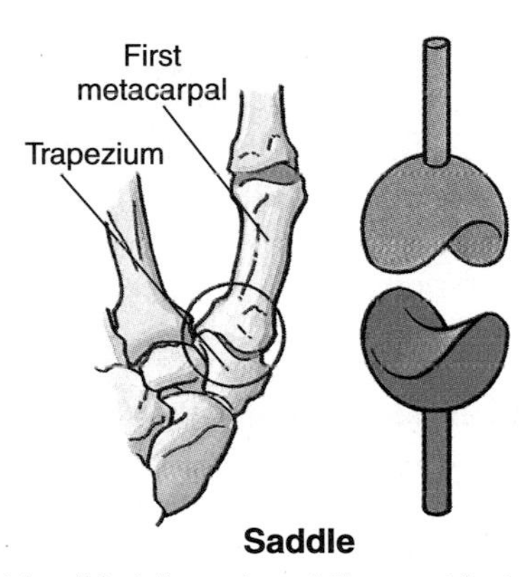

Saddle

Saddle joints (biaxial) are shaped like a saddle; i.e., they are concave and convex where bones articulate.

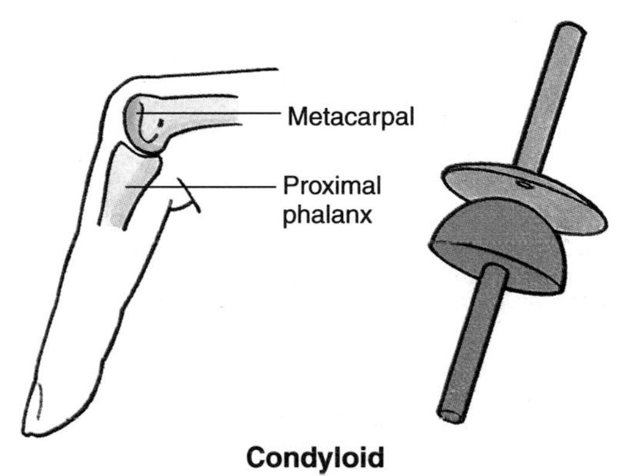

Condyloid

Condyloid joints (biaxial) permit flexion and extension, abduction and adduction, and circumduction [e.g., metacarpophalangeal ("knuckle") joints of digits].

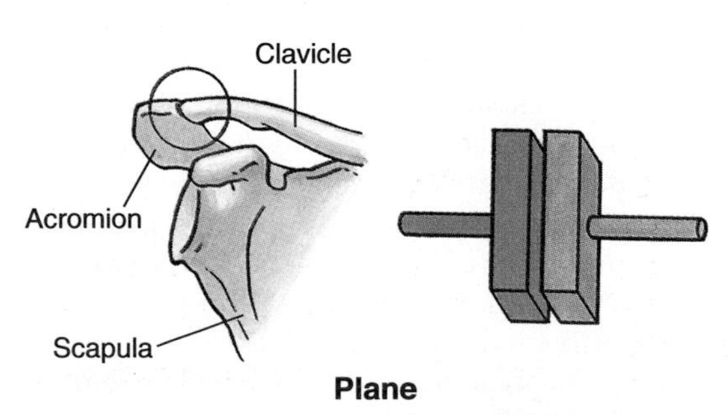

Plane

Plane joints permit gliding or sliding movements (e.g., acromioclavicular joint).

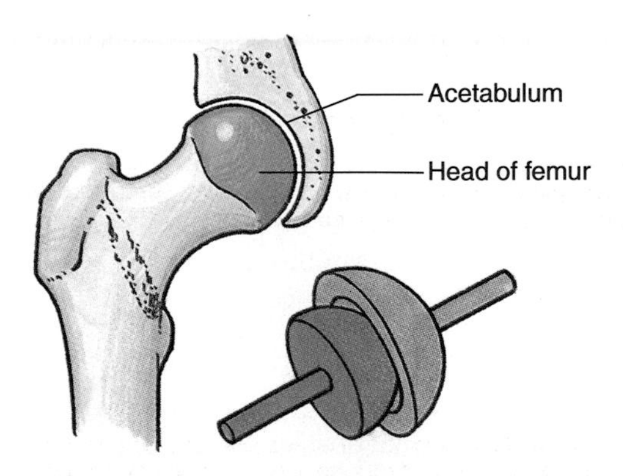

Ball and socket

Ball and socket joints (multiaxial) permit movement in several axes (e.g., flexion-extension, abduction-adduction, medial and lateral rotation, and circumduction). A rounded head fits into a concavity.

- Synovial joints united by cartilage with a synovial membrane enclosing a joint cavity

Synovial joints are the most common type of joint. They provide free movement between the bones they join and are typical of nearly all limb joints. They are called synovial joints because they contain a lubricating substance called *synovial fluid* and are lined with a *synovial membrane*. This vascular connective tissue membrane produces the synovial fluid. There are three distinguishing features of a synovial joint:

- There is a joint cavity
- Bone ends are covered with articular cartilage
- Joints are surrounded by an articular capsule (a fibrous capsule lined with synovial membrane)

Synovial joints are usually reinforced by *accessory ligaments* that are either separate (extrinsic) or attached to the articular capsule (intrinsic). Some synovial joints have other distinguishing features such as fibrocartilaginous *articular discs*, which are present when the articulating surfaces of the bones are incongruous. There are *six types of synovial joint* (Table 1.5); they are classified according to the shape of the articulating surfaces and/or the type of movement they permit.

Vasculature and Innervation of Joints

Numerous *articular arteries* supply joints; they arise from vessels around the joint. The arteries often anastomose (communicate) to form networks. Veins called *venae communicantes* accompany the arteries and, like arteries, are located in the articular capsule, especially in the synovial membrane. Joints have a rich nerve supply; nerve endings are in the articular capsule. The *articular nerves* are branches of those that supply the overlying skin and the muscles that move the joint. *Hilton's law* states that the nerves supplying a joint also supply the muscles moving the joint and the skin covering their attachments.

The main type of sensation from a joint is *proprioception*. It provides information concerning movement and position of the parts of the body. The synovial membrane is relatively insensitive, but pain fibers are numerous in the fibrous capsule and its associated ligaments. These sensory endings respond to twisting and stretching such as occurs with *synovitis* (inflammation of the synovial membrane).

Beginning early in adult life and progressing slowly thereafter, aging of articular cartilage occurs on the ends of the articulating bones, particularly those of the hip, knee, vertebral column, and hands. These irreversible degenerative changes result in the articular cartilage becoming less effective as a shock absorber and lubricated surface. As a result, the articulation becomes vulnerable to the repeated friction that occurs during joint movements. In some people these changes do not produce significant symptoms, but in others they cause considerable pain. *Degenerative joint disease* called osteoarthritis, or osteoarthrosis, is often accompanied by stiffness, discomfort, and pain. *Osteoarthritis* is common in old people and usually affects joints that support the weight of their bodies.

Muscular System

Muscle cells (fibers) produce contractions that move body parts, including internal organs. The associated connective tissue conveys nerve fibers and capillaries to the muscle fibers. Muscles also give form to the body and provide heat. There are three types of muscle:

- Skeletal muscle moves bones and other structures (e.g., eyes)
- Cardiac muscle forms most of the walls of the heart
- Smooth muscle forms part of the walls of most vessels and hollow organs; it moves substances through viscera (e.g., intestine) and controls movement through blood vessels

SKELETAL MUSCLE

Skeletal muscles produce movements of the skeleton. They are often called voluntary muscles because individuals can control many of them at will (Fig. 1.7); however, some of their actions are automatic. For example, the diaphragm usually contracts automatically, but it can be controlled voluntarily (e.g., when taking a deep breath). Skeletal muscles produce movement by shortening; i.e., they pull and never push. The architecture and shape of muscles vary (Fig. 1.8). Some muscles are attached by sheetlike tendons called *aponeuroses* (membranous expansions). Muscles that resemble feathers are described as pennate muscles. Other muscles are oblique, fusiform, spiral, quadrate, and straplike.

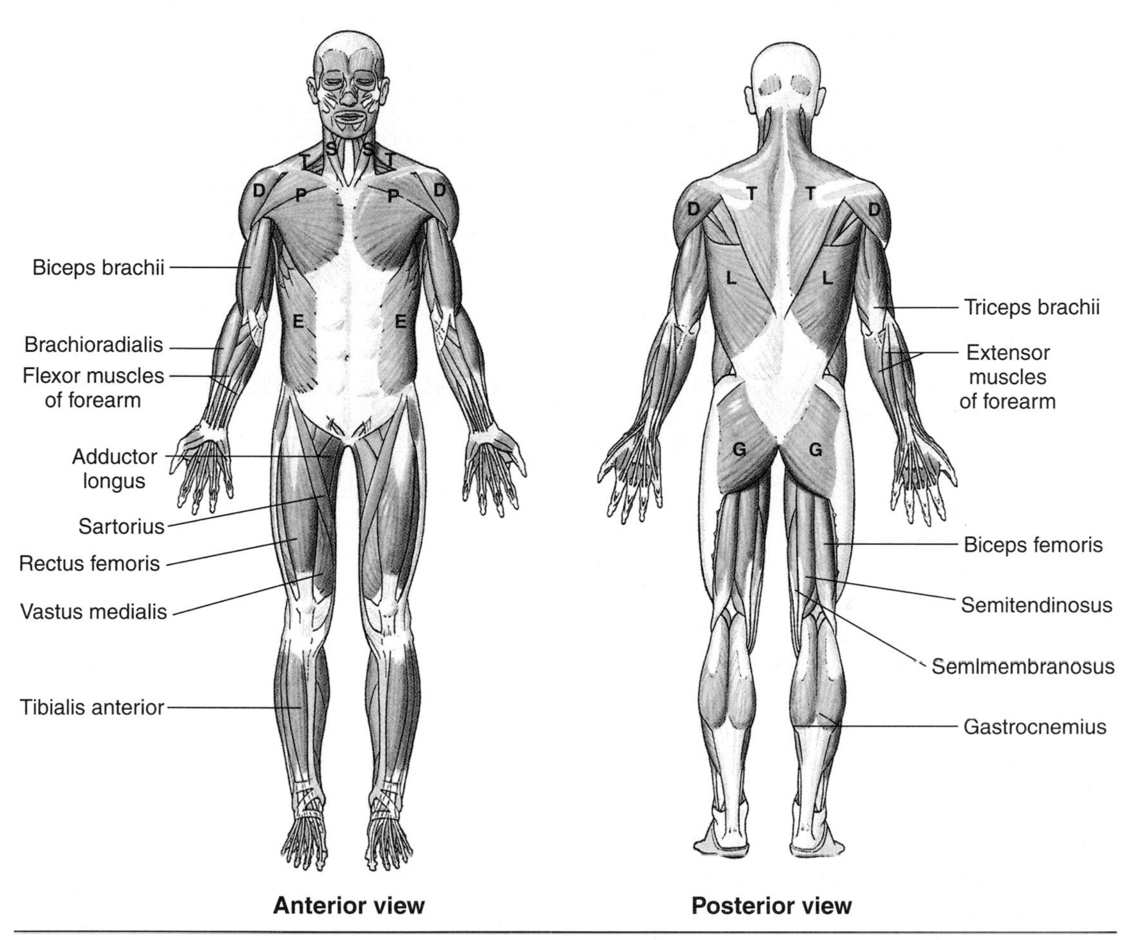

Figure 1.7. Muscles. Some larger muscles are labeled. *S*, sternocleidomastoid; *T*, trapezius; *D*, deltoid; *P*, pectoralis major; *E*, external oblique; *L*, latissimus dorsi; *G*, gluteus maximus.

The structural unit of a muscle is a *muscle fiber*; the functional unit consisting of a motor neuron and the muscle fibers it controls is called a *motor unit*. The number of muscle fibers in a motor unit varies from one to several hundred. The number of fibers varies according to the size and function of the muscle. Large motor units, where one neuron supplies several hundred muscle fibers, are found in the large trunk and thigh muscles, whereas in small eye and hand muscles, where precision movements are required, the motor units contain only a few muscle fibers. Movements result from an increasing number of motor units being activated.

- Agonists or prime movers are the main muscles activated during movements of the body; they contract actively to produce the desired movement
- Antagonists oppose the action of prime movers; as a prime mover contracts the antagonist progressively relaxes, producing a smooth movement
- Synergists prevent movement of the intervening joint when a prime mover passes over more than one joint; they complement the action of the prime movers
- Fixators steady the proximal parts of a limb while movements are occurring in distal parts

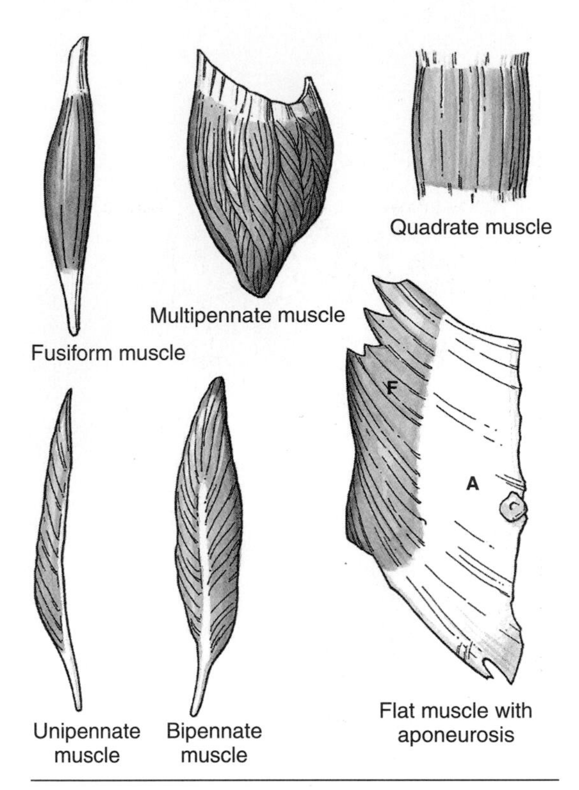

Figure 1.8. Architecture of skeletal muscles. *F*, flat muscle; *A*, aponeurosis.

The same muscle may act as a prime mover, antagonist, synergist, or fixator under different conditions.

A skeletal muscle has at least two attachments, usually to bone. However,

- Some muscles are attached to skin (e.g., facial muscles)
- Muscles are attached to mucous membrane (e.g., intrinsic tongue muscles)
- A few muscles are attached to fascia
- Other muscles form circular bands (e.g., external anal sphincter)

Muscles are commonly described as having an origin and an insertion; however, some muscles can act in both directions under different circumstances. Therefore, it is usually best to use the terms proximal and distal, or medial and lateral, attachments instead of origin and insertion.

Muscle testing is usually performed when nerve injuries are suspected. There are two common testing methods: the patient performs movements against resistance produced by the examiner, or the examiner performs movements against resistance produced by the patient. For example, when testing flexion of the forearm (Table 1.2), patients are asked to flex their forearms while the examiner resists the efforts. The other method is to ask patients to keep their forearms flexed while the examiner attempts to extend them. The latter technique enables the examiner to gauge the power of the movements.

Electromyography is another method for testing muscle action. Surface electrodes are attached, and the patient is asked to perform certain movements. The differences in electrical action potentials of the muscles are amplified and recorded. A resting normal muscle shows no activity. By using this technique, it is possible to analyze the activity of an individual muscle during different movements. *Electrical stimulation of muscles* may also be used as part of the treatment program for restoring the action of muscles.

CARDIAC MUSCLE

Cardiac muscle forms most of the walls of the heart, i.e., the *myocardium*. Although composed of striated fibers, cardiac muscle contractions are not under voluntary control. Heart rate is regulated by a *pacemaker* composed of special cardiac muscle fibers that are innervated by the autonomic nervous system .

The myocardium responds to increased demands by increasing the size of its fibers (cells). This is called *compensatory hypertrophy*. When cardiac muscle fibers are damaged during a heart attack, fibrous scar tissue is formed that produces a *myocardial infarct*.

SMOOTH MUSCLE

Smooth muscle forms a large part of the tunica media (middle coat) of the walls of most blood vessels (Fig. 1.9, *A* and *B*) and the muscular part of the wall of the digestive tract. Like cardiac muscle, smooth muscle is innervated by the autonomic nervous system. Hence it is *involuntary muscle* that can undergo partial contraction for long periods. This is important in regulating the size of the lumen of tubular structures. In the walls of the digestive tract, uterine tubes, and ureters, the smooth muscle cells undergo rhythmic contractions (peristaltic waves). This process, known as *peristalsis*, propels the contents along these tubular structures.

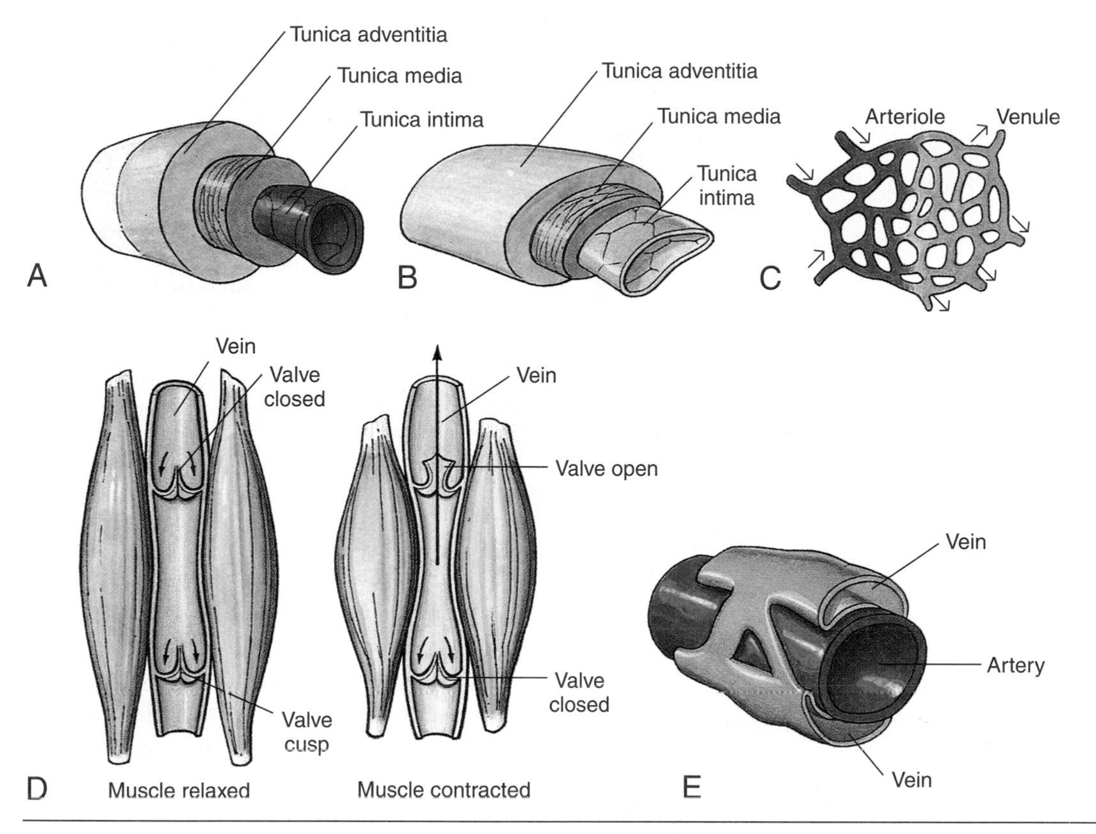

Figure 1.9. Blood vessels. **A.** Muscular (distributing) artery. **B.** Medium-sized vein. Note that its diameter is greater than that of artery. **C.** Capillary bed. **D.** Muscular contractions (e.g., in limbs) function with venous valves to move blood toward heart. *Arrows*, direction of blood flow. **E.** Muscular artery embraced by its accompanying communicating veins (venae communicantes).

> Smooth muscle cells also undergo compensatory hypertrophy in response to increased demands. During pregnancy the smooth muscle cells in the wall of the uterus increase not only in size (*hypertrophy*) but in number (*hyperplasia*).

Cardiovascular System

There are three types of blood vessel in the cardiovascular system:

- Arteries
- Veins
- Capillaries

ARTERIES

Arteries carry blood away from the heart and distribute it to the body (Figs. 1.9 and 1.10). Blood passes from the heart through arteries of ever decreasing caliber. Their walls are divided into three layers: tunica intima, tunica media, and tunica adventitia. The different types of artery are distinguished from each other on the basis of the thickness and differences in the makeup of the layers, especially the tunica media. There are three types of artery: elastic arteries, muscular arteries, and arterioles.

Elastic Arteries

Elastic arteries are the largest type of artery (e.g., aorta). The maintenance of blood pressure in the arterial system between contractions of

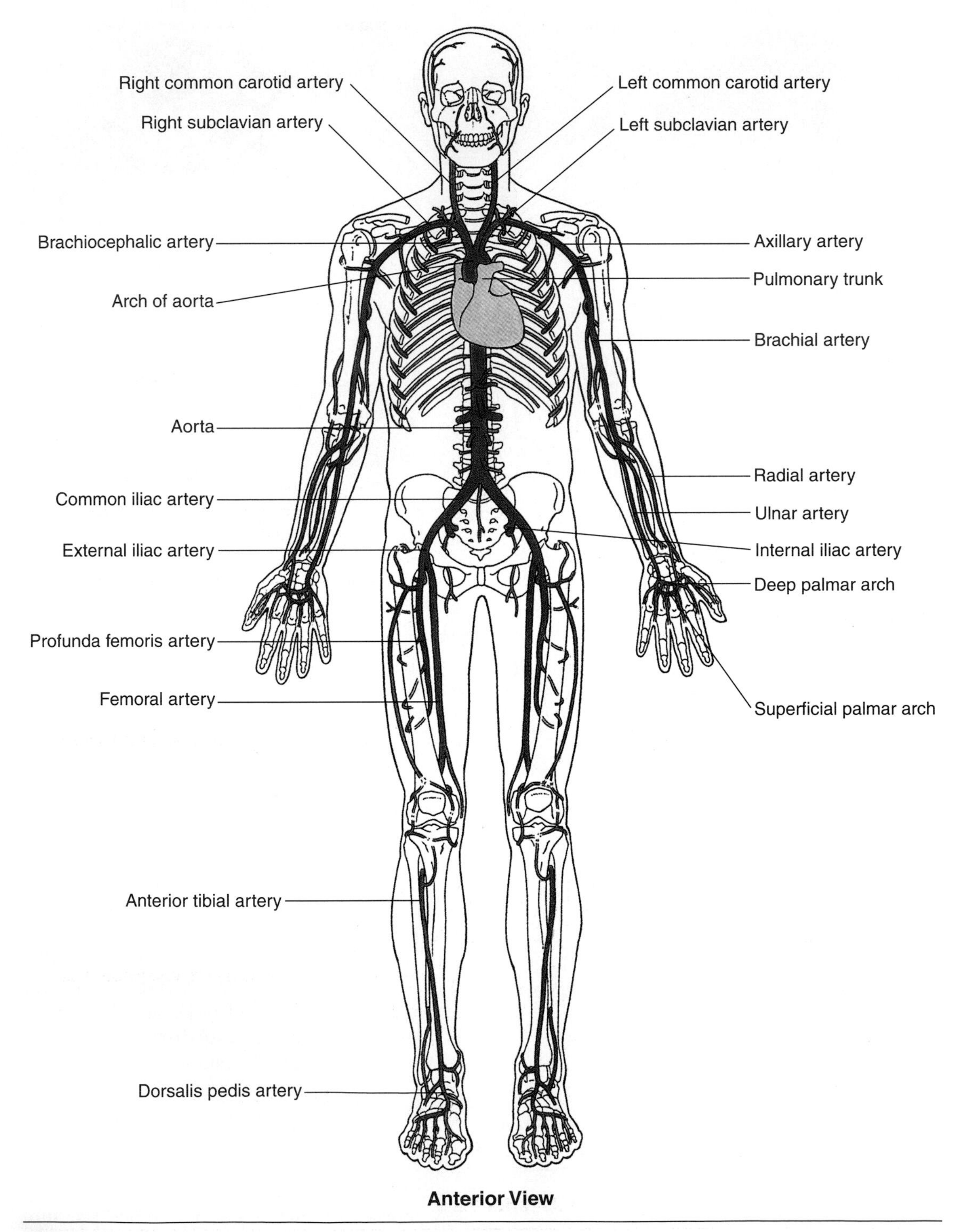

Figure 1.10. Arteries.

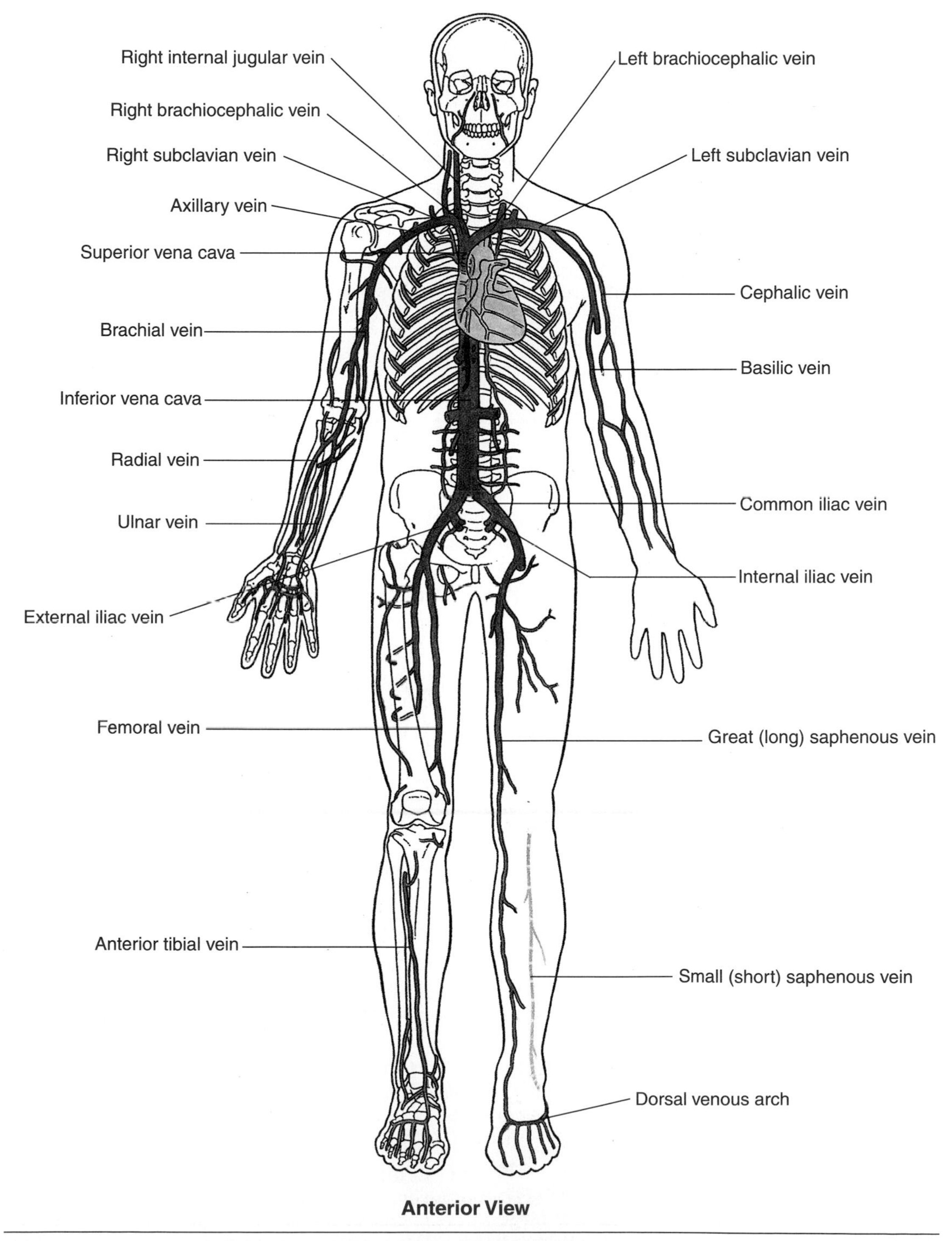

Figure 1.11. Veins. Superficial veins are shown in left limbs and deep veins in right limbs.

the heart results from the elasticity of these arteries. This quality allows them to expand when the heart contracts and to return to normal between cardiac contractions.

Muscular Arteries

Muscular arteries distribute blood to various parts of the body; hence, they are often referred to as *distributing arteries*. Their walls consist chiefly of circularly disposed smooth muscle fibers, which constrict their lumina when they contract. They regulate the flow of blood to different parts of the body as required.

Arterioles

Arterioles are the smallest type of artery. They have relatively narrow lumina and thick muscular walls. The degree of pressure within the arterial system (*arterial pressure*) is mainly regulated by the degree of tonus (firmness) in the smooth muscle in the walls of the arterioles. If the tonus is increased above normal, hypertension results.

The most commonly acquired disease of arteries is *arteriosclerosis* (hardening of arteries), a group of diseases characterized by thickening and loss of elasticity of arterial walls. *Atherosclerosis*, a common form of arteriosclerosis, is associated with the buildup of fat (mainly cholesterol) in the arterial walls. Calcium is deposited, and plaques form (*atheromas*). The complications of atherosclerosis are ischemic heart disease, myocardial infarction (heart attack), stroke, and gangrene (e.g., in parts of the limbs). Expansion of an atherosclerotic lesion in the tunica intima of muscular and elastic arteries may result in *thrombosis* (clotting) and occlusion of arteries.

VEINS

Veins return blood to the heart from the capillary beds (Fig. 1.9, *B* and *C*). The large pulmonary veins are atypical in that they carry well-oxygenated blood from the lungs to the heart. Walls of veins are thinner than those of their companion arteries (Fig. 1.9*E*), because of the lower blood pressure in the venous system.

There are three types of vein: small, medium, and large. Medium-sized veins have *valves* that permit blood to flow toward the heart but not in the reverse direction (Fig. 1.9*D*). Wide bundles of longitudinal smooth muscle characterize the well-developed adventitia of large veins [e.g., superior vena cava (Fig. 1.11)].

Contracting skeletal muscles (e.g., in limbs) compress veins, "milking" the blood superiorly along the vessels toward the heart (Fig. 1.9*D*). The smallest veins are called *venules* (Fig. 1.9*C*). These tributaries unite to form larger veins, which commonly join to form *venous plexuses* [e.g., dorsal venous arch of the foot (Fig. 1.11)]. Veins tend to be double or multiple (Fig. 1.9*E*). Those that accompany deep arteries are called *venae comitantes* (accompanying veins). Systemic veins are more variable than arteries, and anastomoses occur more often between them.

When veins lose their elasticity, they are weak. A weakened vein dilates under the pressure of supporting a column of blood against gravity. The resulting *varicose veins* (e.g., in legs) have a caliber greater than normal, and their valve cusps do not meet or have been destroyed by inflammation. These veins are said to have incompetent valves.

CAPILLARIES

Capillaries are simple endothelial tubes that connect the arterial and venous sides of the circulation (Fig. 1.9*C*). They are generally arranged in networks called *capillary beds*. The blood flowing through a capillary bed is brought to it by arterioles and carried away from it by venules. As blood is forced through the capillary bed by the hydrostatic pressure in the arterioles, nutrients and other cellular materials are exchanged with the surrounding tissue (e.g., oxygen passes to the tissues).

Lymphatic System

The lymphatic system is a vast network of lymphatic vessels (*lymphatics*) that are connected with small masses of lymphatic tissue called *lymph nodes* (Fig. 1.12). The lymphatic system also includes lymphatic organs (e.g., spleen). When tissue fluid enters a lymphatic vessel it is called *lymph*. It is usually clear and watery and has the same constituents as blood plasma. The lymphatic system consists of

- Lymphatic plexuses that are networks of very small lymphatic vessels called *lymphatic capillaries*, which originate in the intercellular spaces of most tissues

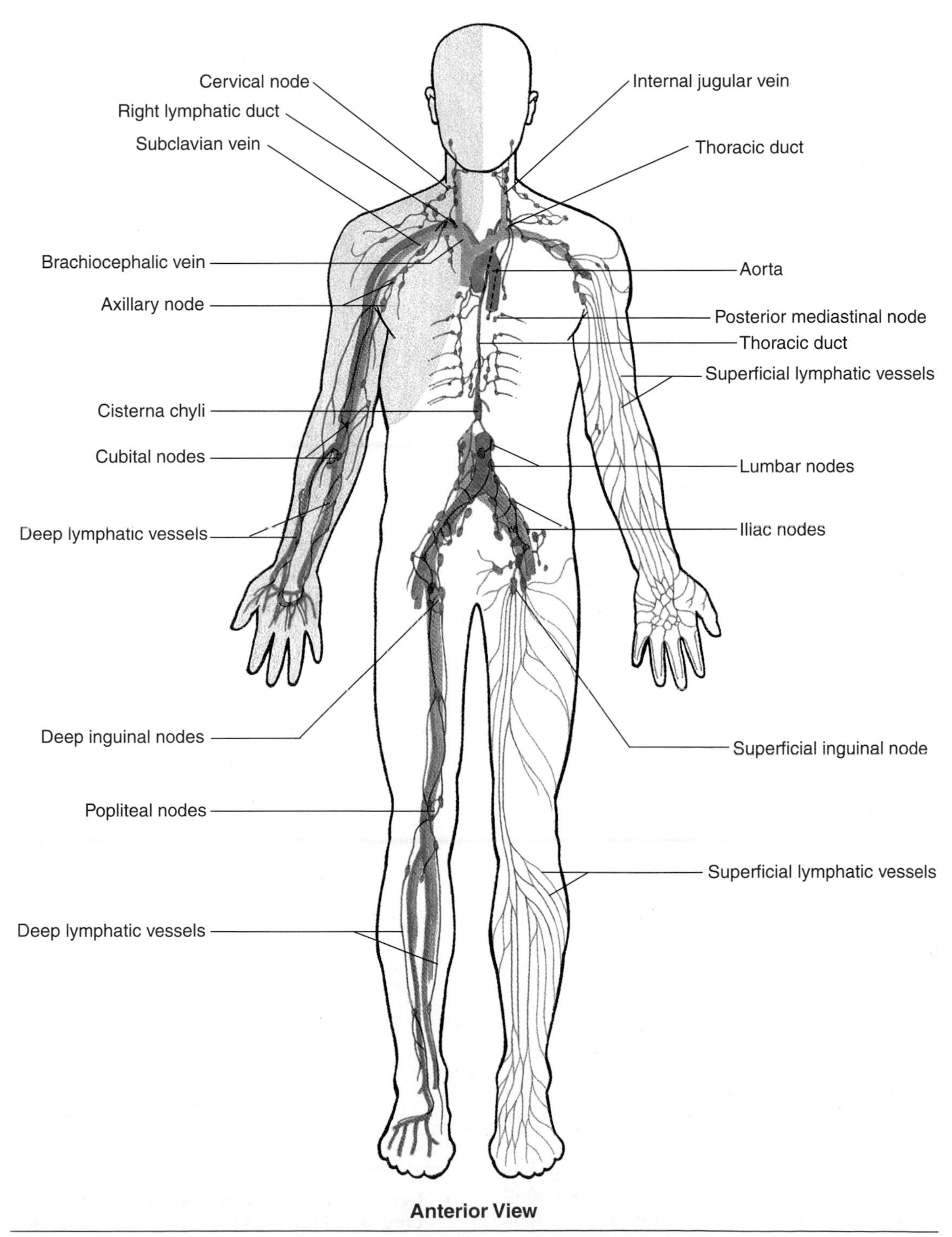

Figure 1.12. Lymphatic system. Right lymphatic duct drains lymph from right side of head and neck and right upper limb (*shaded*). Thoracic duct drains remainder of the body.

- Lymph nodes composed of small masses of lymphatic tissue through which lymph passes on its way to the venous system
- Aggregations of lymphoid tissue located in the walls of the alimentary canal (e.g., tonsils) and in the spleen and thymus
- Circulating lymphocytes that are formed in *lymphoid tissue,* (e.g., in lymph nodes and spleen) and in myeloid tissue in red bone marrow

After traversing one or more lymph nodes, lymph enters larger lymphatic vessels called *lymphatic trunks,* which unite to form either (*a*) the thoracic duct, which begins in the abdomen as a sac called the cisterna chyli and ascends to enter the junction of the left internal jugular and left subclavian veins, or (*b*) the right lymphatic duct, which drains the right side of the head and neck, the right upper limb, and the right half of the thoracic cavity. The thoracic duct drains the remainder of the body.

Superficial lymphatic vessels are located in the skin and superficial fascia. These vessels eventually drain into *deep lymphatic vessels* located in the deep fascia between the muscles and superficial fascia. They accompany the major blood vessels of the region concerned.

Functions of lymphatics follow:

- Drainage of tissue fluid, i.e., the collection of plasma from the tissue spaces and the transport of plasma to the venous system
- Absorption and transport of fat; i.e., the lymphatic capillaries drain fat from the intestine and convey it via the thoracic duct to the left subclavian vein
- Formation of a defense mechanism for the body (e.g., when foreign protein is drained from an infected area, immunologically competent cells produce a specific antibody to the protein, and/or lymphocytes are dispatched to the infected area)

Lymphatic vessels and lymph nodes may become inflamed; this is called lymphangitis and lymphadenitis, respectively. The lymphatic system is commonly involved in the spread (metastasis) of cancer cells. This is referred to as lymphogenous dissemination of cancer cells. *Lymphedema,* accumulation of interstitial fluid, occurs when lymph is not drained from an area of the body (e.g., if cancerous lymph nodes are removed from the axilla, lymphedema of the limb may result).

Nervous System

The nervous system provides the mechanism that enables the body to react to continuous changes in its external and internal environments. It also controls and integrates the various activities of the body (e.g., heart and lungs). For descriptive purposes, the nervous system is divided structurally into the central nervous system and the peripheral nervous system (Fig. 1.13*A*) and functionally into the somatic nervous system and the autonomic nervous system.

Nervous tissue consists of two main types of cell: neurons (nerve cells) and their supporting cells. *Neurons* are the structural and functional units of the nervous system that are specialized for rapid communication. The neuron is composed of a cell body and its processes, dendrites and axons, that carry impulses to and away from the cell body, respectively (Fig. 1.13*B*). Layers of lipid and protein substances called *myelin* form a sheath around some axons, greatly increasing the velocity of impulse conduction. Neurons communicate with each other at *synapses* (points of contact between neurons). *Neuroglia* (glia cells) are nonexcitable cells that form a major component of nervous tissue. Neuroglia provide support and nourishment for neurons.

When the brain or spinal cord is damaged, the injured axons in most circumstances do not recover. Proximal stumps of the axons begin to regenerate, sending sprouts into the area of the lesion, but this growth stops in about 2 weeks. As a result, permanent disability follows destruction of a tract in the brain or spinal cord. For a discussion of axonal degeneration and regeneration in the CNS, see Barr and Kiernan (1993, listed under "Suggested Readings").

CENTRAL NERVOUS SYSTEM

The central nervous system (CNS) consists of the brain and spinal cord (Figs. 1.13*A* and 1.14). The principal roles of the CNS are to

- Integrate and coordinate incoming and outgoing neural signals
- Carry out higher mental functions such as thinking and learning

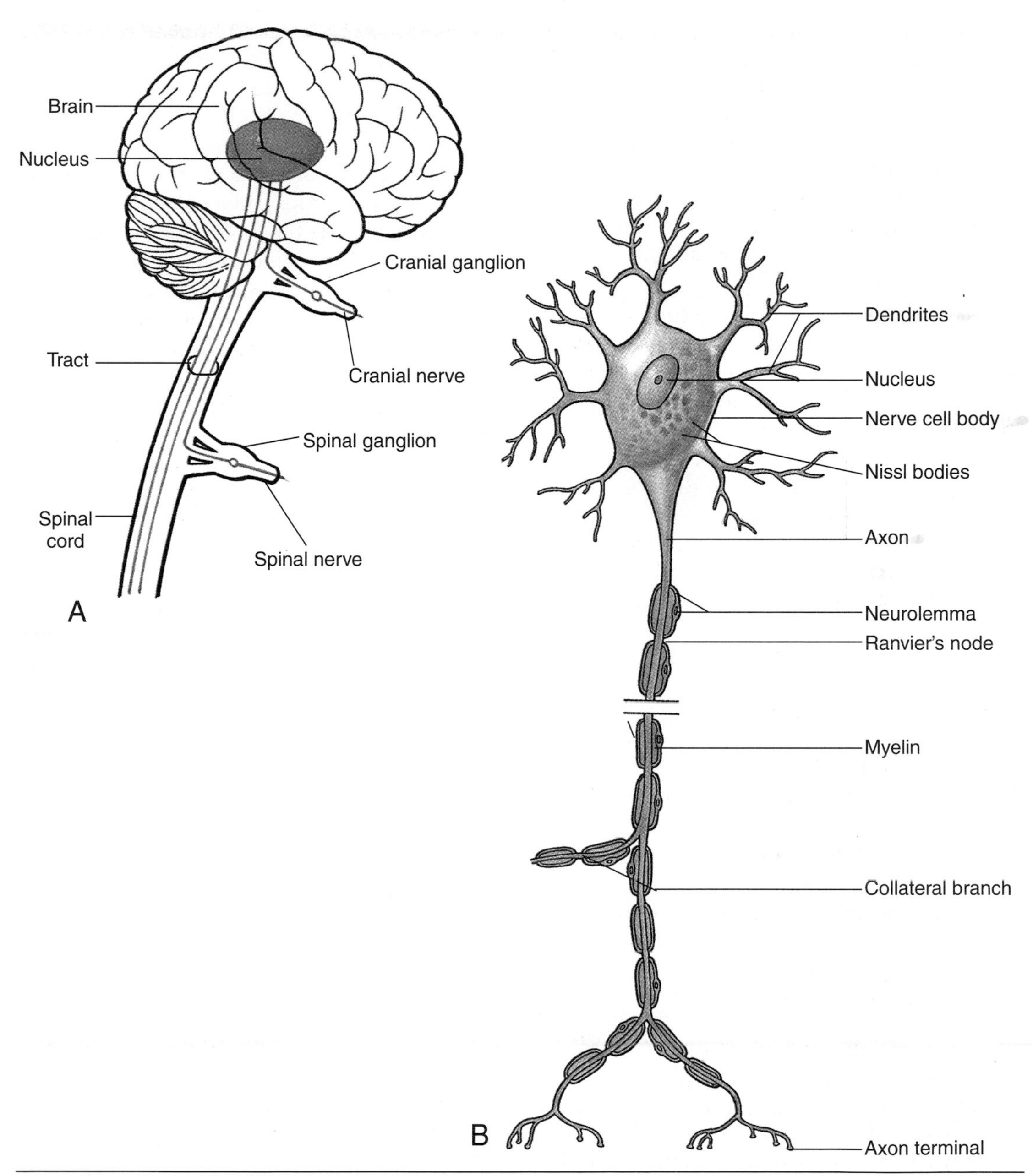

Figure 1.13. **A.** Tract arising from neurons in a nucleus of the brain and passing into spinal cord. **B.** Structure of a typical motor neuron.

A bundle of nerve fibers (axons) connecting neighboring or distant parts of the CNS is called a *tract*, and a collection of nerve cell bodies is called a *nucleus* (Fig. 1.13*A*). *Cerebrospinal fluid* and meninges (membranes enveloping the brain and spinal cord: dura mater, pia mater, and arachnoid mater) surround and protect the CNS.

PERIPHERAL NERVOUS SYSTEM

The peripheral nervous system (PNS) consists of the cranial and spinal nerves. The 12 pairs of *cranial nerves* arise from the brain, and the 31 pairs of *spinal nerves* emerge from the spinal cord (Fig. 1.14). The PNS is anatomically and operationally continuous with the CNS.

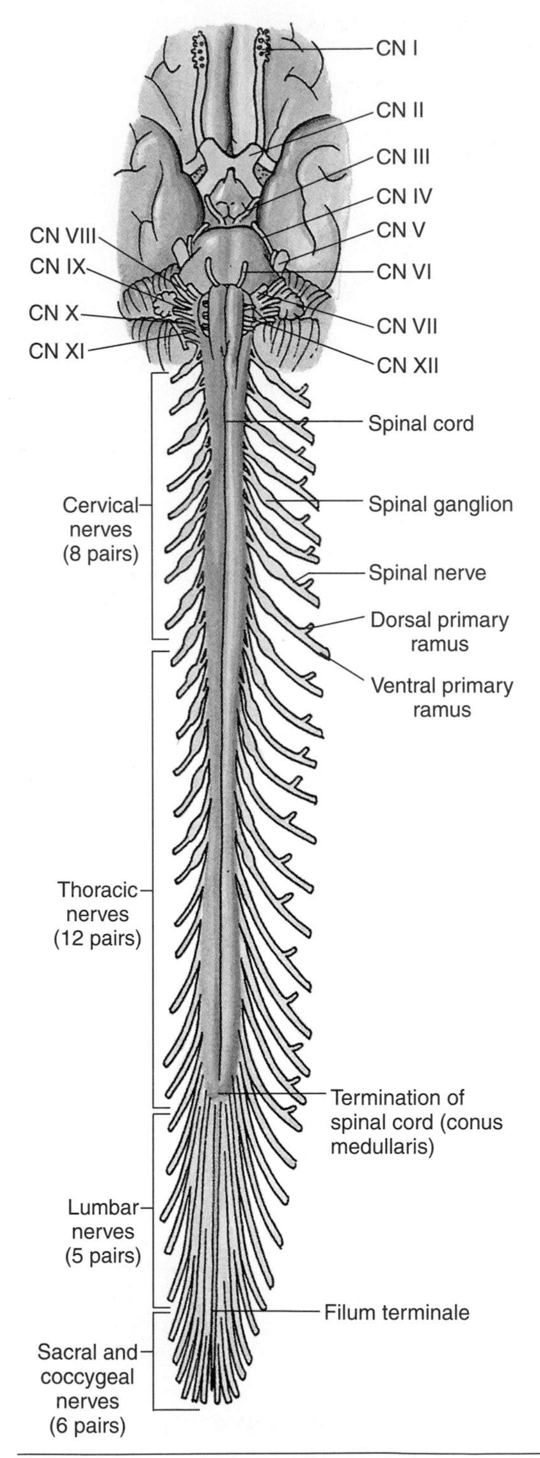

Figure 1.14. Ventral aspect of brainstem: brain and cranial nerves (CN) and spinal cord and spinal nerves. CN I, olfactory nerve; CN II, optic nerve; CN III, oculomotor nerve; CN IV, trochlear nerve; CN V, trigeminal nerve; CN VI, abducent (abducens) nerve; CN VII, facial nerve; CN VIII, vestibulocochlear nerve; CN IX, glossopharyngeal nerve; CN X vagus nerve; CN XI, accessory nerve; CN XII, hypoglossal nerve.

The PNS conveys neural impulses to the CNS from the sense organs (e.g., the eyes) and from sensory receptors in various parts of the body (e.g., skin). The PNS also conveys neural impulses from the CNS to the muscles and glands. A bundle of nerve fibers (axons) in the PNS is called a *nerve*, and a collection of nerve cell bodies outside the CNS is called a *ganglion* (e.g., a spinal ganglion). A network of nerves is called a *nerve plexus*. A peripheral nerve fiber consists of an axon, a myelin sheath and a neurolemma, or sheath of Schwann cells (Fig. 1.13*B*). Not all axons are myelinated (e.g., most fibers in cutaneous nerves) (Fig. 1.2).

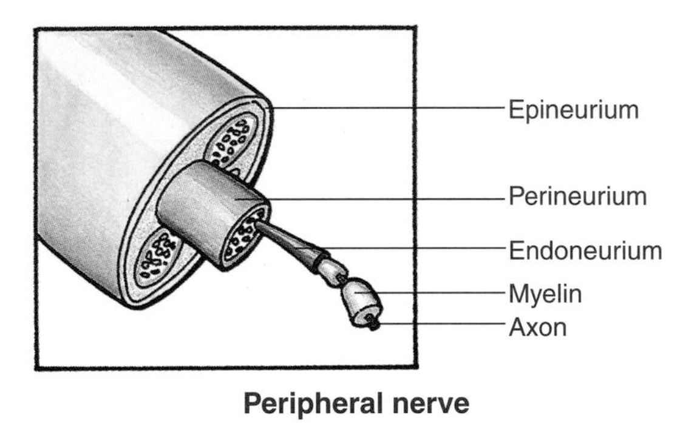

Peripheral nerve

Peripheral nerves are fairly strong and resilient because the delicate nerve fibers in them are strengthened and protected by the following three connective tissue coverings:

- Epineurium, a connective tissue sheath, surrounds the nerve
- Perineurium, a more delicate cellular and connective tissue sheath than the epineurium, encloses a small bundle (fasciculus) of nerve fibers
- Endoneurium, a delicate connective tissue sheath, surrounds individual nerve fibers

A typical spinal nerve arises from the spinal cord by two roots (Fig. 1.15*A*):

- Ventral (anterior) roots contain mainly motor (efferent) fibers from nerve cell bodies in the ventral horn of the spinal cord
- Dorsal (posterior) roots carry sensory (afferent) fibers to the dorsal horn of the spinal cord

The dorsal and ventral roots unite to form a *spinal nerve* that divides into two branches, a dorsal (posterior) primary ramus and a ventral (anterior) primary ramus. The dorsal and ventral rami carry both motor and sensory nerves.

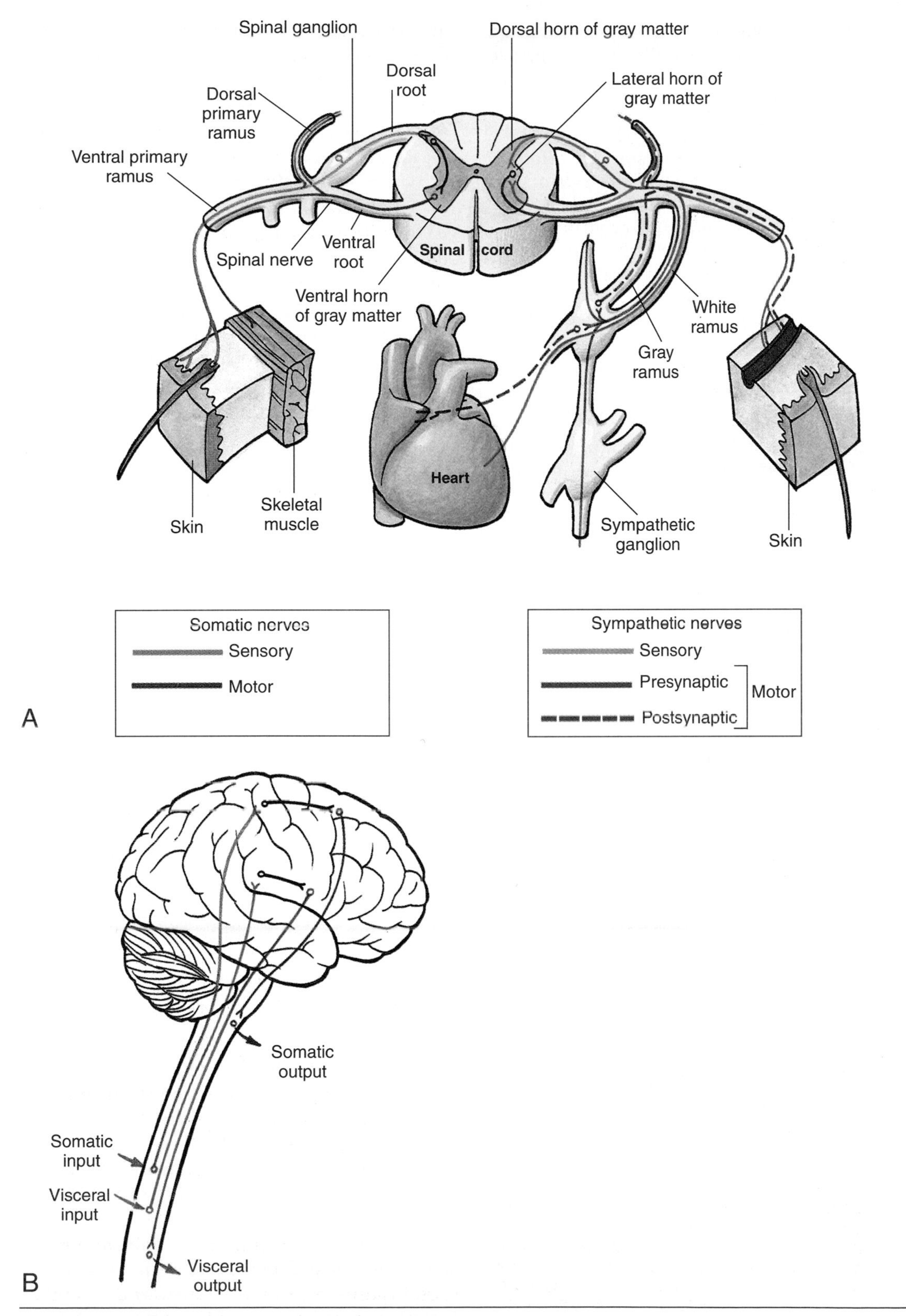

Figure 1.15. Organization of nervous system. **A.** Components of a typical spinal nerve. **B.** Somatic and visceral input and output of CNS.

The dorsal rami supply nerve fibers to the back, whereas the ventral rami supply nerve fibers to the limbs and anterior and lateral regions of the trunk.

Components of a typical spinal nerve follow:

- General sensory (general somatic afferent) fibers that transmit sensations from the body to the spinal cord; they may be exteroceptive sensations (pain, temperature, touch, and pressure) from the skin or proprioceptive sensations from muscles, tendons, and joints that convey information on joint position and the tension of tendon and muscle fibers
- Visceral sensory (general visceral afferent) fibers that transmit sensations from mucous membranes, glands, and blood vessels
- Somatic motor (general somatic efferent) fibers that transmit motor impulses to skeletal muscles
- Visceral motor (general visceral efferent) fibers that transmit impulses to smooth muscle and glandular tissues

When peripheral nerves are crushed or severed, their axons degenerate distal to the lesion because they are dependent on their cell bodies for survival. A crushing nerve injury damages the axons but leaves the connective tissue coverings of the nerve intact. No surgical repair is needed for this type of nerve injury because the connective tissue sheaths are intact to guide the growing axons to their destinations. If the nerve is cut, surgical intervention is necessary because the regeneration of axons requires apposition of the cut ends by sutures through the epineurium. The individual bundles of nerve fibers are realigned as accurately as possible.

SOMATIC NERVOUS SYSTEM

The somatic nervous system is composed of the somatic portions of the CNS and PNS, which consist of sensory and motor components. The somatic sensory system transmits sensations of touch, pain, temperature, and position from sensory receptors. The somatic motor system permits voluntary movement by contraction of skeletal muscle (Fig. 1.15).

AUTONOMIC NERVOUS SYSTEM

The autonomic nervous system is composed of efferent and afferent nerves and ganglia. The efferent nerves are concerned with the distribution of impulses to the heart (cardiac muscle), smooth muscle, and glands. The afferent nerves are concerned with the conduction of visceral pain stimuli and the afferent component of autonomic reflexes. The autonomic nervous system consists of two parts:

- Parasympathetic system, which stimulates activities that conserve and restore body resources (e.g., the heart beats slowly)
- Sympathetic system, which stimulates activities that are performed during emergency and stressful situations when the heart beats quickly and the blood pressure rises

Each part of the autonomic nervous system consists of two neuron chains. The cell body of the first neuron is in the visceral efferent column of the brain or spinal cord, whereas the cell body of the second neuron is in an autonomic ganglion outside the CNS. The axon of the first neuron is called a *presynaptic* or *preganglionic fiber* and that of the second neuron a *postsynaptic* or *postganglionic fiber* (Fig. 1.15).

Parasympathetic Division of Autonomic Nervous System

In the parasympathetic division, the cell bodies of the preganglionic neurons are located in the nuclei of cranial nerves III, VII, IX, and X in the brainstem and in the second, third, and fourth sacral segments of the spinal cord (Fig. 1.14). The preganglionic fiber synapses with the cell body of a postganglionic neuron in a parasympathetic ganglion close to or in the wall of the target organ. The postganglionic fiber innervates the muscle or gland in the organ.

Sympathetic Division of Autonomic Nervous System

In the sympathetic division, the cell bodies of the preganglionic neurons are located in the gray matter (lateral horn) of the spinal cord, beginning at the first thoracic level and ending at the second or third lumbar level. The nerve cell bodies of postganglionic neurons are located in the paravertebral and prevertebral ganglia (Figs. 1.16 and 1.17).

Paravertebral ganglia are located on each side of the vertebral column and are linked by sympathetic trunks (chains) that extend from the base of the skull to the coccyx where they unite at the ganglion impar. *Prevertebral ganglia* are located in plexuses that surround the main branches of the abdominal aorta [e.g., celiac ganglion (Fig. 1.16*A*)].

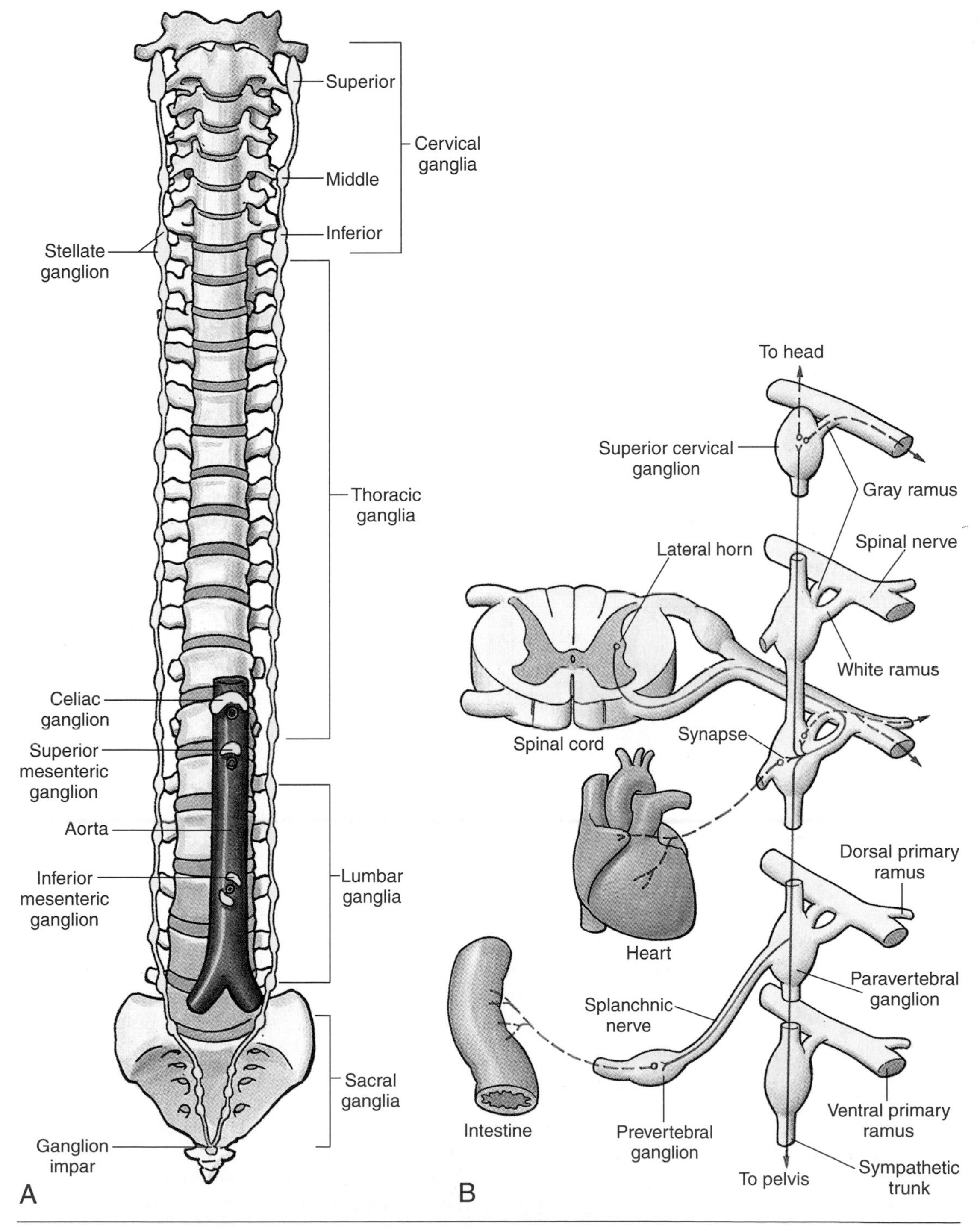

Figure 1.16. **A.** Paravertebral and prevertebral sympathetic ganglia. **B.** Origin and distribution of sympathetic motor fibers. Solid lines, preganglionic fibers; *broken lines*, postganglionic fibers.

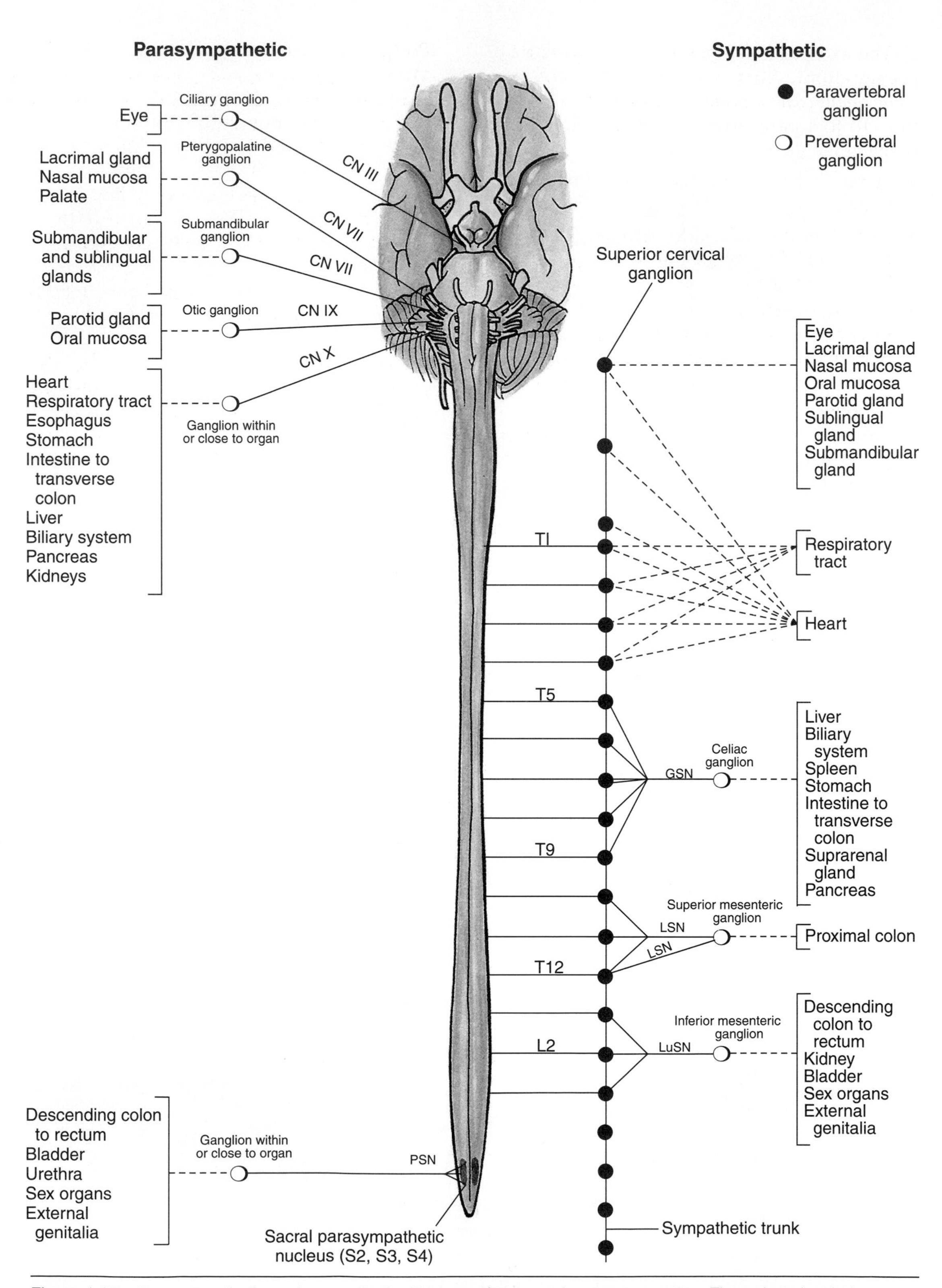

Figure 1.17. Parasympathetic and sympathetic divisions of autonomic nervous system. Thoracic splanchnic nerves [greater splanchnic nerve (*GSN*)], lesser and least splanchnic nerves (*LSN*), and lumbar splanchnic nerves (*LuSN*) provide sympathetic innervation for most abdominopelvic organs. Pelvic splanchnic nerves (*PSN*) arise from S2–S4 segments of spinal cord and provide parasympathetic innervation for descending colon and pelvic organs. *CN*, cranial nerve.

The axon of a sympathetic neuron sends a preganglionic fiber via the ventral root and *white ramus communicans* (white communicating branch) to a paravertebral ganglion. Here, the preganglionic fiber may

- Synapse immediately with a postganglionic neuron
- Ascend or descend in the sympathetic trunk to synapse in a higher or lower paravertebral ganglion
- Pass through the paravertebral ganglia without synapsing to form splanchnic nerves (preganglionic fibers of splanchnic nerves synapse in one of the prevertebral ganglia)
- Pass directly to the suprarenal (adrenal) gland without synapsing in a paravertebral or prevertebral ganglion

Postganglionic fibers are distributed

- Through the *gray ramus communicans* (gray communicating branch) into the spinal nerve, where they run to blood vessels, sweat glands, and the arrector pili muscles associated with hairs
- To the head and neck by synapsing in the *superior cervical ganglion* and then forming a plexus around the blood vessels supplying these regions
- To the thoracic viscera (e.g., heart, lungs) via cardiac, pulmonary, and esophageal plexuses
- From the prevertebral ganglia to the abdominal and pelvic viscera (e.g., intestine and urinary bladder) via nerve plexuses that surround the branches of the abdominal aorta (e.g., celiac plexus surrounds celiac trunk)

2 / THORAX

The thorax (chest) is the superior part of the trunk between the neck and abdomen. The *thoracic cavity*, surrounded by the *thoracic wall*, contains the thymus, heart, lungs, distal part of the trachea, and most of the esophagus. To perform a physical examination of the thorax, a working knowledge of its structure and vital organs is required.

Thoracic Wall

The thoracic wall consists of skin, fasciae, nerves, vessels, muscles, and bones.

SKELETON OF THORACIC WALL

The skeleton of the thoracic wall (Fig. 2.1) forms an *osteocartilaginous thoracic cage* that protects the heart, lungs, and some abdominal organs (e.g., liver). The thoracic skeleton (bony thorax) includes

- Thoracic vertebrae (12) and intervertebral discs
- Ribs (12 pairs) and costal cartilages
- Sternum

The *thoracic vertebrae* and fibrocartilaginous *intervertebral discs* are described in Chapter 5. The special features of thoracic vertebrae include

- Costal facets on their bodies for articulation with the heads of ribs
- Costal facets on their transverse processes for articulation with the tubercles of ribs, except for the inferior two or three
- Long spinous processes

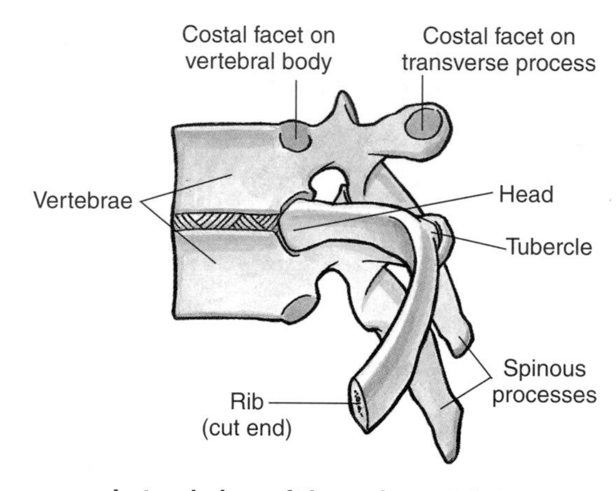

Lateral view of thoracic vertebrae

Ribs

Ribs are narrow, curved, flat bones that form most of the thoracic cage.

- The first seven ribs (sometimes eight) are *true (vertebrosternal) ribs* because they attach the vertebrae to the sternum through their costal cartilages
- The 8th to 10th ribs are *false (vertebrochondral) ribs* because their cartilages are joined to the cartilage of the rib just superior to them
- The 11th and 12th ribs are free or *floating ribs* because their cartilages end in the posterior abdominal musculature

Costal cartilages prolong the ribs anteriorly and contribute to the elasticity of the thoracic wall. The cartilages of the 7th to 10th ribs ascend and join to form an *infrasternal angle* and costal margin on each side. Ribs with their cartilages are separated by *intercostal spaces* that are occupied by intercostal muscles, vessels, and nerves.

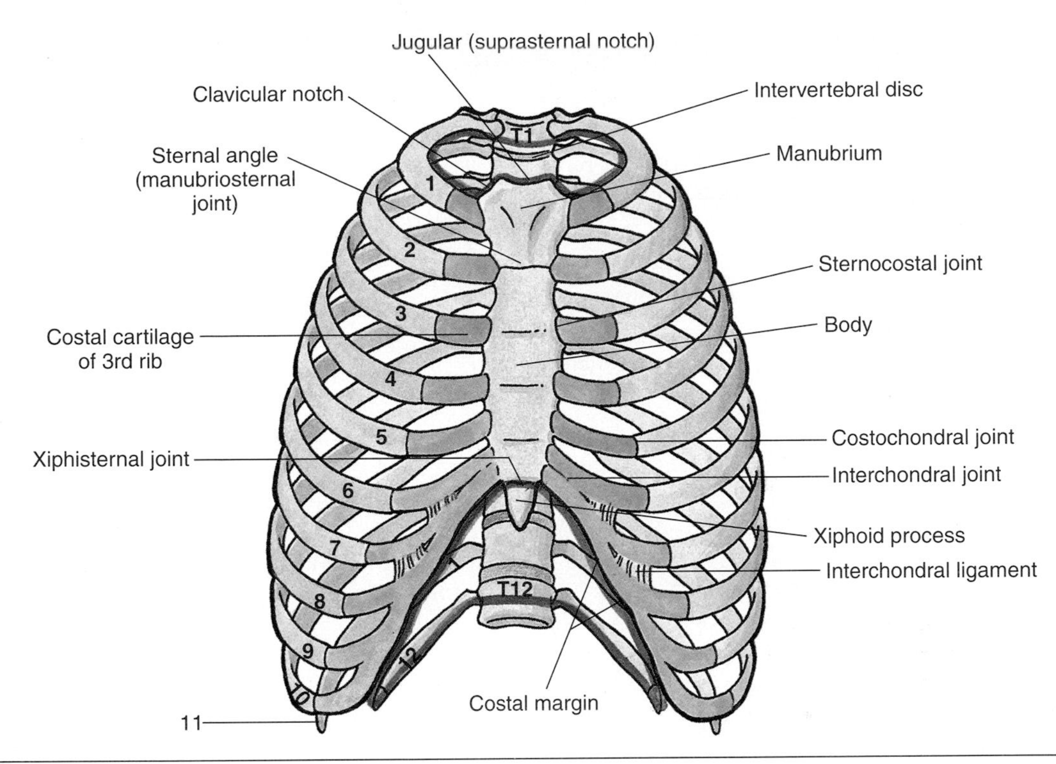

Figure 2.1. Anterior view of osteocartilaginous thoracic cage showing joints of thoracic wall. Ribs are numbered 1 to 12. *T1*, body of first thoracic vertebrae; *T12*, body of twelfth thoracic vertebrae. Superior and inferior thoracic apertures are outlined in *pink*.

Costal cartilages provide resilience to the thoracic cage, preventing many blows from fracturing the sternum and/or ribs. In elderly people, costal cartilages undergo calcification, making the cartilages radiopaque.

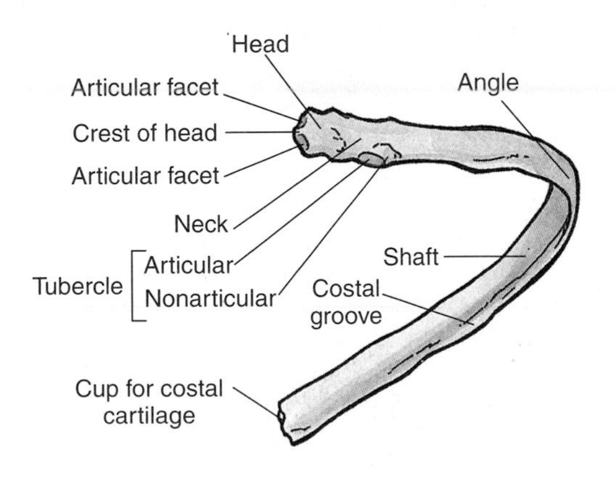

Posterior view of typical rib

Typical ribs (3rd to 9th) have the following:

- Head that is wedge-shaped and has two facets, separated by a crest, for articulation with the numerically corresponding vertebra and the vertebra superior to it
- Neck that separates the head and tubercle
- Tubercle at the junction of the neck and shaft (body) that has a smooth facet for articulation with the corresponding transverse process of the vertebra and a rough nonarticular area for the attachment of the costotransverse ligament
- Shaft that is thin, flat, and curved with an angle at its point of greatest change in curvature

Rib excision is performed by surgeons desiring to access the thoracic cavity. A incision is made through the periosteum along the curve of the rib, and a piece of rib is removed. After the operation, the rib regenerates from the osteogenic layer of the periosteum.

Atypical ribs are dissimilar:

- The 1st rib is the broadest, shortest, and most sharply curved of the seven true ribs
- The 2nd rib is thinner, less curved, much longer than the 1st rib, and has two facets

on its head for articulation with the bodies of T1 and T2 vertebrae

- The 10th rib has only one facet on its head for articulation with T10 vertebra
- The 11th and 12th ribs are short and have one large facet on their heads but no necks or tubercles

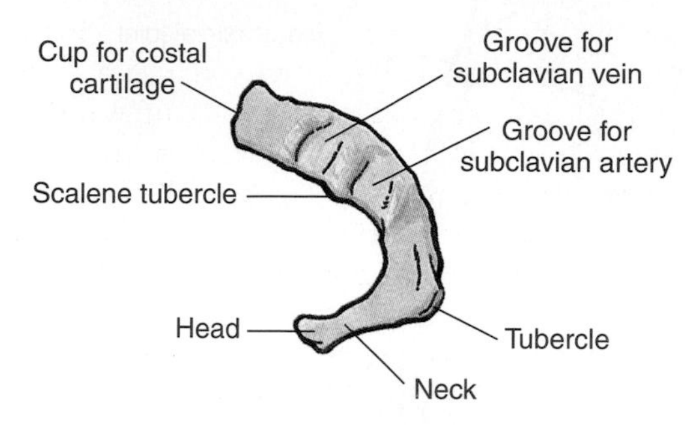

Superior view of 1st rib

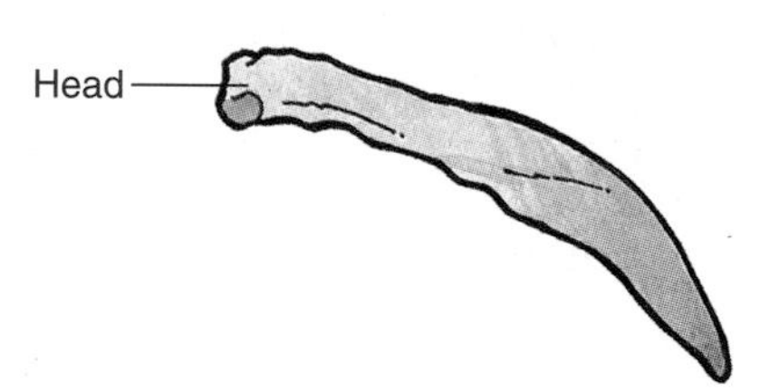

Posterior view of 12th rib

The weakest part of a rib is just anterior to its angle. *Rib fractures* commonly result directly from blows or indirectly from crushing injuries. Direct violence may fracture a rib anywhere, and its broken end may injure internal organs (e.g., lung and/or spleen).

There are usually 12 pairs of ribs, but the number may be increased by the presence of cervical and/or lumbar ribs or decreased by failure of the 12th pair to form. Cervical ribs (incidence 0.5–1%) articulate with C7 vertebra and are clinically significant because they may compress the inferior trunk of the *brachial plexus* of nerves and cause tingling and numbness along the medial border of the forearm. They may also compress the subclavian artery, resulting in *ischemic muscle pain* in the upper limb, i.e., pain caused by poor blood supply.

Sternum

The sternum is the flat, elongated bone that forms the middle of the anterior part of the thoracic cage (Fig. 2.1). The sternum consists of three parts: manubrium, body, and xiphoid process.

The *manubrium,* a roughly triangular bone, lies at the level of T3 and T4 vertebrae. Its thick superior border is indented by the *jugular (suprasternal) notch.* On each side of this notch is a *clavicular notch* that articulates with the medial end of the clavicle. Just inferior to this notch, the costal cartilage of the 1st rib fuses with the lateral border of the manubrium. At the *manubriosternal joint* the manubrium and body lie in slightly different planes; hence their junction forms a projecting *sternal angle* (angle of Louis). This readily palpable clinical landmark is located opposite the second pair of costal cartilages at the level of the disc between T4 and T5 vertebrae.

The *body* is long, narrow, and thinner than the manubrium and is located at the level of T5–T9 vertebrae. The body's nearly flat anterior surface is marked by three transverse ridges in adults that represent the lines of fusion of its four originally separate segments (sternebrae).

The *xiphoid process,* the smallest and most variable part of the sternum, is cartilaginous in young people, but it is more or less ossified in adults older than 40 years.

Sternal fractures are common after traumatic compression of the thoracic wall (e.g., in automobile accidents when the driver's chest is forced into the steering column). The body is commonly fractured, and it is usually a *comminuted fracture,* i.e., broken into several pieces. The installation of air bags in vehicles has reduced the number of sternal and facial fractures.

To access the thoracic cavity for operations on the heart and great vessels, the sternum is divided in the median plane. The sternal body is often used for a *bone marrow needle biopsy* because of its breadth and subcutaneous position.

THORACIC APERTURES

The thoracic cavity communicates with the neck through the kidney-shaped *superior thoracic aperture* (thoracic inlet). Structures entering and leaving the thoracic cavity through this oblique aperture include the trachea, esophagus, vessels, and nerves. The superior thoracic aperture (Fig. 2.1) is bounded by

- First thoracic vertebra
- First pair of ribs and cartilages
- Superior border of manubrium

The thoracic cavity communicates with the abdomen through the *inferior thoracic aperture* (thoracic outlet) that is closed by the diaphragm.

Structures passing to or from the thorax to the abdomen pass through the diaphragm (e.g., inferior vena cava) or posterior to it (e.g., aorta). The inferior thoracic aperture is bounded by

- The 12th thoracic vertebra
- The 12th pair of ribs
- Costal cartilages of ribs 7–12
- Xiphisternal joint

The apices of the lungs and their coverings project through the superior thoracic aperture into the root of the neck (p. 49). On its way to the upper limb, the subclavian artery also passes through the superior thoracic aperture and produces a groove in the 1st rib. Sometimes this artery is compressed between the clavicle superiorly and the 1st rib inferiorly, producing vascular symptoms [e.g., pallor and coldness of the skin (costoclavicular syndrome)].

Clinicians may refer to the superior thoracic aperture as the thoracic outlet, emphasizing that important nerves and vessels pass from the thorax through the aperture into the upper limb. Hence clinicians refer to various types of thoracic outlet syndromes, such as the costoclavicular syndrome just mentioned.

BREASTS

The breasts containing *mammary glands* are located in the superficial fascia of the anterior thoracic wall (Fig. 2.2). Both sexes have breasts, but mammary glands usually develop only in women. They are rudimentary and functionless in males. At the greatest prominence of the breast there is a *nipple,* surrounded by a circular pigmented area called the *areola.* The breast contains up to 20 glands, each of which is drained by a *lactiferous duct* that opens on the nipple. The roughly circular base of the female breast usually lies vertically from the the 2nd to 6th ribs and transversely from the lateral border of the sternum to the midaxillary line.

A small part of the mammary gland extends superolaterally along the inferior border of the pectoralis major muscle toward the axilla to form an *axillary tail.* Two-thirds of the breast rests on the deep (pectoral) fascia covering the pectoralis major muscle; the other third rests on the fascia covering the serratus anterior muscle. Between the gland and the deep fascia, there is loose connective tissue containing little fat

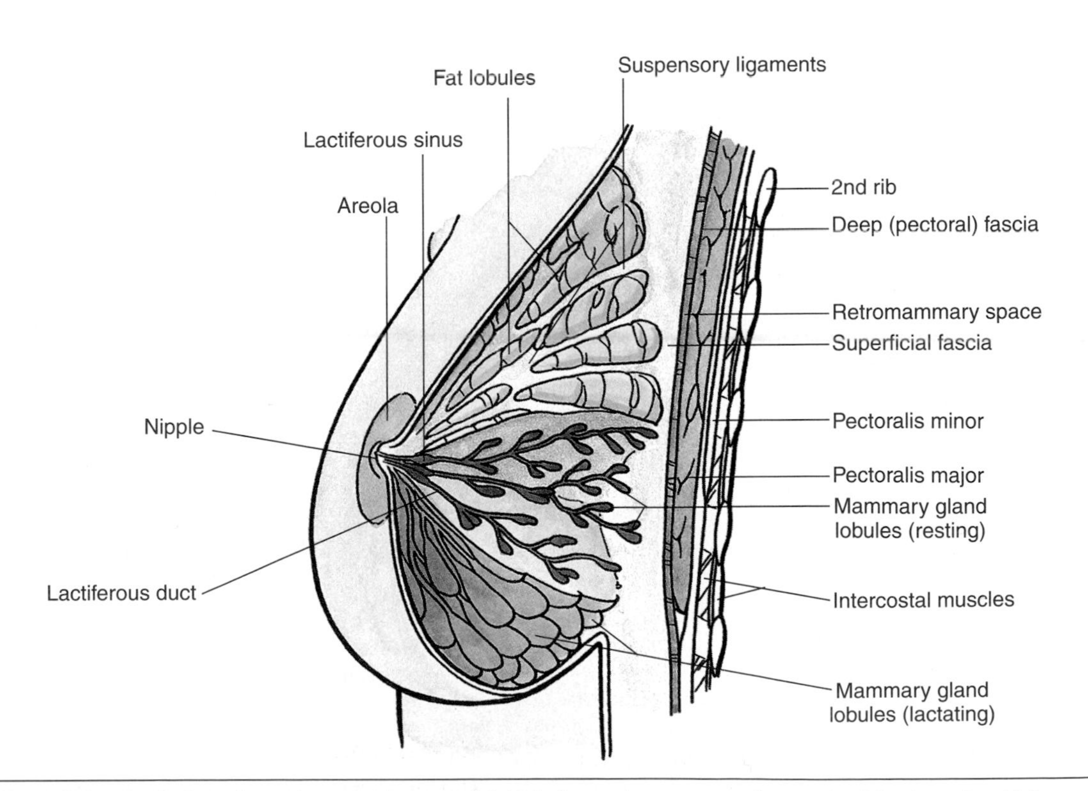

Figure 2.2. Sagittal section of breast. *Upper part,* fat lobules and suspensory ligaments of the breast; *middle part,* appearance of glandular tissue in nonlactating (resting) breast; *lower part,* appearance of glandular tissue in lactating breast.

known as the *retromammary space* (bursa), which allows the breast some degree of movement. The gland is firmly attached to the dermis of the overlying skin by fibrous septa (bands) called the *suspensory ligaments* (Cooper's ligaments). These ligaments are particularly well developed in the superior part of the gland and help support the glandular tissue.

Vasculature

Arterial supply of the breast (Figs. 2.3–2.5) is derived from

- Anterior intercostal branches of the internal thoracic artery, a branch of the subclavian artery
- Lateral thoracic and thoracoacromial arteries, branches of the axillary artery
- Posterior intercostal arteries, branches of the thoracic aorta in the second, third, and fourth intercostal spaces

Venous drainage of the breast is to the axillary (mainly) and internal thoracic veins (Figs. 2.3 and 2.4*B*).

Lymphatic drainage of the breast is important because of its role in the metastasis (spread) of cancer cells. The lymph passes to the subareolar plexus (Fig. 2.3), and from it

- Most lymph (about 75%) drains to the *axillary lymph* nodes, mainly to the pectoral group, but some lymph drains to the apical, subscapular, lateral, and central groups
- Most of the remaining lymph drains to the infraclavicular, supraclavicular, and parasternal (internal thoracic) lymph nodes

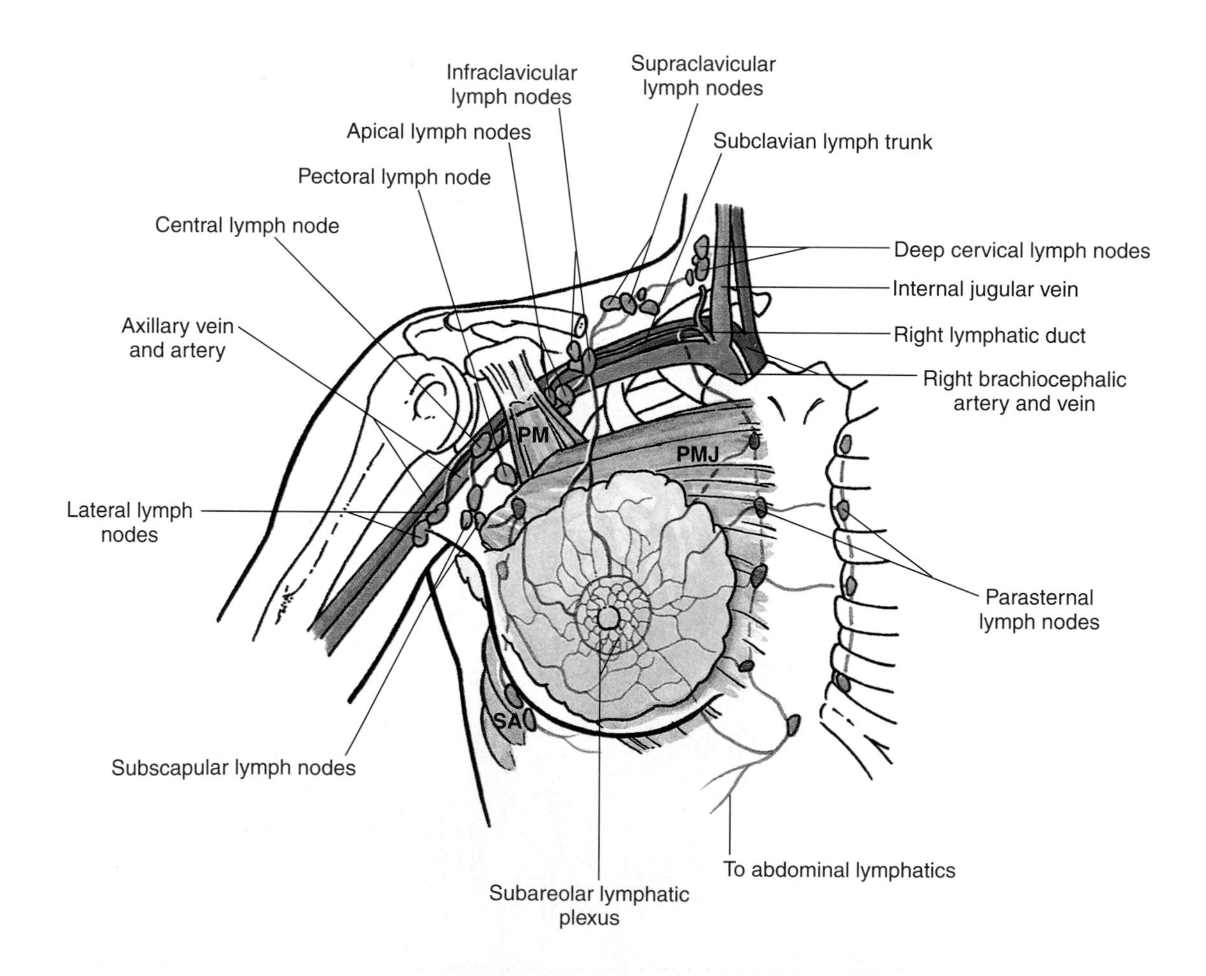

Figure 2.3. Axillary lymph nodes and lymphatic drainage of right breast. *PM*, pectoralis minor muscle; *PMJ*, pectoralis major muscle; *SA*, serratus anterior muscle.

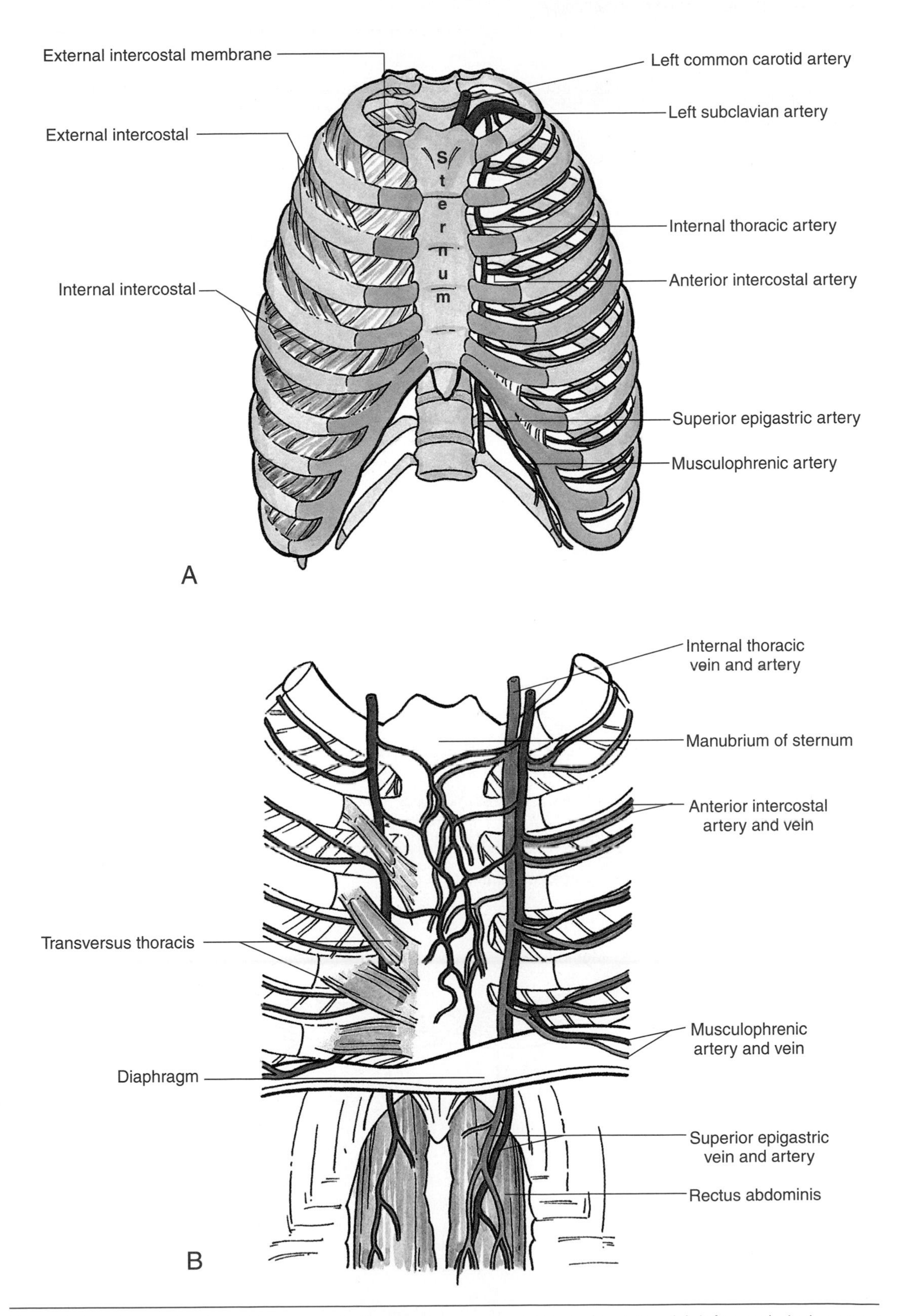

Figure 2.4. **A.** Anterior view of thoracic wall. On left, observe internal thoracic artery arising from subclavian artery and running about 1 cm lateral to border of sternum. On right, observe intercostal muscles. **B.** Posterior view of anterior thoracic wall, transversus thoracis muscle, and blood vessels.

Understanding the lymphatic drainage of the breasts is of practical importance in predicting the metastasis of *breast cancer.* Interference with the lymphatic drainage by cancer produces a leathery thickening of the skin. Often it is dimpled and has prominent pores that give the skin the appearance of orange peel (*peau d'orange sign*). The dimples and pores result from shortening of the suspensory ligaments because of cancerous invasion.

Mastectomy (excision of a breast) is now uncommon. Often only the tumor and surrounding tissues are removed during an operation called a lumpectomy. Although cancer of the breast is uncommon in males, the tumor is often undetected until invasion of lymph nodes has occurred.

- A small amount of lymph enters the lymphatics that drain the opposite breast and anterior abdominal wall

Nerves

The nerves of the breast are derived from anterior and lateral cutaneous branches of the fourth to sixth thoracic nerves (Fig. 2.5). These nerves convey sensory fibers to the skin of the breast and sympathetic fibers to the smooth muscle in the dermis of the areolae and nipples and in the blood vessels.

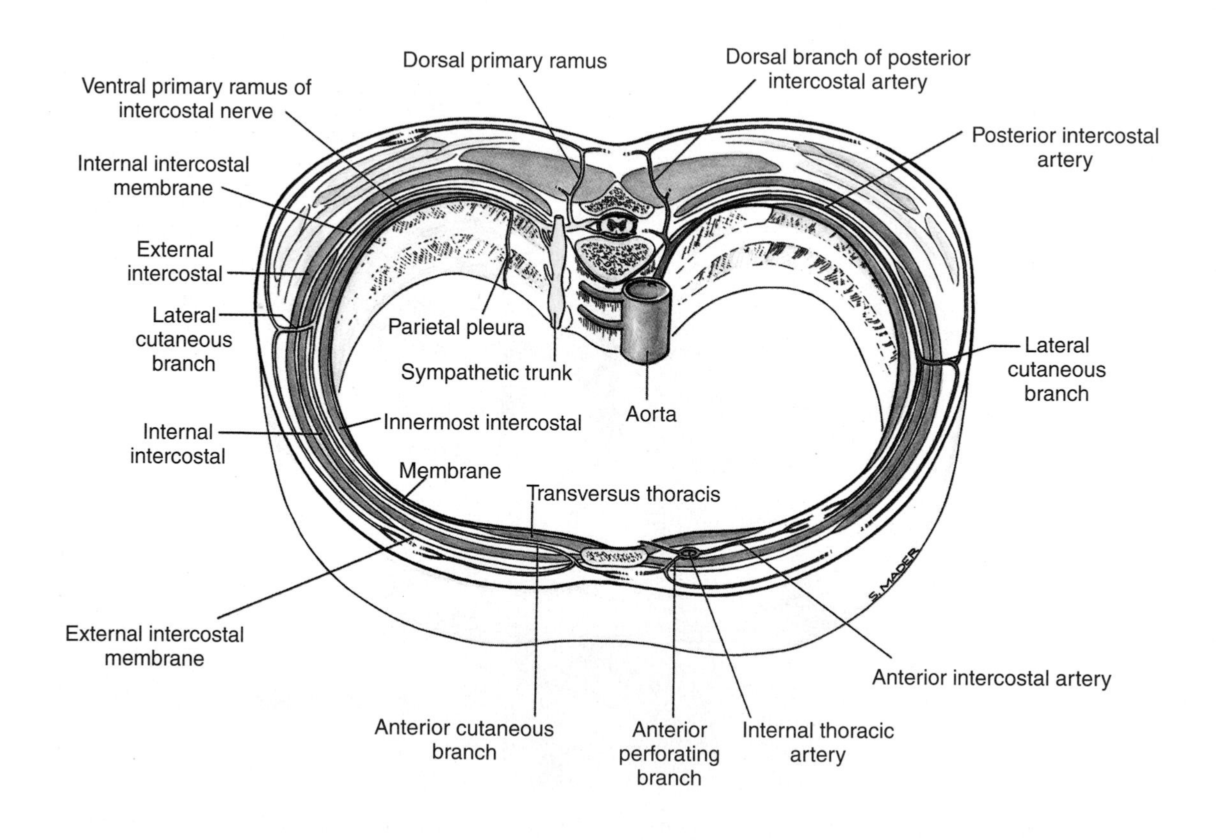

Figure 2.5. Transverse section of thorax showing contents of an intercostal space. Diagram is simplified by showing nerves on right and arteries on left.

Surface Anatomy of Thoracic Wall

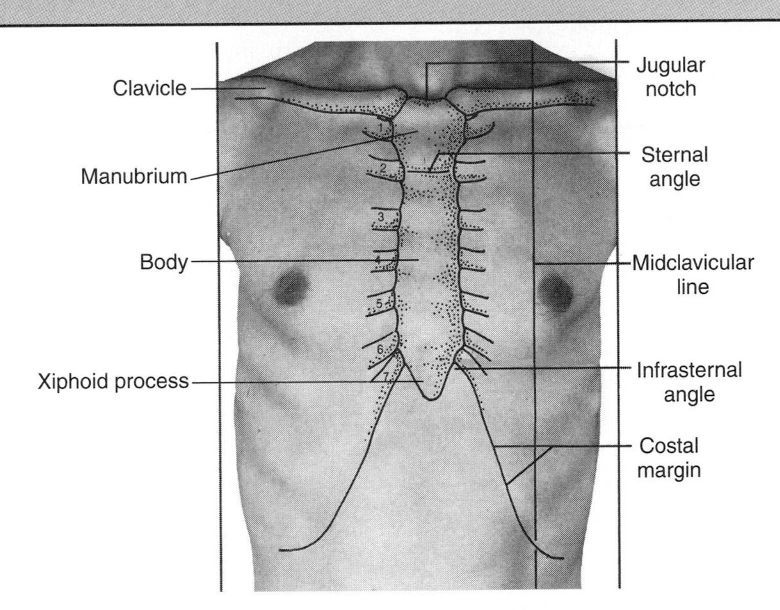

The *clavicles* lie subcutaneously at the junction of the thorax and neck. They can be palpated easily, especially where they articulate with the manubrium.

The *sternum* also lies subcutaneously and is palpable throughout its length. The *jugular notch* in the manubrium is easily felt between the prominent medial ends of the clavicles. The *sternal angle* at the manubriosternal joint is palpable and often visible because this joint between the manubrium and body moves during respiration. The sternal angle, an important landmark, lies at the level of the second pair of costal cartilages. To count the ribs and *intercostal spaces*, slide a digit laterally from the sternal angle onto the second costal cartilage and count the ribs and spaces by moving inferolaterally. The first intercostal space is inferior to the 1st rib; likewise the other spaces are inferior to the similarly numbered ribs. The *xiphoid process* lies in a slight depression where the converging costal margins form the infrasternal angle. This angle is used in *cardiopulmonary resuscitation* for locating the proper hand position on the body of the sternum. The costal margins are palpable with ease, because they extend inferolaterally from the xiphisternal joint. The superior part of the costal margin is formed by the seventh costal cartilage, and its inferior part is formed by the eighth to tenth costal cartilages (Fig. 2.1).

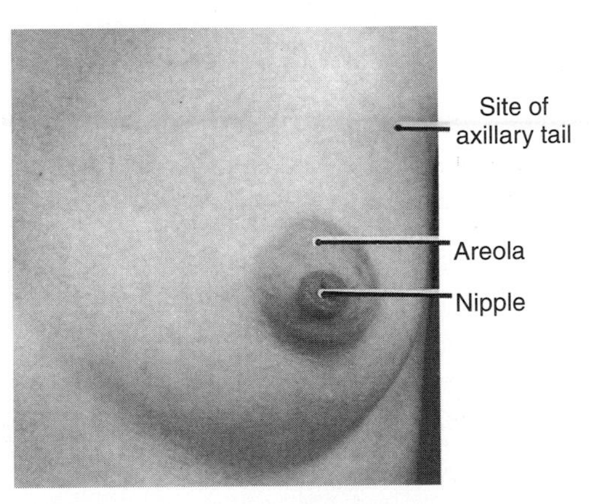

Breasts are prominent features of the anterior thoracic wall. Their flattened superior surfaces show no sharp demarcation from the anterior surface of the wall, but laterally and inferiorly their borders are well defined. The nipple is surrounded by a slightly raised and circular pigmented area called the areola. Its color varies with the woman's complexion, but it darkens during pregnancy and retains that color thereafter.

In men the nipple lies anterior to the fourth intercostal space, about 10 cm from the median plane. In women the position of the nipple is inconstant because the shape of the breast varies markedly among individuals and in the same individual at different ages.

JOINTS AND MOVEMENTS OF THORACIC WALL

Joints of the thoracic wall (Table 2.1) occur between the

- Vertebrae
- Ribs and vertebrae
- Ribs and costal cartilages
- Sternum and costal cartilages
- Parts of the sternum

Table 2.1
Joints of Thoracic Wall

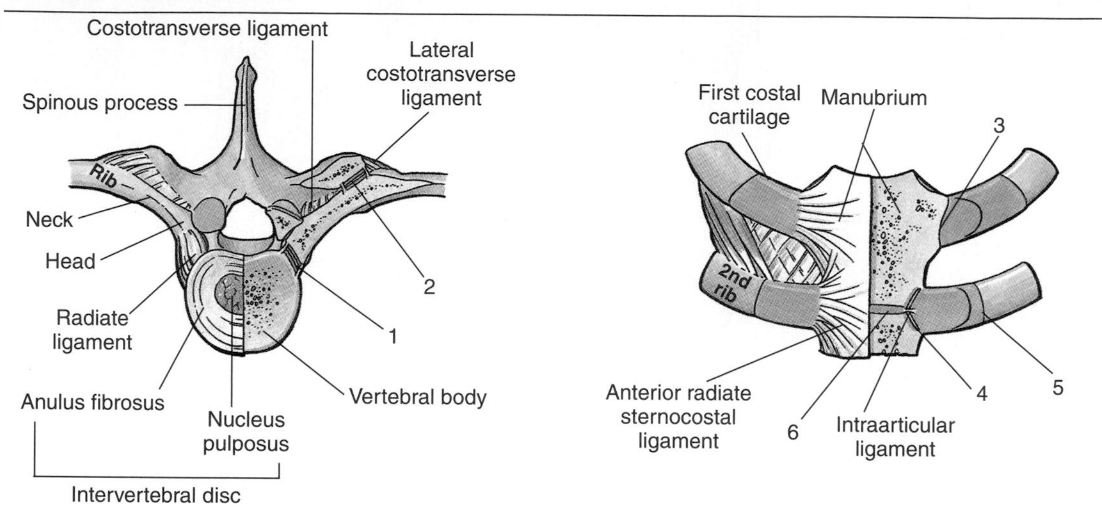

Joint	Type	Articulations	Ligaments	Comments
Costovertebral Joints of heads of ribs (1)	Synovial plane joint	Head of each rib articulates with superior demifacet of corresponding vertebral body and inferior demifacet of vertebral body superior to it	Radiate and intraarticular	Heads of lst, 11th, and 12th ribs (sometimes 10th) articulate only with corresponding vertebral body
Costotransverse (2)	Synovial plane joint	Articulation of tubercle of rib with transverse process of corresponding vertebra	Lateral and superior costotransverse	11th and 12th ribs do not articulate with transverse process of corresponding vertebrae
Sternocostal (3 and 4)	First: primary cartilaginous joint	Articulation of first costal cartilages with manubrium of sternum		
	Second to seventh: synovial plane joints	Articulation of second to seventh pairs of costal cartilages with sternum	Anterior and posterior radiate sternocostal	
Costochondral (5)	Primary cartilaginous joint	Articulation of lateral end of costal cartilage with sternal end of rib	Cartilage and bone are bound together by periosteum	No movement normally occurs at this joint
Interchondral (Fig. 2.1)	Synovial plane joint	Articulation between costal cartilages of 6th–7th, 7th–8th, and 8th–9th ribs	Interchondral ligaments	Articulation between costal cartilages of 9th and 10th ribs is fibrous
Manubriosternal (6)	Secondary cartilaginous joint (symphysis)	Articulation between manubrium and body of sternum		This joint often fuses and becomes a synostosis in older persons
Xiphisternal (Fig. 2.1)	Primary cartilaginous joint (synchondrosis)	Articulation between xiphoid process and body of sternum		This joint often becomes a synostosis in older persons

A *rib dislocation* refers to displacement of a costal cartilage from the sternum; i.e., the sternocostal joint is dislocated. This causes severe pain, particularly during deep respiratory movements. The injury produces a lumplike deformity at the dislocation site. Rib dislocations are common in body contact sports, and possible complications are pressure on or damage to nearby nerves, vessels, and muscles.

A *rib separation* refers to dislocation of a costochondral junction, i.e., between the rib and its costal cartilage. In separations of the 3rd to 10th ribs, tearing of the perichondrium and periosteum usually occurs.

During inspiration movements of the thoracic wall and diaphragm result in an increase in the vertical, transverse, and anteroposterior diameters of the thorax and in the *intrathoracic volume.* The changes in pressure result in air being alternately drawn into the lungs (inspiration) through the nose, mouth, larynx, and trachea and expelled from the lungs (expiration) through the same passages. *During expiration* the diaphragm, intercostal, and other muscles relax; hence the intrathoracic volume is decreased and the intrathoracic pressure is increased. The stretched elastic tissue of the lungs recoils, and much of the air is expelled. Concurrently the intraabdominal pressure is decreased.

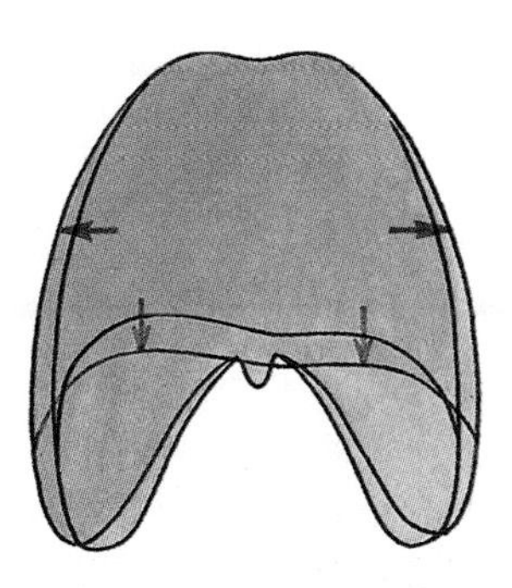

Increased vertical and transverse diameters

The *vertical diameter of the thorax* is increased during inspiration (*green arrows*) as the diaphragm is lowered by contraction. During expiration the vertical diameter is returned to normal by the subatmospheric pressure produced in the pleural cavities by the elastic recoil of the lungs. As a result, the domes of the diaphragm move superiorly, diminishing the vertical diameter. The *transverse diameter of the thorax* is increased slightly (*red arrows*) when the intercostal muscles contract, raising the ribs. As the ribs ascend, they move superolaterally like bucket handles.

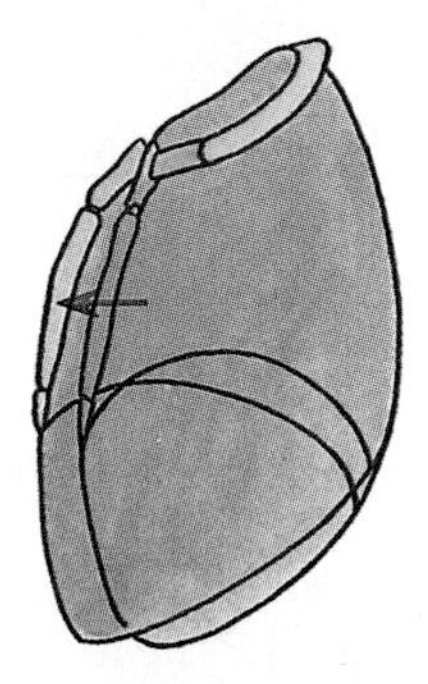

Increased anteroposterior diameter

The *anteroposterior diameter of the thorax* is increased (*blue arrows*) when the intercostal muscles contract.

Increased anteroposterior and transverse diameters

Movement of the ribs (primarily 2nd to 6th) at the costovertebral joints causes the sternum and the anterior ends of the ribs to rise like pump handles. Because the ribs slope inferiorly, their elevation results in movement of the sternum at the manubriosternal joint and an increase in the anteroposterior diameter of the thorax.

Paralysis of half of the diaphragm because of injury to its motor supply from the *phrenic nerve* does not affect the other half because each dome has a separate nerve supply. One can detect paralysis of the diaphragm radiographically by noting its paradoxical movement. Instead of descending on inspiration, the paralyzed dome is pushed superiorly by the abdominal viscera.

MUSCLES, NERVES, AND VASCULATURE OF THORACIC WALL

Muscles

Several muscles are attached to the ribs (e.g., anterolateral muscles of the abdomen and some back and neck muscles). The pectoral muscles covering the anterior thoracic wall usually act on the upper limbs, but the *pectoralis major* (Fig. 2.3) also functions as an accessory muscle of respiration and helps to expand the thoracic cavity when inspiration is deep and forceful (e.g., after a 100-m dash). The *scalene muscles* passing from the neck to the 1st or 2nd ribs also function as accessory respiratory muscles by elevating the 1st and 2nd ribs during forced inspiration. The serratus posterior, levator costarum, intercostal, subcostal, and transversus thoracis are muscles of the thorax proper (Table 2.2).

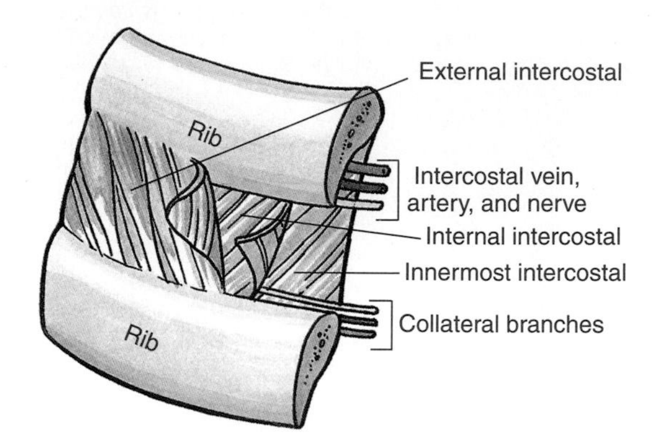

Intercostal space

Typical intercostal spaces contain three layers of *intercostal muscles*. The superficial layer is the external intercostal muscle, the middle layer is the internal intercostal muscle, and the

Table 2.2
Muscles of Thoracic Wall

Muscles	Superior Attachment	Inferior Attachment	Innervation	Action[a]
External intercostal	Inferior border of ribs	Superior border of ribs below	Intercostal n.	Elevate ribs
Internal intercostal	Inferior border of ribs	Superior border of ribs below	Intercostal n.	Depress ribs
Innermost intercostal	Inferior border of ribs	Superior border of ribs below	Intercostal n.	Probably elevate ribs
Transversus thoracis	Posterior surface of lower sternum	Internal surface of costal cartilages 2–6	Intercostal n.	Depress ribs
Subcostal	Internal surface of lower ribs near their angles	Superior borders of 2nd or 3rd ribs below	Intercostal n.	Elevate ribs
Levator costarum	Transverse processes of T7–T11	Subjacent ribs between tubercle and angle	Dorsal primary rami of C8–T11 nn.	Elevate ribs
Serratus posterior superior	Ligamentum nuchae, spinous processes of C7 to T3 vertebrae	Superior borders of 2nd to 4th ribs	Second to fifth intercostal nn.	Elevate ribs
Serratus posterior inferior	Spinous processes of T11 to L2 vertebrae	Inferior borders of of 8th to 12th ribs near their angles	Ventral rami of ninth to twelfth thoracic spinal nn.	Depress ribs

[a] All intercostal muscles keep intercostal spaces rigid, thereby preventing them from bulging out during expiration and from being drawn in during inspiration. Role of individual intercostal muscles and accessory muscles of respiration in moving the ribs is difficult to interpret despite many electromyographic studies.

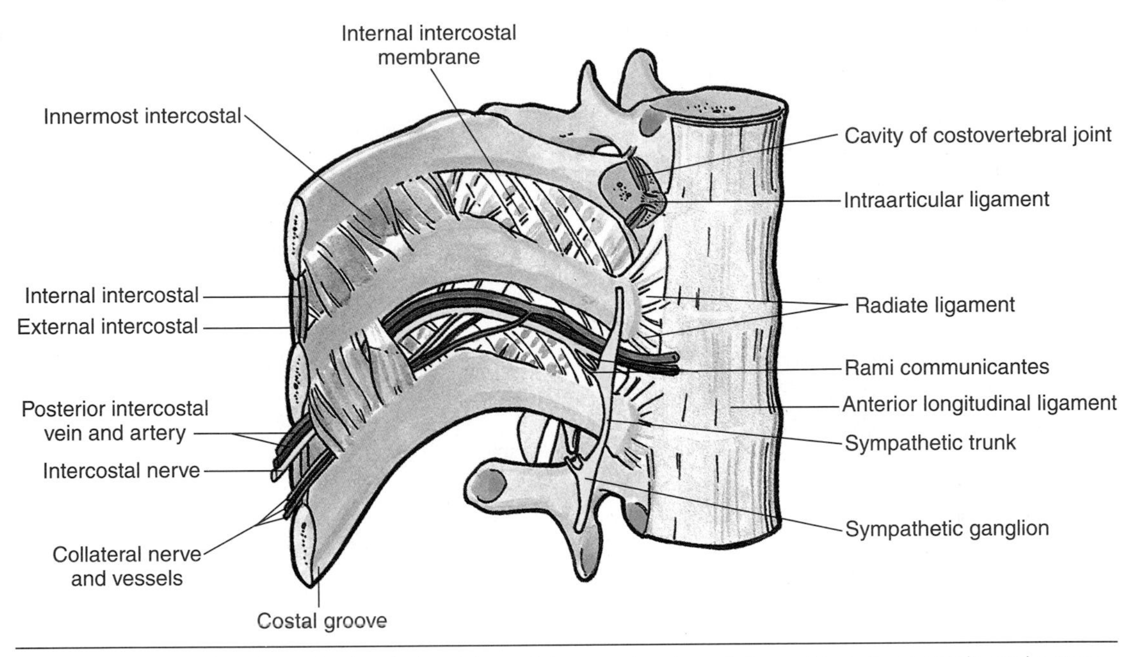

Figure 2.6. Dissection of vertebral end of an intercostal space showing relations of muscles, vessels, and nerves.

deepest layer is the innermost intercostal muscle. Anteriorly, the external intercostal muscles are replaced by *external intercostal membranes* (Fig. 2.5), and posteriorly the internal intercostal muscles are replaced by *internal intercostal membranes.*

Nerves

There are 12 pairs of thoracic spinal nerves; as soon as they pass through the intervertebral foramina, they divide into ventral and dorsal primary rami. The ventral rami of T1–T11 form the *intercostal nerves* that enter the intercostal spaces. The ventral ramus of T12, located inferior to the 12th rib, is called the *subcostal nerve.* The dorsal rami pass posteriorly, immediately lateral to the articular processes of the vertebrae, to supply the muscles, bones, joints, and skin of the back.

Typical intercostal nerves (third to sixth) enter the intercostal spaces posteriorly, between the parietal pleura (the serous membrane lining the thoracic cavity) and the internal intercostal membrane. At first they run across the internal surface of the internal intercostal membrane and muscle near the middle of the intercostal space. Near the angle of the rib, the nerves pass between the internal intercostal and innermost intercostal muscles. Here the nerves enter and are sheltered by the *costal groove,* where they lie just inferior to the intercostal arteries (Fig. 2.6). Collateral branches of these nerves arise near the angles of the ribs and supply the intercostal muscles. The nerves continue anteriorly between the internal and innermost intercostal muscles, giving branches to these and other muscles and forming a lateral cutaneous branch. Anteriorly the nerves appear on the internal surface of the internal intercostal muscle. Near the sternum the nerves turn anteriorly and end as anterior cutaneous branches (Fig. 2.5).

Through the cutaneous branches of the dorsal and ventral rami, each spinal nerve supplies a striplike area of skin extending from the posterior to the anterior median line. These areas are called dermatomes. A *dermatome* is the area of skin supplied by the sensory fibers of a single dorsal root through the dorsal and ventral rami of its spinal nerve.

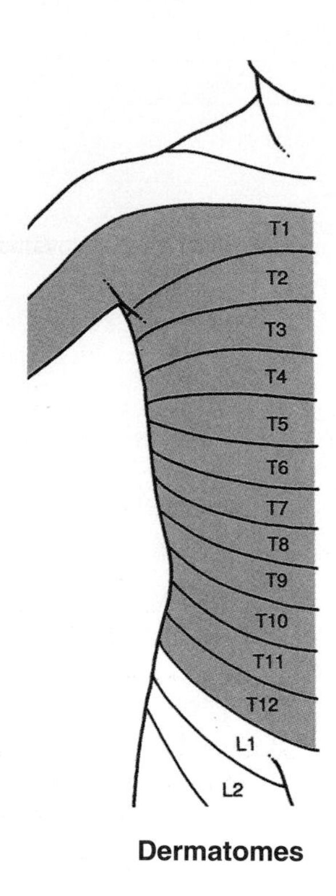

Dermatomes

The dermatomes are arranged in a segmental fashion because the thoracoabdominal nerves arise from segments of the spinal cord. There is considerable overlapping of closely related dermatomes (e.g., C4, C5, and T1). Physicians need a working knowledge of the segmental, or dermatomal, innervation of the skin so they can determine (e.g., with a pin) whether a particular segment of the spinal cord is functioning normally. The group of muscles supplied by a pair of intercostal nerves is known as a *myotome*. Muscular branches of a typical intercostal nerve also supply the subcostal, transversus thoracis, levator costarum, and serratus posterior muscles (Table 2.2).

Rami communicantes connect each intercostal nerve to a sympathetic trunk (Fig. 2.6). Preganglionic fibers leave each nerve as a white ramus and pass to a ganglion of the sympathetic trunk, and postganglionic fibers leave the ganglion as a gray ramus to rejoin the nerve. *Sympathetic nerve fibers* are distributed through all branches of the intercostal nerve to blood vessels, sweat glands, and smooth muscle (Fig. 1.15).

Herpes zoster (*shingles*) is a viral disease of the spinal ganglia. The term means a "creeping, girdle-shaped cutaneous eruption." When the virus invades the ganglia, a sharp burning pain is produced in the dermatomes supplied by the nerves involved. A few days later the involved dermatomes become red, and vesicular eruptions appear.

Sometimes it is necessary to insert a hypodermic needle through an intercostal space to obtain a sample of pleural fluid or to remove blood or pus from the pleural cavity, a procedure called *thoracocentesis*. In other cases it may be necessary to anesthetize an intercostal nerve. To avoid damage to the intercostal nerve and vessels, the needle is inserted superior to the rib, high enough to avoid the collateral branches.

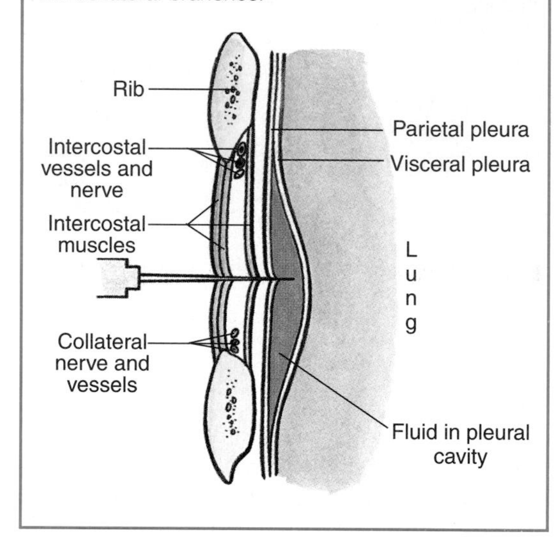

Vasculature

Arterial supply of the thoracic wall (Figs. 2.3–2.6, Table 2.3) is derived from the

- Subclavian artery via the internal thoracic and highest intercostal arteries
- Axillary artery
- Aorta by way of the posterior intercostal and subcostal arteries

Veins accompany the intercostal arteries and nerves and lie deepest (most superior) in the costal grooves. There are 11 *posterior intercostal veins* and one *subcostal vein* on each side. The posterior intercostal veins anastomose with the *anterior intercostal veins* that are tributaries of the internal thoracic veins. Most intercostal veins end in the *azygos vein* (see Fig. 2.11*B*) that conveys venous blood to the superior vena cava.

Table 2.3
Arterial Supply to Thoracic Wall

Artery	Origin	Course	Distribution
Posterior intercostals	Superior intercostal artery (intercostal spaces 1 and 2) and thoracic aorta (remaining intercostal spaces)	Pass between internal and innermost intercostal muscles	Intercostal muscles and overlying skin, parietal pleura
Anterior intercostals	Internal thoracic (intercostal spaces 1–6) and musculophrenic arteries (intercostal spaces 7–9)		
Internal thoracic	Subclavian artery	Passes inferiorly and lateral to sternum between costal cartilages and internal intercostal muscles to divide into superior epigastric and musculophrenic arteries	By way of anterior intercostal arteries to intercostal spaces 1–6
Subcostal	Thoracic aorta	Courses along inferior border of 12th rib	Muscles of anterolateral abdominal wall

Thoracic Cavity

The thoracic cavity is divided into two lateral compartments that contain the pleurae and lungs and a central compartment called the mediastinum that contains all other thoracic structures (Fig. 2.7*A*).

PLEURAE AND LUNGS

Pleurae

Each lung is surrounded by a *pleural sac* that consists of two serous membranes called pleurae: the *parietal pleura* lines the thoracic wall, and the *visceral pleura* (pulmonary pleura) invests the lung, including surfaces within its fissures .

The *pleural cavities* are potential spaces between the layers of pleura and contain a capillary layer of serous *pleural fluid* that lubricates the pleural surfaces and allows the layers of pleura to slide smoothly over each other during respiration.

The *parietal pleura* is adherent to the thoracic wall, mediastinum, and diaphragm. Parts of the parietal pleura include

- Costal pleura covers the internal surface of the thoracic wall (sternum, costal cartilages, ribs, intercostal muscles, intercostal membranes, and sides of thoracic vertebrae)
- Mediastinal pleura covers the mediastinum
- Diaphragmatic pleura covers the thoracic surface of the diaphragm
- Cervical pleura (pleural cupula) extends about 3 cm into the neck, and its apex forms a cup-shaped dome over the apex of the lung

The relatively abrupt lines along which the parietal pleura changes direction from one wall of the pleural cavity to another are known as *pleural reflections*. They occur where the costal pleura becomes continuous with the mediastinal pleura anteriorly and posteriorly and with the diaphragmatic pleura inferiorly. At the root of the lung the visceral and parietal layers of pleura are continuous; a double layer of parietal pleura called the *pulmonary ligament* (p. 48) hangs inferiorly from this region.

The lungs do not completely occupy the pleural cavities during expiration; thus, the diaphragmatic pleura is in contact with the costal pleura, and the potential pleural cavity here is called the *costodiaphragmatic recess* (Fig. 2.7*B*). A similar but smaller pleural recess is located posterior to the sternum where the costal pleura is in contact with the mediastinal pleura. The potential pleural cavity here is called the *costomediastinal recess*; the left recess is larger because of the cardiac notch in the left

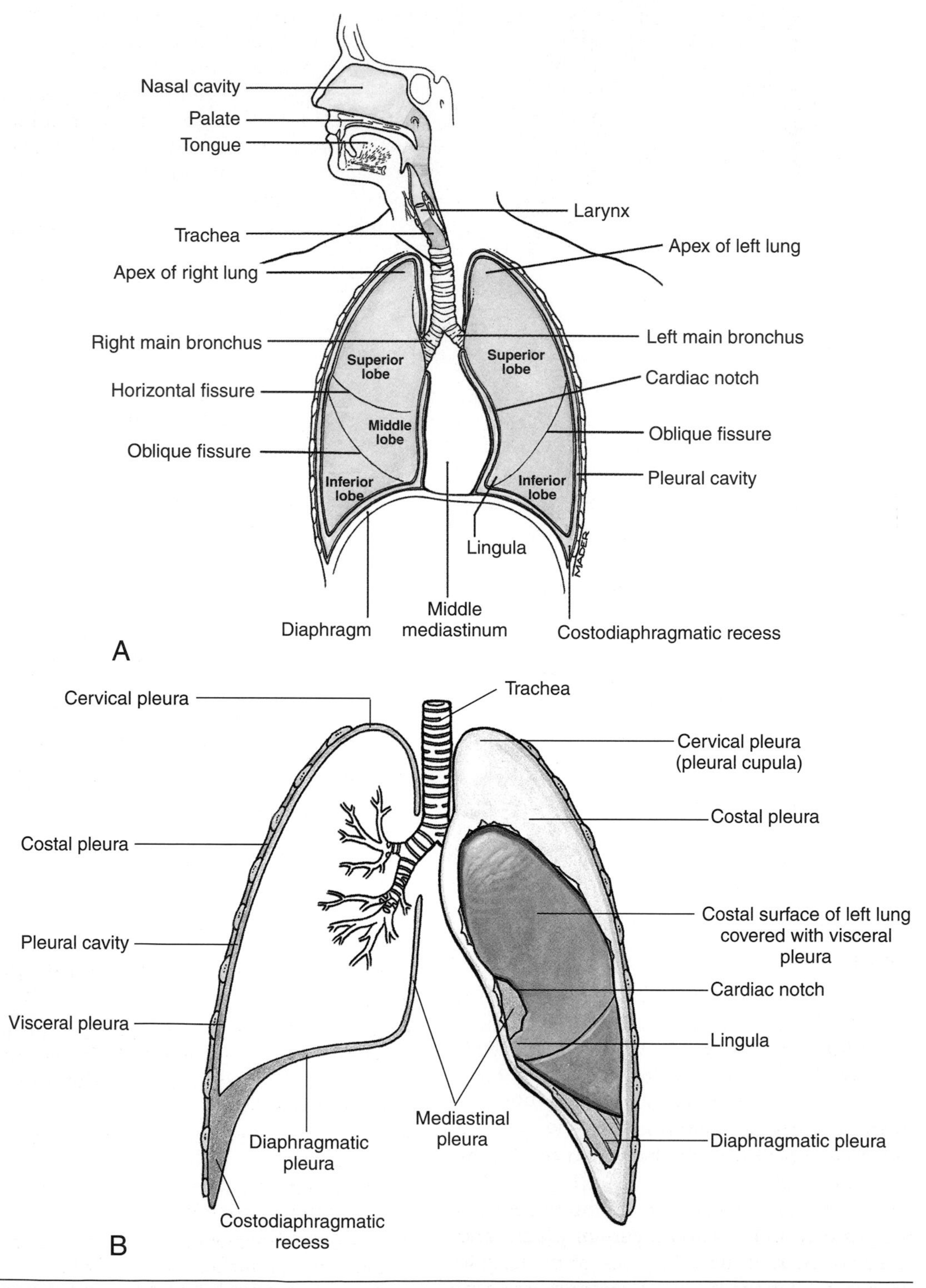

Figure 2.7. **A.** Overview of respiratory system showing relationship of lungs to upper respiratory organs. **B.** Pleurae and pleural cavities.

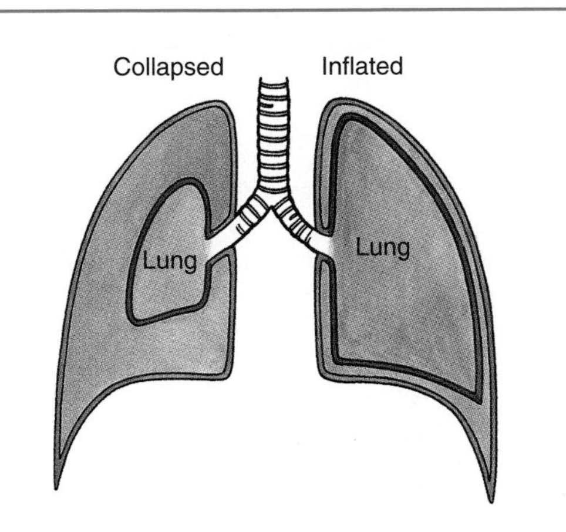

When a lung collapses, the pleural cavity, a potential space (*purple*), becomes an actual space. The pleural cavity is between the parietal pleura (*blue*) and the visceral pleura (*red*). One lung may be collapsed (e.g., after surgery) without collapsing the other because the pleural sacs are separate.

Pneumothorax (entry of air into the pleural cavity) caused by a penetrating wound of the parietal pleura or rupture of a lung results in partial collapse of the lung. Fractured ribs may also produce a pneumothorax. The accumulation of a significant amount of fluid in the pleural cavity (hydrothorax) may result from various causes. If there is a chest wound, blood may also enter the pleural cavity (*hemothorax*).

During inspiration and expiration the normally moist, smooth pleurae make no detectable sound during *auscultation* (listening to breath sounds); however, inflammation of the pleurae called *pleuritis* (pleurisy) makes the lung surfaces rough. The resulting friction (*pleural rub*) may be heard with a stethoscope.

lung. The lungs move into the pleural recesses (not completely filling them) during deep inspiration and out of them during expiration.

Lungs

The lungs are normally light, soft, and spongy. They are also elastic and recoil to about one-third their size when the thoracic cavity is opened. The right and left lungs are separated from each other by the heart and great vessels in the middle mediastinum (Fig. 2.7*A*). The lungs are attached to the heart and trachea by structures in the roots of the lung (Fig. 2.8). The *root of the lung* is the area of continuity between the parietal and visceral layers of pleura and connects the mediastinal surface of the lung to the heart and trachea. The *hilum of the lung* contains the main bronchus, pulmonary vessels, bronchial vessels, lymph vessels, and nerves entering and leaving the lung.

The horizontal and oblique fissures of the visceral pleura divide the lungs into lobes (Table 2.4). Each lung has an apex, three surfaces (costal, mediastinal, and diaphragmatic), and three borders (anterior, inferior, and posterior). The *apex* is the blunt superior end of the lung that is covered by cervical pleurae. The apex and pleurae project superiorly (2–3 cm) through the superior thoracic aperture into the root of the neck (p. 49). Consequently they may be injured by wounds of the neck, producing pneumothorax.

Surfaces of Lung. Each lung has the following surfaces:

- Costal surface, adjacent to the sternum, costal cartilages, and ribs
- Mediastinal surface, related medially to mediastinum and posteriorly to sides of vertebrae
- Diaphragmatic surface (base) rests on convex dome of diaphragm; concavity is deeper in the right lung because of the higher position of the right dome

Auscultation of the lungs (listening to their sounds with a stethoscope) and *percussion of the lungs* (tapping the chest over the lungs) always include the root of the neck to detect sounds in the apices of the lungs. When physicians refer to the base of a lung, they are usually not referring to its diaphragmatic surface (base). They mean the inferior part of the posterior costal surface of the inferior lobe. To auscultate this area, physicians apply a stethoscope to the inferoposterior aspect of the thoracic wall at the level of T10 vertebra.

Borders of Lung. Each lung has the following borders:

- Anterior border is where costal and mediastinal surfaces meet anteriorly and overlap the heart; the *cardiac notch* indents this border of the left lung
- Inferior border circumscribes diaphragmatic surface of lung and separates the diaphragmatic surface from the costal and mediastinal surfaces
- Posterior border is where the costal and mediastinal surfaces meet posteriorly; it is broad and rounded and lies in the cavity at the side of the vertebral column

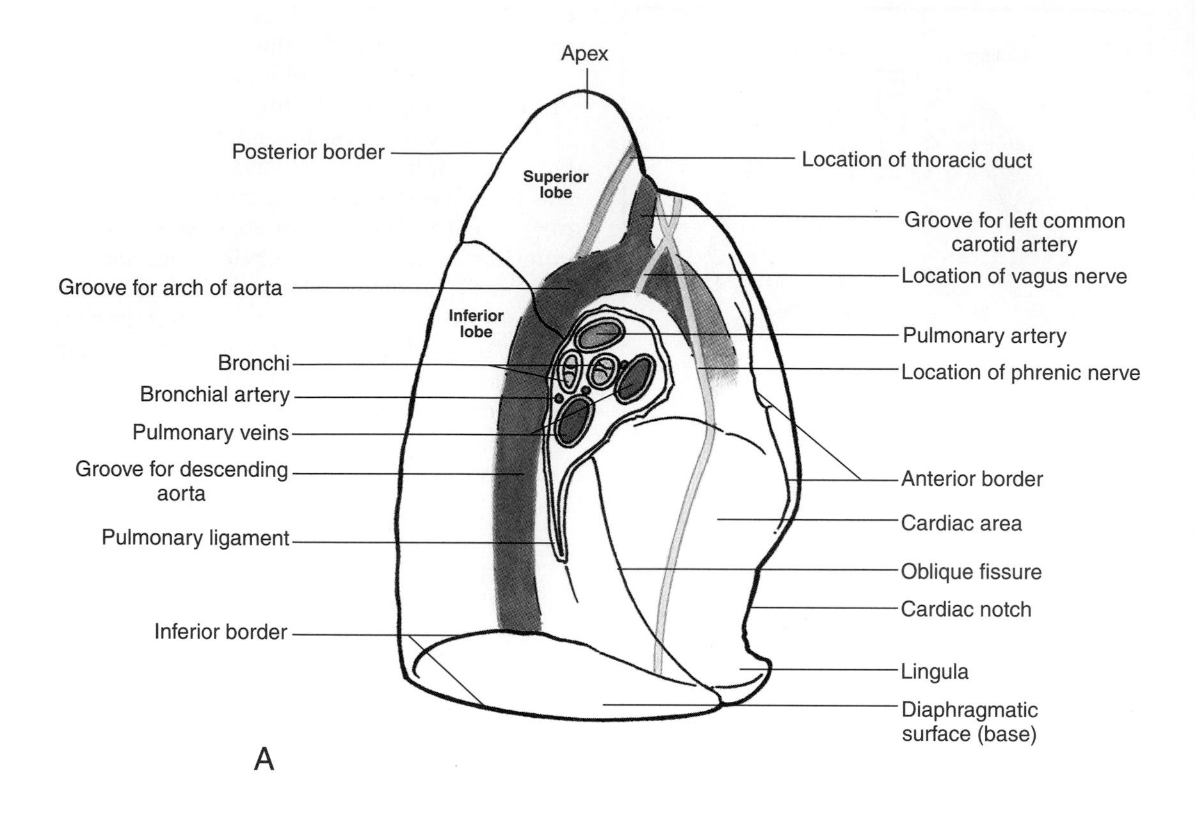

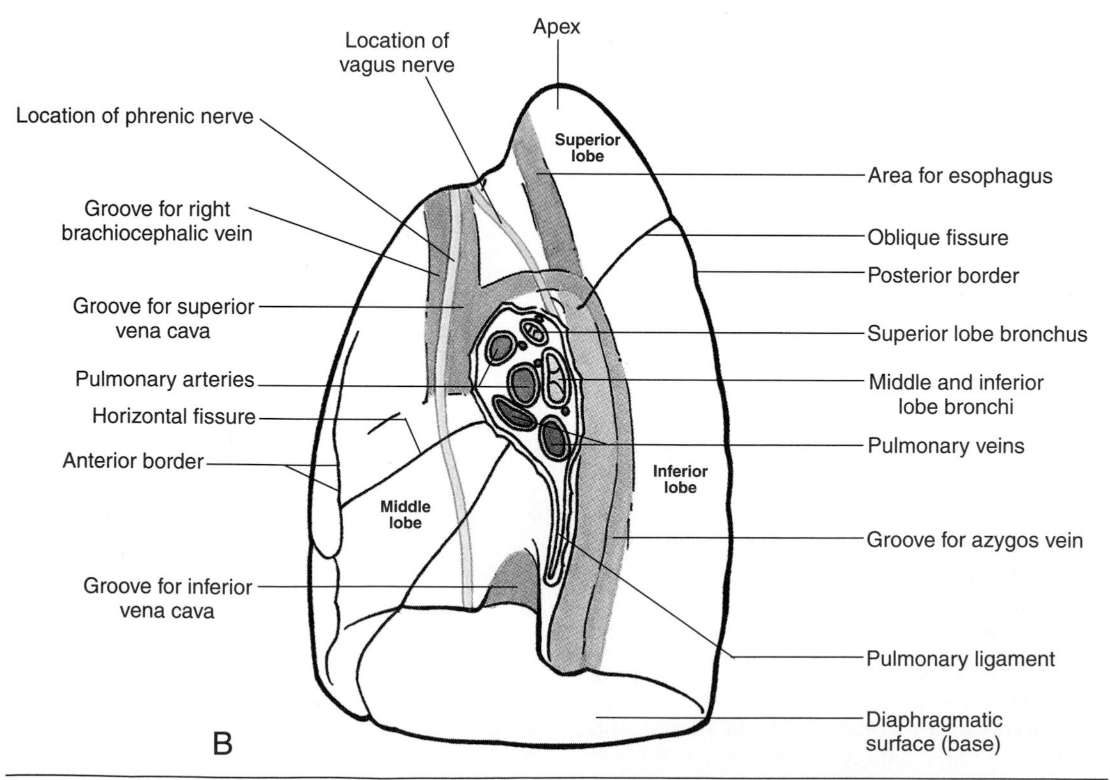

Figure 2.8. Mediastinal surfaces of lungs. Lungs have impressions of structures in contact with them (e.g., aorta, superior vena cava). **A.** Left lung. **B.** Right lung.

Table 2.4
Lobes and Fissures of Lungs

Right lung (three lobes)		Left lung (two lobes)
Superior (upper) lobe		Superior (upper) lobe
separated by horizontal fissure from the		separated by oblique fissure from the
Middle lobe		Inferior (lower) lobe
separated by oblique fissure from the		
Inferior (lower) lobe		

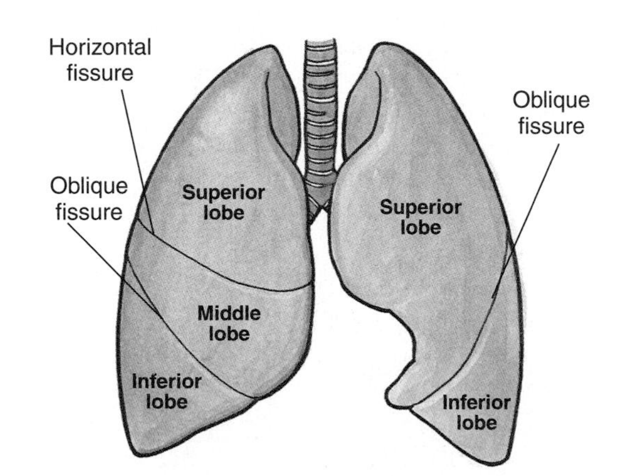

Surface Anatomy of Pleurae and Lungs

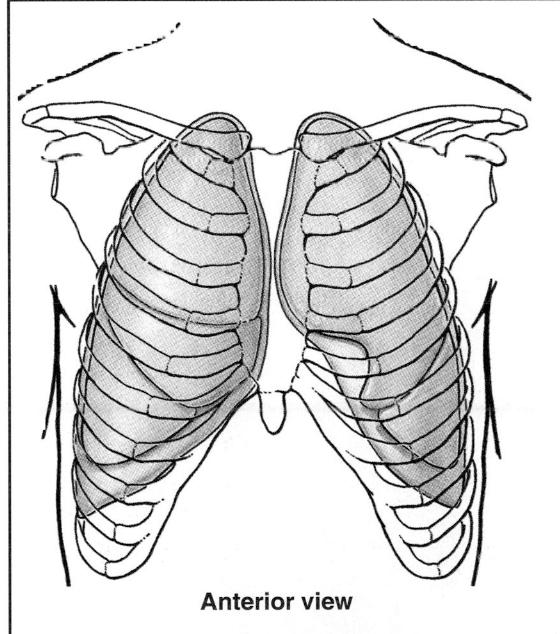
Anterior view

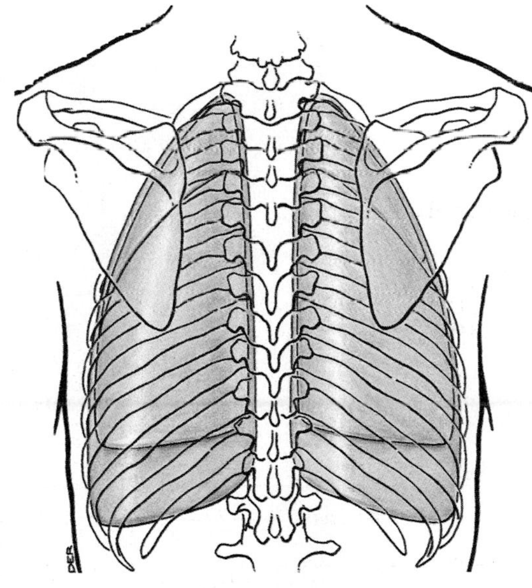

Posterior view

The cervical pleurae and apices of the lungs rise to the level of the neck of the 1st rib and pass through the superior thoracic aperture.

The lungs lie adjacent to the parietal pleura as far distally as the fourth costal cartilages. Here the margin of the left pleural reflection moves laterally and then inferiorly at the cardiac notch to reach the sixth costal cartilage. On the right side, the pleural reflection continues inferiorly from the fourth to the sixth costal cartilage. At the sixth costal cartilages both pleural reflections pass laterally and reach the midclavicular line at the level of the eighth costal cartilage, the 10th rib at the midaxillary line, and the 12th rib at the midscapular line. They then proceed toward the spinous process of T12 vertebra. Generally the parietal pleura extends about two ribs inferior to the lung.

The *oblique fissure* of the right and left lungs extends from the level of the spinous process of the T2 vertebra posteriorly to the sixth costal cartilage anteriorly. The *horizontal fissure* of the right lung extends from the oblique fissure along the 4th rib and costal cartilage anteriorly.

Bronchi. The right and left main bronchi pass inferolaterally from the bifurcation of the trachea to the lungs (Figs. 2.9 and 2.10). The bronchi are supported by C-shaped rings of cartilage.

- Right main bronchus is wider, shorter, and more vertical than left main bronchus and passes directly to the root of the lung
- Left main bronchus passes inferolaterally, inferior to the arch of the aorta and anterior to the esophagus and thoracic aorta, to reach the root of the lung

Because the right main bronchus is wider and shorter and runs more vertically than the left main bronchus, foreign bodies and aspirated material are more likely to enter and lodge in it or one of its branches than in the left bronchus or one of its branches.

The main bronchi accompany the pulmonary arteries into the hila of the lungs and branch in a constant fashion within the lungs to form the bronchial tree (Fig. 2.9). Each main bronchus divides into *secondary bronchi* (*lobar bronchi*), two on the left and three on the right, each of which supplies a lobe of the lung. Each secondary bronchus divides into *tertiary (segmental) bronchi* that supply the bronchopulmonary segments (Fig. 2.10). Each *bronchopulmonary segment* is pyramidal, with its apex facing the root of the lung and its base at the pleural surface. Each segment is named according to the segmental bronchus that supplies it.

A working knowledge of the anatomy of the bronchopulmonary segments is essential for the interpretation of radiographs and other diagnostic images of the thorax. Knowledge of these segments is also essential for surgical resection of diseased segments. Bronchial and pulmonary disorders (e.g., a tumor or abscess) often localize in a bronchopulmonary segment, which may be surgically resected. During the treatment of *lung cancer*, the surgeon may remove a whole lung (*pneumonectomy*), a lobe (*lobectomy*), or one or more bronchopulmonary segments (*segmentectomy*).

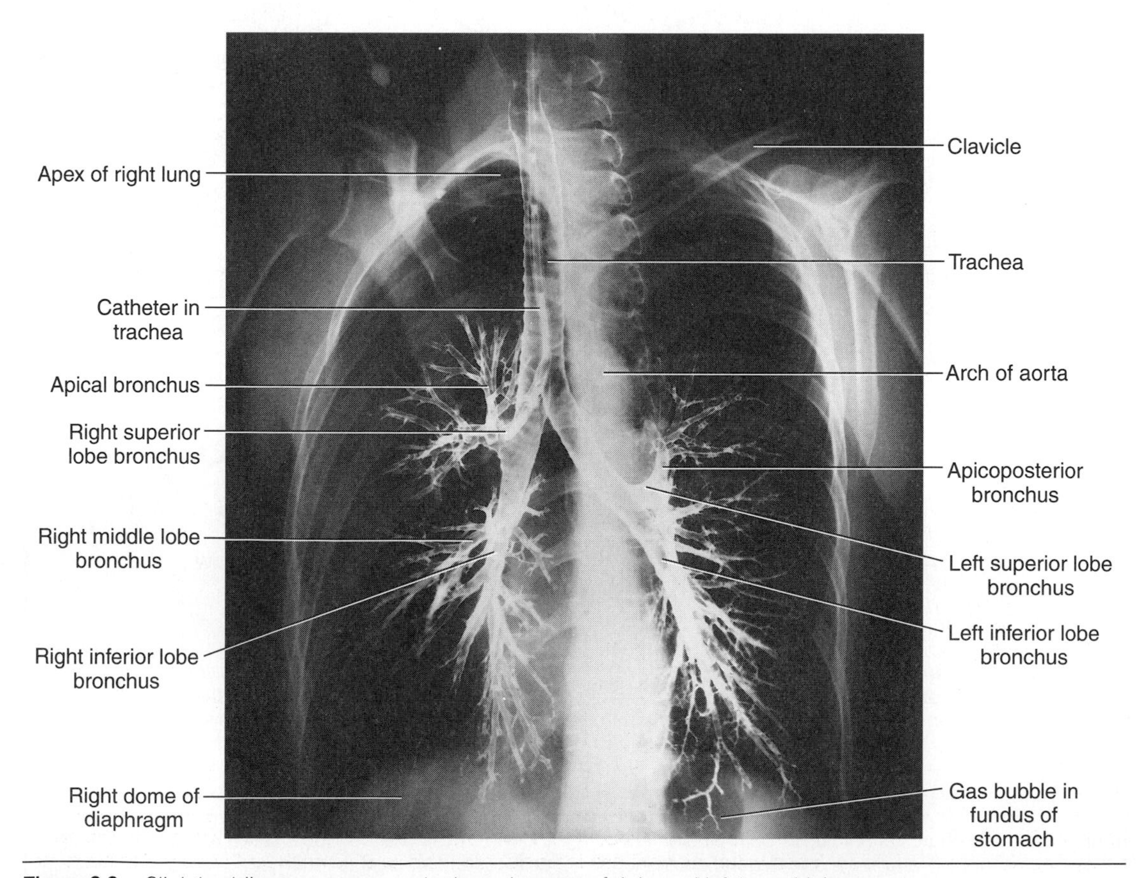

Figure 2.9. Slightly oblique, posteroanterior bronchogram of right and left bronchial tree.

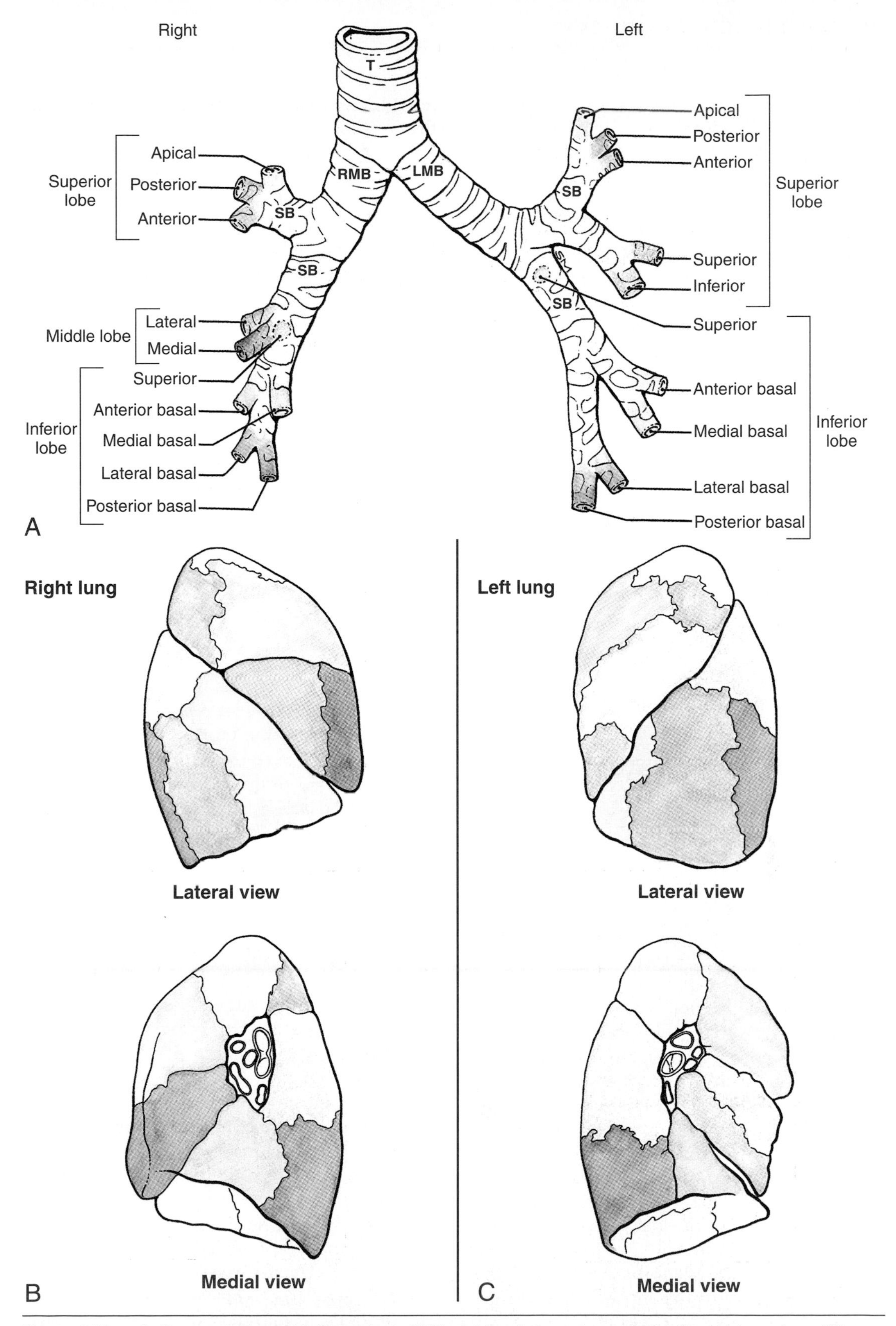

Figure 2.10. **A.** Segmental bronchi. *T*, trachea; *RMB*, right main bronchus; *LMB*, left main bronchus; *SB*, secondary (lobar) bronchi. Tertiary (segmental) bronchi are indicated by *colors*. **B.** Bronchopulmonary segments of right lung. **C.** Bronchopulmonary segments of left lung.

Vessels and Nerves of Lungs and Pleurae

Each lung has a large pulmonary artery supplying blood to it and two pulmonary veins draining blood from it. The right and left *pulmonary arteries* arise from the pulmonary trunk at the level of the sternal angle and carry poorly oxygenated blood to the lungs for oxygenation (Fig. 2.8). The pulmonary arteries pass to the corresponding root of the lung and give off a branch to the superior lobe before entering the hilum. Within the lung each artery descends posterolateral to the main bronchus and divides into lobar and tertiary (segmental) arteries. Hence, there is an arterial branch to each lobe and bronchopulmonary segment of the lung.

Pulmonary thromboembolism is a common cause of morbidity (sickness) and mortality (death). An *embolus* (plug) in a pulmonary artery forms when a *thrombus* (blood clot), fat globule, or air bubble is carried in the blood to the lungs (e.g., from a leg vein after a tibial fracture). The thrombus passes through the right side of the heart and is carried to a lung via the pulmonary artery. The thrombus may block a pulmonary artery (pulmonary thromboembolism) or one of its branches. The immediate result is partial or complete obstruction of blood flow to the lung. The obstruction results in a sector of lung that is ventilated but not perfused with blood. When a large embolus occludes a pulmonary artery, the patient suffers *acute respiratory distress* because of a major decrease in the oxygenation of blood. The patient may die in a few minutes. A medium-sized embolus may block an artery supplying a bronchopulmonary segment, producing a *thrombotic infarct* that seriously compromises the pulmonary air.

The *bronchial arteries* supply blood for the nutrition of the lungs and visceral pleura (Fig. 2.11*A*). They usually arise from the thoracic aorta, but the right artery may arise from a superior posterior intercostal artery. The small bronchial arteries pass along the posterior aspects of the bronchi and supply them and their branches as far distally as the respiratory bronchioles. The arteries anastomose with branches of the pulmonary arteries in the walls of the small bronchi and in the visceral pleura. The parietal pleura is supplied by the arteries that supply the thoracic wall (Figs. 2.3–2.6).

The *pulmonary veins* carry well-oxygenated blood from the lungs to the left atrium of the heart. Beginning in the pulmonary capillaries, the veins unite into larger and larger vessels. A main vein drains each bronchopulmonary segment, usually on the anterior surface of the corresponding bronchus. The *bronchial veins* (Fig. 2.11*B*) drain only part of the blood supplied to the lungs by the bronchial arteries; some blood is drained by the pulmonary veins. The right bronchial vein drains into the *azygos vein*, and the left bronchial vein drains into the *accessory hemiazygos vein* or the *left superior intercostal vein*. The veins from the parietal pleura join the systemic veins in adjacent parts of the thoracic wall. The veins from the *visceral pleura* drain into the pulmonary veins.

There are two lymphatic plexuses in the lung that communicate freely (Fig. 2.11C). The *superficial lymphatic plexus* lies deep to the visceral pleura, and lymphatic vessels from it drain into the *bronchopulmonary lymph nodes* located in the hilum of the lung. From them, lymph drains to the superior and *inferior tracheobronchial lymph nodes* located superior and inferior to the bifurcation of the trachea, respectively. This plexus drains the lung and visceral pleura. The *deep lymphatic plexus* is located in the submucosa of the bronchi and in the peribronchial connective tissue. Lymphatic vessels from this plexus drain into the *pulmonary lymph nodes* located along the large branches of the main bronchi. Lymphatic vessels from these nodes follow the bronchi and pulmonary vessels to the hilum of the lung, where they drain into the *bronchopulmonary lymph nodes*. Lymphatic vessels then pass to *tracheobronchial lymph nodes* around the trachea and main bronchi.

Lymph from the superficial and deep lymphatic plexuses passes to the right and left *bronchomediastinal lymph trunks*. These trunks usually terminate on each side at the junction of the subclavian and internal jugular veins, but the right bronchomediastinal trunk may empty into the right *lymphatic duct*, and the left bronchomediastinal trunk may terminate in the *thoracic duct* (Fig. 2.11C).

Lymphatic vessels from the parietal pleura drain into the lymph nodes of the thoracic wall (intercostal, parasternal, mediastinal, and phrenic) (Fig. 2.3). A few vessels from the cervical pleura drain into the axillary lymph nodes. Lymphatic vessels from the visceral pleura drain into nodes at the hilum of the lung.

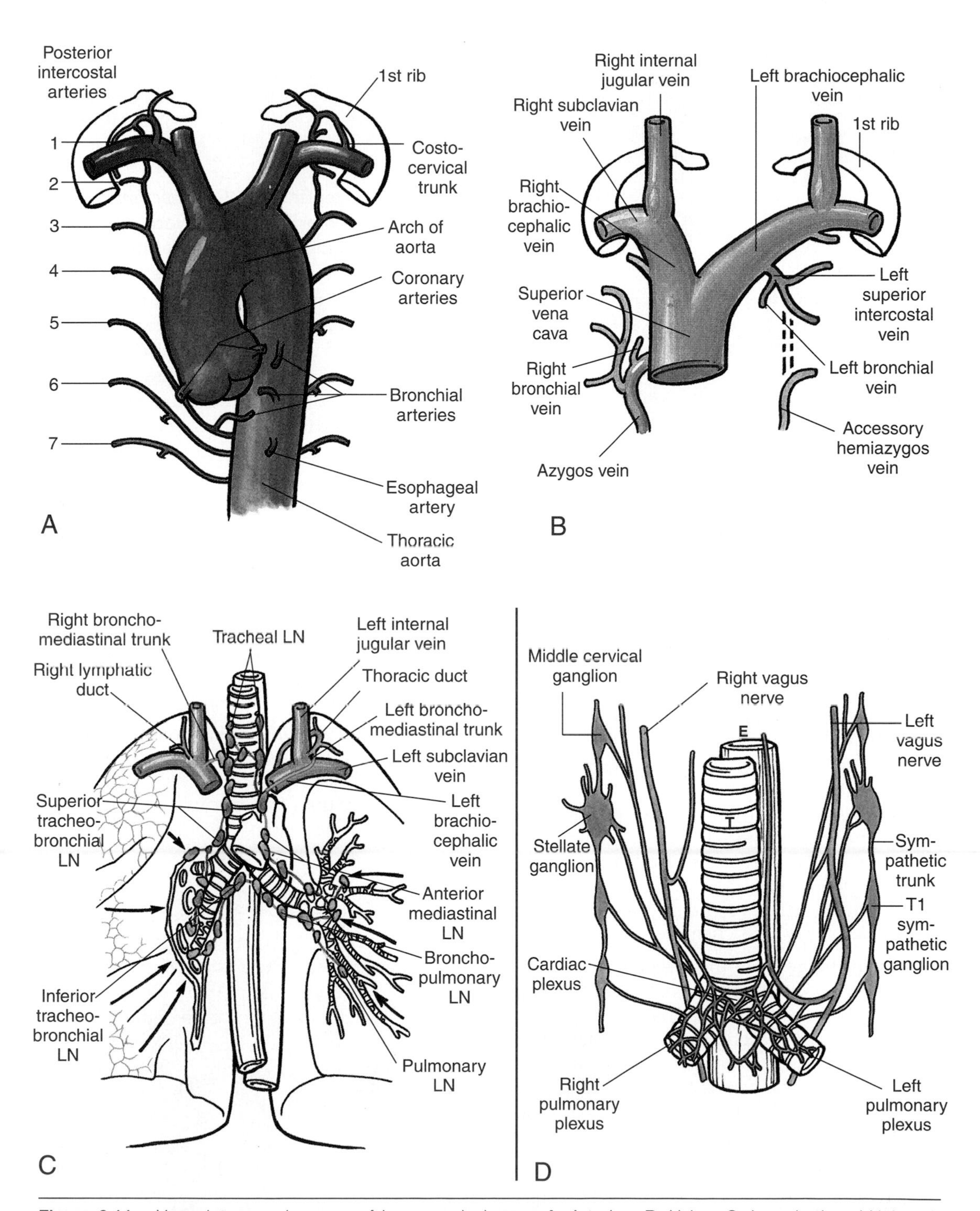

Figure 2.11. Vasculature and nerves of lungs and pleurae. **A.** Arteries. **B.** Veins. **C.** Lymphatics. *LN*, lymph node(s). *Arrows* indicate direction of lymph flow: those on right indicate flow from superficial pulmonary lymphatic plexuses; those on left indicate flow from deep pulmonary plexuses. **D.** Nerves. *E*, esophagus; *T*, trachea. *Orange*, sympathetic; *green*, parasympathetic; *blue*, plexus.

Lymph from the lungs carries *phagocytes* containing ingested carbon particles from inspired air. In many people, especially cigarette smokers and/or city dwellers, these particles color the surface of the lungs and the associated lymph nodes a mottled gray to black.

Bronchogenic carcinoma (*cancer of lung*) is a common type of cancer that is mainly caused by cigarette smoking. These tumors usually metastasize widely because of the arrangement of the lymphatics. Tumor cells may enter the systemic circulation by invading the wall of a sinusoid or venule in the lung. They then pass via the pulmonary veins, left heart, and aorta to all parts of the body, especially the skull and brain.

Nerves of the lungs and visceral pleura are derived from the *pulmonary plexuses* located anterior and posterior to the roots of the lungs (Fig. 2.11*D*). These nerve networks contain *parasympathetic fibers* from the vagus nerve (CN X) and *sympathetic fibers* from the sympathetic trunks. The parasympathetic ganglion cells are located in the pulmonary plexuses and along the branches of the bronchial tree. Sympathetic ganglion cells are located in the *paravertebral ganglia* of the sympathetic trunks. The parasympathetic fibers from the vagus (CN X) are motor to the smooth muscle of the bronchial tree (*bronchoconstrictor*), inhibitor to the pulmonary vessels (*vasodilator*), and secretor to the glands of the bronchial tree (*secretomotor*). The visceral afferent fibers of CN X are distributed to the

- Bronchial mucosa and are probably concerned with cough reflexes
- Bronchial muscles and are involved in stretch reception
- Interalveolar connective tissue and are involved in Hering-Breuer reflexes, the mechanism that tends to limit respiratory excursions
- Pulmonary arteries as pressor receptors and pulmonary veins as chemoreceptors

Afferent fibers from the visceral pleura and bronchi may accompany sympathetic fibers, mediating nociceptive responses to injurious stimuli. The sympathetic fibers are inhibitor to the bronchial muscle (*bronchodilator*), motor to the pulmonary vessels (*vasoconstrictor*), and inhibitor to the glands of the bronchial tree. Nerves of the parietal pleura are derived from the intercostal and phrenic nerves.

Irritation of the parietal pleura can produce local pain and referred pain to the areas supplied by the same segments of the spinal cord. Irritation of the costal and peripheral parts of the diaphragmatic pleura results in local pain and referred pain along the intercostal nerves to the thoracic and abdominal walls. Irritation of the mediastinal and central diaphragmatic areas of the parietal pleura results in pain that is referred to the root of the neck and over the shoulder (C3–C5 dermatomes).

MEDIASTINUM

The central portion of the thoracic cavity between the pleural sacs is called the mediastinum. It extends from the *superior thoracic aperture* to the diaphragm inferiorly and from the sternum and costal cartilages anteriorly to the bodies of the thoracic vertebrae posteriorly. The structures in the mediastinum are surrounded by connective tissue, nerves, blood and lymphatic vessels, lymph nodes, and fat. The looseness of the connective tissue and the elasticity of the lungs and parietal pleura enable the mediastinum to accommodate movement and volume changes in the thoracic cavity.

The mediastinum is divided into superior and inferior parts. The superior mediastinum (*green*) extends inferiorly from the superior thoracic aperture to the plane passing through the sternal angle and the inferior border of T4 vertebra. The inferior mediastinum between this plane and the diaphragm is further subdivided by the pericardium into anterior (*purple*), middle (*yellow*), and posterior (*blue*) parts. The middle mediastinum contains the heart and great vessels. Some structures pass vertically through the mediastinum (e.g., esophagus) and therefore lie in more than one subdivision.

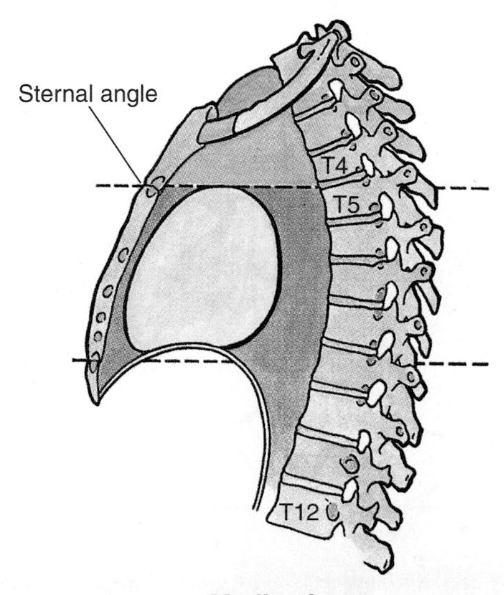

Mediastinum

Much of the mediastinum can be visualized and minor surgical procedures conducted with a *mediastinoscope*. This tubular, lighted instrument is inserted through a small incision in the neck, just superior to the manubrium. *Mediastinoscopy* may be performed to obtain mediastinal lymph nodes to determine if cancer cells have metastasized to them from a bronchogenic carcinoma. The mediastinum can also be explored and biopsies taken by removing part of a costal cartilage.

Pericardium

The pericardium is a double-walled fibroserous sac that encloses the heart and the roots of its great vessels (Table 2.5). The conical *pericardial sac* lies posterior to the body of the sternum and the second to sixth costal cartilages at the level of T5–T8 vertebrae. The pericardial sac is influenced by movements of the heart and great vessels, sternum, and diaphragm because the fibrous pericardium is

- Fused with the tunica adventitia of the great vessels entering and leaving the heart
- Attached to the posterior surface of the sternum by sternopericardial ligaments
- Fused with the central tendon of the diaphragm, both of which are pierced on the right side by the inferior vena cava (IVC) posteriorly

The inelastic *fibrous pericardium* (Fig. 2.12*B*) protects the heart against sudden overfilling. It is pierced superiorly by the aorta, pulmonary trunk, and superior vena cava (SVC). The ascending aorta carries the pericardium superiorly beyond the heart to the level of the sternal angle.

The potential space between the parietal and visceral layers of serous pericardium is called the *pericardial cavity*. It normally contains a thin film of fluid that enables the heart to move and beat in a frictionless environment. The parietal pericardium is fused to the internal surface of the fibrous pericardium. The visceral pericardium forms the *epicardium*, the external layer of the heart wall. The visceral pericardium is reflected from the heart and great vessels and becomes continuous with the parietal pericardium, where

- The aorta and pulmonary trunk leave the heart (a digit can be inserted into the *transverse pericardial sinus* located posterior to these large vessels and anterior to the SVC)
- The superior and inferior venae cavae and pulmonary veins enter the heart; these vessels

Table 2.5
Layers of Pericardium and Heart

Pericardium

External sac called fibrous pericardium
Internal sac called serous pericardium
Parietal layer
Visceral layer (becomes outermost layer of wall of heart, the epicardium)
Thin film of fluid in pericardial cavity between visceral and parietal layers of serous pericardium allows heart to move freely within pericardial sac

Heart

Wall of heart is composed of three layers; from superficial to deep,
Epicardium
Myocardium
Endocardium

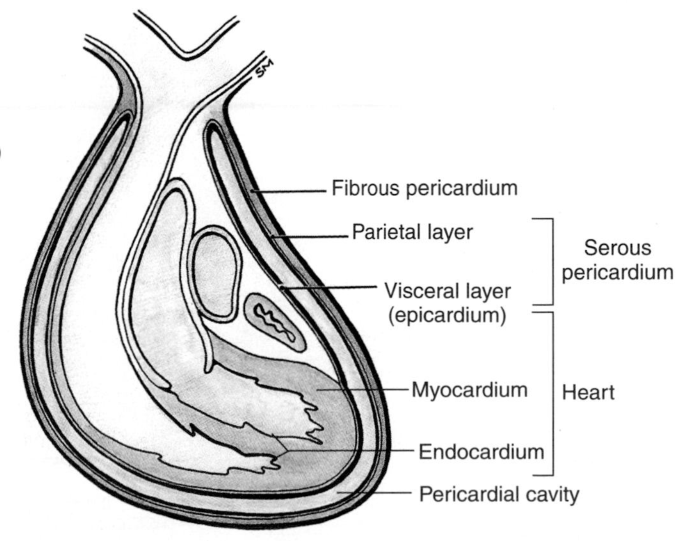

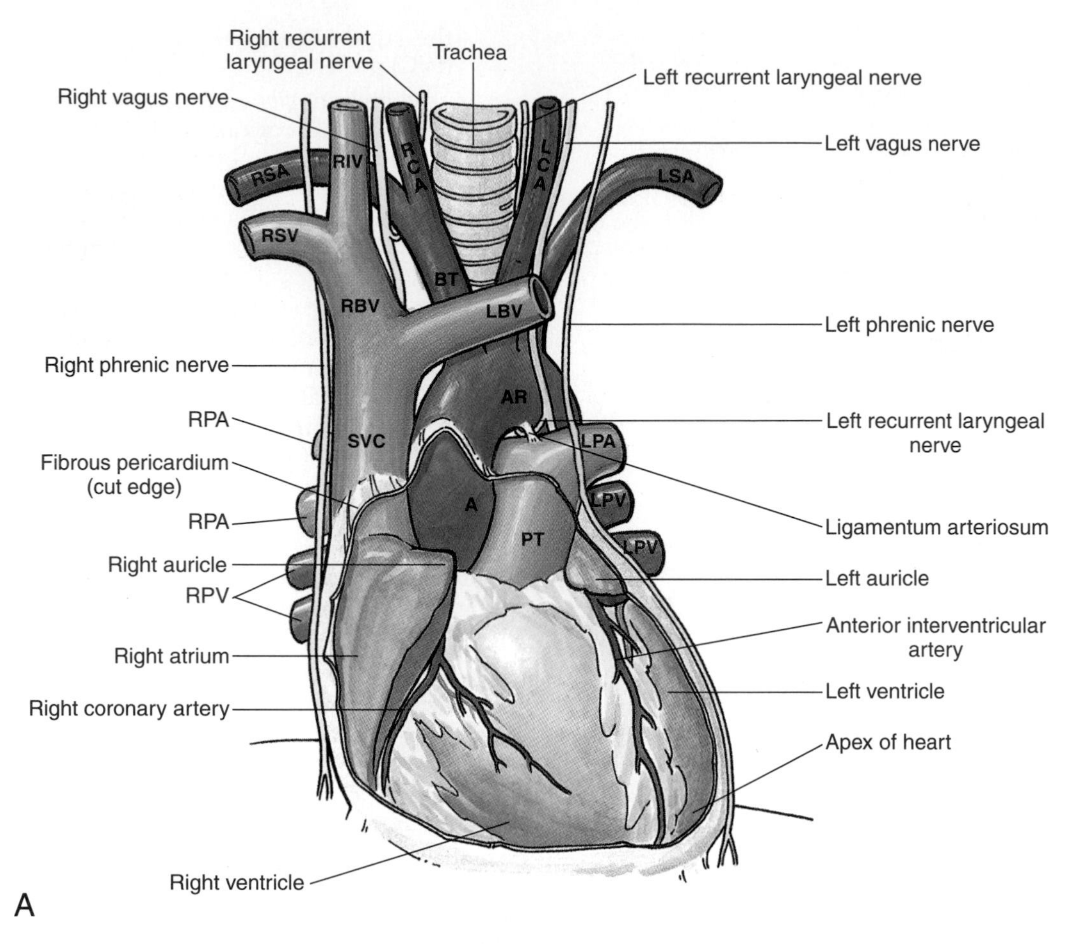
Right recurrent laryngeal nerve
Trachea
Left recurrent laryngeal nerve
Right vagus nerve
Left vagus nerve
RSA
RIV
RCA
LCA
LSA
RSV
BT
RBV
LBV
Left phrenic nerve
Right phrenic nerve
AR
RPA
SVC
LPA
Left recurrent laryngeal nerve
Fibrous pericardium (cut edge)
A
LPV
RPA
PT
Ligamentum arteriosum
LPV
Right auricle
Left auricle
RPV
Anterior interventricular artery
Right atrium
Left ventricle
Right coronary artery
Apex of heart
Right ventricle
A

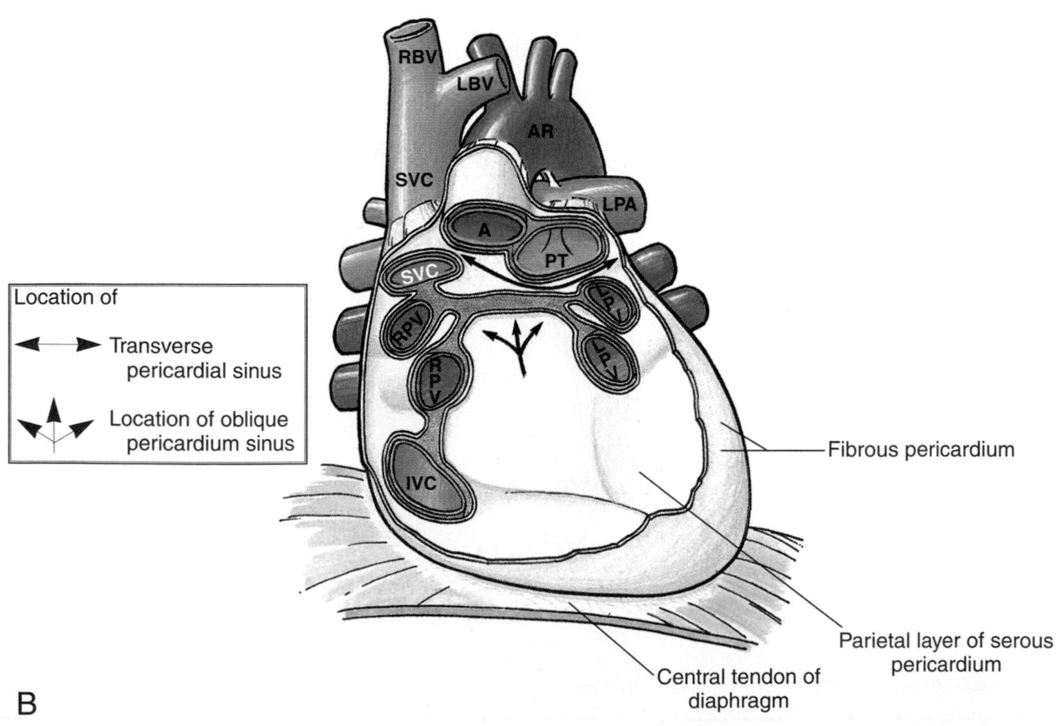
RBV
LBV
AR
SVC
LPA
A
PT
SVC
LPV
RPV
RPV
LPV
IVC
Location of
Transverse pericardial sinus
Location of oblique pericardium sinus
Fibrous pericardium
Parietal layer of serous pericardium
Central tendon of diaphragm
B

are partly covered by serous pericardium which forms the *oblique pericardial sinus*, a wide, slitlike recess posterior to the heart [the oblique pericardial sinus can be entered inferiorly and will admit several digits; however, they cannot pass around any of the vessels because this sinus is a blind recess (cul-de-sac)]

The transverse pericardial sinus is especially important in *cardiac surgery*. After the pericardial sac has been opened, a digit and ligature can be passed through the transverse pericardial sinus. By tightening the ligature, the surgeon can stop the circulation of blood through the aorta or pulmonary trunk while cardiac surgery is performed.

After piercing the diaphragm the entire thoracic part of the IVC (about 2 cm) is within the pericardium. Consequently, the pericardial sac must be opened to expose this large vein. The same is true for the terminal part of the SVC that is partly inside and partly outside the pericardial sac.

Pericarditis (inflammation of the pericardium) causes substernal pain and *pericardial effusion* (passage of fluid from the pericardial capillaries into the pericardial cavity). Usually the layers of serous pericardium make no detectable sound during auscultation. However, pericarditis makes the surfaces rough, and the resulting friction called pericardial friction rub sounds like the rustle of silk.

If there is extensive pericardial effusion, the excess fluid in the pericardial cavity interferes with the action of the heart because the fibrous pericardium is inelastic. This results in *cardiac tamponade* (compression of the heart). Stab wounds that pierce the heart result in blood entering the pericardial cavity and cardiac tamponade. As it accumulates, the heart is compressed and circulation fails. The veins of the face and neck become engorged because of compression of the SVC where it enters the inelastic pericardium. *Pericardiocentesis* (drainage of blood or serous fluid from the pericardial cavity) is usually necessary to relieve pressure on the heart.

Arterial supply of the pericardium is from the

- Pericardiacophrenic artery, a slender branch of the internal thoracic artery that accompanies the phrenic nerve to the diaphragm
- Musculophrenic artery, a terminal branch of the internal thoracic artery (Fig. 2.4*B*)

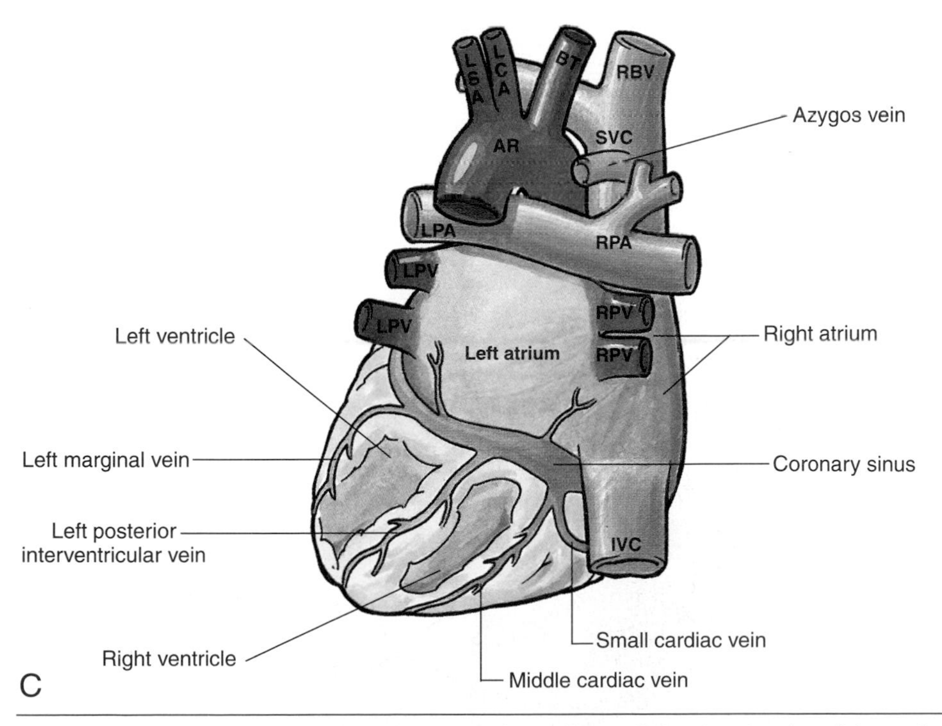

Figure 2.12. **A.** Anterior view of sternocostal surface of heart. **B.** Interior of pericardial sac after removal of heart. **C.** Posteroinferior view of base and diaphragmatic surface of heart. *A*, aorta; *AR*, arch of aorta; *BT*, brachiocephalic trunk; *IVC*, inferior vena cava; *LBV*, left brachiocephalic vein; *LCA*, left common carotid artery; *LPA*, left pulmonary artery; *LPV*, left pulmonary vein; *LSA*, left subclavian artery; *PT*, pulmonary trunk; *RBV*, right brachiocephalic vein; *RCA*, right common carotid artery; *RIV*, right internal jugular vein; *RPA*, right pulmonary artery; *RPV*, right pulmonary vein; *RSA*, right subclavian artery; *RSV*, right subclavian vein; *SVC*, superior vena cava.

- Branches from the thoracic aorta (bronchial, esophageal, and superior phrenic)
- Coronary arteries (visceral layer only)

Venous drainage of the pericardium is from the

- Pericardiacophrenic veins, tributaries of the internal thoracic veins
- Tributaries of azygos venous system (Fig. 2.11*B*).

Nerve supply of the pericardium is from the

- Phrenic nerves (C3–C5)
- Vagus nerves (CN X)
- Sympathetic trunks

Heart and Great Vessels

The heart, slightly larger than a clenched fist (Fig. 2.12*A*), is a double self-adjusting muscular pump, the parts of which work in unison to propel the blood to various parts of the body. The right side of the heart receives poorly oxygenated blood (*blue*) from the body through the SVC and IVC and pumps it to the lungs through the pulmonary trunk for oxygenation, whereas the left side receives well-oxygenated blood (*red*) from the lungs and pumps it into the aorta for distribution to the body. The heart has four chambers: right and left atria and right and left ventricles. The atria are receiving chambers that pump blood into the ventricles, the discharging chambers.

The wall of each heart chamber consists of three layers (Table 2.5):

- Endocardium, an internal layer that lines the chambers of the heart and covers its valves
- Myocardium, a middle layer composed of cardiac muscle fibers
- Epicardium, an external layer formed by the visceral pericardium

The heart and the roots of the great vessels occupy the *pericardial sac*, which is related anteriorly to the sternum, costal cartilages, and medial ends of the 3rd to 5th ribs on the left side. The heart, situated obliquely about two-thirds to the left and one-third to the right of the median plane, has a base, apex, three surfaces, and four borders.

The base of the heart (Fig. 2.12*C*) is

- Located posteriorly
- Formed mainly by the left atrium
- Where the ascending aorta and pulmonary trunk emerge and the SVC enters

The apex of the heart (Fig. 2.12*A*) is

- Formed by the left ventricle
- Located posterior to the left fifth intercostal space in adults, 7–9 cm from the median plane
- Where maximal pulsation of the heart occurs, i.e., the location of the *apex beat* ("heartbeat")

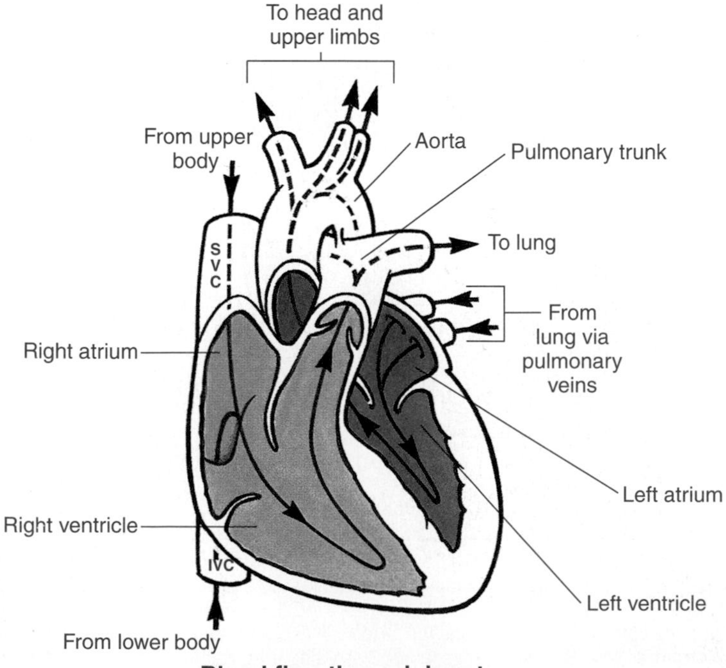

Blood flow through heart

Surfaces of the heart (Fig. 2.12, *A* and *C*) are the following:

- Sternocostal surface, formed mainly by the right ventricle
- Diaphragmatic surface, formed mainly by the left ventricle and partly by the right ventricle; it is usually horizontal or slightly concave and related to the central tendon of the diaphragm
- Pulmonary surface, formed mainly by the left ventricle; it occupies the cardiac notch of the left lung

Borders of the heart (Fig. 2.13) are the following:

- Right border, slightly convex, formed by the right atrium and located between the superior and inferior venae cavae
- Inferior border, nearly horizontal, formed mainly by the right ventricle and only slightly by the left ventricle
- Left border, formed mainly by the left ventricle and slightly by the left auricle
- Superior border, formed by the right and left auricles, which is where the great vessels enter and leave the heart

Percussion [tapping the thorax with short sharp blows (e.g., with a digit)] is a commonly used diagnostic procedure for determining the density of the heart. The character of the sound changes as different areas of the chest are tapped.

Anteroposterior chest films show the contour of the heart and great vessels, i.e., the *cardiovascular silhouette* (cardiac shadow). The silhouette contrasts with the clearer areas occupied by the air-filled lungs because the heart and great vessels are full of blood. The silhouette becomes longer and narrower during inspiration because the fibrous pericardium is attached to the diaphragm that descends during inspiration.

Right Atrium. This chamber forms the right border of the heart and receives venous blood from the superior and inferior venae cavae and the coronary sinus. The *right auricle* is a small, conical muscular pouch that projects from the right atrium and overlaps the ascending aorta. The *coronary sinus* lies in the posterior part of the *atrioventricular (coronary) groove* and receives blood from the cardiac veins (Figs. 2.12C and Fig. 2.14).

The interior of the right atrium has a smooth, thin-walled posterior part called the *sinus venarum*, which receives the venae cavae

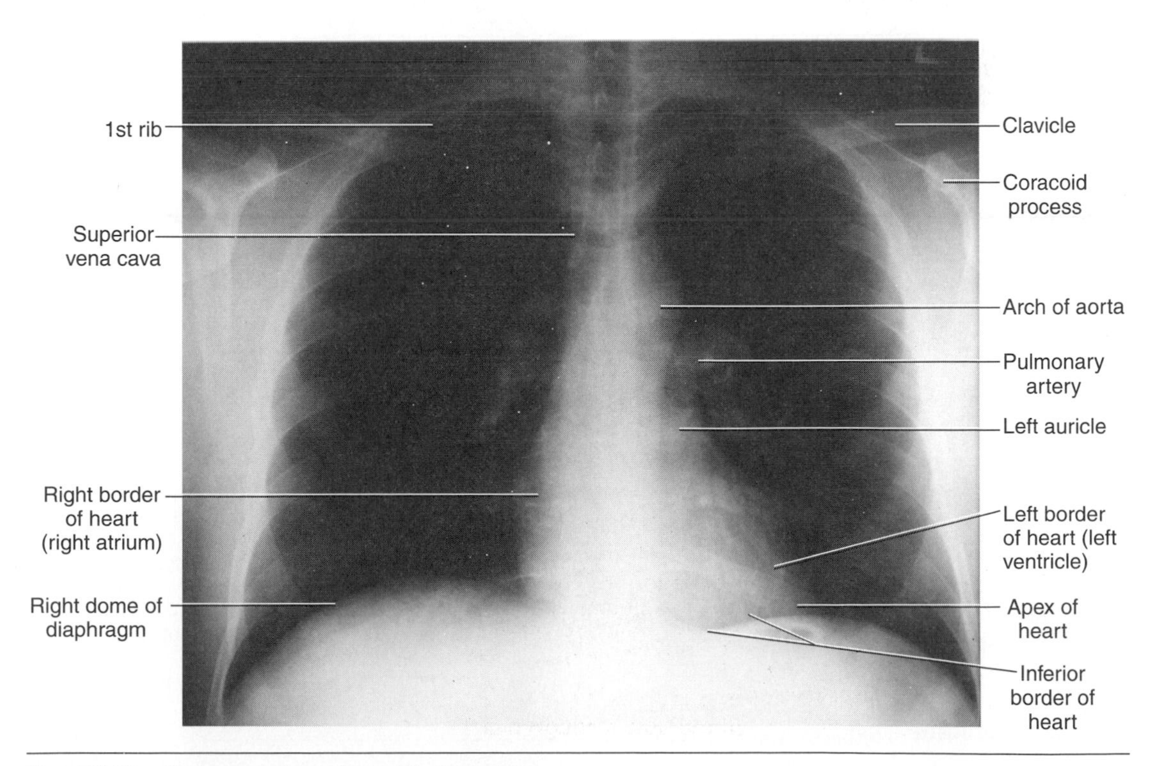

Figure 2.13. Posteroanterior radiograph of chest.

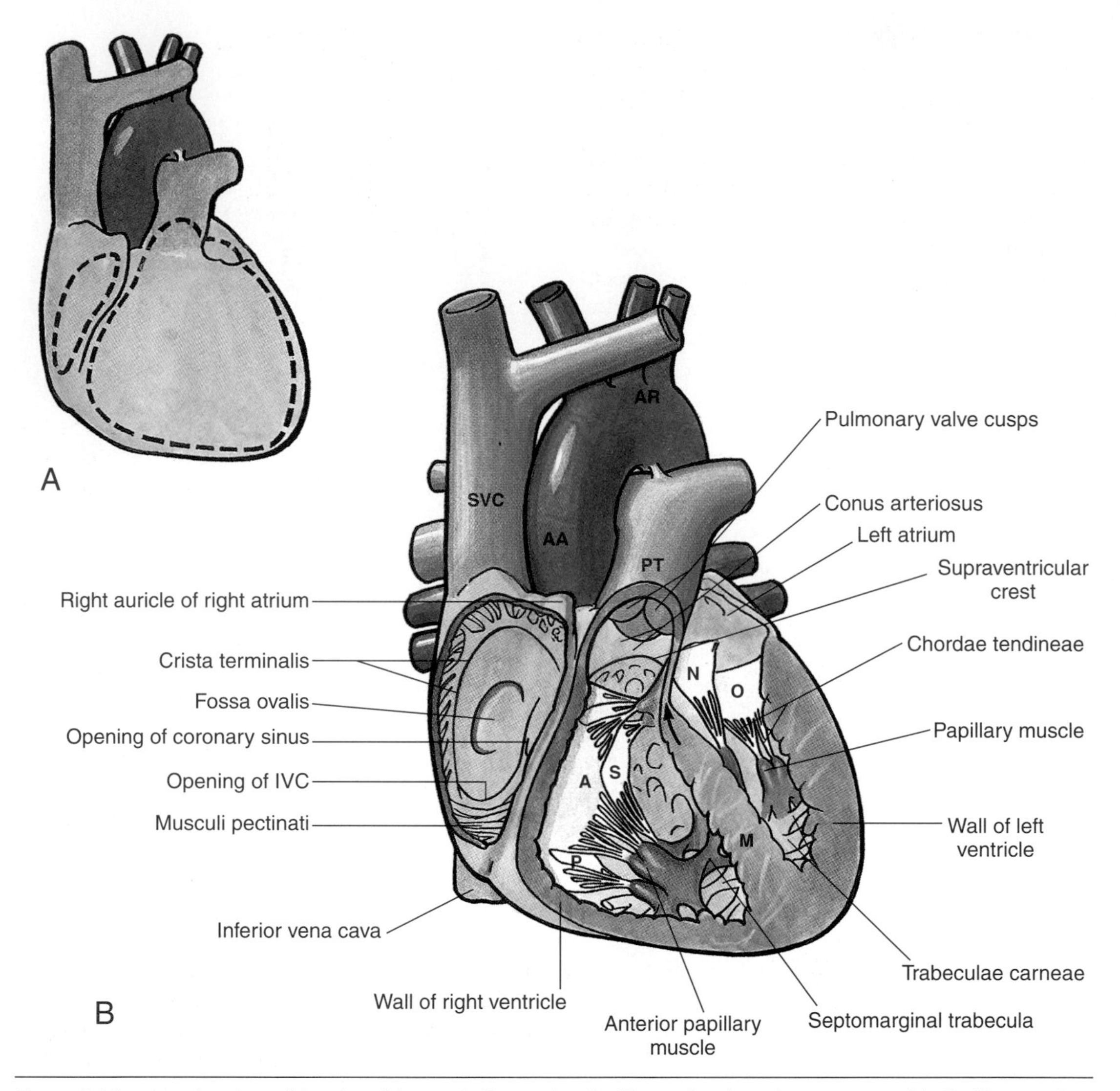

Figure 2.14. Anterior view of interior of heart. **A.** Parts of wall of heart that have been removed in **B.** Observe features of each chamber in **B.** Note three cusps of tricuspid valve [anterior (*A*), posterior (*P*), septal (*S*)] and two cusps of mitral valve [anterior (*N*) and posterior (*O*)]. *AA*, ascending aorta; *AR*, arch of aorta; *M*, muscular part of interventricular septum; *PT*, pulmonary trunk; *SVC*, superior vena cava. *Arrow*, membranous part of interventricular septum.

and coronary sinus, and a rough, more muscular anterior part with muscular ridges (*musculi pectinati*). The two parts are separated externally by a shallow vertical groove called the *sulcus terminalis* and internally by a vertical ridge called the *crista terminalis*. The SVC opens into its superior part at the level of the right third costal cartilage. The IVC opens into its inferior part almost in line with the SVC. The opening of the coronary sinus is between the right atrioventricular orifice and the orifice of the IVC. An *interatrial septum* separates the atria and has an oval thumbprint-sized *fossa ovalis*, a remnant of the *foramen ovale* and its valve in the fetus [see Moore and Persaud (1993) listed under "Suggested Readings"].

A probe-sized *atrial septal defect* appears in the superior part of the fossa ovalis in up to 25% of people. Small atrial septal defects are usually of no clinical significance, but large ones allow oxygenated blood from the lungs to be shunted from the left atrium through the defect into the right atrium. This overloads the pulmonary system and causes enlargement of the right atrium and ventricle and dilation of the pulmonary trunk.

Right Ventricle. This chamber forms the largest part of the sternocostal surface of the heart, a small part of the diaphragmatic surface, and almost the entire inferior border of the heart. Superiorly it tapers into a cone-shaped pouch called the *conus arteriosus* (infundibulum) that leads into the pulmonary trunk (Figs. 2.12 and 2.14).

The interior of the right ventricle has irregular muscular elevations called trabeculae carneae and a thick muscular ridge (*supraventricular crest*) that separates the ridged muscular part of the wall from the smooth-walled conus arteriosus.

The inflow part of the right ventricle receives blood from the right atrium through the right atrioventricular orifice, which is located posterior to the body of the sternum at the level of the fourth and fifth intercostal spaces; the right atrioventricular orifice is surrounded by a fibrous ring that is part of the *fibrous skeleton of the heart* which provides attachment for the valves and muscular fibers (the strong rings around the four major outlets of the heart resist the dilation that would result from blood being forced through them).

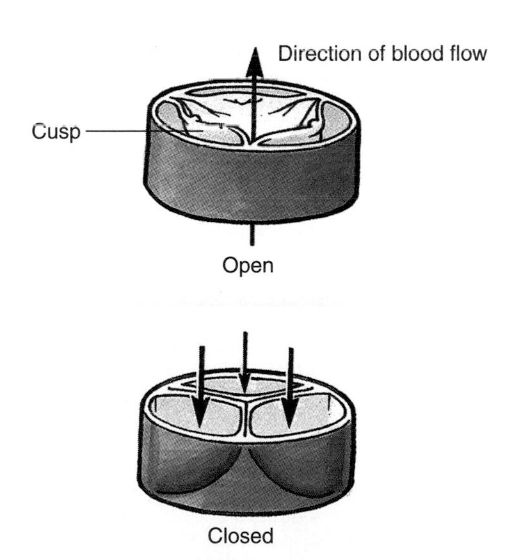

Pulmonary valve

The pulmonary valve at the apex of the conus arteriosus is at the level of the left third costal cartilage; the pulmonary valve consists of three semilunar cusps (anterior, right, and left), each of which is concave when viewed superiorly.

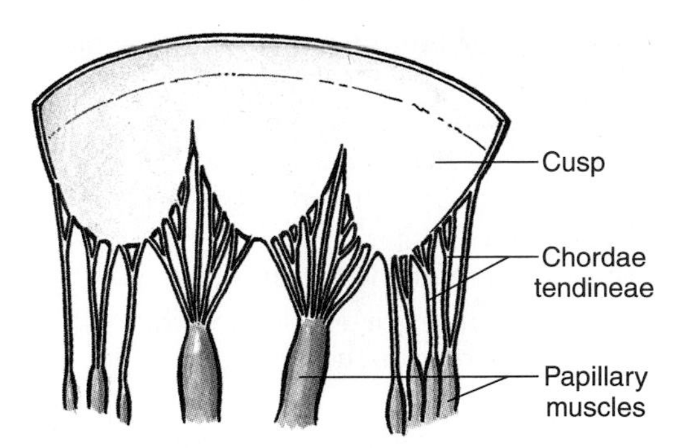

Right atrioventricular valve spread out

The right atrioventricular valve guards the *atrioventricular orifice* [this heart valve is often called the *tricuspid valve* because it has three cusps (anterior, posterior, and septal); the bases of cusps are attached to the fibrous ring of the atrioventricular orifice].

Chordae tendineae are attached to the free edges and ventricular surfaces of the cusps of the right atrioventricular valve; these tendinous cords prevent the cusps from being driven into the right atrium as ventricular pressure rises. *Papillary muscles* form conical projections with their bases attached to the ventricular wall, and chordae tendineae arise from their apices. There are usually three papillary muscles in the right ventricle (anterior, posterior, and septal) that correspond to the cusps.

The *interventricular septum*, composed of membranous and muscular parts, is a strong, obliquely placed partition between the right and left ventricles (the superior and posterior part of the septum is thin and membranous; the longer muscular part of the interventricular septum is thick and bulges to the right because of high blood pressure in the left ventricle).

The septomarginal trabecula, a curved muscular band, runs from the inferior part of the interventricular septum to the base of the anterior papillary muscle; this trabecula is important because it carries the right branch of the

When there is *pulmonary valve stenosis* (narrowing), the cusps of the valve are fused together, forming a dome with a narrow central opening. In *infundibular pulmonary stenosis*, the conus arteriosus is underdeveloped, producing a restriction of right ventricular outflow. Both types of pulmonary stenosis may occur together. The degree of enlargement (hypertrophy) of the right ventricle is variable.

atrioventricular bundle, a part of the conducting system of the heart (p. 66).

Left Atrium. This chamber forms most of the base of the heart. The *left auricle* forms the superior part of the left border of the heart (Fig. 2.12, *A* and *C*).

The interior of the left atrium has

- A smooth interior except for the auricle, which has muscular ridges (*musculi pectinati*)
- Four pulmonary veins (two superior and two inferior) that enter its posterior wall
- A slightly thicker wall than that of the right atrium
- The interatrial septum sloping posteriorly and to the right
- A left atrioventricular orifice that allows oxygenated blood from the left atrium to pass into the left ventricle

Thrombi (blood clots) form on the walls of the left atrium in certain types of heart disease. If these break off, they pass into the systemic circulation and occlude peripheral arteries. Arterial occlusion of an artery in the brain results in a *stroke* that produces paralysis of the parts of the body previously controlled by the damaged area of the brain.

Left Ventricle. This chamber forms the apex of the heart, nearly all of its left surface and border, and the diaphragmatic surface. Because arterial pressure is much higher in the systemic than in the pulmonary circulation, the left ventricle performs more work than the right ventricle (Fig. 2.12, *A* and *C*).

The interior of the left ventricle has

- A wall that is twice as thick as that of the right ventricle
- The ascending aorta, about 2.5 cm in diameter, arising from its superior part
- A conical cavity that is longer than that of the right ventricle
- A superoanterior part that is formed by the smooth walled aortic vestibule opening into the ascending aorta
- An aortic orifice that lies in the right posterosuperior part of the left ventricle and is surrounded by a fibrous ring to which the three semilunar cusps (right, posterior, and left) of the aortic valve are attached
- An interior that is mostly covered with a mesh of trabeculae carneae that is finer and more numerous than in the right ventricle

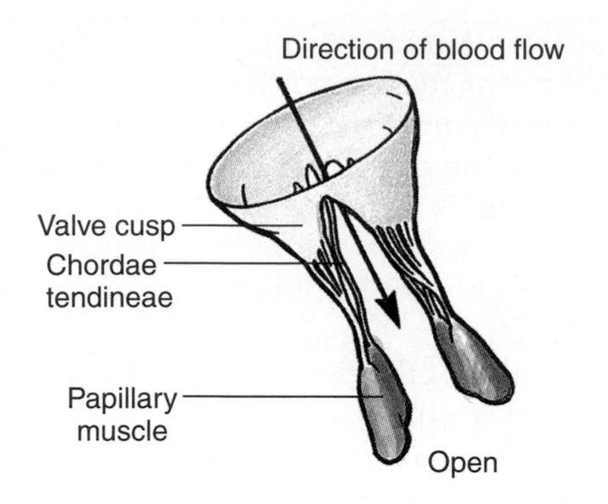

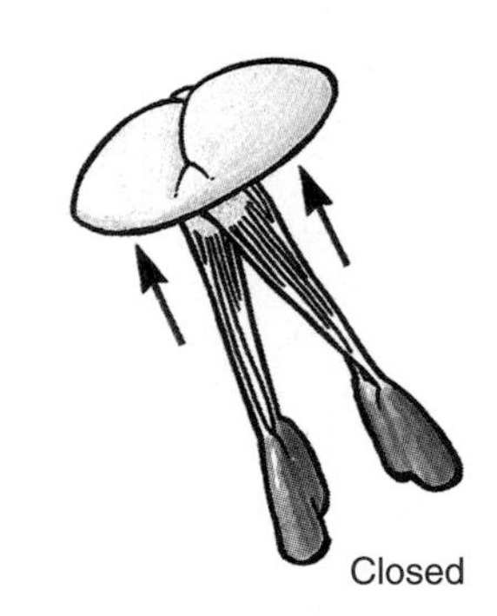

Left atrioventricular valve

- A left atrioventricular (mitral) valve that has two obliquely set cusps, anterior and posterior, and is located posterior to the sternum at the level of the fourth left costal cartilage, where it guards the orifice between the left atrium and ventricle
- Anterior and posterior papillary muscles that are larger than those in the right ventricle because of the greater work done by this ventricle

The left atrioventricular (mitral) valve is the most frequently diseased of the heart valves. Nodules form on the valve cusps, causing irregular blood flow. Later the diseased cusps undergo scarring and shortening, resulting in *atrioventricular valvular incompetence.* Blood regurgitates into the left atrium when the left ventricle contracts, producing a characteristic *heart murmur.*

Aortic valvular incompetence results in aortic regurgitation, a backrush of blood into the left ventricle. This produces a heart murmur and a *collapsing pulse* (a forcible impulse that rapidly disappears). In *aortic valve stenosis* (narrowing), the edges of the valve are usually fused to form a dome with a small opening. Aortic valvular stenosis causes extra work for the heart, resulting in *left ventricular hypertrophy.*

The distal part of the ascending aorta receives a strong thrust of blood from the left ventricle. Because its wall is not reinforced by fibrous pericardium, an *aortic aneurysm* may develop.

Surface Anatomy of Heart

The outline of the heart can be traced on the anterior surface of the thorax by using the following guidelines.

- The superior border corresponds to a line connecting the inferior border of the second left costal cartilage to the superior border of the third right costal cartilage
- The right border corresponds to a line drawn from the third right costal cartilage to the sixth right costal cartilage; this border is slightly convex to the right
- The inferior border corresponds to a line drawn from the inferior end of the right border to a point in the fifth intercostal space close to the left midclavicular line; the left end of this line corresponds to the location of the apex beat
- The left border corresponds to a line connecting the left ends of the lines representing the superior and inferior borders

The pulmonary, aortic, mitral, and tricuspid valves are located posterior to the sternum; however, the sounds produced by them are best heard at the illustrated *auscultatory areas* [pulmonary (*P*), aortic (*A*), mitral (*M*), tricuspid (*T*)]. Clinicians' interest in the surface anatomy of the cardiac valves arises from the need to listen to the valve sounds. The auscultatory areas are as wide apart as possible so that the sounds produced at any given valve may be clearly distinguished from those produced at other valves. Blood tends to carry the sound in the direction of its flow; consequently each area is situated superficial to the chamber or vessel into which the blood has passed and in a direct line with the valve orifice.

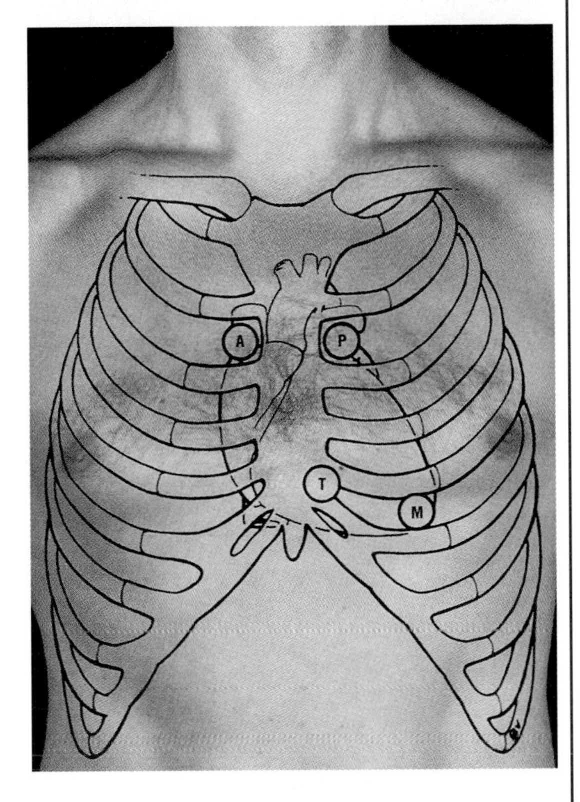

Auscultatory areas

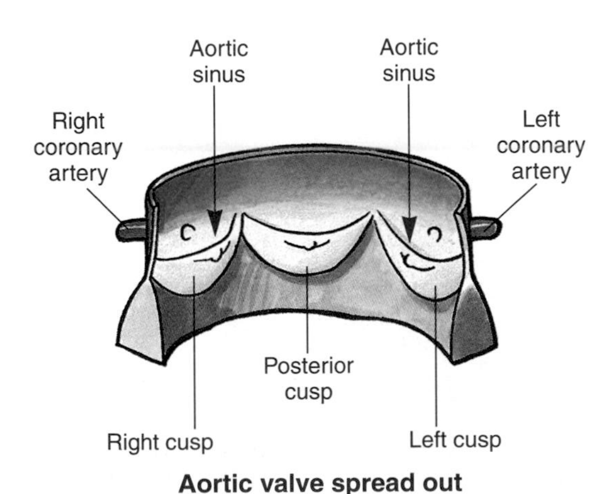

Aortic valve spread out

The *aortic valve* is located obliquely, posterior to the left side of the sternum at the level of the third intercostal space. Aortic sinuses are formed superior to each valve by dilations of the aortic wall. Note that the mouth of the *right coronary artery* is in the right aortic sinus; the mouth of the *left coronary artery* is in the left aortic sinus; and no artery arises from the posterior aortic sinus (noncoronary sinus).

Arterial Supply of Heart. The coronary arteries supply the myocardium and epicardium of the heart. Both vessels arise from the ascending aorta just superior to the aortic valve. They supply both the atria and the ventricles, but the atrial branches are usually small (Fig. 2.12*A*, Table 2.6).

The *right coronary artery* arises from the right aortic sinus and runs in the atrioventricular (coronary) groove. Near its origin the right coronary artery usually (about 60%) gives off a *sinuatrial nodal artery* that supplies the sinuatrial node. The right coronary artery then passes toward the inferior border of the heart and gives off the *right marginal artery* that runs toward the apex. After giving off this branch,

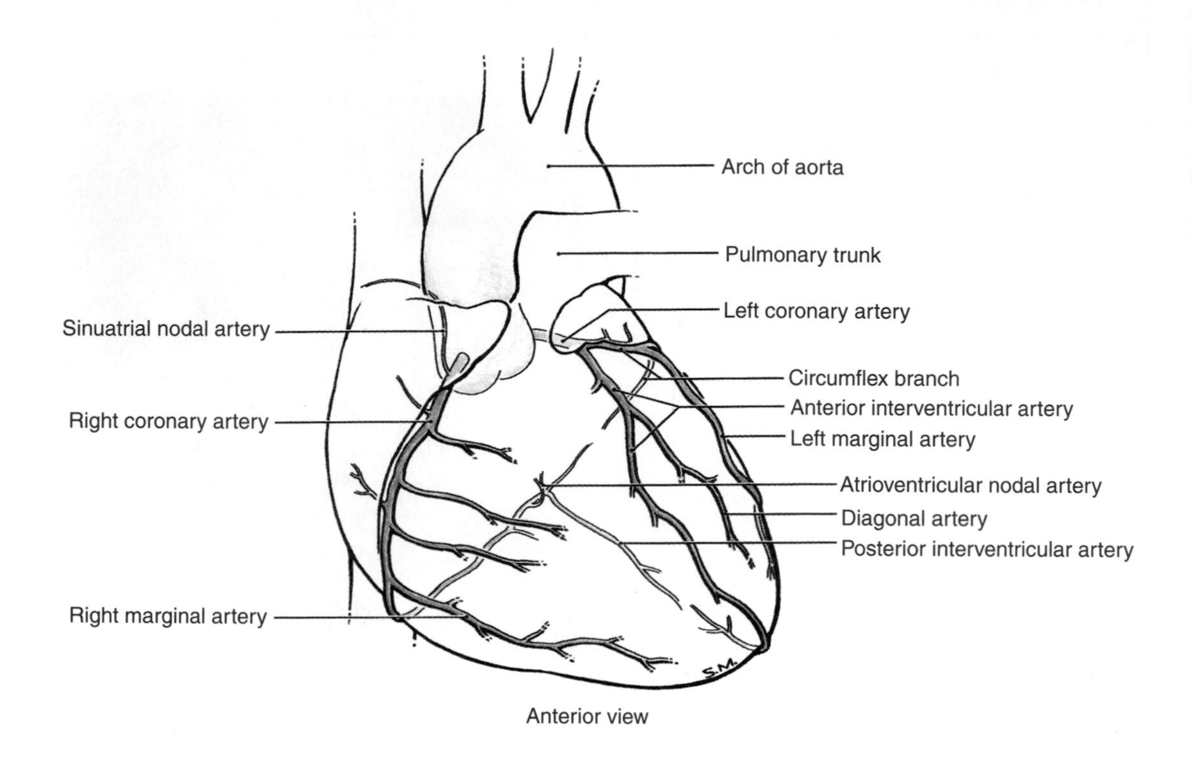

Anterior view

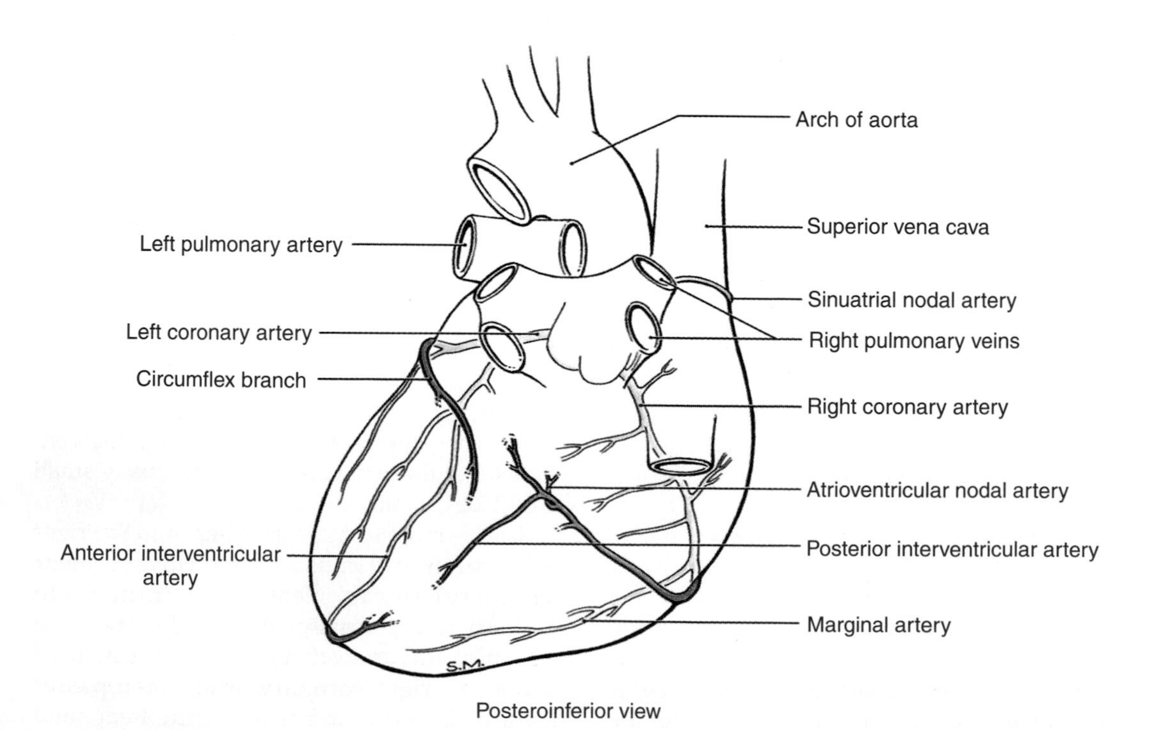

Posteroinferior view

Arterial supply

Table 2.6
Arterial Supply of Heart

Artery	Origin	Course	Distribution	Anastomoses
Right coronary	Right aortic sinus	Follows AV groove between the atria and ventricles	Right atrium, SA and AV nodes, and posterior part of IV septum	Circumflex and anterior IV branches of left coronary
SA nodal	Right coronary a. near its origin (about 60%)	Ascends to SA node	Pulmonary trunk and SA node	
Marginal	Right coronary a.	Passes to inferior margin of heart and apex	Right ventricle and apex	IV branches
Posterior interventricular	Right coronary a.	Posterior IV groove to apex	Right and left ventricles and IV septum	Circumflex and anterior IV branches of left coronary a.
AV nodal	Right coronary a. near its termination	Passes to AV node	AV node	
Left coronary	Left aortic sinus	Runs in AV groove and gives off anterior interventricular and circumflex branches	Most of left atrium and ventricle, IV septum, and AV bundles; may supply AV node	Right coronary a.
SA nodal	Left coronary a. (about 40%)	Ascends on posterior surface of left atrium to SA node	Left atrium and SA node	
Anterior interventricular	Left coronary a.	Passes along anterior IV groove to apex	Right and left ventricles and IV septum	Posterior IV branch of right coronary a.
Circumflex	Left coronary a.	Passes to left in AV groove and runs to posterior surface of heart	Left atrium and left ventricle	Right coronary a.
Marginal	Circumflex a.	Follows left border of heart	Left ventricle	IV branches

AV, atrioventricular; IV, interventricular; SA sinuatrial.

the right coronary artery turns to the left and continues in the posterior atrioventricular groove. At the posterior interventricular groove, it gives off the *posterior interventricular artery* (posterior descending artery) that descends toward the apex. This artery supplies both ventricles, and near the apex it anastomoses with the circumflex and anterior interventricular branches of the left coronary artery. Near its termination the right coronary artery gives rise to the *atrioventricular nodal artery* that supplies the atrioventricular node and bundle.

The *left coronary artery* arises from the *left aortic sinus* and passes between the left auricle and pulmonary trunk to reach the anterior atrioventricular groove. In about 40% of hearts the sinuatrial nodal artery arises from the left coronary artery and ascends on the posterior surface of the left atrium to the sinuatrial node. The large branch of the left coronary artery, the *anterior interventricular artery* (anterior descending branch), passes along the interventricular groove to the apex. Here it turns around the inferior border of the heart and anastomoses with the posterior interventricular branch of the right coronary artery. The anterior interventricular artery supplies both ventricles and the interventricular septum. The smaller *circumflex branch* of the left coronary artery follows the atrioventricular groove around the left border of the heart to the posterior surface of the heart. The circumflex artery has a *marginal branch* that follows the left margin of the heart. The circumflex artery terminates by supplying branches to the left ventricle and atrium. In many people the anterior interventricular artery has a third branch called the diagonal artery that descends on the sternocostal surface of the heart.

The branches of coronary arteries are end arteries; i.e., they supply regions of the myocardium without overlap from other large branches. However, there are anastomoses between the small branches of the coronary arteries. When there is a sudden occlusion of a major artery by an embolus, the region of myocardium supplied by the occluded vessel becomes infarcted (rendered virtually bloodless) and soon degenerates. An area of myocardium that has undergone necrosis is called a *myocardial infarct*. The most common cause of *ischemic heart disease* is coronary insufficiency, resulting from atherosclerosis of the coronary arteries. For a fuller discussion of the clinical anatomy of coronary artery disease, angina, and coronary bypass surgery, see Moore (1992) listed under "Suggested Readings." A patient with coronary arterial disease often suffers from *angina pectoris* (usually pain in the substernal region and down the medial side of the left arm and forearm).

Venous Drainage of Heart. The heart is drained mainly by veins that empty into the coronary sinus and partly by small veins (venae cordis minimae and anterior cardiac veins) that empty into the right atrium. The *coronary sinus*, the main vein of the heart, is a wide venous channel that runs from left to right in the posterior part of the atrioventricular groove. It receives the great cardiac vein at its left end and the middle and small cardiac veins at its right end. The left posterior ventricular vein and left marginal vein also open into the coronary sinus (Fig. 2.12C).

Lymphatic Drainage of Heart. The lymphatic vessels in the myocardium and subendocardial connective tissue pass to the *subepicardial lymphatic plexus*. Vessels from this plexus pass to the atrioventricular groove and follow the coronary arteries. A single lymphatic vessel, formed by the union of various vessels from the heart, ascends between the pulmonary trunk and left atrium and ends in the inferior *tracheobronchial lymph nodes* (Fig. 2.11C), usually on the right side.

Conducting System of Heart. The impulse-conducting system consists of cardiac muscle cells and conducting fibers that are specialized for initiating impulses and conducting them rapidly through the heart. They initiate the normal heartbeat and coordinate contractions of the four heart chambers (Fig. 2.15).

The *sinuatrial node* is a small collection of specialized cardiac muscle fibers that initiates and regulates the impulses for contraction. It gives off an impulse about 70 times per minute in most people. The sinuatrial node is located anterolaterally immediately deep to the epicardium at the junction of the SVC and right atrium. The sinuatrial node is supplied by both divisions of the autonomic nervous system.

The *atrioventricular node* is a smaller collection of nodal tissue located in the posteroinferior region of the interatrial septum near the opening of the coronary sinus. Impulses from cardiac muscle fibers in both atria converge on the atrioventricular node, which distributes them to the ventricles via the *atrioventricular bundle (of His)*. Sympathetic stimulation speeds up conduction and parasympathetic stimulation slows it down.

When there is an atrial septal defect, the atrioventricular bundle usually lies in the margin of the defect. Obviously this vital part of the conducting system of the heart must be preserved during surgical repair of the atrial septal defect because its destruction would cut the only physiological link between the atrial and ventricular musculature.

The passage of impulses over the heart from the sinuatrial node can be amplified and recorded as an *electrocardiogram*. Because many heart problems involve abnormal functioning of the conducting system, electrocardiograms are of considerable importance in detecting the cause of irregularities of the heartbeat. Deficiencies in the stimulating action of the sinuatrial node may result in a *heart block* (delay or interruption of impulse propagation). Heart blocks can result from myocardial infarction or ischemia resulting from coronary atherosclerosis. In some people with heart blocks, a cardiac *pacemaker* is inserted subcutaneously with its electrode introduced into the right ventricle via the SVC.

The *atrioventricular bundle*, the only bridge between the atrial and ventricular myocardium,

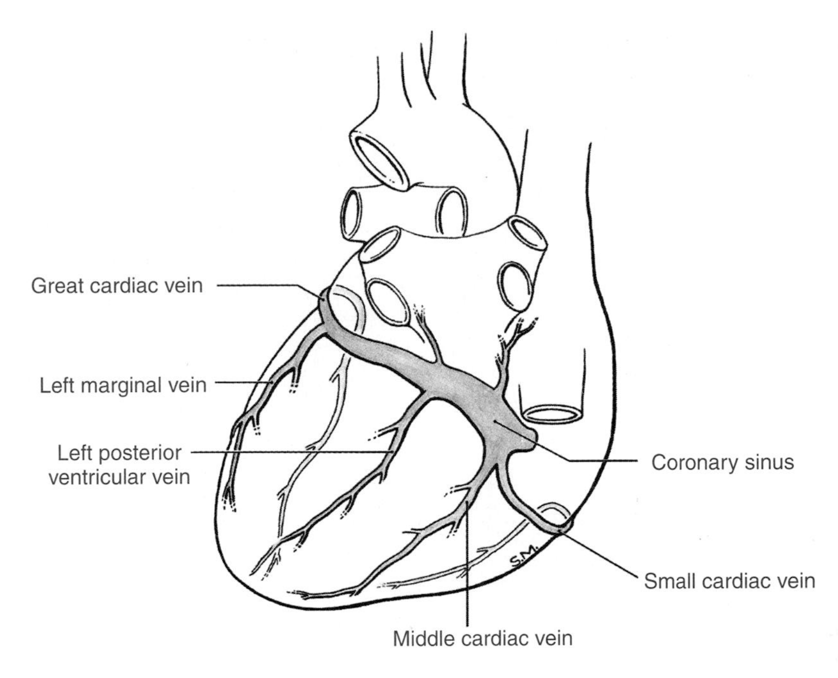

Posteroinferior view

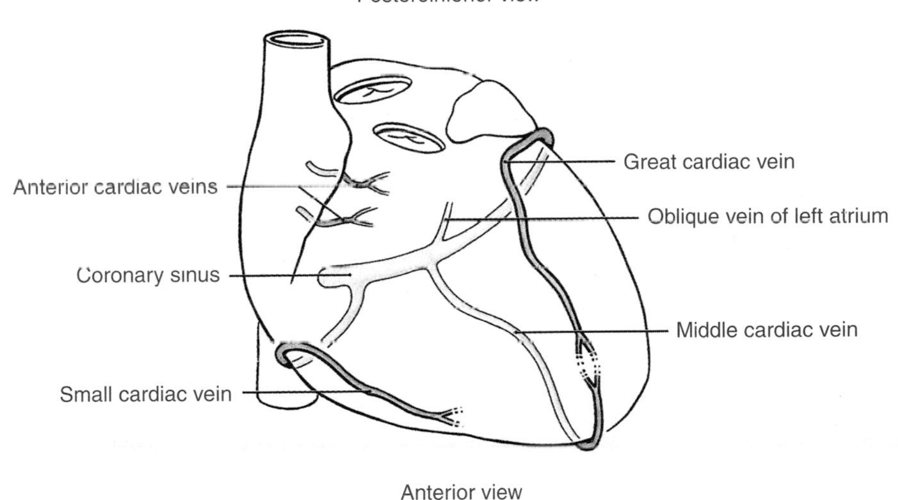

Anterior view

Venous drainage

passes from the atrioventricular node and runs through the membranous part of the interventricular septum. At the junction of the membranous and muscular parts of the septum, the atrioventricular bundle divides into right and left bundle branches. Each branch passes deep to the endocardium into the walls of the ventricles. The *right bundle branch* innervates the muscle of the interventricular septum, the anterior papillary muscle through the septomarginal band, and the wall of the right ventricle. The *left bundle branch* supplies the interventricular septum, the anterior and posterior papillary muscles, and the wall of the left ventricle.

Innervation of Heart. The heart is supplied by autonomic nerve fibers from the *cardiac plexuses*. These nerve networks lie anterior to the bifurcation of the trachea, posterior to the arch of the aorta, and superior to the bifurcation of the pulmonary trunk. The *sympathetic supply* comes from the cervical and superior thoracic parts of the sympathetic trunks. The

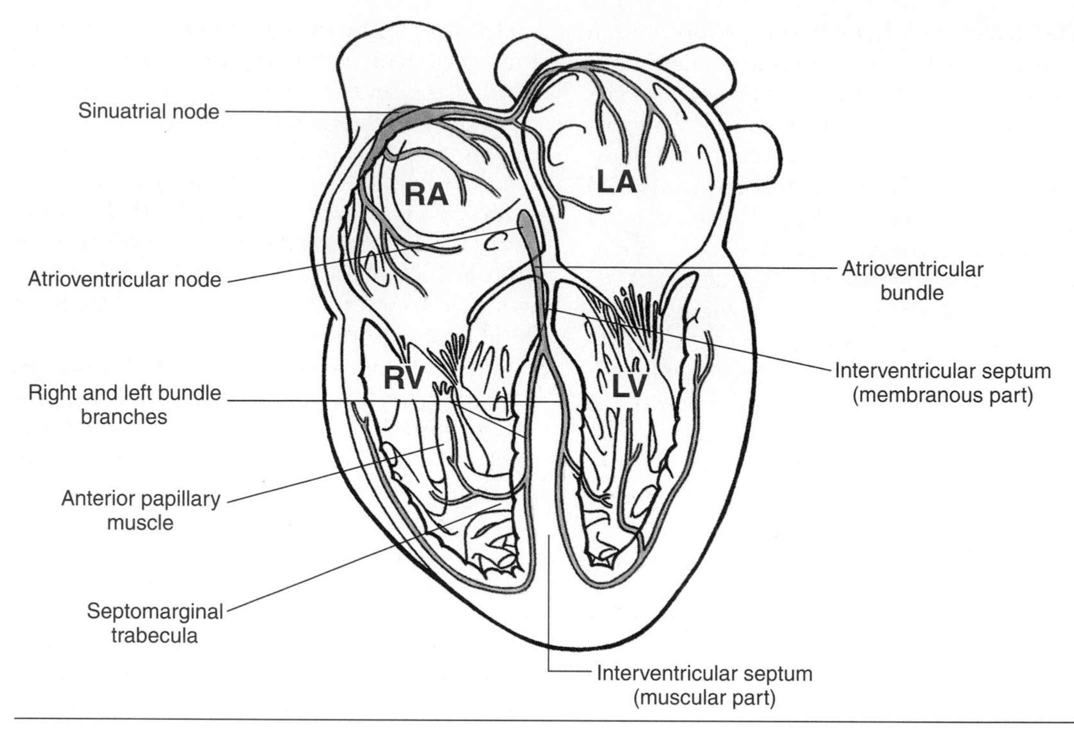

Figure 2.15. Conducting system of heart. *LA*, left atrium; *LV*, left ventricle; *RA*, right atrium; *RV*, right ventricle.

cell bodies of the preganglionic sympathetic fibers are in the lateral horn of the superior five or six thoracic segments of the spinal cord. The *parasympathetic supply* comes from the vagus nerves (CN X).

The postganglionic sympathetic fibers end in the sinuatrial and atrioventricular nodes, on cardiac muscle fibers, and on the coronary arteries (Fig. 2.11*D*). Stimulation of sympathetic nerves increases the heart rate and the force of the heartbeat and causes dilation of the coronary arteries. This supplies more oxygen and nutrients to the myocardium. The postganglionic parasympathetic fibers end in the sinuatrial and atrioventricular nodes and in the coronary arteries. Stimulation of parasympathetic nerves slows the heart rate, reduces the force of the heartbeat, and constricts the coronary arteries.

Superior Mediastinum

The superior mediastinum (*green*) is located superior to the horizontal plane passing through the sternal angle and the inferior border of T4 vertebra. From anterior to posterior, the main contents of the superior mediastinum are

- Thymus
- Great vessels related to heart and pericardium: brachiocephalic veins, SVC, and arch of aorta
- Phrenic and vagus nerves
- Cardiac plexus of nerves
- Trachea
- Left recurrent laryngeal nerve
- Esophagus
- Thoracic duct
- Prevertebral muscles

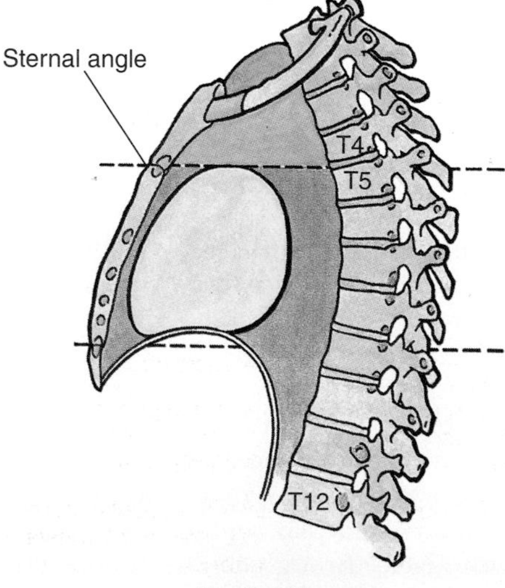

Mediastinum

The *thymus* is a lymphoid organ located in the anterior part of the superior mediastinum. It lies posterior to the manubrium and extends into the anterior mediastinum (*purple*), anterior to the pericardium. After puberty the thymus undergoes gradual involution and is composed largely of fat. For a discussion of the function of the thymus, see Cormack (1993).

The *brachiocephalic veins* (right and left) are formed posterior to the sternoclavicular joints by the union of the internal jugular and subclavian veins (Fig. 2.11*B*). At the level of the inferior border of the first right costal cartilage, the brachiocephalic veins unite to form the SVC. The left brachiocephalic vein is over twice as long as the right brachiocephalic vein because it passes from the left to the right side, shunting blood from the head, neck, and left upper limb to the right atrium. The right brachiocephalic vein receives lymph from the right lymphatic duct, and the left brachiocephalic vein receives lymph from the thoracic duct (Fig. 2.11*C*).

The SVC returns blood from all structures superior to the diaphragm, except the lungs and heart. It forms posterior to the first right costal cartilage by the union of the right and left brachiocephalic veins. It passes inferiorly and ends at the level of the third costal cartilage, where it enters the right atrium. The SVC lies in the right side of the superior mediastinum, anterolateral to the trachea and posterolateral to the ascending aorta (Fig. 2.12*A*). The right phrenic nerve lies between the SVC and the mediastinal pleura. The terminal half of the SVC is in the middle mediastinum, where it lies beside the ascending aorta.

The *arch of the aorta*, the curved continuation of the ascending aorta (Fig. 2.12, Table 2.7), begins posterior to the second right sternocostal joint at the level of the sternal angle and arches superoposteriorly and to the left (its main direction is posterior). The arch passes anterior to the trachea to reach the left side of the trachea and esophagus. It then passes over the root of the left lung as it descends on the left side of the body of T4 vertebra. The aortic arch ends by becoming the thoracic aorta posterior to the second left sternocostal joint. The *ligamentum arteriosum* (the remnant of the fetal ductus arteriosus) passes from the root of the left pulmonary

Table 2.7
Aorta and Its Branches in Thorax

Artery	Origin	Course	Branches
Ascending aorta	Aortic orifice of left ventricle	Ascends about 5 cm to sternal angle where it becomes arch of aorta	Right and left coronary arteries
Arch of aorta	Continuation of ascending aorta	Arches posteriorly on left side of trachea and esophagus and superior to left main bronchus	Brachiocephalic, left common carotid, left subclavian
Thoracic aorta	Continuation of arch of aorta	Descends in posterior mediastinum to left of vertebral column; gradually shifts to right to lie in median plane at aortic hiatus posterior to diaphragm	Posterior intercostal arteries, subcostal, some phrenic arteries and visceral branches (e.g., esophageal)
Esophageal (4–5 branches)	Anterior aspect of thoracic aorta	Run anteriorly to esophagus	To esophagus
Superior phrenic (branches vary in number)	Anterior aspect of thoracic aorta of diaphragm	Arise at aortic hiatus and pass to superior aspect	To diaphragm
Bronchial (1–2 branches)	Anterior aspect of aorta or posterior intercostal artery	Run with tracheobronchial tree	Bronchial and peribronchial tissue, visceral pleura

artery to the inferior surface of the arch. The *left recurrent laryngeal* nerve hooks around the arch and ligamentum arteriosum and then ascends between the trachea and esophagus.

The branches of the arch of the aorta are the

- Brachiocephalic trunk
- Left common carotid artery
- Left subclavian artery

The *brachiocephalic trunk*, the first and largest of the branches, arises posterior to the manubrium where it is anterior to the trachea and posterior to the left brachiocephalic vein. It ascends superolaterally to reach the right side of the trachea and the right sternoclavicular joint, where it divides into the right common carotid and right subclavian arteries.

The *left common carotid artery*, the second branch of the arch, arises posterior to the manubrium and slightly posterior and to the left of the brachiocephalic trunk. It ascends anterior to the left subclavian artery and is at first anterior to the trachea and then is to its left. It enters the neck by passing posterior to the left sternoclavicular joint.

The *left subclavian artery*, the third branch of the arch, arises from the posterior part of the arch, just posterior to the left common carotid artery. It ascends with the left common carotid artery through the superior mediastinum: it has no branches in the mediastinum. As it leaves the thorax and enters the root of the neck, it passes posterior to the left sternoclavicular joint and to the left of the left common carotid artery.

The *vagus nerves* (CN X) arise bilaterally from the medulla of the brain and descend through the neck, posterolateral to the common carotid arteries. Each nerve enters the superior mediastinum posterior to the respective sternoclavicular joint and brachiocephalic vein (Figs. 2.12*A*, 2.16, and 2.17, Table 2.8).

The *right vagus nerve* enters the thorax anterior to the right subclavian artery, where it gives rise to the *right recurrent laryngeal nerve*. This nerve hooks around the right subclavian artery and ascends between the trachea and esophagus to supply the larynx. The right vagus nerve runs posteroinferiorly through the superior mediastinum on the right side of the trachea. It then passes posterior to the right brachiocephalic vein, the SVC, and the root of the right lung. Here it breaks into many branches that contribute to the *pulmonary plexus* (Fig. 2.18). Usually the nerve leaves this plexus as a single nerve and passes to the esophagus, where it again breaks up and contributes fibers to the *esophageal plexus*. The right vagus nerve also gives rise to nerves that contribute to the *cardiac plexus*.

The *left vagus nerve* descends in the neck posterior to the left common carotid artery. It enters the mediastinum between the left common carotid artery and the left subclavian artery. When it reaches the left side of the arch of the aorta, the left vagus nerve diverges posteriorly from the left phrenic nerve. It is separated laterally from the phrenic nerve by the left superior intercostal vein. As the left vagus nerve curves medially at the inferior border of the arch of the aorta, it gives off the *left recurrent laryngeal nerve*. This nerve passes inferior to the arch of the aorta just lateral to the ligamentum arteriosum and ascends to the larynx in the groove between the trachea and esophagus. The left vagus nerve passes posterior to the root of the left lung, where it breaks into many branches that contribute to the *pulmonary plexus*. The nerve leaves this plexus as a single trunk and passes to the esophagus, where it joins fibers from the right vagus in the *esophageal plexus*.

The recurrent laryngeal nerves supply all intrinsic muscles of the larynx, except one. Consequently, any investigative procedure or disease process in the superior mediastinum may involve these nerves and affect the voice. Because the left recurrent laryngeal nerve winds around the arch of the aorta and ascends between the trachea and esophagus, it may be involved when there is a bronchial or *esophageal carcinoma*, enlargement of mediastinal lymph nodes, or an aneurysm of the arch of the aorta. In the latter condition the nerve may be stretched by the dilation of the aorta.

Each phrenic nerve enters the superior mediastinum between the subclavian artery and the origin of the brachiocephalic vein (Figs. 2.12*A*, 2.16, and 2.17, Table 2.8). Phrenic nerves are the sole motor supply to the diaphragm; about one-third of their fibers are sensory to the diaphragm.

The *right phrenic nerve* passes along the right side of the right brachiocephalic vein, SVC, and the pericardium over the right atrium. It also passes anterior to the root of the right lung and descends on the right side of the IVC to the

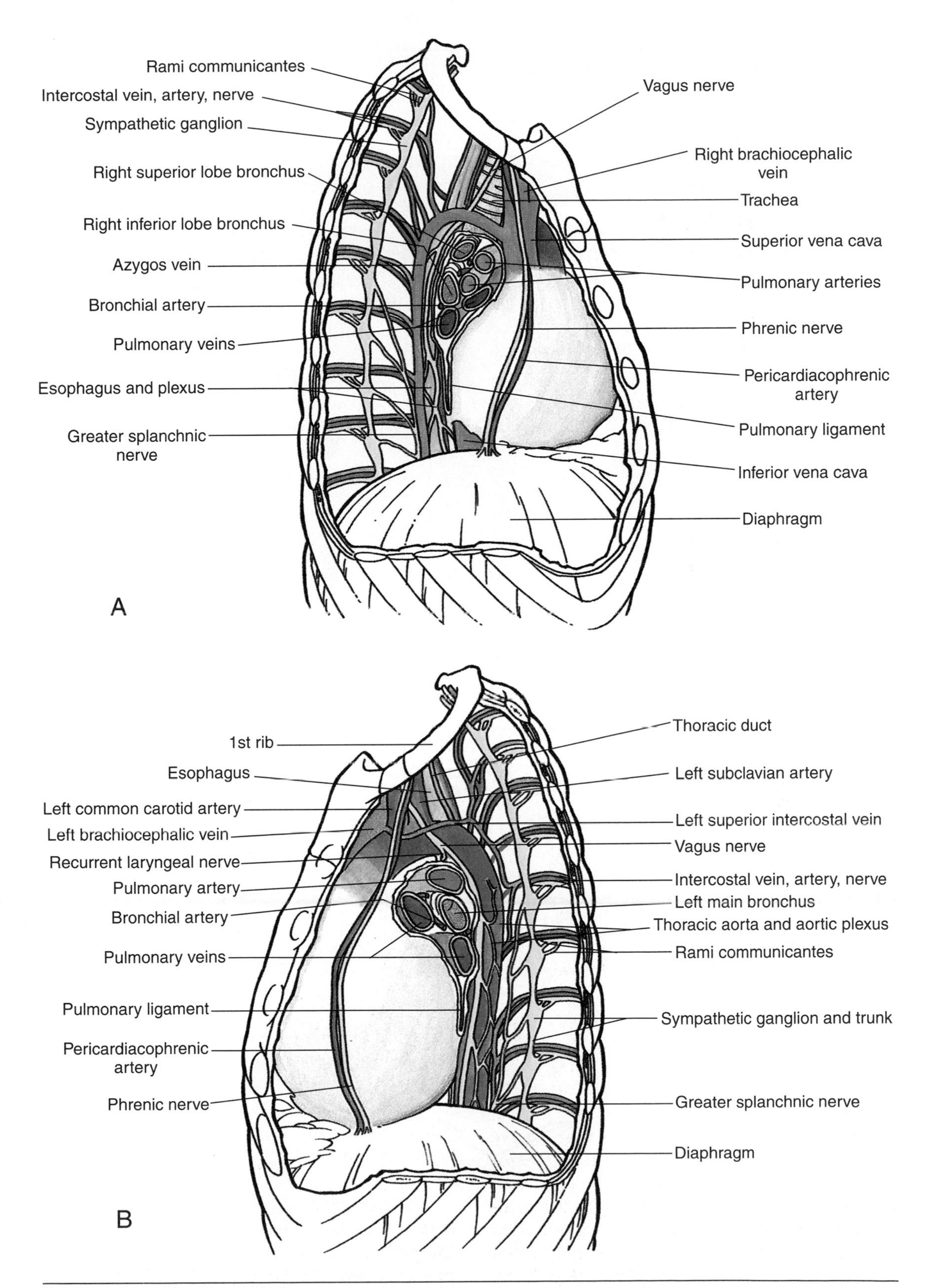

Figure 2.16. Mediastinum. **A.** Right side. **B.** Left side.

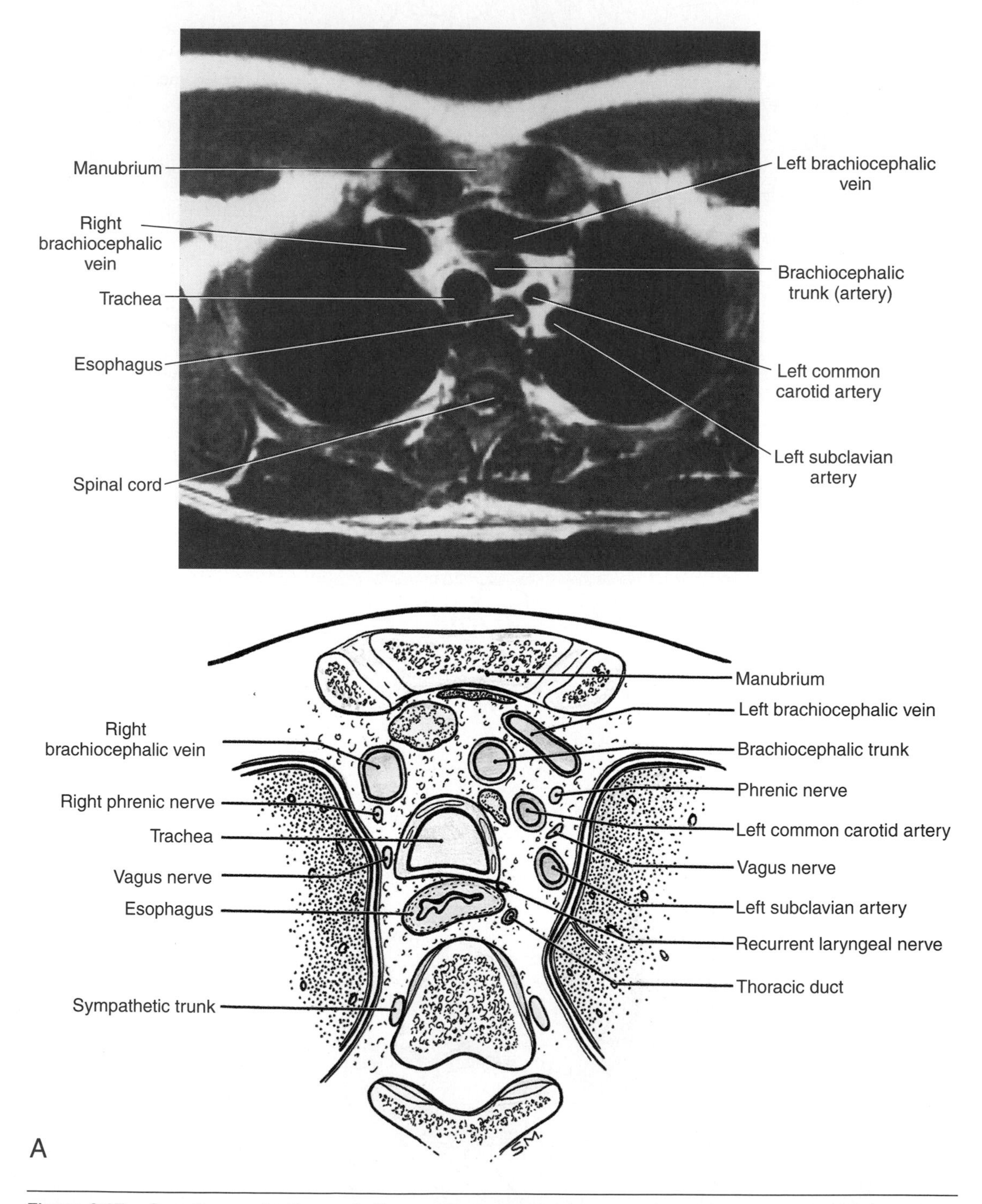

Figure 2.17. Superior mediastinum. **A.** Transverse magnetic resonance image superior to arch of aorta. **B.** Transverse magnetic resonance image at level of arch of aorta. **C.** *Broken lines* indicate level of magnetic resonance imaging in **A** and **B.** *BT*, brachiocephalic trunk; *E*, esophagus; *LB*, left brachiocephalic vein; *LC*, left common carotid; *LS*, left subclavian; *RB*, right brachiocephalic vein; *T*, trachea.

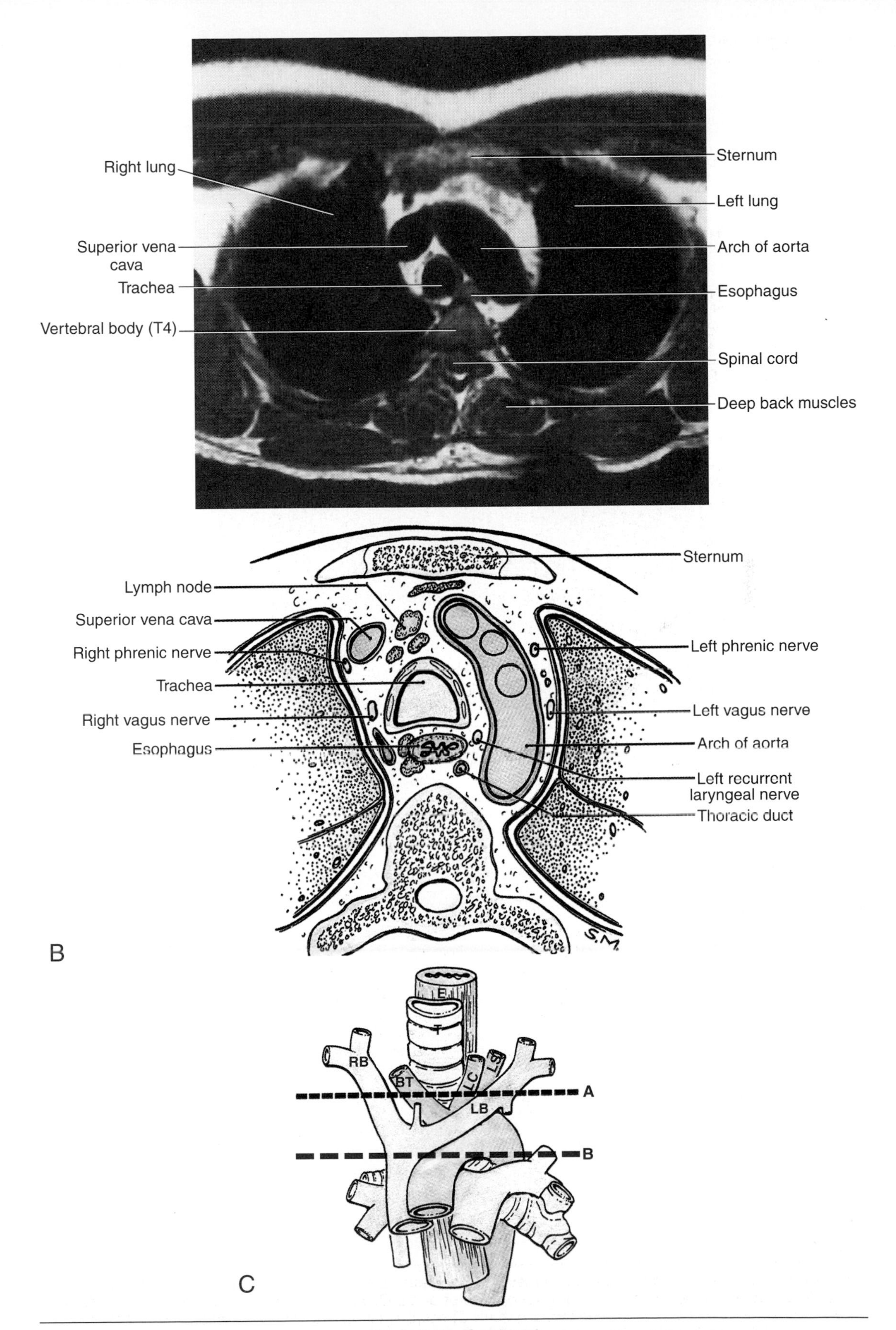

Figure 2.17. *Continued.*

Table 2.8
Nerves of Thorax

Nerve	Origin	Course	Distribution
Vagus (CN X)	Eight to ten rootlets from medulla of brainstem	Enters superior mediastinum posterior to sternoclavicular joint and brachiocephalic vein; gives rise to recurrent laryngeal n.; continues into abdomen	Pulmonary plexus, esophageal plexus, and cardiac plexus
Phrenic	Ventral rami of C3–C5 nn.	Passes through superior thoracic aperture and runs between mediastinal pleura and pericardium	Central portion of diaphragm
Intercostals	Ventral rami of T1–T11 nn.	Run in intercostal spaces between internal and innermost layers of intercostal muscles	Muscles in and skin over intercostal space lower nerves supply muscles and skin of anterolateral abdominal wall
Subcostal	Ventral ramus of T12 n.	Follows inferior border of 12th rib and passes into abdominal wall	Abdominal wall and skin of gluteal region
Recurrent laryngeal	Vagus n.	Loops around subclavian on right; on left runs around arch of aorta and ascends in tracheoesophageal groove	Intrinsic muscles of larynx (except cricothyroid); sensory inferior to level of vocal folds
Cardiac plexus	Cervical and thoracic cardiac branches of vagus nn. and sympathetic trunks	From arch of aorta and and posterior surface of heart, fibers extend along coronary arteries and to SA node	Impulses pass to SA node; parasympathetic fibers slow rate; reduce force of heartbeat, and constrict coronary arteries; sympathetic fibers have opposite effect
Pulmonary plexus	Vagus nn. and sympathetic trunks	Plexus forms on root of lung and extends along bronchial subdivisions	Parasympathetic fibers constrict bronchioles; sympathetic fibers dilate them
Esophageal plexus	Vagus nn., sympathetic ganglia and greater splanchnic nn.	Distal to tracheal bifurcation, the vagus and sympathetic nerves form a plexus around the esophagus	Vagal and sympathetic fibers to smooth muscle and glands of inferior two-thirds of the esophagus

SA, sinuatrial; AV, atrioventricular.

diaphragm, which it pierces near the vena caval foramen. The *left phrenic nerve* descends between the left subclavian and left common carotid arteries. It crosses the left surface of the arch of the aorta anterior to the left vagus nerve and passes over the left superior intercostal vein. It then descends anterior to the root of the left lung and runs along the pericardium, superficial to the left atrium and ventricle of the heart, where it pierces the diaphragm to the left of the pericardium.

The *trachea* descends in the neck anterior to the esophagus and enters the superior mediastinum, inclining a little to the right of the median plane (Fig. 2.19). The posterior surface of the trachea is flat where it is related to the esophagus. The trachea ends at the level of the sternal angle by dividing into the right and left main bronchi.

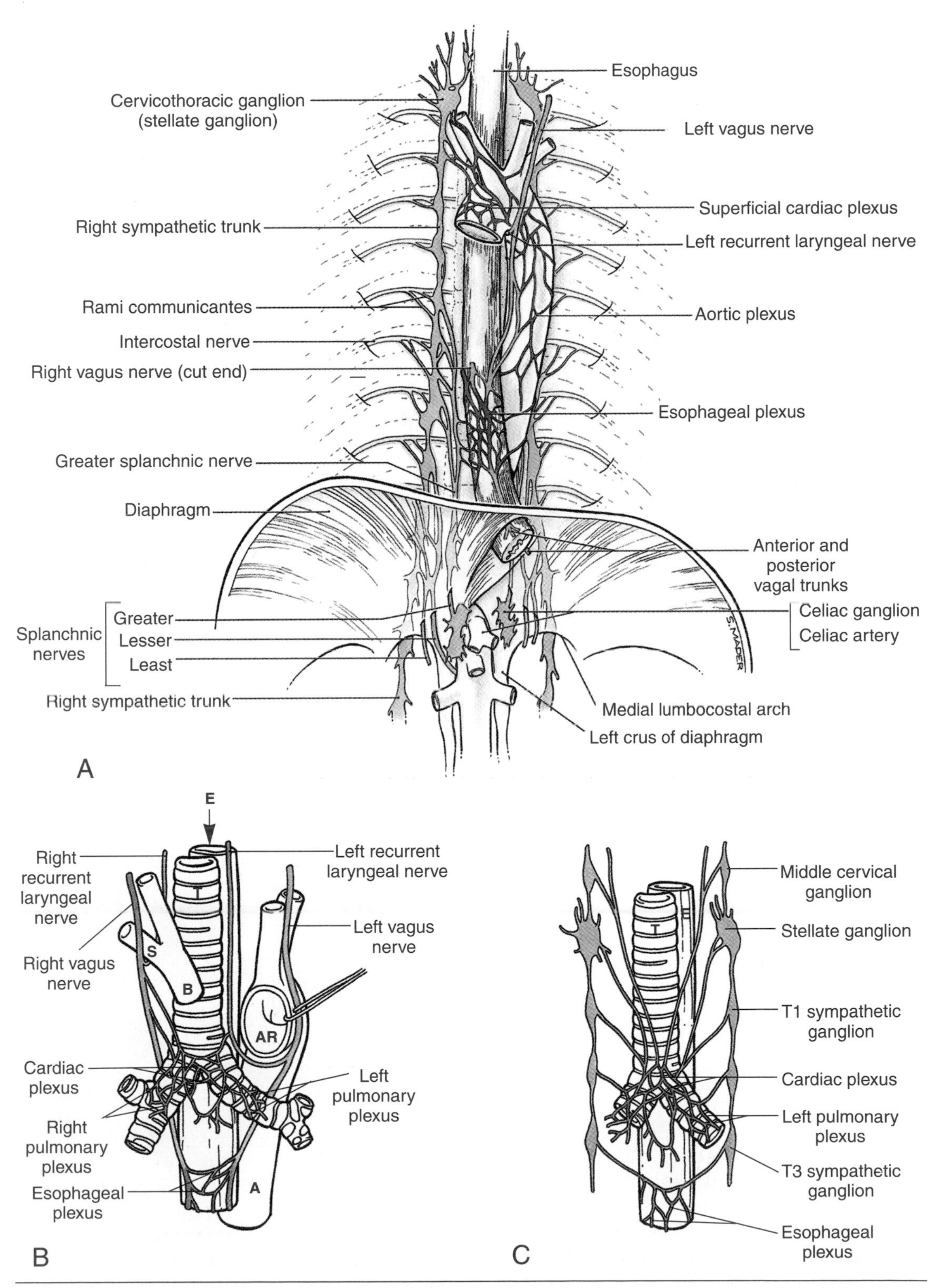

Figure 2.18. **A.** Autonomic nerves in superior and posterior parts of mediastinum. *Orange*, sympathetic; *green*, parasympathetic; *blue*, plexus. **B.** Parasympathetic nerves. *A*, aorta; *AR*, arch of aorta; *B*, right brachiocephalic artery; *E*, esophagus; *S*, right subclavian artery; *T*, trachea. **C.** Sympathetic nerves.

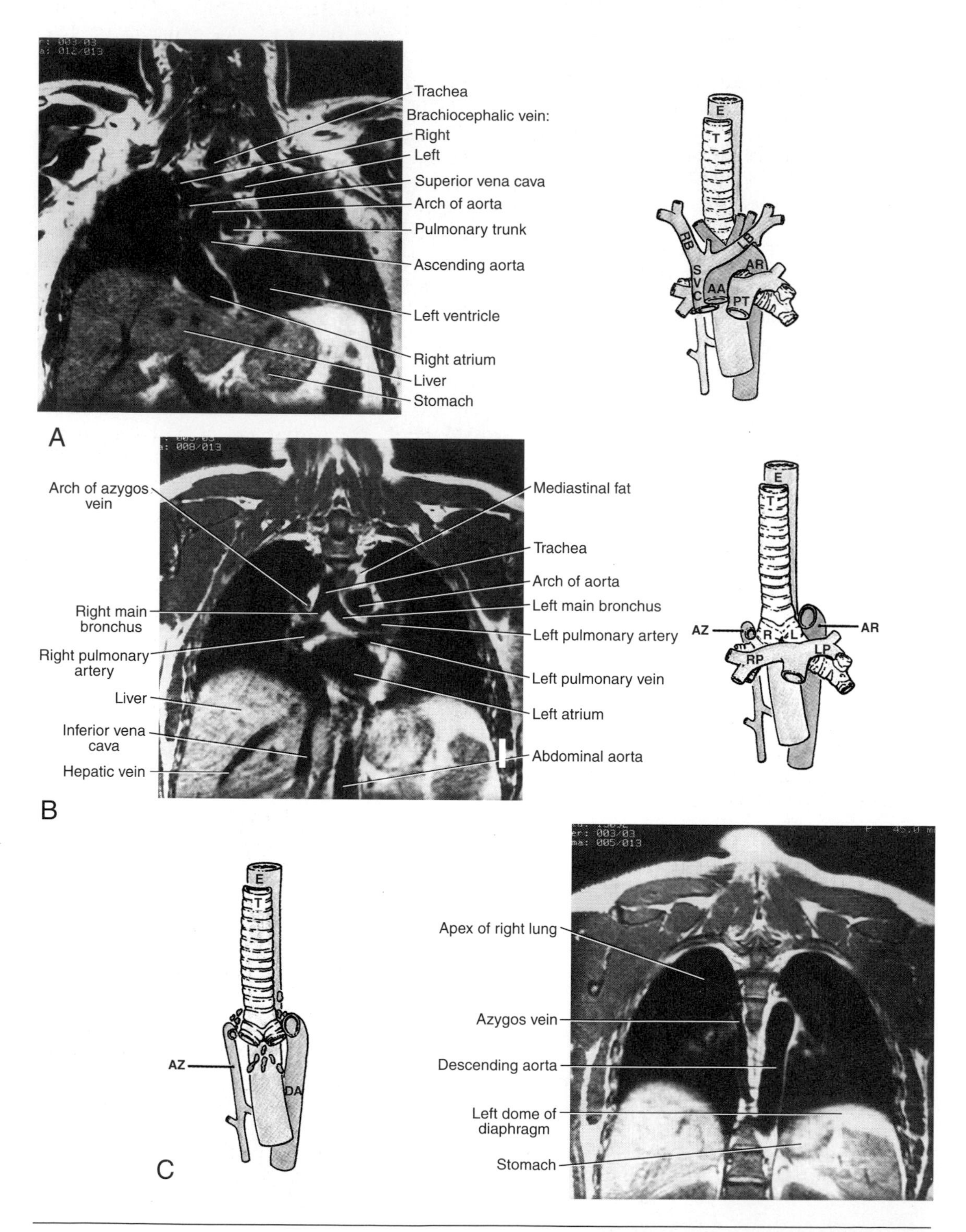

Figure 2.19. Coronal magnetic resonance images of thorax and upper abdomen, with orientation drawings of main structures. *AA*, ascending aorta; *AR*, arch of aorta; *AZ*, azygos vein; *DA*, descending aorta; *E*, esophagus; *L*, left main bronchus; *LB*, left brachiocephalic vein; *LP*, left pulmonary artery; *R*, right main bronchus; *RB*, right brachiocephalic vein; *RP*, right pulmonary vein; *T*, trachea.

Numerous tracheal and *tracheobronchial lymph nodes* are associated with the trachea (Fig. 2.11*C*). They are clinically important because lymph from the lungs drains into them. Consequently they enlarge when *bronchogenic carcinoma* develops. Widening or distortion of the carina of the trachea results from invasion and enlargement of these nodes by cancer cells. The *carina* is a ridge produced by the last tracheal cartilage. It runs anteroposteriorly between the orifices of the two main bronchi.

The *esophagus* is a fibromuscular tube that is usually flattened anteroposteriorly. It enters the superior mediastinum between the trachea and vertebral column, where it lies anterior to T1–T4 vertebrae (Fig. 2.17). Initially it inclines to the left, but it is moved by the arch of the aorta to the median plane opposite the root of the left lung. Inferior to the arch, the esophagus again inclines to the left as it approaches and passes through the diaphragm. The thoracic duct usually lies on the left side of the esophagus and deep to the arch of the aorta.

Posterior Mediastinum

The posterior mediastinum (*blue*) is located anterior to T5–T12 vertebrae, posterior to the pericardium and diaphragm, and between the parietal pleura of the two lungs. The posterior mediastinum contains the thoracic aorta, thoracic duct, posterior mediastinal lymph nodes, azygos and hemiazygos veins, esophagus, esophageal plexus, thoracic sympathetic trunks, and thoracic splanchnic nerves.

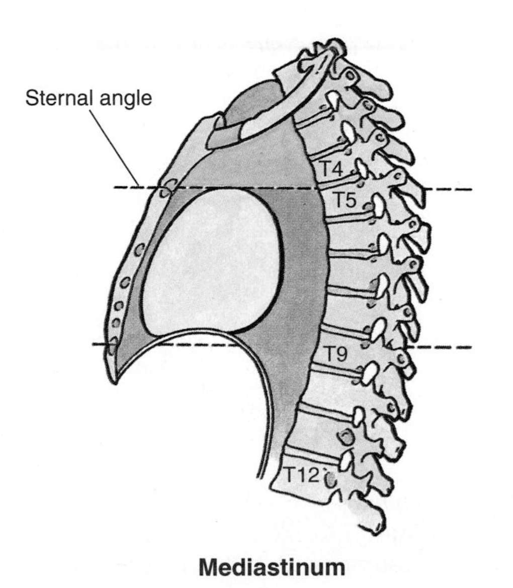

The *thoracic aorta* is the continuation of the arch of the aorta (Figs. 2.19C and 2.20, Table 2.7). It begins on the left side of the inferior border of the body of T4 vertebra and descends in the posterior mediastinum on the left sides of T5–T12 vertebrae. As it descends it approaches the median plane and displaces the esophagus to the right. The thoracic aorta descends through the posterior mediastinum against the left parietal pleura, with the thoracic duct and azygos vein to its right (Fig. 2.16*B*). The *thoracic aortic plexus*, an autonomic nerve network, surrounds it. The thoracic aorta lies posterior to the root of the left lung, the pericardium, and esophagus. It terminates anterior to the inferior border of T12 vertebra and enters the abdomen through the *aortic hiatus* (opening) in the diaphragm. The thoracic duct and azygos vein descend on its right side and accompany it through this hiatus. The thoracic aorta has the following branches (Figs. 2.11*A* and 2.20):

- Bronchial
- Esophageal

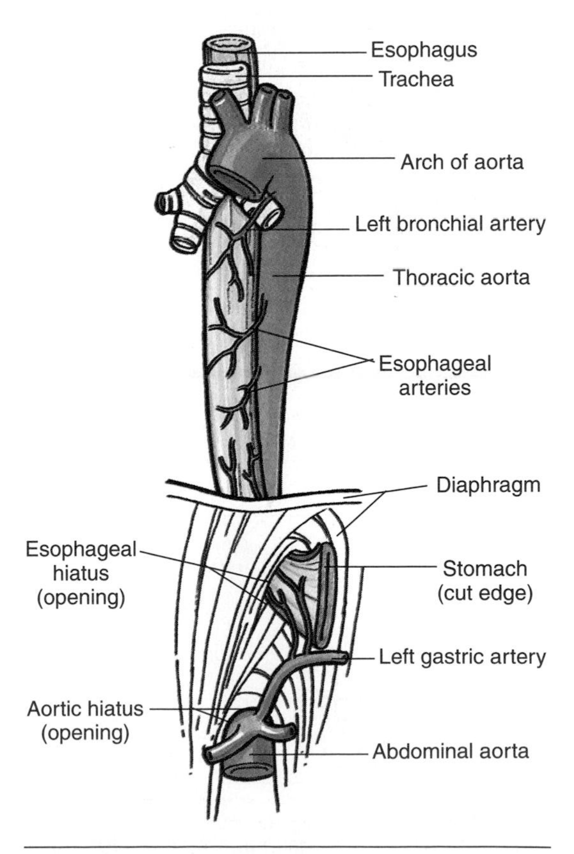

Figure 2.20. Anterior view of trachea, esophagus, and aorta.

- Pericardial
- Mediastinal
- Posterior intercostal
- Subcostal
- Superior phrenic

The *posterior intercostal arteries* (nine pairs) pass into the third to eleventh intercostal spaces. The *superior phrenic arteries* pass to the posterior surface of the diaphragm, where they anastomose with the musculophrenic and pericardiacophrenic branches of the internal thoracic artery.

The *thoracic duct,* the main lymphatic duct, lies on the bodies of the inferior seven thoracic vertebrae (Fig. 2.21). It conveys most lymph of the body to the venous system, i.e., from the lower limbs, pelvic cavity, abdominal cavity, left side of the thorax, and left side of the head, neck, and left upper limb. The thoracic duct ascends from the *cisterna chyli* through the aortic hiatus in the diaphragm. It is usually thin-walled and dull white; often it is beaded owing to its numerous valves. The thoracic duct ascends in the posterior mediastinum between the thoracic aorta and the azygos vein. At the level of T4, T5, or T6 vertebrae, the thoracic duct crosses to the left, posterior to the esophagus, and ascends into the superior mediastinum. The thoracic duct usually empties into the venous system near the union of the left internal jugular and subclavian veins.

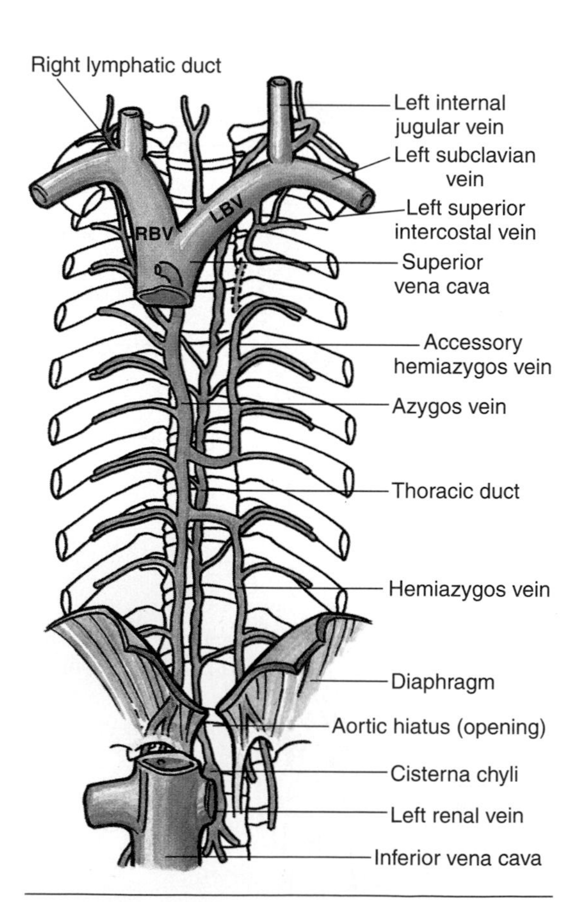

Figure 2.21. Azygos venous system and thoracic duct. *RBV,* right brachiocephalic vein; *LBV,* left brachiocephalic vein.

Because the thoracic duct is not easily identified, it may be inadvertently injured during investigative and/or surgical procedures in the posterior mediastinum. *Laceration of the thoracic duct* results in chyle escaping from it.

The *posterior mediastinal lymph nodes* lie posterior to the pericardium where they are related to the esophagus and thoracic aorta. There are several nodes posterior to the inferior part of the esophagus and more (up to eight) anterior and lateral to it. The posterior mediastinal lymph nodes receive lymph from the esophagus, the posterior aspect of the pericardium and diaphragm, and the middle posterior intercostal spaces.

The *azygos venous system,* on each side of the vertebral column, drains the back and thoracic and abdominal walls (Fig. 2.21). It exhibits much variation, not only in its origin but in its course, tributaries, anastomoses, and termination. The *azygos vein* and its main tributary, the *hemiazygos vein,* usually arise from the posterior aspect of the IVC and the left renal vein, respectively. The azygos vein connects the superior and inferior venae cavae and drains blood from the posterior walls of the thorax and abdomen. It ascends in the posterior mediastinum, passing close to the right sides of the bodies of the inferior eight thoracic vertebrae. It arches over the superior aspect of the root of the lung to join the SVC (Fig. 2.16*B*).

The azygos, hemiazygos, and accessory hemiazygos veins offer an alternate means of venous drainage from the thoracic, abdominal, and back regions when there is obstruction of the IVC.

The *esophagus* descends from the superior mediastinum into the posterior mediastinum, passing posterior and to the right of the arch of the aorta and posterior to the pericardium and left atrium. It then deviates to the left and passes through the *esophageal hiatus* in the diaphragm at about the level of T10 vertebra, anterior to the descending aorta (Fig. 2.20).

The *thoracic sympathetic trunks* and their associated ganglia represent a major portion of the autonomic nervous system (Fig. 2.18C). They are in continuity with the cervical and lumbar sympathetic trunks. The thoracic trunks lie against the heads of the ribs in the superior part of the chest, the costovertebral joints at midthoracic level, and the sides of the vertebral bodies in the inferior part of the thorax.

The *thoracic splanchnic nerves,* consisting of preganglionic fibers from the fifth to twelfth sympathetic ganglia, pass through the diaphragm and synapse in prevertebral ganglia in the abdomen. They supply sympathetic innervation for most of the abdominal viscera. The greater, lesser, and least splanchnic nerves (Fig. 2.18*A*) are discussed with the abdomen.

Anterior Mediastinum

The anterior mediastinum, the smallest subdivision of the mediastinum, lies between the body of the sternum and the transversus thoracis muscles anteriorly and the pericardium posteriorly. It is continuous with the superior mediastinum at the sternal angle and is limited inferiorly by the diaphragm. The anterior mediastinum consists of loose connective tissue, fat, lymphatic vessels, some lymph nodes, and branches of the internal thoracic vessels (Fig. 2.4*B*).

3 / ABDOMEN

The abdomen is between the thorax and pelvis. The abdominal cavity is

- Surrounded by the abdominal wall
- Separated from the thoracic cavity by the diaphragm
- Under cover of the thoracic cage superiorly
- Continuous inferiorly with the pelvic cavity

The *abdominal cavity* contains the peritoneum, most of the digestive organs (i.e., stomach, intestine, liver, gallbladder, and pancreas), as well as the spleen, kidneys, suprarenal (adrenal) glands, and parts of the ureters. To describe the location of an abdominal organ or the distribution of pain, the abdominal cavity is usually divided either into *nine regions* (areas), which are defined by four planes—two horizontal (subcostal and transtubercular) and two vertical (midclavicular)—or more simply into *four quadrants*, which are defined by two planes—one horizontal (transumbilical) and one vertical (median).

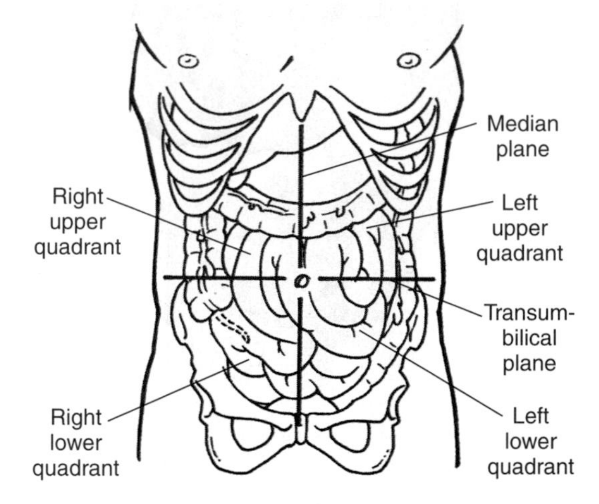

Four abdominal quadrants

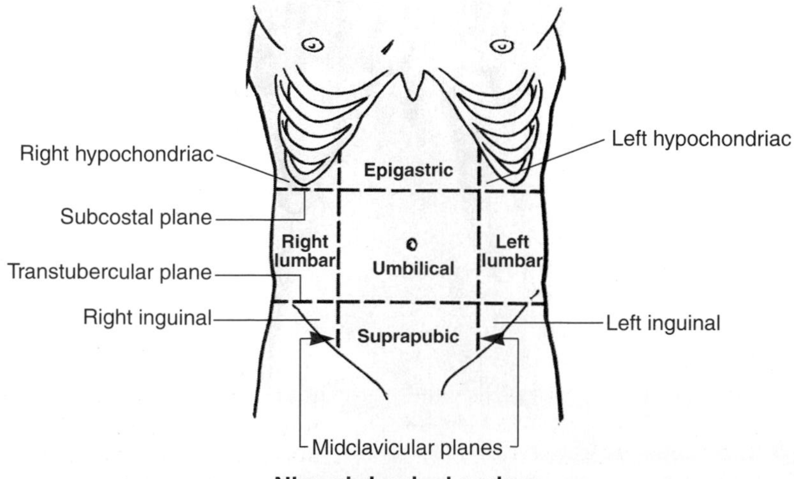

Nine abdominal regions

The horizontal planes are the

- Subcostal plane passing through the inferior border of the tenth costal cartilage on each side and the body of L3 vertebra
- Transumbilical plane passing through the umbilicus and the disc between L3 and L4 vertebrae
- Transtubercular plane passing through the iliac tubercles [tubercles of iliac crests (p. 85)] and the body of L5 vertebra

The vertical planes are the

- Median plane passing longitudinally through the body, dividing it into right and left halves
- Midclavicular plane passing from the midpoint of the clavicle to the midinguinal point [middle of the line joining the anterior superior iliac spine and pubic symphysis (p. 85)]

Anterolateral Abdominal Wall

The anterolateral abdominal wall is bounded superiorly by the cartilages of the 7th to 12th ribs and the xiphoid process and inferiorly by the inguinal ligament and pelvic bones. The wall consists of skin, subcutaneous connective tissue, muscles, fasciae, and peritoneum (Figs. 3.1 and 3.2).

FASCIAE OF ANTEROLATERAL ABDOMINAL WALL

The *superficial fascia* (subcutaneous connective tissue) over most of the wall consists of one layer that contains a variable amount of fat. In the inferior part of the wall, the superficial fascia can be divided into two layers:

- A fatty superficial layer (Camper's fascia)
- A membranous deep layer (Scarpa's fascia)

A thin layer of *investing* or *deep fascia* invests the external oblique (Fig. 3.1*A*) and cannot be separated easily from this muscle. A firm membranous sheet called the *transversalis fascia* lines most of the abdominal wall (Fig. 3.2*B*). This fascia covers the deep surface of the transversus abdominis and its aponeurosis, and its right and left sides are continuous deep to the *linea alba*. The parietal peritoneum is internal to the transversalis fascia and is separated from it by a variable amount of extraperitoneal fat.

Surgeons use the membranous deep layer of superficial fascia for holding sutures when closing abdominal skin incisions. Between this layer and the deep fascia covering the rectus abdominis and external oblique muscles, there is a potential space where fluid may accumulate (e.g., urine from a ruptured urethra). The fluid cannot spread into the thigh because the deep layer of fascia fuses with the deep fascia of the thigh along a line inferior and parallel to the inguinal ligament.

MUSCLES OF ANTEROLATERAL ABDOMINAL WALL

There are four important muscles in the anterolateral abdominal wall (Figs. 3.1 and 3.2): three *flat muscles* (external oblique, internal oblique, and transversus abdominis) and one straplike *vertical muscle* (rectus abdominis). Their attachments, nerve supply, and main actions are described in Table 3.1.

- The *external oblique* is the superficial flat muscle; its fibers pass inferomedially, and slips of the serratus anterior, a pectoral (chest) muscle, interdigitate with those of the external oblique
- The *internal oblique* is the intermediate flat muscle; its fibers run at right angles to those of the external oblique
- The *transversus abdominis* is the innermost flat muscle; its fibers, except for the most inferior ones, run more or less horizontally
- The *rectus abdominis* is the prominent, straplike vertical muscle, most of which is enclosed in the *rectus sheath*; the anterior layer of this sheath is firmly attached to the rectus at three *tendinous intersections*, which are located at the level of the xiphoid process, umbilicus, and halfway between these structures

All three flat muscles end anteriorly in a strong sheetlike aponeurosis. The fibers of each aponeurosis interlace at the *linea alba* with their fellows of the opposite side to form the sheath of the rectus muscle.

The rectus sheath has

- An anterior layer consisting of the aponeuroses of the external oblique and internal oblique
- A posterior layer consisting of the fused aponeuroses of the internal oblique and transversus abdominis (the inferior one-fourth of this layer is deficient because the aponeuroses of the three flat muscles pass anterior to the rectus abdominis, leaving the

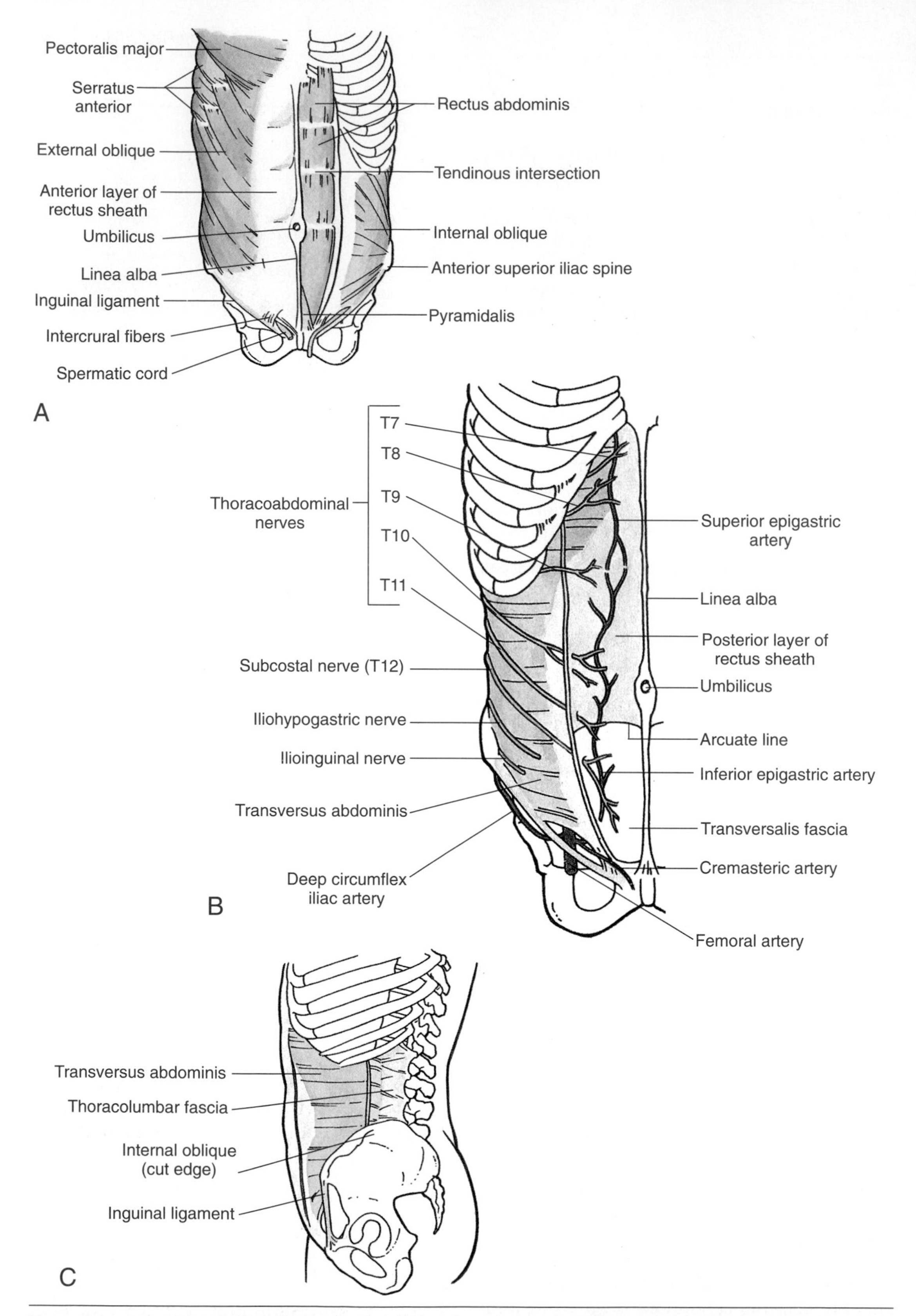

Figure 3.1. Anterolateral abdominal wall. **A.** Anterior view of external oblique muscle and rectus sheath on right and internal oblique and rectus abdominis muscle on left. **B.** Anterior view of transversus abdominis muscle and cutaneous nerves. Rectus abdominis has been removed to show extent of posterior layer of rectus sheath and epigastric vessels. **C.** Lateral view showing transversus abdominis muscle and thoracolumbar fascia.

Table 3.1
Principal Muscles of Anterolateral Abdominal Wall

Muscles	Origin	Insertion	Innervation	Action(s)
External oblique	External surfaces of 5th to 12th ribs	Linea alba, pubic tubercle, and anterior half of iliac crest	Inferior six thoracic nn. and subcostal n.	Compress and support abdominal viscera; flex and rotate trunk (external and internal oblique)
Internal oblique	Thoracolumbar fascia, anterior two-thirds of iliac crest, and lateral half of inguinal ligament	Inferior borders of 10th–12th ribs, linea alba,and pubis via conjoint tendon	Ventral rami of inferior six thoracic and first lumbar nn. (internal oblique and transversus abdominis)	
Transversus abdominis	Internal surfaces of seventh to twelth costal cartilages, thoracolumbar fascia, iliac crest, and lateral third of inguinal ligament	Linea alba with aponeurosis of internal oblique, pubic crest, and pecten pubis via conjoint tendon		Compresses and supports abdominal viscera
Rectus abdominis	Pubic symphysis and pubic crest	Xiphoid process and fifth to seventh costal cartilages	Ventral rami of inferior six thoracic nn.	Flexes trunk and compresses abdominal viscera

posterior surface of the muscle in contact with the transversalis fascia)

- A crescentic line of transition called the *arcuate line* between the transversalis fascia and the aponeurotic posterior wall of the rectus sheath

The contents of the rectus sheath are the rectus abdominis; a small inconstant muscle, the pyramidalis; the superior and inferior epigastric vessels; lymphatic vessels; and the ventral primary rami of T7–T12 nerves.

The three-ply structure of the flat muscles and their aponeuroses form a strong expandable support for the anterolateral abdominal wall and provide considerable protection for the viscera (e.g., stomach and intestine). When the diaphragm contracts, the anterolateral abdominal wall expands as its muscles relax. When the diaphragm relaxes, the wall passively sinks in as the muscles contract. The muscles also compress the viscera and increase the intraabdominal pressure. The combined actions of the four anterolateral muscles assist in expelling air during expiration and produce the force required for defecation (bowel movement), micturition (urination), and parturition (childbirth). These muscles are also involved in movements of the trunk and pelvis and in the maintenance of posture. The rectus muscle is the most powerful flexor of the thoracic and lumbar regions of the vertebral column. The rectus can be seen and palpated when a supine patient is asked to raise the head and shoulders against resistance. The flat abdominal muscles also assist in movements of the trunk, especially lateral flexion and rotation, but their principal role is controlling intraabdominal pressure.

NERVES OF ANTEROLATERAL ABDOMINAL WALL

The skin and muscles of the anterolateral abdominal wall are supplied mainly by the thoracoabdominal *intercostal nerves* (Fig. 3.1*B*, Table 3.2), formed by the ventral rami of the *inferior six thoracic nerves* (T7–T11) and by the *subcostal nerves* (T12). The intercostal nerves pass inferoanteriorly from the intercostal spaces and run between the internal oblique and transversus abdominis to supply the abdominal skin and muscles.

The *cutaneous nerves* pierce the rectus sheath a short distance from the median plane. Branches of cutaneous nerve(s)

- T7–T9 supply the skin superior to the umbilicus
- T10 innervates the skin around the umbilicus
- T11, T12, and L1 supply the skin inferior to the umbilicus

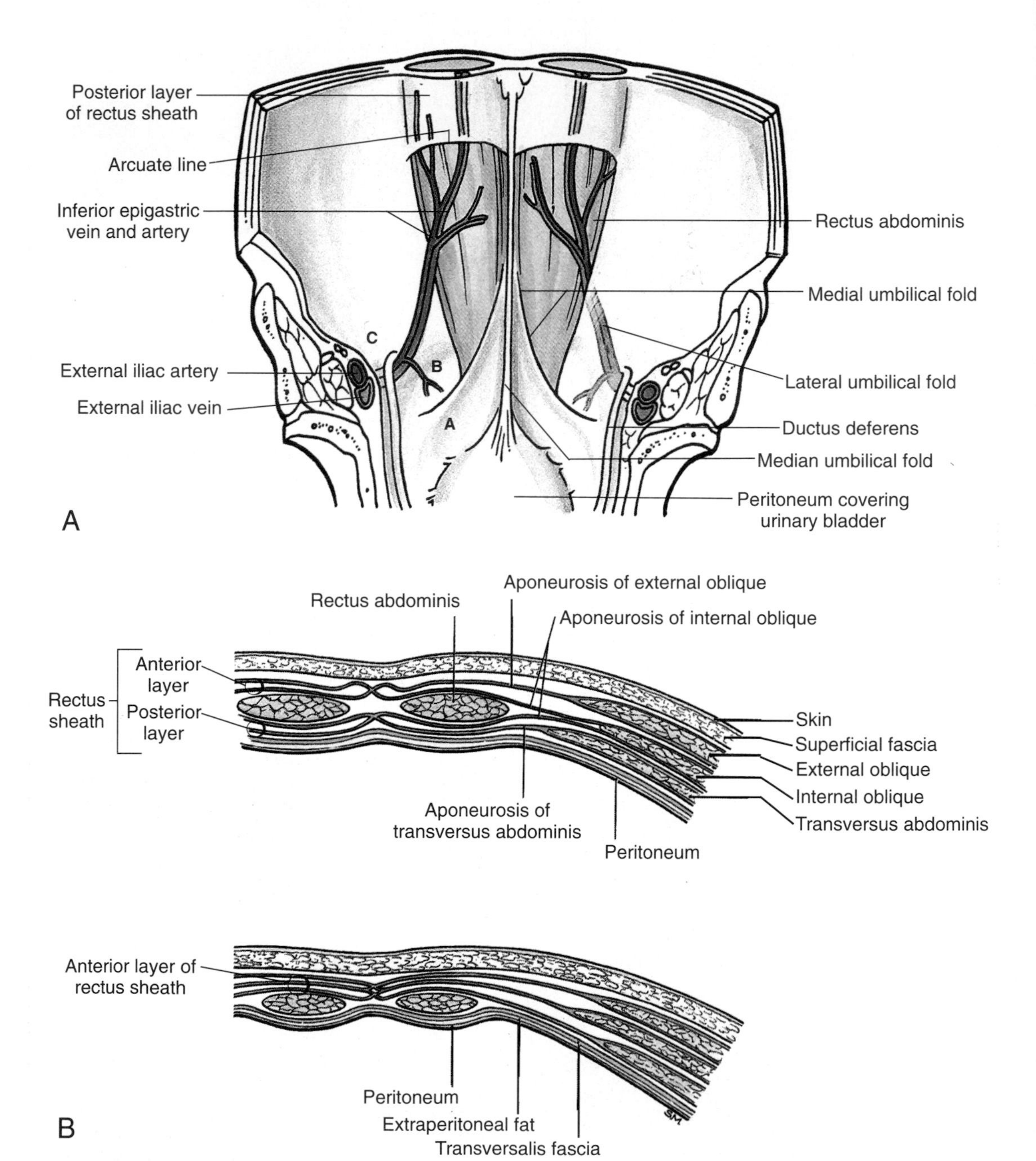

Figure 3.2. **A.** Internal view of infraumbilical part of anterior abdominal wall showing umbilical peritoneal folds and fossae: supravesical fossa (*A*), medial inguinal fossa (*B*), and lateral inguinal fossa (*C*). **B.** Rectus sheath. Transverse sections of anterolateral abdominal wall superior to umbilicus (*upper*) and inferior to umbilicus (*lower*).

Surface Anatomy of Anterolateral Abdominal Wall

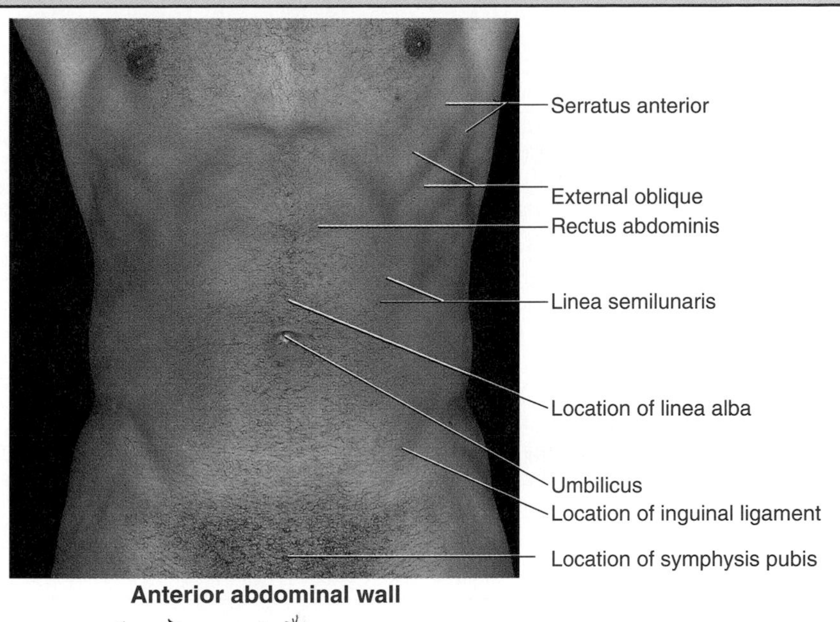

Anterior abdominal wall

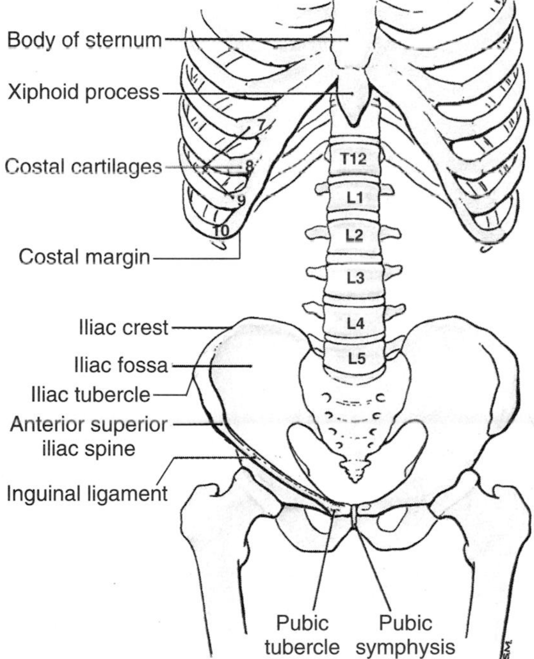

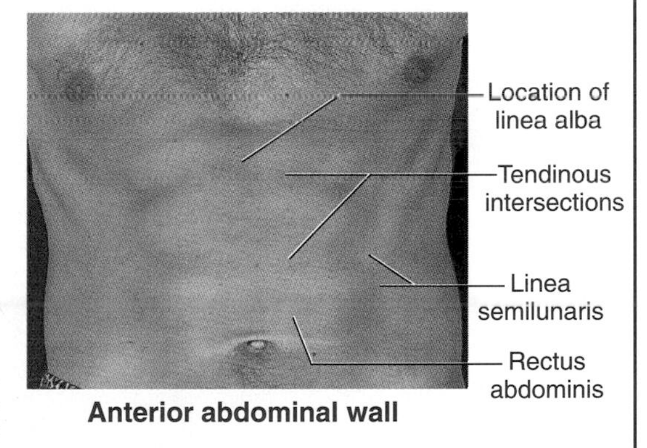

Anterior abdominal wall

The *umbilicus* is an obvious feature of this wall and is the reference point for the transumbilical plane. The *linea alba*, often indicated by a vertical skin groove, is a subcutaneous fibrous band extending from the xiphoid process to the pubic symphysis (Fig. 3.1*B*). This symphysis is a cartilaginous joint that can be felt as a firm resistance in the median plane distal to the linea alba. The bony *iliac crest* at the level of L4 vertebra can be easily palpated as it extends posteriorly from the *anterior superior iliac spine*. Curved skin grooves called *lineae semilunaris* extend from the inferior tips of the costal margins near the ninth costal cartilages to the *pubic tubercles*. These semilunar lines indicate the lateral borders of the rectus abdominis. Three transverse skin grooves overlie the tendinous intersections of the rectus muscle. The interdigitating bellies of the serratus anterior and external oblique muscles are usually visible. The location of the *inguinal ligament* is indicated by a skin crease just inferior and parallel to the ligament. This crease marks the division between the anterolateral abdominal wall and thigh.

Table 3.2
Principal Nerves of Anterolateral Abdominal Wall

Nerve	Origin	Course	Distribution
Thoracoabdominal (T7–T11)	Continuations of lower intercostal nn.	Run between second and third layers of abdominal mm.	Anterior abdominal mm. and overlying skin; periphery of diaphragm
Subcostal (T12)	Ventral ramus of twelfth thoracic n.	Runs along inferior border of 12th rib	Lowest slip of external oblique m. and skin over anterior superior iliac spine and hip
Iliohypogastric [L1 (T12)]	Chiefly from ventral ramus of first lumbar n.	Pierces transversus abdominis m.; branches pierce external oblique aponeurosis	Skin of hypogastric region and over iliac crest; internal oblique and transversus abdominis
Ilioinguinal (L1)	Ventral ramus of first lumbar n.	Passes between second and third layers of abdominal mm. and passes through inguinal canal	Skin of scrotum or labium majus, mons pubis, and adjacent medial aspect of thigh; internal oblique and transversus abdominis

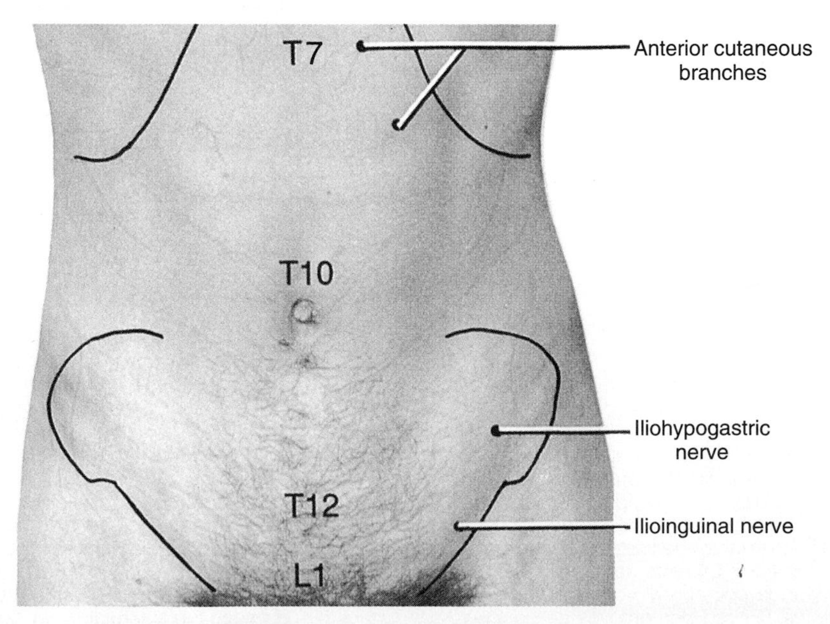

Nerves of anterolateral abdominal wall

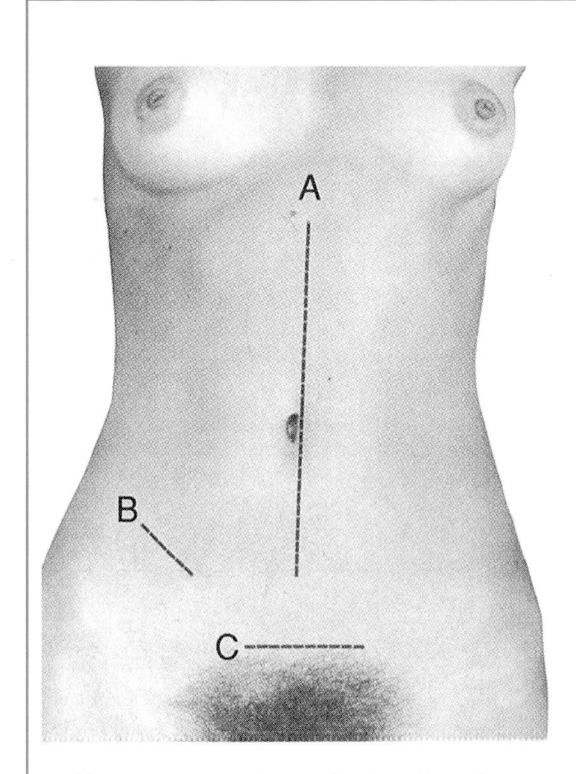

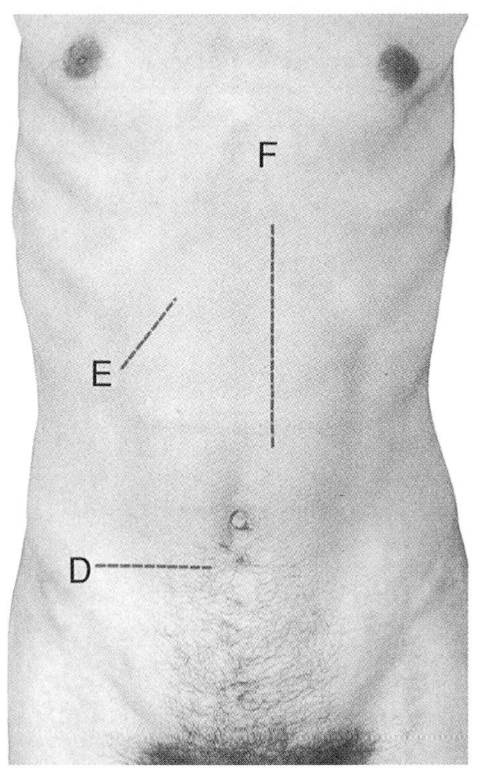

The cutaneous and muscular branches of nerves run inferoanteriorly and parallel to each other. For this reason, transverse incisions (e.g., *D*) through the rectus abdominis provide good access to the viscera (internal organs) and cause the least possible damage to the nerve supply of the muscle. Transverse incisions are not made through the tendinous intersections because cutaneous nerves and branches of the superior epigastric vessels pierce them (Fig. 3.1*B*). These incisions can be made without serious damage because a new transverse band similar to a tendinous intersection forms when parts of the muscle are rejoined.

Median (midline) incisions (*A*) are often used because they can be made rapidly through the linea alba without cutting major vessels or nerves. When making *paramedian (parasagittal) incisions* (e.g., *F*), the anterior layer of the rectus sheath is incised, the rectus is pulled laterally, and the posterior layer of the sheath is incised. The rectus is not pulled medially because its nerves entering the sheath laterally may be torn.

The common incision for an appendectomy is the muscle-splitting (gridiron) incision (*B*). A subcostal incision (*E*) is used to expose the gallbladder, and a suprapubic incision (*C*) is used for exposure of the uterus and tubes.

VESSELS OF ANTEROLATERAL ABDOMINAL WALL

The arteries and veins of the anterolateral abdominal wall (Figs. 3.1, *A* and *B*, and 3.2*A*, Table 3.3) are the

- Superior epigastrics from the internal thoracic vessels
- Inferior epigastrics and deep circumflex iliacs from the external iliac vessels
- Small vessels from anterior and collateral branches of the *posterior intercostal vessels* in the tenth and eleventh intercostal spaces and from anterior branches of the *subcostal vessels*

The superficial *lymphatic vessels* superior to the umbilicus drain to the axillary lymph nodes, whereas those inferior to it drain to the superficial inguinal lymph nodes. The deep lymphatic vessels drain to the lumbar (lateral aortic) and common and external iliac lymph nodes.

INTERNAL SURFACE OF ANTEROLATERAL ABDOMINAL WALL

The internal surface of the anterolateral abdominal wall is covered with *parietal peritoneum*. The infraumbilical part of the wall exhibits several *peritoneal folds*, some of which contain remnants of vessels that carried blood

Table 3.3
Arterial Supply to Anterolateral Abdominal Wall

Artery	Origin	Course	Distribution
Superior epigastric	Internal thoracic a.	Descends in rectus sheath deep to rectus abdominis	Rectus abdominis and anterolateral abdominal wall
Inferior epigastric	External iliac a.	Runs superiorly and enters rectus sheath; runs deep to rectus abdominis	Rectus abdominis and anterolateral abdominal wall
Deep circumflex iliac	External iliac a.	Runs on deep aspect of anterior abdominal wall, parallel to inguinal ligament	Iliacus muscle and inferior part of anterolateral abdominal wall
Superficial circumflex iliac	Femoral a.	Runs in superficial fascia along inguinal ligament	Superficial fascia and skin over inferior portion of anterolateral abdominal wall
Superficial epigastric	Femoral a.	Runs in superficial fascia and runs toward umbilicus	Superficial fascia and skin over suprapubic region

to and from the fetus (Fig. 3.2*A*). *Five umbilical folds* (two on each side and one in the median plane) pass superiorly toward the umbilicus.

- Two lateral umbilical folds covering the inferior epigastric vessels run superomedially on each side (because these folds contain blood vessels, they bleed if cut)
- Two medial umbilical folds cover the medial umbilical ligaments, the remnants of the fetal umbilical arteries
- One median umbilical fold extending from the apex of the urinary bladder to the umbilicus covers the median umbilical ligament, the remnant of the *urachus* that joined the apex of the fetal bladder to the umbilicus

The fossae between the umbilical folds are the

- Supravesical fossae between the median and medial umbilical folds
- Medial inguinal fossae between the medial and lateral umbilical folds
- Lateral inguinal fossae lateral to the lateral umbilical folds, which are useful landmarks during laparoscopic repair of inguinal hernias

INGUINAL CANAL

The inguinal canal is an oblique, inferomedially directed passage for the spermatic cord through the inferior part of the anterior abdominal wall (Fig. 3.3, *A* and *B*). It lies parallel and just superior to the medial half of the *inguinal ligament*. The contents of the inguinal canal are the spermatic cord in males, the round ligament of the uterus in females, and the ilioinguinal nerve in both sexes. The inguinal canal has an

- Anterior wall formed mainly by the external oblique aponeurosis that is reinforced laterally by internal oblique fibers
- Posterior wall formed by transversalis fascia that is reinforced medially by the *conjoint tendon*, the common tendon of the internal oblique and transversus abdominis
- Roof formed by arching fibers of the internal oblique and transversus abdominis
- Floor formed by the superior surface of the inguinal ligament that is reinforced medially by the *lacunar ligament*, an extension of the inguinal ligament
- Superficial inguinal ring, a triangular aperture in the external oblique aponeurosis; its margins are called *crura* (the lateral crus is attached to the pubic tubercle and the medial crus to the body of the pubis; *intercrural fibers* from the inguinal ligament arch across the superficial ring, helping to prevent the crura from spreading apart)
- Deep inguinal ring, an outpouching of the transversalis fascia located just superior to the midpoint of the inguinal ligament and lateral to the inferior epigastric artery

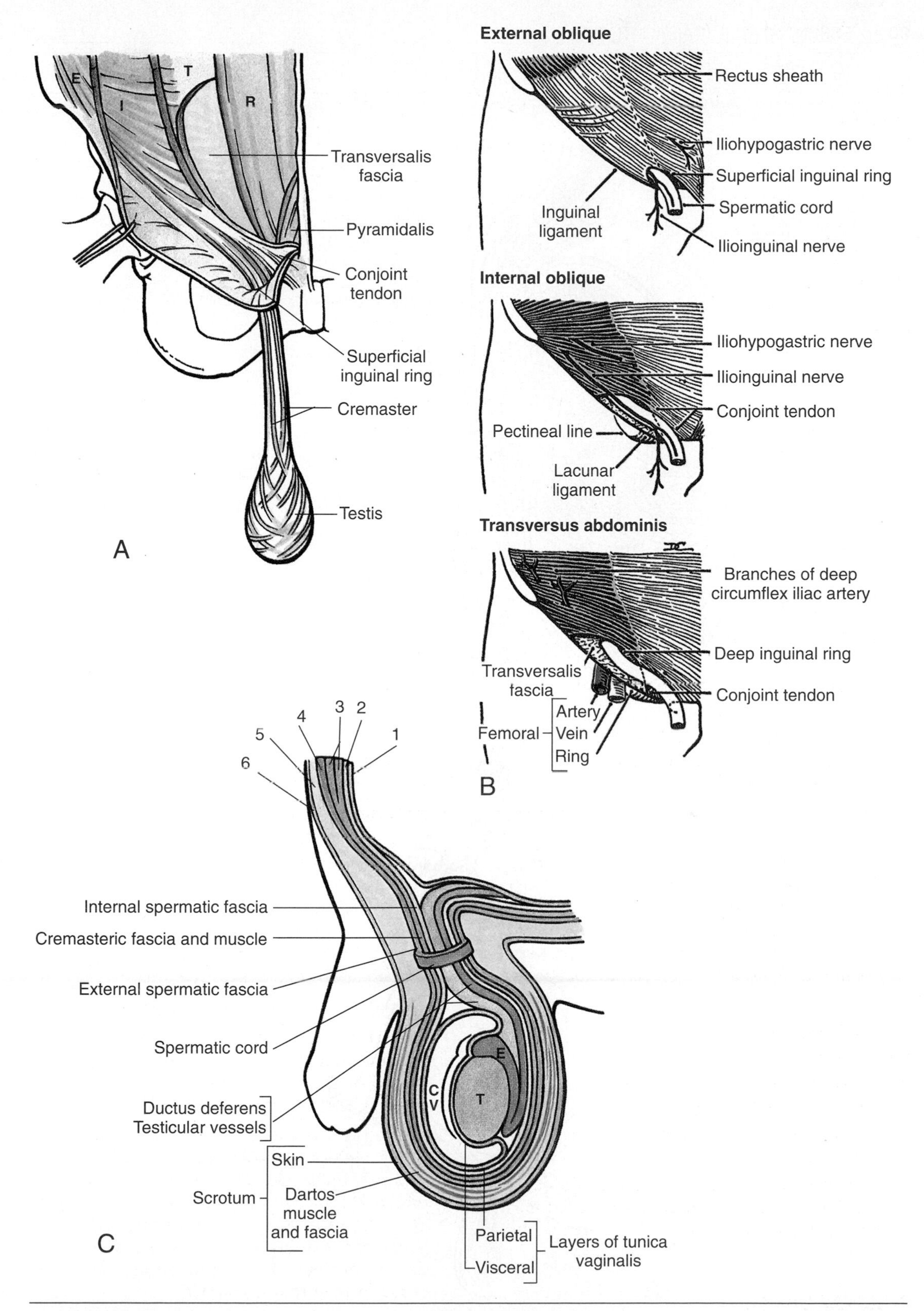

Figure 3.3. Inguinal region. **A.** Anterior view of a dissection showing spermatic cord in inguinal canal and emerging from superficial ring. *E*, external oblique; *I*, internal oblique; *T*, transversus abdominis; *R*, rectus abdominis. **B.** Progressive dissections of flat abdominal muscles showing walls of inguinal canal and superficial and deep inguinal rings. **C.** Coverings of spermatic cord and testis (*T*). *E*, epididymis; *CV*, cavity of tunica vaginalis; *1*, peritoneum; *2*, transversalis fascia; *3*, transversus abdominis, internal oblique; *4*, external oblique; *5*, subcutaneous fat; *6*, skin

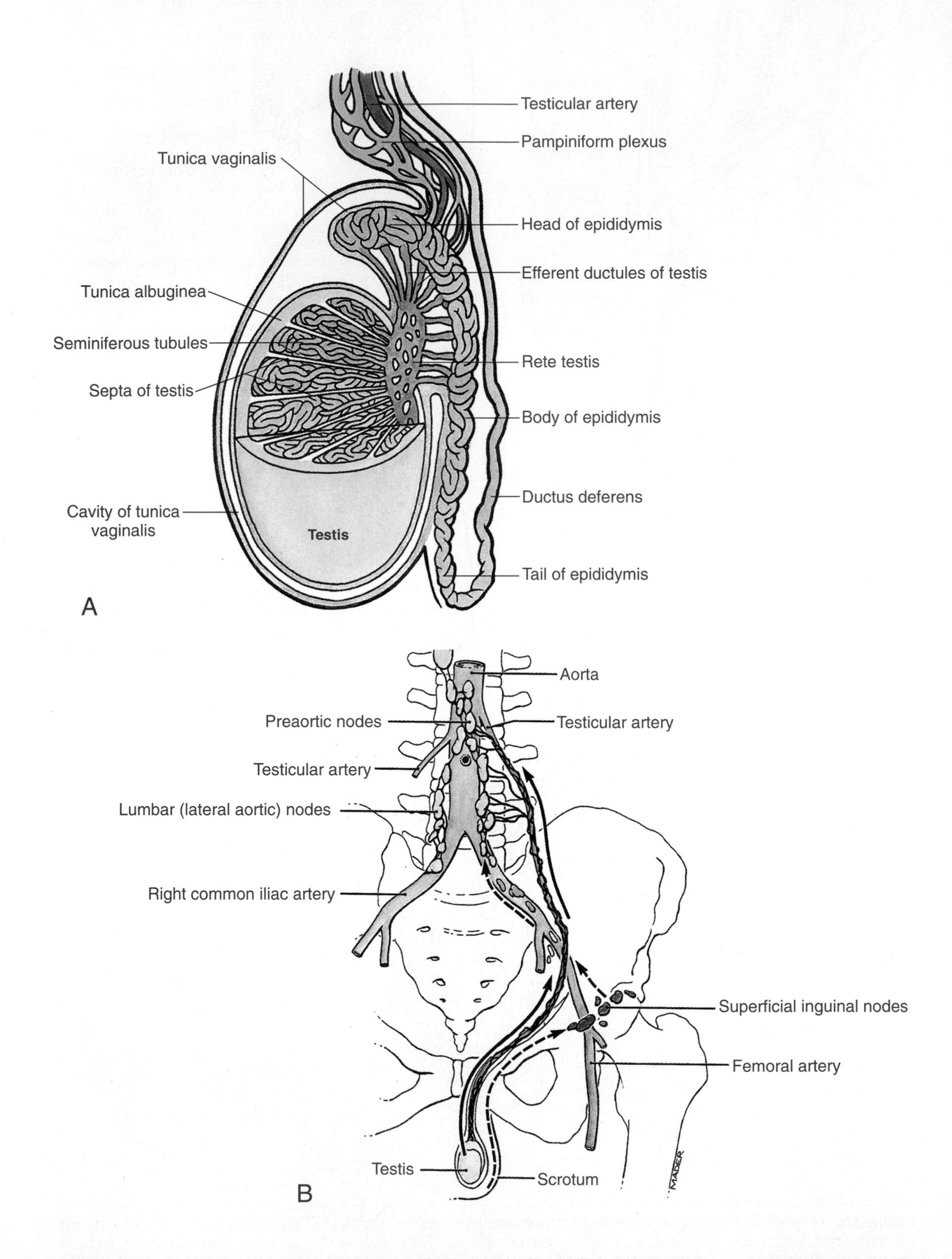

Figure 3.4. **A.** Dissection of testis, epididymis, and spermatic cord. **B.** Lymphatic drainage of testis and scrotum. *Arrows*, flow of lymph to lymph nodes.

The deep and superficial inguinal rings do not overlap because of the oblique path of the inguinal canal (Fig. 3.3*B*). Consequently, increases in intraabdominal pressure act on the deep ring, forcing the posterior wall of the canal against the anterior wall and strengthening this weak part of the abdominal wall. Contraction of the external oblique approximates the anterior wall of the canal to the posterior wall. Contraction of the internal oblique and transversus abdominis makes the roof of the canal descend, which constricts the canal. The superficial inguinal ring is palpable superolateral to the pubic tubercle and can be examined by invaginating the scrotum superior to the testis with the index finger and probing gently superolaterally along the spermatic cord. If the ring is dilated, it may admit the digit without causing pain.

SPERMATIC CORD

The spermatic cord (Figs. 3.3 and 3.4*A*) suspends the testis in the scrotum and contains structures running to and from the testis. It begins at the deep inguinal ring lateral to the inferior epigastric artery, passes through the inguinal canal, and ends at the posterior border of the testis in the scrotum. The spermatic cord is surrounded by fascial coverings derived from the anterior abdominal wall (Fig. 3.3C).

The *coverings of the spermatic cord* are formed by three layers of fascia derived from the anterior abdominal wall during the fetal period:

- The internal spermatic fascia from the transversalis fascia
- The cremasteric fascia from the fascia covering the internal oblique
- The external spermatic fascia from the external oblique aponeurosis

The cremasteric fascia contains loops of the *cremaster muscle* that reflexly draws the testis superiorly in the scrotum, particularly when it is cold. The cremaster, derived from the internal oblique, is innervated by the genital branch of the genitofemoral nerve (L1, L2).

The *constituents of the spermatic cord* are

- Ductus deferens (vas deferens), a muscular tube about 45 cm long that conveys sperms from the epididymis
- Testicular artery arising from the lateral aspect of the aorta that supplies the testis and epididymis
- Artery of the ductus deferens arising from the inferior vesical artery
- Cremasteric artery arising from the inferior epigastric artery
- Pampiniform plexus, a venous network formed by the anastomosis of up to 12 veins
- Sympathetic nerve fibers on the arteries and sympathetic and parasympathetic fibers on the ductus deferens
- Genital branch of the genitofemoral nerve supplying the cremaster muscle
- Lymphatic vessels draining the testis and closely associated structures and passing to the lumbar and preaortic lymph nodes

The vinelike pampiniform plexus of veins may dilate, becoming varicose and producing a *varicocele*. These dilated and tortuous vessels often result from defective valves in the testicular vein. The wormlike enlargement usually disappears when the person lies down.

The testes descend from the posterior abdominal wall during development. As a result, their lymphatic drainage differs from that of the scrotum, an outpouching of the abdominal skin (Fig. 3.4*B*). Consequently,

- *Cancer of the testis* metastasizes to the lumbar and preaortic lymph nodes
- *Cancer of the scrotum* metastasizes to the superficial inguinal nodes

Contraction of the cremaster muscle can be produced by lightly stroking the skin on the medial aspect of the superior part of the thigh with an applicator stick or tongue depressor. This area is supplied by the *ilioinguinal nerve*. The rapid elevation of the testis on the same side is called the *cremasteric reflex*.

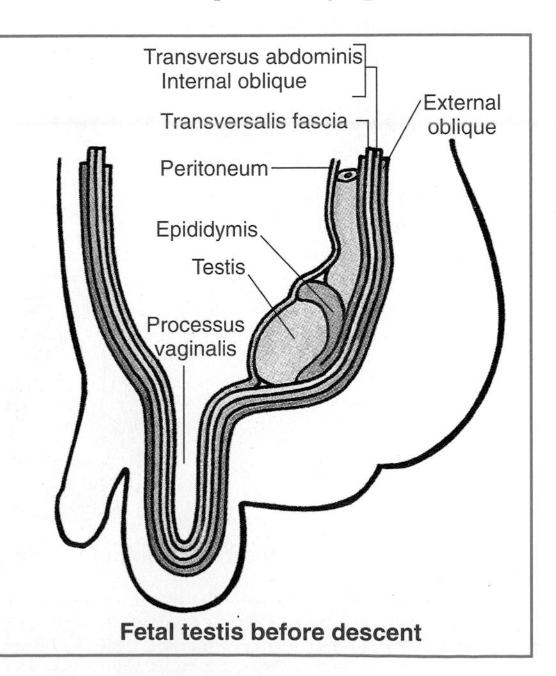

Fetal testis before descent

A *hernia* is a protrusion of a structure, viscus (organ), or part of it through a normal or abnormal opening from the cavity in which it belongs. About 90% of hernias are located in the inguinal region. There are two types of inguinal hernia: indirect and direct.

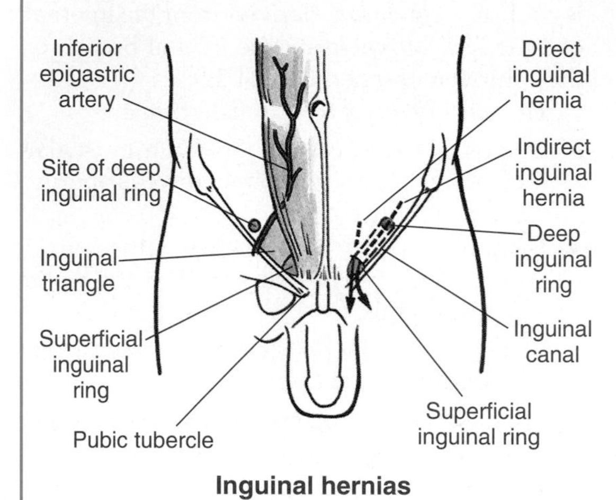

Inguinal hernias

An indirect inguinal hernia

- Leaves the abdominal cavity lateral to the inferior epigastric vessels to enter the deep inguinal ring
- Passes through the inguinal canal
- Exits the superficial inguinal ring and enters the scrotum
- Has a hernial sac formed by the persistent processus vaginalis
- Is covered by all three fascial coverings of the spermatic cord.

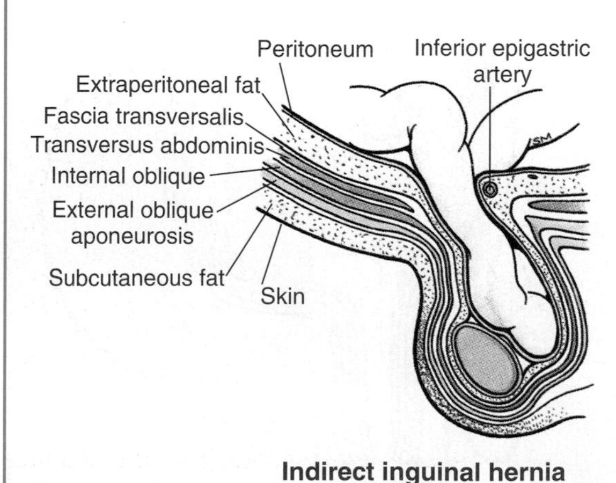

Indirect inguinal hernia

An indirect inguinal hernia can occur in women, but it is about 20 times more common in males of all ages.

A direct inguinal hernia

- Has a peritoneal covering
- Leaves the abdominal cavity medial to the inferior epigastric vessels (usually passes through the inferior part of the inguinal triangle that lies between the inferior epigastric artery laterally, the rectum abdominis medially, and the inguinal ligament inferiorly)

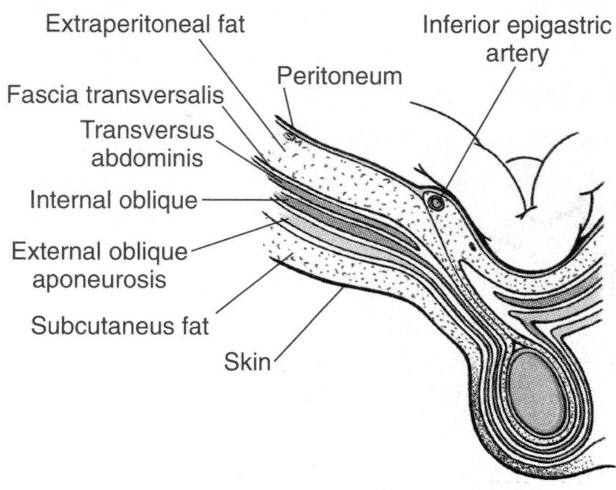

Direct inguinal hernia

- Does not enter the deep inguinal ring (protrudes through posterior wall of inguinal canal)
- Lies outside the coverings of the spermatic cord
- Emerges through or around the conjoint tendon to reach the superficial inguinal ring

In the area of the *inguinal triangle* (Hesselbach triangle), the transversalis fascia is covered only by the external oblique. Hence this triangle is a weak area of the abdominal wall. Direct inguinal hernias are much less common than indirect inguinal hernias, and they are usually acquired in middle-aged and elderly men. Direct inguinal hernias are uncommon in women.

SCROTUM

The scrotum is a cutaneous sac consisting of two layers: skin and superficial fascia (Figs. 3.3C and 3.4). The superficial fascia is devoid of fat but contains a thin sheet of smooth muscle called the *dartos* that contracts in response to cold, reducing the surface area of the skin. The fascia is continuous anteriorly with the membranous deep layer of superficial fascia of the anterolateral abdominal wall and posteriorly with the superficial fascia of the perineum.

The arterial supply of the scrotum (Figs. 3.1*B*, 3.3*B*, and 3.4) is from the

- Perineal branch of the *internal pudendal artery*
- External pudendal branches of the *femoral artery*
- Cremasteric branch of the *inferior epigastric artery*

Scrotal veins accompany the arteries.

Lymphatic vessels drain into the superficial inguinal lymph nodes.

Nerves of the scrotum (Figs. 3.1*B* and 3.3*B*) are

- The genital branch of the *genitofemoral nerve* (L1, L2) that sends sensory branches to the anterior and lateral surfaces of the scrotum
- The anterior surface of the scrotum is also supplied by branches of the *ilioinguinal nerve* (L1)
- The posterior surface of the scrotum is supplied by the perineal branch of the *pudendal nerve* (S2–S4)
- The inferior surface of the scrotum is supplied by perineal branches of the *posterior femoral cutaneous nerve* (S2, S3)

TESTES

The testes, located in the scrotum (Fig. 3.4*A*), form sperms (spermatozoa) and hormones, principally testosterone. The surface of each testis is covered by the visceral layer of the tunica vaginalis, except where the testis is attached to the epididymis and spermatic cord. The *tunica vaginalis* (Fig. 3.3C) is a peritoneal sac surrounding the testis and is a derivative of the embryonic *processus vaginalis* (p. 91). Its *parietal layer* is adjacent to the internal spermatic fascia, and its *visceral layer* is adherent to the testis and epididymis. A small amount of fluid in the cavity of the tunica vaginalis separates the visceral and parietal layers and enables the testis to move freely in the scrotum.

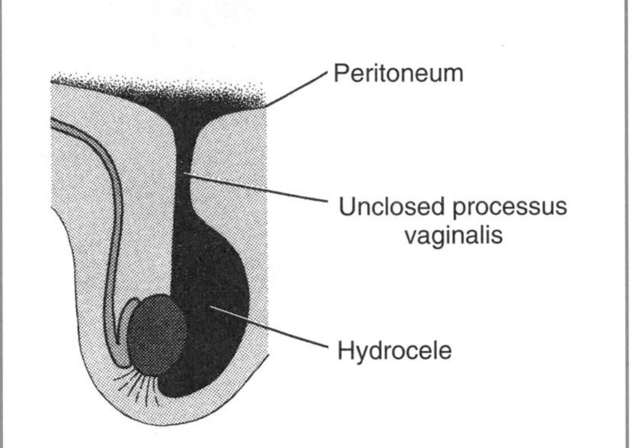

The presence of excess fluid in a persistent processus vaginalis is called a *hydrocele*. Certain pathological conditions (e.g., injury and/or inflammation of the epididymis) may also produce a hydrocele of the testis and spermatic cord. A *hematocele* of the testis is a collection of blood in the tunica vaginalis.

The *epididymis* is a convoluted duct that is applied to the superior and posterolateral surfaces of the testis (Fig. 3.4*A*).

- The superior expanded part called the head of the epididymis is composed of lobules formed by the coiled ends of the efferent ductules
- The efferent ductules transmit sperms from the testis to the epididymis where they are stored
- The body of the epididymis consists of the convoluted duct of the epididymis
- The tail of the epididymis is continuous with the *ductus deferens* that transports sperms from the epididymis to the *ejaculatory duct* for expulsion into the prostatic part of the urethra

The *ductus deferens* (vas deferens) is ligated bilaterally when sterilizing a man. To perform this operation called a *deferentectomy*, or more often a *vasectomy*, the duct is isolated on each side by incising the superoanterior scrotal wall. The duct is sectioned, and both ends are ligated. Sperms can no longer pass to the urethra; they degenerate in the epididymis and proximal end of the ductus deferens. However, the secretions of the auxiliary genital glands (e.g., seminal vesicles and prostate) can still be ejaculated.

The *testicular artery* arises from the abdominal aorta just inferior to the renal artery (Fig. 3.4). Veins emerge from the testis and join the *pampiniform plexus* from which the *testicular vein* arises in the inguinal canal. Lymphatic drainage of the testis is to the lumbar and preaortic lymph nodes. The autonomic nerves of the testis arise from the *testicular plexus* on the testicular artery. It contains vagal parasympathetic fibers and sympathetic fibers from T7 segment of the spinal cord.

Peritoneum and Peritoneal Cavity

The peritoneum (Fig. 3.5C) is a transparent, continuous serous membrane that consists of two layers:

- Parietal peritoneum lining the abdominal wall
- Visceral peritoneum covering the viscera (e.g., stomach and intestine)

The peritoneal cavity, the space between the layers of peritoneum, is a potential space because the organs are packed so closely. The peritoneal cavity contains a thin film of fluid that lubricates the peritoneal surfaces, enabling the viscera to move over each other without friction. The peritoneal cavity is closed in males, but there is a communication in females with the exterior of the body through the uterine tubes, uterus, and vagina.

The peritoneum and all viscera are within the *abdominal cavity* (Fig. 3.5). The relationship of the viscera to the peritoneum is as follows:

- Intraperitoneal organs (e.g., stomach) are viscera that are covered with visceral peritoneum
- Extraperitoneal (retroperitoneal) organs (e.g., kidneys, pancreas, and ascending and descending colon) are viscera that are located between the parietal peritoneum and the posterior abdominal wall

If the peritoneum is damaged (e.g., by a stab wound), the parietal and visceral layers of peritoneum may adhere, forming an *adhesion* that interferes with the movement of the viscera. The surgical separation of adhesions is called an *adhesiotomy*.

Under certain pathological conditions the peritoneal cavity may be distended with several liters of fluid (*ascites*). Widespread metastases of cancer cells cause exudation (escape) of fluid that is often bloodstained and contains cancer cells. *Paracentesis* of the abdomen may have to be performed to remove the excess fluid. A needle or cannula is inserted through the anterior wall into the peritoneal cavity (e.g., through the linea alba).

If the intestine is ruptured after a penetrating wound (e.g., by a knife), gas and intestinal material enter the peritoneal cavity and cause peritonitis (inflammation of the peritoneum), a very painful condition.

TERMS DESCRIBING PARTS OF PERITONEUM

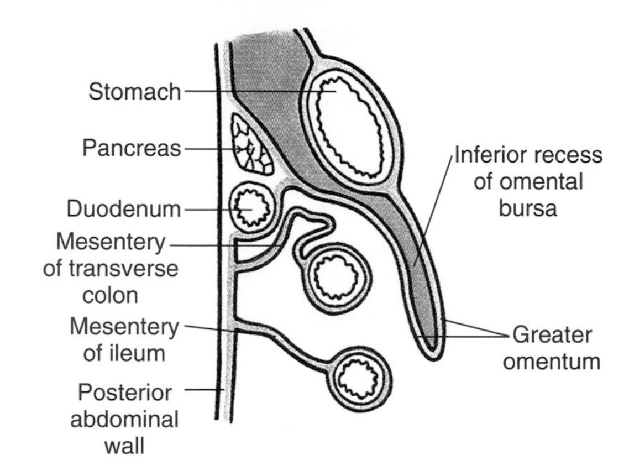

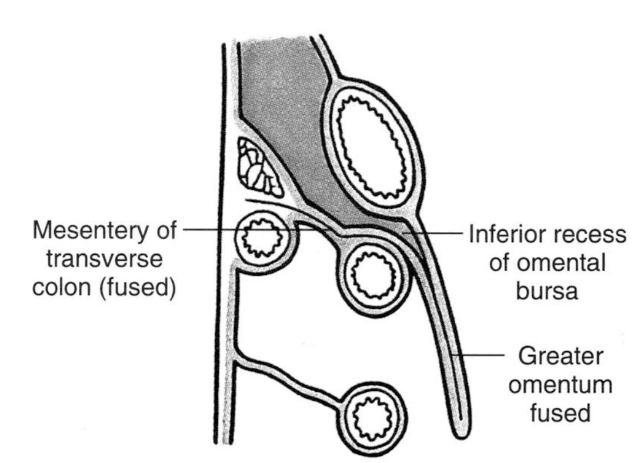

Greater omentum and omental bursa

Various terms are used to describe the parts of the peritoneum connecting organs with other organs or the abdominal wall.

A *mesentery* is a double layer of peritoneum that begins as an extension of the visceral peritoneum covering an organ. A mesentery connects the organ to the body wall (e.g., mesentery of the small intestine). Mesenteries have a core of connective tissue containing blood vessels, lymphatic vessels, nerves, fat, and lymph nodes. Viscera with a mesentery are mobile; the degree of mobility depends on the length of the mesentery.

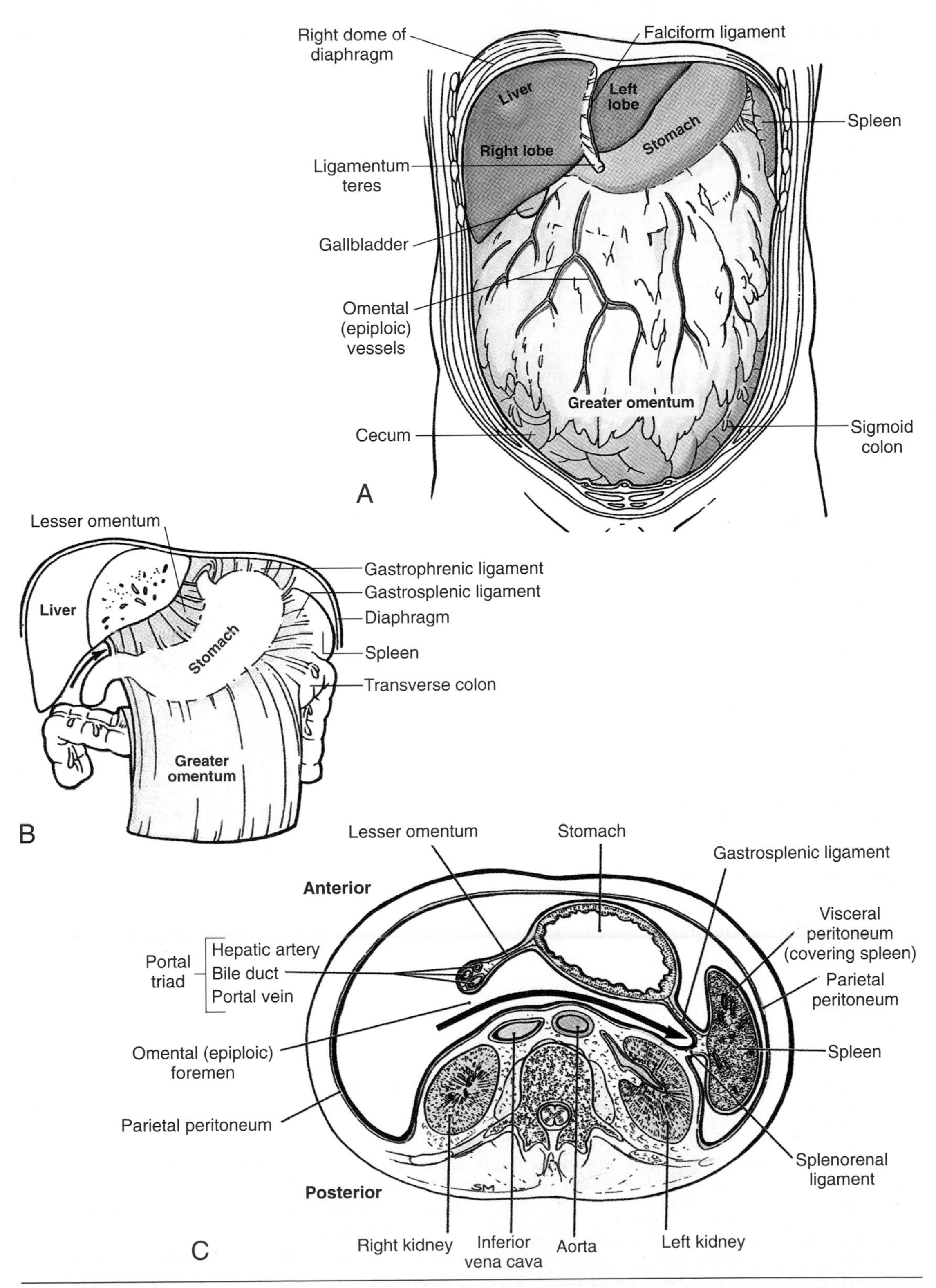

Figure 3.5. Abdominal contents. **A.** Anterior thoracic and abdominal walls are cut away to show undisturbed contents. **B.** Stomach and lesser and greater omenta. *Arrow*, site of omental (epiploic) foramen. **C.** Transverse section of abdomen at level of omental foramen to show horizontal extent of omental bursa (lesser sac). *Arrow*, path taken by a digit passed from greater peritoneal sac through omental foramen into omental bursa toward spleen.

An *omentum* is a double-layered extension of visceral peritoneum that passes from the stomach and the proximal part of the duodenum to another organ or structure.

- The lesser omentum connects the lesser curvature of the stomach and the proximal part of the duodenum to the liver
- The greater omentum, large and fat-laden, arises from the greater curvature of the stomach and the inferior border of the proximal half of the superior (first) part of the duodenum; it descends and then folds back to attach to the transverse colon (the *inferior recess of the omental bursa* is between the layers of the greater omentum; usually the anterior descending layer fuses with the posterior ascending layer, obliterating most of the inferior recess of the bursa).

The greater omentum prevents the visceral peritoneum from adhering to the parietal peritoneum lining the abdominal wall. It has considerable mobility and can migrate throughout the peritoneal cavity and wrap itself around an inflamed organ such as the appendix; i.e., it "walls off" and protects other viscera from the infected organ.

Peritoneal ligaments (Fig. 3.5) also consist of a double layer of peritoneum. The liver is connected to the anterior abdominal wall by the *falciform ligament*, and the stomach is connected to the

- Inferior surface of the diaphragm by the gastrophrenic ligament
- Spleen by the gastrosplenic ligament (gastrolienal ligament) that reflects onto the hilum of the spleen
- Transverse colon by the gastrocolic ligament (part of greater omentum)

A *peritoneal fold* is a reflection of peritoneum that is raised from the body wall by underlying blood vessels, ducts, and obliterated fetal vessels [e.g., medial and lateral umbilical folds (Fig. 3.2*A*)].

A *peritoneal recess* is a pouch of peritoneum that is formed by a peritoneal fold [e.g., subphrenic recess between the diaphragm and liver (Fig. 3.6*B*)].

Recesses of the peritoneal cavity are of clinical importance in connection with the spread of pathological fluids (e.g., pus). The recesses determine the extent and direction of the spread of fluids that may enter the peritoneal cavity when an organ is diseased or injured.

SUBDIVISIONS OF PERITONEAL CAVITY

As the fetal organs assume their final positions, the peritoneal cavity is divided into two peritoneal sacs called the *greater and lesser sacs*. A surgical incision through the anterior abdominal wall enters the greater peritoneal sac.

The *transverse mesocolon*, or mesentery of the transverse colon (Fig. 3.6), divides the greater sac into a

- Supracolic compartment containing the stomach, liver, and spleen
- Infracolic compartment containing the small intestine and ascending and descending colon

The infracolic compartment lies posterior to the greater omentum and is divided into right and left infracolic spaces by the mesentery of the small intestine (Fig. 3.6). There is free communication between the supracolic and infracolic compartments by way of the right *paracolic gutters* (fossae) on the medial and lateral sides of the descending colon.

The *omental bursa* (lesser sac) lies posterior to the stomach and lesser omentum (Figs. 3.5*C* and 3.6). The omental bursa has

- A superior recess that is limited superiorly by the diaphragm and the posterior layers of the coronary ligament of the liver
- An inferior recess between the superior part of the layers of the greater omentum

Most of the inferior recess of the bursa is a potential space because it is usually shut from the main part of the omental bursa because of adhesion of the anterior and posterior layers of the greater omentum (Fig. 3.6*B*). The omental bursa communicates with the greater peritoneal sac through the *omental foramen* (epiploic foramen). This opening is located posterior to the free edge of the lesser omentum (Fig. 3.5, *B* and *C*) and will usually admit two digits.

The boundaries of the omental foramen are

- Anteriorly, the portal vein, hepatic artery, and (common) bile duct
- Posteriorly, the inferior vena cava (IVC) and right crus of the diaphragm
- Superiorly, the caudate lobe of the liver
- Inferiorly, the superior part of the duodenum, portal vein, hepatic artery, and bile duct

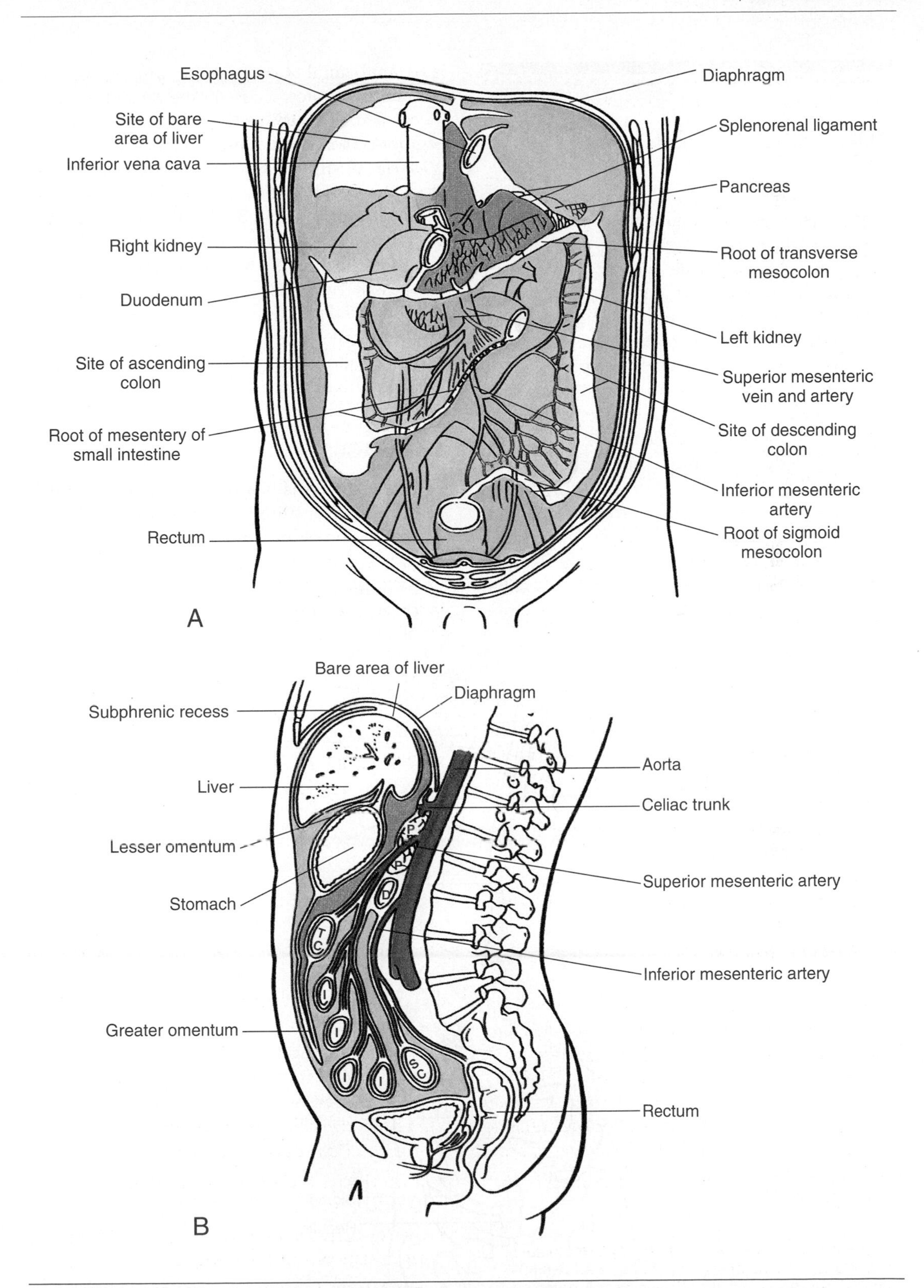

Figure 3.6. Mesenteries and vessels. A. Peritoneum covering posterior abdominal wall after removal of some viscera. *Light blue*, greater peritoneal sac; *dark blue*, omental bursa (lesser peritoneal sac). **B.** Sagittal section of abdomen and pelvis showing viscera, arrangement of peritoneum, and blood supply of stomach and intestine. *P*, pancreas; *D*, duodenum; *TC*, transverse colon; *I*, jejunum and ileum; *SC*, sigmoid colon.

Abdominal Viscera

The principal viscera of the abdomen are the esophagus, stomach, small and large intestines, spleen, pancreas, liver, biliary ducts and gallbladder, portal vein and portal-systemic anastomoses, renal fascia and fat, and kidneys, ureters, and suprarenal (adrenal) glands. Food passes from the mouth and pharynx through the esophagus to the stomach where it mixes with gastric secretions. Digestion occurs in the stomach and duodenum (Fig. 3.7). Absorption occurs principally in the small intestine, a coiled tube 5–6 m long consisting of the duodenum, jejunum, and ileum. The stomach continues into the duodenum, which receives the openings of the pancreas and the biliary tract from the liver. The large intestine consists of the cecum, which receives the ileum of the small intestine, vermiform appendix, colon (ascending, transverse, and descending), rectum, and anal canal ending at the anus. Most reabsorption of water occurs in the ascending colon. Feces (excrement) forms in the descending and sigmoid colon and is stored in the rectum.

Arterial supply to the digestive tract, spleen, pancreas, gallbladder, and liver is from the *abdominal aorta* (Figs. 3.6*B* and 3.8*A*). The three major branches of the aorta supplying the gut are the celiac trunk and the superior and inferior mesenteric arteries.

The *portal vein* (Fig. 3.8*B*) is the main channel of the portal system of veins that collects blood from the abdominal part of the gastrointestinal tract, gallbladder, pancreas, and spleen and carries it to the liver.

ESOPHAGUS

The esophagus is a muscular tube (about 25 cm long) extending from the pharynx to the stomach. The esophagus

- Follows the curve of the vertebral column as it descends through the neck and posterior mediastinum

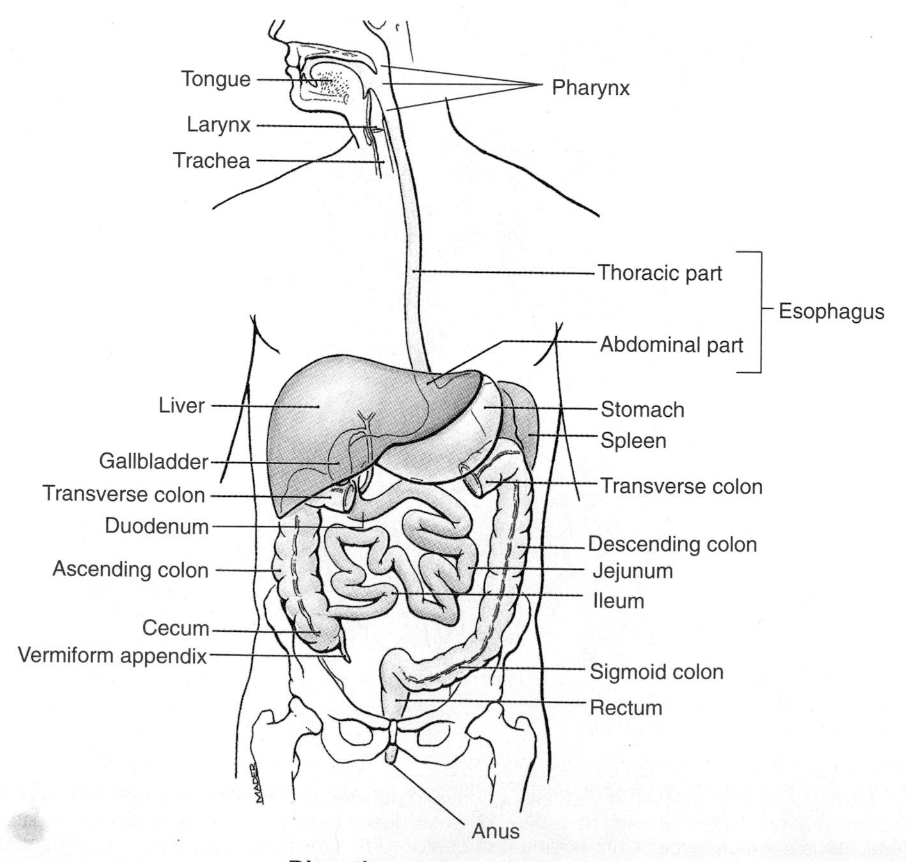

Digestive system

- Pierces the diaphragm just to the left of the median plane
- Enters the cardia of the stomach at the level of the seventh left costal cartilage and T10 (or T11) vertebrae
- Is encircled by the esophageal plexus of nerves distally
- Is covered anteriorly and laterally in the abdomen by peritoneum; i.e., it is retroperitoneal

The short abdominal part of the esophagus receives its arterial supply through the left gastric branch of the *celiac trunk* and the left *inferior phrenic artery*. Venous drainage is into the portal venous system via the *left gastric vein* and into the systemic venous system via the *azygos vein*. Lymphatic drainage is into the *left gastric lymph nodes* (Fig. 3.8C); efferent lymphatic vessels from these nodes drain mainly to the *celiac lymph nodes*. Innervation is from the *vagal trunks* (anterior and posterior gastric nerves), the thoracic *sympathetic trunks*, the greater and lesser *splanchnic nerves*, and the plexus of nerves around the left gastric and inferior phrenic arteries (Fig. 3.8*D*)

STOMACH

The stomach (Fig. 3.7) has a

- Lesser curvature forming its concave border
- Greater curvature forming its longer convex border
- Sharp indentation about two-thirds of the distance along the lesser curvature called the *angular notch*, which indicates the junction of the body and pyloric part
- Cardia around the opening of the esophagus
- Fundus, its dilated superior part that is related to the left dome of the diaphragm
- Body that lies between the fundus and pyloric antrum
- Pyloric part, its funnel-shaped part; its wide portion, the *pyloric antrum*, leads into the *pyloric canal*, its narrow portion
- Pylorus, the distal sphincteric region, that is thickened to form the *pyloric sphincter*, which controls discharge of the stomach contents through the *pyloric orifice* into the duodenum

Surface Anatomy of Stomach

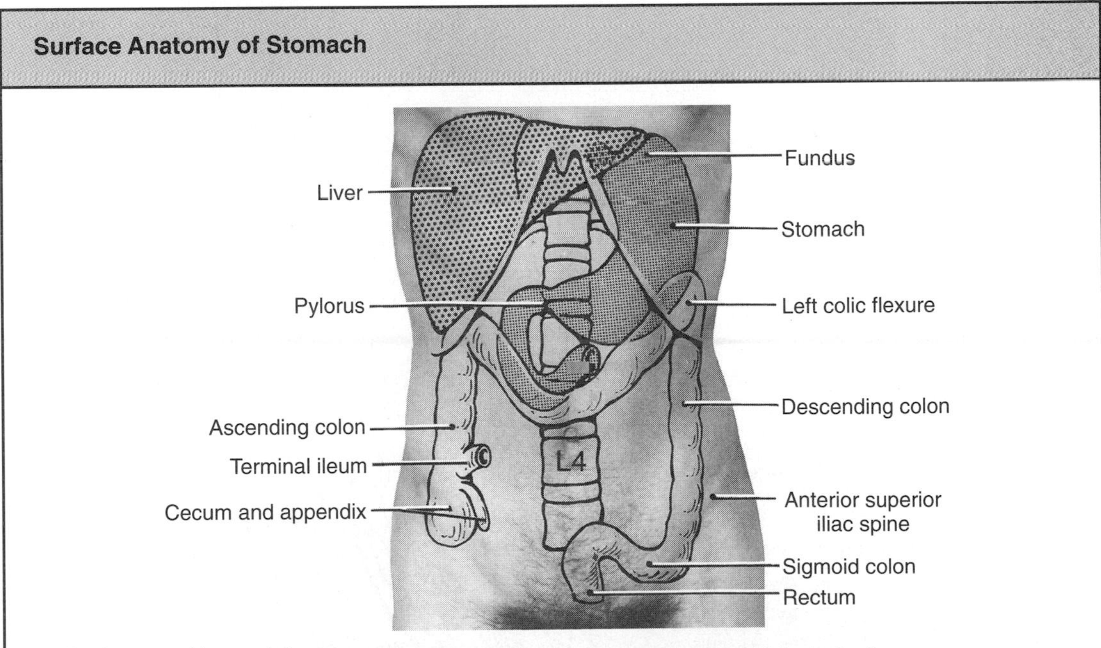

Surface markings of the stomach vary greatly because its size and position change under various circumstances (e.g., after a heavy meal).

- The cardiac orifice is usually located posterior to the seventh left costal cartilage, 2–4 cm from the median plane at the level of T10 or T11 vertebra
- The fundus is usually located posterior to the left 5th rib in the midclavicular line
- The pyloric part usually lies in the transpyloric plane that passes through the ninth costal cartilages at the level of L1 vertebra
- In the erect position the location of the pylorus varies from L2 to L4 vertebrae; it is usually on the right side

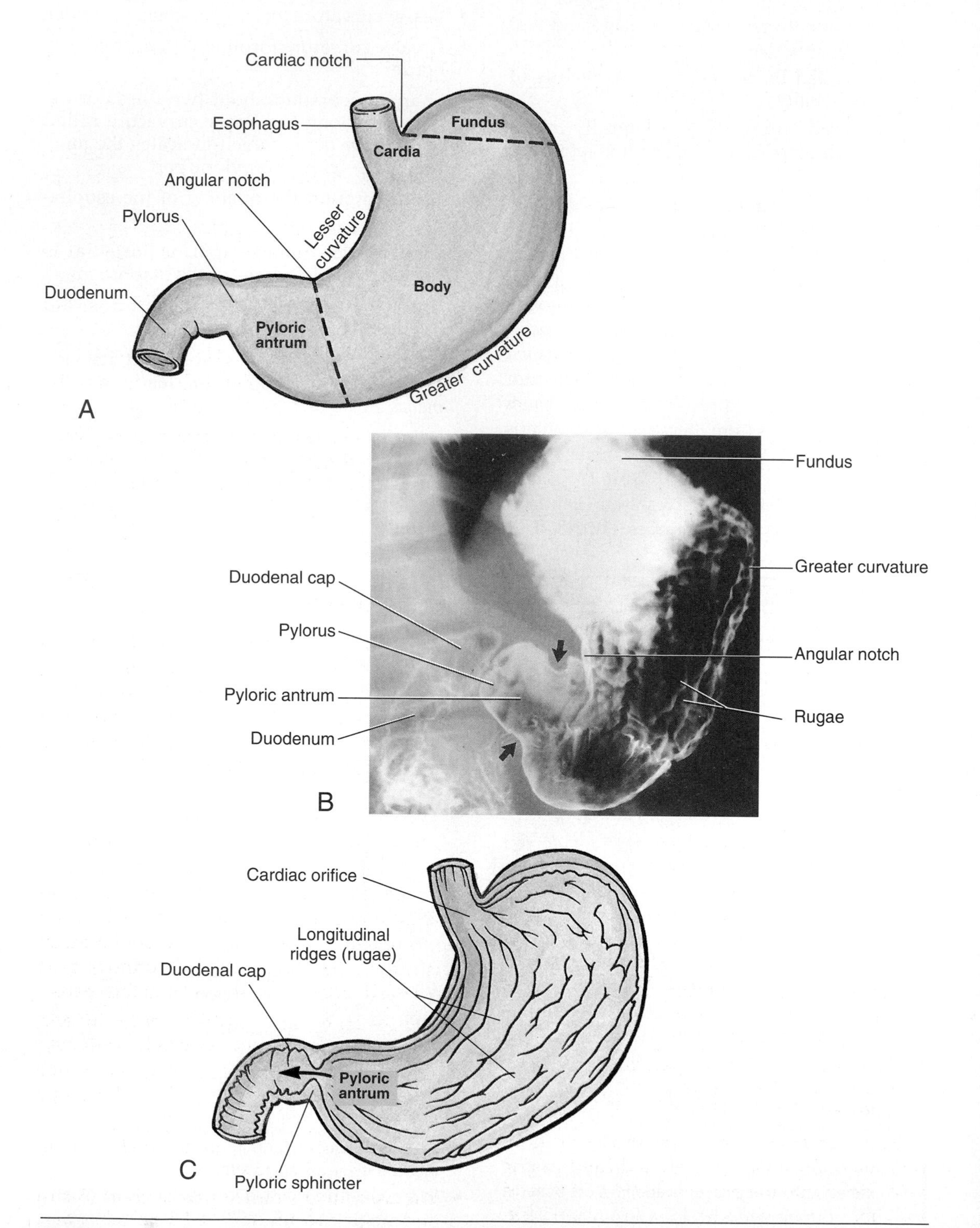

Figure 3.7. Stomach and proximal duodenum. **A.** External surface. **B.** Radiograph after a barium meal. *Arrows*, peristaltic wave. **C.** Internal surface. *Arrow* passes through pyloric canal.

Relations of Stomach

The stomach is covered by peritoneum (Figs. 3.5 and 3.6*B*), except where blood vessels run along its curvatures and in a small area posterior to the cardiac orifice. The two layers of the lesser omentum extend around the stomach and leave its greater curvature as the greater omentum. The anterior surface of the stomach is in contact with the

- Diaphragm
- Left lobe of the liver
- Anterior abdominal wall

The *stomach bed* on which the stomach rests in the supine position (see Fig. 3.11*A*) is formed by the posterior wall of the omental bursa and the structures between it and the posterior abdominal wall:

- Diaphragm
- Transverse colon, transverse mesocolon, pancreas, spleen, and celiac trunk and its three branches
- Left suprarenal (adrenal) gland and superior part of left kidney

Vessels and Nerves of Stomach

The *gastric arteries* arise from the celiac trunk and its branches (Fig. 3.8*A*, Table 3.4):

- Left gastric artery arises from the celiac trunk and runs in the lesser omentum to the cardia and then turns abruptly to course along the lesser curvature to anastomose with the right gastric artery
- Right gastric artery arises from the hepatic artery (common hepatic artery) and runs to the left along the lesser curvature to anastomose with the left gastric artery
- Right gastroomental (gastroepiploic) artery arises from the gastroduodenal artery and runs to the left along the greater curvature to anastomose with the left gastroomental artery
- Left gastroomental artery arises from the splenic artery and anastomoses with the right gastroomental artery
- Short gastric arteries arise from the distal end of the splenic artery and pass to the fundus

The *gastric veins* parallel the arteries in position and course. The left and right gastric veins drain into the portal vein (Fig. 3.8*B*), and the short gastric veins and the left *gastroomental (gastroepiploic) vein* drain into the splenic vein, which joins the superior mesenteric vein to form the portal vein. The right gastroomental vein empties in the superior mesenteric vein.

The *gastric lymphatic vessels* accompany the arteries along the greater and lesser curvatures. They drain lymph from its anterior and posterior surfaces toward its curvatures, where many of the *gastroomental (gastroepiploic) lymph nodes* are located (Fig. 3.8C). The efferent vessels from these nodes accompany the large arteries to the *celiac lymph nodes*.

The parasympathetic *nerve supply of the stomach* is from the anterior and posterior *vagal trunks* and their branches (Fig. 3.8*D*). The sympathetic nerve supply from T6 to T9 segments of the spinal cord passes to the *celiac plexus* and is distributed through the plexuses around the gastric and gastroomental arteries.

Hiatal hernias occur most often in people after middle age, possibly because of weakening of the muscle of the diaphragm around the esophageal hiatus (opening). The cardia and part of the fundus of the stomach may herniate through this hiatus into the thorax.

SMALL INTESTINE

The small intestine extends from the pylorus to the ileocecal junction, where it joins the large intestine. The pylorus empties the contents of the stomach into the duodenum, the first part of the small intestine (Fig. 3.7); its other two parts are the jejunum and ileum.

Duodenum

The duodenum is the shortest, widest, and most fixed part of the small intestine. It pursues a C-shaped course around the head of the pancreas. The duodenum begins at the pylorus on the right side and ends at the duodenojejunal junction on the left side. For descriptive purposes the duodenum is divided into four parts:

- The superior (first) part is short (5 cm) and lies anterolateral to the body of L1 vertebra
- The descending (second) part is longer (7–10 cm) and descends along the right sides of L1–L3 vertebrae
- The horizontal (third) part is 6–8 cm long and crosses L3 vertebra
- The ascending (fourth) part is short (5 cm) and begins to the left of L3 vertebra and rises superiorly as far as the superior border of L2 vertebra

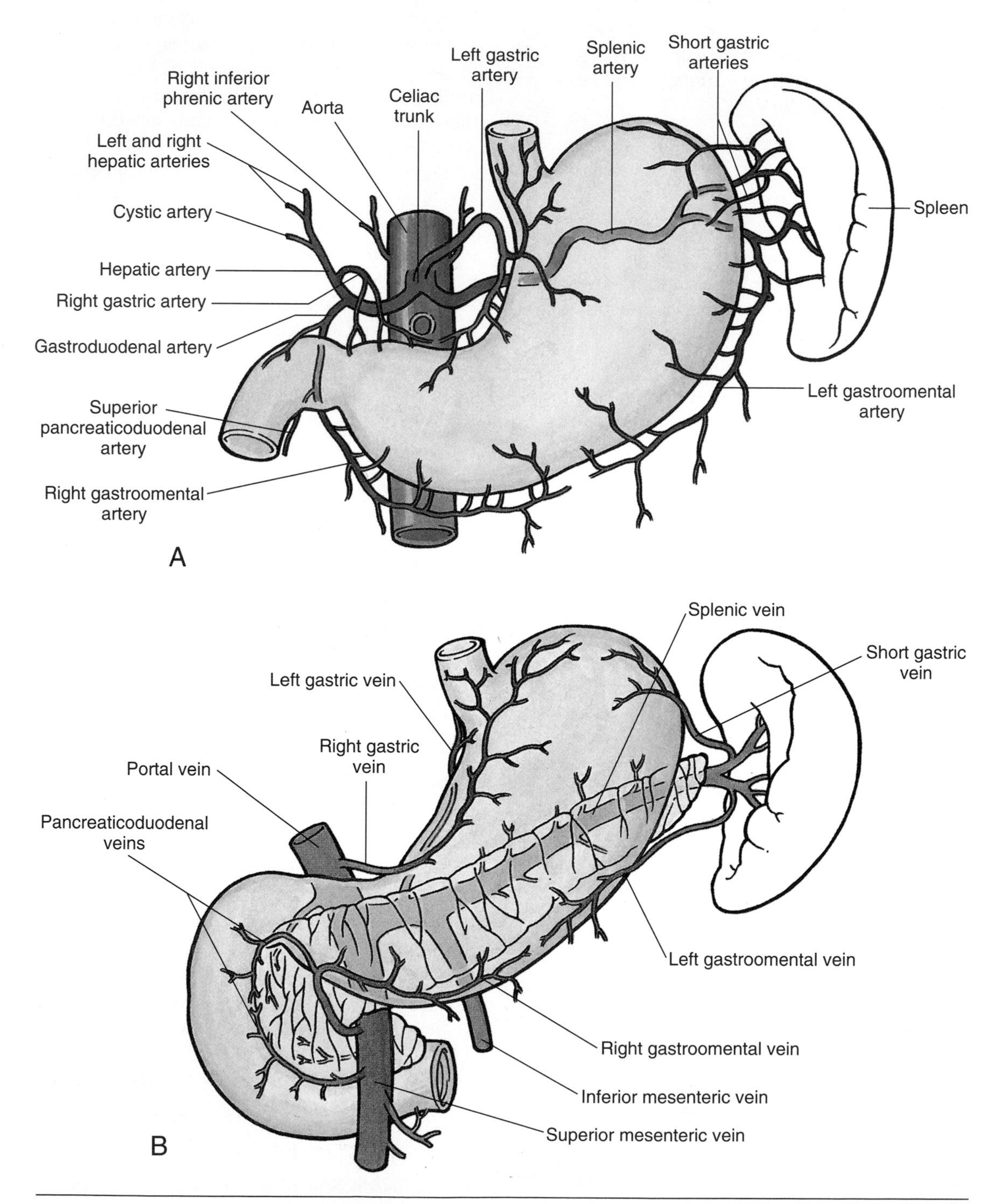

Figure 3.8. **A.** Arteries of stomach and spleen. **B.** Portal venous drainage of stomach and spleen. **C.** Lymphatic drainage of stomach and small intestine. *Arrows*, lymph flow to lymph nodes. **D.** Innervation of stomach and small intestine. *Arrows*, afferent and efferent nerves.

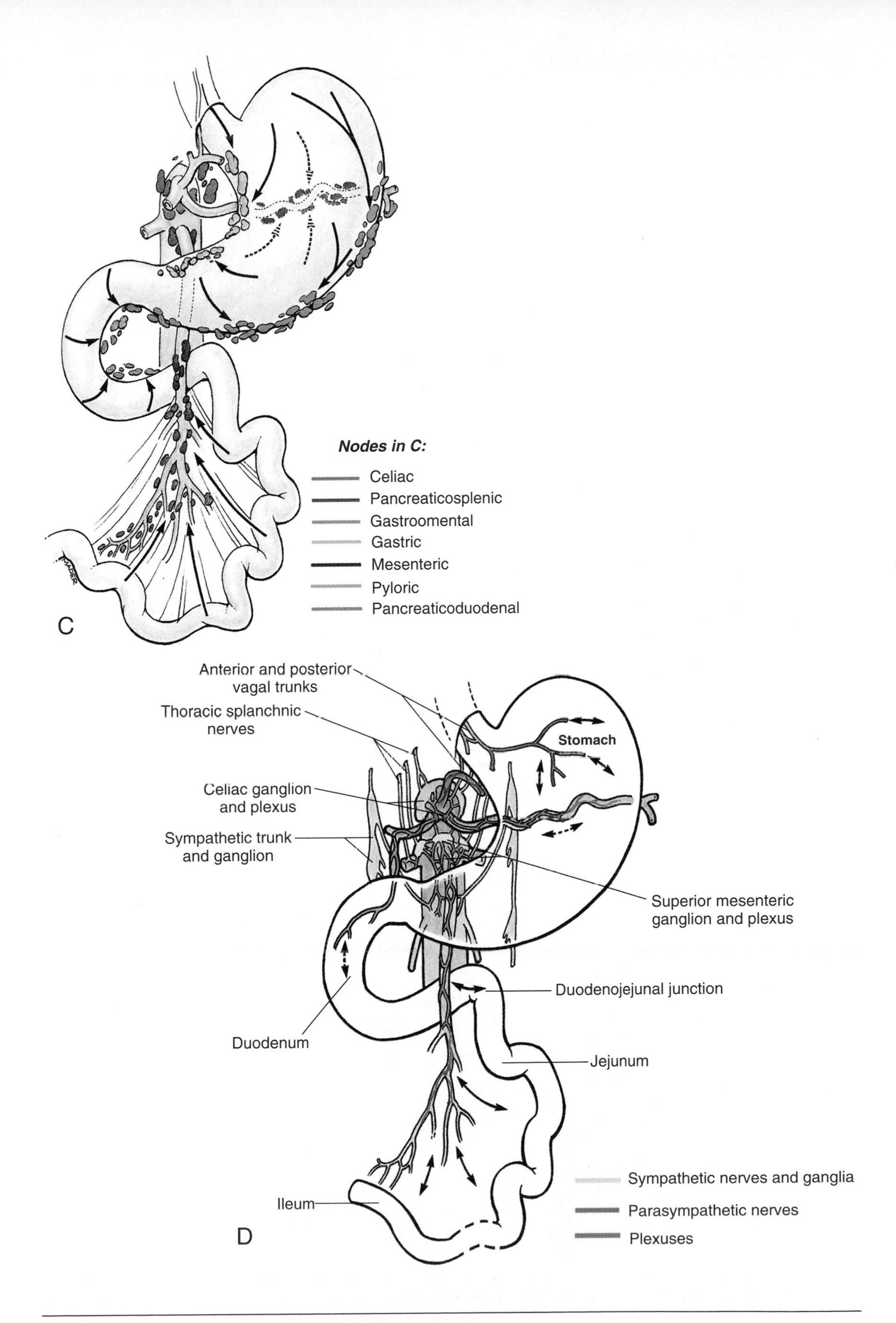

Figure 3.8. *Continued.*

Table 3.4
Arterial Supply to Esophagus, Stomach, Duodenum, Liver, Gallbladder, Pancreas, and Spleen

Artery	Origin	Course	Distribution
Celiac	Abdominal aorta just distal to aortic hiatus of diaphragm	Soon divides into left gastric, splenic, and common hepatic aa.	Supplies esophagus, stomach, duodenum (proximal to bile duct), liver and biliary apparatus, and pancreas
Left gastric	Celiac trunk	Ascends retroperitoneally to esophageal hiatus where it passes between layers of hepatogastric ligament	Distal portion of esophagus and lesser curvature of stomach
Splenic	Celiac trunk	Runs retroperitoneally along superior border of pancreas; it then passes between layers of of splenorenal ligament to hilum of spleen	Body of pancreas, spleen, and greater curvature of stomach
Left gastroomental (gastroepiploic)	Splenic a. in hilum of spleen	Passes between layers of gastrosplenic ligament to greater curvature of stomach	Left portion of greater curvature of stomach
Short gastric (n = 4–5)	Splenic a. in hilum of spleen	Pass between layers of gastrosplenic ligament to fundus of stomach	Fundus of stomach
Hepatic[a]	Celiac trunk	Passes retroperitoneally to reach hepatoduodenal ligament and passes between its layers to porta hepatis; divides into right and left hepatic aa.	Liver, gallbladder, stomach, pancreas, duodenum, and respective lobes of liver
Cystic	Right hepatic a.	Arises within hepatoduodenal ligament	Gallbladder and cystic duct
Right gastric	Hepatic a.	Runs between layers of hepatogastric ligament	Right portion of lesser curvature of stomach
Gastroduodenal	Hepatic a.	Descends retroperitoneally, posterior to gastroduodenal junction	Stomach, pancreas, first part of duodenum, and distal part of bile duct
Right gastroomental (gastroepiploic)	Gastroduodenal a.	Passes between layers of greater omentum to greater curvature of stomach	Right portion of greater curvature of stomach
Anterior and posterior superior pancreaticoduodenal	Gastroduodenal a.	Descends on head of pancreas	Proximal portion of duodenum and head of pancreas
Anterior and posterior inferior pancreaticoduodenal	Superior mesenteric a.	Ascends retroperitoneally on head of pancreas	Distal portion of duodenum and head of pancreas

[a] For descriptive purposes, hepatic artery is often divided into common hepatic artery from its origin to origin of gastroduodenal artery, and remainder of vessel is called hepatic artery proper.

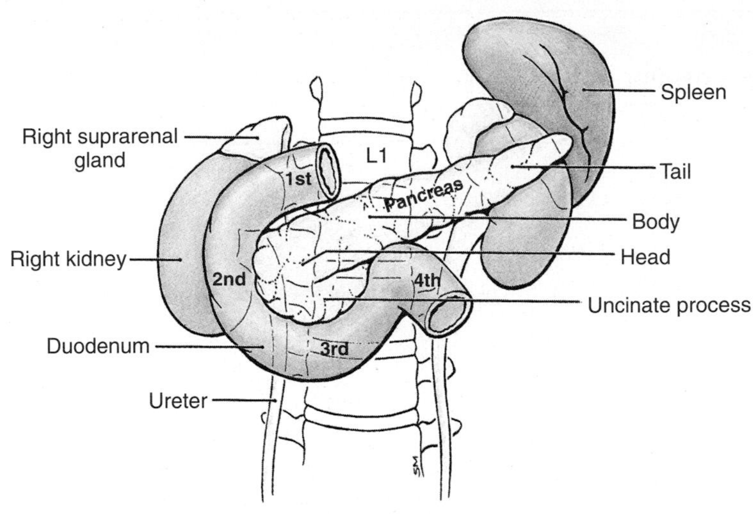

Relations of duodenum

The first 2 cm of the duodenum has a mesentery and is mobile. Radiologists refer to this free part as the *duodenal cap* (Fig. 3.7*B*). The distal 3 cm of the superior part and the other three parts of the duodenum have no mesentery and are immobile because they are retroperitoneal. The principal relations of the duodenum are listed in Table 3.5. The descending part of the duodenum runs inferiorly, initially to the right and parallel to the IVC. The *bile and pancreatic ducts* enter its posteromedial wall (see Fig. 3.11). These ducts usually unite to form the *hepatopancreatic ampulla*, which opens on the summit of the *major duodenal papilla*. The horizontal part of the duodenum is crossed by the superior mesenteric vessels and the root of the mesentery of the jejunum and ileum. The ascending part of the duodenum runs superiorly on the left side of the aorta to reach the inferior border of the pancreas. Here it curves anteriorly to join the jejunum at the *duodenojejunal flexure* (Fig. 3.8*C*). This curvature is supported by a fibromuscular band called the *suspensory muscle of the duodenum* (ligament of Treitz). Contraction of this muscle also widens the angle of the flexure, facilitating movement of its contents.

The *duodenal arteries* arise from the celiac trunk and superior mesenteric artery (Figs. 3.6*B* and 3.8*A*, Table 3.4). The *celiac trunk*, via the gastroduodenal artery and its branch, the superior pancreaticoduodenal artery, supplies the duodenum proximal to the entry of the bile duct, i.e., the part derived from the embryonic foregut. The *superior mesenteric artery*, via its branch, the inferior pancreaticoduodenal artery, supplies the duodenum distal to the entry of the bile duct, i.e., the part derived from the embryonic midgut

The *duodenal veins* follow the arteries and drain into the *portal vein* (Fig. 3.8*B*); some drain directly and others indirectly via the superior mesenteric and splenic veins.

The anterior *lymphatic vessels of the duodenum* follow the arteries and drain into the *pancreaticoduodenal lymph nodes* located along the splenic artery and the pyloric lymph nodes along the gastroduodenal artery (Fig. 3.8*C*). The posterior lymphatic vessels pass posterior to the head of the pancreas and drain into the superior mesenteric lymph nodes. Efferent lymphatic vessels from the duodenal lymph nodes drain into the *celiac lymph nodes*.

The duodenum is supplied by the vagus and sympathetic nerves by way of plexuses on the pancreaticoduodenal arteries (Fig. 3.8*D*).

Most *duodenal ulcers* (inflammatory erosions of the duodenal wall) are located in the superior (first) part of the duodenum (95%). Occasionally an ulcer perforates the duodenal wall, permitting its contents to enter the peritoneal cavity and produce *peritonitis*. Because the superior part of the duodenum is closely related to the liver and gallbladder, either may adhere to or be ulcerated by a duodenal ulcer. Erosion of the gastroduodenal artery, a posterior relation of the superior part of the duodenum (Fig. 3.8*A*), by a duodenal ulcer results in severe hemorrhage into the peritoneal cavity.

Table 3.5
Relations of Duodenum

Part	Anterior	Posterior	Medial	Superior	Inferior
Superior (first part)	Peritoneum; gallbladder; quadrate lobe of liver	Bile duct; portal vein; IVC; gastroduodenal a.		Neck of gallbladder	Neck of pancreas
Descending (second part)	Transverse colon; transverse mesocolon; coils of small intestine	Hilum of right kidney; renal vessels; ureter; psoas major	Head of pancreas; pancreatic duct; bile duct		
Horizontal (third part)	Superior mesenteric artery; superior mesenteric vein; coils of small intestine	Right psoas major IVC; aorta; right ureter		Head and uncinate process of pancreas; superior mesenteric vessels	
Ascending (fourth part)	Beginning of root of mesentery; coils of jejunum	Left psoas major; left margin of aorta	Head of pancreas	Body of pancreas	

Jejunum and Ileum

The jejunum begins at the duodenojejunal junction, and the ileum ends at the *ileocecal junction*, the union of the ileum with the cecum (Figs. 3.8*C* and 3.9). Together the jejunum and ileum are 6–7 m long; the jejunum constitutes about two-fifths and the ileum the remainder. Most of the jejunum lies in the umbilical region, whereas most of the ileum is in the suprapubic and right inguinal regions. The terminal part of the ileum is usually in the pelvis from which it ascends to end in the medial aspect of the cecum. Although there is no clear line of demarcation between the jejunum and ileum, they have distinctive characteristics that are of surgical importance (Fig. 3.9*C*, Table 3.6).

The *mesentery* attaches most of the small intestine to the posterior abdominal wall (Fig. 3.6*B*). The root of the mesentery (about 15 cm long) is directed obliquely, inferiorly, and to the right from the left side of L2 vertebra to the right sacroiliac joint (Fig. 3.6*A*). The root of the mesentery crosses the

- Horizontal part of the duodenum
- Abdominal aorta
- Inferior vena cava
- Right psoas major muscle
- Right ureter
- Right testicular or ovarian vessels

The *superior mesenteric artery* supplies the jejunum and ileum (Figs. 3.6 and 3.9, Table 3.7). It runs between the layers of the mesentery and sends 15–18 branches to the intestine. The arteries unite to form loops or arches called *arterial arcades*, from which straight arteries called *vasa recta* arise. The *superior mesenteric vein* drains the jejunum and ileum. It lies anterior and to the right of the superior mesenteric artery in the root of the mesentery. The superior mesenteric vein ends posterior to the neck of the pancreas where it unites with the splenic vein to form the portal vein (Fig. 3.8*B*).

The lymphatic vessels of the jejunum and ileum pass between the layers of the mesentery to the mesenteric lymph nodes (Fig. 3.8*C*), which are located

- Close to the wall of the intestine
- Among the arterial arcades
- Along the proximal part of the superior mesenteric artery

The lymphatic vessels from the terminal ileum follow the ileal branch of the ileocolic artery to the *ileocolic lymph nodes*. Efferent lymphatic vessels from the mesenteric lymph nodes drain into the *superior mesenteric lymph nodes*.

The sympathetic nerves to the jejunum and ileum originate in segments T5–T9 of the spinal cord and reach the celiac plexus through the sym-

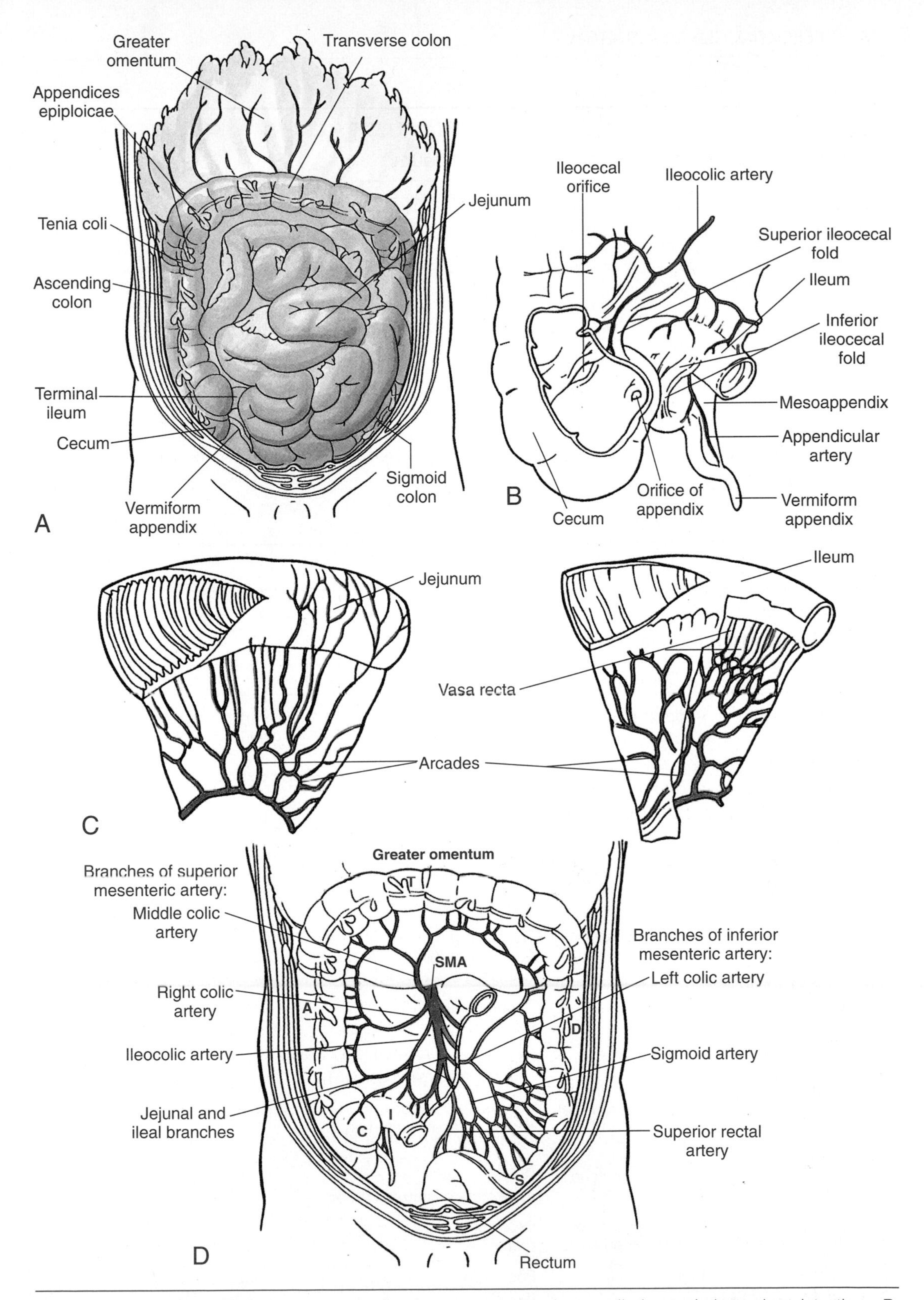

Figure 3.9. Intestine and mesenteries. **A.** Greater omentum has been pulled superiorly to show intestine. **B.** Cecum and vermiform appendix showing their blood supply. A window has been cut in wall of cecum to show ileocecal orifice and orifice of appendix. **C.** Blood supply and interior of jejunum and ileum. **D.** Most of small intestine has been removed to show blood supply. *T*, transverse colon; *SMA*, superior mesenteric artery; *A*, ascending colon; *D*, descending colon; *I*, terminal ileum; *C*, cecum; *S*, sigmoid colon.

Table 3.6
Distinguishing Characteristics of Jejunum and Ileum

Characteristic	Jejunum	Ileum
Color	Deeper red	Paler pink
Caliber	2–4 cm	2–3 cm
Wall	Thick and heavy	Thin and light
Vascularity	Greater	Less
Vasa recta	Long	Short
Arcades	A few large loops	Many short loops
Fat in mesentery	Less	More
Circular folds (plicae circulares)	Large, tall, and closely packed	Low and sparse; absent in distal part
Lymphoid nodules (Peyer's patches)	Few	Many

Table 3.7
Arterial Supply to Intestines[a]

Artery	Origin	Course	Distribution
Superior mesenteric	Abdominal aorta	Runs in root of mesentery to ileocecal junction	Part of gastrointestinal tract derived from midgut
Intestinal (n = 15–18)	Superior mesenteric a.	Pass between the two layers of mesentery	Jejunum and ileum
Middle colic	Superior mesenteric a.	Ascends retroperitoneally and passes between layers of transverse mesocolon	Transverse colon
Right colic	Superior mesenteric a.	Passes retroperitoneally to reach ascending colon	Ascending colon
Ileocolic	Terminal branch of superior mesenteric a.	Runs along root of mesentery and divides into ileal and colic branches	Ileum and cecum
Appendicular	Ileocolic a.	Passes between layers of mesoappendix	Vermiform appendix
Inferior mesenteric	Abdominal aorta	Descends retroperitoneally to left of abdominal aorta	Supplies part of gastrointestinal tract derived from hindgut
Left colic	Inferior mesenteric a.	Passes retroperitoneally toward left to descending colon	Descending colon
Sigmoid (n = 3–4)	Inferior mesenteric a.	Run between layers of mesentery of sigmoid colon	Sigmoid colon
Superior rectal	Terminal branch of inferior mesenteric a.	Descends retroperitoneally to rectum	Proximal part of rectum
Middle rectal	Internal iliac a.	Passes retroperitoneally to rectum	Midpart of rectum
Inferior rectal	Internal pudendal a.	Crosses ischioanal fossa to reach rectum	Distal part of rectum and anal canal

[a] See Table 3.4 for arterial supply to duodenum.

sympathetic trunks and the greater splanchnic nerves (Fig. 3.8*D*). The preganglionic sympathetic fibers synapse in the *celiac and superior mesenteric ganglia*. Parasympathetic nerves are derived from the posterior vagal trunks. The postganglionic sympathetic fibers and the preganglionic parasympathetic fibers synapse in the myenteric and submucous plexuses in the intestinal wall. In general, sympathetic stimulation reduces motility and secretion and acts as a vasoconstrictor, whereas parasympathetic stimulation increases motility of the intestine and secretion. There are also sensory fibers. The intestine is insensitive to most pain stimuli, including cutting and burning, but is sensitive to distention, which is perceived as colic (cramps).

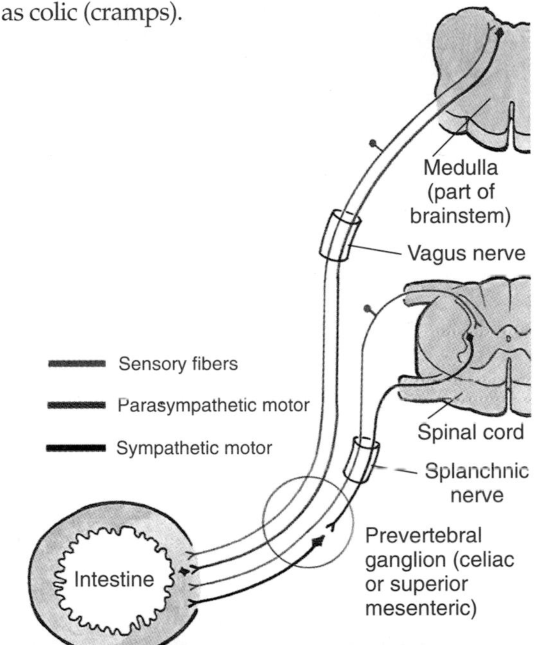

Nerve supply to intestine

Occlusion of the vasa recta by an embolus results in reduced blood supply (ischemia) for the part of the intestine concerned. If the ischemia is severe, *necrosis* of that segment results, and ileus of the paralytic type occurs. *Ileus* is a severe colicky pain, accompanied by vomiting and often fever and dehydration. If the condition is diagnosed early enough (e.g., using a *superior mesenteric arteriogram*), the obstructed portion of the vessel may be cleared surgically.

An *ileal (Meckel) diverticulum* is a congenital anomaly that occurs in 1–2% of people. It is a pouchlike (3–6 cm long) remnant of the proximal part of the yolk stalk [see Moore and Persaud (1993) listed under "Suggested Readings"]. It sometimes becomes inflamed and produces pain that may mimic that produced by appendicitis.

LARGE INTESTINE

The large intestine consists of the cecum, vermiform appendix, colon, rectum, and anal canal (Fig. 3.9, *A*, *B*, and *D*). The large intestine can be distinguished from the small intestine by

- Three thickened bands of muscle called teniae coli
- Sacculations between the teniae called haustra
- Small pouches of omentum filled with fat called omental (epiploic) appendages

Cecum

The cecum is the first part of the large intestine and is continuous with the ascending colon. The cecum is located in the right lower quadrant, where it lies in the iliac fossa. Usually it is almost entirely enveloped by peritoneum and can be lifted freely, but the cecum does not have a mesentery. The ileum enters the cecum obliquely and partly invaginates into it, forming folds (lips) superior and inferior to the *ileocecal orifice*. These folds form the *ileocecal valve*.

Vermiform Appendix

The vermiform appendix is a worm-shaped blind tube that joins the cecum inferior to the ileocecal junction. It has a short triangular mesentery called the *mesoappendix* that suspends it from the mesentery of the terminal ileum. The position of the appendix is variable, but it is usually retrocecal. The base of the appendix lies deep to a point (McBurney's point) that is one-third of the way along the oblique line joining the right anterior superior iliac spine to the umbilicus.

The cecum is supplied by the *ileocolic artery*, a branch of the superior mesenteric artery, and the appendix is supplied by the *appendicular artery*, a branch of the ileocolic artery (Fig. 3.9*A*, Table 3.7). The *ileocolic vein*, a tributary of the superior mesenteric vein, drains blood from the cecum and appendix. Lymphatic vessels from the cecum and appendix pass to lymph nodes in the mesoappendix and to the *ileocolic lymph nodes* that lie along the ileocolic artery (Fig. 3.10*A*). Efferent lymph vessels pass to the *superior mesenteric lymph nodes*.

The nerve supply to the cecum and appendix is derived from the sympathetic and parasympathetic nerves from the superior mesenteric plexus (Fig. 3.10*B*). The sympathetic

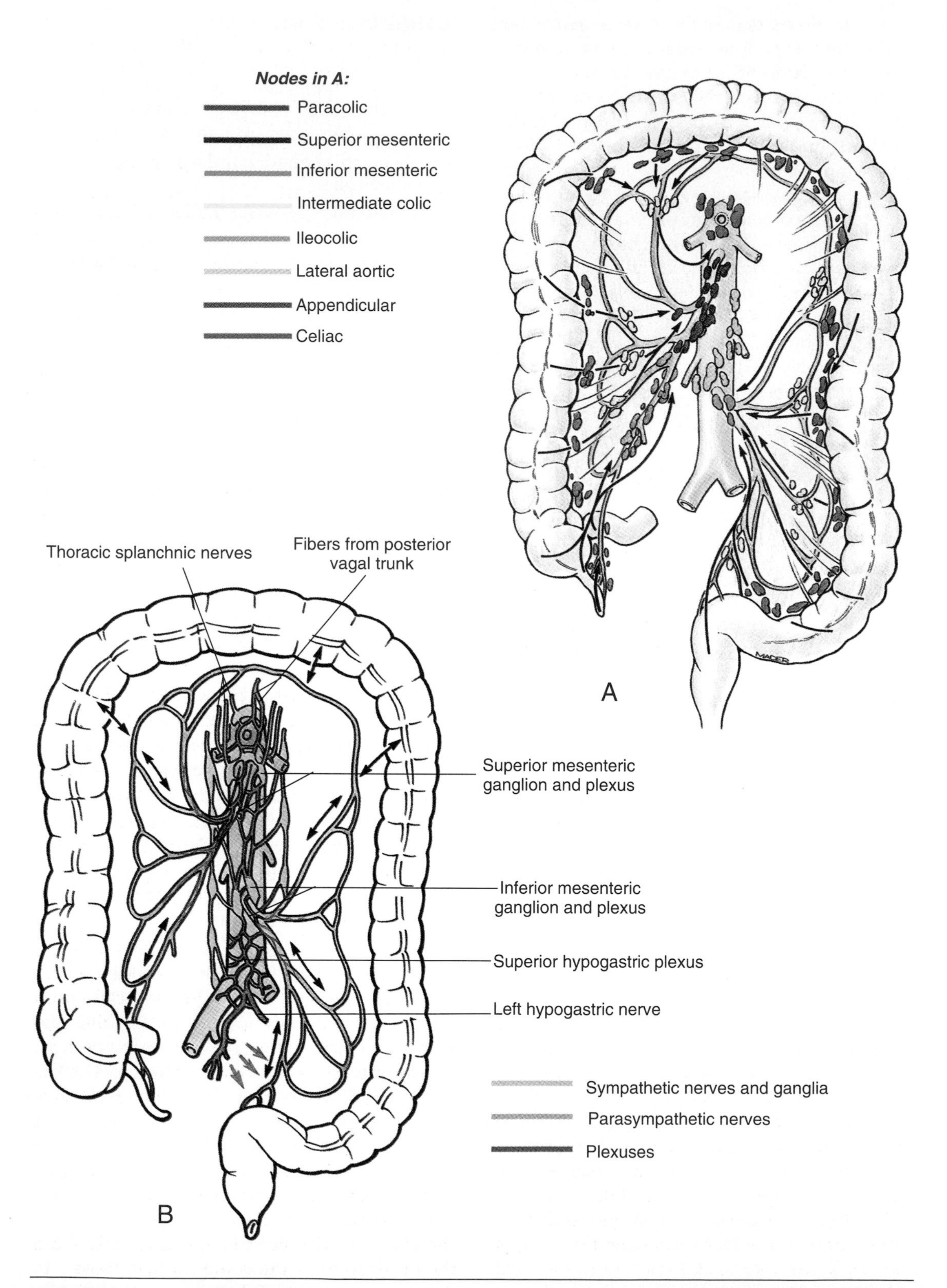

Figure 3.10. **A.** Lymphatic drainage of large intestine. *Arrows*, direction of lymph flow. **B.** Innervation of large intestine. *Black arrows*, afferent and efferent nerves; *green arrows*, pelvic splanchnic nerves (S2,3,4).

nerve fibers originate in the lower thoracic part of the spinal cord, and the parasympathetic fibers are derived from the vagus nerves. Afferent nerve fibers from the appendix accompany the sympathetic nerves to the T10 segment of the spinal cord.

Acute inflammation of the appendix is a common cause of acute abdominal pain (acute abdomen). Digital pressure over McBurney's point registers the maximum abdominal tenderness. *Appendicitis* is usually caused by obstruction of the appendix, most often by fecal material. When its secretions cannot escape, it swells and stretches the visceral peritoneum. The pain of acute appendicitis usually commences as a vague (dull) pain in the periumbilical region because afferent pain fibers enter the spinal cord at the T10 level. Later severe pain develops in the right lower quadrant; it is caused by irritation of the parietal peritoneum on the posterior abdominal wall. Pain can be elicited by extending the thigh at the hip joint. *Thrombosis* (clotting of the blood) in the appendicular artery often results in perforation of an acutely inflamed appendix.

Colon

The *ascending colon* passes superiorly from the cecum on the right side of the abdominal cavity to the liver, where it turns to the left as the *right colic flexure* (Fig. 3.11*D*). The ascending colon lies retroperitoneally along the right side of the posterior abdominal wall, but it is covered by peritoneum anteriorly and on its sides. On the medial and lateral sides of the ascending colon, the peritoneum forms *paracolic gutters*. The ascending colon is usually separated from the anterior abdominal wall by coils of small intestine and the greater omentum.

Arterial supply to the ascending colon and right colic flexure is through the *ileocolic and right colic arteries*, branches of the superior mesenteric artery (Fig. 3.9*D*). The ileocolic and right colic veins, tributaries of the superior mesenteric vein, drain blood from the ascending colon. The lymphatic vessels pass to the *paracolic and epicolic lymph nodes* and from them to the *superior mesenteric nodes* (Fig. 3.10*A*). The nerves to the ascending colon are derived from the *superior mesenteric plexus*, as described for the cecum and appendix (Fig. 3.10*B*).

The *transverse colon* is the largest and most mobile part of the large intestine (Fig. 3.9, *A* and *D*). It crosses the abdomen from the *right colic flexure to the left colic flexure*, where it bends inferiorly to become the descending colon. The left colic flexure lies on the inferior part of the left kidney and is attached to the diaphragm by the *phrenicocolic ligament*. The *transverse mesocolon* is the mobile mesentery of the transverse colon. The root of this mesentery is located along the inferior border of the pancreas and becomes continuous with the parietal peritoneum posteriorly (Fig. 3.6). Because it is freely movable, the transverse colon is variable in position. It usually hangs down to the level of the umbilicus. In tall, thin people the transverse colon may reach the level of the pelvis.

Arterial supply of the transverse colon is mainly from the *middle colic artery* (Fig. 3.9*D*, Table 3.7), a branch of the superior mesenteric artery, but it also receives blood from the right and left colic arteries. Venous drainage is via the *superior mesenteric vein*. Lymphatic drainage is to the *middle colic lymph nodes* that in turn drain to the *superior mesenteric lymph nodes* (Fig. 3.10*A*). The nerves are derived from the superior mesenteric plexus and follow the right and middle colic arteries (Fig. 3.10*B*). They transmit sympathetic and parasympathetic (vagal) nerve fibers. The nerves that follow the left colic artery are derived from the *inferior mesenteric plexus*.

The *descending colon* passes retroperitoneally from the left colic flexure into the left iliac fossa (Fig. 3.9*D*), where it is continuous with the sigmoid colon. The peritoneum covers the colon anteriorly and laterally and binds it to the posterior abdominal wall. As it descends, the colon passes anterior to the lateral border of the left kidney. As with the ascending colon, there are *paracolic gutters* on the medial and lateral sides of the descending colon.

The *sigmoid colon*, an S-shaped loop that is variable in length, links the descending colon and rectum (Fig. 3.9*D*). The sigmoid colon extends from the pelvic brim to the third segment of the sacrum where it joins the rectum. The termination of the teniae coli indicates the beginning of the rectum. The *rectosigmoid junction* is about 15 cm from the anus. The sigmoid colon usually has a long mesentery (*sigmoid mesocolon*) and therefore has considerable freedom of movement. The root of this mesentery has a V-shaped attachment, superiorly along the external iliac vessels and inferiorly from the bifurcation of the common iliac vessels to the anterior aspect of the sacrum. Posterior to the apex of the sigmoid mesocolon (i.e.,

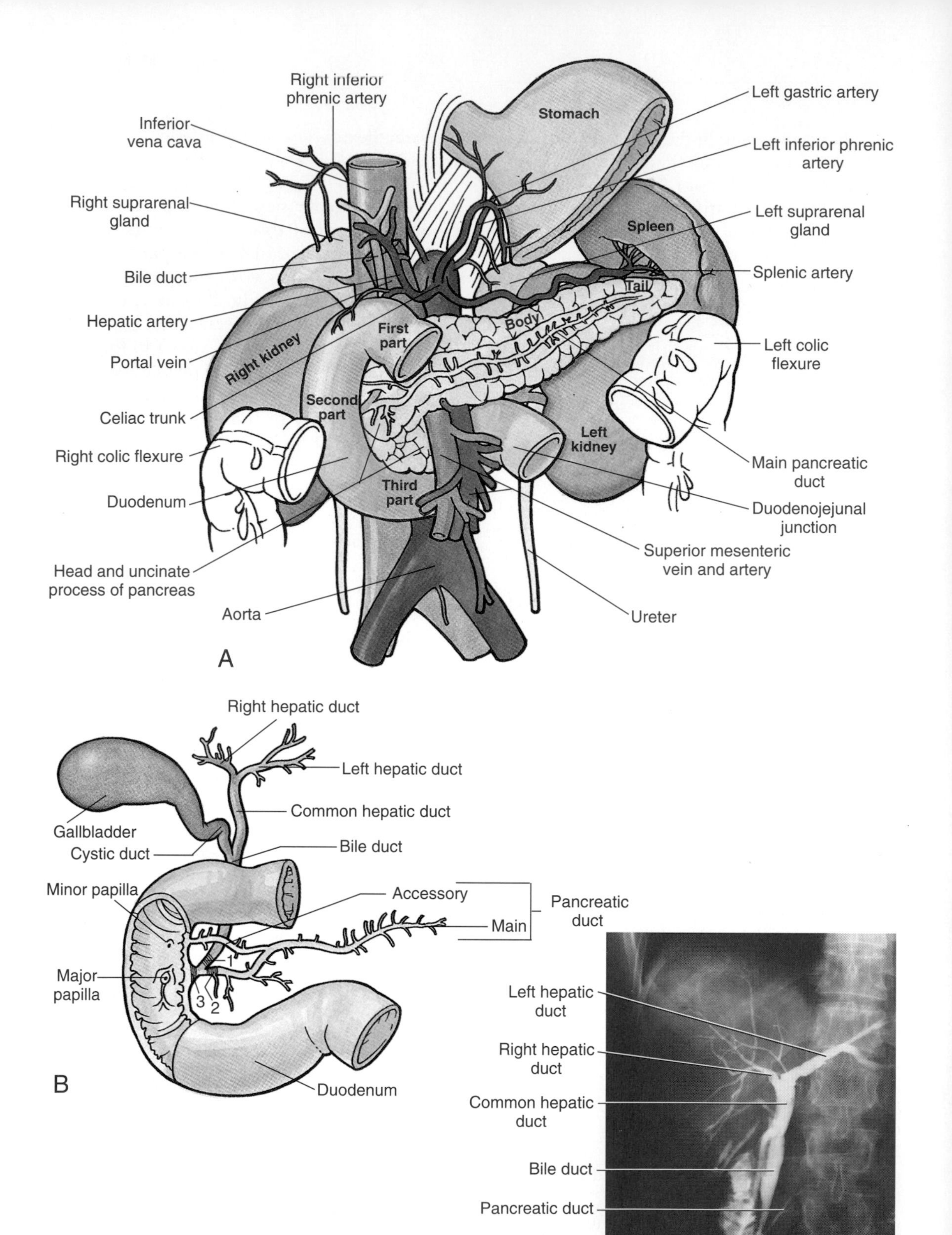

Figure 3.11. **A.** Stomach bed. **B.** Extrahepatic bile passages. Choledochal sphincter (*1*), pancreatic duct sphincter (*2*), and hepatopancreatic sphincter in wall of hepatopancreatic ampulla (*3*). **C.** Endoscopic retrograde cholangiography and pancreatography of bile and pancreatic ducts.

retroperitoneally) lies the left ureter and the division of the left common iliac artery. The *omental appendages* are long in the sigmoid colon.

The arterial supply of the descending colon is from the left colic and superior sigmoid arteries (Fig. 3.9*D*, Table 3.7). The *sigmoid arteries*, branches of the inferior mesenteric artery, descend obliquely to the left where they divide into ascending and descending branches that supply the sigmoid colon. The most superior sigmoid artery anastomoses with the descending branch of the left colic artery. The inferior mesenteric vein returns blood from the sigmoid and descending colon.

Lymphatic vessels from the descending and sigmoid colon pass to the intermediate colic lymph nodes along the left colic artery (Fig. 3.10*A*). From them the lymph passes to *inferior mesenteric lymph nodes* around the inferior mesenteric artery. However, lymph vessels from the left colic flexure also drain to the *superior mesenteric lymph nodes*. The descending and sigmoid colon receive their sympathetic nerve supply from the lumbar part of the sympathetic trunk and the *superior hypogastric plexus* by means of plexuses on the branches of the inferior mesenteric artery (Fig. 3.10*B*). The parasympathetic supply is derived from the *pelvic splanchnic nerves* (Table 3.8).

Rectum and Anal Canal

The rectum is the fixed terminal part of the large intestine (Fig. 3.9*D*). It is continuous inferiorly with the anal canal. The rectum and anal canal are described with the pelvis in Chapter 4.

Chronic disease of the colon (ulcerative colitis) is characterized by severe inflammation and ulceration of the colon and rectum. In some patients, a *total colectomy* is performed during which the terminal ileum and colon, as well as the rectum and anal canal, are removed. An *ileostomy* is then constructed to establish an opening between the ileum and the skin of the anterior abdominal wall.

Most tumors of the large intestine occur in the rectum; about 12% of them appear near the rectosigmoid junction. The interior of the sigmoid colon can be observed with a fiberoptic instrument known as a *sigmoidoscope*. This instrument is inserted into the colon via the anus.

SPLEEN

The spleen is the largest of the lymphatic organs and is located in the left upper quadrant (Fig. 3.11*A*). It varies considerably in size and shape but is usually about 12 cm long and 7 cm wide (roughly the size and shape of a clenched fist). Its diaphragmatic surface is convexly curved to fit the concavity of the diaphragm. The anterior and superior borders of the spleen are sharp and often notched, whereas its posterior and inferior borders are rounded. The spleen normally contains a large amount of blood that is expelled periodically into the circulation by the action of the smooth muscle in its capsule and trabeculae. The spleen contacts the posterior wall of the stomach and is connected to its greater curvature by the *gastrosplenic ligament* and to the left kidney by the *splenorenal ligament*. These ligaments are attached to the hilum of the spleen on its medial aspect, where the branches of the splenic artery enter and the tributaries of the splenic vein leave. Except at the hilum, the spleen is completely enclosed by peritoneum. The hilum

Table 3.8
Splanchnic Nerves

Name	Type	Origin
1. Thoracic (greater, lesser, and least) splanchnic nn.	Sympathetic	Greater (T5–T9) or T10 Lesser (T10–T11) Least (T12)
2. Lumbar splanchnic nn.	Sympathetic	Branches of four lumbar sympathetic ganglia
3. Sacral splanchnic nn.	Sympathetic	Branches of second and third sacral sympathetic ganglia
4. Pelvic splanchnic nn.	Parasympathetic	Branches of ventral rami of sacral spinal nerves [S2, S3, (S4)]

of the spleen is related to the tail of the pancreas.

The *splenic artery*, the largest branch of the celiac trunk, follows a tortuous course posterior to the omental bursa, anterior to the left kidney, and along the superior border of the pancreas (Fig. 3.11*A*). Between the layers of the splenorenal ligament, the splenic artery divides into five or more branches that enter the hilum of the spleen. The *splenic vein* is formed by several tributaries that emerge from the hilum of the spleen (Fig. 3.8*B*). It is joined by the inferior mesenteric vein and runs posterior to the body and tail of the pancreas throughout most of its course. The splenic vein unites with the superior mesenteric vein posterior to the neck of the pancreas to form the *portal vein*.

The splenic lymphatic vessels leave the lymph nodes in the hilum of the spleen and pass along the splenic vessels to the *pancreaticosplenic lymph nodes* (Fig. 3.12*A*). These nodes are related to the posterior surface and superior border of the pancreas. Nerves of the spleen are derived from the *celiac plexus* (Fig. 3.12*B*). They are distributed mainly to branches of the splenic artery and are vasomotor in function.

Although well protected by the ribs, the spleen is the most frequently injured organ in the abdomen when severe blows to the ribs are received on the left side. If ruptured, the spleen bleeds profusely because its capsule is thin and its parenchyma is soft and pulpy. *Rupture of the spleen* causes severe intraperitoneal hemorrhage and shock. Repair of a ruptured spleen is difficult; consequently *splenectomy* is often performed to prevent the patient from bleeding to death.

PANCREAS

The pancreas is an elongated digestive gland that lies transversely across the posterior abdominal wall, posterior to the stomach (Fig. 3.11*A*). The transverse mesocolon extends to its anterior margin (Fig. 3.6*A*). The pancreas produces

- An exocrine secretion (*pancreatic juice*) that enters the duodenum via the pancreatic duct
- Endocrine secretions (*glucagon* and *insulin*) that enter the blood

The *head of the pancreas* is in the curve of the duodenum. It has a prolongation called the *uncinate process* that extends superiorly and to the left and lies posterior to the superior mesenteric vessels. The head rests posteriorly on the IVC, the right renal artery and vein, and the left renal vein. The *bile duct*, on its way to the duodenum, lies in a groove on the posterosuperior surface of the head (Fig. 3.11*A*).

The *neck of the pancreas* is grooved posteriorly by the superior mesenteric vessels. Its anterior surface is covered with peritoneum and is adjacent to the pylorus. The superior mesenteric vein joins the splenic vein posterior to the neck of the pancreas to form the *portal vein*.

The *body of the pancreas* extends to the left across the aorta and L2 vertebra, posterior to the omental bursa. The body is intimately related to the splenic vessels. The anterior surface of the pancreas is covered with peritoneum and forms part of the bed of the stomach. Its posterior surface is devoid of peritoneum where it is in contact with the aorta, superior mesenteric artery, left suprarenal gland, and the left kidney and its vessels.

The *tail of the pancreas* passes between the layers of the splenorenal ligament with the splenic vessels. The tip of the tail usually contacts the hilum of the spleen.

The *pancreatic duct* begins in the tail of the pancreas and runs through the substance of the gland to the head, where it turns inferiorly and comes into close relationship with the bile duct. Usually the pancreatic and bile ducts unite to form a short, dilated *hepatopancreatic ampulla*, which opens by a common duct into the duodenum at the summit of the *major duodenal papilla* (Fig. 3.11, *B* and *C*). There is a sphincter around the terminal part of the main duct called the *pancreatic duct sphincter*. There is also one around the hepatopancreatic ampulla called the *hepatopancreatic sphincter* (of Oddi). These sphincters control the flow of bile and pancreatic juice into the duodenum.

The *accessory pancreatic duct* drains the uncinate process and the inferior part of the head of the pancreas. Usually this duct communicates with the main pancreatic duct, but in about 9% of people it is a separate duct. Typically it opens into the duodenum at the *minor duodenal papilla*.

The pancreatic arteries are derived from the splenic and pancreaticoduodenal arteries (Fig. 3.8*A*, Table 3.4). Up to 10 branches of the splenic artery supply the body and tail. The anterior and posterior superior *pancreaticoduodenal arteries*, branches of the gastroduodenal artery, and

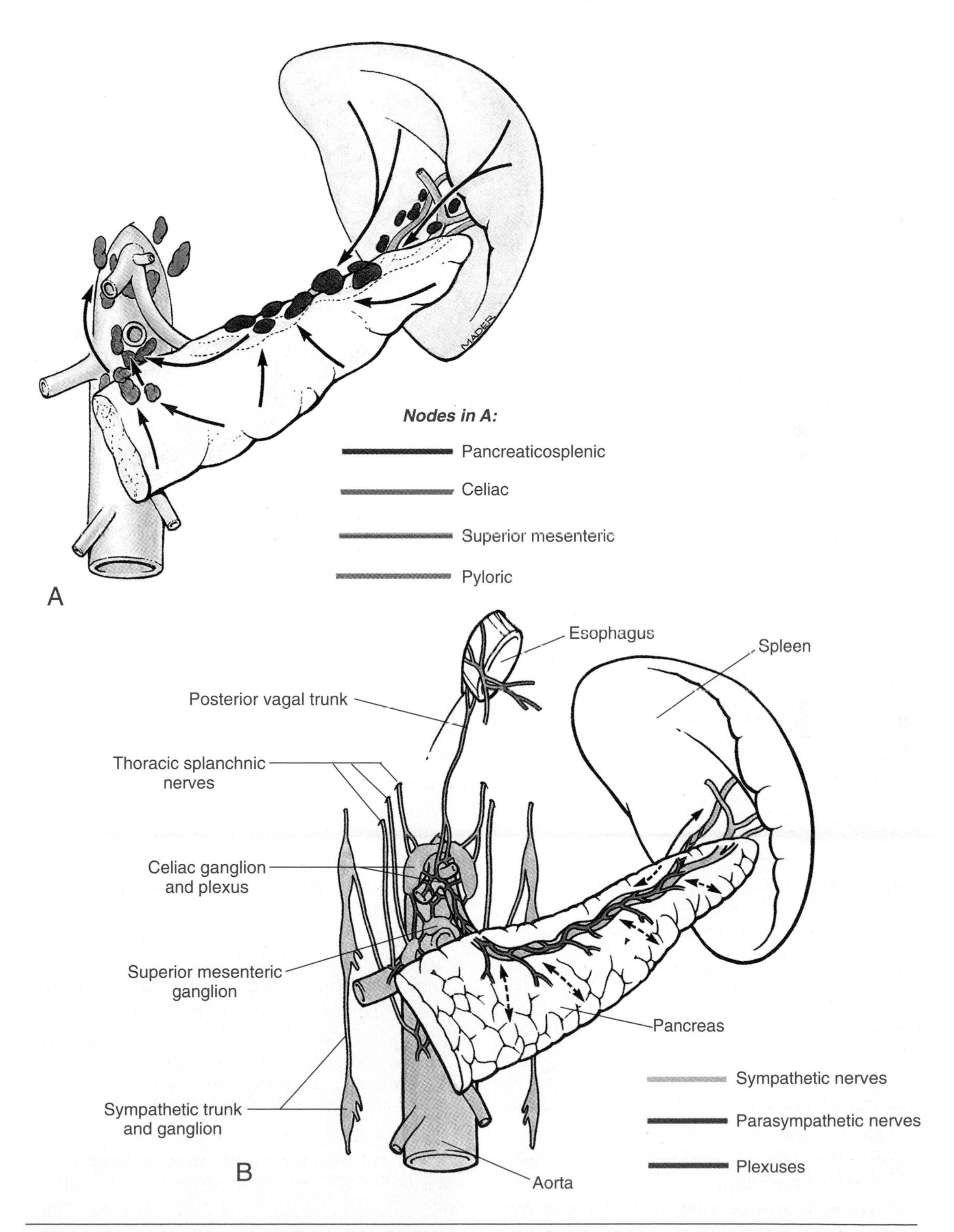

Figure 3.12. Spleen and pancreas. **A.** Lymphatic drainage. *Arrows*, lymph flow to lymph nodes. **B.** Innervation. *Arrows*, afferent and efferent nerves.

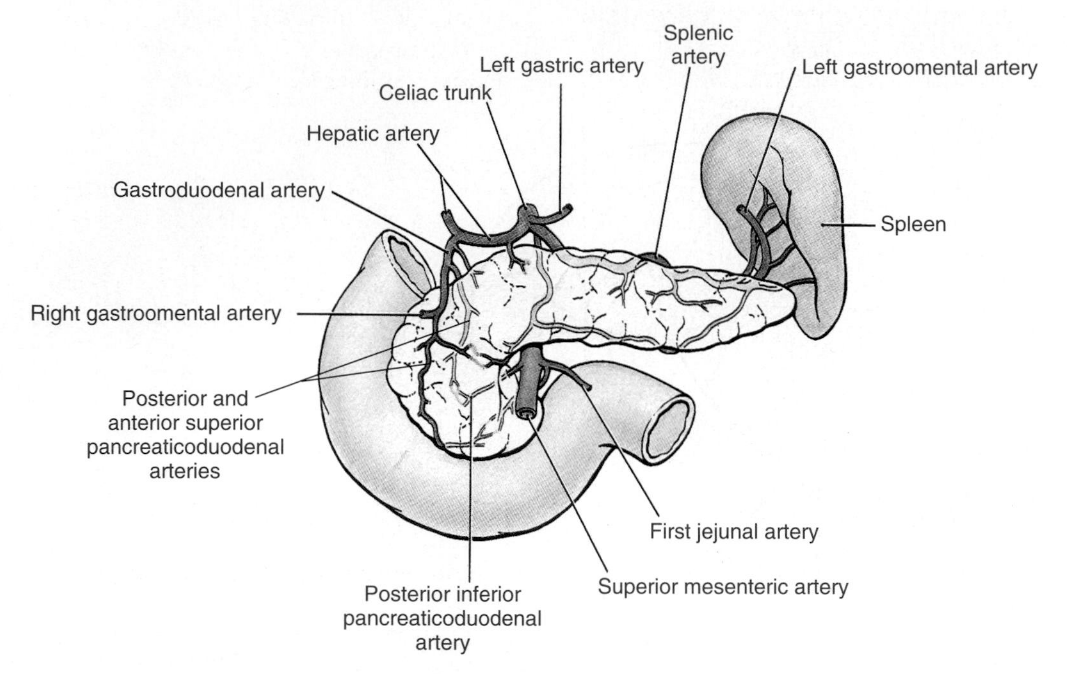

Pancreatic arteries

Surface Anatomy of Spleen and Pancreas

Xiphisternal joint
Midclavicular line
Spleen
Suprarenal gland
Transpyloric plane
Head of pancreas
Left kidney
Duodenum
Transumbilical plane
Anterior superior iliac spine

The spleen lies on the left side between the 9th and 11th ribs, and its costal surface is convex to fit these bones. The long axis of the spleen normally lies in the line of the 10th rib, where it rests on the left colic flexure. Normally the spleen does not extend inferior to the left costal margin. The tail of the pancreas touches the hilum of the spleen, and its body lies at the level of the disc between T12 and L1 vertebrae.

the anterior and posterior inferior pancreaticoduodenal arteries, branches of the superior mesenteric artery, supply the head. The pancreatic veins drain into the portal, splenic, and superior mesenteric veins, but most of them empty into the *splenic vein* (Fig. 3.8*B*).

The pancreatic lymphatic vessels follow the blood vessels (Fig. 3.12*A*). Most of them end in the *pancreaticosplenic nodes* that lie along the splenic artery, but some vessels end in the *pyloric lymph nodes*. Efferent vessels from these nodes drain to the celiac, hepatic, and superior mesenteric lymph nodes. Nerves of the pancreas are derived from the *vagus and thoracic splanchnic nerves* (Fig. 3.12*B*, Table 3.8). The parasympathetic and sympathetic fibers reach it by passing along the arteries from the celiac and superior mesenteric plexuses.

Pancreatic injury may occur when there is sudden, severe forceful compression of the abdomen, as may occur in an automobile accident. Because the pancreas lies transversely, the vertebral column acts like an anvil, and the traumatic force may rupture the pancreas. *Rupture of the pancreas* frequently tears its duct system, allowing pancreatic juice to enter the substance of the gland and invade adjacent tissues. Digestion of pancreatic and other tissues by pancreatic juice is painful.

Cancer of the pancreas usually involves the head and accounts for most cases of extrahepatic obstruction of the biliary system. Cancer of the head often results in obstruction of the bile duct and/or the hepatopancreatic ampulla. This causes retention of bile pigments, enlargement of the gallbladder, and yellow staining of most body tissues. *Jaundice* is the name given to the yellow color of the skin, mucous membranes, and conjunctiva.

Surface Anatomy of Liver

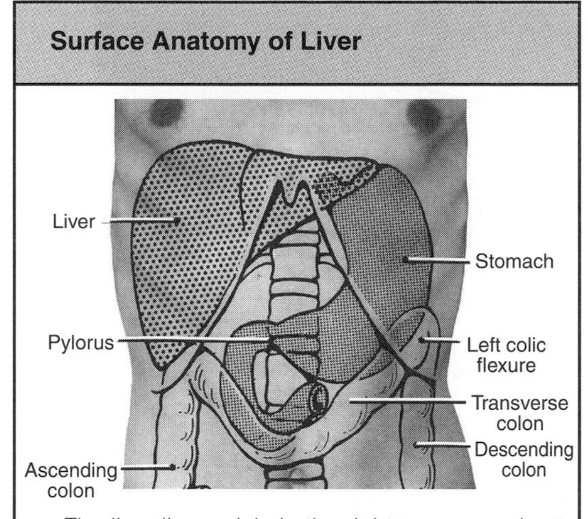

The liver lies mainly in the right upper quadrant of the abdomen where it is hidden and protected by the thoracic cage. The liver lies more inferiorly when one is erect because of gravity. The liver is pyramidal, with its base to the right and its apex to the left. Normally the liver extends inferiorly as far as the right costal margin. It is easily palpated when the patient is asked to inspire deeply because of the inferior movement of the diaphragm and liver.

LIVER

The liver is the largest gland in the body (Fig 3.13). In addition to its many metabolic activities, the liver stores glycogen and secretes bile. Bile from the liver passes through the hepatic ducts and cystic duct to the gallbladder, where it is concentrated by absorption of water.

The liver has diaphragmatic and visceral (posteroinferior) surfaces that are separated by its inferior border. The *diaphragmatic surface* is smooth and dome-shaped where it conforms to the concavity of the inferior surface of the diaphragm, but it is largely separated from the diaphragm by the *subphrenic recesses* of the peritoneal cavity. The liver is covered with peritoneum except posteriorly in the *bare area*, where it lies in direct contact with the diaphragm. The bare area is demarcated by the reflection of peritoneum from the diaphragm to it as the anterior (upper) and posterior (lower) layers of the *coronary ligament* (Fig. 3.13*B*). These layers meet at the right to form the right triangular ligament. The layers of the coronary ligament diverge toward the left and enclose the triangular bare area. The anterior layer of the coronary ligament is continuous on the left with the right layer of the falciform ligament, and the posterior layer is continuous with the right layer of the lesser omentum. The left layers of the falciform ligament and lesser omentum meet to form the left triangular ligament.

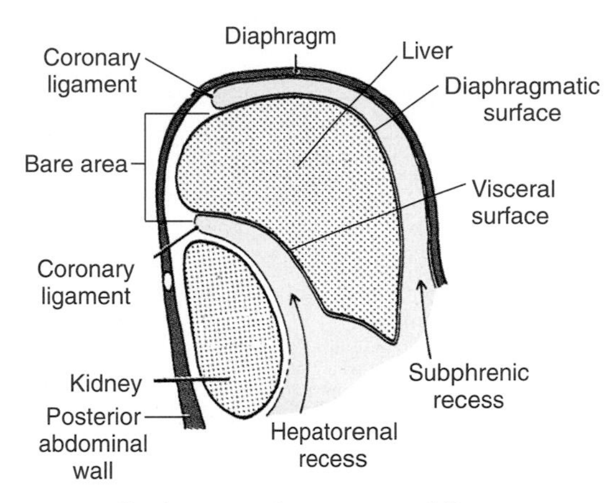

Surfaces and recesses of liver

The visceral surface is covered with peritoneum, except at the gallbladder and porta hepatis. The visceral surface of the liver is related to the

- Right side of the stomach (gastric area)
- Superior part of the duodenum (duodenal area)
- Lesser omentum
- Gallbladder
- Right colic flexure (colic area)
- Right kidney and suprarenal gland (renal area)

The *hepatorenal recess* is located between the visceral surface of the right lobe of the liver and the right kidney.

Lobes of Liver

The liver is divided into functionally independent right and left lobes (Fig. 3.13). Each lobe has its own blood supply from the hepatic artery and portal vein and its own venous and biliary drainage.

The *right lobe* of the liver is demarcated from the left lobe by the gallbladder fossa and the fossa for the IVC on the visceral surface of the liver and by an imaginary line over the diaphragmatic surface that runs from the fundus of the gallbladder to the IVC.

The *left lobe* of the liver includes the caudate lobe and most of the quadrate lobe (Fig. 3.13*B*). It is separated from these lobes by the fissure for the ligamentum teres and the fissure for the

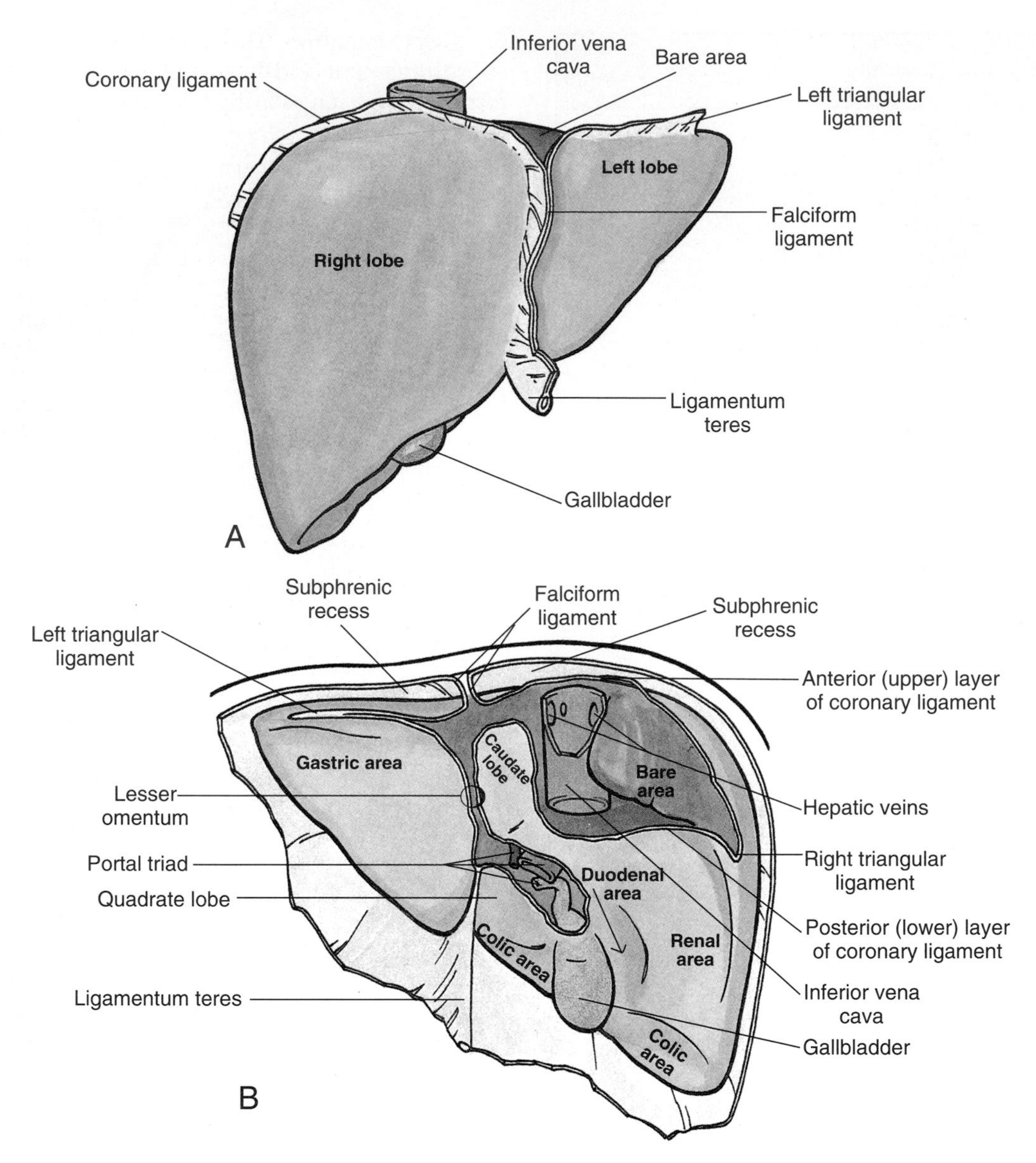

Figure 3.13. Liver and gallbladder. **A.** Anterior surface. **B.** Visceral surface. Gastric, duodenal, colic, and renal areas indicate where these organs are related to liver.

ligamentum venosum on the visceral surface and by the attachment of the ligamentum teres on the diaphragmatic surface.

The *ligamentum teres* is the obliterated remains of the umbilical vein that carried well-oxygenated blood from the placenta to the fetus. The *ligamentum venosum* is the fibrous remnant of the fetal ductus venosus that shunted blood from the umbilical vein to the IVC, short-circuiting the liver.

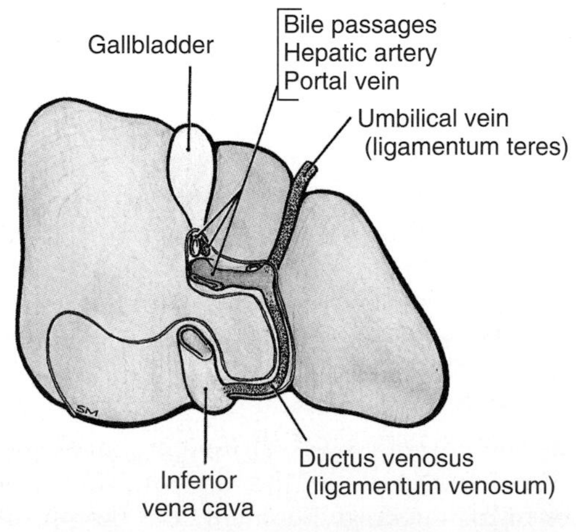

Peritoneal Relations of Liver

The lesser omentum, enclosing the *portal triad* (portal vein, bile duct, and hepatic artery) at the porta hepatis, passes to the lesser curvature of the stomach and the first 2 cm of the superior part of the duodenum. The part of the lesser omentum extending between the liver and stomach is called the *hepatogastric ligament*, and the part between the liver and duodenum is called the *hepatoduodenal ligament* (Fig. 3.14*A*). The free edge of the lesser omentum encloses the portal triad, a few lymph nodes and lymphatic vessels, and the *hepatic plexus of nerves* (Figs. 3.5*C* and 3.13*B*).

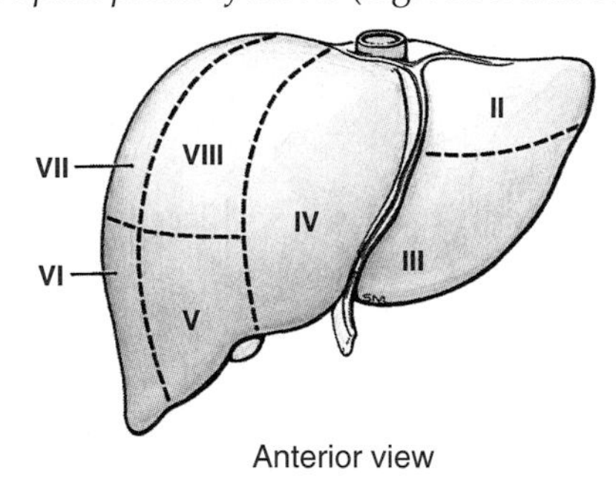

Anterior view

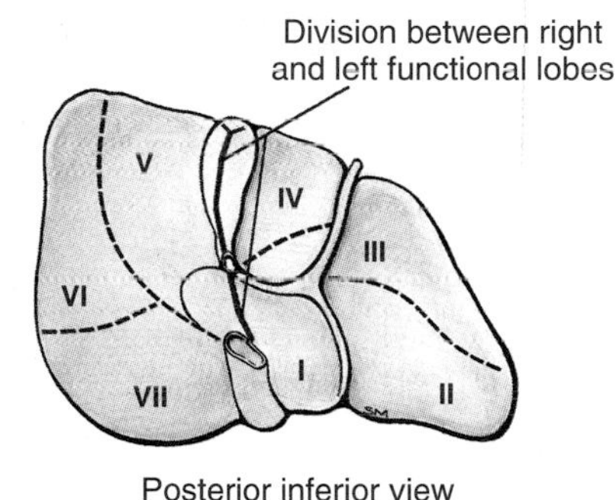

Posterior inferior view

Segments of liver

Vessels and Nerves of Liver

The liver receives blood from two sources: the hepatic artery (30%) and portal vein (70%). The *hepatic artery* carries oxygenated blood from the aorta, and the *portal vein* carries poorly oxygenated blood from the gastrointestinal tract, except the inferior part of the anal canal (Figs. 3.8*B* and 3.14*B*). At the porta hepatis, the hepatic artery and portal vein terminate by dividing into right and left branches to supply the right and left lobes of the liver, respectively. These lobes function separately. Within each lobe the primary branches of the portal vein and hepatic artery are consistent enough to describe *vascular segments*. A horizontal plane through each lobe divides the liver into eight vascular segments. Between the segments are the hepatic veins that drain adjacent segments.

The *hepatic veins*, formed by the union of the central veins of the liver, open into the IVC just inferior to the diaphragm (Fig. 3.13*B*). The attachment of these veins to the IVC helps to hold the liver in position.

When it was discovered that the right and left hepatic arteries and ducts, as well as tributaries of the right and left branches of the portal vein, do not communicate, it became possible to perform *hepatic lobectomies* and to remove segments of the liver without excessive bleeding. Surgeons still must contend with intersegmental hepatic veins that are a source of bleeding.

The right branch of the hepatic artery usually runs posterior to the common hepatic duct (Fig. 3.14*B*). Sometimes the hepatic artery arises from the superior mesenteric artery or the aorta instead of from the celiac trunk. In these cases the artery usually passes posterior to the portal vein. In about 20% of people the left hepatic artery arises from the left gastric artery and may replace the normal artery or exist as an accessory hepatic artery.

The lymphatics of the liver are superficial and deep (Fig. 3.15*A*). Most superficial lymphatic vessels join those in the porta hepatis and enter the *hepatic lymph nodes*. Most deep lymphatic vessels also converge at the porta hepatis and end in the hepatic lymph nodes scattered along the hepatic vessels and ducts in the lesser omentum. Efferent lymphatic vessels from the hepatic lymph nodes drain into the *celiac lymph nodes*, and from them lymph enters the *thoracic duct*. Some deep lymphatic vessels follow the hepatic veins to the *vena caval foramen* in the diaphragm and end in the *phrenic lymph nodes*. Lymphatic vessels from the bare area pass through the vena caval foramen and enter the phrenic and mediastinal lymph nodes. Lymph from these nodes empties into the right lymphatic duct and thoracic duct. From the posterior surface of the left lobe, a few lymphatic vessels pass to the esophageal hiatus of the diaphragm and end in the left gastric lymph nodes.

Nerves of the liver are derived from the hepatic plexus (Fig. 3.15*B*), the largest derivative of the celiac plexus. The hepatic plexus accompanies the branches of the hepatic artery and portal vein to the liver. It consists of sympathetic fibers from the celiac plexus and parasympathetic fibers from the anterior and posterior vagal trunks.

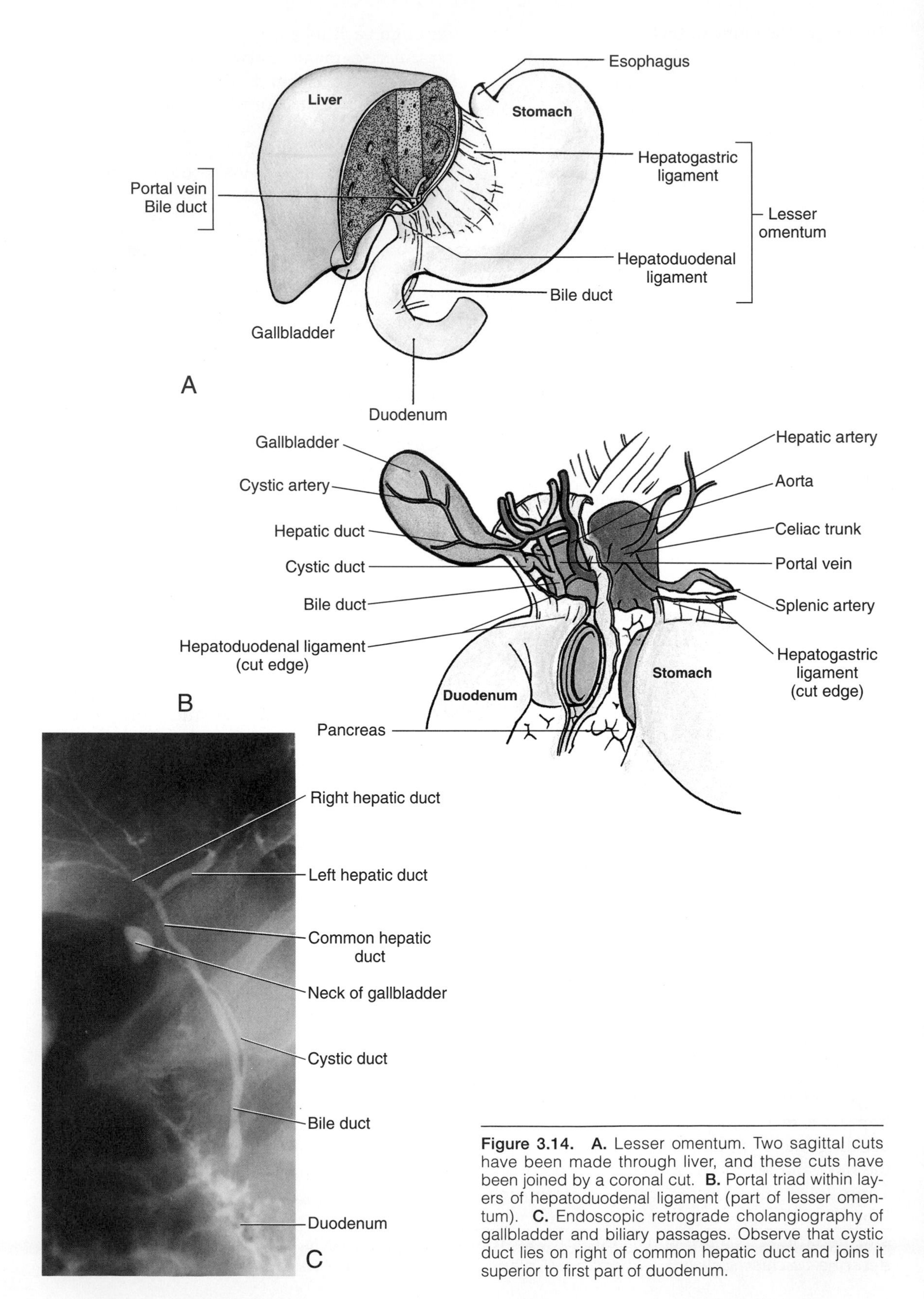

Figure 3.14. **A.** Lesser omentum. Two sagittal cuts have been made through liver, and these cuts have been joined by a coronal cut. **B.** Portal triad within layers of hepatoduodenal ligament (part of lesser omentum). **C.** Endoscopic retrograde cholangiography of gallbladder and biliary passages. Observe that cystic duct lies on right of common hepatic duct and joins it superior to first part of duodenum.

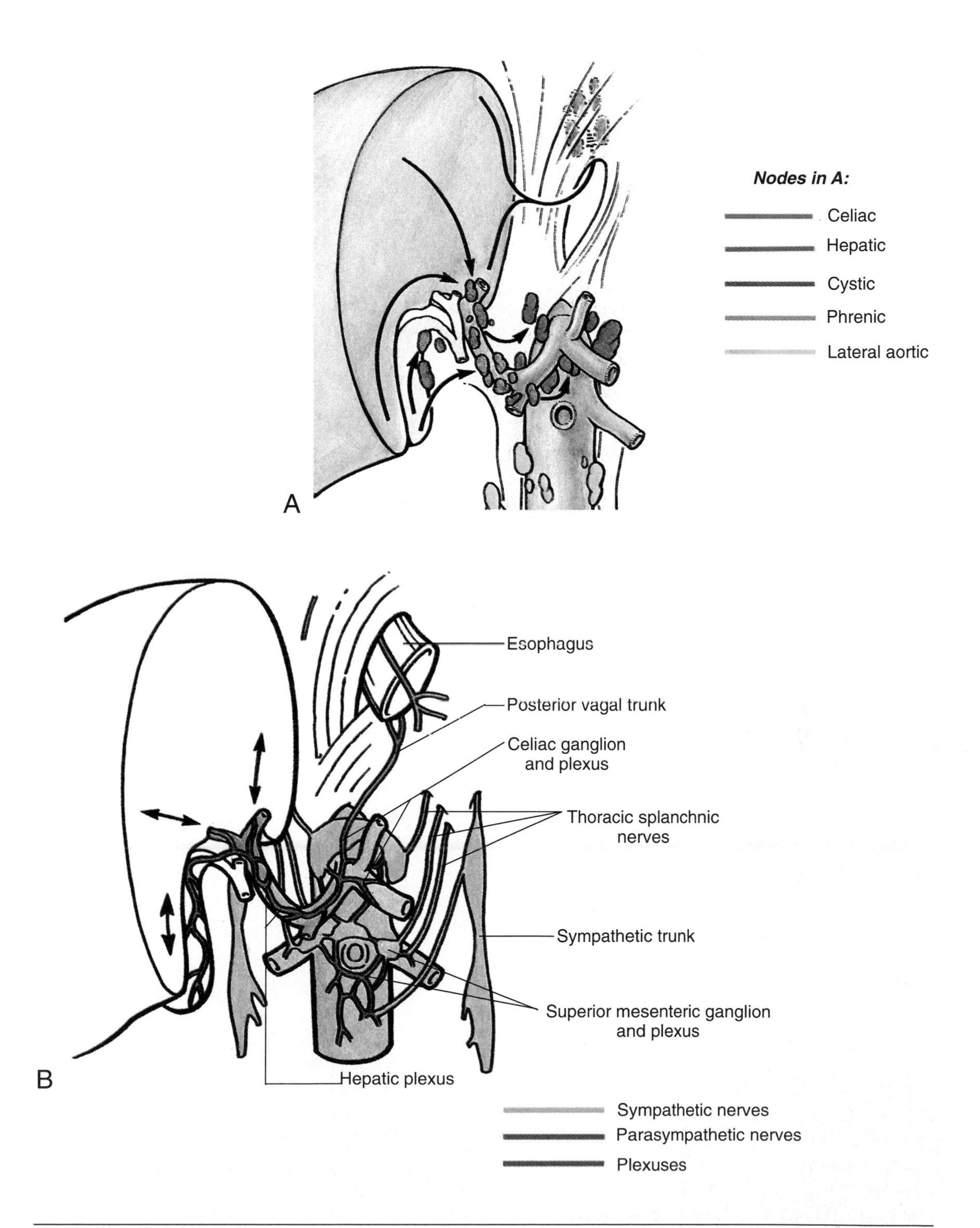

Figure 3.15. Liver. **A.** Lymphatic drainage. *Arrows*, lymph flow to lymph nodes. **B.** Innervation. *Arrows*, afferent and efferent nerves.

Hepatic tissue may be obtained for diagnostic purposes by *liver biopsy*. The needle puncture is commonly made through the right tenth intercostal space in the midaxillary line. The biopsy is taken while the patient is holding his or her breath in full expiration to reduce the *costodiaphragmatic recess* and to lessen the possibility of damaging the lung and contaminating the pleural cavity.

The liver is easily ruptured because it is large, fixed in position, and friable. Often the liver is torn by a fractured rib that perforates the diaphragm. Owing to the liver's great vascularity and friability, liver lacerations often cause considerable hemorrhage and right upper quadrant pain. The liver is enlarged in many diseases (e.g., heart failure). When there is massive liver enlargement, its inferior edge may reach the right lower quadrant of the abdomen.

The liver is also a common site of metastatic carcinoma from organs drained by the portal system of veins. Carcinoma may also pass to the liver from the thorax, especially the breast, because of communications between thoracic lymph nodes and lymphatic vessels draining the bare area of the liver.

In *cirrhosis of the liver*, there is progressive destruction of hepatocytes and replacement of them by fibrous tissue. This tissue surrounds the intrahepatic blood vessels and biliary ducts, making the liver very firm and impeding circulation of blood through it.

BILIARY DUCTS AND GALLBLADDER

Bile is secreted by hepatic cells into *bile canaliculi* that drain into small interlobular biliary ducts which join to form the right and left hepatic ducts (Figs. 3.11, *B* and *C*, and 3.14). The right hepatic duct drains the right lobe of the liver, and the left hepatic duct drains the left lobe, which includes the caudate lobe and most of the quadrate lobe. Shortly after leaving the porta hepatis, the hepatic ducts unite to form the *common hepatic duct*. This duct is joined on the right side by the *cystic duct* from the gallbladder to form the bile duct, which conveys bile to the duodenum.

The *bile duct* (common bile duct) begins in the free edge of the lesser omentum by the union of the cystic and common hepatic ducts. The bile duct descends posterior to the superior part of the duodenum and lies in a groove on the posterior surface of the head of the pancreas. On the left side of the descending part of the duodenum, the bile duct comes into contact with the pancreatic duct. The two of them run obliquely through the wall of this part of the duodenum, where they unite to form the *hepatopancreatic ampulla*. The distal end of the ampulla opens into the duodenum through the *major duodenal papilla* (Fig. 3.11). The muscle around the distal end of the bile duct is thickened to form the *choledochal sphincter*. When this sphincter contracts, bile cannot enter the ampulla and/or the duodenum; hence bile backs up and passes along the cystic duct to the gallbladder for concentration and storage.

The arterial supply of the bile duct is as follows (Fig. 3.8*A*):

- The proximal part is supplied by the cystic artery
- The middle part is supplied by the right hepatic artery
- The retroduodenal part is supplied by the posterior superior pancreaticoduodenal artery and the gastroduodenal artery

The veins from the proximal part of the bile duct and the hepatic ducts generally enter the liver directly. The *posterior superior pancreaticoduodenal vein* drains the distal part of the bile duct (Fig. 3.8*B*) and empties into the portal vein or one of its tributaries. Lymphatic vessels from the bile duct pass to the *cystic lymph node* near the neck of the gallbladder, the node of the omental foramen, and the *hepatic lymph nodes* (Fig. 3.15*A*). Efferent lymphatic vessels pass to the *celiac lymph nodes*.

The *gallbladder* (7–10 cm long) lies in the gallbladder fossa on the visceral surface of the liver (Figs. 3.13*B* and 3.14*A*). The posterior surface of the pear-shaped gallbladder is covered by visceral peritoneum, and its anterior surface adheres to the liver. Peritoneum completely surrounds its fundus and binds its body and neck to the liver.

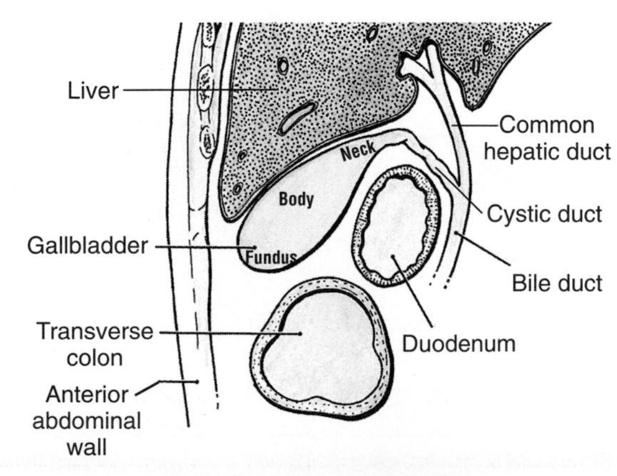

Relations of gallbladder

There are three parts of the gallbladder.

- The fundus is the wide end that projects from the inferior border of the liver and is usually located at the tip of the right ninth costal cartilage in the midclavicular line
- The body contacts the visceral surface of the liver, the transverse colon, and the superior part of the duodenum
- The neck is narrow, tapered, and directed toward the porta hepatis

The neck of the gallbladder makes an S-shaped bend and joins the cystic duct. The mucosa of the neck is shaped into a spiral fold called the *spiral valve* that keeps the cystic duct open so that (*a*) bile can easily be diverted into the gallbladder when the distal end of the bile duct is closed by the choledochal sphincter and/or the hepatopancreatic sphincter or (*b*) bile can pass to the duodenum as the gallbladder contracts.

The *cystic duct* (about 4 cm long) connects the neck of the gallbladder to the common hepatic duct. The cystic duct passes between the layers of the lesser omentum, usually parallel to the *common hepatic duct*, which it joins to form the bile duct.

The *cystic artery* (Fig. 3.14*B*) supplies the gallbladder and cystic duct. The artery commonly (72%) arises from the right hepatic artery in the angle between the common hepatic duct and the cystic duct. The *cystic vein* (Fig. 3.16) draining the biliary ducts and the neck of the gallbladder may pass to the liver directly or drain through the portal vein to the liver. The veins of the fundus and body pass directly into the visceral surface of the liver.

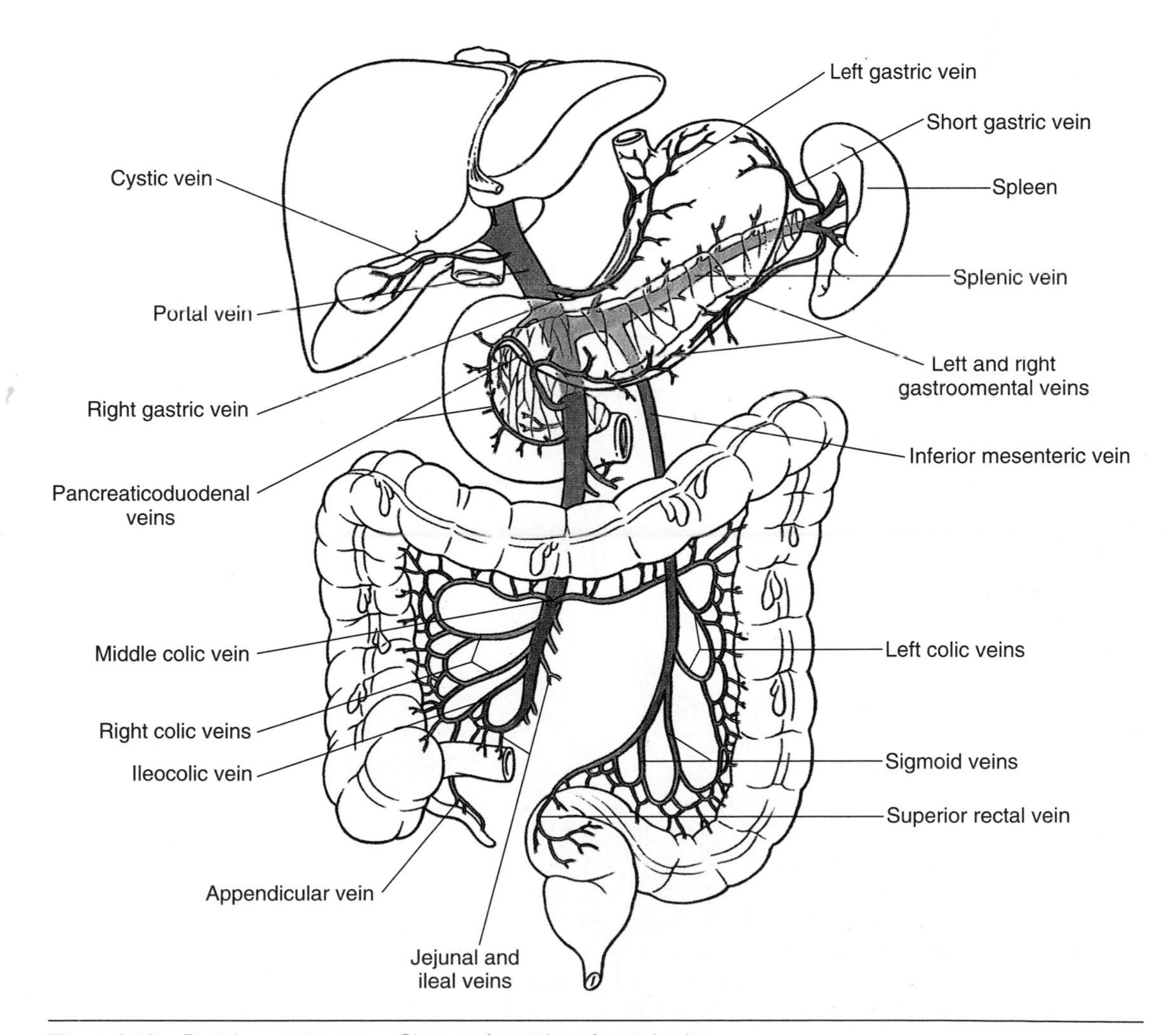

Figure 3.16. Portal venous system. Observe formation of portal vein.

Lymphatic drainage of the gallbladder is to the *hepatic lymph nodes* (Fig. 3.15*A*), often by way of the *cystic lymph node* located near the neck of the gallbladder. Efferent lymphatic vessels from these nodes pass to the *celiac lymph nodes*. The nerves to the gallbladder and cystic duct (Fig. 3.15*B*) pass along the cystic artery from the celiac plexus (sympathetic), the vagus nerve (parasympathetic), and the right phrenic nerve (sensory).

The distal end of the hepatopancreatic ampulla is the narrowest part of the biliary passages and is the common site for *impaction of a gallstone*. A stone may also lodge in the cystic duct, causing pain (*biliary colic*) in the epigastric region. When the gallbladder relaxes, the stone may pass back into the gallbladder. If the stone blocks the cystic duct, inflammation of the gallbladder (*cholecystitis*) occurs, and the pain shifts to the right hypochondriac region. If bile cannot leave the gallbladder it enters the blood and causes jaundice.

Errors in gallbladder surgery can result from failure to appreciate the common variations in the anatomy of the biliary system. Before dividing any structure and removing the gallbladder, surgeons identify all three biliary ducts, as well as the cystic and hepatic arteries.

PORTAL VEIN AND PORTAL-SYSTEMIC ANASTOMOSES

The portal vein is the main channel of the portal system of veins (Fig. 3.16). It collects blood from the abdominal part of the gastrointestinal tract (gallbladder, pancreas, and spleen) and carries it to the liver. There it branches to end in expanded capillaries known as *sinusoids*. The portal venous system communicates with the systemic venous system in the following locations:

- Between esophageal veins draining into either the azygos vein (systemic) or the left gastric vein (portal); when dilated these are *esophageal varices* (*A*)
- Between rectal veins, the inferior and middle draining into the IVC (systemic) and the superior rectal vein continuing as the inferior mesenteric vein (portal); when dilated these are *hemorrhoids* (*B*)
- Paraumbilical veins (portal) anastomosing with small epigastric veins of the anterior abdominal wall (systemic); when dilated these veins produce "caput medusae," so named because of their resemblance to the

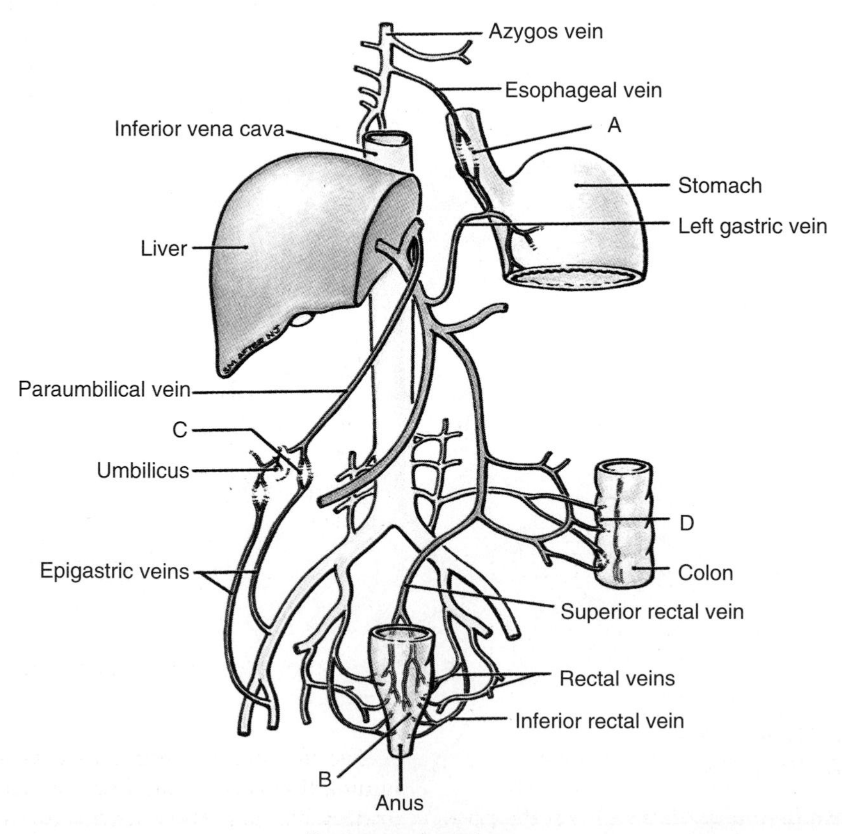

Portacaval system

serpents on the head of Medusa, a character in Greek mythology (*C*)

- Twigs of colic veins (portal) anastomosing with systemic retroperitoneal veins (*D*)

The portal-systemic anastomoses are important clinically. When the portal circulation is obstructed (e.g., because of liver disease), blood from the gastrointestinal tract can still reach the right side of the heart through the IVC via several collateral routes. These alternate routes are available because the portal vein and its tributaries have no valves; hence blood can flow in a reverse direction to the IVC.

When scarring and fibrosis from cirrhosis obstruct the portal vein in the liver, pressure rises in the portal vein and its tributaries (*portal hypertension*). At the sites of anastomoses between portal and systemic veins, portal hypertension produces enlarged *varicose veins* and blood flow from the portal to the systemic system of veins. The veins may become so dilated that their walls rupture, resulting in hemorrhage. Bleeding from *esophageal varices* at the distal end of the esophagus is often severe and may be fatal.

RENAL FASCIA AND FAT

The fibrous tissue surrounding the kidney called renal fascia is separated from the fibrous capsule of the kidney by *perirenal fat* (Fig. 3.17), which is continuous at the hilum of the kidney with the fat in the renal sinus. External to the renal fascia is *pararenal fat* that is most obvious posterior to the kidney. Movement of the kidneys during respiration is accommodated by the perirenal and pararenal fat. The renal fascia sends bundles of collagen through the fat, which, along with the renal vessels and ureter, hold the kidney in position. The renal fascia ascends to envelop the suprarenal glands. Inferior to the kidney the renal fascia is replaced by loose connective tissue that connects the parietal peritoneum to the posterior abdominal wall.

The attachments of the renal fascia are clinically important because they prevent extension of a *perinephric abscess* (collection of pus). For example, fascia at the renal hilum is firmly attached to the renal vessels and the ureter, which usually prevents spread of pus to the opposite side. Similarly, pus does not pass inferiorly because of the attachment of the renal fascia to the posterior abdominal wall.

KIDNEYS, URETERS, AND SUPRARENAL GLANDS

The kidneys lie retroperitoneally on the posterior abdominal wall, one on each side of the vertebral column at the level of T12 to L3 vertebrae (Figs. 3.18 and 3.19). The right kidney lies at a slightly lower level than the left kidney because of the large size of the right lobe of the liver. Each kidney has anterior and posterior surfaces, medial and lateral margins, and superior and inferior poles.

Superiorly each kidney is related to the *diaphragm* that separates it from the pleural cavity and the 12th rib. More inferiorly, the posterior surface of the kidney is related to the quadratus lumborum muscle. The subcostal nerve and vessels and the iliohypogastric and ilioinguinal nerves descend diagonally across the posterior surface of the kidney. The liver, duodenum, and ascending colon are anterior to the right kidney, whereas the left kidney is related to the stomach, spleen, pancreas, jejunum, and descending colon.

At the concave medial margin of each kidney there is a vertical cleft called the *renal hilum* where the renal artery enters and the renal vein and renal pelvis leave. The left hilum lies in the transpyloric plane, about 5 cm from the median plane at the level of L1 vertebra. At the hilum the renal vein is anterior to the renal artery, which is anterior to the renal pelvis. The renal hilum leads into a space within the kidney called the *renal sinus* (Fig. 3.17), which is occupied by the renal pelvis, calices, vessels and nerves, and a variable amount of fat.

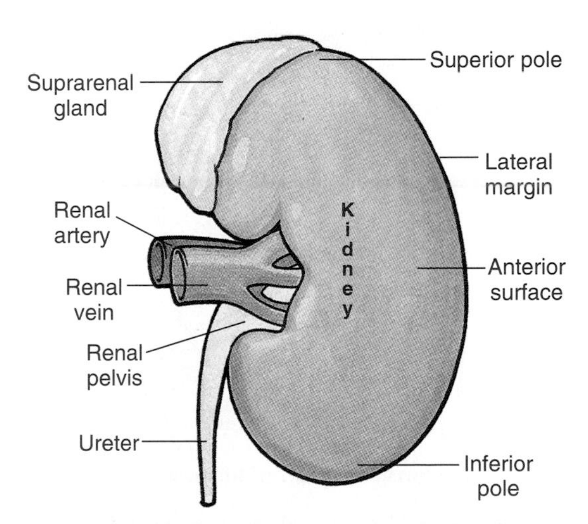

Right kidney and suprarenal gland

The *ureters* are muscular ducts with narrow lumina that carry urine from the kidneys to the urinary bladder. The superior expanded end of the ureter, called the *renal pelvis*, divides into two

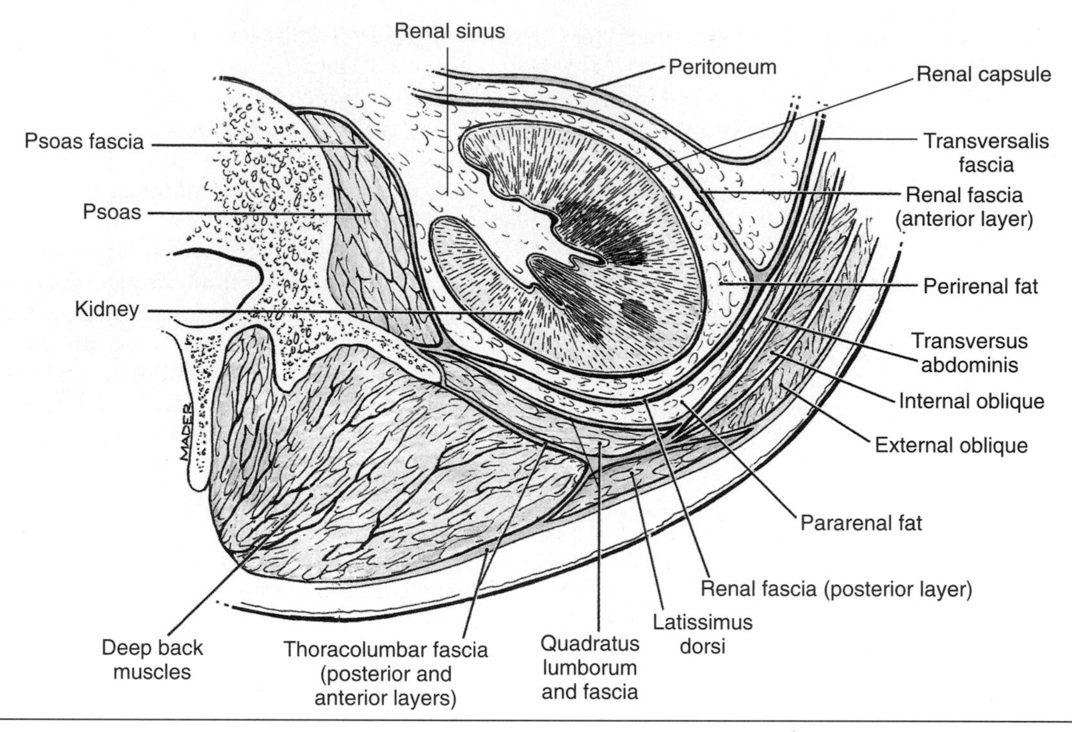

Figure 3.17. Transverse section of kidney showing its relationships to muscles and fasciae.

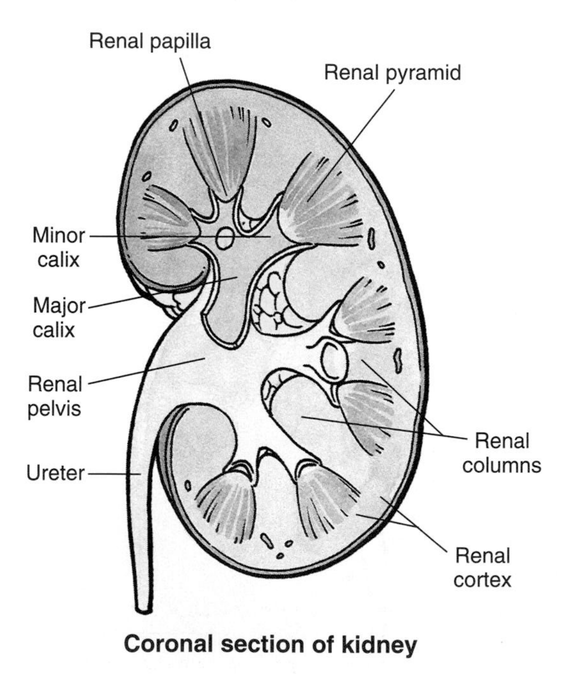

or three *major calices* (calyces), each of which divides into two or three *minor calices*. Each minor calix (calyx) is indented by the apex of the *renal pyramid* called the *renal papilla*. The abdominal part of the ureter adheres closely to the parietal peritoneum and is retroperitoneal throughout its course. The ureters run inferomedially along the transverse processes of the lumbar vertebrae and cross the external iliac artery just beyond the bifurcation of the common iliac artery (Fig. 3.19). They then run along the lateral wall of the pelvis to enter the urinary bladder.

Renal transplantation is now an established operation for the treatment of selected cases of chronic renal failure. The site for the transplanted kidney is the lower abdomen. Its renal artery and vein are joined to the external iliac artery and vein, and the ureter is sutured into the bladder.

Renal cysts, multiple or solitary, are common findings during dissection of cadavers. Adult polycystic disease of the kidneys is an important cause of renal failure that is inherited as an autosomal dominant trait. The kidneys are markedly enlarged and distorted by cysts as large as 5 cm.

The *suprarenal (adrenal) glands* are located on the superomedial aspects of the kidneys (Fig. 3.19). Each gland is enclosed within a fibrous capsule and is enveloped by the *renal fascia*. The shape and relations of the suprarenal glands differ on the two sides. The triangular right gland lies anterior to the diaphragm and makes contact anteriorly with the IVC medially and the liver laterally. The semilunar left gland is related to the spleen, stomach, pancreas, and left crus of the diaphragm.

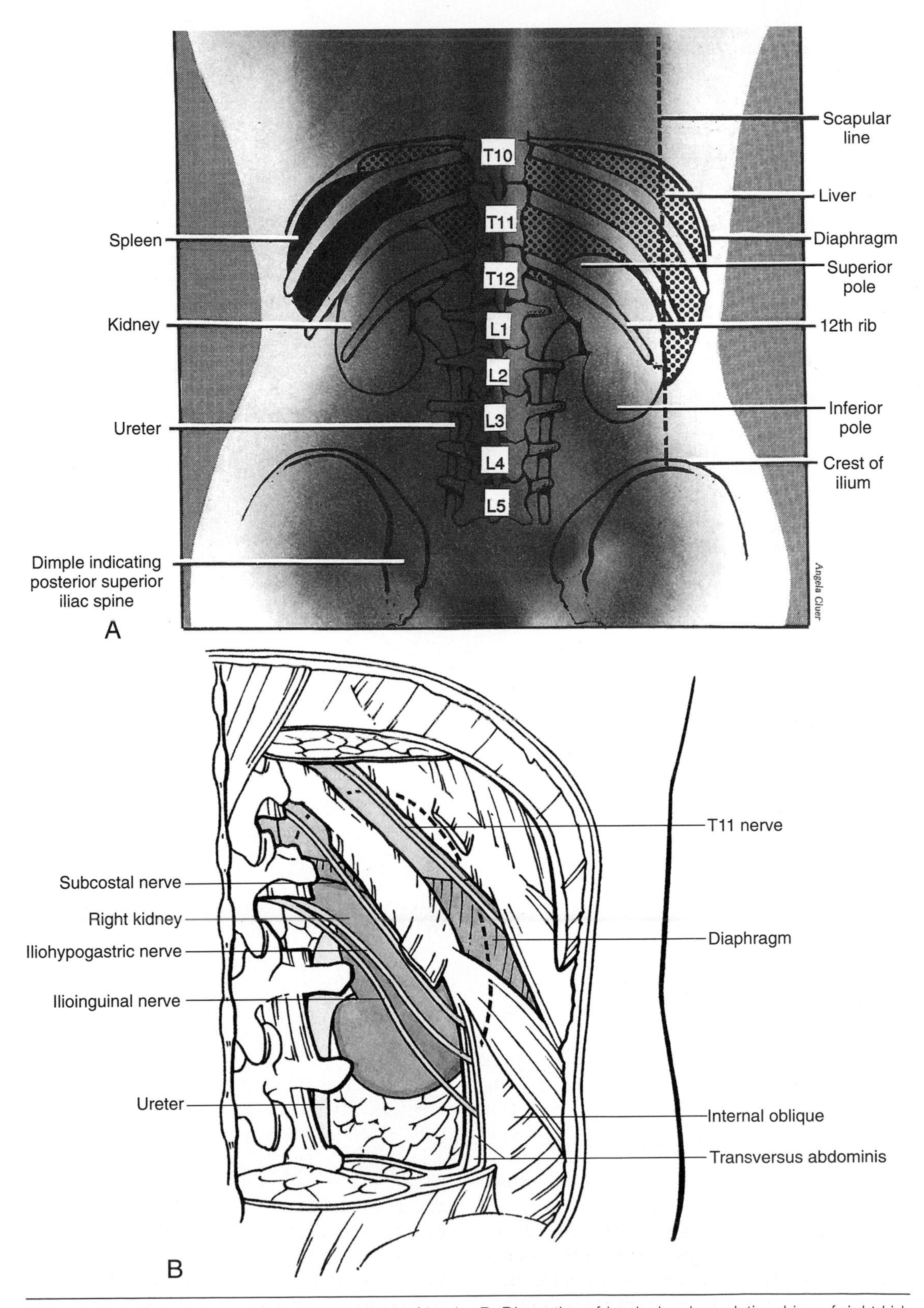

Figure 3.18. **A.** Surface anatomy and markings of back. **B.** Dissection of back showing relationships of right kidney.

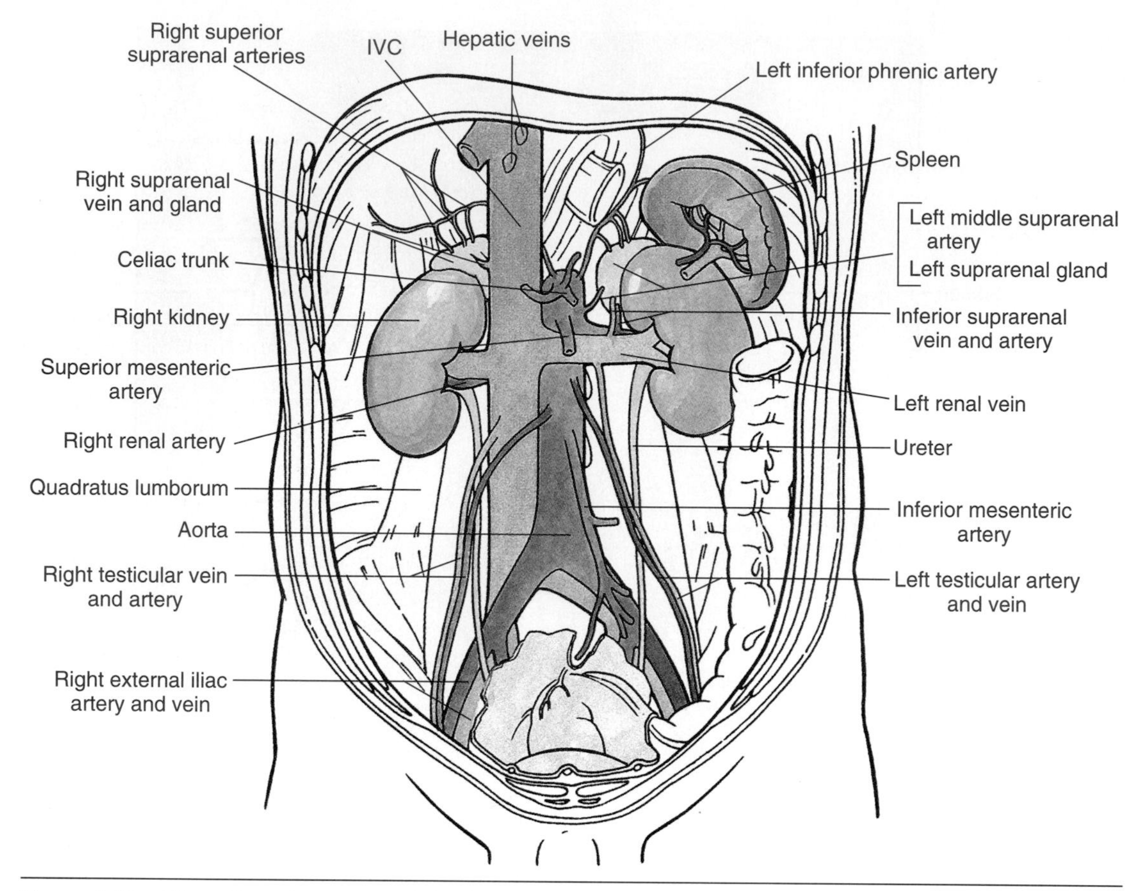

Figure 3.19. Dissection of posterior abdominal wall showing great vessels, spleen, kidneys, and suprarenal glands.

Vessels of Kidneys, Ureters, and Suprarenal Glands

The renal arteries arise at the level of the intervertebral disc between L1 and L2 vertebrae (Fig. 3.19). The longer right renal artery passes posterior to the IVC. Typically each artery divides close to the hilum into five *segmental arteries* that are end arteries; i.e., they do not anastomose. They are distributed to the segments of the kidney. Several veins drain the kidney and unite in a variable fashion to form the renal vein. The renal veins lie anterior to the renal arteries, and the longer left renal vein passes anterior to the aorta. Each renal vein drains into the IVC.

The suprarenal glands have a profuse blood supply from the superior *suprarenal arteries* (*n* = 6–8) from the inferior phrenic artery, the middle suprarenal arteries (one or more) from the abdominal aorta, and the inferior suprarenal arteries from the renal artery. Each suprarenal gland is drained by a large *suprarenal vein* and often many small ones. The short right suprarenal vein drains into the IVC, whereas the longer left one joins the left renal vein.

The arteries to the abdominal part of the ureter usually originate from three sources: renal artery, testicular or ovarian artery, and aorta. Venous drainage from the ureters is into the testicular or ovarian veins.

During their "ascent," the embryonic kidneys receive their blood supply and venous drainage from successively more superior vessels [see Moore and Persaud (1993) listed under "Suggested Readings"]. Usually the inferior vessels degenerate as superior ones take over the blood supply and venous drainage. Failure of some of these vessels to degenerate results in *accessory renal arteries and veins.* Variations in the number and position of these vessels occur in about 25% of people.

The renal lymphatic vessels follow the renal veins and drain into the *lumbar (aortic) lymph nodes* (Fig. 3.20*A*). Lymphatic vessels from the superior part of the ureter may join those from the kidney or pass directly to the lumbar lymph nodes. Lymphatic vessels from the middle part of the ureter usually drain into the common iliac lymph nodes, whereas those from its inferior part drain into the common, external, or internal *iliac lymph nodes* (Fig. 3.4*B*).

The suprarenal lymphatic vessels arise from a plexus deep to the capsule of the gland and from one in the medulla. Many lymphatic vessels leave the suprarenal glands; most of them end in the *lumbar lymph nodes*.

Nerves of Kidneys, Ureters, and Suprarenal Glands

Nerves to the kidneys and ureters arise from the renal plexus and consist of sympathetic and parasympathetic fibers (Fig. 3.20*B*). This plexus is supplied by fibers from the thoracic splanchnic nerves (Table 3.8). The suprarenal glands have a rich nerve supply from the celiac plexus and thoracic splanchnic nerves.

Excessive distention of the ureter from a *ureteric calculus* or stone causes severe rhythmic pain called *ureteric colic*. These stones may cause complete or intermittent obstruction of urinary flow. Ureteric colic is usually a sharp, stabbing pain that follows the course of the ureter.

Thoracic Diaphragm

The thoracic diaphragm is a dome-shaped, musculotendinous partition separating the thoracic and abdominal cavities. The diaphragm is the chief muscle of inspiration and forms the convex floor of the thoracic cavity and the concave roof of the abdominal cavity (Fig. 3.21). Although the diaphragm descends during inspiration, only its dome moves because its peripheral parts are attached to the inferior margin of the thoracic cage and the superior lumbar vertebrae. The diaphragm curves superiorly into right and left domes; the right is higher than the left. During expiration the right dome reaches as high as the 5th rib, and the left dome ascends to the fifth intercostal space. The level of the domes varies according to the

- Phase of respiration
- Posture assumed
- Size and degree of distention of the abdominal viscera

The diaphragm is composed of two portions: a peripheral muscular part and a central aponeurotic part, the central tendon. The muscular fibers of the diaphragm converge radially to the *central tendon* and may be divided into three parts according to their attachments (Fig. 3.21):

- A sternal part consisting of two muscular slips that are attached to the posterior aspect of the xiphoid process
- A costal part consisting of wide muscular slips that arise from the internal surfaces of the inferior six ribs and their cartilages
- A lumbar part arising from the lumbar vertebrae by two crura and from the arcuate ligaments

The *crura of the diaphragm* are musculotendinous bundles that arise from the anterior surfaces of the bodies of the lumbar vertebrae, the anterior longitudinal ligament, and the intervertebral discs (Fig. 3.21*B*). The right crus, longer than the left crus, arises from the first three lumbar vertebrae, whereas the left crus arises from only the first two. The crura are united by the *median arcuate ligament* that passes over the anterior surface of the aorta. The diaphragm is also attached on each side to the *medial and lateral arcuate ligaments*, which are thickenings of the fascia covering the psoas and quadratus lumborum muscles, respectively.

The *central tendon* is the tendon of all muscular fibers of the diaphragm that is fused with the inferior surface of the fibrous pericardium. The central tendon has no bony attachments and is incompletely divided into three leaves, which together resemble a cloverleaf. Vessels and nerves of the diaphragm are described in Table 3.9.

DIAPHRAGMATIC APERTURES

The diaphragmatic apertures permit structures to pass between the thorax and abdomen (Fig. 3.21).

The *vena caval foramen* for the IVC is in the central tendon. It is located to the right of the median plane at the level of the disc between T8 and T9 vertebrae. The IVC is adherent to the margin of the foramen; consequently when the diaphragm contracts during inspiration, it widens the foramen and dilates the IVC. These changes facilitate blood flow through this large

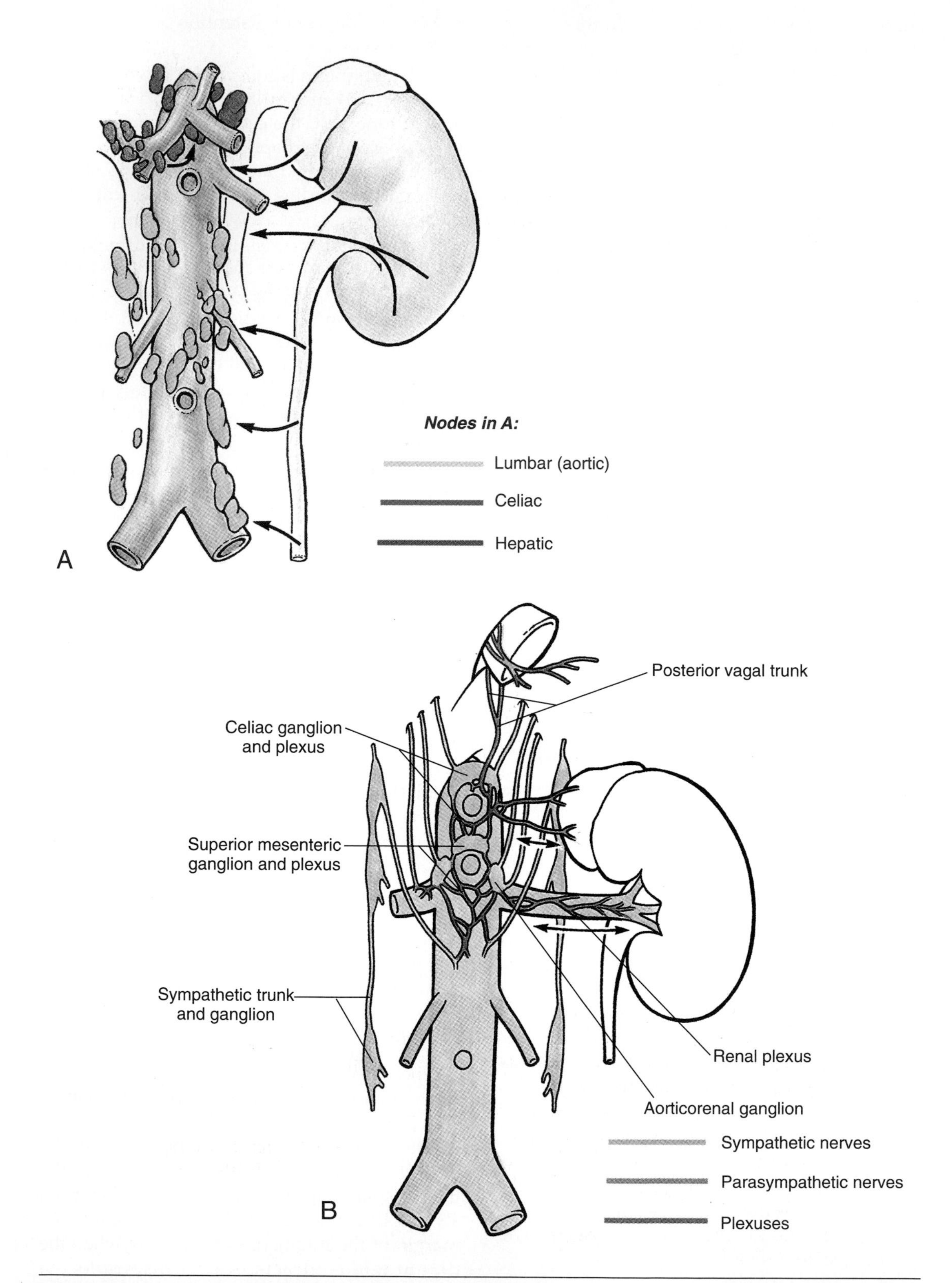

Figure 3.20. Kidney. **A.** Lymphatic drainage. *Arrows*, lymph flow to lymph nodes. **B.** Innervation. *Arrows*, afferent and efferent nerves.

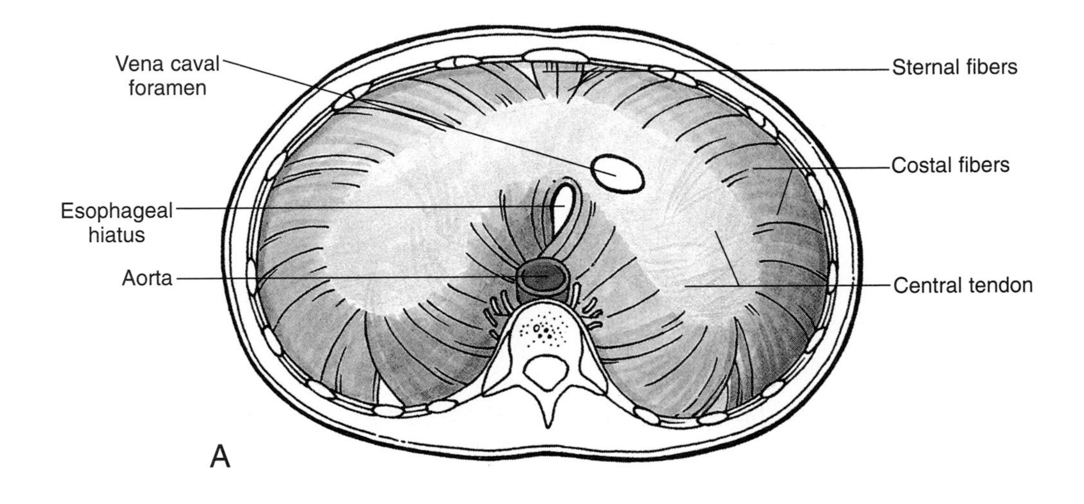

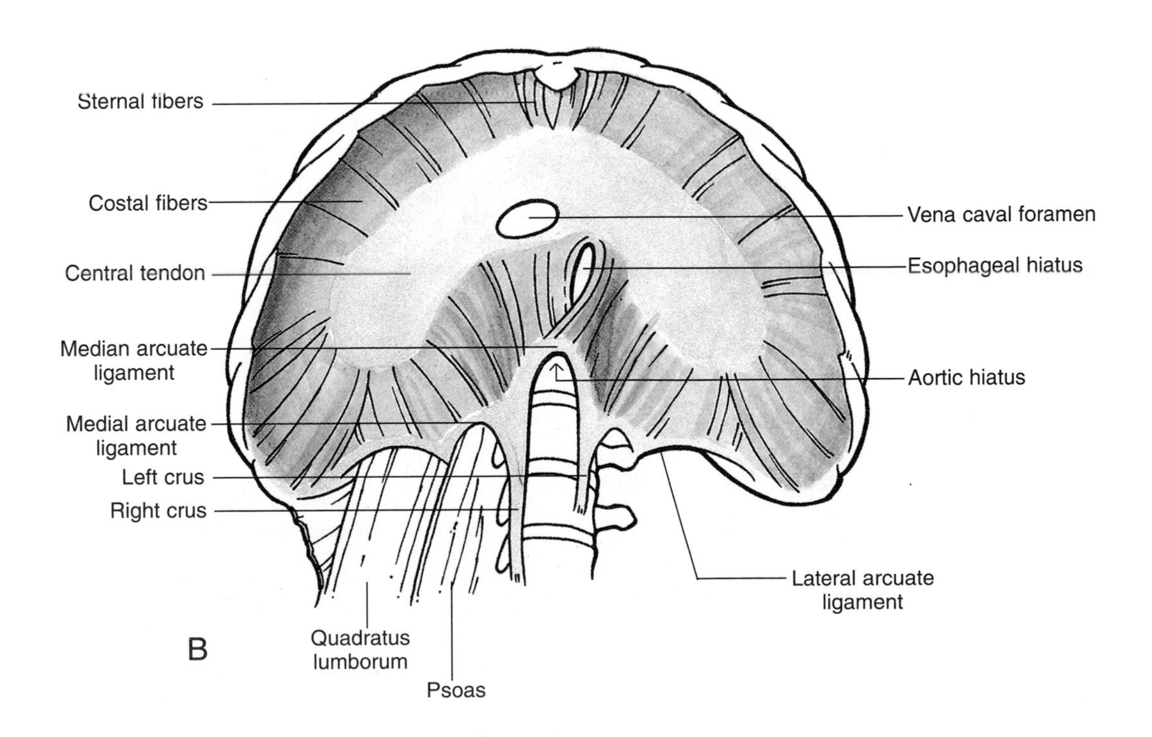

Figure 3.21. Diaphragm. **A.** Superior view. **B.** Inferior view.

Table 3.9
Vessels and Nerves of Diaphragm

	Superior Surface of Diaphragm	Inferior Surface of Diaphragm
Arterial supply	Superior phrenic aa. from thoracic aorta Musculophrenic and pericardiophrenic aa. from internal thoracic aa.	Inferior phrenic aa. from abdominal aorta
Venous drainage	Musculophrenic and pericardiacophrenic vv. drain into internal thoracic vv.	Inferior phrenic vv.: right v. drains into IVC; left v. often drains into left suprarenal v.
Lymphatic drainage	Diaphragmatic lymph nodes to phrenic then to parasternal and posterior mediastinal nodes	Superior lumbar lymph nodes; lymphatic plexuses on superior and inferior surfaces communicate freely
Innervation	Motor supply: phrenic nerves (C3–C5) Sensory supply: centrally by phrenic nn. (C3–C5); peripherally by intercostal nn. (T5–T11) and subcostal nn. (T12)	

vein. Terminal branches of the *right phrenic nerve* and a few lymphatic vessels, passing from the liver to the phrenic and mediastinal lymph nodes, also pass through the vena caval foramen.

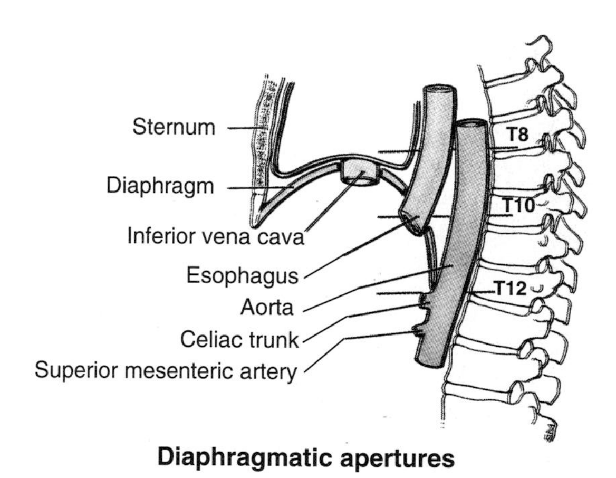

Diaphragmatic apertures

The *esophageal hiatus* is an oval aperture in the muscle of the right crus of the diaphragm at the level of T10 vertebra. Hence the esophagus is constricted when the diaphragm contracts. This opening for the esophagus also transmits the vagal trunks and esophageal branches of the left gastric vessels.

The *aortic hiatus* is posterior to the diaphragm; hence the aorta does not pierce the diaphragm, and blood flow through it is not affected by its movements. The aorta passes between the crura of the diaphragm and posterior to the median arcuate ligament, which is at the level of T12 vertebra. The opening for the aorta also transmits the *thoracic duct* and *azygos vein*.

ACTIONS OF DIAPHRAGM

The diaphragm is the chief muscle of inspiration. When it contracts, it moves inferiorly. As the diaphragm descends, it pushes the abdominal viscera before it. This increases the volume of the thoracic cavity and decreases the intrathoracic pressure, resulting in air being taken into the lungs. In addition, the volume of the abdominal cavity is slightly decreased, and the intraabdominal pressure is somewhat increased. Diaphragm movements are also important in circulation because the increased intraabdominal pressure and decreased intrathoracic pressure help to return blood to the heart. When the diaphragm contracts, compressing the abdominal viscera, blood in the IVC is forced superiorly into the heart.

Referred pain from the diaphragm is felt in two different areas because of the difference in the sensory nerve supply of the diaphragm (Table 3.9). Irritation of the diaphragmatic pleura or diaphragmatic peritoneum is referred to the shoulder region, the area of skin supplied by C5 segment of the spinal cord. This segment also contributes ventral rami to the phrenic nerve. Irritation of peripheral regions of the diaphragm, innervated by the inferior intercostal nerves, is referred to the skin over the costal margin of the anterior abdominal wall.

Posterior Abdominal Wall

The posterior abdominal wall is composed mainly of muscles and fascia attached to the vertebrae, hip bones, and ribs. It also contains fat, nerves, vessels, and lymph nodes.

MUSCLES OF POSTERIOR ABDOMINAL WALL

There are three paired muscles in the posterior abdominal wall (Fig. 3.22, Table 3.10).

- Psoas major muscles pass inferolaterally
- Quadratus lumborum muscles lie adjacent to the transverse processes of the lumbar vertebrae and lateral to the superior parts of the psoas major muscles
- Iliacus muscles lie along the lateral sides of the inferior part of the psoas major muscles

The *iliopsoas* (combined iliacus and psoas major muscles) has extensive and clinically important relations to the kidneys, ureters, cecum, appendix, sigmoid colon, pancreas, lumbar lymph nodes, and nerves of the posterior abdominal wall. When any of these structures is diseased, movements of the iliopsoas may cause pain. Because it lies along the vertebral column and crosses the sacroiliac joint, disease of the intervertebral and sacroiliac joints may cause *spasm of the iliopsoas muscle*, a protective reflex.

Owing to the relationship of the pancreas to the posterior abdominal wall (Fig. 3.11*A*), *cancer of the pancreas* in advanced stages invades the muscles and nerves of the posterior abdominal wall, producing excruciating pain.

FASCIAE OF POSTERIOR ABDOMINAL WALL

The posterior abdominal wall is covered with a continuous layer of fascia that lies between the parietal peritoneum and the muscles. The fascia lining the posterior abdominal

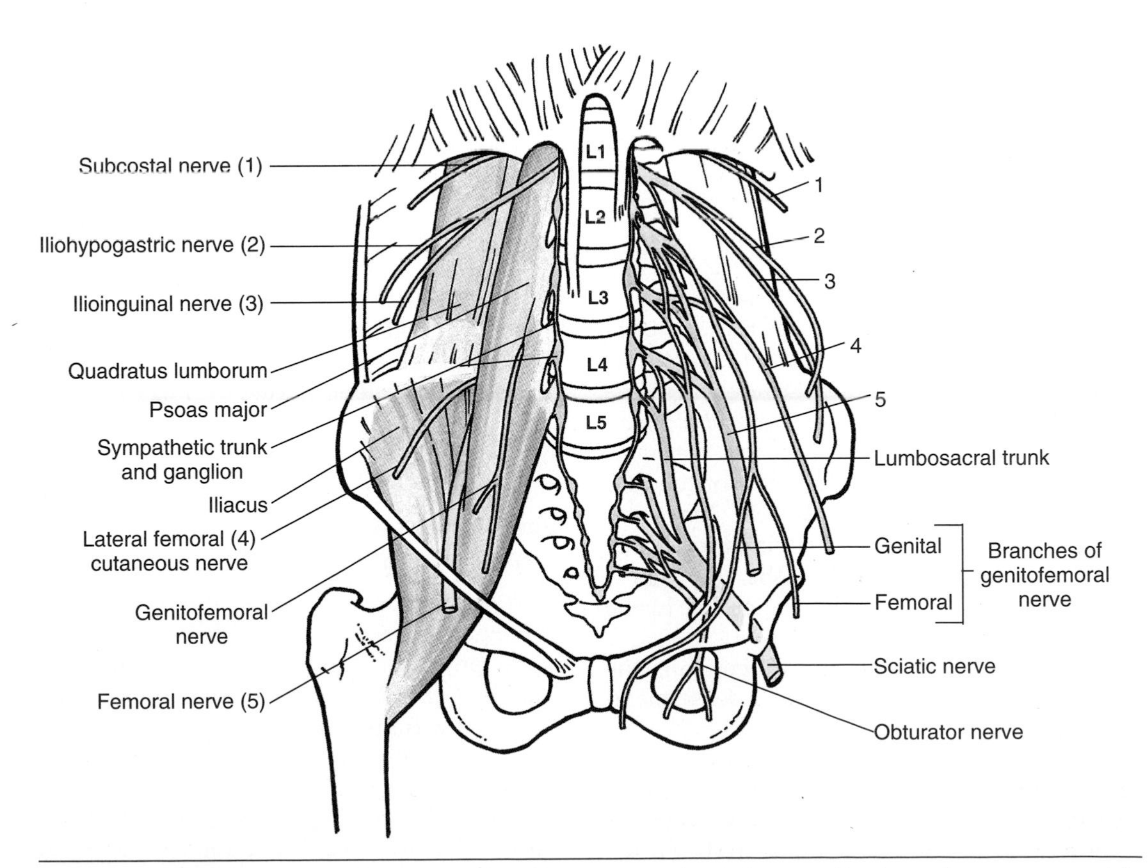

Figure 3.22. Dissection of posterior abdominal wall showing muscles and nerves. *Numbers* refer to nerves on right side.

Table 3.10
Principal Muscles of Posterior Abdominal Wall

Muscle	Superior Attachments	Inferior Attachment(s)	Innervation	Actions
Psoas major[a]	Transverse processes of lumbar vertebrae; sides of bodies of T12–L5 vertebrae and intervening intervertebral discs	By a strong tendon to lesser trochanter of femur	Lumbar plexus via ventral branches of L2–L4 nn.	Acting superiorly with iliacus, it flexes thigh; acting inferiorly it flexes vertebral column laterally; it is used to balance the trunk when sitting; acting inferiorly with iliacus; it flexes trunk
Iliacus[a]	Superior two-thirds of iliac fossa, ala of sacrum, and anterior sacroiliac ligaments	Lesser trochanter of femur and shaft inferior to it, and to psoas major tendon	Femoral n. (L2–L4)	Flexes thigh and stabilizes hip joint; acts with psoas major
Quadratus lumborum	Medial half of inferior border of 12th rib and tips of lumbar transverse processes	Iliolumbar ligament and internal lip of iliac crest	Ventral branches of T12 and L1–L4 nn.	Extends and laterally flexes vertebral column; fixes 12th rib during inspiration

[a] Psoas major and iliacus muscles are often described together as the iliopsoas muscle when flexion of the thigh is discussed (Chapter 6). The iliopsoas is the chief flexor of the thigh, and when thigh is fixed, it is a strong flexor of the trunk (e.g., during situps).

wall is continuous with the transversalis fascia that lines the transversus abdominis muscle (p. 81). It is customary to name the fascia according to the structure it covers (e.g., the diaphragmatic fascia covers the inferior surface of the diaphragm).

The *psoas fascia* covering the psoas major is attached medially to the lumbar vertebrae and pelvic brim. Superiorly the psoas fascia is thickened to form the *medial arcuate ligament* (Fig. 3.21). The psoas fascia is fused laterally with the quadratus lumborum and *thoracolumbar fasciae* (Fig. 3.17). Inferior to the iliac crest, the psoas fascia is continuous with the part of the iliac fascia covering the iliacus muscle. The psoas fascia also blends with the fascia covering the quadratus lumborum muscle. The *iliac fascia* is continuous inferiorly with the fascia in the thigh.

The *quadratus lumborum fascia* covering the quadratus lumborum muscle is a dense membranous layer that is continuous laterally with the anterior layer of the thoracolumbar fascia. The quadratus lumborum fascia is attached to the anterior surfaces of the transverse processes of the lumbar vertebrae, the iliac crest, and the 12th rib and is continuous with the transversalis fascia. It is thickened superiorly to form the *lateral arcuate ligaments* and is adherent inferiorly to the iliolumbar ligaments.

The *thoracolumbar fascia* is an extensive fascial sheet that covers the deep muscles of the back. The *lumbar fascia*, part of the thoracolumbar fascia, extends between the 12th rib and the iliac crest. It is attached laterally to the internal oblique and transversus abdominis muscles. The thoracolumbar fascia splits into two layers. The deep back muscles are enclosed between its anterior and posterior layers.

NERVES OF POSTERIOR ABDOMINAL WALL

Somatic Nerves

The *subcostal nerves*, the ventral rami of T12, arise in the thorax, pass posterior to the lateral arcuate ligaments into the abdomen, and cross the anterior surface of the quadratus lumborum muscle (Fig. 3.22). They disappear by passing through the transversus abdominis and internal oblique muscles to supply the muscles and skin of the anterolateral abdominal wall (Fig. 3.1*B*).

The *lumbar nerves* pass from the spinal cord through the intervertebral foramina inferior to the corresponding vertebrae, where they divide into dorsal and ventral primary rami. Each ramus contains sensory and motor fibers. The dorsal primary rami pass posteriorly to supply the muscles and skin of the back, whereas the ventral primary rami pass into the psoas major muscles and are connected to the *sympathetic trunk* by rami communicantes.

The *lumbar plexus* is a nerve network within the psoas major muscle that is formed by the ventral rami of L1–L4 nerves. All ventral rami receive gray rami communicantes from the sympathetic trunk, and the superior two send white rami communicantes to the sympathetic trunk. The following nerves are branches of the lumbar plexus; the three largest are listed first (Fig. 3.22).

- The obturator nerve (L2–L4) emerges from the medial border of the psoas to supply the adductor muscles of the thigh
- The femoral nerve (L2–L4) emerges from the lateral border of the psoas and innervates the iliacus muscle and the extensor muscles of the knee
- The lumbosacral trunk (L4, L5) passes over the ala (wing) of the sacrum and descends into the pelvis to take part in the formation of the sacral plexus along with the ventral rami of S1–S4 nerves
- The ilioinguinal and iliohypogastric nerves (L1) arise from the ventral ramus of L1 and enter the abdomen posterior to the *medial arcuate ligaments* and pass inferolaterally, anterior to the quadratus lumborum muscle; they pierce the transversus abdominis muscle near the anterior superior iliac spine and pass through the internal and external oblique muscles to supply the skin of the suprapubic and inguinal regions (both nerves also supply branches to the abdominal musculature)
- The genitofemoral nerve (L1, L2) pierces the anterior surface of the psoas major muscle and runs inferiorly on it deep to the psoas fascia; it divides lateral to the common and external iliac arteries into femoral and genital branches
- The lateral femoral cutaneous nerve (L2, L3) runs inferolaterally on the iliacus muscle and enters the thigh posterior to the inguinal ligament, just medial to the anterior superior iliac spine; it supplies the skin on the anterolateral surface of the thigh

Autonomic Nerves

The autonomic nerves consist of sympathetic and parasympathetic nerves that are distributed to the abdominal viscera by a rich tangle of nerve plexuses and ganglia along the anterior surface of the abdominal aorta (Fig. 3.23, Table 3.8).

The sympathetic portion consists of

- Thoracic splanchnic nerves from the thoracic sympathetic trunk
- Lumbar splanchnic nerves from the lumbar sympathetic trunk

The parasympathetic portion consists of

- Anterior and posterior vagal trunks
- Pelvic splanchnic nerves

Sympathetic Nerves. The *thoracic splanchnic nerves* are the main source of sympathetic nerves in the abdomen. The greater (T5–T9 or T10), lesser (T10 and T11), and least (T12) thoracic splanchnic nerves are composed of preganglionic fibers that come from the spinal cord via white rami communicantes and pass through the sympathetic ganglia without synapsing. They pierce the corresponding crus of the diaphragm to synapse in the celiac and aorticorenal ganglia that convey both sympathetic and parasympathetic (vagal) fibers, branches from the first (sometimes second) lumbar splanchnic nerve, renal branches from the intermesenteric nerves, and contributions from the superior hypogastric plexus. The postganglionic fibers are relayed through autonomic plexuses to the viscera.

The *lumbar splanchnic nerves* arise from the abdominal portion of the sympathetic trunk. The abdominal part of the sympathetic trunk on each side enters the abdomen by passing posterior to the medial arcuate ligament (Figs. 3.22 and 3.23). It is usually composed of *four lumbar sympathetic ganglia* with interconnecting fibers. The sympathetic trunks lie in a groove along the medial border of the psoas major muscle. The lumbar sympathetic trunks receive white rami communicantes from L1, L2, and possibly L3 spinal nerves. Medially the lumbar sympathetic trunk gives off three to four lumbar splanchnic nerves to end in the intermesenteric, inferior mesenteric, and superior hypogastric plexuses. Postganglionic fibers of the inferior mesenteric plexus supply the distal part of the large intestine.

Although sympathetic nerves are motor, they also carry sensory fibers from sense organs in the viscera. These fibers pass toward the spinal cord in the splanchnic nerves as far as the sympathetic trunk. They then leave the

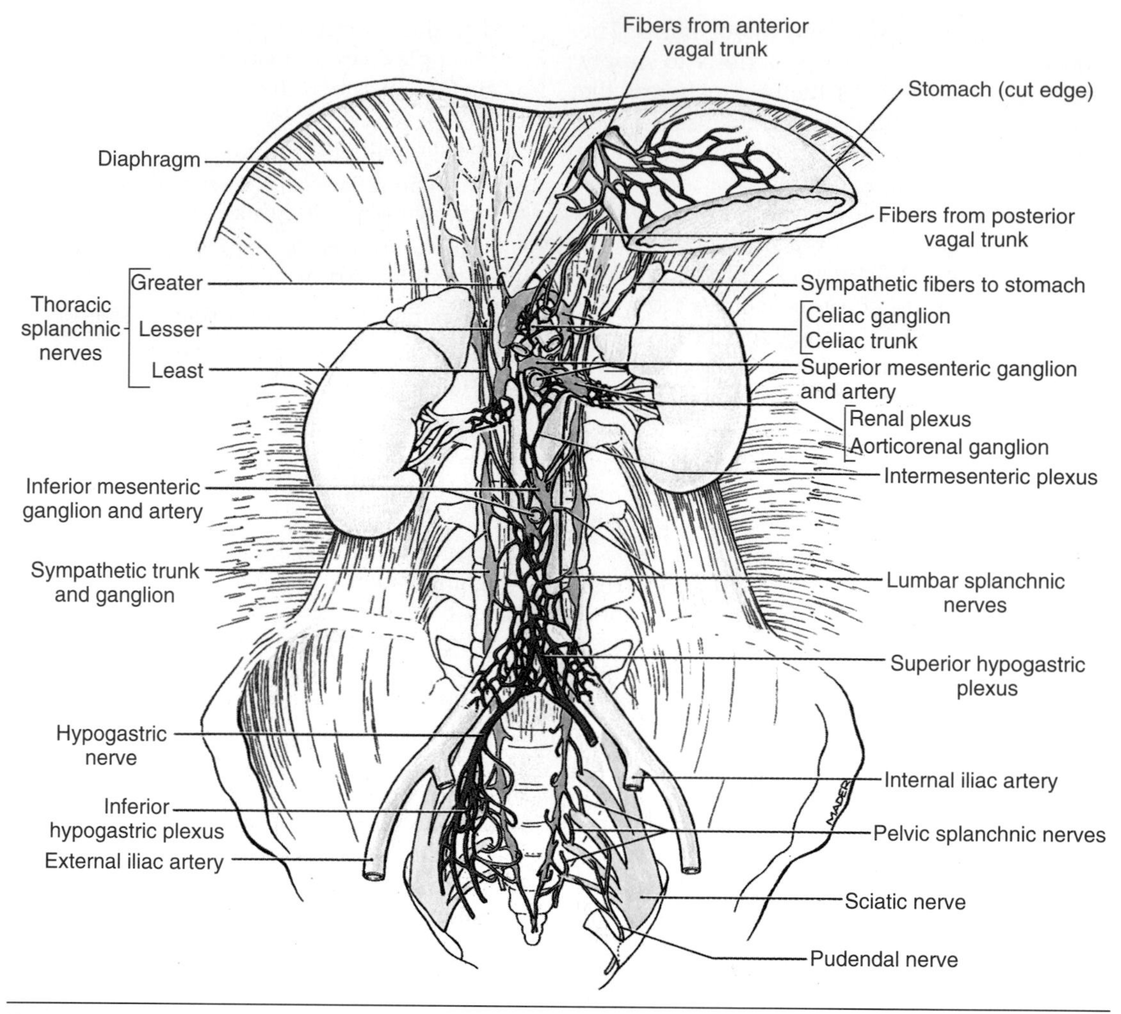

Figure 3.23. Autonomic nerve supply of abdomen. *Orange*, sympathetic; *green*, parasympathetic; *blue*, plexus; *yellow*, nerves of sacral plexus.

sympathetic trunk in a *white ramus communicans* and enter a spinal nerve to reach the spinal cord through its ventral root (Fig. 1.16*B*).

Parasympathetic Nerves. Abdominal branches arise from the *anterior and posterior vagal trunks* and are conveyed by the abdominal autonomic plexuses where they synapse with ganglion cells in the walls of the viscera. The postganglionic fibers supply smooth muscle in the digestive tract as far as the left colic flexure. The *pelvic splanchnic nerves* are derived from S2–S4 levels of the spinal cord and accompany the inferior hypogastric plexus to supply the descending and sigmoid colon, rectum, and pelvic organs.

Abdominal Autonomic Plexuses. These nerve networks consist of both sympathetic and parasympathetic fibers that surround the abdominal aorta and its major branches. The celiac, superior mesenteric, and inferior mesenteric plexuses are interconnected (Fig. 3.23). The *sympathetic ganglia* are scattered among the celiac and mesenteric plexuses. The *parasympathetic ganglia* are located in the walls of the viscera [e.g., the myenteric plexus (Auerbach's plexus) is in the muscular coat of the stomach and intestine].

The *intermesenteric plexus* is located between the superior and inferior mesenteric arteries and gives rise to renal, testicular or ovarian, and ureteric plexuses.

The *superior hypogastric plexus* is continuous with the intermesenteric plexus and the inferior mesenteric plexus and lies anterior to the inferior part of the aorta at its bifurcation. Right and left hypogastric nerves join the superior

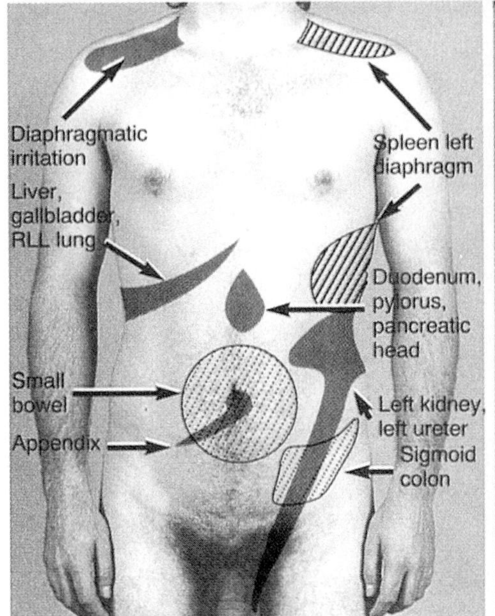

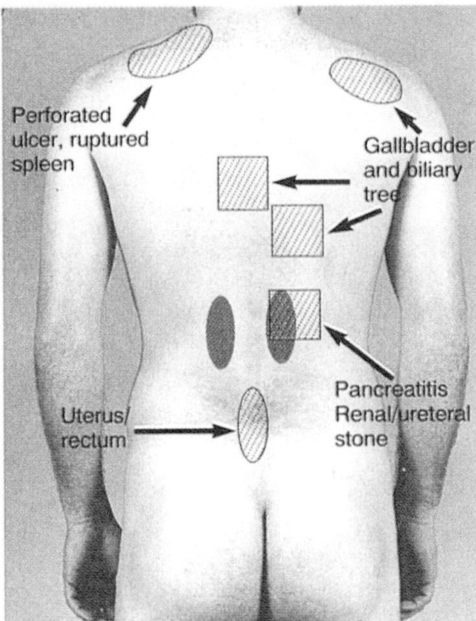

Pain arising from a viscus (e.g., stomach) varies from dull to very severe, but it is poorly localized. It radiates to the part of the body supplied by somatic sensory fibers associated with the same segment of the spinal cord that receives visceral sensory fibers from the viscus concerned. This is called *visceral referred pain* (e.g., pain from a gastric ulcer is referred to the epigastric region because the stomach is supplied by pain afferents that reach T7 and T8 segments of the spinal cord via the *greater splanchnic nerve*). The pain is interpreted by the brain as though the irritation occurred in the area of skin supplied by the dorsal roots of T7–T9 nerves. (*RLL*, right lower lobe.)

hypogastric plexus to the inferior hypogastric plexus. The superior hypogastric plexus supplies ureteric and testicular or ovarian plexuses and a plexus on each common iliac artery.

The *inferior hypogastric plexus* is formed on each side by a hypogastric nerve from the superior hypogastric plexus. The right and left plexuses are situated on the sides of the rectum, uterine cervix, and urinary bladder. The plexuses receive small branches from the superior sacral sympathetic ganglia and the sacral parasympathetic outflow from S2 to S4 (*pelvic splanchnic nerves*). Extensions of the inferior hypogastric plexus send autonomic fibers along the blood vessels, which form visceral plexuses on the walls of the pelvic viscera (e.g., *rectal and vesical plexuses*).

ARTERIES OF POSTERIOR ABDOMINAL WALL

Most arteries of the posterior abdominal wall arise from the *abdominal aorta* (Figs. 3.24–3.26). Only the subcostal arteries arise from the thoracic aorta. The abdominal aorta begins at the aortic hiatus in the diaphragm at the level of T12 vertebra and ends at the level of L4 vertebra by dividing into two common iliac arteries. The *aortic bifurcation* is just to the left of the midpoint of the line joining the highest points of the iliac crests. Branches of the aorta may be described as visceral or parietal, paired or unpaired.

The unpaired visceral branches arise at the following vertebral levels:

- Celiac trunk (T12)
- Superior mesenteric artery (L1)
- Inferior mesenteric artery (L3)

The paired visceral branches are the

- Suprarenal arteries (L1)
- Renal arteries (L1)
- Gonadal arteries, the ovarian or testicular arteries (L2)

The paired parietal branches are the

- Subcostal arteries that enter the abdomen posterior to the lateral arcuate ligaments with the subcostal nerves (T12)
- Inferior phrenic arteries that arise from the aorta just inferior to the diaphragm

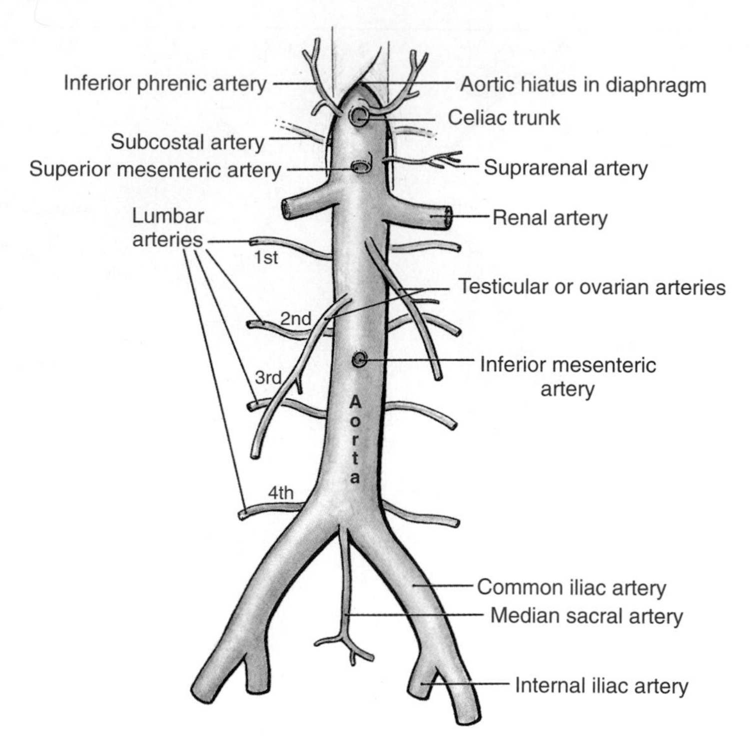

Arteries of posterior abdominal wall

- Lumbar arteries that pass around the sides of the superior four lumbar vertebrae

The unpaired parietal branch is the *median sacral artery* that arises from the aorta at its bifurcation.

> An aneurysm of the abdominal aorta (localized enlargement) distal to the renal arteries may be repaired by opening the aneurysm and inserting and sewing a Dacron graft into position.

VEINS OF POSTERIOR ABDOMINAL WALL

The veins of the posterior abdominal wall are tributaries of the IVC, except for the left testicular or ovarian vein that enters the renal vein. The IVC, the largest vein in the body, has no valves except for a variable, nonfunctional one that is located at its orifice in the right atrium of the heart. The IVC returns blood from the lower limbs, most of the abdominal wall, and the abdominopelvic viscera. Blood from the viscera passes through the *portal venous system* and liver before entering the IVC via the hepatic veins. The IVC begins anterior to L5 vertebra by the union of the common iliac veins. This union occurs about 2.5 cm to the right of the median plane, inferior to the bifurcation of the aorta and posterior to the proximal part of the right common iliac artery. The IVC ascends on the right psoas major muscle to the right of the median plane and aorta and passes through the vena caval foramen in the diaphragm to enter the thorax.

> Three collateral routes are available for venous blood to pass to the right side of the heart when the IVC is obstructed or ligation of it is necessary.
>
> - The first route is via various anastomoses in the abdomen and pelvis that enable blood to reach the superior and inferior *epigastric veins*; blood ascends in them to the thoracoepigastric veins, superior epigastric veins, and superior vena cava
> - The second route is via tributaries of the IVC that anastomose with the vertebral and azygos systems of veins
> - The third route is via the *lateral thoracic veins* that connect the circumflex iliac veins with the axillary veins

Tributaries of the IVC correspond to branches of the aorta.

- Common iliac veins
- Third and fourth lumbar veins
- Right testicular or ovarian vein
- Renal veins

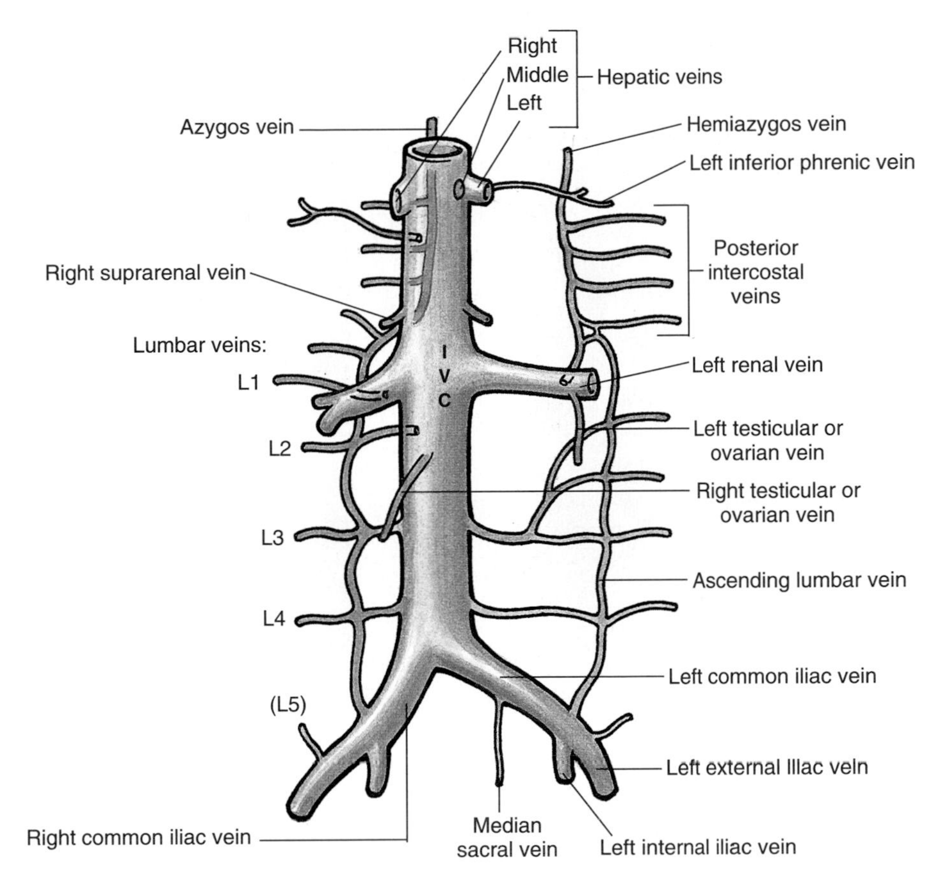

Veins of posterior abdominal wall

- Azygos vein
- Right suprarenal vein
- Inferior phrenic veins
- Hepatic veins

LYMPHATICS OF POSTERIOR ABDOMINAL WALL

The lymph nodes lie along the aorta, IVC, and iliac vessels (Figs. 3.4B and 3.20A). The *common iliac lymph nodes* receive lymph from the external and internal iliac lymph nodes. Lymph from the common iliac lymph nodes passes to the lumbar lymph nodes. Lymph from the digestive tract, liver, spleen, and pancreas passes along the celiac and superior and inferior mesenteric arteries to *preaortic lymph nodes* around the origins of these arteries from the aorta (Figs. 3.12*A* and 3.15*A*).

The *lumbar (lateral aortic) lymph nodes* lie on both sides of the aorta and IVC. They receive lymph directly from the posterior abdominal wall, kidneys, ureters, testes or ovaries, uterus, and uterine tubes (Fig. 3.20*A*). They also receive lymph from the descending colon, pelvis, and lower limbs through the inferior mesenteric and common iliac lymph nodes. Efferent lymphatic vessels from these large lymph nodes form the right and left *lumbar lymph trunks* that terminate in the cisterna chyli.

The *cisterna chyli* (about 5 cm long and 6 mm wide) is a thin-walled sac at the inferior end of the *thoracic duct*, located anterior to the bodies of L1 and L2 vertebrae between the right crus of the diaphragm and the aorta. The cisterna chyli receives lymph from the right and left lumbar lymphatic trunks, the intestinal lymphatic trunks, and a pair of lymphatic vessels that descend from the inferior intercostal lymph nodes.

The *thoracic duct* begins at the cisterna chyli and ascends through the *aortic hiatus* in the diaphragm to the thorax. The thoracic duct receives all lymph that forms inferior to the diaphragm and empties into the junction of the left subclavian and left internal jugular veins (Fig. 1.12).

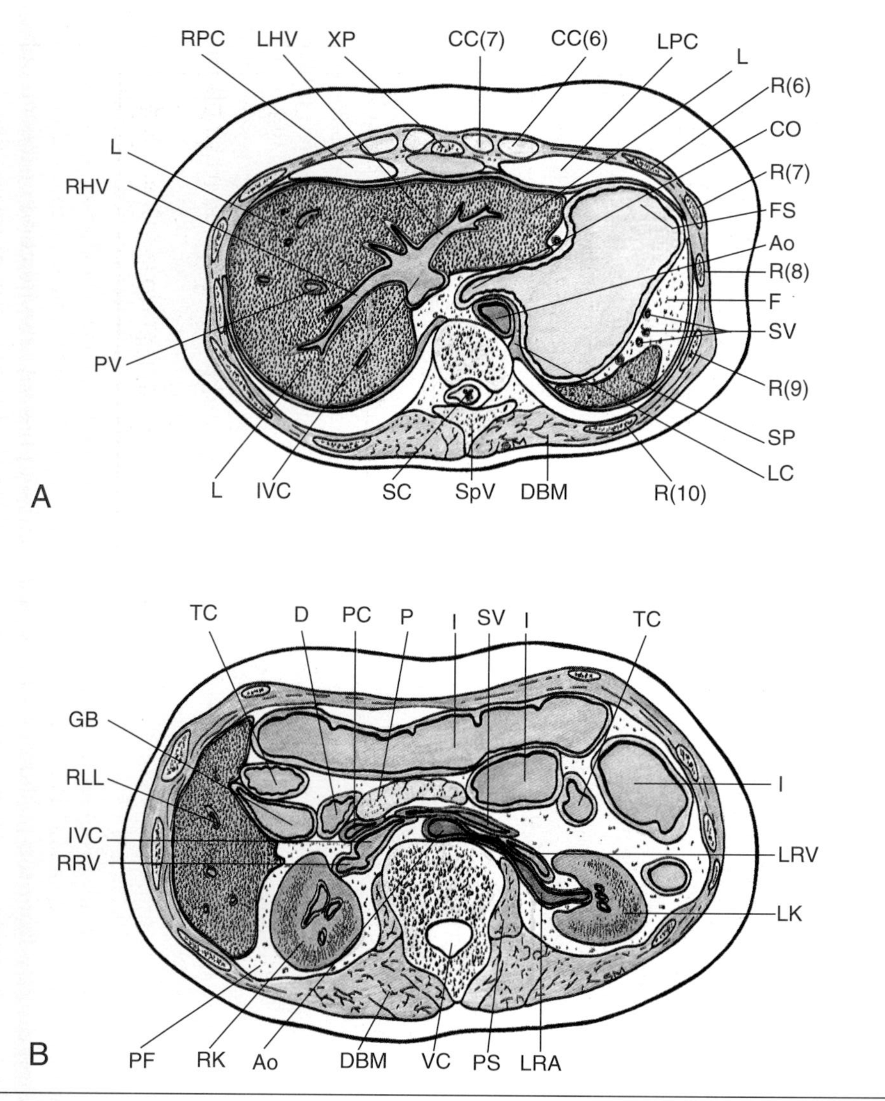

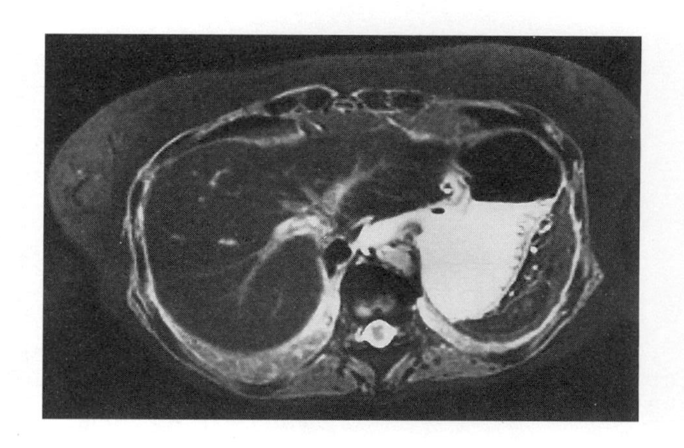

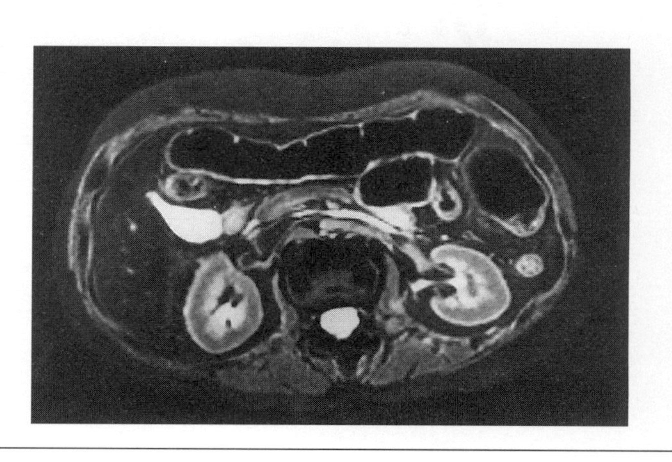

Figure 3.24. Transverse magnetic resonance images of abdomen.

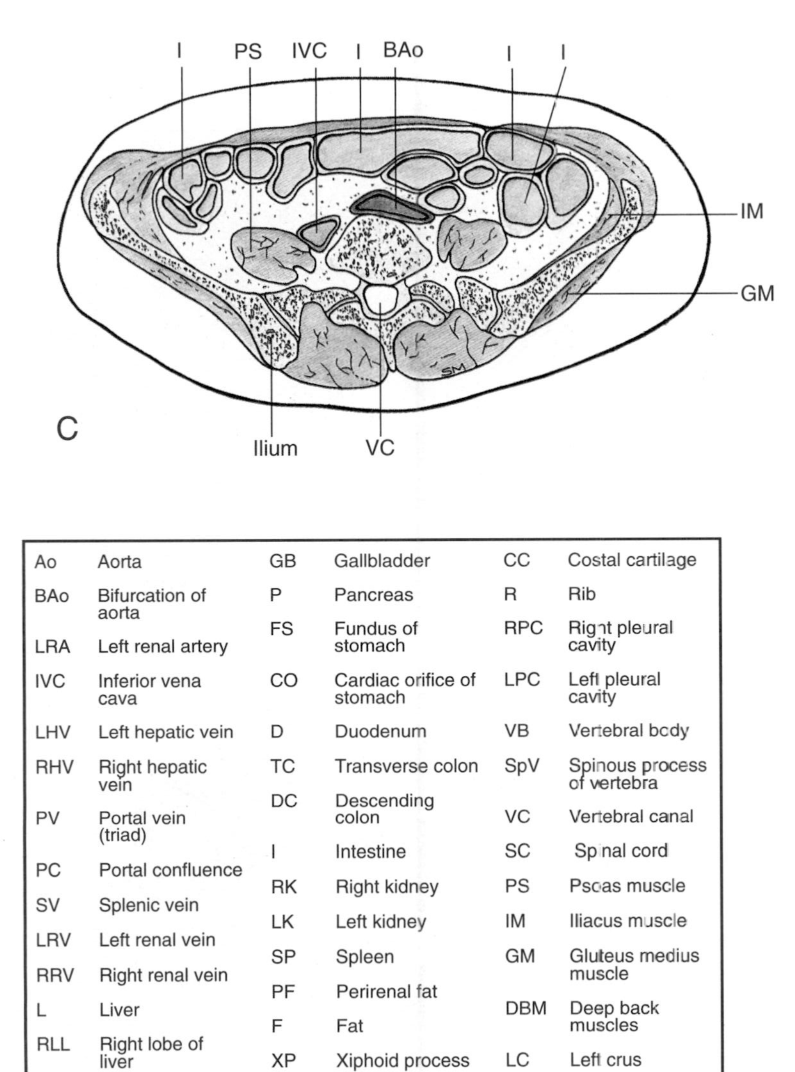

Ao	Aorta	GB	Gallbladder	CC	Costal cartilage
BAo	Bifurcation of aorta	P	Pancreas	R	Rib
LRA	Left renal artery	FS	Fundus of stomach	RPC	Right pleural cavity
IVC	Inferior vena cava	CO	Cardiac orifice of stomach	LPC	Left pleural cavity
LHV	Left hepatic vein	D	Duodenum	VB	Vertebral body
RHV	Right hepatic vein	TC	Transverse colon	SpV	Spinous process of vertebra
PV	Portal vein (triad)	DC	Descending colon	VC	Vertebral canal
PC	Portal confluence	I	Intestine	SC	Spinal cord
SV	Splenic vein	RK	Right kidney	PS	Psoas muscle
LRV	Left renal vein	LK	Left kidney	IM	Iliacus muscle
RRV	Right renal vein	SP	Spleen	GM	Gluteus medius muscle
L	Liver	PF	Perirenal fat	DBM	Deep back muscles
RLL	Right lobe of liver	F	Fat		
		XP	Xiphoid process	LC	Left crus

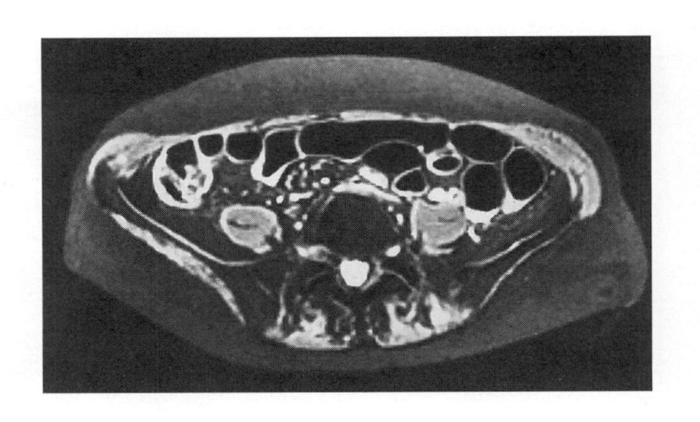

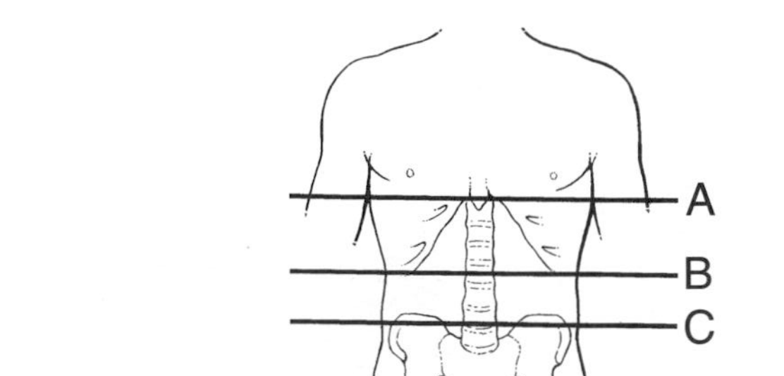

Figure 3.24. *Continued*

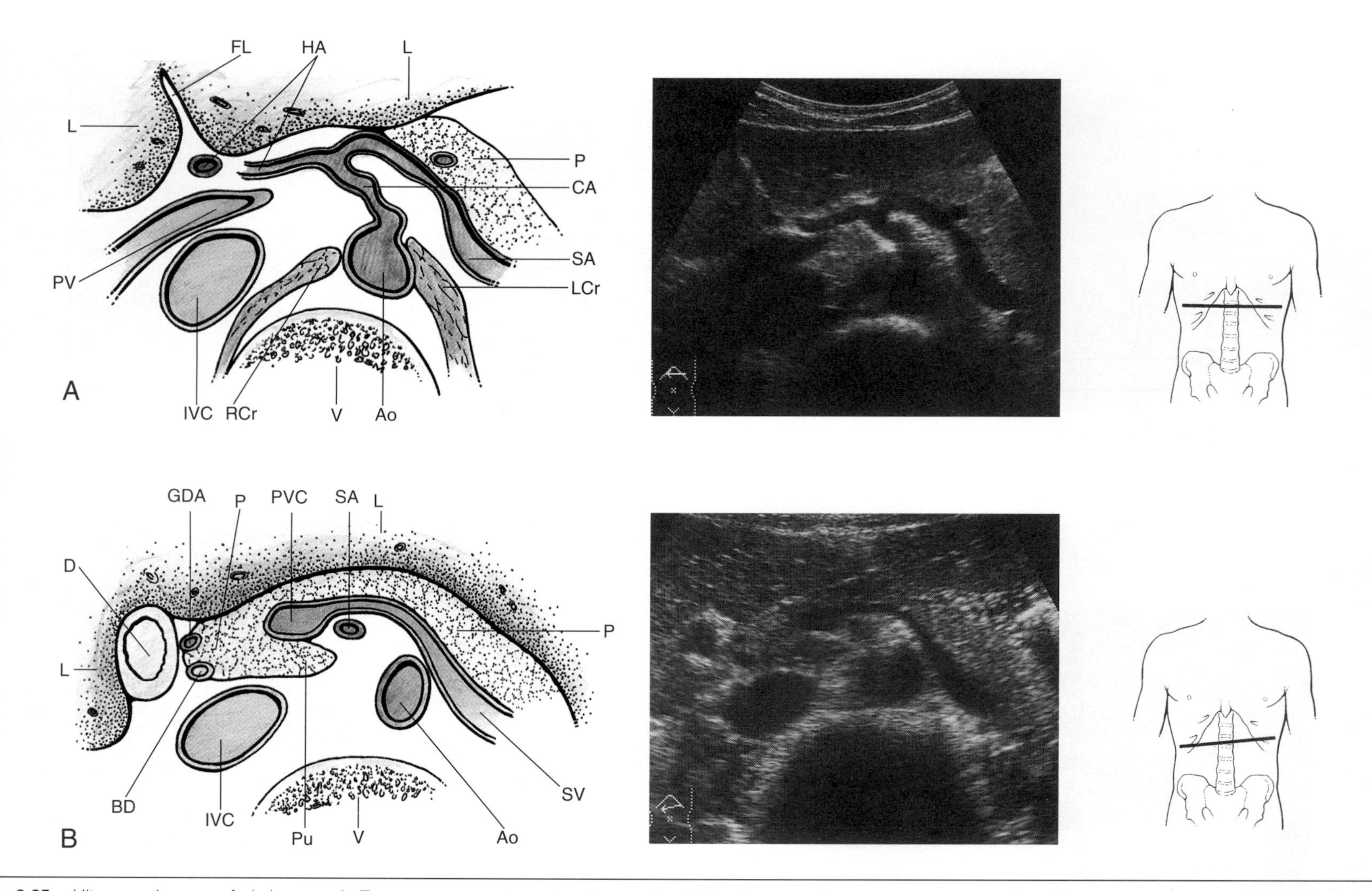

Figure 3.25. Ultrasound scans of abdomen. **A.** Transverse scan through celiac trunk. **B.** Transverse scan through pancreas. **C.** Sagittal scan through aorta. *Ao*, aorta; *BD*, bile duct; *CA*, celiac artery; *D*, duodenum; *FL*, falciform ligament; *GDA*, gastroduodenal artery; *GE*, gastroesophageal junction; *HA*, hepatic artery; *IVC*, inferior vena cava; *L*, liver; *LCA*, left gastric artery; *LCr*, left crus of diaphragm; *LGA*, left gastric artery; *LRV*, left renal vein; *P*, pancreas; *Pu*, uncinate process of pancreas; *PV*, portal vein; *PVC*, portal venous confluence; *RCr*, right crus of diaphragm; *SA*, splenic artery; *SMA*, superior mesenteric artery; *SMV*, superior mesenteric vein; *SV*, splenic vein; *V*, vertebra.

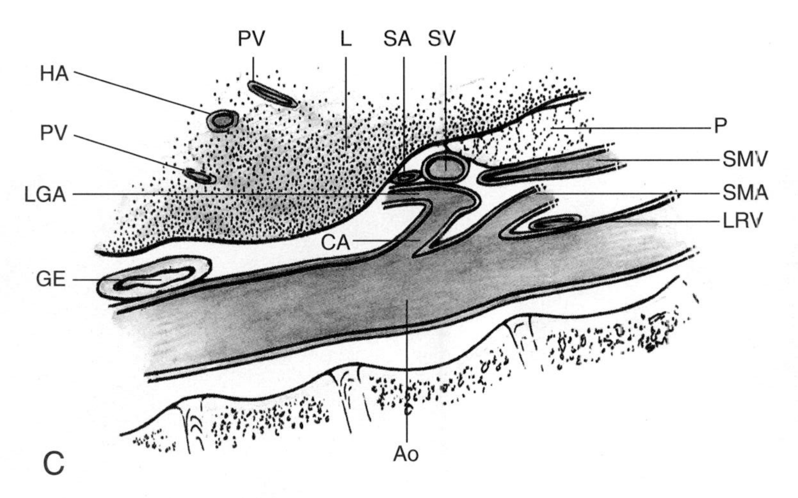

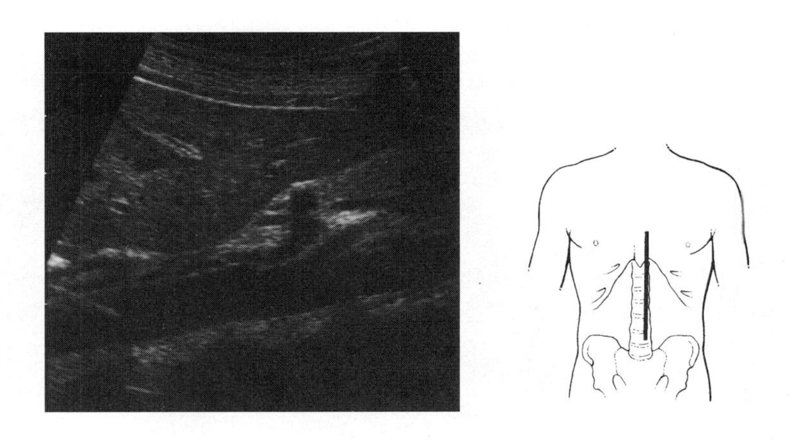

Figure 3.25. *Continued.*

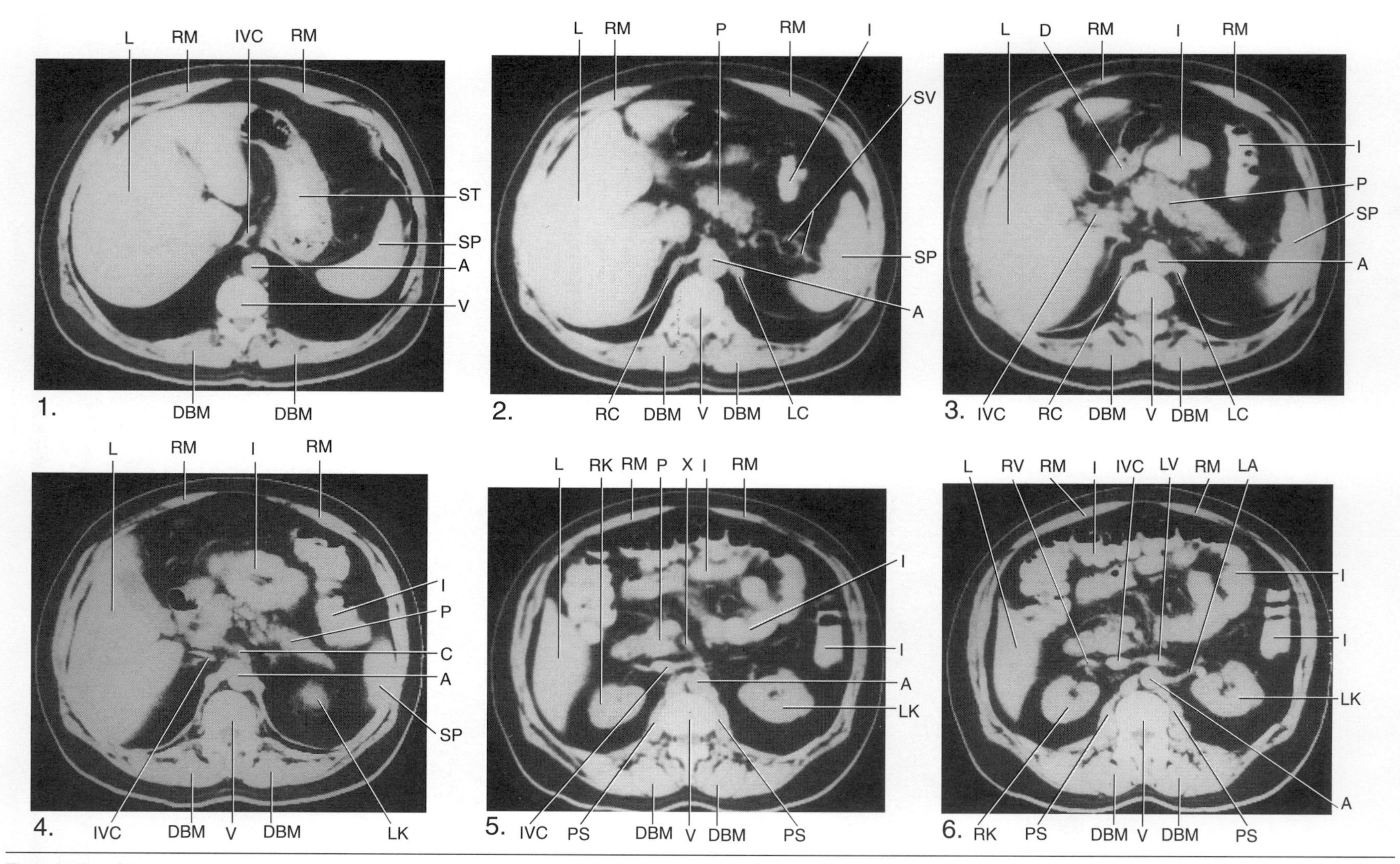

Figure 3.26. Computerized tomography scans of abdomen at progressively lower levels showing viscera and blood vessels. *A*, aorta; *C*, celiac trunk; *D*, duodenum; *DBM*, deep back muscles; *I*, intestine; *IVC*, inferior vena cava; *L*, liver; *LA*, left renal artery; *LC*, left crus of diaphragm; *LK*, left kidney; *LV*, left renal vein; *P*, pancreas; *PS*, psoas major; *RA*, renal artery; *RC*, right crus of diaphragm; *RK*, right kidney; *RM*, rectus abdominis; *RV*, right renal vein; *SP*, spleen; *ST*, stomach; *SV*, splenic vessels; *V*, vertebral body; *X*, superior mesenteric artery.

4 / PELVIS AND PERINEUM

The *pelvis* is the region of the trunk inferoposterior to the abdomen and the area of transition from the trunk to the lower limbs (Fig. 4.1). The *perineum* is the region of the trunk inferior to the pelvic diaphragm between the thighs and buttocks.

Pelvis

The pelvis is enclosed by bony, ligamentous, and muscular walls. The funnel-shaped *pelvic cavity* contains the urinary bladder, terminal parts of the ureters, pelvic genital organs, rectum, blood vessels, lymphatics, and nerves.

BONY PELVIS

The bony pelvis [skeleton of pelvis (Table 4.1)] is formed by the

- Two hip bones (each of which is composed of three bones: ilium, ischium, and pubis)
- Sacrum
- Coccyx

The hip bones are joined anteriorly at the pubic symphysis to form a *pelvic girdle* that is firmly attached to the sacrum for support of the lower limbs.

The *pelvis major* (greater pelvis, false pelvis) is

- Superior to the superior pelvic aperture
- The location of some abdominal viscera (e.g., sigmoid colon)
- Bounded by the abdominal wall anteriorly, the iliac fossae laterally, and L5 and S1 vertebrae posteriorly

The *pelvis minor* (lesser pelvis, true pelvis) is

- Between the superior pelvic aperture (pelvic inlet) and the inferior pelvic aperture (pelvic outlet)
- The location of the pelvic viscera (e.g., urinary bladder)
- Bounded by the pelvic surfaces of the hip bones, sacrum, and coccyx
- Limited inferiorly by the pelvic diaphragm

The superior pelvic aperture separates the pelvis major from the pelvis minor. The edge of this aperture is sometimes referred to as the pelvic brim. The *superior pelvic aperture* is bounded by the

- Superior margin of the pubic symphysis
- Posterior border of the pubic crest
- Pecten of the pubis
- Arcuate line of the ilium
- Anterior border of the winglike ala of the sacrum
- Promontory of the sacrum

The *inferior pelvic aperture* is bounded

- Anteriorly by the inferior margin of the pubic symphysis
- Anterolaterally on each side by the inferior ramus of the pubis and the ischial tuberosities
- Posterolaterally on each side by the sacrotuberous ligament
- Posteriorly by the tip of the coccyx

The pelves of males and females differ in several respects. These sex differences are related mainly to the heavier build and larger muscles of men and to the adaptation of the pelves of women for childbearing.

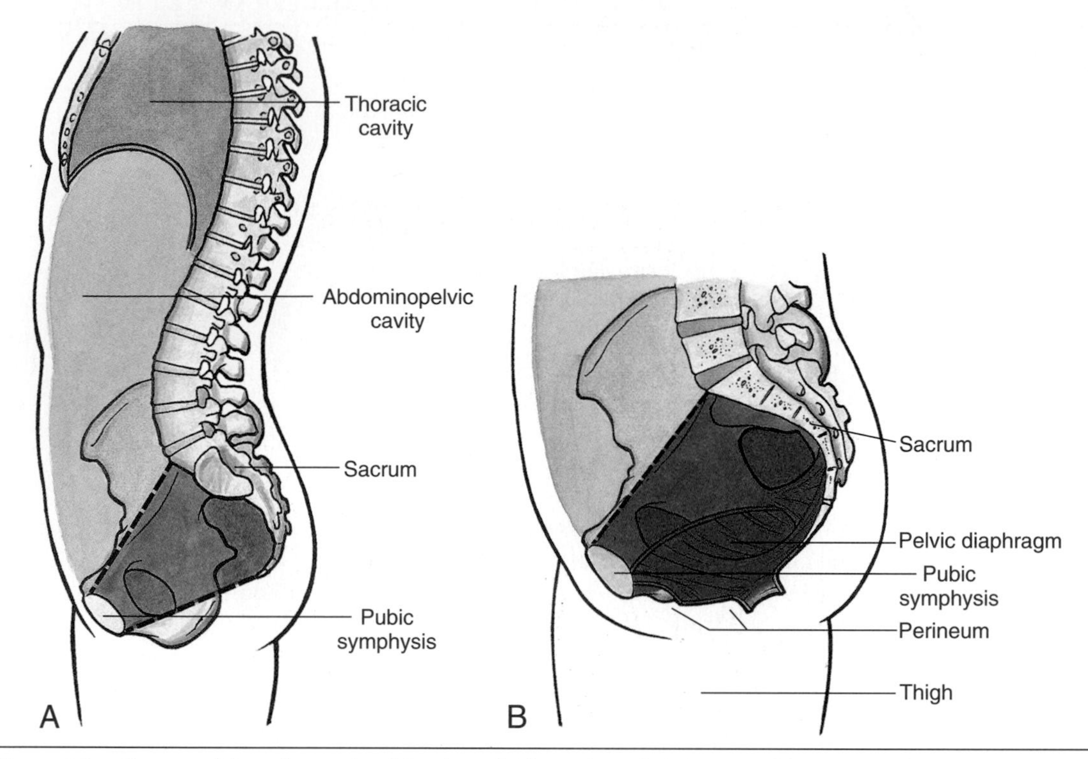

Figure 4.1. *Green*, pelvis major; *red*, pelvis minor. **A.** Thoracic and abdominopelvic cavities. *Superior broken line*, plane of pelvic brim surrounding superior pelvic aperture. *Inferior broken line*, plane of inferior pelvic aperture. **B.** Pelvic diaphragm separating pelvic cavity from perineum.

The pelvis minor is important in obstetrics because it is the bony canal through which the fetus passes during birth. The shape of the superior pelvic aperture is significant, for it is through this inlet that the fetal head enters the pelvic cavity during labor. To make a decision concerning the capacity of the female pelvis for childbirth, its diameters may be determined radiographically. The ischial spines face each other, and the distance between them is the narrowest part of the pelvic cavity.

Fractures of the pelvic girdle may result from direct trauma to the pelvic bones (e.g., during an automobile accident) or be caused by forces transmitted to these bones from the lower limbs. *Pelvic fractures* may be associated with injury to pelvic soft tissues, blood vessels, nerves, and organs.

PELVIC JOINTS

Joints of the pelvis include the lumbosacral joints, the sacrococcygeal joint, the sacroiliac joints, and the pubic symphysis (Fig. 4.2).

Lumbosacral Joints

L5 and S1 vertebrae articulate at the *anterior intervertebral joint* formed by the intervertebral disc between their bodies and at two *posterior zygapophyseal (facet) joints* between their articular processes. Facets on S1 vertebra face posteriorly and medially, thereby preventing L5 vertebra from sliding anteriorly. The *iliolumbar ligaments* unite the ilia and L5 vertebra.

Sacrococcygeal Joint

In this secondary cartilaginous joint, fibrocartilage and ligaments join the apex of the sacrum to the base of the coccyx. The *sacrococcygeal ligaments* correspond to the anterior and posterior longitudinal ligaments of the other intervertebral joints. The sacral and coccygeal horns (cornua) are also united by intercornual ligaments.

Sacroiliac Joints

These articulations are strong, weightbearing synovial joints between the articular surfaces of the sacrum and ilium. These surfaces have irregular elevations and depressions that produce some interlocking of the bones. The sacrum is suspended between the iliac bones and is firmly attached to them by interosseous and posterior *sacroiliac ligaments*. Sacroiliac

Table 4.1
Differences between Male and Female Pelves

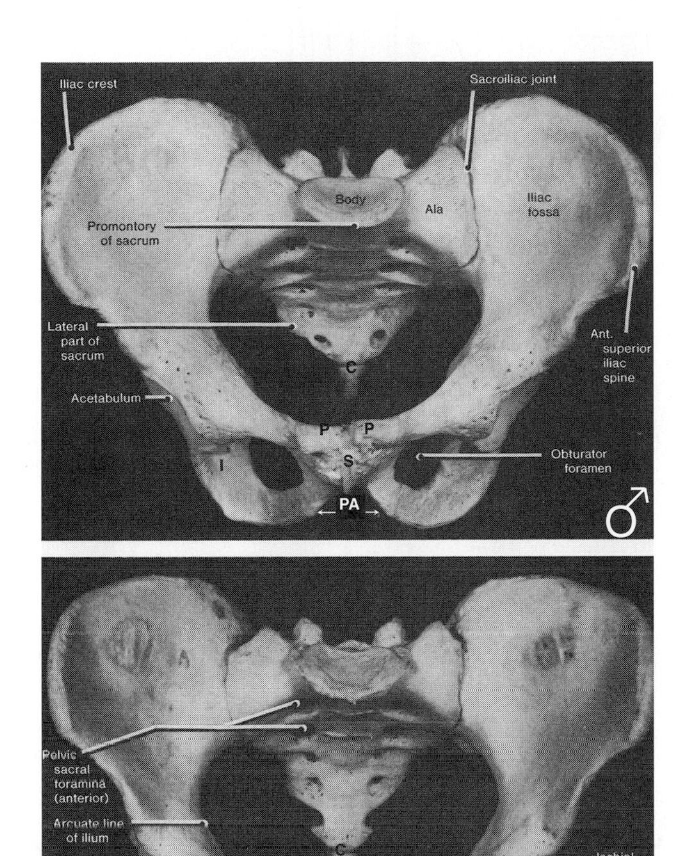

P, pubis; *S*, pubic symphysis; *C*, coccyx; *I*, ramus of ischium; *PA*, pubic arch.

	Male	Female
General structure	Thick and heavy	Thin and light
Pelvis major	Deep	Shallow
Pelvis minor	Narrow and deep	Wide and shallow
Superior pelvic aperture	Heart-shaped	Oval or rounded
Inferior pelvic aperture	Comparatively small	Comparatively large
Pubic arch	Narrow	Wide
Obturator foramen	Round	Oval
Acetabulum	Large	Small

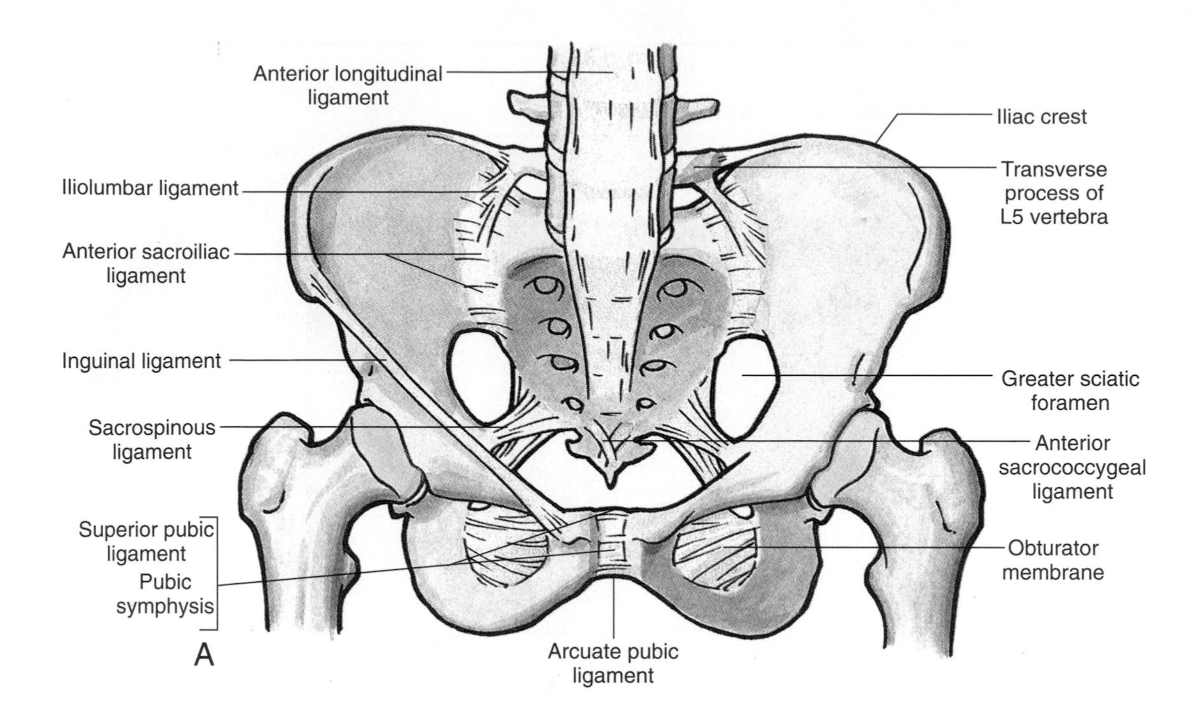

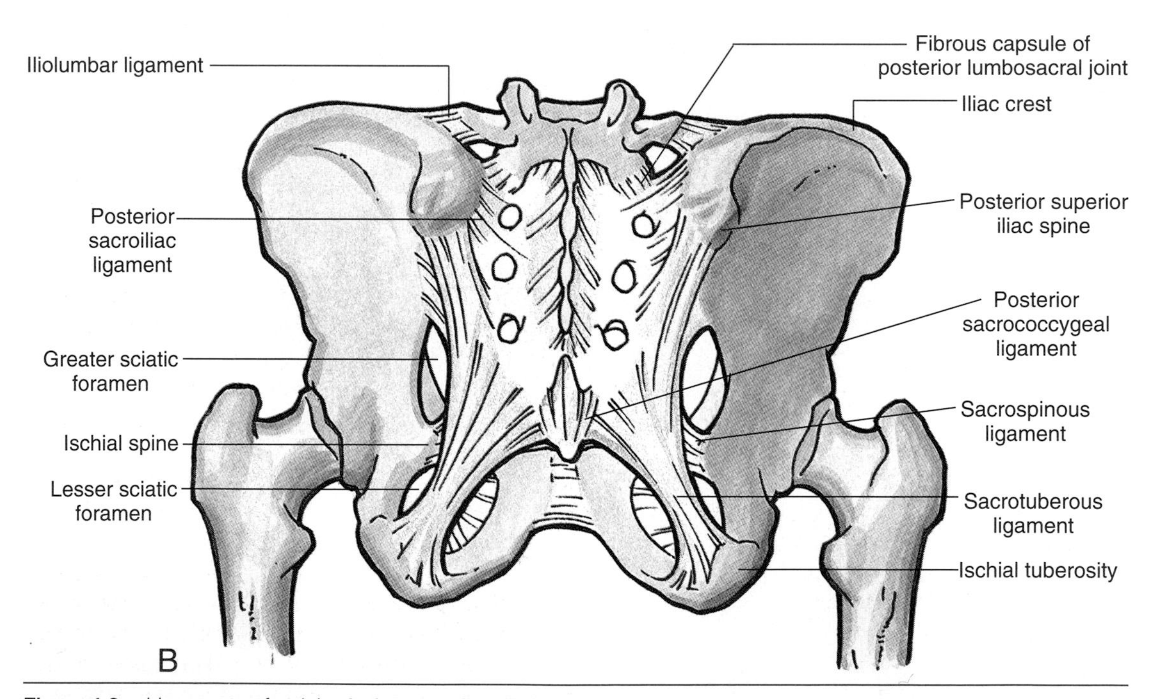

Figure 4.2. Ligaments of pelvis. **A.** Anterior view. **B.** Posterior view.

joints differ from most synovial joints in that they possess very little mobility because of their role in transmitting the weight of most of the body to the hip bones. Movement of the sacroiliac joints is limited to gliding and rotary movements, except when subject to considerable force such as occurs during a high jump. In this case the force is transmitted through the lumbar vertebrae to the superior end of the sacrum, which tends to rotate anteriorly. This rotation is counterbalanced by the interlocking articular surfaces and the strong supporting ligaments,

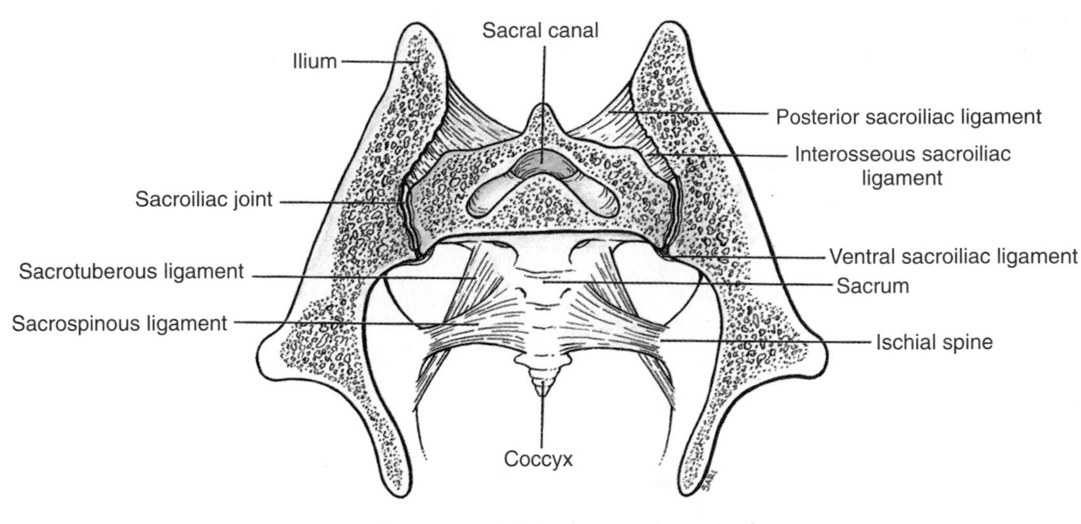

Sacroiliac joints (coronal section)

especially the sacrotuberous and sacrospinous ligaments, which join the sacrum to the ischium. The *sacrotuberous and sacrospinous ligaments* allow only limited movement of the inferior end of the sacrum, thereby giving resilience to this region when the vertebral column must sustain sudden weight increases.

Pubic Symphysis

This secondary cartilaginous joint is formed by the union of the bodies of the pubic bones in the median plane. The fibrocartilaginous *interpubic disc* of the symphysis is generally thicker in women than in men. The ligaments joining the bones are thickened superiorly and inferiorly to form the *superior pubic ligament* and the *arcuate pubic ligament*, respectively.

During pregnancy the vertebropelvic ligaments relax because of the influence of hormones, making freer movements between the inferior part of the vertebral column and the pelvis. Loosening of the interpubic disc also occurs, resulting in an increase in the distance between the pubic bones. The coccyx also moves posteriorly during childbirth. All these changes facilitate passage of the fetus through the pelvis.

PELVIC WALLS

The pelvic walls are divided into an anterior wall, two lateral walls, a posterior wall, and a floor (Fig. 4.3, Table 4.2).

Anterior Pelvic Wall. The anterior pelvic wall is formed primarily by the bodies and rami of the pubic bones and the pubic symphysis.

Lateral Pelvic Walls. The lateral pelvic walls have a bony framework formed by parts of the hip bones. The *obturator internus muscles* cover most of these walls. Medial to these muscles are the *obturator nerves and vessels* and other branches of the internal iliac vessels. Each obturator internus passes from the pelvis through the *lesser sciatic foramen* and attaches to the femur.

Posterior Pelvic Wall. The posterior pelvic wall is formed by the sacrum, adjacent parts of the ilia, and the sacroiliac joints and their associated ligaments. The *piriformis muscles* line this wall laterally. Each muscle leaves the pelvis minor through the *greater sciatic foramen*. Medial to the piriformis muscles are the nerves of the *sacral plexus* (see Table 4.3) and the *internal iliac vessels* and their branches.

Pelvic Floor. The pelvic floor is formed by the *pelvic diaphragm*, which consists of the levator ani and coccygeus and the fasciae that cover the superior and inferior aspects of these muscles. The pelvic diaphragm stretches between the pubis anteriorly and the coccyx posteriorly and from one lateral pelvic wall to the other. This gives the pelvic diaphragm the appearance of a funnel suspended from these attachments.

The *levator ani* is the larger and more important muscle in the pelvic floor. Although the levator ani is a paired muscle, it is usually described as a single structure. The levator ani

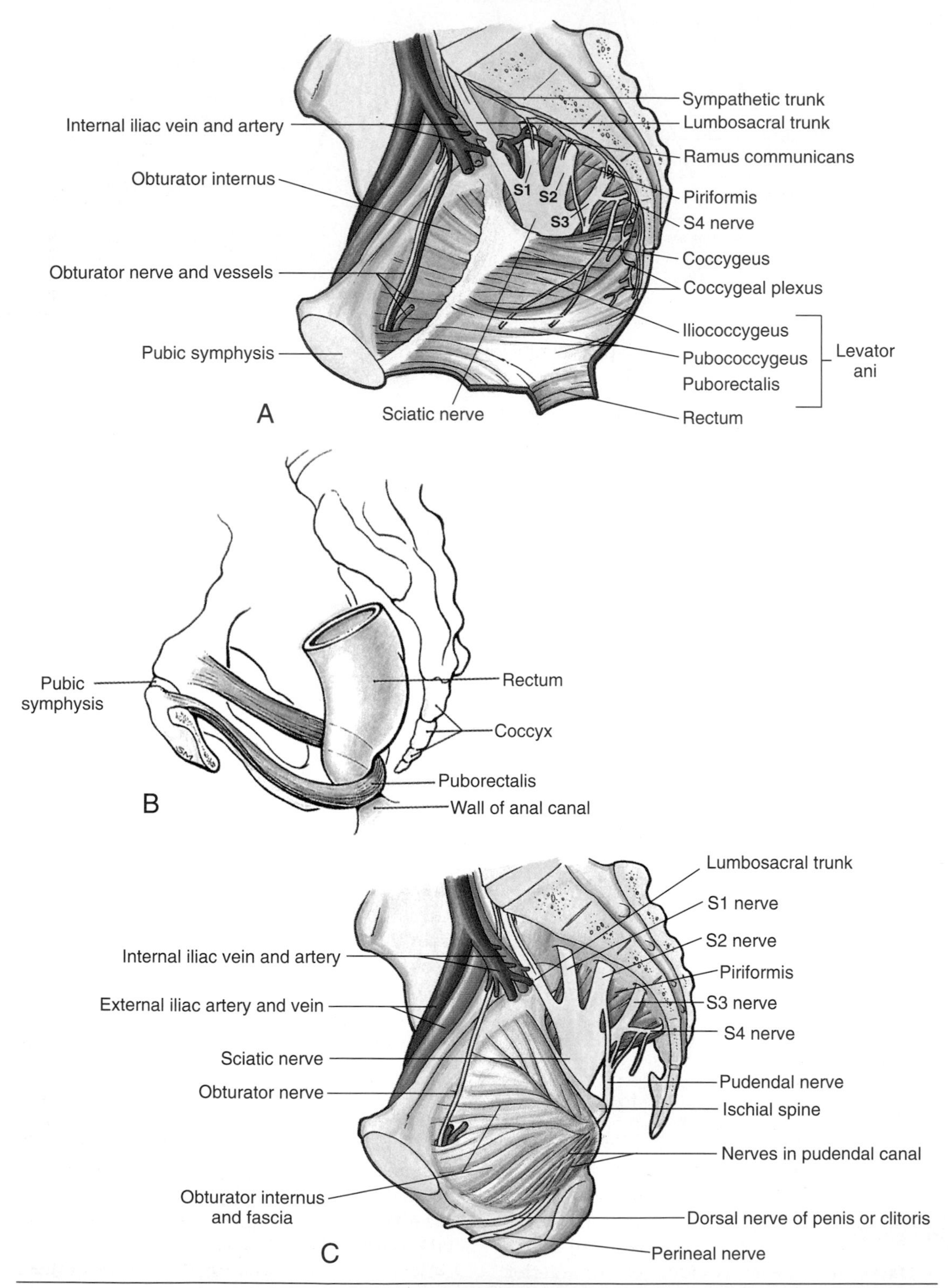

Figure 4.3. **A.** Lateral wall of pelvis minor showing pelvic diaphragm and its relationship to sacral and coccygeal plexuses. **B.** Puborectalis muscle, part of levator ani. **C.** Lateral wall of pelvis minor showing obturator internus and piriformis muscles, sacral plexus, obturator nerve and vessels, and pudendal canal.

Table 4.2
Muscles of Pelvic Walls

Muscle	Proximal Attachment	Distal Attachment	Innervation	Main Action
Obturator internus	Pelvic surfaces of ilium and ischium; obturator membrane	Greater trochanter of femur	Nerve to obturator internus (L5, S1, and S2)	Rotates thigh laterally; assists in holding head of femur in acetabulum
Piriformis	Pelvic surface of second to fourth sacral segments: superior margin of greater sciatic notch, and sacrotuberous ligament	Greater trochanter of femur	Ventral rami of S1 and S2	Rotates thigh laterally; abducts thigh; assists in holding head of femur in acetabulum
Levator ani (pubococcygeus, puborectalis, and iliococcygeus)	Body of pubis, tendinous arch of obturator fascia, and ischial spine	Perineal body, coccyx, anococcygeal ligament, walls of prostate or vagina, rectum, and anal canal	Branches of S4 and pudendal	Helps to support the pelvic viscera and resists increases in intraabdominal pressure
Coccygeus (ischiococcygeus)	Ischial spine	Inferior end of sacrum	Branches of S4 and S5	Forms small part of pelvic diaphragm that supports pelvic viscera; flexes coccyx

consists of three parts, designated according to the direction and attachment of the fibers.

- The puborectalis unites with its partner to form a U-shaped muscular sling around the anorectal junction
- The pubococcygeus is the main part of the levator ani
- The iliococcygeus, the most posterior part of the levator ani, is thin and often poorly developed

The levator ani forms a muscular sling that supports the abdominopelvic viscera, resists increases in intraabdominal pressure, and helps to hold the pelvic viscera in position. Acting together, the parts of the levator ani raise the pelvic floor, thereby assisting the anterolateral abdominal muscles in compressing the abdominal and pelvic contents. This action is an important part of forced expiration, coughing, vomiting, urinating, defecating, and fixation of the trunk during strong movements of the upper limbs (e.g., when lifting heavy objects). The levator ani also has important functions in the voluntary control of urination and in the support of the uterus.

During childbirth the levator ani supports the fetal head while the cervix of the uterus is dilating to permit delivery of the fetus. The levator ani may be injured during difficult deliveries; the pubococcygeus is the part that is usually damaged. This part is important because it encircles and supports the urethra, vagina, and anal canal. Weakening of the levator ani from stretching or tearing during childbirth may cause *urinary stress incontinence*, which is characterized by dribbling of urine when the intraabdominal pressure is raised (e.g., during coughing and lifting). In addition, prolapse of the urinary bladder may occur; prolapse of the uterus through the vagina may also occur.

PELVIC NERVES

The pelvis is innervated mainly by the *sacral and coccygeal nerves* and the pelvic part of the autonomic nervous system. The piriformis muscles form a bed for the sacral and coccygeal nerve plexuses. The ventral rami of S2 and S3 nerves emerge between the digitations of these muscles. The descending part of L4 nerve unites with the ventral ramus of L5 nerve to form the thick, cordlike *lumbosacral trunk*. It

Table 4.3
Nerves of Sacral Plexus

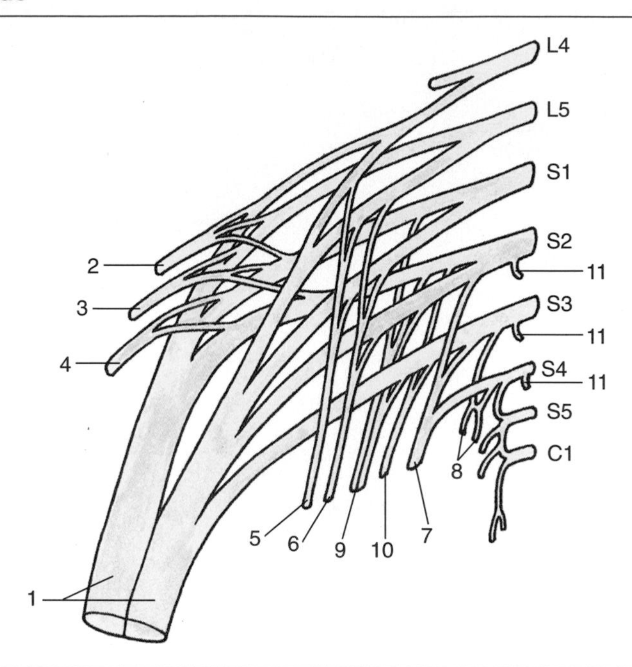

Nerve	Segmental Origin	Distribution
1. Sciatic	L4, L5, S1, S2, S3	Articular branches to hip joint and muscular branches to flexors in thigh and all muscles in leg and foot
2. Superior gluteal	L4, L5, S1	Gluteus medius and gluteus minimus mm.
3. Inferior gluteal	L5, S1, S2	Gluteus maximus m.
4. Nerve to piriformis	S1, S2	Piriformis m.
5. Nerve to quadratus femoris and inferior gemellus	L4, L5, S1	Quadratus femoris and inferior gemellus mm.
6. Nerve to obturator internus and superior gemellus	L5, S1, S2	Obturator internus and superior gemellus mm.
7. Pudendal	S2, S3, S4	Structures in perineum: sensory to genitalia, muscular branches to sphincter urethrae and external anal sphincter
8. Nerves to levator ani and coccygeus	S3, S4	Levator ani and coccygeus mm.
9. Posterior femoral cutaneous	S2, S3	Cutaneous branches to buttock and posterior surface of thigh
10. Perforating cutaneous	S2, S3	Cutaneous branches to medial part of buttock
11. Pelvic splanchnic	S2, S3, S4	Pelvic viscera via inferior hypogastric plexus

passes inferiorly, anterior to the ala of the sacrum, where it joins the sacral plexus.

Sacral Plexus

The sacral plexus (Fig. 4.3, *A* and *C*, Table 4.3) is located on the posterior wall of the pelvis minor where it is closely related to the anterior surface of the *piriformis* muscle. The two main nerves of the plexus are the sciatic and pudendal. The major branches of the sacral plexus leave the pelvis through the *greater sciatic foramen*.

The *sciatic nerve*, the large nerve to the lower limb, is formed by the ventral rami of L4–S3 that converge on the anterior surface of the piriformis. The sciatic nerve passes through the greater sciatic foramen inferior to the piriformis to enter the gluteal (buttock) region.

The *pudendal nerve* arises from the sacral plexus by separate branches of the ventral rami of S2–S4. It accompanies the internal pudendal artery and leaves the pelvis through the greater sciatic foramen between the piriformis and coccygeus muscles. The pudendal nerve hooks around the sacrospinous ligament and enters the perineum through the *lesser sciatic foramen*. It supplies the muscles of the perineum and ends as the dorsal nerve of the penis or clitoris. The pudendal nerve is also sensory to the external genitalia.

Obturator Nerve

The obturator nerve arises in the abdomen from the *lumbar plexus* (L2–L4) and enters the pelvis on the deep medial aspect of the psoas major. It runs in the extraperitoneal fat along the lateral wall of the pelvis to the *obturator foramen*, where it divides into anterior and posterior parts that leave the pelvis through this foramen and supply the medial thigh muscles (Chapter 6).

Coccygeal Plexus

The coccygeal plexus is a small network of nerves that is formed by the ventral rami of S4 and S5 nerves and the coccygeal nerves. It lies on the pelvic surface of the coccygeus and supplies this muscle, part of the levator ani, and the sacrococcygeal joint. The anococcygeal nerves arising from this plexus pierce the sacrotuberous ligament and supply a small area of skin in the region of the coccyx.

> During birth the fetal head may compress the nerves of the sacral plexus, producing pain in the mother's lower limbs. The obturator nerve is also vulnerable to injury during surgery (e.g., during removal of cancerous lymph nodes from the lateral pelvic wall). Injury to this nerve may cause painful spasms of the adductor muscles of the thigh.

Pelvic Autonomic Nerves

The sacral sympathetic trunks are the inferior continuation of the lumbar sympathetic trunks (Fig. 4.4). Each sacral trunk is smaller than the lumbar trunks and usually has four ganglia. The trunks descend on the pelvic surface of the sacrum just medial to the pelvic sacral foramina and converge to form the small median *ganglion impar* anterior to the coccyx. The sympathetic trunks descend posterior to the rectum in the extraperitoneal connective tissue and send gray rami communicantes to each ventral ramus of the sacral and coccygeal nerves. They also send small branches to the median sacral artery and the inferior hypogastric plexus.

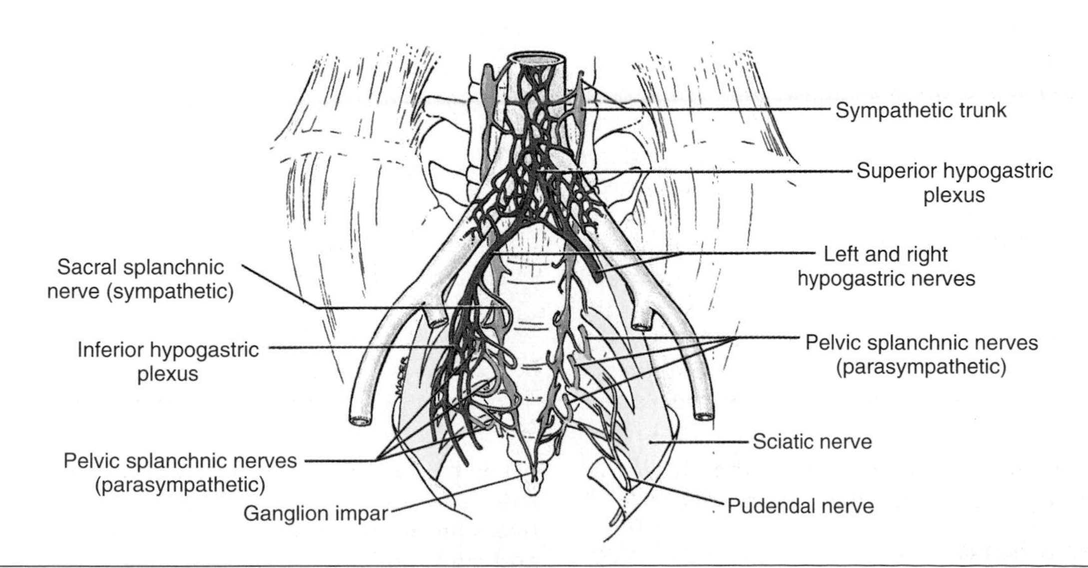

Figure 4.4. Autonomic nerves of pelvis. *Orange*, sympathetic trunk and nerves; *green*, parasympathetic nerves; *purple*, plexuses; *yellow*, sciatic and pudendal nerves.

The *hypogastric plexuses*—superior and inferior—are autonomic nerves. The superior hypogastric plexus descends into the pelvis and lies just inferior to the bifurcation of the aorta. This plexus is the inferior prolongation of the *intermesenteric plexus* that also receives the L3 and L4 splanchnic nerves. Branches from the superior hypogastric plexus enter the pelvis and descend anterior to the sacrum as the right and left *hypogastric nerves*. In males these nerves descend lateral to the rectum, prostate, seminal vesicles, and against the inferolateral surfaces of the bladder. In females the cervix of the uterus and the lateral fornices of the vagina take the place of the seminal vesicles and prostate. The inferior hypogastric plexuses are fanlike expansions of the hypogastric nerves.

The *pelvic splanchnic nerves* containing parasympathetic and visceral afferent fibers are derived from S2–S4 spinal cord segments. Hence, the inferior hypogastric plexuses contain both sympathetic and parasympathetic fibers. These fibers pass along the branches of the internal iliac artery to the pelvic viscera.

PELVIC FASCIAE

The pelvic fasciae comprise the visceral pelvic fascia and the parietal pelvic fascia. The *visceral pelvic fascia* (endopelvic fascia) surrounds the pelvic viscera and binds them to each other and to the parietal pelvic fascia. The *parietal pelvic fascia* is part of the general layer that lines the internal aspect of the abdominal and pelvic walls. This fascia also forms part of the pelvic floor (superior and inferior fasciae of the pelvic diaphragm) and is separated from the parietal peritoneum by extraperitoneal fat (Fig. 4.5). The parietal pelvic fascia covers the pelvic surfaces of the obturator internus, piriformis, coccygeus, sphincter urethrae, and levator ani muscles. The name of the fascia is derived from the muscle it encloses (e.g., obturator fascia). The parietal pelvic fascia attaches to the periosteum of the ilium just inferior to the pelvic brim. In females the parietal pelvic fascia attaches to the posterior aspect of the body of the pubis, urinary bladder, cervix of the uterus, vagina, and rectum to form the *pubovesical*, *transverse cervical* (cardinal), and *uterosacral ligaments*. In males the parietal pelvic fascia is attached to the rectum, prostate, urinary bladder, and pubis. The fascia attached to the prostate and bladder forms the medial and lateral *puboprostatic (pubovesical) ligaments* (Fig. 4.6).

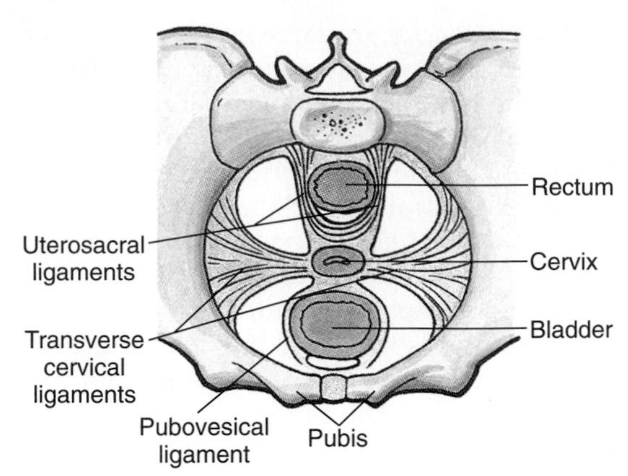

Ligaments of pelvic floor (female)

The *retropubic space* (Figs. 4.6 and 4.7) is between the parietal pelvic fascia and the anterior surface of the urinary bladder. It contains loose connective tissue, fat, vessels, and nerves. The space accommodates the expansion of the urinary bladder as urine accumulates.

Pelvic Viscera

The pelvic viscera include the inferior part of the digestive tract (rectum), the urinary bladder, and parts of the ureters and reproductive system.

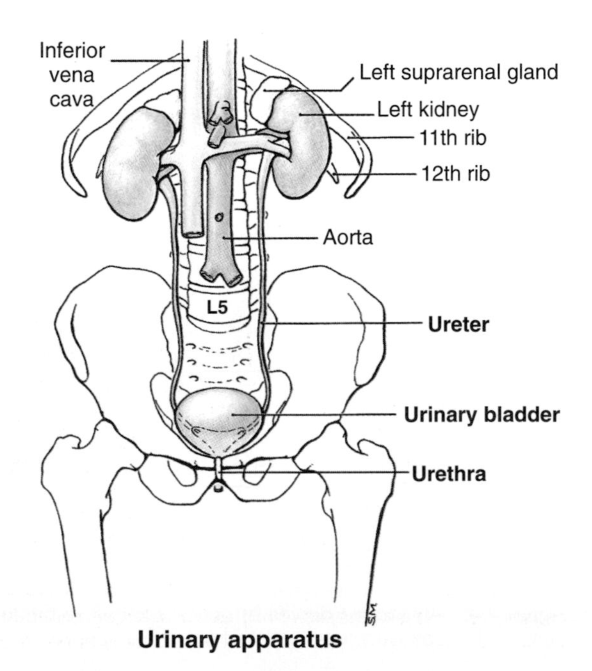

Urinary apparatus

URINARY ORGANS

The pelvic urinary organs are the

- Ureters that carry urine from the kidneys
- Urinary bladder that stores urine
- Urethra that conducts urine to the exterior

Ureters

The ureters pass over the pelvic brim (edge of superior pelvic aperture), anterior to the origins of the external iliac arteries. They then run posteroinferiorly on the lateral walls of the pelvis, external to the parietal peritoneum and anterior to the internal iliac arteries. They then curve anteromedially, superior to the levator ani, to enter the bladder. In males the only structure that passes between the ureter and the peritoneum is the ductus deferens. The ureter lies posterolateral to this duct and enters the posterosuperior angle of the bladder, just superior to the seminal vesicle. In females the ureter passes medial to the origin of the uterine artery and continues to the level of the ischial spine, where it is crossed superiorly by the uterine artery. It then passes close to the lateral portion of the fornix of the vagina and enters the posterosuperior angle of the bladder.

Arterial Supply. The external and internal *iliac arteries* and *vesical arteries* supply the pelvic part of the ureter. The most constant arteries

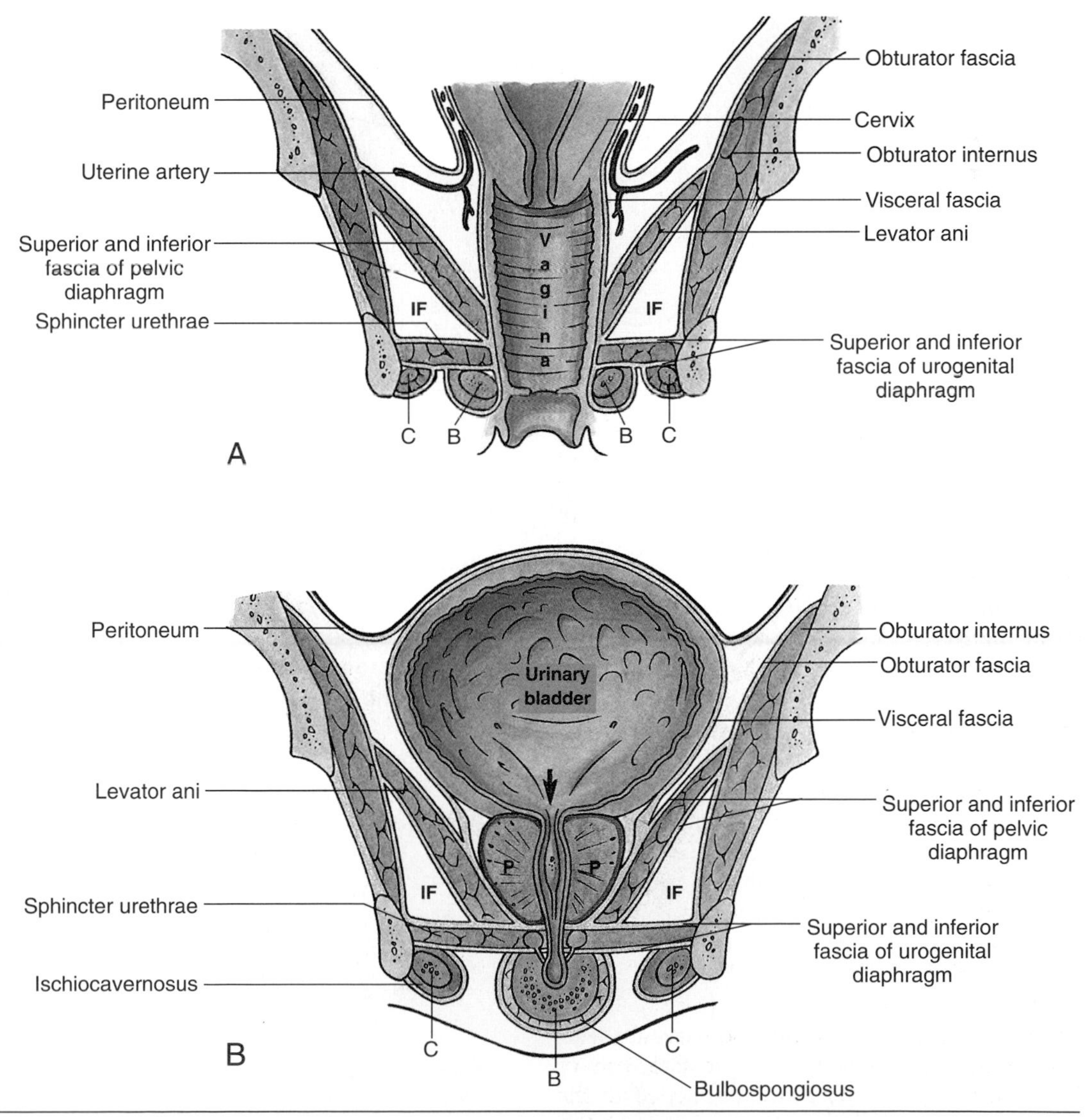

Figure 4.5. Coronal sections of pelvis illustrating pelvic fasciae. *IF*, ischioanal fossa. **A.** Female. *C*, crus of clitoris; *B*, bulb of vestibule. **B.** Male. *P*, prostate; *C*, crus of penis; *B*, bulb of penis. *Arrow*, internal urethral orifice.

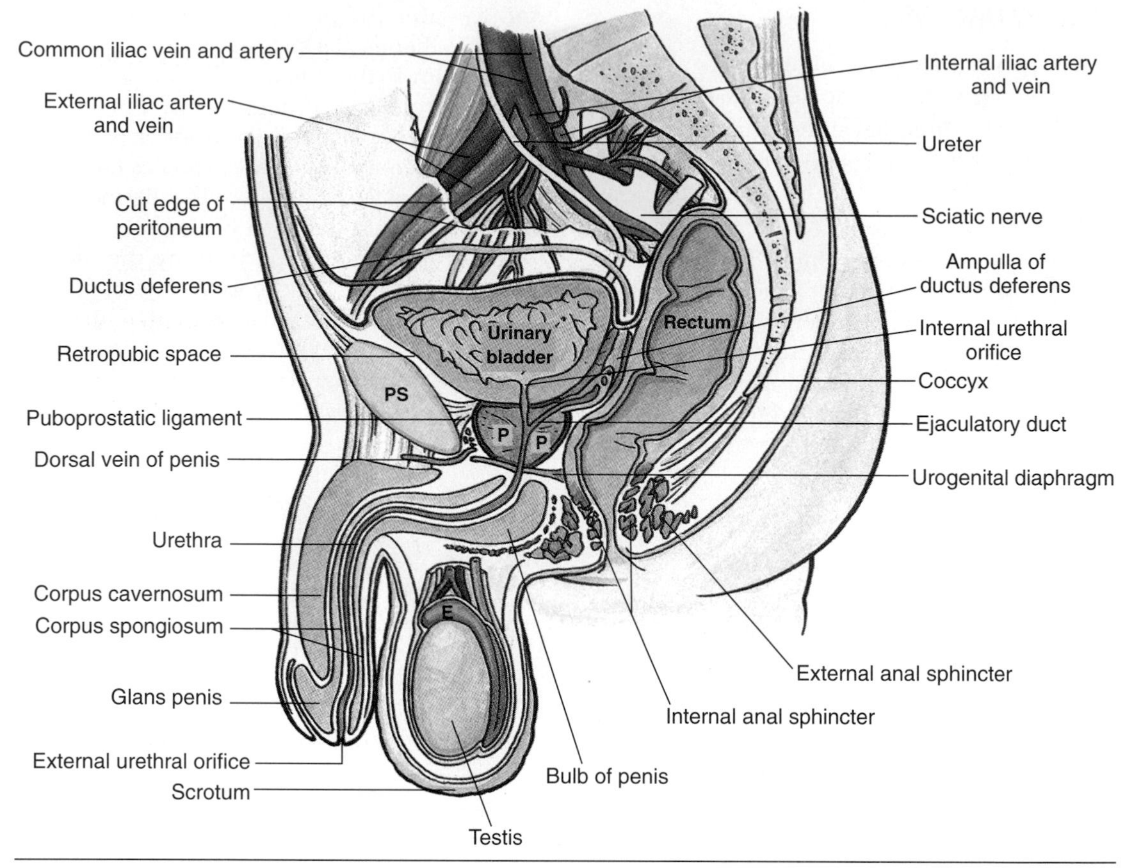

Figure 4.6. Median section of male pelvis. *PS*, pubic symphysis; *P*, prostate; *E*, epididymis.

supplying this part of the ureter in females are branches of the *uterine arteries*. The sources of similar branches in males are the *inferior vesical arteries* (Fig. 4.8, Table 4.4).

Venous and Lymphatic Drainage. Veins from the ureters accompany the arteries and have corresponding names. Lymph drains into the lumbar (lateral aortic), common iliac, external iliac, and internal iliac lymph nodes. (Fig. 4.9*A*).

Innervation. The nerves derive from adjacent *autonomic plexuses* (renal, testicular or ovarian, and inferior hypogastric). Afferent fibers (pain) from the ureters reach the spinal cord through the dorsal roots of T11, T12, and L1 nerves (Fig. 4.9*B*).

Ureteric calculi (stones) may cause complete or intermittent *obstruction of urinary flow*. The obstruction may occur anywhere along the ureter, but it occurs most often where the ureter crosses the external iliac artery and the pelvic brim and where it passes through the wall of the bladder. The severity of the pain associated with calculi depends on the location, type, size, and texture of the stone.

Urinary Bladder

The urinary bladder is in the pelvis minor when empty, posterior and slightly superior to the pubic bones. It is separated from these bones by the *retropubic space* (Figs. 4.6 and 4.7)

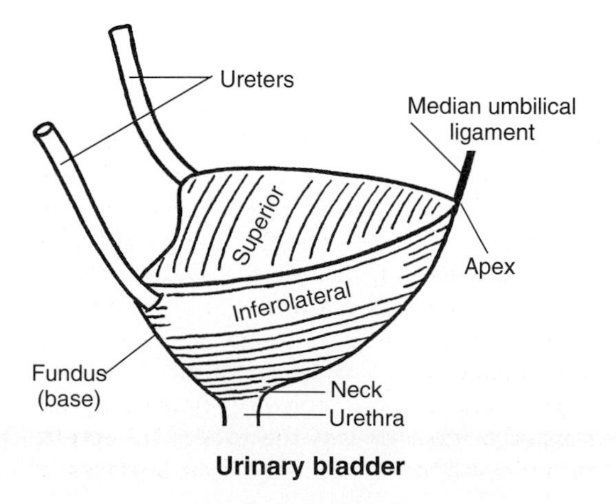

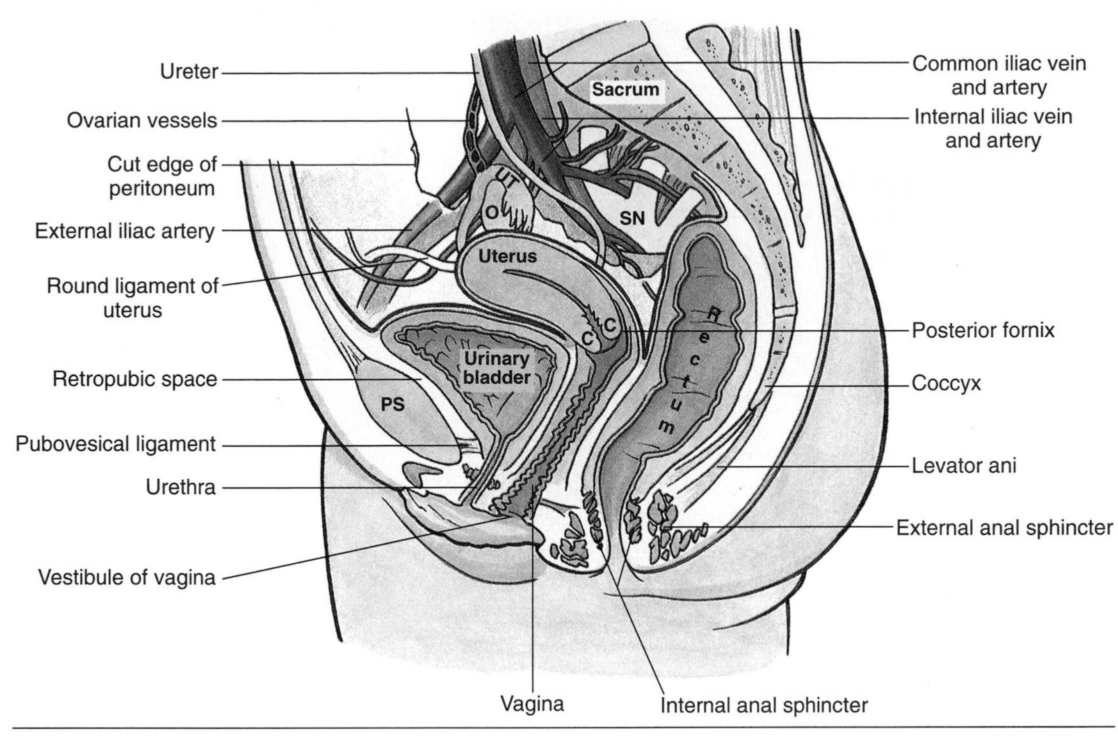

Figure 4.7. Median section of female pelvis. *UT*, uterine tube; *O*, ovary; *SN*, sciatic nerve; *C*, cervix of uterus; *PS*, pubic symphysis.

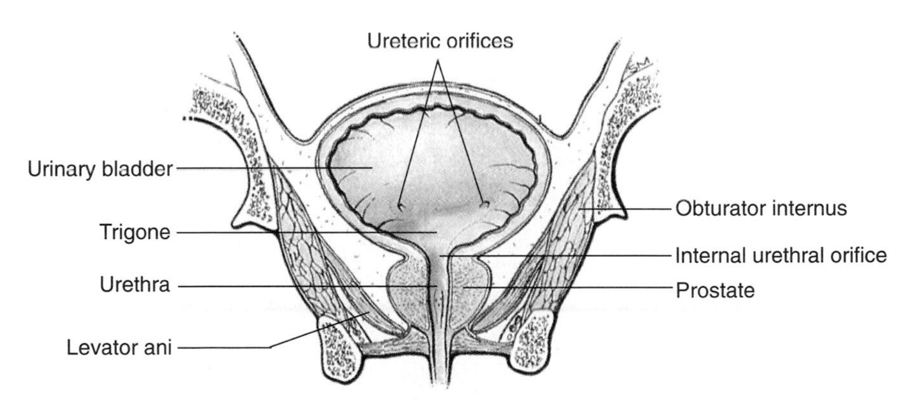

and lies inferior to the peritoneum where it rests on the pelvic floor. The bladder is relatively free within the extraperitoneal fatty tissue, except for its neck, which is held firmly by the *pubovesical ligaments* in females and the *puboprostatic ligaments* in males. As the bladder fills, it expands superiorly into the extraperitoneal fatty superficial layer of fascia of the anterior abdominal wall.

The bladder always contains some urine and is usually more or less rounded. The empty pyramid-shaped bladder has four surfaces: a superior surface, two inferolateral surfaces, and a posterior surface. The inferolateral surfaces are in contact with the fascia covering the levator ani. The posteroinferior surface of the bladder is the *fundus* (base). In females the fundus is closely related to the anterior wall of the vagina; in males it is related to the rectum. The *apex of the bladder* (anterior end) points toward the superior edge of the pubic symphysis. The *neck of the bladder* (inferior part) is where the fundus and inferolateral surfaces converge.

The *bladder bed* is formed on each side by the

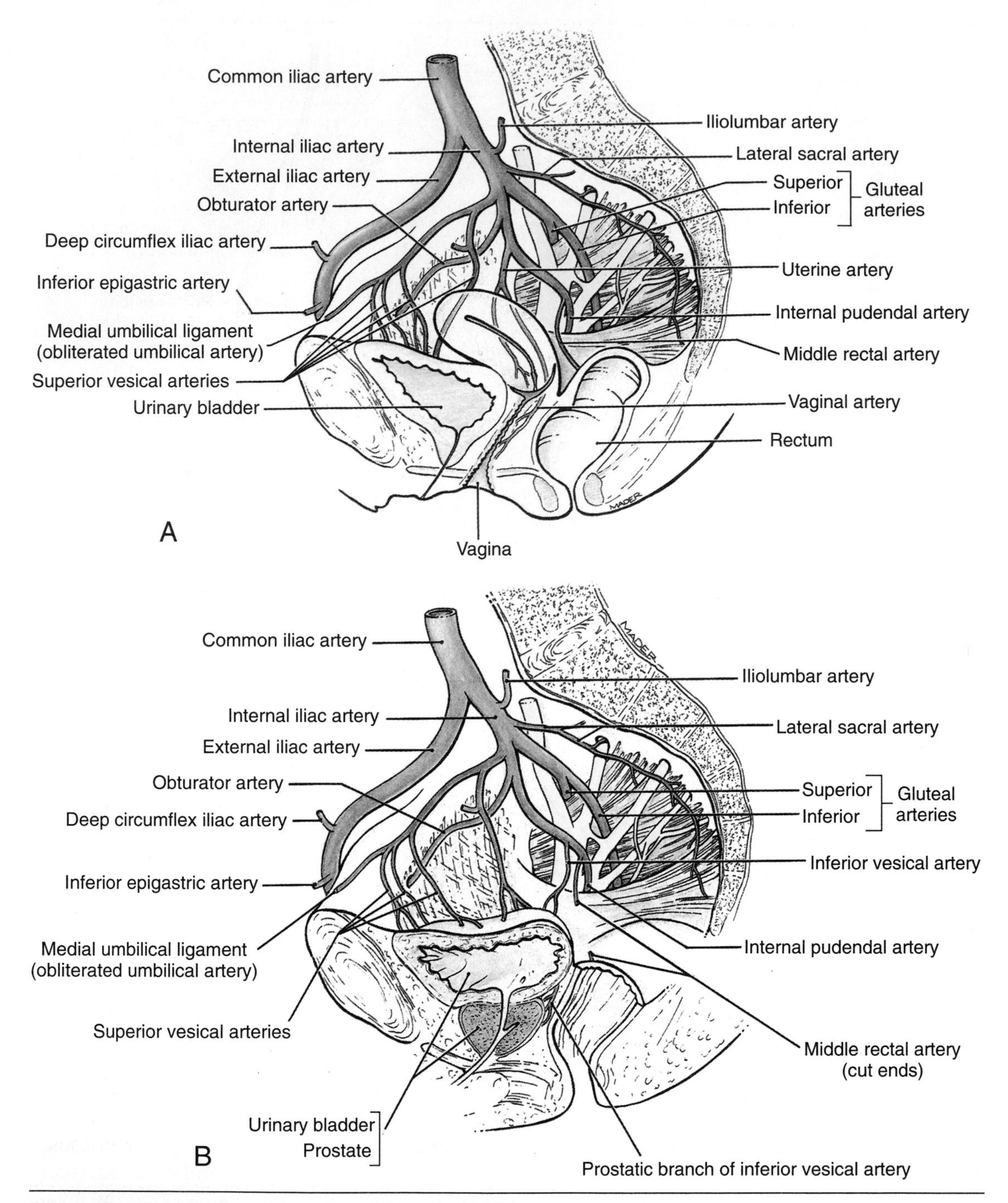

Figure 4.8. Iliac arteries and branches. **A.** Female. **B.** Male.

pubic bones, the obturator internus and levator ani, and posteriorly by the rectum. The entire bladder is enveloped by loose connective tissue called vesical fascia that contains the *vesical venous plexus*. The wall of the bladder is composed chiefly of the *detrusor muscle*. Toward the neck of the bladder, its muscle fibers form the involuntary *internal sphincter*. Some of its fibers run radially and assist in opening the *internal urethral orifice*. In males the muscle fibers in the neck of the bladder are continuous with the fibromuscular tissue of the prostate, whereas in females these fibers are continuous with muscle fibers in the wall of the urethra. The *ureteric ori-*

Table 4.4
Arteries of Pelvis

Artery	Origin	Course	Distribution
Internal iliac	Common iliac a.	Passes over brim of pelvis to reach pelvic cavity	Main blood supply to pelvic organs, gluteal mm. and perineum
Anterior divison of internal iliac a.	Internal iliac a.	Passes anteriorly and divides into visceral branches and obturator a.	Pelvic viscera and muscles in medial compartment of thigh
Umbilical	Anterior division of internal iliac a.	Short pelvic course and ends as superior vesical a. in females	Superior aspect of urinary bladder in females; ductus deferens in males
Obturator	Anterior division of internal iliac a.	Runs anteroinferiorly on lateral pelvic wall	Pelvic mm., nutrient a. to ilium, and head of femur
Superior vesical	Remnant of proximal part of umbilical a.	Passes to superior aspect of urinary bladder	Superior aspect of urinary bladder
Artery to ductus deferens	Inferior vesical a.	Runs retroperitoneally to ductus deferens	Ductus deferens
Inferior vesical	Anterior division of internal iliac a.	Passes retroperitoneally to inferior portion of urinary bladder in males	Urinary bladder, seminal vesicle, and prostate
Middle rectal	Anterior division of internal iliac a.	Descends in pelvis to rectum	Seminal vesicle, prostate, and rectum
Internal pudendal	Anterior division of internal iliac a.	Leaves pelvis through greater sciatic foramen and enters perineum by passing through lesser sciatic foramen	Muscles of anal canal and perineum; skin and structures in urogenital triangle
Inferior gluteal	Anterior division of internal iliac a.	Leaves pelvis through greater sciatic foramen	Piriformis, coccygeus, levator ani, and gluteal mm.
Uterine	Anterior division of internal iliac a.	Runs medially on levator ani; crosses ureter to reach base of broad ligament	Uterus, ligament of uterus, uterine tube, and vagina
Vaginal	Uterine a.	At junction of body and cervix of uterus, it descends to vagina	Vagina and branches to inferior part of urinary bladder
Gonadal (testicular and ovarian)	Abdominal aorta	Descends retroperitoneally; testicular a. passes into deep inguinal ring; ovarian a. crosses brim of pelvis and runs medially in suspensory ligament to ovary	Testis and ovary, respectively
Posterior division of internal iliac a.	Internal iliac a.	Passes posteriorly and gives rise to parietal branches	Pelvic wall and gluteal region
Iliolumbar	Posterior division of internal iliac a.	Ascends anterior to sacroiliac joint and posterior to common iliac vessels and psoas major	Iliacus, psoas major, quadratus lumborum mm., and cauda equina in vertebral canal
Lateral sacral (superior and inferior)	Posterior division of internal iliac a.	Run on superficial aspect of piriformis	Piriformis and vertebral canal

fices and the internal urethral orifice are at the angles of the *trigone of the bladder*. The ureters pass obliquely through the bladder wall in an inferomedial direction. An increase in bladder pressure presses the walls of the ureters together, preventing pressure in the bladder from forcing urine into the ureters.

In females the peritoneum passes

- From the anterior abdominal wall (*1*)
- Superior to the pubic bone (*2*)
- On the superior surface of the urinary bladder (*3*)
- From the bladder to the uterus, forming the *vesicouterine pouch* (*4*)

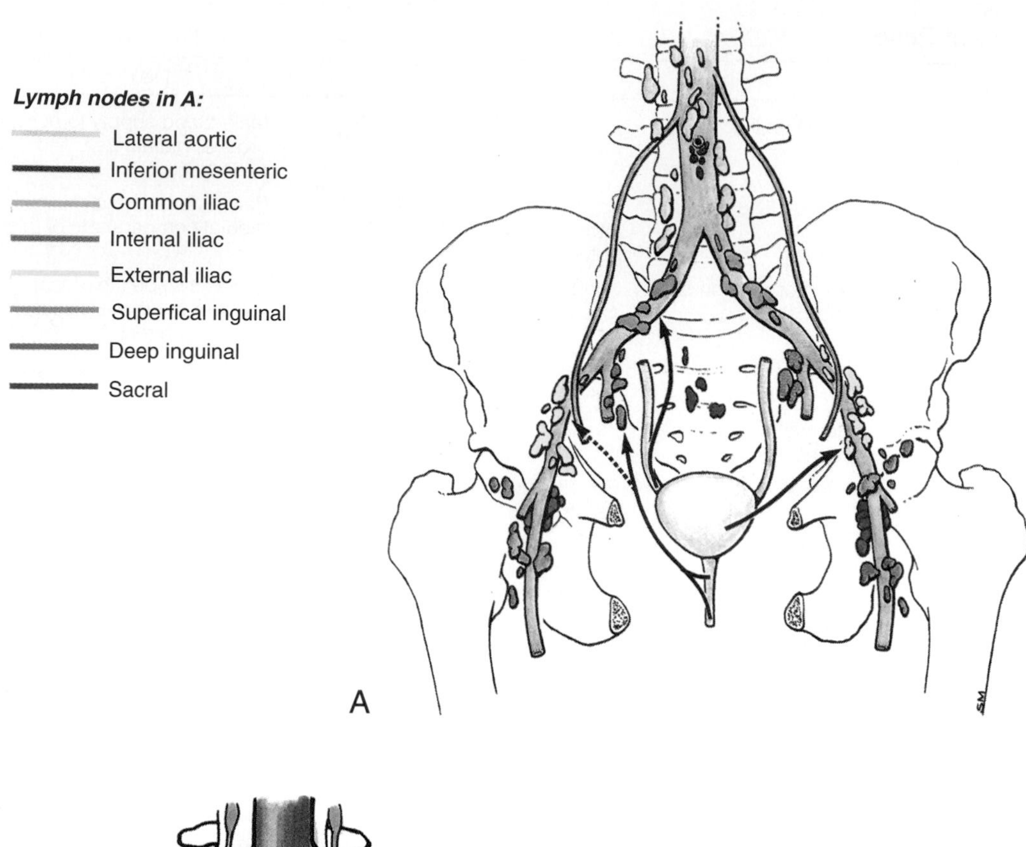

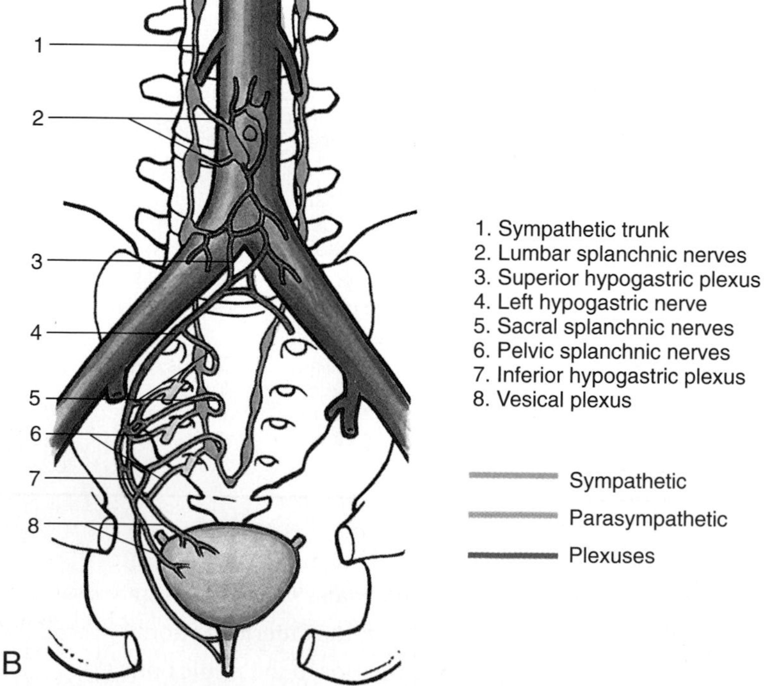

Figure 4.9. Urinary bladder and urethra. **A.** Lymphatic drainage. *Arrows*, lymph flow to lymph nodes. **B.** Autonomic innervation.

- On the fundus and body of the uterus, posterior fornix, and wall of the vagina (*5*)
- Between the rectum and uterus, forming the *rectouterine pouch* (*6*)
- On the anterior and lateral sides of the rectum (*7*)
- Posteriorly to become the sigmoid mesocolon (*8*)

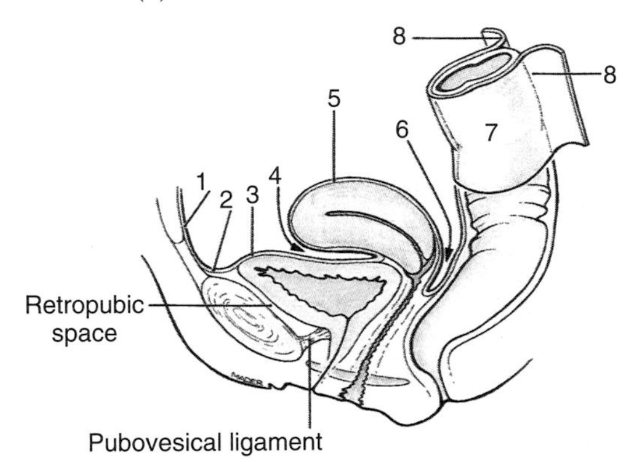

Peritoneal reflections in female pelvis (median section)

In males the peritoneum passes

- From the anterior abdominal wall (*1*)
- Superior to the pubic bone (*2*)
- On the superior surface of the urinary bladder (*3*)
- 2 cm inferiorly on the posterior surface of the urinary bladder (*4*)
- On the superior ends of the seminal vesicles (*5*)
- Posteriorly to line the *rectovesical pouch* (*6*)
- To cover the rectum (*7*)
- Posteriorly to become the sigmoid mesocolon (*8*)

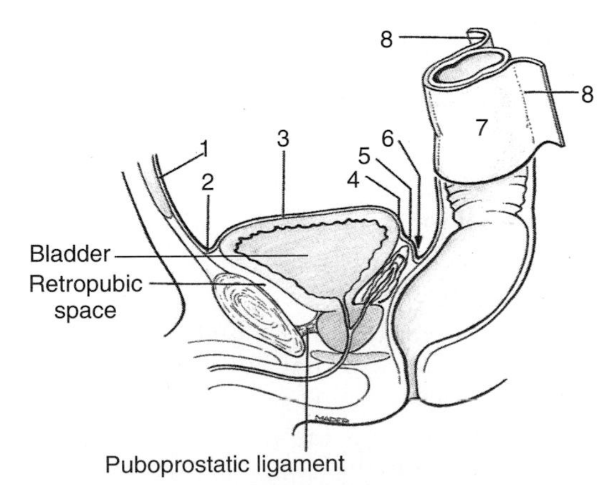

Peritoneal reflections in male pelvis (median section)

Arterial Supply. The main arteries supplying the bladder are branches of the internal iliac arteries (Table 4.4). The superior vesical arteries supply anterosuperior parts of the bladder. In males the inferior vesical arteries supply the fundus of the bladder. In females the vaginal arteries replace the inferior vesical arteries and send small branches to posteroinferior parts of the bladder. The obturator and inferior gluteal arteries also supply small branches to the bladder.

Venous and Lymphatic Drainage. The names of the veins correspond to the arteries and are tributaries of the internal iliac veins. In males the *vesical venous plexus*, which combines with the *prostatic venous plexus*, envelops the base of the bladder and prostate, the seminal vesicles, the ductus deferentes (plural of deferens), and the inferior ends of the ureters. The prostatic venous plexus, a dense network of veins, receives blood from the dorsal vein of the penis. The vesical venous plexus mainly drains through the inferior vesical veins into the internal iliac veins, but it may drain via the sacral veins into the *vertebral venous plexuses*. In females the vesical venous plexus envelops the pelvic part of the urethra and the neck of the bladder and receives blood from the dorsal vein of the clitoris and communicates with the *vaginal venous plexus*.

In both sexes lymphatic vessels leave the superior surface of the bladder and pass to the external iliac lymph nodes (Fig. 4.9*A*), whereas those from the posteroinferior surface pass to the internal iliac lymph nodes. Some vessels from the neck of the bladder drain into the sacral or common iliac lymph nodes.

Innervation. Parasympathetic fibers to the bladder are derived from the *pelvic splanchnic nerves* (Fig. 4.9*B*). They are motor to the detrusor muscle and inhibitory to the internal sphincter. Hence, when these fibers are stimulated by stretching, the bladder contracts, the internal sphincter relaxes, and urine flows into the urethra. *Sympathetic fibers* to the bladder are derived from T11, T12, L1, and L2 nerves. The nerves supplying the bladder form the *vesical nerve plexus*, which consists of both sympathetic and parasympathetic fibers. This plexus is continuous with the *inferior hypogastric plexus*. Sensory fibers from the bladder are visceral and transmit pain sensations such as those from overdistention.

When excessively distended, the bladder rises to the level of the umbilicus. In so doing, it lifts the parietal peritoneum from the anterior abdominal wall (Fig. 4.10*A*). The bladder then lies adjacent to this wall without the intervention of peritoneum. Consequently the distended bladder may be punctured (*suprapubic cystostomy*) or approached surgically superior to the pubic symphysis for the introduction of instruments without traversing the peritoneum and entering the peritoneal cavity. Because of the superior position of the distended bladder, it may be ruptured by injuries to the inferior part of the anterior abdominal wall or by fractures of the pelvis. The rupture may result in the escape of urine extraperitoneally or intraperitoneally.

Male Urethra

The urethra conveys urine from the urinary bladder to the exterior through the external urethral orifice at the tip of the glans penis. The urethra also provides an exit for semen [sperms and secretions from glands (e.g., prostate)]. For descriptive purposes the urethra is divided into three parts: prostatic, membranous, and spongy. The last two parts are described under "Perineum."

The *prostatic urethra* begins at the internal urethral orifice at the apex of the trigone of the bladder and descends through the prostate, forming a gentle curve that is concave anteriorly (Fig. 4.10*A*). It ends by piercing the superior fascia of the urogenital diaphragm. The prostatic part is the widest and most dilatable part of the urethra. It receives its blood supply from the urethral artery, a branch of the internal pudendal artery. The internal surface of the posterior wall of the prostatic urethra has notable features (p. 165). The most prominent structure is the *urethral crest*, a median ridge that has a groove on each side called a *prostatic sinus*. Most prostatic ducts open into these sinuses. In the middle part of this crest is the *seminal colliculus*, a rounded eminence with a slitlike orifice. This opening leads into a small, vestigial cul-de-sac called the *prostatic utricle*. On each side of this orifice is the opening of an ejaculatory duct.

Arterial Supply. The prostatic urethra is supplied by arteries of the prostate, which are branches of the inferior vesical and middle rectal arteries (Fig. 4.8*B*).

Venous and Lymphatic Drainage. The veins follow the arteries and have similar names. The lymphatic vessels pass mainly to the internal iliac lymph nodes; a few drain into the external iliac lymph nodes (Fig. 4.9*A*).

Innervation. The nerves are derived from the pudendal nerve and the *prostatic plexus* of the autonomic nervous system. This plexus arises from the inferior part of the inferior hypogastric plexus (Fig. 4.9*B*).

Female Urethra

The female urethra passes anteroinferiorly from the urinary bladder, posterior and then inferior to the pubic symphysis (Fig. 4.7). The *external urethral orifice* is in the vestibule of the vagina (p. 187). The urethra lies anterior to the vagina and passes with it through the pelvic and urogenital diaphragms and the perineal membrane. Its inferior end is surrounded by the sphincter urethrae muscle, and some of these muscle fibers enclose both the urethra and vagina.

Arterial Supply. Blood is supplied by the internal pudendal and vaginal arteries (Fig. 4.8*A*).

Venous and Lymphatic Drainage. The veins follow the arteries and have similar names. Most lymphatic vessels from the urethra pass to the *sacral and internal iliac lymph nodes*. A few vessels drain into the inguinal lymph nodes (Fig. 4.9*A*).

Innervation. Nerves to the urethra arise from the *pudendal nerve*. Most afferents from the urethra run in the pelvic splanchnic nerves.

MALE INTERNAL GENITAL ORGANS

The male internal genital organs include the testes, ductus deferentes, seminal vesicles, ejaculatory ducts, prostate, and bulbourethral glands (Fig. 4.10). The testes are described in Chapter 3.

Ductus Deferens

The ductus deferens is the continuation of the duct of the epididymis. The ductus deferens

- Begins in the tail of the epididymis
- Ascends in the spermatic cord
- Passes through the inguinal canal
- Crosses over the external iliac vessels to enter the pelvis
- Passes along the lateral wall of the pelvis where it lies external but adherent to the parietal peritoneum
- Ends by joining the duct of the seminal vesicle to form the ejaculatory duct

During its course no other structure intervenes between the ductus deferens and the

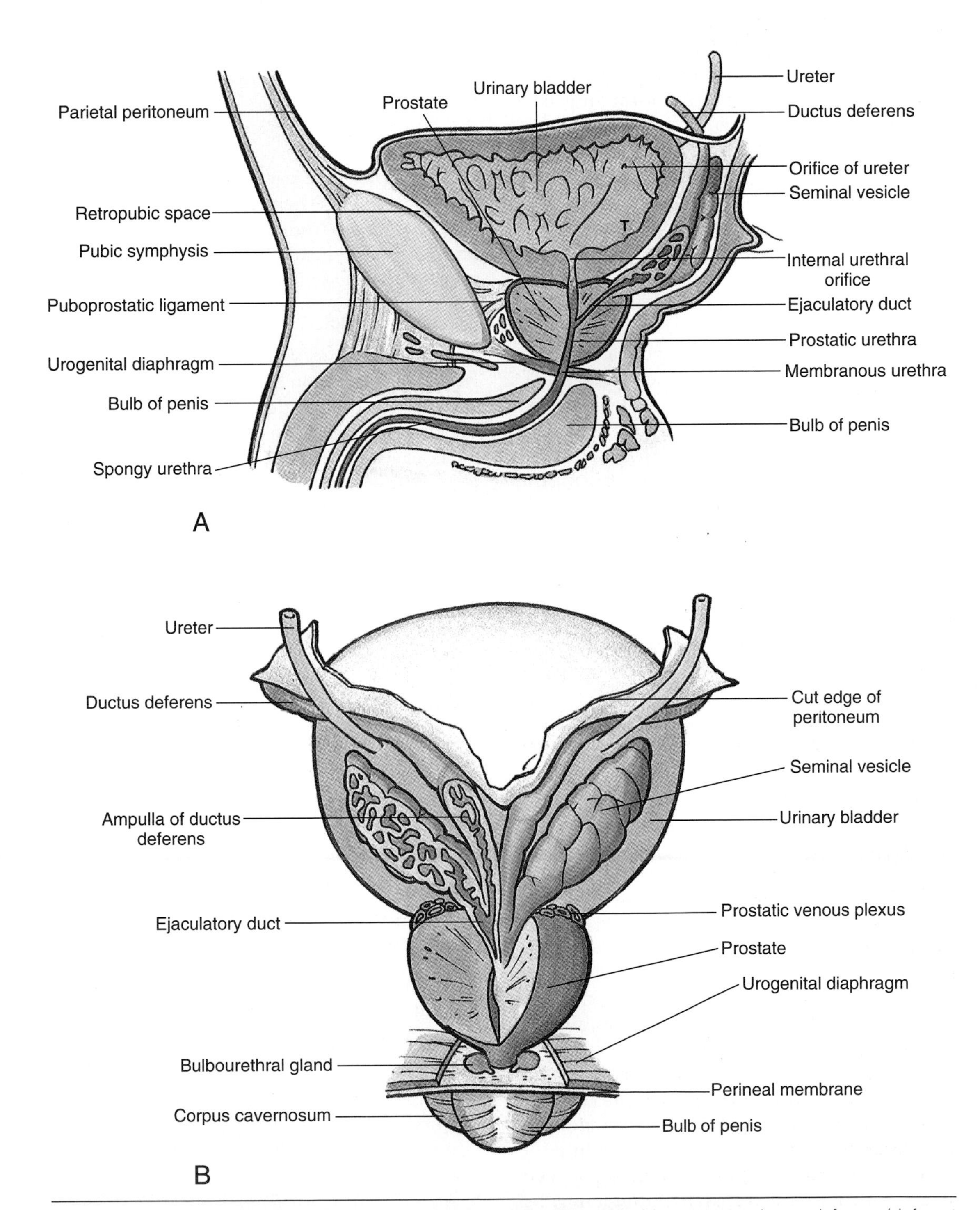

Figure 4.10. **A.** Sagittal section of male pelvis showing relationship of bladder, prostate, ductus deferens (deferent duct), and ejaculatory duct. *T*, trigone of bladder. **B.** Posterior view of urinary bladder, deferent ducts, and prostate. Left seminal vesicle and ampulla of ductus deferens have been opened, and prostate has been cut away to expose ejaculatory duct. A piece of the urogenital diaphragm has been removed to show the bulbourethral glands.

peritoneum. The ductus deferens crosses the ureter near the posterolateral angle of the bladder, running between the ureter and peritoneum to reach the fundus of the bladder. At first it lies superior to the seminal vesicle; then it descends medial to the ureter and this vesicle. The ductus deferens enlarges to form the *ampulla of the ductus deferens* as it passes posterior to the bladder. It then narrows and joins the duct of the seminal vesicle to form the *ejaculatory duct*.

Arterial Supply. The tiny *artery of the ductus deferens* usually arises from the inferior vesical artery and terminates by anastomosing with the *testicular artery*, posterior to the testis.

Venous and Lymphatic Drainage. The veins accompany the arteries and have similar names. Like the prostate and seminal vesicle, lymphatic vessels from the ductus deferens end in the *external iliac lymph nodes* (Fig. 4.12*A*).

Innervation. The nerves of the ductus deferens are derived from the *inferior hypogastric plexus*. The ductus is richly innervated by autonomic nerve fibers, thereby facilitating its rapid contraction for expulsion of sperms during ejaculation.

A method of sterilizing males is *deferentectomy*, popularly called a *vasectomy*. During this procedure part of the ductus deferens is ligated and/or excised. Hence, the ejaculated fluid from the seminal vesicles, prostate, and bulbourethral glands contains no sperms. The unexpelled sperms degenerate in the epididymis and the proximal part of the ductus deferens.

Seminal Vesicles

Each seminal vesicle is an elongated structure that lies between the fundus of the bladder and the rectum (Fig. 4.10). The seminal vesicles, obliquely placed superior to the prostate, *do not store sperms*. They secrete a thick alkaline fluid that mixes with the sperms as they pass into the ejaculatory ducts and urethra. The superior ends of the seminal vesicles are covered with peritoneum and lie posterior to the ureters, where the peritoneum of the *rectovesical pouch* separates them from the rectum. The inferior ends of the seminal vesicles are closely related to the rectum and are separated from it only by the *rectovesical septum*. The duct of each seminal vesicle joins the ductus deferens to form the ejaculatory duct.

Arterial Supply. The arteries to the seminal vesicles derive from the inferior vesical and middle rectal arteries (Fig. 4.8*B*).

Venous and Lymphatic Drainage. The veins accompany the arteries and have similar names. The iliac lymph nodes, especially the *internal iliac lymph nodes*, receive lymph from the seminal vesicles (Fig. 4.12*A*).

Innervation. The walls of these vesicles contain a plexus of nerve fibers and some sympathetic ganglia. The preganglionic sympathetic fibers emerge from the *superior lumbar nerves*, and the parasympathetic fibers derive from the *pelvic splanchnic nerves*.

Ejaculatory Ducts

Each ejaculatory duct is a slender tube that is formed by the union of the ducts of the seminal vesicle and ductus deferens. The ejaculatory ducts, formed near the neck of the bladder, run close together as they pass anteroinferiorly through the posterior part of the prostate and along the sides of the prostatic utricle. The ejaculatory ducts traverse the main part of the prostate and converge to open by slitlike apertures into the posterior wall of the prostatic urethra, one on each side of the orifice of the prostatic utricle, a vestigial nonfunctional structure (p. 165).

Arterial Supply. The arteries of the ductus deferens supply the ejaculatory ducts.

Venous and Lymphatic Drainage. The veins join the prostatic and vesical venous plexuses. The lymphatic vessels drain into the external iliac lymph nodes.

Prostate

The prostate is the walnut-sized, fibromuscular accessory glandular organ that surrounds the prostatic urethra (Figs. 4.10 and 4.11). The prostate has a dense *fibrous capsule* that is surrounded by a fibrous *prostatic sheath*, part of the visceral layer of pelvic fascia. The relationships of the prostate follow:

- Its base is related to the neck of the bladder
- Its apex rests on the urogenital diaphragm
- Its anterior surface is separated from the pubic symphysis by retroperitoneal fat in the retropubic space
- Its posterior surface is related to the ampulla of the rectum
- Its inferolateral surfaces are related to the levator ani

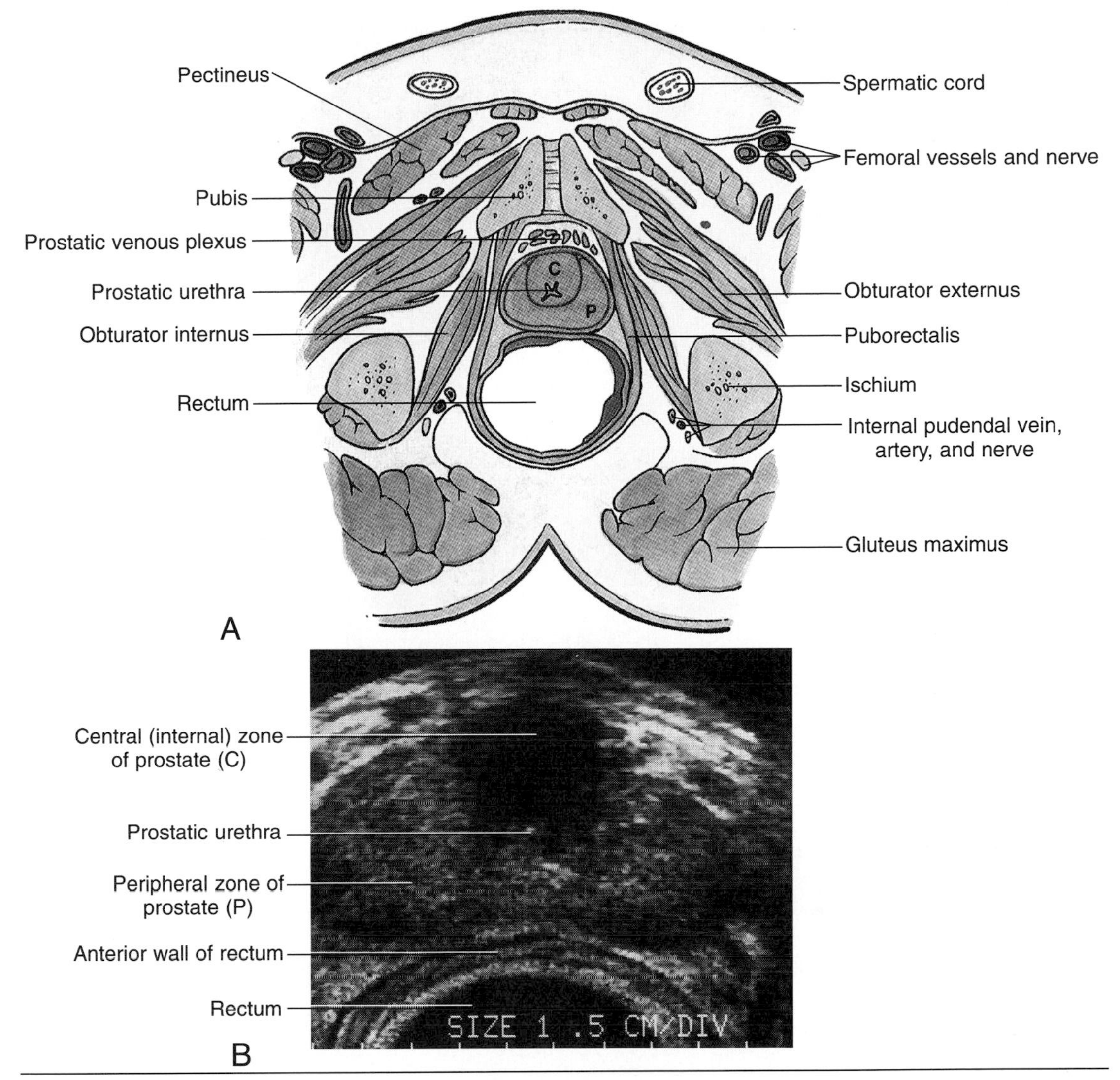

Figure 4.11. **A.** Transverse section of male pelvis. *C* and *P*, zones of prostate shown in ultrasound scan. **B.** Transverse (transrectal) ultrasound scan primarily to show zones of prostate.

Although not clearly separated anatomically, several lobes of the prostate are described.

- The anterior lobe, or isthmus, lies anterior to the urethra; it is fibromuscular and contains little, if any, glandular tissue
- The posterior lobe lies posterior to the urethra and inferior to the ejaculatory ducts; it is readily palpable by digital rectal examination
- The lateral lobes lie on either side of the urethra; they form the major part of the prostate
- The middle (median) lobe lies between the urethra and the ejaculatory ducts and is closely related to the neck of the bladder

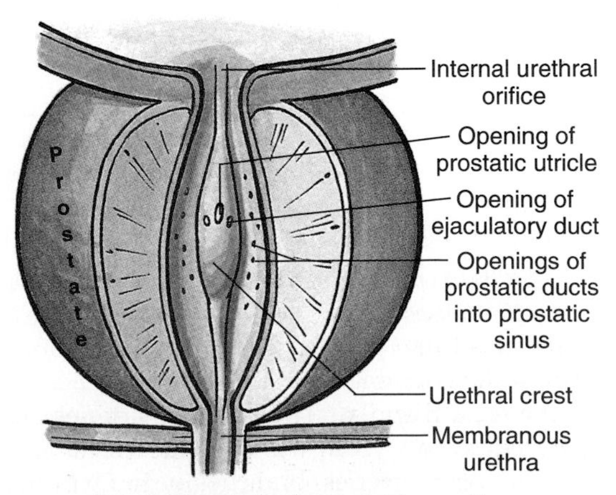

Posterior wall of prostatic urethra

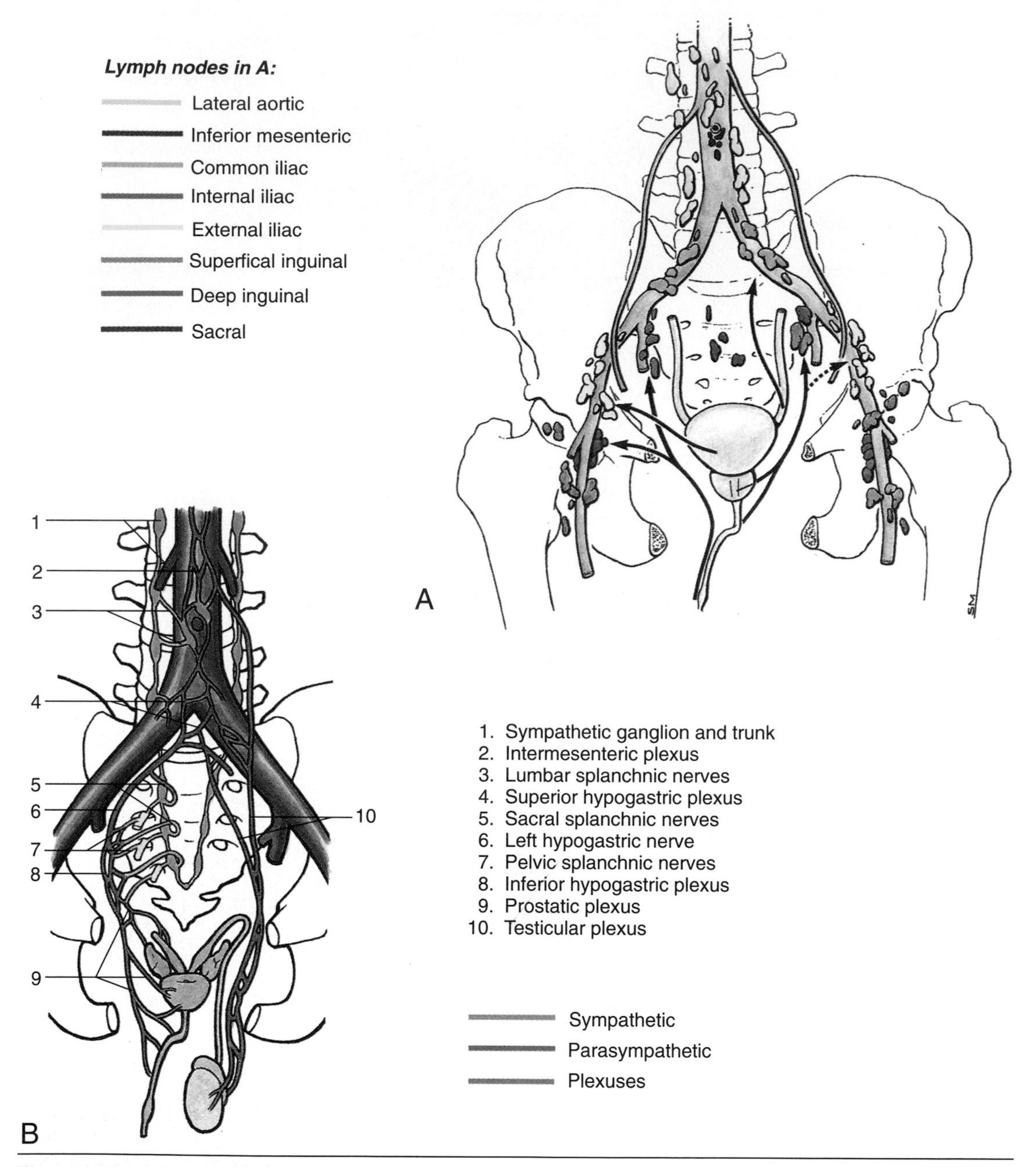

Figure 4.12. **A.** Lymphatic drainage of testis, ductus deferens, prostate, and seminal vesicles. *Arrows*, lymph flow to lymph nodes. **B.** Autonomic innervation of these structures.

The 20–30 *prostatic ducts* open chiefly into the *prostatic sinuses* on the posterior wall of the prostatic urethra. Prostatic fluid provides about 20% of the volume of seminal fluid.

Arterial Supply. The prostatic arteries are derived mainly from the inferior vesical and middle rectal arteries, branches of the internal iliac artery.

Venous and Lymphatic Drainage. The veins join to form the *prostatic venous plexus* (Fig. 4.11*A*) around the sides and base of the prostate. This plexus, between the fibrous capsule and the prostatic sheath, drains into the *internal iliac veins*. The prostatic venous plexus also communicates with the vesical venous plexus and the vertebral venous plexuses. The

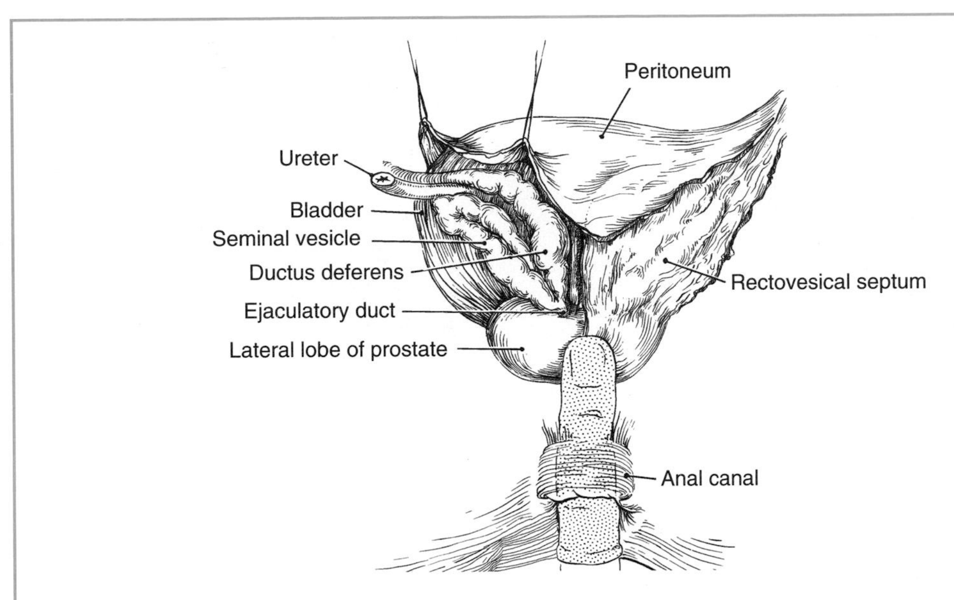

The prostate is of medical interest because enlargement of the gland is common after age 40. An enlarged prostate compresses the prostatic urethra as it passes through the prostate and interferes with the passage of urine. *Prostatic cancer* is a common tumor in men older than 55. During a rectal examination the malignant prostate feels hard and often irregular. In advanced stages cancer cells metastasize to the iliac and sacral lymph nodes and then to distant nodes and bone.

lymphatic vessels (Fig. 4.12*A*) terminate chiefly in the internal iliac and sacral lymph nodes.

Innervation. Parasympathetic fibers arise from the pelvic splanchnic nerves (S2–S4). The sympathetic fibers derive from the *inferior hypogastric plexus* (Fig. 4.12*B*).

Bulbourethral Glands

The pea-sized bulbourethral glands (Fig. 4.10*B*) lie posterolateral to the membranous urethra. Their ducts pass through the inferior fascia of the urogenital diaphragm (perineal membrane) with the urethra and open through minute apertures into the proximal part of the spongy urethra in the bulb of the penis. Their mucuslike secretion enters the urethra during sexual arousal.

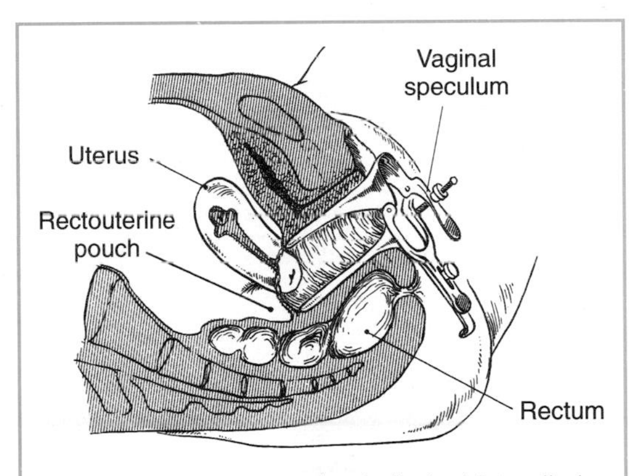

Distention of the vagina is limited laterally by the presence of the *ischial spines* and sacrospinous ligaments. The interior of the vagina can be examined with a *vaginal speculum*. The cervix can also be palpated with the digits in the vagina or rectum.

A pelvic abscess (collection of pus) in the rectouterine pouch can be drained through an incision made in the posterior vaginal fornix (*colpotomy*). Similarly, fluid in the peritoneal cavity (e.g., blood) can be aspirated by this technique (*culdocentesis*).

FEMALE INTERNAL GENITAL ORGANS

The female internal genital organs include the vagina, uterus, uterine tubes, and ovaries (Figs. 4.13 and 4.14).

Vagina

The vagina serves as the excretory passage for menstrual fluid, forms the inferior portion of the "birth canal," and receives the penis during sexual intercourse. The vagina communicates superiorly with the cavity of the cervix (neck) of the uterus and inferiorly with the *vestibule of the vagina* (Fig. 4.7). Its anterior and posterior walls are in contact except at its supe-

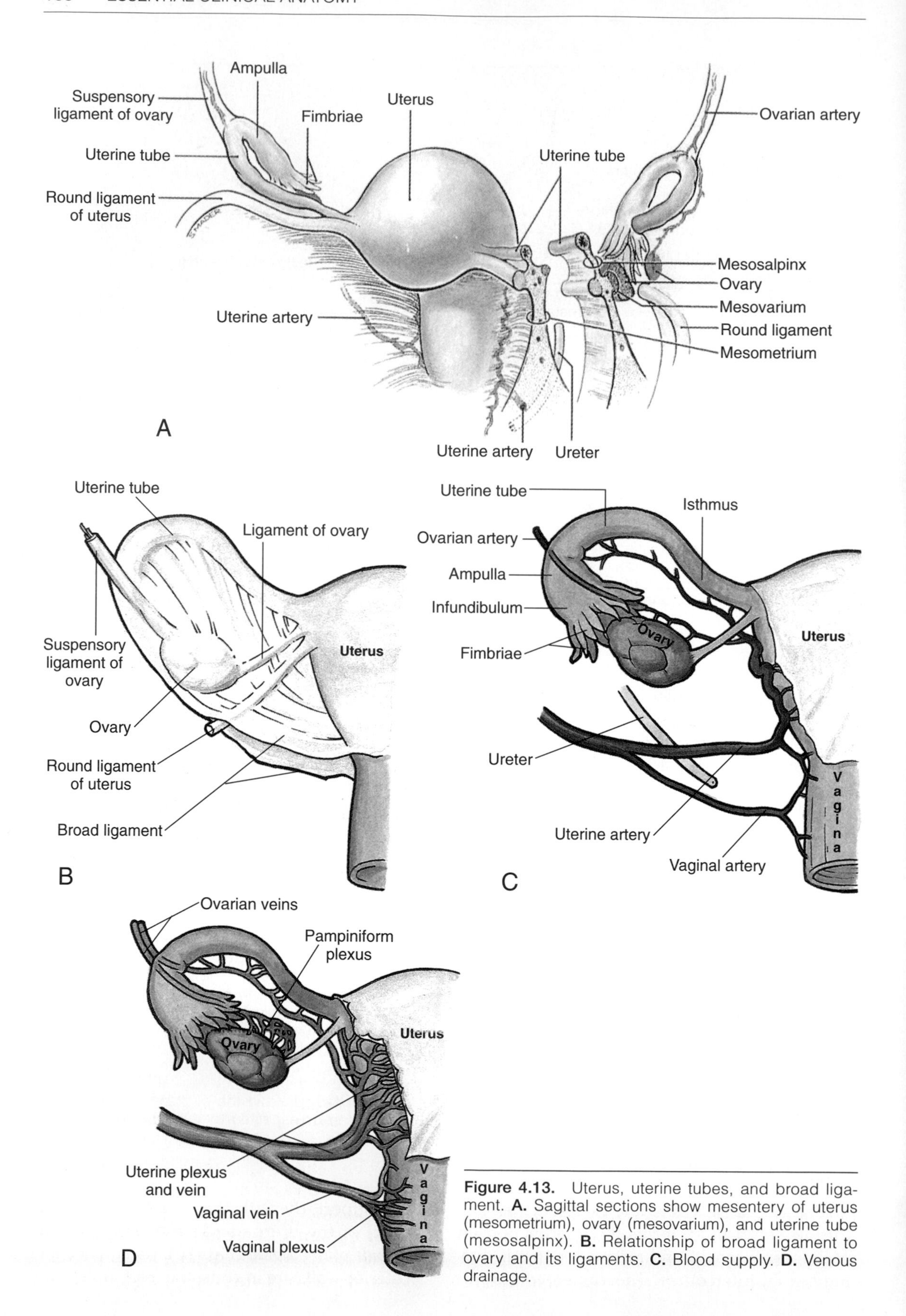

Figure 4.13. Uterus, uterine tubes, and broad ligament. **A.** Sagittal sections show mesentery of uterus (mesometrium), ovary (mesovarium), and uterine tube (mesosalpinx). **B.** Relationship of broad ligament to ovary and its ligaments. **C.** Blood supply. **D.** Venous drainage.

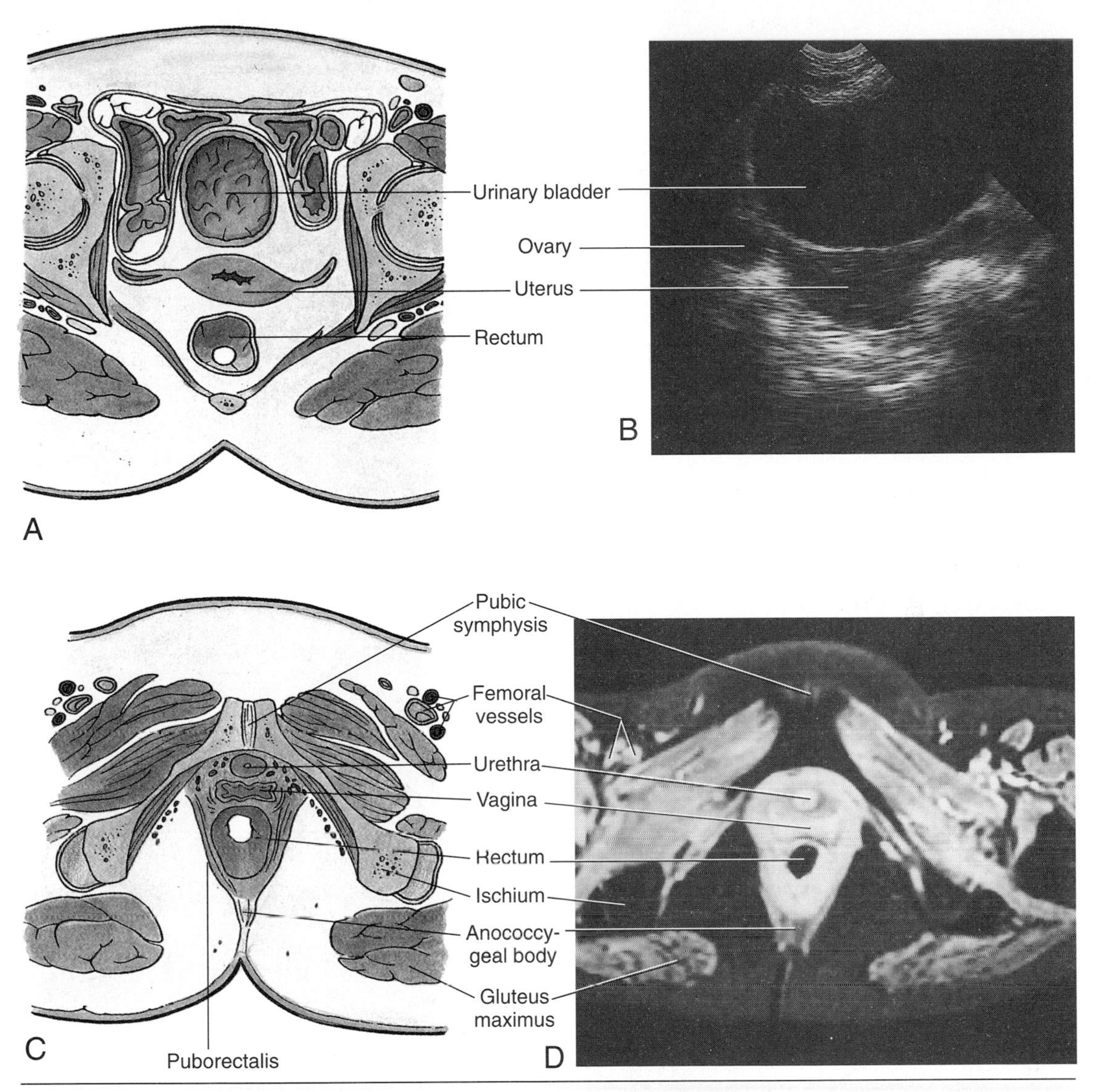

Figure 4.14. Transverse sections of female pelvis. **A.** Through urinary bladder, uterus, and rectum. **B.** Transverse ultrasound scan. **C.** Through urethra, vagina, and rectum. **D.** Transverse magnetic resonance image.

rior end where the cervix holds them apart. The vagina is posterior to the urinary bladder and anterior to the rectum. It passes between the medial margins of the levator ani muscles and pierces the urogenital diaphragm. The *fornix,* the vaginal recess around the cervix, is divisible into anterior, posterior, and lateral regions. The posterior fornix is the deepest and is closely related to the *rectouterine pouch* (of Douglas). Three muscles compress the vagina and act like sphincters: pubovaginalis, urogenital diaphragm (sphincter urethrae and deep transverse perineus muscles), and bulbospongiosus.

Arterial Supply. The blood vessels supplying the superior part of the vagina derive from the *uterine arteries*. The *vaginal arteries* supplying the middle and other parts of the vagina derive from the middle rectal artery and the internal pudendal artery.

Venous and Lymphatic Drainage. The vaginal veins form vaginal venous plexuses along the sides of the vagina and within the

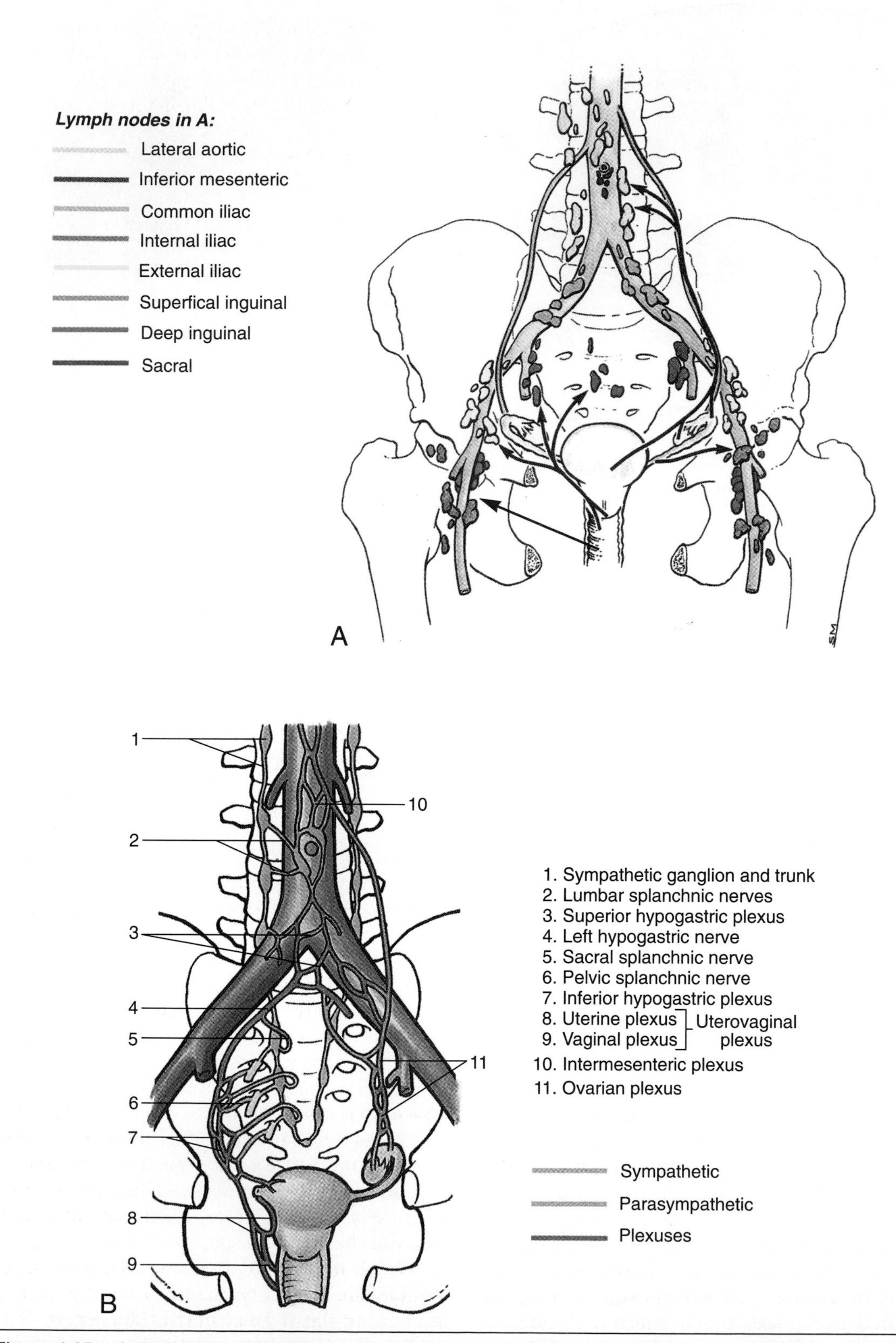

Figure 4.15. **A.** Lymphatic drainage of uterus, vagina, and ovaries. *Arrows*, lymph flow to lymph nodes. **B.** Autonomic innervation of these structures.

vaginal mucosa. These veins drain into the internal iliac veins and communicate with the vesical, uterine, and rectal venous plexuses. The *vaginal lymphatic vessels* (Fig. 4.15*A*) from the

- Superior part drain into the internal and external iliac lymph nodes
- Middle part drain into the internal iliac lymph nodes
- Inferior part drain into the sacral and common iliac nodes, as well as the superficial inguinal lymph nodes

Innervation. The vaginal nerves derive from the *uterovaginal plexus* that lies with the uterine artery between the layers of the broad ligament. The uterovaginal plexus is an extension of the *inferior hypogastric plexus* (Fig. 4.15*B*). Sympathetic, parasympathetic, and afferent fibers pass through this plexus. Most afferent fibers ascend through the hypogastric plexus to the spinal cord via T10–T12 thoracic nerves and the subcostal nerve (T12).

Uterus

The uterus is a thick-walled, pear-shaped muscular organ between the bladder and rectum (Figs. 4.7, 4.13, and 4.14*A*). The uterus is usually bent anteriorly (anteflexed) over the bladder, but its position changes with the degree of fullness of the bladder and rectum. The uterus consists of two major parts:

- Body, the expanded superior two-thirds
- Cervix, the cylindrical inferior third

The *body of the uterus* lies between the layers of the broad ligament and is freely movable. The *cervix* is divided into vaginal and supravaginal parts. The rounded vaginal part communicates with the vagina via the external ostium of the uterus. The isthmus is the constricted zone between the body and cervix, and the fundus is the rounded superior part of the body. The superolateral regions where the uterine tubes enter are the *horns* (cornua). The *ligament of the ovary* is attached to the uterus posteroinferior to the uterotubal junction (Fig. 4.13*B*). The *round ligament of the uterus* is attached anteroinferiorly to this junction.

The *principal supports of the uterus* are the pelvic floor formed by the pelvic diaphragm and the pelvic organs surrounding the uterus. The cervix is not very mobile because it is held in position by ligaments that are condensations of the pelvic fascia in the pelvic floor that contains smooth muscle (p. 154).

- Transverse cervical ligaments extend from the cervix and lateral parts of the fornix of the vagina to the lateral walls of the pelvis
- Uterosacral ligaments pass superiorly and slightly posteriorly from the sides of the cervix to the middle of the sacrum and are palpable on rectal examination

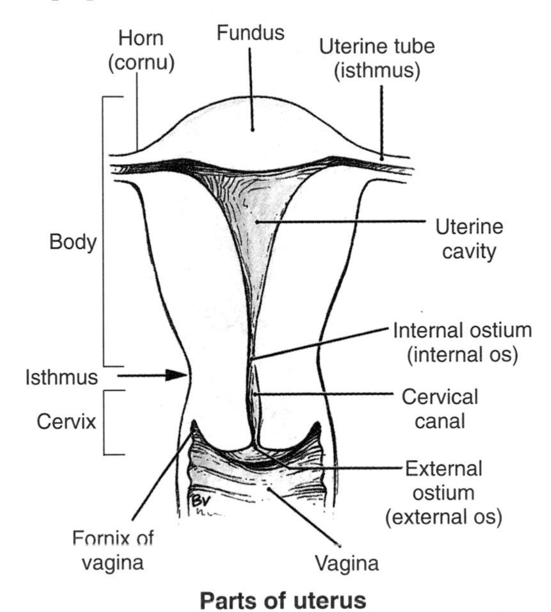

Parts of uterus

The *broad ligaments* are double layers of peritoneum that extend from the sides of the uterus to the lateral walls and floor of the pelvis. These ligaments assist in keeping the uterus in position. The two layers of the broad ligament are continuous with each other at a free edge that surrounds the uterine tube. Laterally the peritoneum of the broad ligament is prolonged superiorly over the vessels as the *suspensory ligament of the ovary*. The ligament of the ovary lies posterosuperiorly, and the *round ligament* of the uterus lies anteroinferiorly between the layers of the broad ligament. The broad ligament gives attachment to the ovary through the *mesovarium*. The part of the broad ligament between the ligament of the ovary, the ovary, and the uterine tube is the *mesosalpinx*. The major part of the broad ligament, the *mesometrium*, is attached to the uterus. Some clinicians use this term to denote the myometrium, the muscular layer of the uterus.

Relationships of Uterus. The *peritoneum* covers the uterus anteriorly and superiorly, except for the vaginal part of the cervix. It is reflected anteriorly from the uterus onto the bladder and posteriorly over the posterior part of the fornix of the vagina onto the rectum (p. 161).

Anteriorly, the uterine body is separated from the urinary bladder by the *vesicouterine pouch,* where the peritoneum is reflected from the uterus onto the posterior margin of the superior surface of the bladder. Posteriorly, the body and supravaginal part of the cervix are separated from the *sigmoid colon* by a layer of peritoneum and the peritoneal cavity and from the rectum by the *rectouterine pouch.* Laterally, the uterine artery crosses the ureter superiorly, near the cervix.

Arterial Supply. The blood supply derives mainly from the uterine arteries with an additional supply from the ovarian arteries (Fig. 4.13*C*).

Venous and Lymphatic Drainage. The uterine veins enter the broad ligaments with the arteries and form a *uterine venous plexus* on each side of the cervix. Veins from the plexus drain into the *internal iliac veins.* Each venous plexus is connected with a superior rectal vein, forming a *portal-systemic anastomosis* (p. 124). The uterine lymphatic vessels follow three main routes (Fig. 4.15*A*): (*a*) most vessels from the fundus pass to the lumbar (lateral aortic) lymph nodes, but some vessels pass to the external iliac lymph nodes or run along the round ligament of the uterus to the superficial inguinal lymph nodes; (*b*) vessels from the body pass within the broad ligament to the external iliac lymph nodes; (*c*) vessels from the cervix pass to the internal iliac and sacral lymph nodes.

Innervation. The nerves arise from the inferior hypogastric plexus, largely through the *uterovaginal plexus.* Parasympathetic fibers are from the *pelvic splanchnic nerves* (S2–S4), and sympathetic fibers derive from the uterovaginal plexus. Most visceral afferents ascend through the hypogastric plexus and enter the spinal cord through the T10–T12 and subcostal nerves (L1) (Fig. 4.15*B*).

> The uterine artery crosses anterior to the ureter near the lateral fornix of the vagina (Fig. 4.13*C*). Hence, the ureter is in danger of being inadvertently clamped or severed when the uterine artery is tied off during a hysterectomy (excision of the uterus). The point of crossing of the artery and the ureter is about 2 cm superior to the ischial spine. The left ureter is particularly vulnerable because it runs close to the lateral aspect of the cervix.

Uterine Tubes

The uterine tubes extend laterally from the horns of the uterine body and open into the peritoneal cavity near the ovaries. For descriptive purposes, each uterine tube is divided into four parts.

- The infundibulum is the funnel-shaped distal end that opens into the peritoneal cavity through the *abdominal ostium* (the fingerlike processes of the infundibulum—*fimbriae*—spread over the medial surface of the ovary; one large *ovarian fimbria* is attached to the superior pole of the ovary)
- The ampulla, the widest and longest part, begins at the medial end of the infundibulum
- The isthmus, the thick-walled part, enters the horn of the uterine body
- The uterine part, the short proximal segment that passes through the wall of the uterus

The uterine tubes lie in the *mesosalpinx* formed by the free edges of the broad ligaments. The tubes extend posterolaterally to the lateral pelvic walls where they ascend and arch over the ovaries.

Arterial Supply. The tubal arteries derive from the uterine and ovarian arteries (Fig. 4.13*C*).

Venous and Lymphatic Drainage. The tubal veins drain into the uterine and ovarian veins (Fig. 4.13*D*). The lymphatic vessels drain to the lumbar lymph nodes (Fig. 4.15*A*).

Innervation. The nerve supply derives partly from the ovarian plexus and partly from the uterine plexus (Fig. 4.15*B*). Afferent fibers are contained in T11, T12, and L1 nerves.

> Because the female genital tract communicates with the peritoneal cavity through the abdominal ostia located deep in the infundibula of the tubes, infections of the vagina, uterus, and tubes may result in peritonitis. Conversely, inflammation of the tube (*salpingitis*) may result from infections that spread from the peritoneal cavity. A major cause of infertility in women is blockage of the uterine tubes, often the result of infection. Patency of a uterine tube may be determined by *hysterosalpingography,* a radiographic procedure that involves injection of a radiopaque material into the uterus and tubes. *Ligation of the uterine tubes* is one method of birth control.

Ovaries

The ovaries are almond-shaped glands located close to the lateral pelvic walls and are attached to the mesovarium of the broad ligament (Fig. 4.13, *A* and *B*). The distal end of the ovary connects to the lateral wall of the pelvis by the *suspensory ligament of the ovary*. This ligament contains the ovarian vessels, lymphatics, and nerves that pass through the mesovarium to the ovary. Each ovary also attaches to the uterus by the *ligament of the ovary* or ovarian ligament that also runs within the mesovarium. It connects the proximal (uterine) end of the ovary to the lateral angle of the uterus, just inferior to the entrance of the tube.

Arterial Supply. The ovarian arteries from the abdominal aorta descend along the posterior abdominal wall. At the pelvic brim, they cross over the external iliac vessels and enter the suspensory ligaments (Fig. 4.13, *A* and *C*). The ovarian artery sends branches through the mesovarium to the ovary and continues medially in the broad ligament to supply the uterine tube and uterus. Both branches anastomose with the uterine artery.

Venous and Lymphatic Drainage. The veins leave the ovary and form a vinelike *pampiniform plexus* of vessels in the broad ligament near the ovary and uterine tube (Fig. 4.13*D*). Each ovarian vein arises from the pampiniform plexus and leaves the pelvis minor with the ovarian artery. The right ovarian vein ascends to the inferior vena cava; the left vein drains into the left renal vein. The lymphatic vessels follow the ovarian blood vessels and join those from the uterine tubes and fundus as they ascend to the *lumbar lymph nodes* (Fig. 4.15*A*).

Innervation. The nerves descend along the ovarian vessels from the *ovarian plexus* that communicates with the uterine plexus (Fig. 4.15*B*). The parasympathetic fibers in the plexus are derived from the vagus nerves. Afferent fibers from the ovary enter the spinal cord through T10 nerve.

The ureter is vulnerable to injury when the ovarian vessels are being tied off (e.g., during an *ovariectomy*) because these structures lie close to each other when they cross the pelvic brim; the ureter is medial to the ovarian vessels. On the right side the vermiform appendix often lies close to the ovary and uterine tube. This close relationship explains why a *ruptured tubal pregnancy* may be misdiagnosed as acute appendicitis. In both cases the parietal peritoneum is inflamed in the same general area, and the pain is referred to the right lower quadrant of the abdomen.

RECTUM

The rectum is continuous proximally with the sigmoid colon and distally with the anal canal (Fig. 4.16*A*). The rectum begins anterior to the level of S3 vertebra, follows the curve of the sacrum and coccyx, and ends anteroinferior to the tip of the coccyx by turning posteroinferiorly and becoming the anal canal. The dilated terminal part of the rectum is the *rectal ampulla*, which supports and holds the fecal mass before it is expelled during defecation. The rectum is S-shaped and has three sharp flexures as it follows the sacrococcygeal curve. Its terminal part bends sharply in a posterior direction (*anorectal flexure*) where it joins the anal canal. At each of three concavities formed by the flexures are infoldings (*transverse rectal folds*) of the mucous and submucous coats and most of the circular muscle layer of the rectal wall.

Peritoneum covers the anterior and lateral surfaces of the superior third of the rectum, only the anterior surface of the middle third, and no surface of the inferior third. In males the peritoneum reflects from the rectum to the posterior wall of the bladder where it forms the floor of the *rectovesical pouch* (p. 161). In females the peritoneum reflects from the rectum to the posterior fornix of the vagina where it forms the floor of the *rectouterine pouch*. In both sexes, lateral reflections of peritoneum from the upper one-third of the rectum form *pararectal fossae*, which permit the rectum to distend as it fills with feces.

The rectum rests posteriorly on the inferior three sacral vertebrae and on the coccyx, anococcygeal ligament, median sacral vessels, and inferior ends of the sympathetic trunks and sacral plexuses. In males the rectum is related anteriorly to the fundus of the urinary bladder, terminal parts of the ureters, ductus deferentes, seminal vesicles, and prostate (Fig. 4.6). The *rectovesical septum* lies between the fundus of the bladder and the ampulla of the rectum and is closely associated with the seminal vesicles and prostate. In females the rectum is related anteriorly to the vagina (Fig. 4.7) and is separated from its posterior fornix and the cervix by the *rectouterine pouch* (p. 161). Inferior to this pouch, the weak rectovaginal septum separates the superior half of the posterior wall of the vagina from the rectum.

Arterial Supply. The superior rectal artery, the continuation of the inferior mesenteric artery, supplies the proximal part of the rectum. The two middle rectal arteries supply the middle and inferior parts of the rectum, and the inferior rectal arteries supply the distal part of the rectum (Fig. 4.16*B*).

Venous and Lymphatic Drainage. Blood from the rectum drains via superior, middle, and inferior rectal veins (Fig. 4.16*B*). Because the superior rectal vein drains into the portal venous system and the middle and inferior rectal veins drain into the systemic system, this communication is an important area of portacaval anastomosis (p. 124). The submucosal rectal venous plexus surrounds the rectum and communicates with the vesical venous plexus in males and the uterovaginal venous plexus in females. The rectal venous plexus consists of two parts, the internal rectal venous plexus just

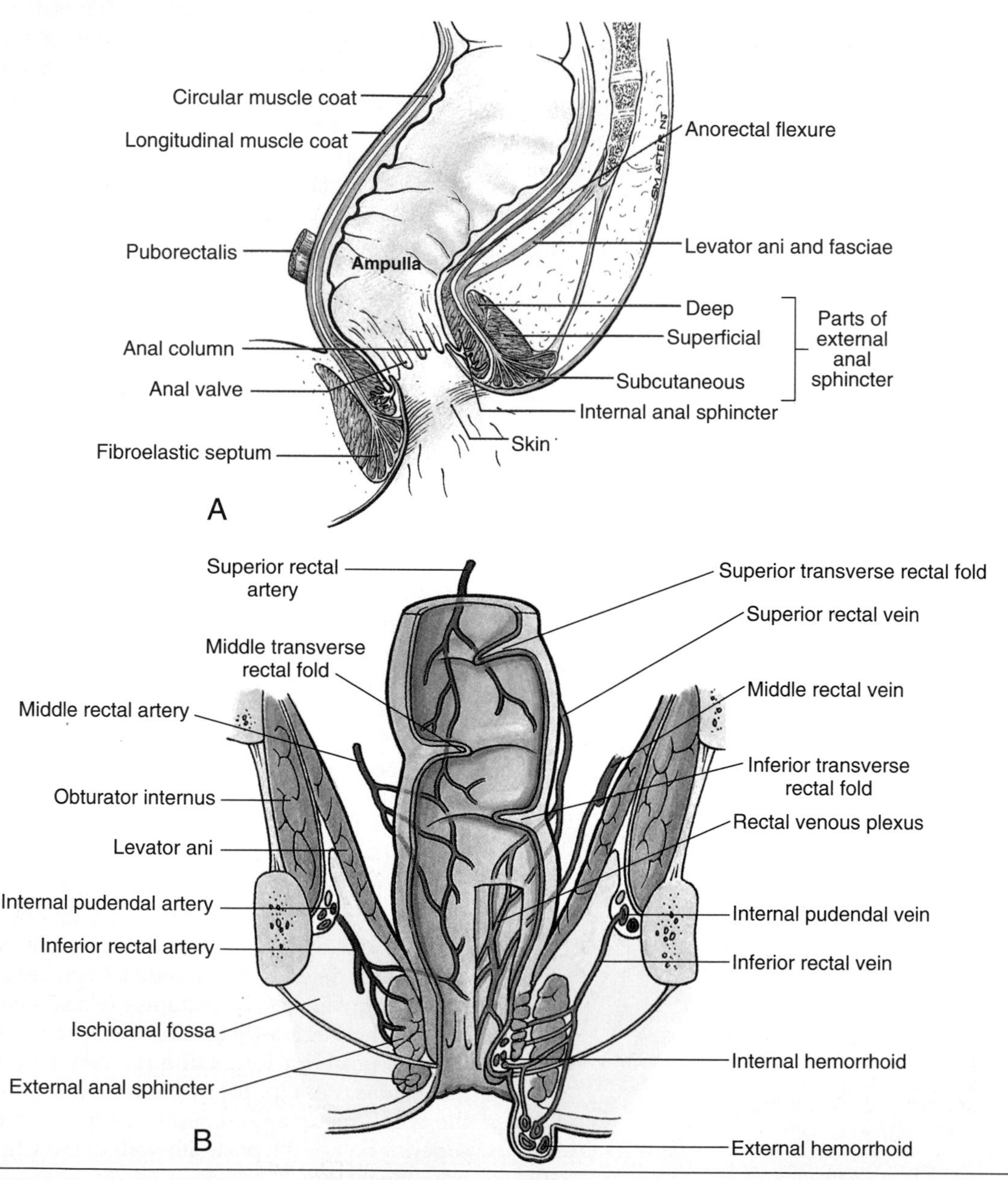

Figure 4.16. Rectum and anal canal. **A.** Median section showing anal sphincters. **B.** Coronal section showing arterial supply (right side) and venous drainage (left side). Note presence of hemorrhoids. **C.** Lymphatic drainage. *Arrows*, lymph flow to lymph nodes. **D.** Innervation.

deep to the epithelium of the rectum and the external rectal venous plexus external to the muscular wall of the rectum. Lymphatic vessels from the *superior half of the rectum* ascend along the superior rectal vessels to the pararectal lymph nodes (Fig. 4.16C); then, they pass to lymph nodes in the inferior part of the mesentery of the sigmoid colon and from them to the inferior mesenteric and lumbar lymph nodes. Lymphatic vessels from the *inferior half of the rectum* pass superiorly with the middle rectal arteries and drain into the *internal iliac lymph nodes.*

Innervation. Nerve supply to the rectum derives from the sympathetic and parasympathetic systems. The rectum derives its sympathetic supply from the lumbar part of the sympathetic trunk and the superior hypogastric

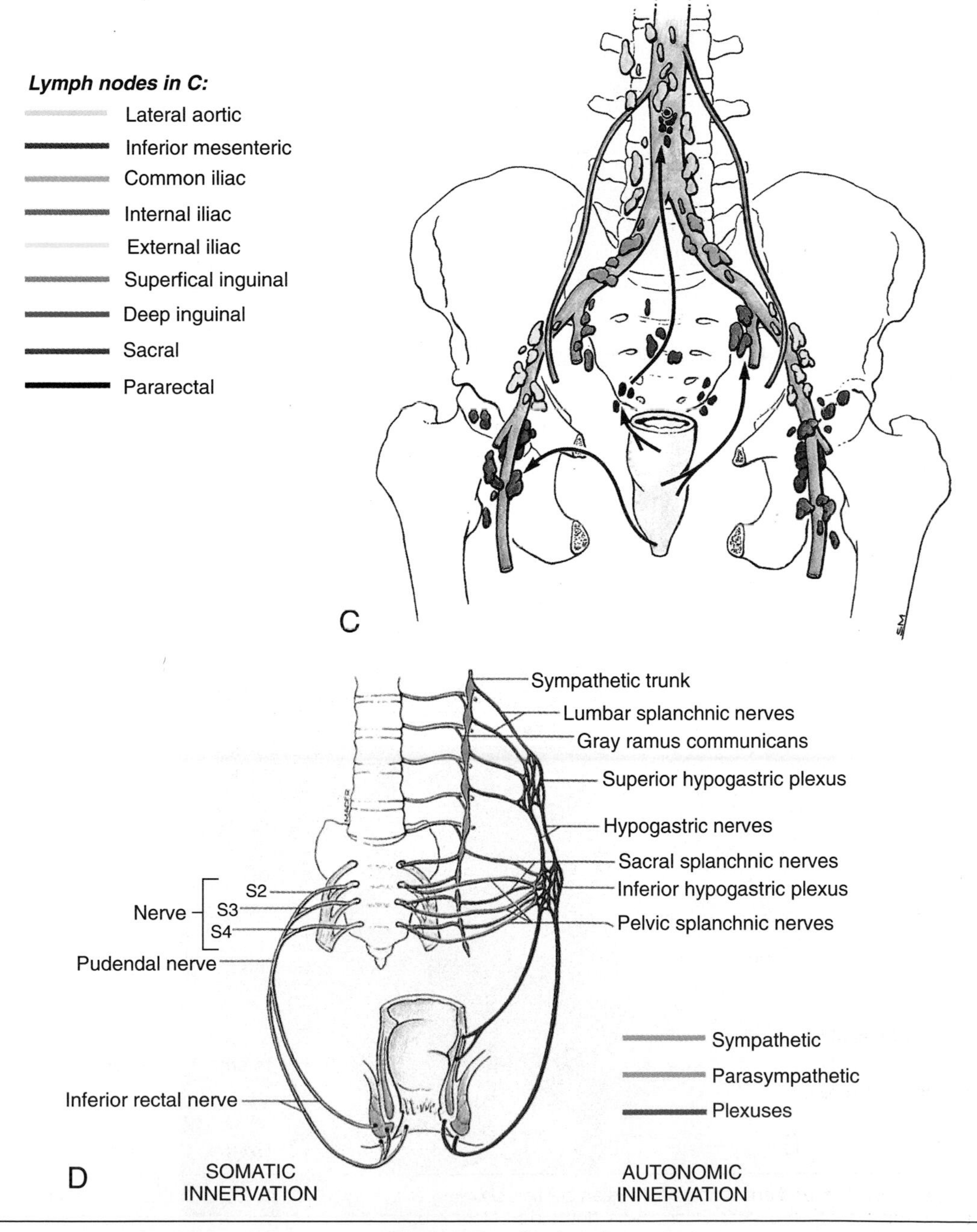

Figure 4.16. *Continued.*

plexus through plexuses on the branches of the inferior mesenteric artery. The parasympathetic supply derives from the pelvic splanchnic nerves. Fibers pass from these nerves to the left and right inferior hypogastric plexuses to supply the rectum. Visceral afferent or sensory fibers also join these plexuses and reach the spinal cord through the pelvic splanchnic or lumbar splanchnic nerves (Fig. 4.16*D*).

When resecting the rectum in males (e.g., in cancer treatment), the plane of the rectovesical septum is located so that the prostate and urethra can be separated from the rectum. In this way these organs are not damaged during excision of the rectum.

Structures related to the posteroinferior part of the rectum may be palpated through its walls (e.g., prostate and seminal vesicles in males and cervix in females). Tenderness of an inflamed vermiform appendix can also be detected rectally if it hangs into the pelvis.

Perineum

The perineum lies inferior to the inferior pelvic aperture and is separated from the pelvic cavity by the pelvic diaphragm (Figs. 4.1 and 4.5). In the anatomical position the perineum is the narrow region between the proximal parts of the thighs; however, when they are spread apart, the perineum is a diamond-shaped area extending from the pubic symphysis anteriorly to the ischial tuberosities laterally and the tip of the coccyx posteriorly (Fig. 4.17). Obstetricians apply the term perineum to a more restricted region—the area between the vagina and anus. Structures that mark the boundaries of the perineum are the

- Pubic symphysis (*S*)
- Inferior pubic rami (*P*)
- Ischial rami (*I*)
- Ischial tuberosities (*T*)
- Sacrotuberous ligaments (*L*)
- Coccyx (*C*)

A transverse line joining the anterior ends of the ischial tuberosities divides the perineum into two unequal triangles. The *anal triangle* (*AT*), containing the anus, is posterior to this line, and the *urogenital triangle* (*UT*), containing the root of the scrotum and penis or the external genitalia in females, is anterior to this line. The midpoint of the line is the central point of the perineum that overlies the *perineal body*, the central attachment for the perineal muscles.

UROGENITAL DIAPHRAGM

The urogenital diaphragm is a thin sheet of striated muscle that stretches between the two sides of the pubic arch and covers the anterior part of the inferior pelvic aperture (Figs. 4.5, 4.6, and 4.17). The *deep transverse perineus muscles*, the most anterior and posterior fibers of the urogenital diaphragm, run transversely,

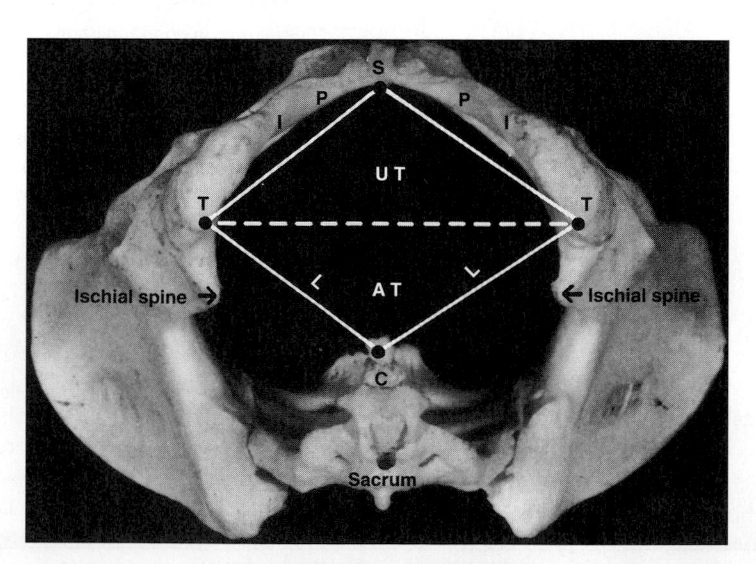

Urogenital and anal triangles

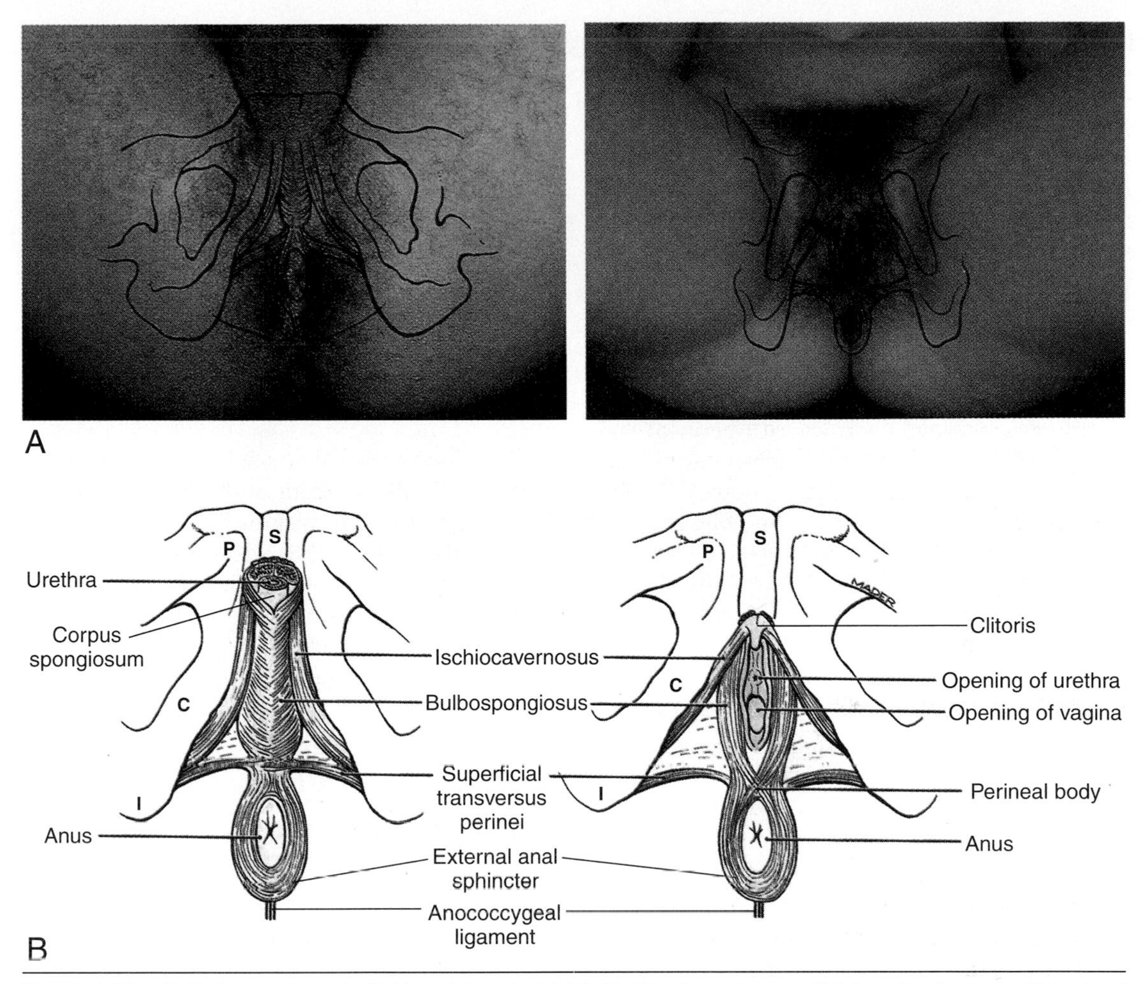

Figure 4.17. **A.** Perineum in male (*left*) and female (*right*). **B.** Structures in superficial perineal space. *P*, body of pubis; *S*, pubic symphysis; *C*, conjoint ramus formed by inferior rami of ischium and pubis; *I*, ischial tuberosity.

whereas the *sphincter urethrae muscle,* its middle fibers, surrounds the male urethra inferior to the prostate. The *perineal body* is the site of convergence of several muscles: the bulbospongiosus, external anal sphincter, and superficial and deep perineus muscles. The perineal body lies anterior to the anus and attaches to the posterior border of the perineal membrane (inferior fascia of urogenital diaphragm).

> The perineal body is an especially important structure in women because it is the final support of the pelvic viscera. Stretching or tearing of this central attachment for the perineal muscles can occur during childbirth, removing support from the inferior part of the posterior wall of the vagina. As a result, *prolapse of the vaginal wall* through the vaginal orifice may occur.
>
> During labor an *episiotomy* (a clean surgical incision) is often made in the perineum to enlarge the vaginal orifice and to prevent a jagged tear.

The *perineal fascia* consists of superficial and deep layers. The *superficial perineal fascia* (subcutaneous connective tissue) consists of a fatty superficial layer and a membranous deep layer (Colles' fascia). The fatty superficial layer is continuous between the scrotum and thighs with the superficial fascia of the abdomen and posteriorly with a similar layer in the anal region. In females the fatty superficial layer continues anteriorly into the labia majora and from there into the mons pubis and the fatty superficial layer of the abdomen. The membranous deep layer is attached posteriorly to the posterior margin of the deep transverse perineus muscle and the perineal body. Laterally it is attached to the ischiopubic ramus where the fascia lata of the thigh also attaches. Anteriorly the deep layer of fascia is continuous with the dartos muscle in the scrotum, but laterally the deep layer becomes continuous with the membranous deep layer of the abdomen. In females the membranous deep layer passes through the labia majora and becomes continuous with the membranous deep layer of subcutaneous fascia in the abdomen.

The *deep perineal fascia* is attached to the posterior margin of the deep transverse perineus muscle. It is also attached laterally to the ischiopubic ramus superior to the attachment of the membranous deep layer of perineal fascia. Anteriorly it is fused to the suspensory ligament of the penis and is continuous with the deep fascia covering the external oblique muscle of the abdomen and the rectus sheath. In females the deep perineal fascia is fused with the suspensory ligament of the clitoris and the deep fascia of the abdomen.

SUPERFICIAL PERINEAL SPACE

The superficial perineal space (pouch) lies between the superficial perineal fascia and the perineal membrane (Figs. 4.17 and 4.18).

Rupture of the spongy urethra results in urine passing into the superficial perineal space. The attachments of the perineal fascia determine the direction of flow of the extravasated urine. Hence, the urine may pass into the loose connective tissue in the scrotum, around the penis, and superiorly into the fatty layer of subcutaneous connective tissue of the lower anterior abdominal wall. Urine cannot pass far into the thighs because the superficial perineal fascia blends with the fascia lata enveloping the thigh muscles, just distal to the inguinal ligament. In addition, urine cannot pass posteriorly into the anal triangle because the superficial and deep layers of perineal fascia are continuous with each other around the superficial perineus muscles. A rupture of a blood vessel into the superficial perineal space from trauma would result in a similar containment of blood in the superficial perineal space.

In males the superficial perineal space contains the

- Root of the penis and muscles associated with it (ischiocavernosus and bulbospongiosus)
- Proximal part of the spongy urethra
- Superficial transverse perineus muscles
- Branches of the internal pudendal vessels
- Pudendal nerves

In females the superficial perineal space contains the

- Root of the clitoris and muscles associated with it (ischiocavernosus and bulbospongiosus)
- Bulbs of the vestibule
- Superficial perineus muscles
- Related vessels and nerves
- Greater vestibular glands (p. 187)

DEEP PERINEAL SPACE

The deep perineal space is the fascial space enclosed by the superior and inferior fasciae of the urogenital diaphragm. The two layers of fascia attach laterally to the pubic arch and blend with each other anteriorly at the apex and posteriorly at the base of the urogenital diaphragm.

In males the deep perineal space contains the

- Membranous urethra
- Sphincter urethrae muscle
- Bulbourethral glands
- Deep transverse perineus muscles
- Related vessels and nerves

In females the deep perineal space contains the

- Proximal part of the urethra
- Sphincter urethrae muscle
- Deep transverse perineus muscles
- Related vessels and nerves

PELVIC DIAPHRAGM

The pelvic diaphragm (consisting of the levator ani and coccygeus and the fascia covering these muscles) divides the pelvic cavity from the perineum. The pelvic diaphragm forms the funnel-shaped floor of the pelvic cavity and the inverted V-shaped roof of each ischioanal fossa.

Ischioanal Fossae

The ischioanal fossae (ischiorectal fossae) at the sides of the anal canal are large fascia-lined, wedge-shaped spaces between the skin of the anal region and the pelvic diaphragm. The apex of each fossa lies superiorly where the levator ani arises from the obturator fascia. The ischioanal fossae, wide inferiorly and narrow

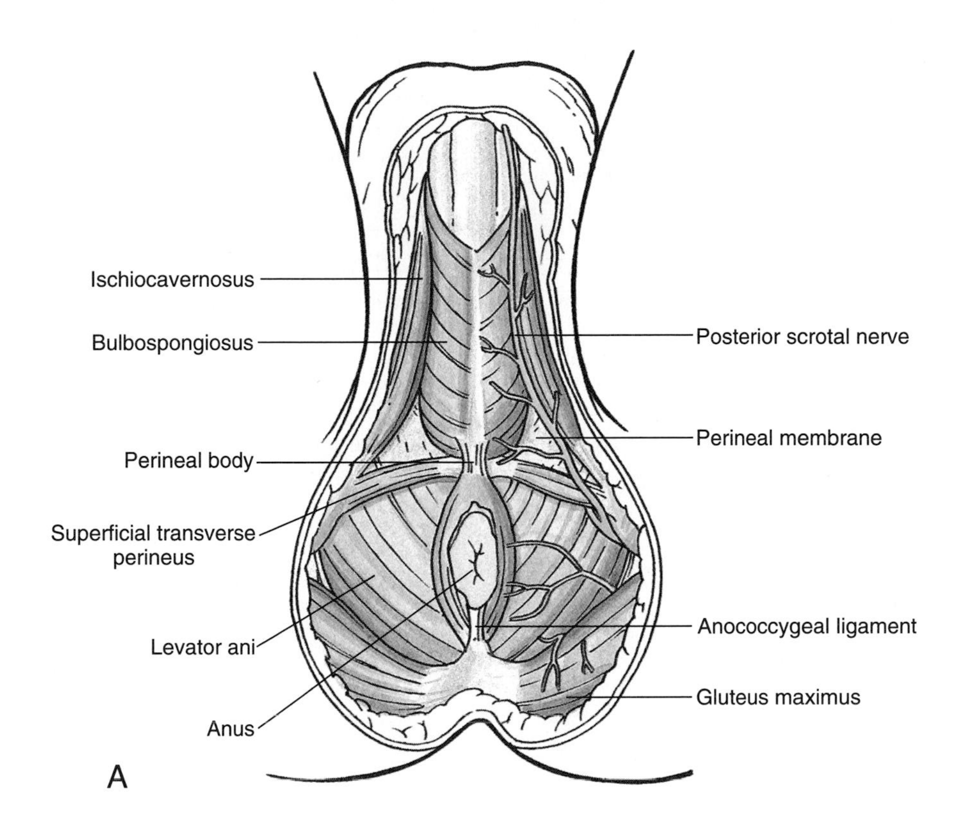

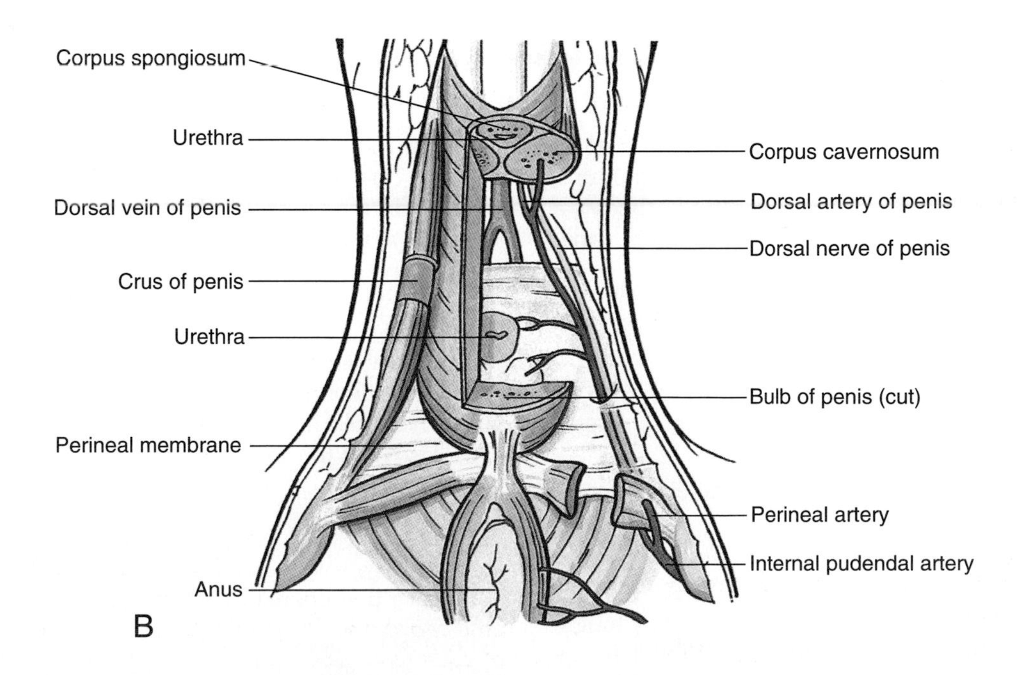

Figure 4.18. Dissections of male perineum. **A.** Superficial dissection. **B.** Deep dissection.

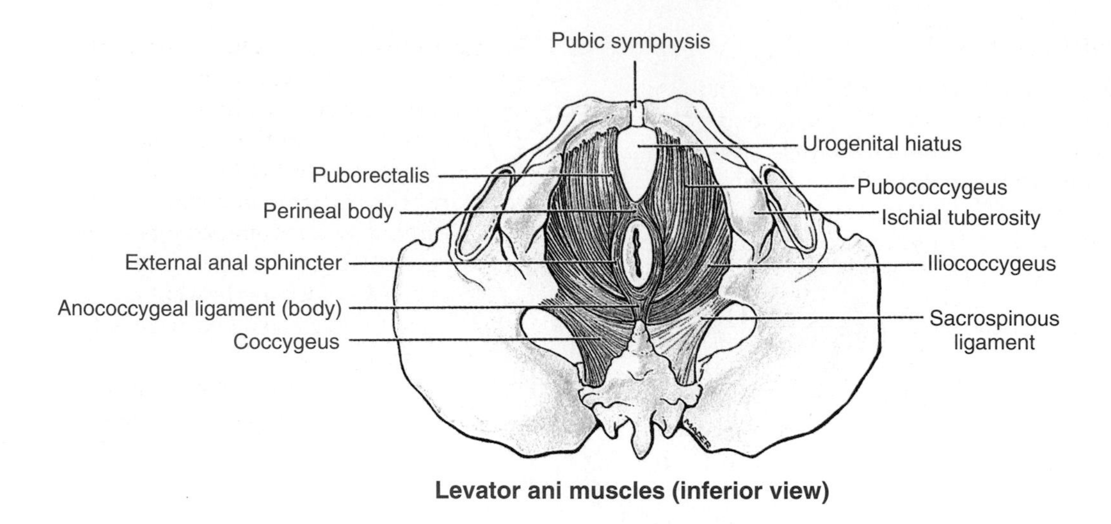

Levator ani muscles (inferior view)

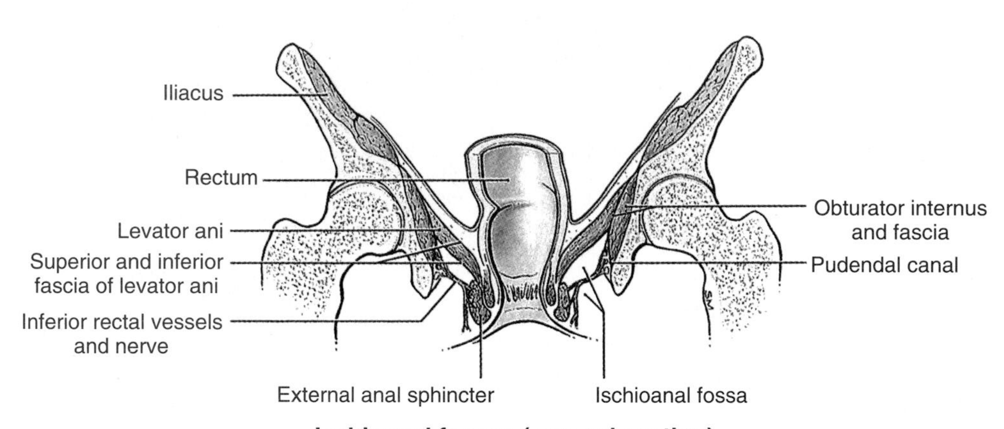

Ischioanal fossae (coronal section)

superiorly, are filled with fat and loose connective tissue. The two ischioanal fossae communicate over the anococcygeal ligament (body), a fibrous mass located between the anal canal and the tip of the coccyx.

Each ischioanal fossa is bounded

- Laterally by the ischium and the inferior part of the obturator internus
- Medially by the anal canal to which the levator ani and external anal sphincter are applied
- Posteriorly by the sacrotuberous ligament and gluteus maximus
- Anteriorly by the base of the urogenital diaphragm and its fasciae

The ischioanal fossae are traversed by tough, fibrous bands and filled with fat that forms *ischioanal pads of fat*. These fat pads support the anal canal, but they can readily displace to permit expansion of the anal canal when feces is present. The lateral walk of the ischioanal fossae contain the *internal pudendal vessels* and the *pudendal nerve*. Posteriorly these vessels and the nerve branch into the inferior rectal vessels and nerves, respectively, which cross the ischioanal fossae and become superficial as they supply the external anal sphincter and the perianal skin. Two other cutaneous nerves, the perforating branch of S2 and S3 and the perineal branch of S4, also pass through the ischioanal fossae.

The ischioanal fossae are occasionally the sites of infection that may result in the formation of *ischioanal abscesses*. These collections of pus are annoying and painful. Diagnostic signs of an ischioanal abscess are fullness and tenderness between the anus and ischial tuberosity. An ischioanal abscess may spontaneously open into the anal canal, rectum, or perianal skin.

Pudendal Canal

The pudendal canal is a fibrous tunnel in the obturator internus fascia (Fig. 4.3C) that lies in the lateral wall of the ischioanal fossa. The pudendal canal begins at the posterior border of the ischioanal fossa and runs from the lesser sciatic notch adjacent to the ischial spine to the posterior edge of the urogenital diaphragm. The internal pudendal artery and vein, the pudendal nerve, and the nerve to the obturator internus enter this canal at the lesser sciatic notch, inferior to the ischial spine.

The *pudendal nerve* supplies most innervation to the perineum. Toward the distal end of the pudendal canal, the pudendal nerve splits to form the *dorsal nerve of the penis or clitoris* and the *perineal nerve*. These nerves run anteriorly on each side of the internal pudendal artery. The *perineal nerve* gives off scrotal or labial branches and continues to supply the muscles of the deep and superficial perineal spaces. The dorsal nerve of the penis or clitoris, a sensory nerve, runs through the deep perineal space to reach its area of supply. The *inferior rectal nerve* arises from the pudendal nerve at the start of the pudendal canal. After leaving the canal it crosses the ischioanal fossa to the anus. It supplies the external anal sphincter and perianal skin and communicates with the posterior scrotal or labial and perineal nerves.

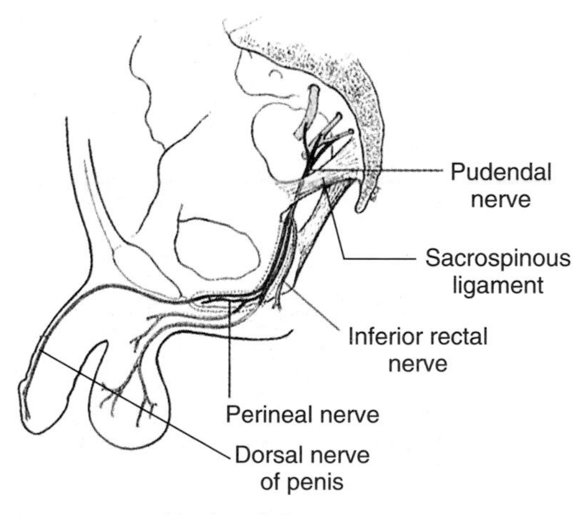

Pudendal nerve

Anal Canal

The anal canal, the terminal part of the large intestine, lies inferior to the pelvic diaphragm and opens on the surface of the perineum as the anus (Figs. 4.16*A* and 4.17). The anal canal (2.5–3.5 cm long) begins where the rectal ampulla narrows at the level of the U-shaped sling formed by the puborectalis muscle (Fig. 4.3*B*). The anal canal ends at the *anus*, the external outlet of the gastrointestinal tract. The anal canal, surrounded by internal and external anal sphincters, descends posteroinferiorly between the anococcygeal ligament and the perineal body. The anal canal is collapsed except during passage of feces. Both sphincters must relax before defecation can occur.

The *external anal sphincter* is a voluntary sphincter that forms a broad band on each side of the inferior two-thirds of the anal canal (Fig. 4.16*A*). This sphincter blends superiorly with the puborectalis muscle. The sphincter is supplied mainly by S4 through the inferior rectal nerve.

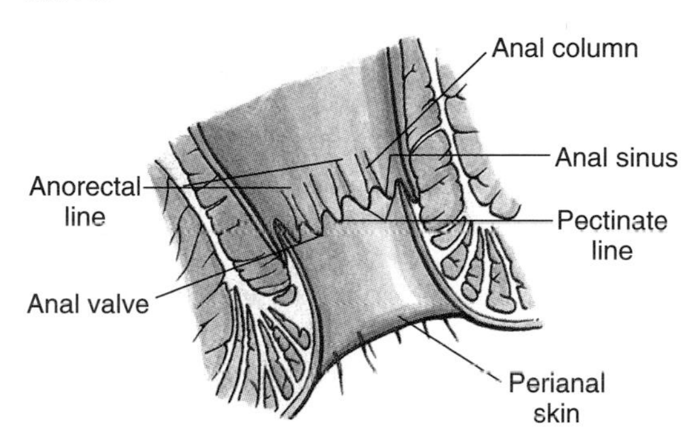

Anal canal

The *internal anal sphincter* is an involuntary sphincter surrounding the superior two-thirds of the anal canal. It forms from a thickening of the circular muscle layer of the intestine (Fig. 4.16*A*) and is innervated by the parasympathetic *pelvic splanchnic nerves*. This sphincter reacts to the pressure of feces in the rectal ampulla.

In the interior of the anal canal, the superior half of the mucous membrane is characterized by a series of longitudinal ridges, the *anal columns*. These columns contain the terminal branches of the superior rectal artery and vein. The *anorectal line*, indicated by the superior ends of the anal columns, is where the rectum joins the anal canal. The inferior ends of these columns are joined by *anal valves*. Superior to the valves are small recesses called *anal sinuses*. When compressed by feces, the anal sinuses exude mucus, which aids in evacuation of feces from the anal canal. The inferior comb-shaped limit of the anal valves forms an irregular line known as the *pectinate line*, which indicates the junction of the superior part of the anal canal (derived from the hindgut) and the inferior part (derived from the proctodeum). The anal canal superior to the pectinate line differs from the part inferior to the pectinate line in its arterial supply, innervation, and venous and lymphatic drainage. These differences result from their different embryological origins [Moore and Persaud (1993) listed under "Suggested Readings"].

Arterial Supply. The superior rectal artery supplies the anal canal superior to the pectinate line. The two inferior rectal arteries supply the inferior part of the anal canal, as well as the surrounding muscles and perianal skin. The middle rectal arteries assist with the blood supply to the anal canal by forming anastomoses with the superior and inferior rectal arteries (Fig. 4.16*B*).

Internal hemorrhoids are varicosities of the tributaries of the superior rectal veins and are covered by mucous membrane (Fig. 4.16*B*). *External hemorrhoids* are varicosities of the tributaries of the inferior rectal veins and are covered by skin. Hemorrhoids that prolapse through the external anal sphincter are often compressed, impeding blood flow. As a result they tend to strangulate and ulcerate.

The anastomoses between the superior, middle, and inferior rectal veins form clinically important communications between the portal and systemic venous systems (p. 124). The superior rectal vein (portal system) drains into the inferior mesenteric vein, whereas the middle and inferior rectal veins drain through the systemic system into the inferior vena cava. Any abnormal increase in pressure in the valveless portal system may cause enlargement of the superior rectal veins, resulting in internal hemorrhoids. In portal hypertension, as in *hepatic cirrhosis*, the anastomotic veins in the anal canal and elsewhere become varicose and may rupture.

Because autonomic nerves supply the anal canal superior to the pectinate line, an incision or a needle insertion in this region is relatively painless. However, the anal canal inferior to the pectinate line is very sensitive (e.g., to the prick of a hypodermic needle) because it is supplied by the *inferior rectal nerves* containing sensory fibers.

Venous and Lymphatic Drainage. The internal rectal venous plexus drains in both directions from the level of the pectinate line. Superior to the pectinate line the internal rectal plexus drains chiefly into the *superior rectal vein*, a tributary of the inferior mesenteric vein and the portal system (Fig. 4.16*B*). Inferior to the pectinate line the internal rectal plexus drains into *inferior rectal veins* around the margin of the external anal sphincter and a tributary of the systemic venous system. The *middle rectal veins*, tributaries of the internal iliac veins, mainly drain the muscularis externa of the ampulla and form anastomoses with the superior and inferior rectal veins. *Superior to the pectinate line* the lymphatic vessels drain into the *internal iliac lymph nodes* and through them into the common iliac and lumbar lymph nodes (Fig. 4.16*C*). *Inferior to the pectinate line* the lymphatic vessels drain into the *superficial inguinal lymph nodes*.

Innervation. Superior to the pectinate line the nerve supply to the anal canal is from the *inferior hypogastric plexus* (Fig. 4.16*D*). The superior part of the anal canal is sensitive only to stretching. The nerve supply of the anal canal inferior to the pectinate line is derived from the *inferior rectal nerves*, branches of the *pudendal nerve*. This part of the anal canal is sensitive to pain, touch, and temperature.

MALE PERINEUM

The male perineum includes the

- Anal canal
- Membranous and spongy parts of the urethra
- Root of the penis and scrotum

Urethra

The *prostatic urethra*, the first part, is described with the pelvis (p. 162). The *membranous urethra* is the shortest, thinnest, and narrowest portion of the urethra. The membranous urethra begins at the apex of the prostate and ends at the bulb of the penis where it joins the spongy urethra (Fig. 4.10). The membranous urethra traverses the deep perineal space where it is surrounded by the sphincter urethrae muscle and the perineal membrane. Posterolateral to the membranous urethra, on each side, is a small *bulbourethral gland* and its slender duct.

The circular investment of the sphincter urethrae muscle makes the membranous urethra the least distensible part. Because of its thin wall, the inferior part of it is vulnerable to penetration by a urethral catheter or to rupture during an accident.

The *spongy urethra*, the longest part, passes through the bulb and corpus spongiosum of the penis and ends at the *external urethral orifice* (Fig. 4.19, *A* and *B*). There are minute openings

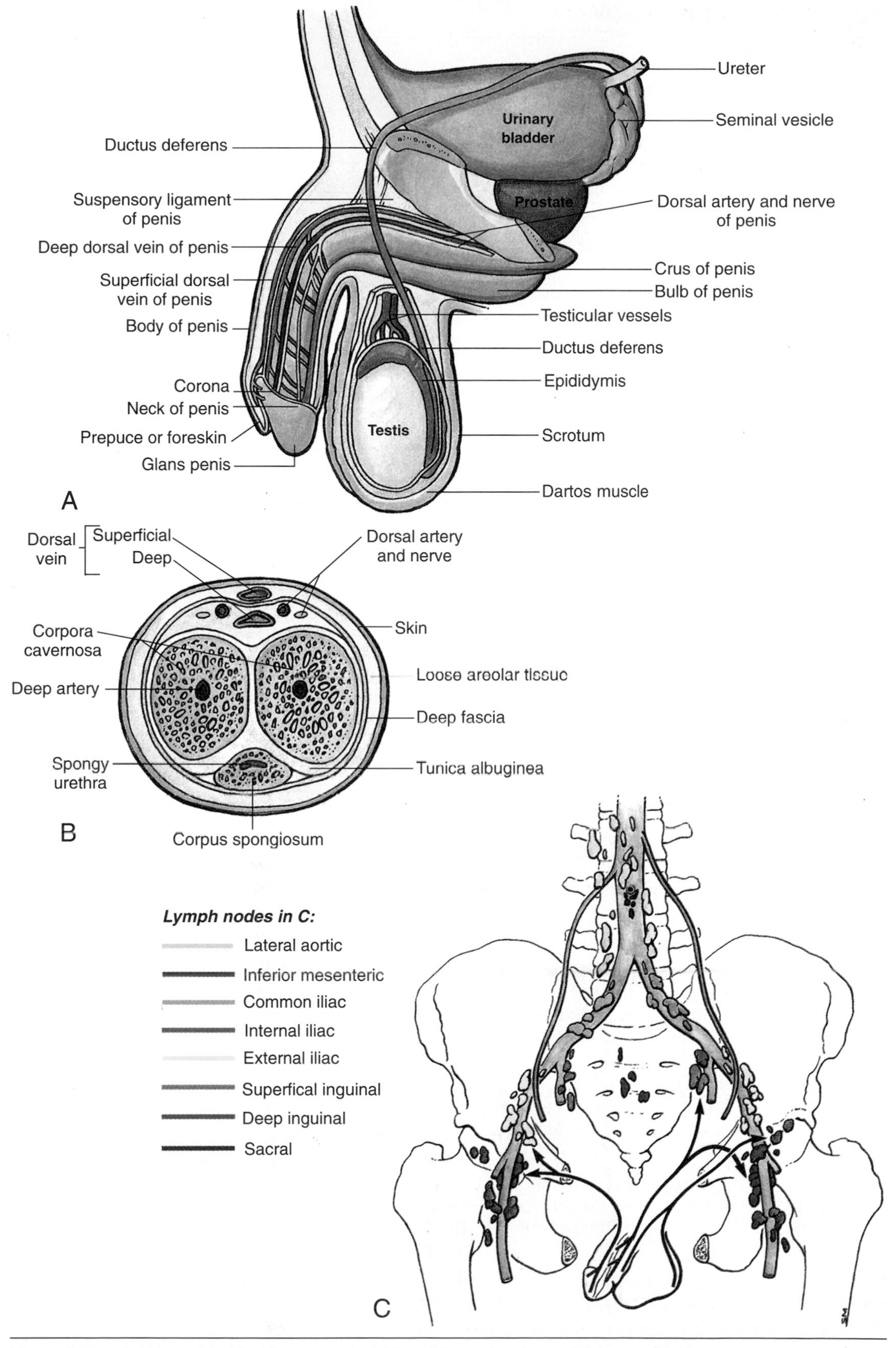

Figure 4.19. **A.** Urinary bladder, prostate, seminal vesicle, and male genital organs. **B.** Transverse section through body of penis. **C.** Lymphatic drainage of penis and scrotum. *Arrows*, lymph flow to lymph nodes.

Table 4.5
Arterial Supply of Perineum

Artery	Origin	Course	Distribution
Internal pudendal	Internal iliac a.	Leaves pelvis through greater sciatic foramen; hooks around ischial spine and enters perineum by way of lesser sciatic foramen and passes to pudendal canal	Perineum and external genital organs
Inferior rectal	Internal pudendal a.	Leaves pudendal canal and crosses ischioanal fossa	Distal portion of anal canal
Perineal	Continuation of internal pudendal a.	Leaves pudendal canal and enters superficial perineal space	Supplies superficial perineal muscles and scrotum
Posterior scrotal or labial	Superficial branch of perineal a.	Runs in superficial fascia of scrotum or labium majus	Skin of scrotum or labium majus
Artery of bulb of penis or vestibule of vagina	Deep branch of perineal a.	Pierces perineal membrane to reach bulb of penis or vestibule of vagina	Supplies bulb of penis or vestibule and bulbourethral gland (male) and greater vestibular gland (female)
Deep artery of penis or clitoris	Deep branch of perineal a.	Pierces perineal membrane to reach penis or clitoris	Supplies erectile tissue of penis or clitoris
Dorsal artery of penis or clitoris	Termination of deep branch of pudendal a.	Pierces perineal membrane and passes through suspensory ligament of penis or clitoris to dorsum of penis or clitoris	Skin of penis and erectile tissue of penis or clitoris

of the ducts of mucus-secreting *urethral glands* into the spongy urethra.

Arterial Supply. The arterial supply of the membranous and spongy parts of the urethra is from branches of the internal pudendal artery (Fig. 4.18*B*, Table 4.5).

Urethral stricture may result from external trauma of the penis or from infection of the urethra. Instruments are used to dilate the urethra (e.g., for insertion of a *cystoscope*). *Urethral catheterization* is done to remove urine from a patient who cannot micturate.

Rupture of the spongy urethra in the bulb of the penis is fairly common in "straddle injuries." The urethra is torn when it is caught between a hard object (e.g., a steel beam) and the person's pubic arch. Urine escapes into the *superficial perineal space* and passes from there inferiorly into the scrotum and superiorly into the fatty layer of subcutaneous connective tissue of the anterior abdominal wall.

Venous and Lymphatic Drainage. Veins accompany the arteries and have similar names. Lymphatic vessels from the membranous part of the urethra drain mainly into the internal iliac lymph nodes (Fig. 4.19*C*). Most lymphatic vessels from the spongy urethra pass to the deep inguinal lymph nodes, but some vessels pass to the external iliac lymph nodes.

Innervation. The nerves are branches of the *pudendal nerve* (p. 181). Most afferent fibers from the urethra run in the *pelvic splanchnic nerves*. Nerves from the *prostatic plexus*, which arise from the inferior hypogastric plexus, are distributed to all parts of the urethra.

Scrotum

The scrotum is a cutaneous fibromuscular sac for the testes and associated structures (Fig. 4.19*A*). It is situated posteroinferior to the penis and inferior to the pubic symphysis. The bilateral embryonic formation of the scrotum is indicated by the midline *scrotal raphe*, which continues on the ventral surface of the penis as the *penile raphe* and posteriorly along the median line of the perineum as the *perineal raphe*. The contents of the scrotum are described with the abdomen (p. 93).

Arterial Supply. The external pudendal arteries supply the anterior aspect of the scrotum, and the internal pudendal arteries supply its posterior aspect. It also receives branches from the testicular and cremasteric arteries.

Venous and Lymphatic Drainage. The scrotal veins accompany the arteries and join the external pudendal veins. Lymphatic vessels from the scrotum drain into the *superficial inguinal lymph nodes.*

Innervation. The anterior part of the scrotum is supplied by the *ilioinguinal nerve* and by the genital branch of the genitofemoral nerve. Its posterior part is supplied by the medial and lateral scrotal branches of the *perineal nerve* and the perineal branch of the *posterior femoral cutaneous nerve.*

Penis

The penis is the male sexual organ and the common outlet for urine and semen (Fig. 4.19, *A* and *B*). It is composed of three cylindrical bodies of erectile cavernous tissue that are enclosed by a fibrous capsule, the *tunica albuginea.* Superficial to this layer is the *deep fascia of the penis* that forms a common covering for the corpus spongiosum and the two corpora cavernosa. The *corpus spongiosum* contains the *spongy urethra.* The corpora cavernosa are fused with each other in the median plane, except posteriorly, where they separate to form two *crura* that are attached on each side to the conjoint rami of the pubis and ischium.

The *root of the penis* consists of the crura, bulb, and ischiocavernosus and bulbospongiosus muscles. The *body of the penis* is the free part that is pendulous in the flaccid condition. Except for fibers of the bulbospongiosus covering the bulb and the ischiocavernosus overlying the crura, the penis has no muscles. It consists of the corpora cavernosa and corpus spongiosum and is covered by skin. Distally the corpus spongiosum penis expands to form the *glans penis.* The margin of the glans, the corona, projects beyond the ends of the corpora cavernosa penis. It overhangs an obliquely grooved constriction, the *neck of the penis,* that separates the glans from the body. The slitlike opening of the spongy urethra, the *external urethral orifice,* is near the tip of the glans. The skin and fasciae of the penis are prolonged as a double layer of skin, the *prepuce,* which covers the glans to a variable extent.

The *suspensory ligament of the penis* is a condensation of superficial fascia that arises from the anterior surface of the pubic symphysis (Fig. 4.19*A*). It passes inferiorly and splits to form a sling that is attached to the deep fascia of the penis at the junction of its fixed and mobile parts.

The *superficial perineal muscles* are the superficial transverse perineus, bulbospongiosus, and ischiocavernosus (Table 4.6). These muscles are in the superficial perineal space, and the perineal nerve supplies all of them.

Arterial Supply. The penis is supplied by branches of the *internal pudendal artery* (Fig. 4.19, *A* and *B*).

- Dorsal arteries run in the interval between the corpora cavernosa on each side of the deep dorsal vein
- Deep arteries pierce the crura and run within the corpora cavernosa
- The artery of the bulb of the penis enters it on each side

The deep arteries and branches of the dorsal artery supply the crura and corpora cavernosa. The artery of the bulb and the dorsal artery supply the bulb and corpus spongiosum. The dorsal artery also supplies the skin and superficial coverings of the penis.

Venous and Lymphatic Drainage. Blood from the cavernous spaces is drained by a venous plexus that joins the *deep dorsal vein* in the deep fascia. This vein passes mostly deep to the arcuate pubic ligament and joins the *prostatic venous plexus.* Blood from the superficial coverings of the penis drain into the *superficial dorsal vein* and then into the superficial external pudendal vein. Some blood passes to the lateral pudendal vein. The lymphatic vessels from most of the penis drain into the superficial inguinal lymph nodes (Fig. 4.19C).

Innervation. The dorsal nerve of the penis is one of the two terminal branches of the *pudendal nerve;* the perineal nerve is the other. The *dorsal nerve of the penis* arises in the pudendal canal and passes anteriorly into the deep perineal space. It then runs to the dorsum of the penis where it passes lateral to the arteries. It supplies both the skin and glans penis. The penis is supplied with sensory nerve endings—especially the glans—thus it is very sensitive. The *cavernous nerves* from the inferior hypogastric plexus pass through the urogenital diaphragm to reach the penis.

Erection. When a male is stimulated erotically, the smooth muscle in the fibrous trabeculae

Table 4.6
Muscles of Perineum

Muscle	Origin	Insertion	Innervation	Action(s)
External anal sphincter	Skin and fascia surrounding anus and coccyx via anococcygeal ligament	Perineal body	Inferior rectal n.	Closes anal canal
Bulbospongiosus	Male: median raphe, ventral surface of bulb of penis, and perineal body	Male: corpus spongiosum and cavernosa and fascia of bulb of penis	Deep branch of perineal n., a branch of pudendal n.	Male: compresses bulb of urethra and assists in erection of penis
	Female: perineal body	Female: fascia of corpus cavernosa		Female: reduces lumen of vagina and assists in erection of clitoris
Ischiocavernosus	Ischial ramus and tuberosity	Crus of penis or clitoris		Maintains erection of penis or clitoris by compression of crura
Superficial transverse perineus	Ischial ramus and tuberosity	Perineal body		Supports perineal body
Deep transverse perineus	Inner aspect of ischiopubic ramus	Median raphe, perineal body, and external anal sphincter		Fixes perineal body
Sphincter urethrae	Inferior pubic ramus	Surrounds urethra; in females some fibers also enclose vagina		Compresses urethra (also compresses vagina in females)

and coiled arteries relaxes because of *parasympathetic stimulation* (S2–S4 nerves). As a result the arteries straighten and their lumina enlarge, allowing blood to flow into the cavernous spaces. Blood engorges and dilates these spaces; the bulbospongiosus and ischiocavernosus muscles compress the venous plexuses at the periphery of the corpora cavernosa and impede the return of venous blood. As a result the corpora cavernosa and the corpus spongiosum become enlarged and rigid, and the penis erects.

Emission. Semen is delivered (sperms and glandular secretions) to the prostatic urethra through the ejaculatory ducts after peristalsis of the ductus deferentes and seminal vesicles. Prostatic fluid is added to the seminal fluid as the smooth muscle in the prostate contracts. Emission is a sympathetic response (L1 and L2 nerves).

Ejaculation. Semen is expelled from the urethra through the external urethral orifice. This process results from

- Closure of the vesical sphincter at the neck of the bladder (sympathetic, L1 and L2 nerves)
- Contraction of urethral muscle, which is a parasympathetic response (S2–S4 nerves)
- Contraction of the bulbospongiosus muscles (pudendal nerves S2–S4)

After ejaculation, the penis gradually returns to its flaccid state. This results from sympathetic stimulation that causes constriction of the smooth muscle in the coiled (helicine) arteries. The bulbospongiosus and ischiocavernosus muscles relax, allowing more blood to flow into the veins. Blood is slowly drained from the cavernous spaces into the deep dorsal vein.

FEMALE PERINEUM

In the female perineum, the vagina pierces the urogenital diaphragm, and the urethra is in the anterior wall of the vagina. The superficial perineal fascia consists of a fatty layer and a membranous deep layer of subcutaneous connective tissue. These layers are continuous over the labia majora. The membranous layer of fascia attaches medially to the pubic symphysis and laterally to the body of the pubis.

The superficial perineal muscles (Fig. 4.17) are the

- Superficial transverse perineus
- Ischiocavernosus
- Bulbospongiosus

The external genital organs known collectively as the *vulva*, or pudendum (Fig. 4.20*A*), consist of the

- Mons pubis
- Labia majora and minora
- Vestibule of the vagina
- Clitoris
- Bulb of the vestibule
- Greater and lesser vestibular glands

Mons Pubis

The mons pubis is the fatty prominence anterior to the pubic symphysis and suprapubic region (Fig. 4.20*A*). The amount of fat increases during puberty and decreases after menopause. After puberty the mons pubis is covered with coarse pubic hairs.

Labia Majora

The labia majora are folds of skin that provide protection for the urethral and vaginal orifices. Each labium majus, largely filled with subcutaneous fat, passes posteriorly from the mons pubis toward the anus. They are situated on both sides of the *pudendal cleft*, the slit between the labia majora.

Labia Minora

The labia minora are folds of smooth hairless skin that lie between the labia majora. They contain a core of spongy connective tissue with many small blood vessels but no fat. Although the internal surface of each labium minus consists of thin moist skin, it has the typical pink color of a mucous membrane and contains many sensory nerve endings.

Vestibule of Vagina

The vestibule of the vagina is the space between the labia minora (Figs. 4.7 and 4.20*A*). The urethra, vagina, and ducts of the greater vestibular glands open into the floor of the vestibule. The *external urethral orifice* is located 2–3 cm posterior to the clitoris and just anterior to the vaginal orifice. On each side of the external urethral orifice are the openings of the ducts of the *paraurethral glands*. The size and appearance of the *vaginal orifice* vary with the condition of the *hymen*, a thin fold of mucous membrane that surrounds the vaginal orifice.

The *greater vestibular glands*, about 0.5 cm in diameter, are on each side of the vestibule, posterolateral to the vaginal orifice. They are round or oval and are partly overlapped posteriorly by the *bulbs of the vestibule*. The slender ducts of these glands pass deep to these bulbs and open into the vestibule of the vagina on each side of the vaginal orifice. These glands secrete mucus into the vestibule of the vagina during sexual arousal. The *lesser vestibular glands* open into the vestibule between the urethral and vaginal orifices. These small glands also secrete mucus into the vestibule to moisten the labia and the vestibule.

Clitoris

The clitoris is an erectile organ located where the labia minora meet anteriorly. The clitoris consists of a *root* and a *body* that is composed of two crura deep to the ischiocavernosus muscles, two corpora cavernosa, and a *glans* (Fig. 4.20, *A* and *B*). The clitoris is not traversed by the urethra. Parts of the labia minora pass anterior to the clitoris and form the *prepuce of the clitoris*, and the parts of the labia minora passing posterior to the clitoris form the *frenulum of the clitoris*. The clitoris enlarges upon tactile stimulation and is a highly sensitive organ that is very important in sexual arousal.

Bulbs of Vestibule

The bulb of the vestibule is split by the vagina so that it appears as two masses of elongated *erectile tissue*, about 3 cm long, that lie along the sides of the vaginal orifice, deep to the bulbospongiosus muscles (Figs. 4.5*A* and 4.20*A*).

Arterial Supply. The arterial supply to the vulva is from two *external pudendal arteries* and an *internal pudendal artery* on each side. The

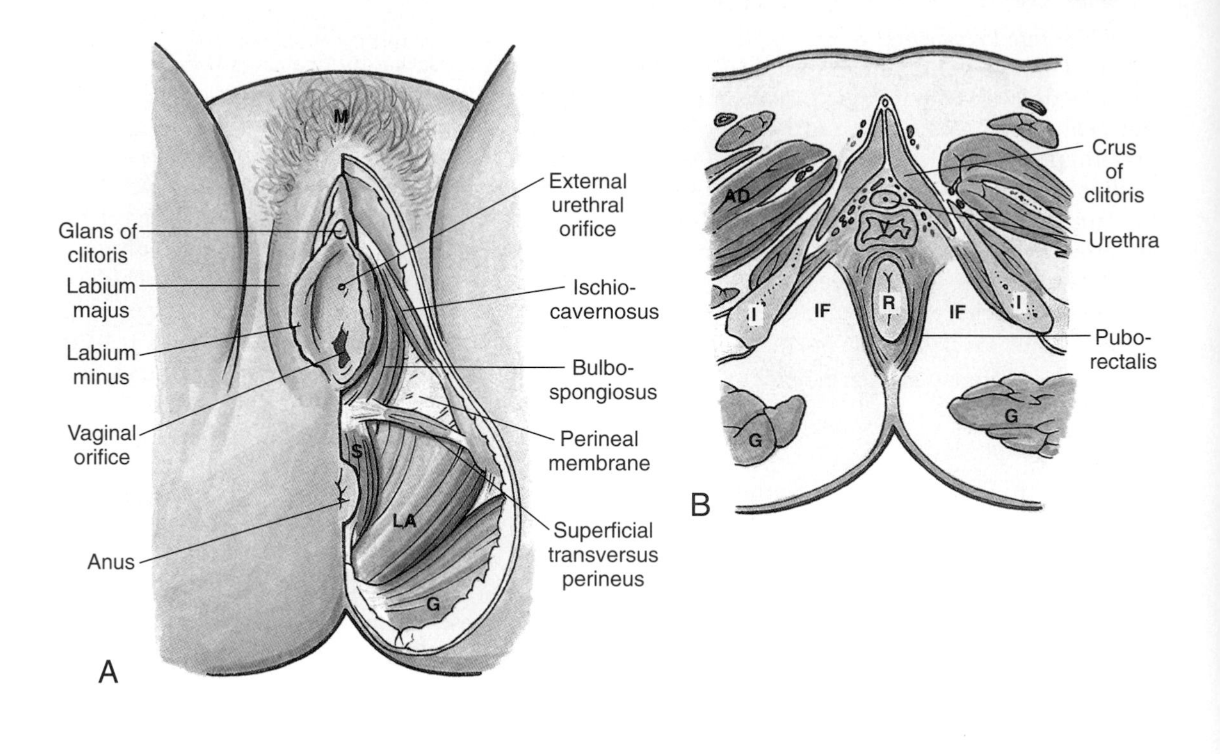

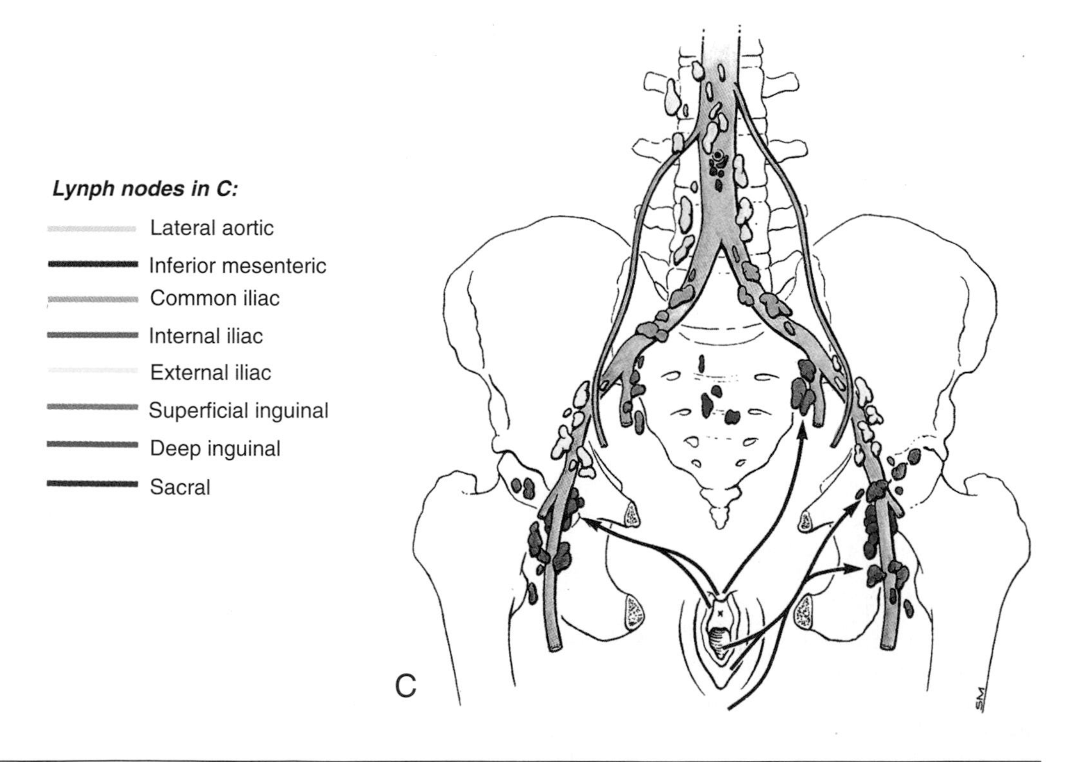

Figure 4.20. **A.** Female perineum. Left side has been dissected to show muscles. *M*, mons pubis; *S*, external anal sphincter; *LA*, levator ani; *G*, gluteus maximus. **B.** Part of a transverse section of female perineum. *AD*, adductor muscles of thigh; *V*, vagina; *I*, ischium; *IF*, ischioanal fossa; *R*, rectum; *G*, gluteus maximus. **C.** Lymphatic drainage of female external genitalia. *Arrows*, lymph flow to lymph nodes.

internal pudendal artery supplies the skin, sex organs, and perineus muscles.

Venous and Lymphatic Drainage. The labial veins are tributaries of the *internal pudendal veins* and venae comitantes of the internal pudendal artery. The vulva contains an exceedingly rich network of lymphatic channels. Lymph vessels pass laterally to the *superficial inguinal lymph nodes* (Fig. 4.20C).

Innervation. Nerves to the vulva are branches of the *ilioinguinal nerve,* the genital branch of the *genitofemoral nerve,* the perineal branch of the *femoral cutaneous nerve of the thigh,* and the *perineal nerve.*

Parasympathetic stimulation produces

- Increased vaginal secretion
- Erection of the clitoris
- Engorgement of erectile tissue in the bulbs of the vestibule

The female urethra is very distensible because it contains considerable elastic tissue as well as smooth muscle. It can easily dilate without injury to it; consequently, the passage of catheters or cystoscopes in females is much easier than it is in males.

The greater vestibular glands are usually not palpable, but they are when infected. *Bartholinitis,* inflammation of the greater vestibular glands (Bartholin's glands), may result from several pathogenic organisms. Infected glands may enlarge to a diameter of 4–5 cm and impinge on the wall of the rectum.

To relieve the pain associated with childbirth, *pudendal block anesthesia* may be performed by injecting a local anesthetic agent into the tissues surrounding the pudendal nerve. The injection may be made where the pudendal nerve crosses the lateral aspect of the sacrospinous ligament, near its attachment to the ischial spine. Although a pudendal nerve block anesthetizes most of the perinuem, it does not abolish sensation from the anterior part of the perineum that is innervated by the ilioinguinal and genitofemoral nerves.

5/ BACK

The back (posterior aspect of neck and trunk) consists of skin, superficial fascia containing a variable amount of fat, deep fascia, muscles, ligaments, vertebral column, ribs (in thoracic region), spinal cord, vessels, and nerves.

Vertebral Column

The vertebral column (spine) extends from the skull to the tip of the coccyx and forms the skeleton of the neck and back and is the main part of the axial skeleton (bones of cranium, vertebral column, ribs, and sternum). The vertebral column protects the spinal cord, supports the weight of the body, and provides a partly rigid and flexible axis for the body and a pivot for the head. Consequently, it has important roles in posture, support of body weight, and locomotion.

The vertebral column consists of 33 vertebrae arranged in five regions, but only 24 of them [7 cervical (*red*), 12 thoracic (*tan*), and 5 lumbar (*yellow*)] are movable in adults. In adults the five sacral vertebrae fuse to form the *sacrum* (*purple*), and the four coccygeal vertebrae fuse to form the *coccyx* (*dark blue*). The vertebral bodies gradually become larger toward the inferior end of the vertebral column and then become progressively smaller toward the end of the coccyx. These structural differences are related to the fact that the lumbar and sacral regions carry more weight than the cervical and thoracic regions. The 24 cervical, thoracic, and lumbar vertebrae articulate at *intervertebral joints* that give the vertebral column considerable flexibility. The vertebral bodies contribute about three-fourths of the length of the vertebral column, and the intervertebral discs (light blue) contribute about one-fourth. The shape and strength of the vertebrae and intervertebral discs, ligaments, and muscles provide stability to the vertebral column.

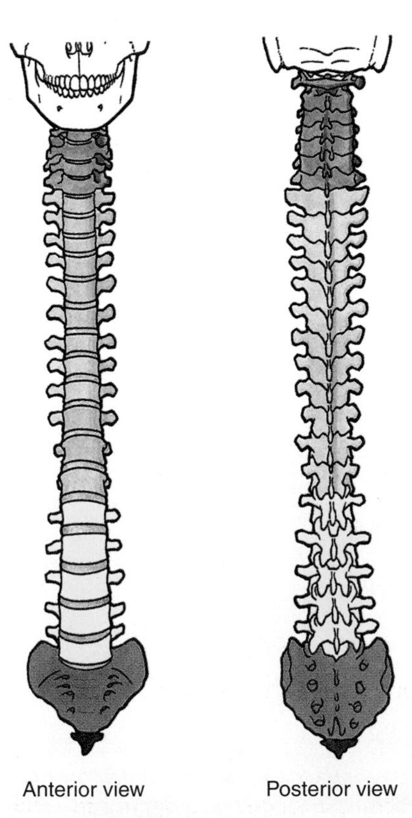

Vertebral column

The number of cervical vertebrae is constant, but numerical variations of thoracic, lumbar, and sacral vertebrae occur in about 5% of otherwise normal people. Although these variations may be clinically important, most of them are detected in diagnostic images (e.g., radiographs) and during dissections and autopsies of persons with no history of back problems. Caution is therefore required in ascribing symptoms (e.g., back pain) to numerical variations of vertebrae. When counting the vertebrae (e.g., to report the site of a fracture), begin at the base of the neck because what may appear to be an extra lumbar vertebra in a radiograph or other image may be an extra thoracic or sacral vertebra.

CURVATURES OF VERTEBRAL COLUMN

There are four curvatures of the vertebral column in adults. The thoracic (*tan*) and sacrococcygeal (*orange*) curvatures are concave anteriorly, whereas the cervical (*pink*) and lumbar (*yellow*) curvatures are concave posteriorly.

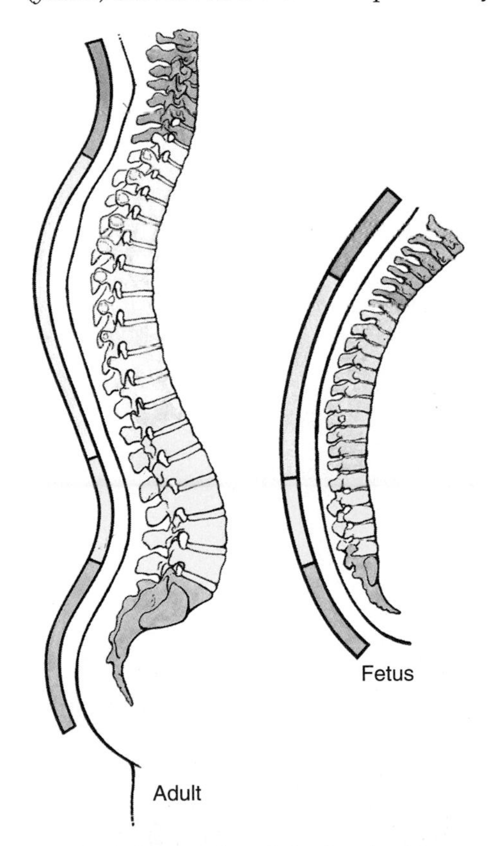

Curvatures of vertebral columns

The thoracic and sacral curvatures are *primary curvatures* that develop during the fetal period. The cervical and lumbar curvatures are *sec-*

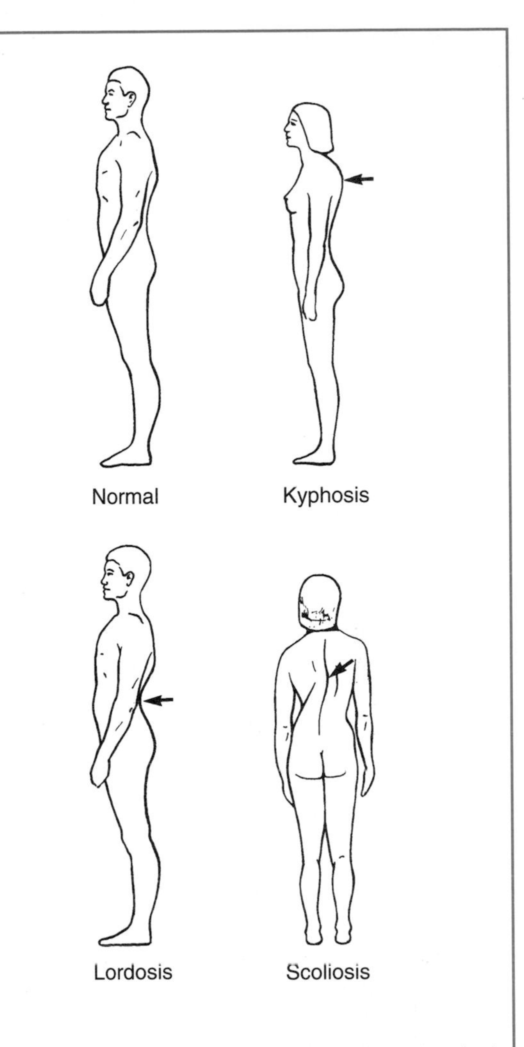

In some people the curvatures of the vertebral column are abnormal because of maldevelopment and/or pathological processes.

Kyphosis is characterized by an abnormal increase in the thoracic curvature. This abnormality results from erosion of the anterior part of one or more vertebrae [e.g., because of demineralization (osteoporosis)]. Progressive erosion and collapse of vertebrae results in an overall loss of height.

Lordosis is characterized by an abnormal increase in the lumbar curvature. It is often associated with weakened trunk musculature, especially of the anterolateral abdominal wall. To restore their line of gravity to the normal position, women develop a temporary lumbar lordosis during late pregnancy.

Scoliosis is characterized by a lateral curvature of the vertebral column, accompanied by rotation of the vertebrae. The spinous processes turn toward the cavity of the abnormal curvature. Asymmetric weakness of the intrinsic back muscles (*myopathic scoliosis*), failure of half of a vertebra to develop (*hemivertebra*), and a difference in the length of the lower limbs are some causes of scoliosis.

ondary curvatures that begin to appear in the cervical and lumbar regions before birth, but they are not obvious until infancy. Secondary curvatures are caused mainly by differences in thickness between the anterior and posterior parts of the intervertebral discs. The *cervical curvature* is accentuated when an infant begins to hold its head erect. The *thoracic curvature* results from the slightly wedge-shaped vertebral bodies. The *lumbar curvature* becomes obvious when an infant begins to walk. This curvature, generally more pronounced in females, ends at the sacrovertebral angle formed at the junction of L5 vertebra with the sacrum. The *sacrococcygeal curvature* is also permanent and differs in males and females; the curve is sharper in females.

STRUCTURE AND FUNCTION OF VERTEBRAE

Vertebrae vary in size and other characteristics from one region to another and to a lesser degree within each region. Typical vertebrae consist of the vertebral body and vertebral arch.

The *vertebral body* is the anterior part that gives strength to the vertebral column and supports body weight. The bodies, especially from T4 inferiorly, become progressively larger in order to bear progressively greater weight.

The *vertebral arch* (neural arch) is the posterior part of a vertebra that is formed by pedicles and laminae. The *pedicles* are short stout processes that join the arch to the vertebral body; *vertebral notches* are indentations of the pedicles. The superior and inferior vertebral notches of adjacent vertebrae form an *intervertebral foramen*. The pedicles project posteriorly to meet two broad, flat plates of bone, the *laminae*. The arch and posterior surface of the body form the walls of the *vertebral foramen*. The succession of vertebral foramina in the articulated column forms the *vertebral canal* (spinal canal), which contains the spinal cord, its meninges (protective membranes), fat, nerve roots, and vessels.

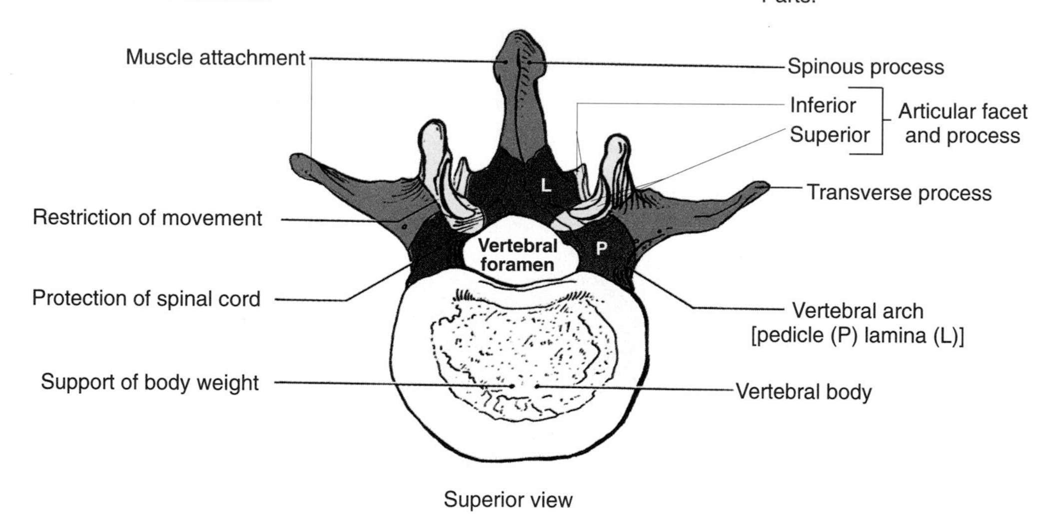

Superior view

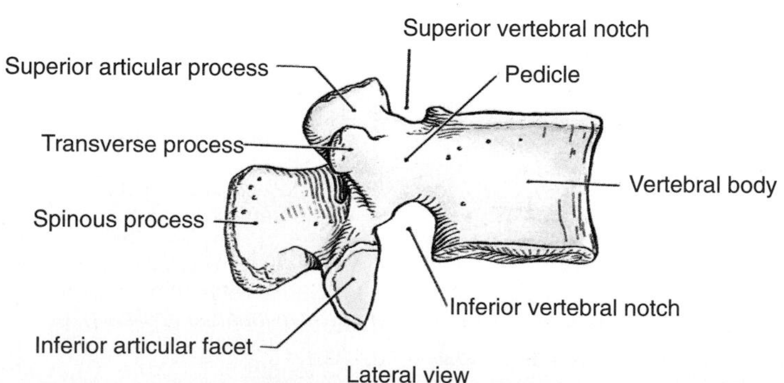

Lateral view

Typical lumbar vertebra

Seven processes arise from the vertebral arch. The

- Spinous process (spine) projects from the junction of the laminae and overlaps the vertebra below
- Two transverse processes project posterolaterally from the junctions of the pedicles and laminae
- Superior and inferior articular processes also arise from the junctions of the pedicles and laminae

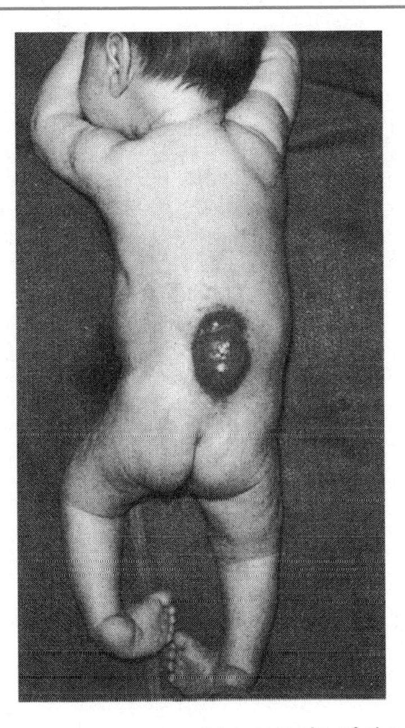

The common congenital anomaly of the vertebral column is *spina bifida occulta*, in which the laminae of L5 and/or S1 fail to develop normally and fuse. This bony defect, usually in the vertebral arch of L5 and/or S1, is concealed by skin, but its location is often indicated by a tuft of hair. In more severe types of spina bifida, *spina bifida cystica*, there may be almost complete failure of the vertebral arch(es) to develop [see Moore and Persaud (1993) listed under "Suggested Readings"]. Spina bifida cystica is associated with herniation of the meninges (meningocele) and/or the spinal cord (meningomyelocele). Usually neurological symptoms are present in these cases (e.g., in the lumbar region paralysis of the limbs and disturbances in bladder and bowel control may be present).

REGIONAL CHARACTERISTICS OF VERTEBRAE

Vertebrae in different regions of the vertebral column show some modification from typical vertebrae. The main regional characteristics of vertebrae are summarized in Tables 5.1–5.4. The direction of the facets on the articular processes determines the nature of the movement between adjacent vertebrae.

JOINTS OF VERTEBRAL COLUMN

The joints of the vertebral column include the joints of the vertebral bodies, joints of the vertebral arches, craniovertebral joints, costovertebral joints (Chapter 2), and sacroiliac joints (p. 146).

Joints of Vertebral Bodies

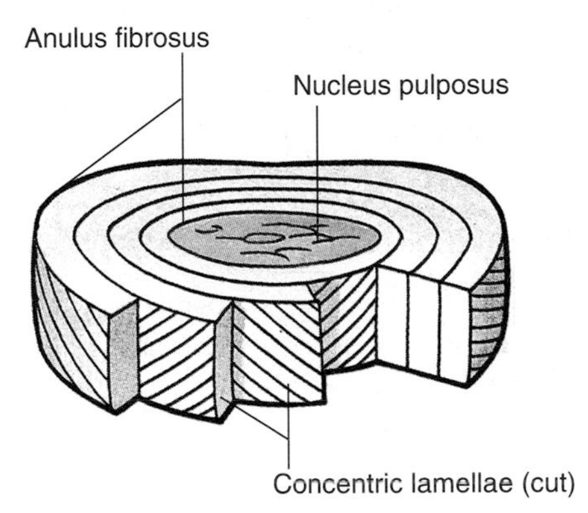

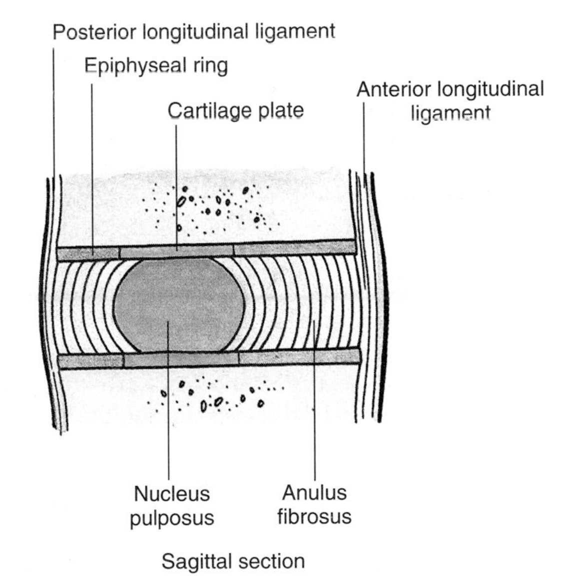

Intervertebral discs and ligaments

The joints of the vertebral bodies are secondary cartilaginous joints (symphyses) designed for weightbearing and strength. The articulating surfaces of adjacent vertebrae are connected by a disc and ligaments. Each *intervertebral disc* consists of an *anulus fibrosus* com-

Surface Anatomy of Vertebrae

Neck

Spinous processes of lumbar vertebrae

Location of sacrum

The tips of the spinous processes of some cervical and all thoracic and lumbar vertebrae are palpable and often visible when the vertebral column is flexed. The spinous processes of C3–C5 are not palpable because they are short and covered by a thick ligament. There is a slight depression posterior to the arch of the atlas (C1 vertebra) because it has no spinous process. C6, C7 (vertebra prominens), and T1 are easily palpable when the neck is fully flexed. The inferior angles of the scapulae lie at the level of the tip of the spinous process of T7 vertebra. In the anatomical position, there is a furrow over the spinous processes of the lower thoracic and lumbar vertebrae.

The spinous processes of lumbar vertebrae are large and easy to observe and palpate. A horizontal line joining the highest points of the iliac crests passes through the tip of the L4 spinous process and the L4/L5 disc. This is a useful landmark when a lumbar puncture is performed (p. 213).

The transverse processes of C1, C6, and C7 vertebrae are palpable. The tubercles of C1 can be palpated by deep pressure posteroinferior to the tips of the mastoid processes of the temporal bones (bony prominences posterior to the ears). The transverse processes of other vertebrae are covered with thick muscles and are palpated with more difficulty. (*Large arrow*, C7 spinous process; *small arrow*, displaced T2 spinous process.)

posed of concentric lamellae of fibrocartilage, which surrounds a gelatinous *nucleus pulposus*. The anuli insert into the smooth, rounded rims (epiphyseal rings) on the articular surfaces of the vertebral bodies. There is no disc between C1 (atlas) and C2 (axis). The most inferior functional disc is between L5 and S1. The discs vary in thickness in different regions; they are thickest in the lumbar region and thinnest in the superior thoracic region. The discs are thicker anteriorly in the cervical and lumbar regions and more uniform in thickness in the thoracic region.

The *uncovertebral joints* (of Luschka) are between the uncinate (uncal) processes of C3–C6 vertebrae and the bevelled surfaces of the vertebral bodies superior to them. These synovial joints (fissures) are at the lateral and posterolateral margins of the discs.

The *anterior longitudinal ligament* is a strong, broad fibrous band that covers and connects the anterior aspects of the vertebral bodies and intervertebral discs. It extends from the pelvic surface of the sacrum to the anterior tubercle of C1 (atlas) and the occipital bone anterior to the foramen magnum. This ligament maintains stability of the joints between the vertebral bodies and helps prevent hyperextension of the vertebral column.

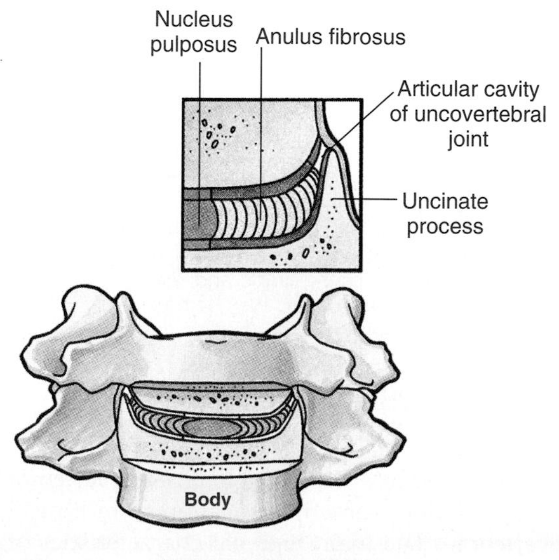

Uncovertebral joint

Table 5.1
Cervical Vertebrae

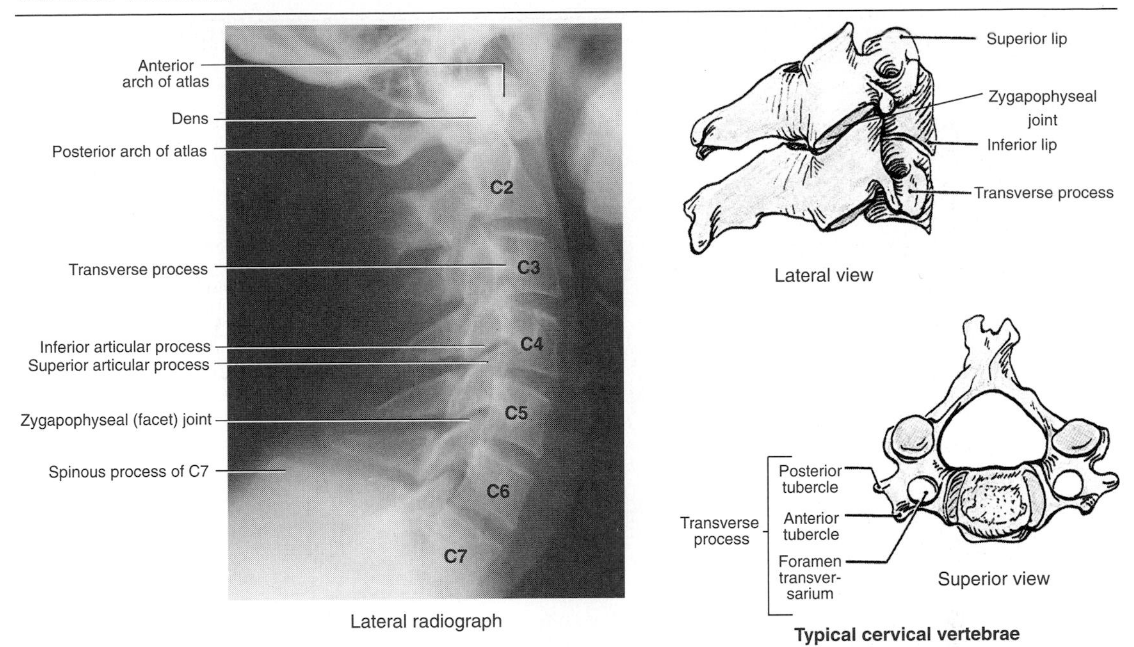

Lateral radiograph

Typical cervical vertebrae

Cervical vertebrae (C1–C7) form skeleton of neck. They are typical vertebrae except for C1 and C2.

Part	Distinctive Characteristics
Body	Small and longer from side to side than anteroposteriorly; superior suface is concave and inferior surface is convex
Vertebral foramen	Large and triangular
Transverse processes	Transverse foramina (foramina transversaria); small or absent in C7; vertebral arteries and accompanying venous and sympathetic plexuses pass through foramina, except C7, which transmits only small accessory vertebral veins; anterior and posterior tubercles
Articular processes	Superior facets directed superoposteriorly; inferior facets directed inferoanteriorly
Spinous process	Short (C3–C5) and bifid (C3–C5); process of C6 is long but that of C7 is longer (because of this C7 is called vertebra prominens)

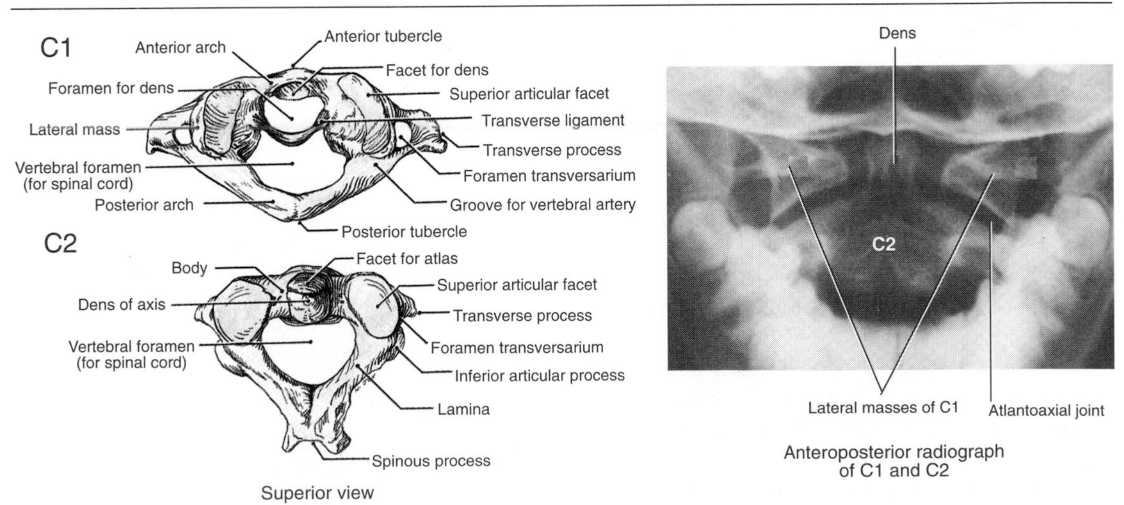

Superior view

Anteroposterior radiograph of C1 and C2

C1 and C2 vertebrae are atypical. The ringlike C1 vertebra, the atlas, is kidney-shaped. Its concave superior articular facets receive the occipital condyles. C1 has no spinous process or body and consists of two lateral masses connected by anterior and posterior arches. C1 carries the skull and rotates on C2's large flat superior articular facets. C2 vertebra, the axis, is the strongest cervical vertebra. Its distinguishing feature is the dens, which projects superiorly from its body.

Table 5.2
Thoracic Vertebrae

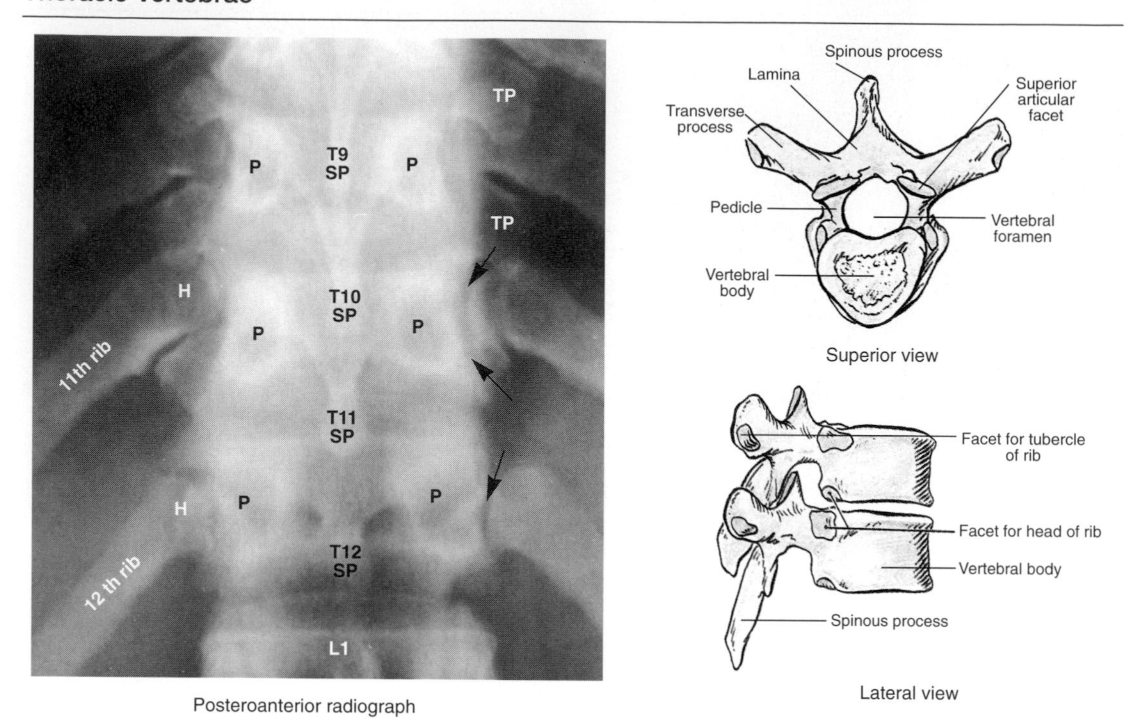

Posteroanterior radiograph

Thoracic vertebrae (T1–T12) form posterior part of skeleton of thorax and articulate with ribs. Space between vertebral bodies is site of intervertebral disc. *P*, pedicle; *arrows*, costovertebral joints.

Part	Distinctive Characteristics
Body	Heart-shaped; has one or two facets for articulation with head of a rib (*H*)
Vertebral foramen	Circular and smaller than in cervical and lumbar regions
Transverse process (*TP*)	Long and strong and extends posterolaterally; length diminishes from T1–T12 (T1–T10 have facets for articulation with tubercle of a rib)
Articular processes	Superior facets directed posteriorly; inferior facets directed anteriorly
Spinous process (*SP*)	Long and slopes posteroinferiorly; tip extends to level of vertebral body below

Table 5.3
Lumbar Vertebrae

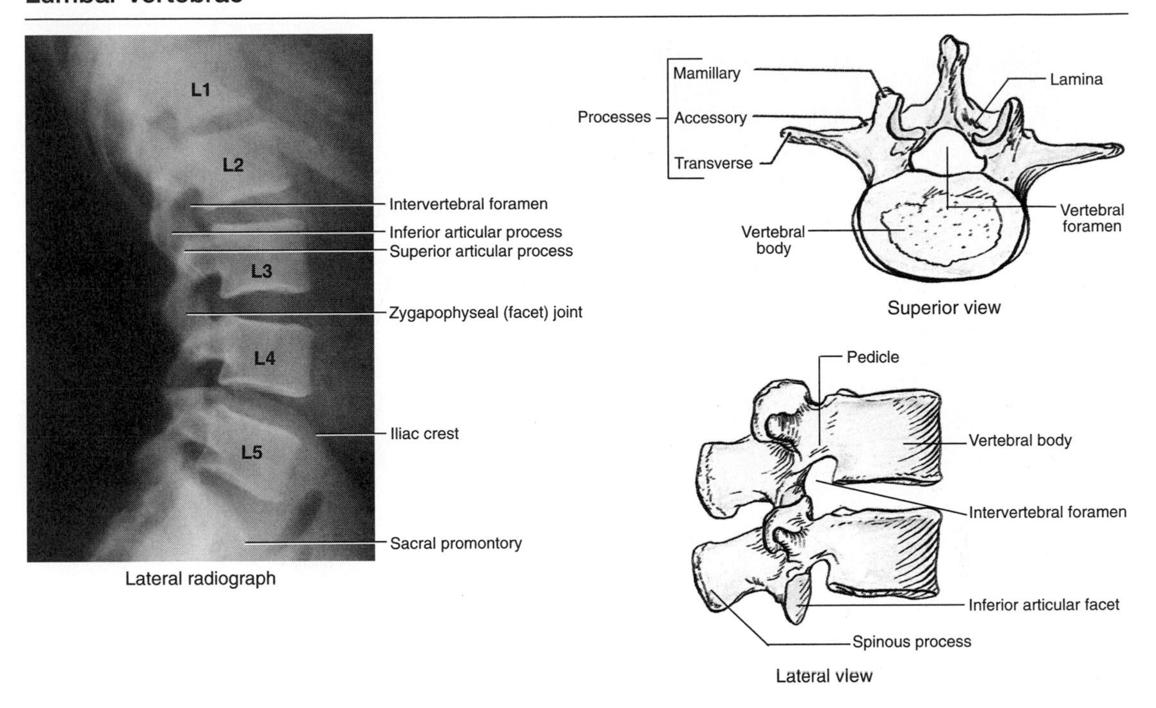

Lateral radiograph

Superior view

Lateral view

Lumbar vertebrae (L1–L5) are larger and heavier than in other regions. Space between vertebral bodies is site of intervertebral disc.

Part	Distinctive Characteristics
Body	Massive; kidney-shaped when viewed superiorly
Vertebral foramen	Triangular; larger than in thoracic region and smaller than in cervical region
Transverse processes	Long and slender; accessory process on posterior surface of base of each process
Articular processes	Superior facets directed posteromedially (or medially); inferior facets directed anterolaterally (or laterally); mamillary process on posterior surface of each superior articular process
Spinous process	Short and sturdy

Table 5.4
Sacrum and Coccyx

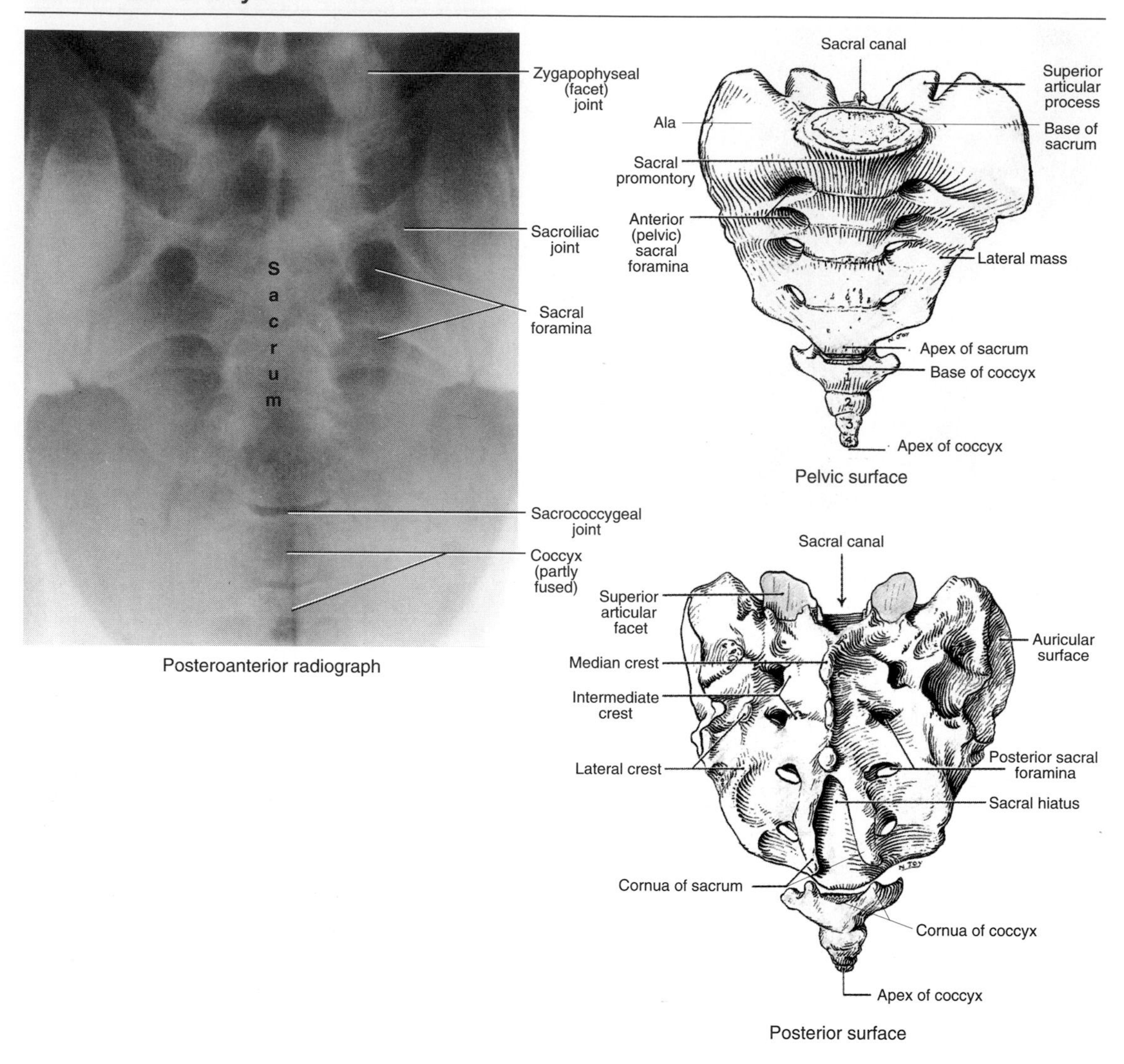

Posteroanterior radiograph

The large, wedge-shaped *sacrum* in adults is composed of five fused sacral vertebrae. The sacrum provides strength and stability to the pelvis and transmits body weight to the pelvic girdle through the sacroiliac joints. The base of the sacrum is formed by the superior surface of S1 vertebra. Its superior articular processes articulate with the inferior articular processes of L5 vertebra. The projecting anterior edge of the body of the first sacral vertebra is the sacral promontory.

On the pelvic and dorsal surfaces are four pairs of sacral foramina for the exit of the rami of the first four sacral nerves and the accompanying vessels. The pelvic surface of the sacrum is smooth and concave. The four transverse lines indicate where fusion of the sacral vertebrae occurred. The posterior surface of the sacrum is rough and convex. The fused spinous processes form the median sacral crest. The inverted U-shaped *sacral hiatus* results from the absence of the laminae and spinous processes of S4 and S5 vertebrae. The hiatus leads into the sacral canal, the inferior end of the vertebral canal. The sacral cornua (L. horns), representing the inferior articular processes of S5 vertebra, project inferiorly on each side of the sacral hiatus and are a helpful guide to its location. The lateral surface of the sacrum has an ear-shaped articular surface that participates in the sacroiliac joint.

The vertebrae of the tapering *coccyx* are remnants of the skeleton of the embryonic tail. These vertebrae are reduced in size and have no pedicles, laminae, or spinous processes. The distal three vertebrae fuse during middle life to form the coccyx, a beaklike bone that articulates with the sacrum.

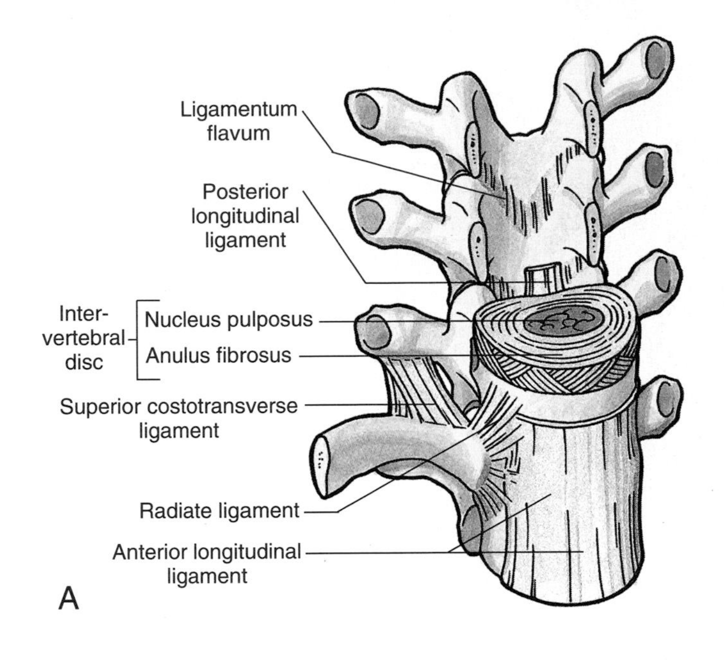

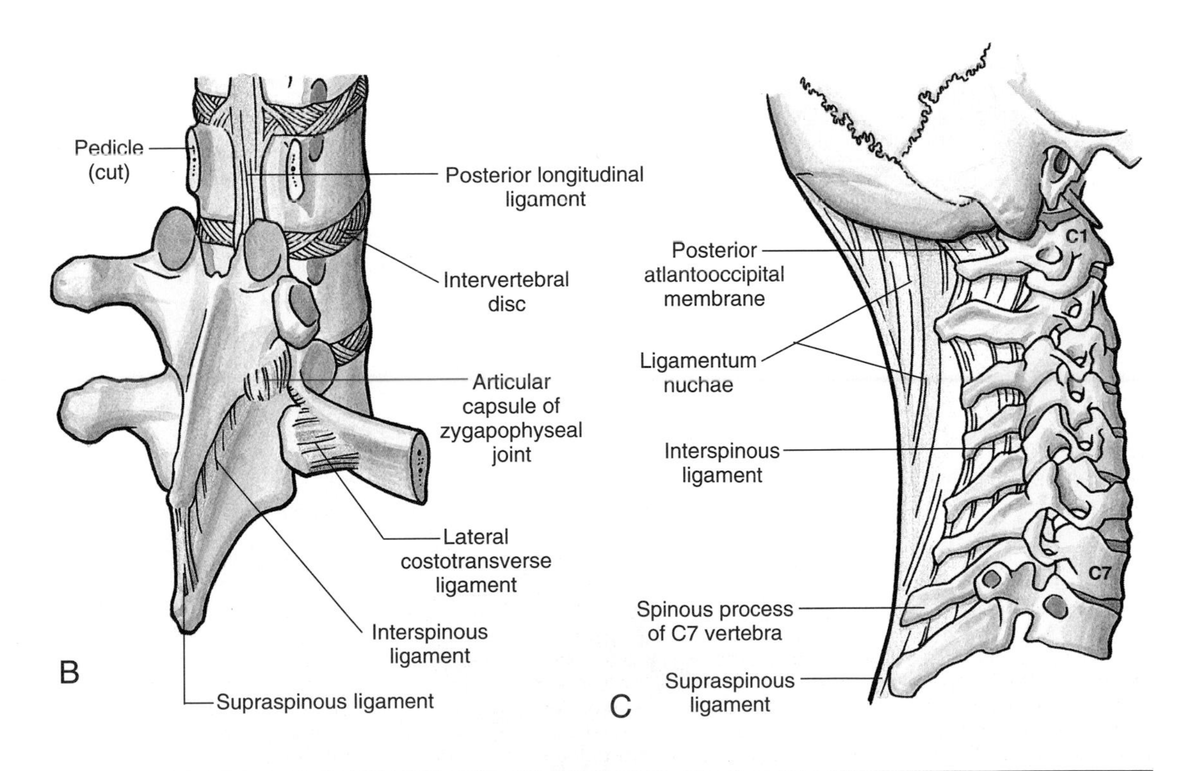

Figure 5.1. Joints and ligaments of vertebral column. **A.** Anterior view. Pedicles of upper vertebrae have been sawn through, and their bodies have been removed. A rib and its costovertebral joint and associated ligaments are also illustrated. **B.** Dorsolateral view of two articulated thoracic vertebrae and their associated ligaments. Vertebral arch of upper vertebra has been removed. **C.** Lateral view of ligaments of cervical region. Note prominent spinous process of C7 (vertebra prominens).

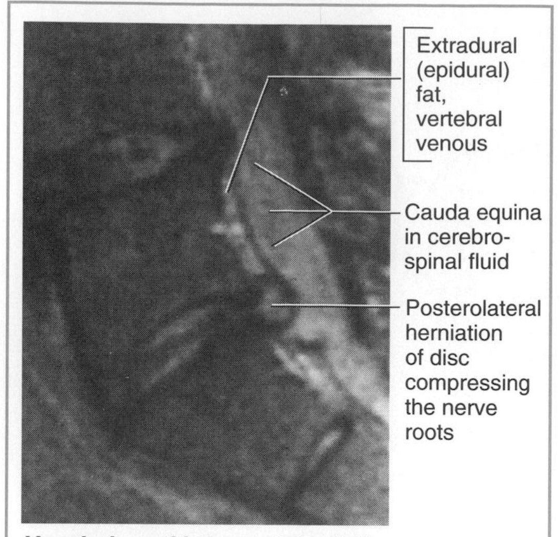

Herniation of intervertebral disc

With increasing age, the nuclei pulposi lose their resiliency and become thin because of dehydration and degeneration. The anuli also undergo degenerative changes, apparently from constant wear and tear. Consequently an anulus, usually lumbar, may bulge or rupture. If the tear is complete, part of the nucleus pulposus may herniate through the tear and compress a spinal nerve. *Disc protrusions* (herniations or "slipped discs") may occur in any direction, but posterolateral is the most common where the anulus is thin and the posterior longitudinal ligament is weak. Because spinal nerves pass over the posterolateral part of the disc, the protruding nucleus pulposus may compress a spinal nerve, causing low back pain and/or leg pain.

As degenerative changes occur, the cervical discs thin out, and the uncinate processes approach the bevelled inferior surfaces of the cervical vertebrae superiorly. This results in encroachment of the intervertebral foramina, pressure on the nerve roots, and neck pain.

The *posterior longitudinal ligament* is a narrower, somewhat weaker band than the anterior longitudinal ligament. It runs within the vertebral canal along the posterior aspect of the vertebral bodies. It is attached to the intervertebral discs and the posterior edges of the vertebral bodies from C2 (axis) to the sacrum. This ligament helps prevent hyperflexion of the vertebral column and posterior protrusion of the discs.

The zygapophyseal joints are of clinical interest because they are close to the intervertebral foramina through which the spinal nerves emerge from the vertebral canal. When these joints are injured or diseased, the spinal nerves may be affected. This disturbance causes pain along the distribution patterns of the dermatomes and spasm in the muscles derived from the associated myotomes. *Denervation of lumbar zygapophyseal joints* is one procedure used for treatment of back pain caused by disease of these joints [see Moore (1992) listed under "Suggested Readings"].

Joints of Vertebral Arches

The joints of the vertebral arches are the *zygapophyseal joints* (facet joints). These articulations are plane synovial joints between the articular processes (zygapophyses) of adjacent vertebrae. Each joint is surrounded by a thin, loose articular capsule (Fig. 5.1*B*), especially in the cervical region. The capsule is attached to the margins of the articular processes of adjacent vertebrae. Accessory ligaments unite the laminae, transverse processes, and spinous processes and help to stabilize the joints. The zygapophyseal joints permit gliding move-

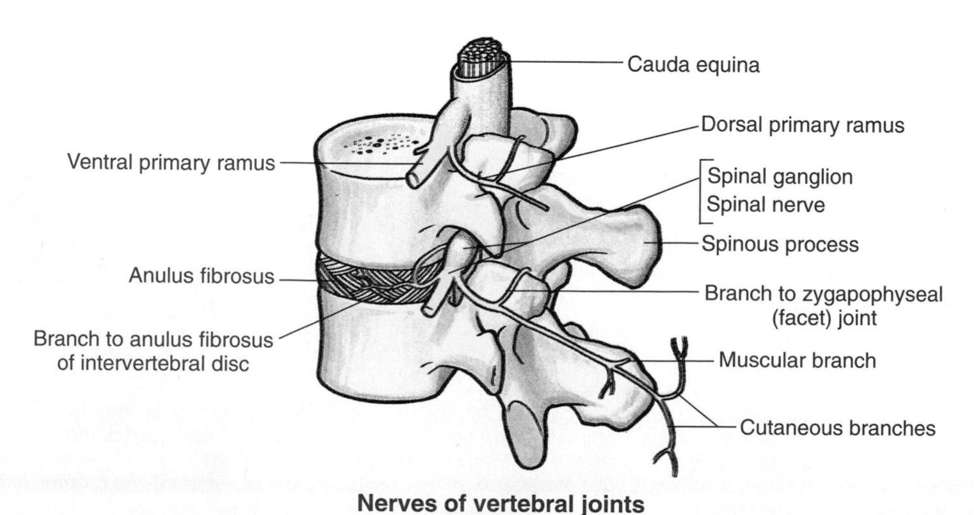

Nerves of vertebral joints

ments between the vertebrae; the shape of the articular surfaces limits the range of movement possible. The zygapophyseal joints are innervated by nerves that arise from branches of the dorsal primary rami of spinal nerves.

Accessory Ligaments of Intervertebral Joints

The laminae of adjacent vertebral arches are joined by broad elastic bands, *ligamenta flava,* that extend almost vertically from the lamina above to the lamina below (Fig. 5.2*B*). These elastic ligaments help to preserve the normal curvature of the vertebral column and to straighten the column after it has been flexed. Adjacent spinous processes are joined by weak *interspinous ligaments* and a strong cordlike *supraspinous ligament*. The latter ligament merges superiorly with the ligamentum nuchae (Fig. 5.1C), the strong median ligament of the neck (nucha refers to the back of the neck). The *ligamentum nuchae* attaches to the skull and spinous processes of the cervical vertebrae. Because of the shortness of the C3–C5 spinous processes, the ligamentum nuchae substitutes for bone in providing muscular attachments. The *intertransverse ligaments*, connecting adjacent transverse processes, consist of a few scattered fibers, except in the lumbar region where they are membranous and more substantial.

Craniovertebral Joints

There are two craniovertebral suboccipital joints: atlantooccipital (between the skull and C1 vertebra) and atlantoaxial [between C1 and C2 (Fig. 5.2)]. The suboccipital joints are synovial joints and have no intervertebral discs.

Atlantooccipital Joints. These articulations between the lateral masses of C1 (atlas) and the occipital condyles permit nodding of the head, i.e., neck flexion and extension that occurs when indicating approval (atlanto refers to the atlas). They are *synovial joints of the condyloid type* and have thin, loose articular capsules. The skull and C1 are also connected by anterior and posterior *atlantooccipital membranes*, which extend from the anterior and posterior arches of C1 to the anterior and posterior margins of the foramen magnum. The atlantooccipital membranes prevent excessive movement of these joints.

The *transverse ligament of the atlas* is a strong band extending between the tubercles on the lateral masses of C1 vertebrae (Fig. 5.2*A*). It holds the dens of C2 (axis) against the anterior arch of C1 (atlas). Vertically oriented superior and inferior bands pass from the transverse ligament to the occipital bone superiorly and to the body of C2 inferiorly. They form the *cruciform ligament*, so named because of its resemblance to a cross (Fig. 5.2C). The *alar ligaments* extend from the sides of the dens to the lateral margins of the foramen magnum. These short, rounded cords attach the skull to C1 vertebra and check rotation and side-to-side movements of the head. The *tectorial membrane* is the superior continuation of the posterior longitudinal ligament. It runs from the body of C1 to the internal surface of the occipital bone and covers the alar and transverse ligaments.

Atlantoaxial Joints. There are two lateral joints and one median joint. These synovial joints are between the lateral masses of C1 and C2 and between the dens of C2 and the anterior arch of the atlas. Movement (rotation) at these joints permits the head to be turned from side to side (e.g., when rotating the head to indicate disapproval). During this movement, the skull and C1 rotate as a unit on C2. Excessive rotation of these joints is prevented by the *alar ligaments*. During rotation of the head, the dens of C2 is held in a collar formed by the anterior arch of the atlas and the transverse ligament of the atlas. The articulation of the dens with C1 is a pivot joint.

If the transverse ligament of the atlas ruptures as the result of trauma or is weakened by disease, the dens is set free and may be driven into the cervical region of the spinal cord, causing *quadriplegia*, or into the medulla of the brainstem, causing death.

MOVEMENTS OF VERTEBRAL COLUMN

Movements of the vertebral column vary according to the region and individual. The mobility of the column results primarily from the compression and elasticity of the intervertebral discs. The following movements of the vertebral column are possible: flexion, extension, lateral bending (lateral flexion), and rotation. The normal range of movement of the vertebral column is limited by the

- Thickness and compressibility of the discs
- Shape and orientation of the zygapophyseal joints

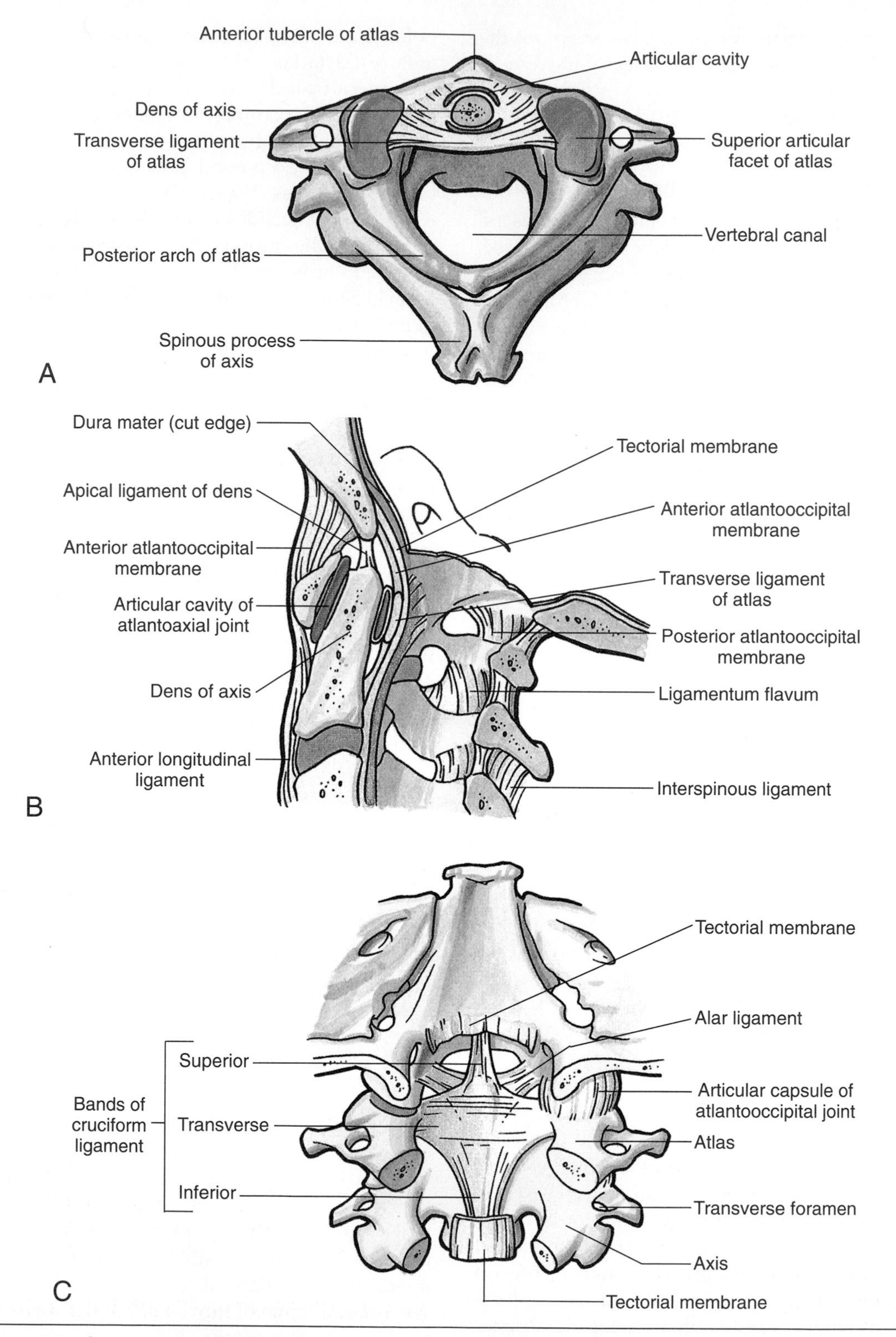

Figure 5.2. Craniovertebral joints. **A.** Superior view of atlantoaxial joint. Observe large vertebral foramen of atlas (C1), which is divided into two foramina by transverse ligament. Larger posterior foramen is for spinal cord, and smaller anterior foramen is for dens of axis (C2). **B.** Median section of craniovertebral region showing ligaments and joints. **C.** Posterior view of craniovertebral joints. Observe bow-shaped transverse ligament of atlas that, by addition of superior and inferior bands, becomes the cross-shaped cruciform ligament.

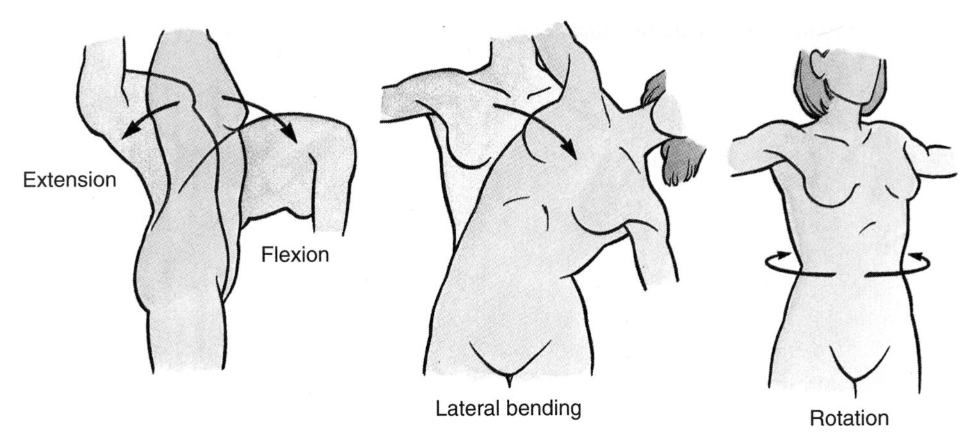

Movements of vertebral column

- Resistance of the back muscles and ligaments
- Tension of capsules of the zygapophyseal joints

The back muscles producing movements of the vertebral column are discussed on pages 207 and 209. However, the movements are not produced exclusively by the back muscles. They are assisted by gravity and the action of the anterolateral abdominal muscles (see Table 5.7). Movements between adjacent vertebrae take place on the resilient nuclei pulposi of the discs and at the zygapophyseal joints. The orientation of these joints permits some movements while restricting others. For example, in the thoracic region, their oblique orientation allows some rotation and lateral bending but prevents flexion and posterior sliding of the vertebrae. Although movements between adjacent vertebrae are relatively small, especially in the thoracic region, the summation of all the small movements produces a considerable range of movement of the vertebral column as a whole.

Movements of the vertebral column are freer in the cervical and lumbar regions than elsewhere. The thoracic region of the column is relatively stable because of its connection to the sternum via the ribs and costal cartilages. Flexion is greatest in the cervical region and is almost nonexistent in the thoracic region. *Lateral bending* is greatest in the cervical and lumbar regions and is restricted in the thoracic region by the ribs. *Extension* is most marked in the lumbar region and is usually more extensive here than flexion. Flexion, extension, and lateral bending of the vertebral column involve compression of the discs at one surface and stretching at the other. Flexion, extension, lateral bending, and rotation of the neck are free because of the thin discs, loose articular capsules, and almost horizontal plane of the articular processes.

Fractures, dislocations, and fracture dislocations of the vertebral column usually result from sudden forceful flexion, as in an automobile accident or from a violent blow to the back of the head. The common injury is a *compression (crush) fracture* of the body of one or more vertebrae. If there is violent anterior movement of the vertebra in addition to compression, a vertebra may be displaced anteriorly on the vertebra inferior to it. Usually this dislocates and fractures the articular facets between the two vertebrae and ruptures the interspinous ligaments. Irreparable injuries to the spinal cord accompany most severe flexion injuries of the vertebral column.

Sudden forceful extension can also injure the vertebral column and spinal cord. Severe hyperextension is most likely to injure posterior parts of the vertebrae, i.e., the vertebral arches and their processes. *Severe hyperextension of the neck* may pinch the posterior arch of C1 between the occipital bone and C2. In these cases C1 usually breaks at one or both grooves for the vertebral arteries. The anterior longitudinal ligament and adjacent anulus fibrosus of the C2/C3 intervertebral disc may also rupture. In these cases the skull, C1, and C2 are separated from the rest of the axial skeleton, and the spinal cord is usually severed. Persons with this injury seldom survive.

VASCULATURE OF VERTEBRAL COLUMN

Spinal arteries supplying the vertebrae are branches of the

- Vertebral and ascending cervical arteries in the neck

- Posterior intercostal arteries in the thoracic region
- Subcostal and lumbar arteries in the abdomen
- Iliolumbar and lateral sacral arteries in the pelvis

Spinal arteries enter the intervertebral foramina and divide into terminal and radicular branches, some of which anastomose with arteries of the spinal cord (see Fig. 5.7).

Spinal veins form venous plexuses extending along the vertebral column both inside (internal vertebral venous plexus) and outside (external vertebral venous plexus) the vertebral canal. The clinical significance of these plexuses is discussed on page 216. The *basivertebral veins* are within the bodies of the vertebrae (see Fig. 5.8).

Muscles of the Back

Because most body weight is anterior to the vertebral column, many strong muscles are attached to the spinous and transverse processes to support the column. Three groups of muscles are in the back: superficial, intermediate, and deep. The superficial and intermediate groups are *extrinsic back muscles* that control limb movements and respiration, respectively. The deep group comprises the *intrinsic back muscles* that act on the vertebral column to produce movements.

EXTRINSIC BACK MUSCLES

The superficial extrinsic back muscles (trapezius, latissimus dorsi, levator scapulae, and rhomboids) connect the upper limbs to the trunk and control limb movements (see Chapter 7). The intermediate extrinsic back muscles (serratus posterior) are superficial respiratory muscles and are described on page 42.

INTRINSIC BACK MUSCLES

The deep intrinsic back muscles (postvertebral muscles) maintain posture and control movements of the vertebral column and head. The muscles are enclosed by fascia that attaches medially to the ligamentum nuchae, the tips of the spinous processes, the supraspinous ligament, and the median crest of the sacrum. The fascia attaches laterally to the cervical and lumbar transverse processes and to the angles of the ribs. The thoracic and lumbar parts of the fascia constitute the *thoracolumbar fascia,* which forms a thin cover for the deep muscles in the thoracic region and a strong thick cover for muscles in the lumbar region. The intrinsic back muscles are grouped according to their relationship to the surface.

Superficial Layer

The *splenius muscles* lie on the lateral and posterior aspects of the neck, somewhat like a bandage, which explains their name (G. splenion, bandage). The muscles arise from the midline and extend superolaterally to the cervical vertebrae (splenius cervicus) and skull (splenius capitis). The splenius muscles cover and hold the deep neck muscles in position (Fig. 5.3, Table 5.5).

Intermediate Layer

The massive *erector spinae* lies in a trough on each side of the spinous processes and forms a prominent bulge on each side of the median plane. The erector spinae is the chief extensor of the vertebral column and can be divided into three vertical columns (Table 5.5):

- Iliocostalis, lateral column
- Longissimus, intermediate column
- Spinalis, medial column

Each muscle can be subdivided into parts according to its superior attachments (e.g., iliocostalis lumborum, iliocostalis thoracis, and iliocostalis cervicis).

Deep Layer

Deep to the erector spinae is an obliquely disposed group of muscles known as the *transversospinal muscle group* (semispinalis, multifidus, and rotatores). The semispinalis is superficial, the multifidus is deeper, and the rotatores are deepest. Minor deep back muscles are the small interspinales and intertransversarii.

Back strain is a common problem that usually results from extreme movements of the vertebral column. The term "strain" is used to indicate some degree of stretching of the muscle and/or ligaments of the back. The muscles usually involved are those producing movements of the lumbar intervertebral joints.

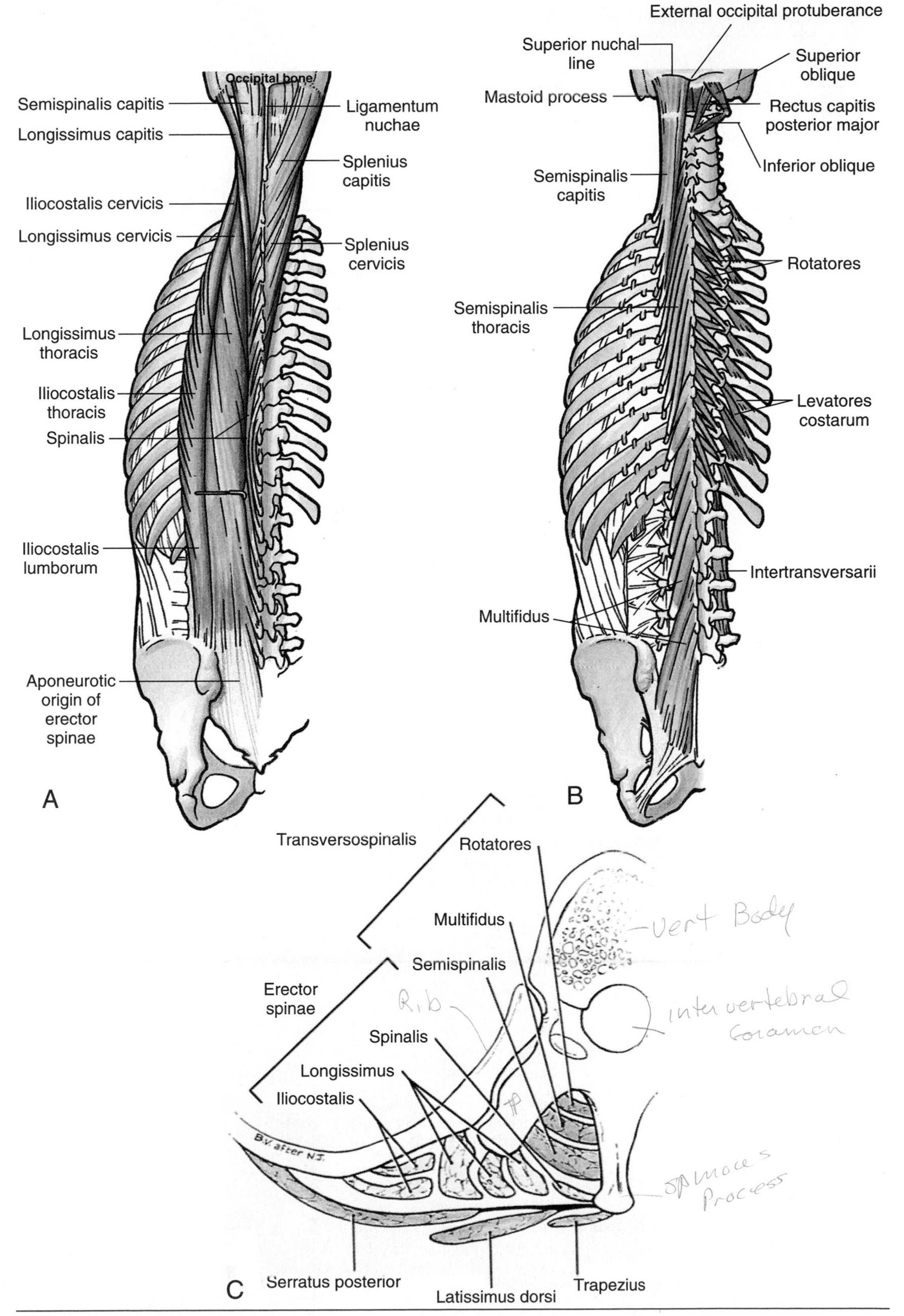

Figure 5.3. Intrinsic back muscles. **A.** Erector spinae, splenius, and semispinalis. **B.** Dissection of back showing transversospinalis. **C.** Partial transverse section of back showing muscles. Note that erector spinae muscle is in three columns, and transversospinalis muscle is in three layers.

Table 5.5
Intrinsic Back Muscles

Muscles	Origin	Insertion	Nerve Supply[a]	Main Actions
Superficial layer				
Splenius	Arises from ligamentum nuchae and spinous processes of C7–T3 or T4 vertebrae	*Spenius capitis*: fibers run superolaterally to mastoid process of temporal bone and lateral third of superior nuchal line of occipital bone *Splenius cervicis*: posterior tubercles of transverse of C1–C3 or C4 vertebrae	Dorsal rami of spinal nn.	*Acting alone*, they laterally bend and rotate head to side of active muscles; *acting together*, they extend head and neck
Intermediate layer				
Erector spinae	Arises by a broad tendon from posterior part of iliac crest, posterior surface of sacrum, sacral and inferior lumbar spinous processes, and supraspinous ligament	*Iliocostalis*: lumborum, thoracis, and cervicis; fibers run superiorly to angles of lower ribs and cervical transverse processes *Longissimus*: thoracis, cervicis, and capitis; fibers run superiorly to ribs between tubercles and angles, to transverse processes in thoracic and cervical regions, and to mastoid process of temporal bone *Spinalis*: thoracis, cervicis, and capitis; fibers run superiorly to spinous processes in the upper thoracic region and to skull	Dorsal rami of spinal nn.	*Acting bilaterally,* they extend vertebral column and head; as back is flexed they control movement by gradually lengthening their fibers; *acting unilaterally*, they laterally bend vertebral column
Deep layer				
Transversospinal	Semispinalis arises from thoracic and cervical transverse processes	*Semispinalis*: thoracis, cervicis, and capitis; fibers run superomedially and attach to occipital bone and spinous processes in thoracic and cervical regions, spanning four to six segments	Dorsal rami of spinal nn.	Extend head and thoracic and cervical regions of vertebral column and rotate them contralaterally
	Multifidus arises from sacrum and ilium, transverse processes of T1–T3, and articular processes of C4–C7	Fibers pass superomedially to spinous processes, spanning two to four segments	Dorsal rami of spinal nn.	Stabilizes vertebrae during local movements of vertebral column
	Rotatores arise from transverse processes of vertebrae; best developed in thoracic region	Pass superomedially and attach to junction of lamina and transverse process of vertebra of origin or into spinous process above their origin, spanning one to two segments	Dorsal rami of spinal nn.	Stabilize vertebrae and assist with local extension and rotary movements of vertebral column

Table 5.5. *Continued*

Muscles	Origin	Insertion	Nerve Supply[a]	Main Actions
Minor deep layer				
Interspinales	Superior surfaces of spinous processes of cervical and lumbar vertebrae	Inferior surfaces of spinous processes of vertebrae superior to vertebrae of origin	Dorsal rami of spinal nn.	Aid in extension and rotation of vertebral column
Intertransversarii	Transverse processes of cervical and lumbar vertebrae	Transverse processes of adjacent vertebrae	Dorsal and ventral rami of spinal nn.	Aid in lateral bending of vertebral column; acting bilaterally, they stabilize vertebral column
Levatores costarum	Tips of transverse processes of C7 and T1–T11 vertebrae	Pass inferolaterally and insert on rib between its tubercle and angle	Dorsal rami of C8–T11 spinal nn.[b]	Elevate ribs, assisting inspiration Assist with lateral bending of vertebral column

[a] Most back muscles are innervated by dorsal rami of spinal nerves, but a few are innervated by ventral rami. Anterior intertransversarii of cervical region are supplied by ventral rami.
[b] Levatores costarum were once said to be innervated by ventral rami, but investigators now agree that they are innervated by dorsal rami.

Muscles Producing Movements of Intervertebral Joints

The principal muscles producing movements of the cervical, thoracic, and lumbar intervertebral joints are summarized in Tables 5.6 and 5.7.

SUBOCCIPITAL REGION

The suboccipital region is the triangular area (*suboccipital triangle*) between the occipital bone and the posterior aspects of C1 (atlas) and C2 (axis) and lies deep to the trapezius and semispinalis capitis muscles. The four small muscles in this region, two rectus capitis posterior and two oblique muscles, are innervated by the dorsal ramus of C1, the *suboccipital nerve*. They are mainly postural muscles, but they also help to move the head.

- Rectus capitis posterior major arises from the spinous process of C2 and inserts into the lateral part of the inferior nuchal line and the occipital bone
- Rectus capitis posterior minor arises from the posterior arch of C1 and inserts into the medial part of the inferior nuchal line
- Inferior oblique (obliquus capitus inferior) arises from the spinous process of C2 and inserts into the transverse process of C1
- Superior oblique (obliquus capitis superior) arises from the transverse process of C1 and inserts into the occipital bone between the superior and inferior nuchal lines

The action of the suboccipital group of muscles is to extend the head on C1 and rotate the head on C1 and C2.

The boundaries and contents of the suboccipital triangle are

- Superomedially—rectus capitis posterior major
- Superolaterally—superior oblique
- Inferolaterally—inferior oblique
- Floor—posterior atlantooccipital membrane and posterior arch of C1
- Roof—semispinalis capitis
- Contents—vertebral artery and suboccipital nerve

The principal muscles producing movements of the craniovertebral joints are summarized in Tables 5.8 and 5.9, and the nerve sup-

Table 5.6
Principal Muscles Producing Movements of Cervical Intervertebral Joints

Flexion	Extension	Lateral Bending	Rotation
Bilateral action of Longus coli Scalene Sternocleidomastoid	Bilateral action of Splenius capitis Semispinalis capitis and cervicis	Unilateral action of Iliocostalis cervicis Longissimus capitis and cervicis Splenius capitis and cervicis	Unilateral action of Rotatores Semispinalis capitis and cervicis Multifidus Splenius cervicis

Table 5.7
Principal Muscles Producing Movements of Thoracic and Lumbar Intervertebral Joints

Flexion	Extension	Lateral Bending	Rotation
Bilateral action of Rectus abdominus Psoas major Gravity	Bilateral action of Erector spinae Multifidus Semispinalis thoracis	Unilateral action of Iliocostalis thoracis and lumborum Longissimus thoracis Multifidus External and internal oblique Quadratus lumborum	Unilateral action of Rotatores Multifidus External oblique acting synchronously with opposite internal oblique Semispinalis thoracis

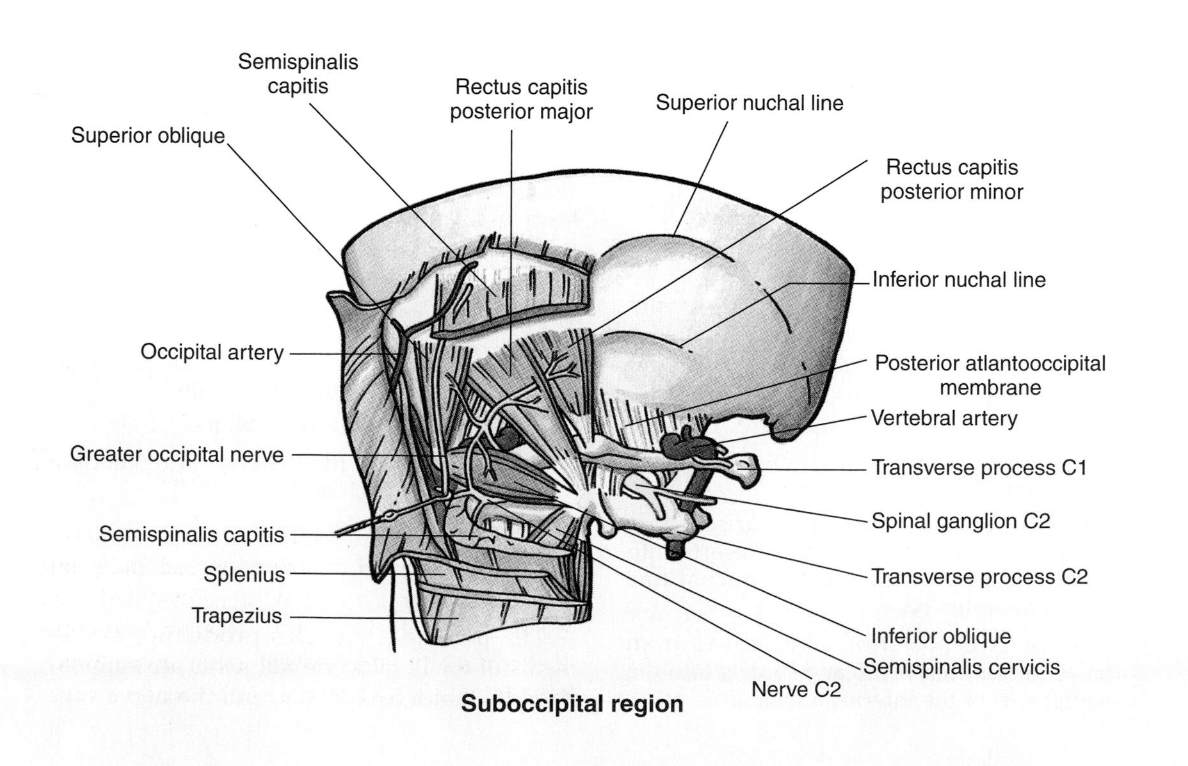

Suboccipital region

Table 5.8
Principal Muscles Producing Movements of Atlantooccipital Joints

Flexion	Extension	Lateral Bending
Longus capitis Rectus capitis anterior Anterior fibers of sternocleidomastoid	Rectus capitis posterior major and minor Obliquus capitis superior Semispinalis capitis Splenius capitis Longissimus capitis Trapezius	Sternocleidomastoid Obliquus capitis superior and inferior Rectus capitis lateralis Longissimus capitis Splenius capitis

Table 5.9
Principal Muscles Producing Rotation at Atlantoaxial Joints[a]

Ipsilateral[b]	Contralateral
Obliquus capitis inferior Rectus capitis posterior, major and minor Longissimus capitis Splenius capitis	Sternocleidomastoid Semispinalis capitis

[a] Rotation is the specialized movement at these joints. Movement of one joint involves the other.
[b] Same side to which head is rotated.

Table 5.10
Nerve Supply of Suboccipital Triangle, Back, and Back of Neck

Nerve	Origin	Course	Distribution
Suboccipital	Dorsal ramus C_1 n.	Runs between skull and first cervical vertebra to reach suboccipital triangle	Muscles of suboccipital triangle
Greater occipital	Dorsal ramus of C_2 n.	Emerges inferior to inferior oblique and ascends to back of scalp	Skin over neck and occipital bone
Lesser occipital	Ventral ramus of C_2 and sometimes C_3 n.	Passes directly to skin	Skin of neck and scalp
Dorsal rami	Spinal nn.	Pass segmentally to muscles and skin	Intrinsic muscles and skin adjacent to vertebral column

ply of the muscles in the suboccipital triangle, back, and back of the neck is summarized in Table 5.10.

The winding course of the vertebral arteries through the suboccipital triangle becomes clinically significant when blood flow through them is reduced (e.g., owing to *arteriosclerosis*). Under these conditions, prolonged turning of the head (as occurs when backing up a motor vehicle) may cause dizziness and other symptoms from the interference with the blood supply to the brainstem.

Spinal Cord and Meninges

The spinal cord and meninges are in the *vertebral canal* (Fig. 5.4). The spinal cord, the major reflex center and conduction pathway between the body and the brain, is a cylindrical structure that is slightly flattened anteriorly and posteriorly. It is protected by the vertebrae, their asso-

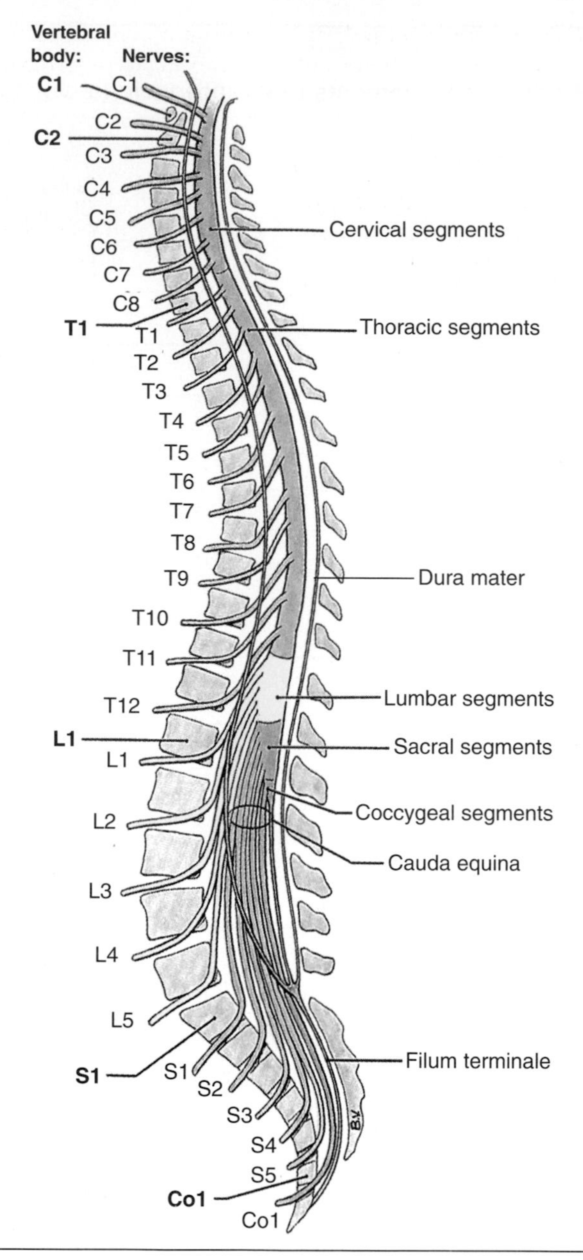

Figure 5.4. Spinal cord in vertebral canal, illustrating relationship of spinal cord segments to vertebrae. Observe that spinal cord ends between bodies of L1 and L2 vertebrae.

ciated ligaments and muscles, the spinal meninges, and the cerebrospinal fluid (CSF). The spinal cord begins as a continuation of the medulla (oblongata), the inferior part of the brainstem. *In adults the spinal cord extends from the foramen magnum in the occipital bone to the intervertebral disc between L1 and L2 vertebrae, but it may terminate at T12 or L3.* Thus, the spinal cord occupies only the superior two-thirds of the vertebral canal. It is enlarged in two regions for innervation of the limbs:

- The cervical enlargement extends from C4 to T1 segments of the spinal cord, and most of the ventral rami arising from it form the *brachial plexus* of nerves that innervates the upper limbs (Chapter 7)
- The lumbosacral enlargement extends from T11 to L1 segments of the spinal cord, and

the corresponding nerves make up the lumbar and sacral plexuses that innervate the lower limbs (Chapter 6)

STRUCTURE OF SPINAL NERVES

Thirty-one pairs of spinal nerves are attached to the spinal cord. Several rootlets emerge from the dorsal and ventral surfaces of the spinal cord and converge to form dorsal and ventral roots (Figs. 5.4 and 5.5). *The part of the spinal cord from which the rootlets of one dorsal and ventral root emerge is a segment of the cord.* The dorsal roots contain afferent or sensory fibers from skin, subcutaneous and deep tissues, and often from viscera. The *ventral roots* contain efferent or motor fibers to skeletal muscle, and many contain preganglionic autonomic fibers. The cell bodies of axons making up the ventral roots are in the ventral gray horns of the spinal cord, whereas the cell bodies of axons making up the dorsal roots are outside the spinal cord in the *spinal ganglion* (dorsal root ganglion) at the distal end of the dorsal root. The dorsal and ventral nerve roots unite at their points of exit from the vertebral canal to form a *spinal nerve*. The nerves are distributed as follows: 8 cervical, 12 thoracic, 5 lumbar, 5 sacral, and 1 coccygeal (Fig. 5.4). The first cervical nerves lack dorsal roots in 50% of people and the coccygeal nerve may be absent. Each spinal nerve divides almost immediately into a *ventral primary ramus* and a *dorsal primary ramus*. The dorsal rami supply the skin and muscles of the back, and the ventral rami supply the limbs and the rest of the trunk.

Because the adult spinal cord is shorter than the vertebral column, there is a progressive obliquity of the spinal nerve roots (Fig. 5.4). Because of the increasing distance between the spinal cord segments and the corresponding vertebrae, the length of the nerve roots increases progressively as the inferior end of the vertebral column is approached. The lumbar and sacral nerve rootlets are the longest. They descend until they reach their intervertebral foramina of exit in the lumbar and sacral regions. The bundle of spinal nerve roots in the subarachnoid space caudal to the termination of the spinal cord is the *cauda equina*.

The inferior end of the spinal cord tapers into the *conus medullaris*. From its inferior end, a threadlike strand of pia mater, the *filum terminale*, descends among the nerve roots of the cauda equina. It leaves the inferior end of the dural sac, passes through the sacral hiatus, and ends by attaching to the dorsum of the coccyx.

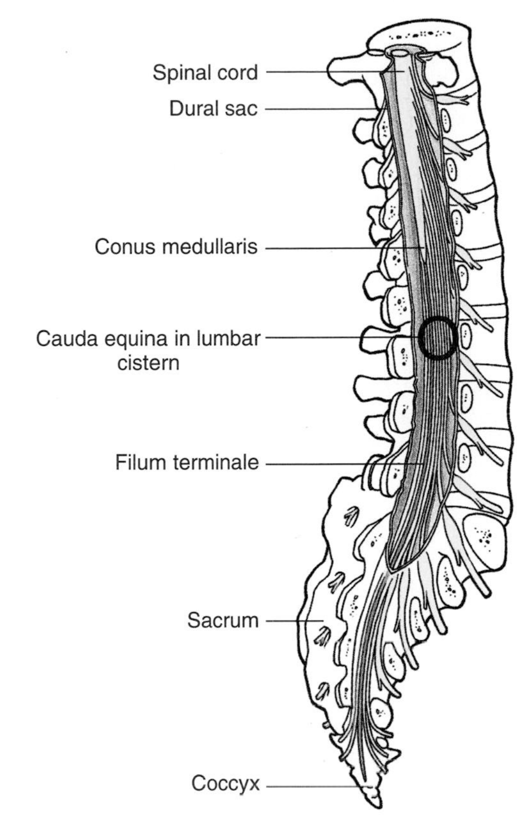

Cauda equina in lumbar cistern

SPINAL MENINGES AND CEREBROSPINAL FLUID

The dura, arachnoid, and pia surrounding the spinal cord are known collectively as the spinal meninges. These membranes and CSF surround, support, and protect the spinal cord and cauda equina.

Dura Mater

The spinal dura mater, composed of tough fibrous and elastic tissue, is the outermost covering membrane of the spinal cord. The spinal dura forms the external layer of the *dural sac*, a long tubular sheath that is free within the vertebral canal (Fig. 5.5). The dural sac adheres to the margin of the foramen magnum of the skull where it is continuous with the cranial dura mater. It is anchored inferiorly to the coccyx by the filum terminale. The dura mater is separat-

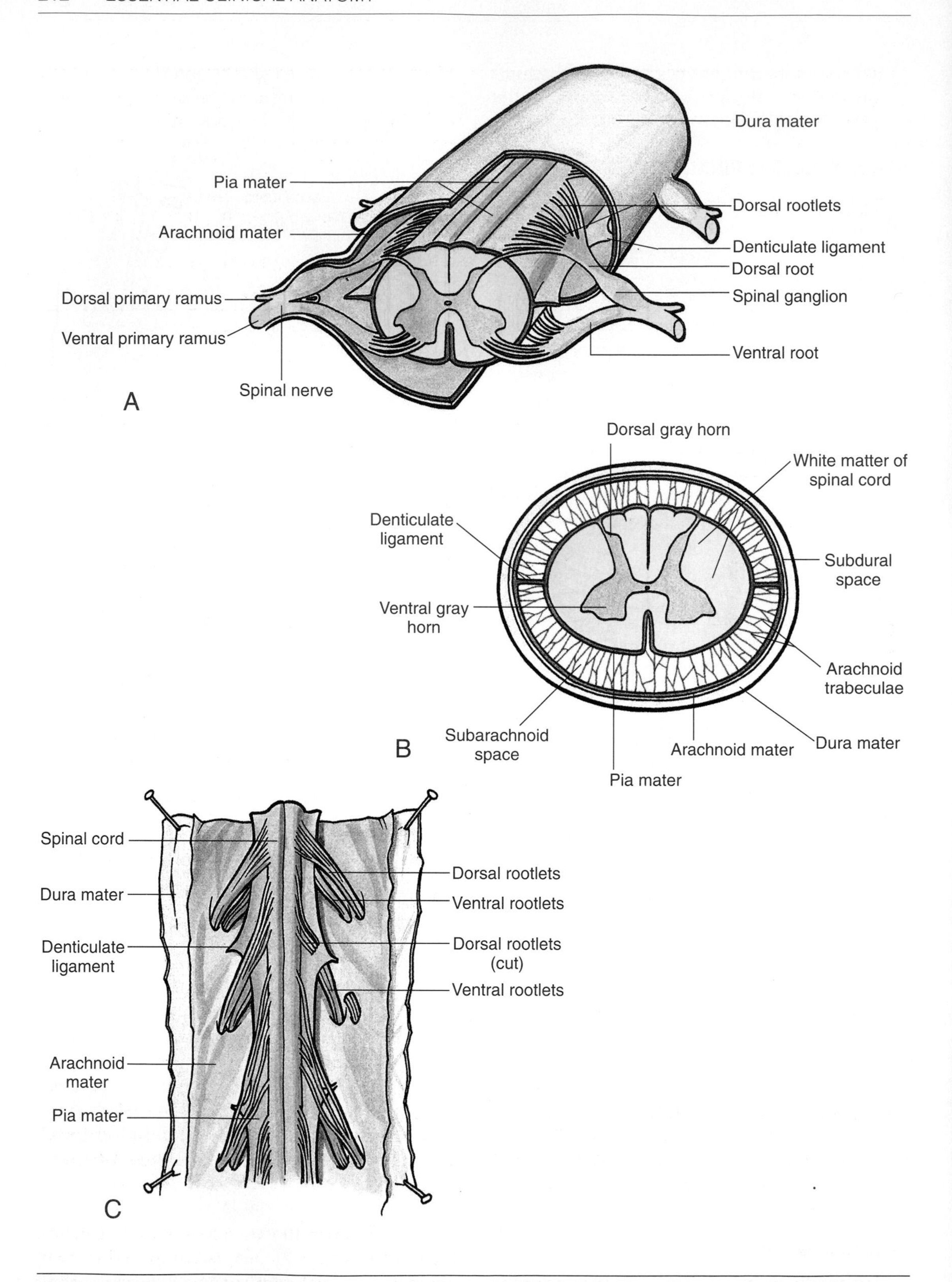

Figure 5.5. **A.** Three-dimensional drawing of spinal cord and meninges. On each side observe rootlets that arise from a segment of the cord and join to form dorsal and ventral roots. **B.** Transverse section of spinal cord and meninges. **C.** Meninges have been cut and spread out. Pia matter (*pink*) covers spinal cord and projects laterally as denticulate ligament.

Table 5.11
Spaces Associated with Spinal Meninges

Space	Location	Contents
Extradural (epidural)	Between wall of vertebral canal and dura mater	Fat, loose connective tissue, internal venous plexuses, and distal to L2 vertebra, the roots of spinal nerves
Subdural[a]	Between dura and arachnoid mater	Capillary layer of serous fluid
Subarachnoid	Between arachnoid and pia mater	CSF, spinal arteries, and veins

[a] Recent evidence indicates that the subdural space is a creation of a cleft in this area as the result of tissue damage [Haines (1991) listed under "Suggested Readings"].

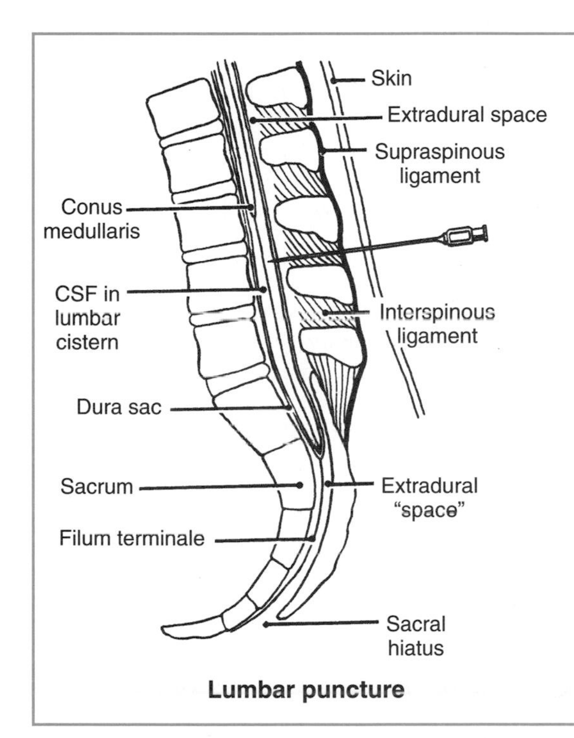

Lumbar puncture

To obtain a sample of CSF (*lumbar puncture*) agent, a needle is inserted into the subarachnoid space. The needle is inserted between the arches of L3 and L4 vertebrae and enters the lumbar cistern. At this level, there is no danger of damaging the spinal cord.

During an *epidural block*, an anesthetic agent is injected into the extradural (epidural) space and diffuses through the dura and arachnoid. It enters the subarachnoid space and acts on the nerve roots in the cauda equina (Fig. 5.6). The patient loses sensation inferior to the level of the block.

An anesthetic agent can also be injected through the sacral hiatus into the extradural space in the sacral canal. The agent spreads superiorly and acts on the spinal nerves (*caudal analgesia*). The distance the agent ascends depends on the amount injected and the position of the patient.

Laminectomy is the surgical procedure used to expose the spinal cord. The laminae of several vertebrae are removed to relieve pressure on neural structures from bony fragments, protruding discs, tumors, hematomas, and other lesions.

ed from the periosteum of the vertebrae by a space, the *extradural* or *epidural space* (Table 5.11). The spinal dura mater extends into the intervertebral foramina and along the dorsal and ventral nerve roots and spinal ganglia to form *dural root sleeves*. These sleeves adhere to the periosteum lining the intervertebral foramina and end by blending with the epineurium of the spinal nerves.

Arachnoid Mater

The arachnoid mater is the delicate, avascular membrane composed of fibrous and elastic tissue that lines the dural sac. The arachnoid mater is separated from the dura mater by a potential "space," the *subdural space*. The arachnoid also covers the spinal nerve roots and spinal ganglia and blends with the sheaths of the spinal nerves. The arachnoid is separated from the pia mater by the *subarachnoid space*, but the two layers are connected by *arachnoid trabeculae* (delicate strands of connective tissue).

Pia Mater

The pia mater, the innermost covering membrane of the spinal cord, is composed of two fused layers of loose connective tissue that adhere closely to the surface of the spinal cord. It also covers the roots of the spinal nerves and the spinal blood vessels and ensheathes the

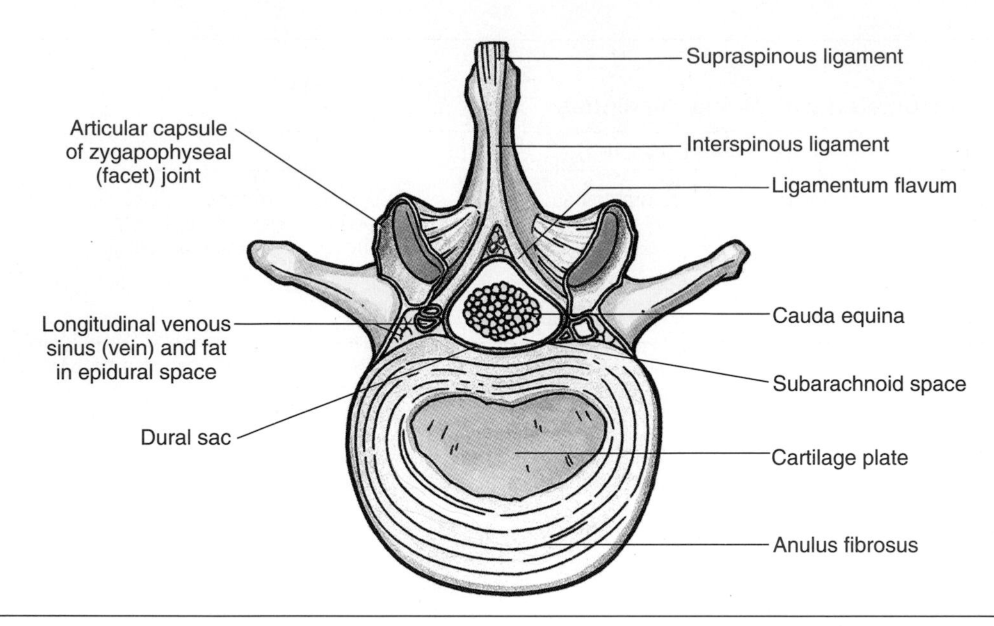

Figure 5.6. Transverse section of an intervertebral disc and associated intervertebral ligaments. Nucleus pulposus has been removed to show hyaline cartilage plate covering superior surface of body of vertebra. Observe cauda equina in dural sac.

anterior spinal artery. Inferior to the conus medullaris, the pia continues as the filum terminale that pierces the dural sac and attaches to the coccyx.

The spinal cord is suspended in the dural sac by a saw-toothed *denticulate ligament* on each side. These ligaments consisting of pia mater extend from the lateral surfaces of the pia mater midway between the dorsal and ventral nerve roots, and attach to the dural sac by 21 toothlike processes. The denticulate ligaments extend from the foramen magnum to the area between T12 and L1 nerve roots.

Subarachnoid Space

The subarachnoid space, between the arachnoid and pia mater, contains CSF, a clear slightly alkaline fluid. This space is largest in the inferior part of the vertebral canal, where CSF forms the *lumbar cistern* that surrounds the nerve roots comprising the cauda equina and filum terminale. CSF is formed by the choroid plexuses in the lateral, third, and fourth ventricles of the brain (Chapter 8).

VASCULATURE OF SPINAL CORD

The vessels supplying the spinal cord derive from branches of the vertebral, deep cervical, intercostal, and lumbar arteries. Three longitudinal arteries supply the cord: an *anterior spinal artery* and *two posterior spinal arteries* (Fig. 5.7). These vessels are reinforced by blood from segmental vessels called *radicular arteries*. The anterior and posterior radicular arteries accompany the dorsal and ventral roots of the spinal nerves. Some of these arteries are small and supply only the nerve roots and pia mater; others are large and join the anterior and posterior spinal arteries. There are about 14 large radicular and 12 anterior radicular arteries that join the spinal arteries.

The *great anterior radicular artery* (spinal artery of Adamkiewicz) supplies the inferior thoracic and superior lumbar regions of the spinal cord, including the lumbosacral enlargement. It is usually much larger than other radicular arteries. The great radicular artery arises more frequently on the left side from an intercostal or lumbar artery. It is clinically important because it makes a major contribution to the anterior spinal artery that provides the main blood supply to the inferior two-thirds of the spinal cord.

There are usually three anterior and three posterior *spinal veins*. They are arranged longitudinally, communicate freely with each other, and are drained by numerous radicular veins. The veins draining the spinal cord and vertebrae form *internal vertebral plexuses* of thin-walled, valveless veins that surround the dura

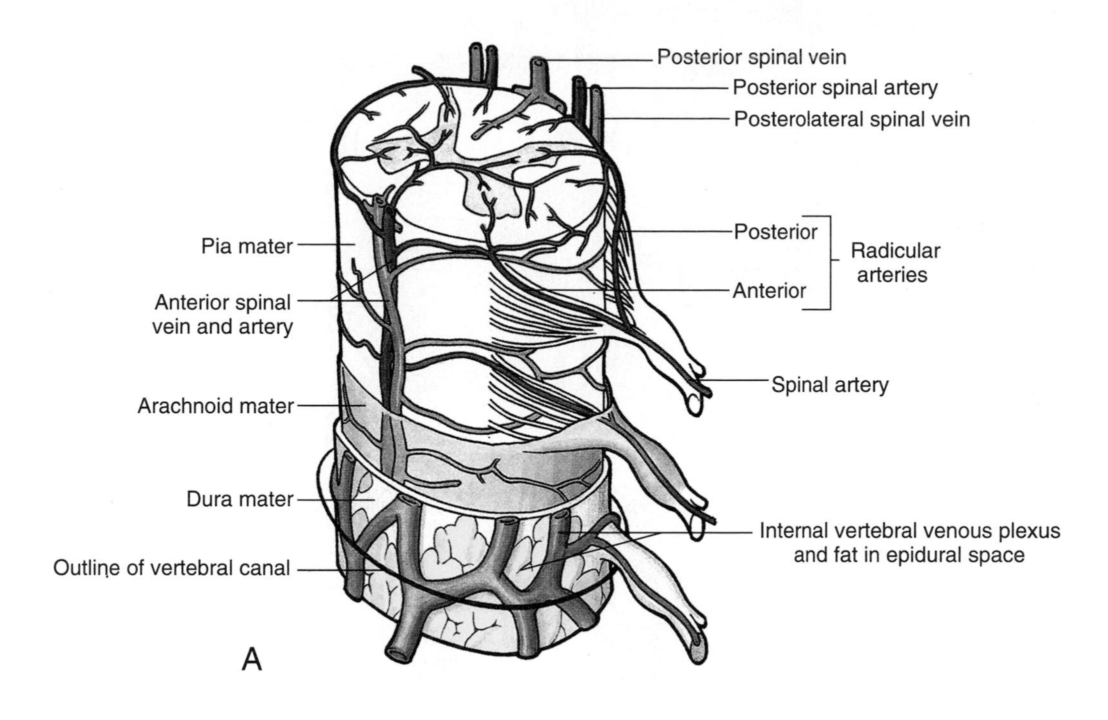

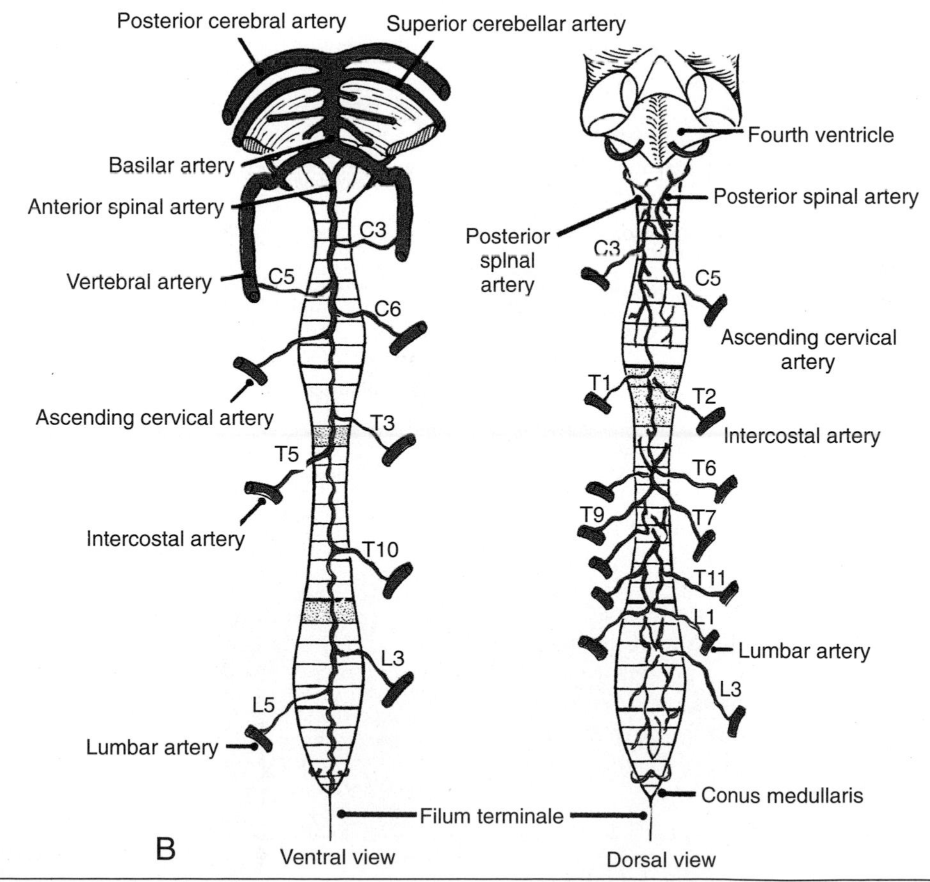

Figure 5.7. **A.** Arterial supply and venous drainage of spinal cord. **B.** Ventral and dorsal views of arteries of spinal cord. *Stippled areas of spinal cord*, regions that are most vulnerable to vascular deprivation when contributing arteries are injured.

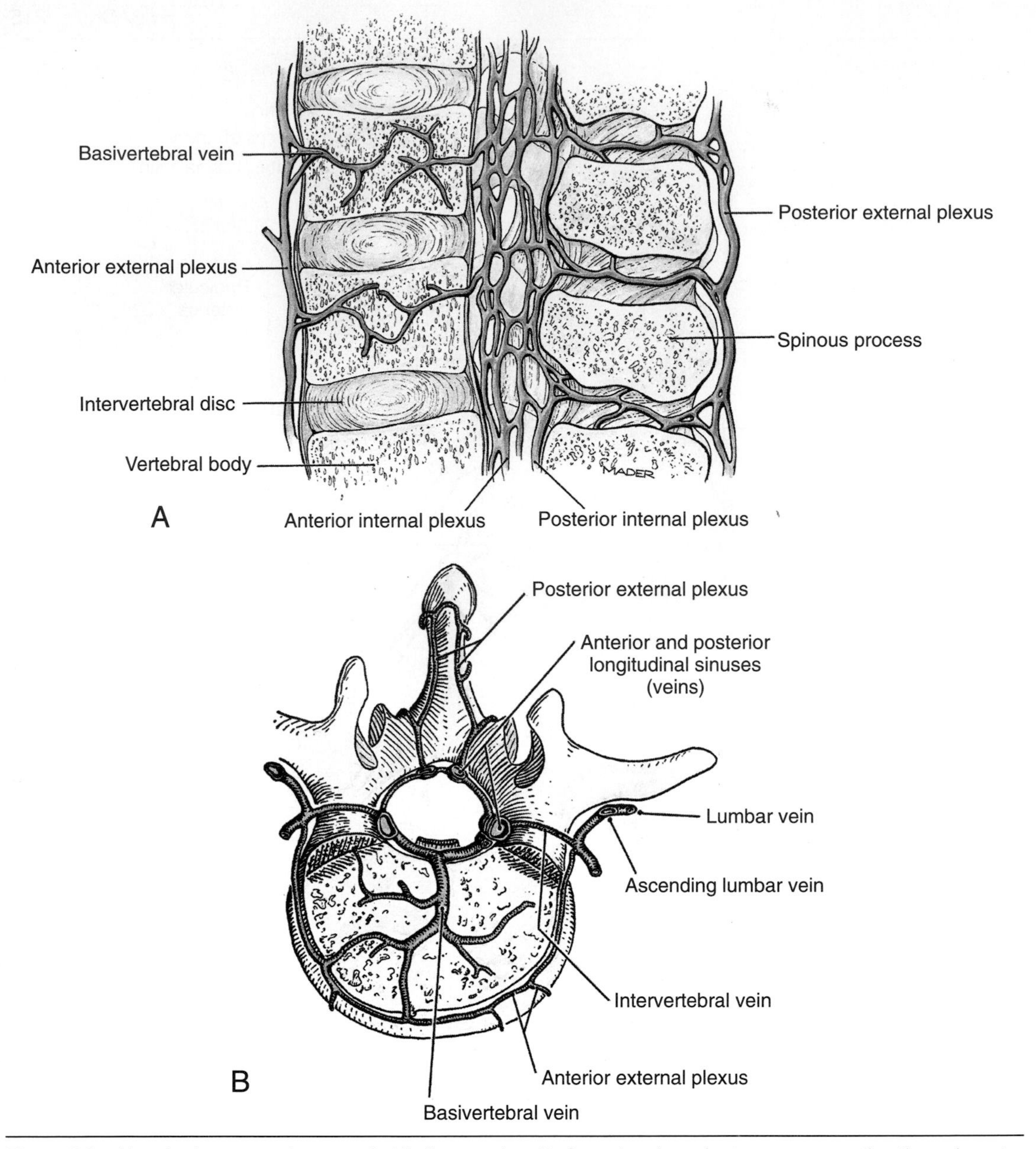

Figure 5.8. Vertebral venous plexuses. **A.** Median section. **B.** Superior view of a transverse section through vertebral body.

mater (Fig. 5.8). These veins communicate via anterior and posterior longitudinal sinuses with the *venous sinuses of the cranial dura mater* (Chapter 8). The anterior and posterior spinal veins and the vertebral venous plexuses drain into intervertebral veins and from them into the vertebral veins, ascending lumbar veins, and *azygos venous system* (p. 78)

The segmental reinforcements of blood supply from the radicular arteries are important in supplying the anterior and posterior spinal arteries.

Fractures, dislocations, and fracture-dislocations may interfere with the blood supply to the spinal cord from the radicular arteries, causing weakness and paralysis of muscles. Patients with ruptured aneurysms of the great radicular artery may suffer from paralysis of the lower limbs because this artery provides the main blood supply to the inferior two-thirds of the spinal cord.

The *vertebral venous plexuses* are important because blood may return from the pelvis or abdomen through them and reach the heart via the superior vena cava. These veins can also provide a route for metastasis of cancer cells to the vertebrae or the brain from an abdominal or pelvic tumor (e.g., prostate cancer).

6 / LOWER LIMB

The lower limb is specialized for locomotion, bearing weight, and maintaining equilibrium. It has four parts:

- Hip (*blue*) containing the hip bone (coxal bone, innominate bone), which connects the skeleton of the limb to the vertebral column
- Thigh (*purple*) containing the femur, which connects hip and knee, and the patella
- Leg (*pink*) containing the tibia and fibula, which connect the knee and ankle
- Foot (*orange*) containing the tarsus, metatarsus, and phalanges, which connect the ankle and foot.

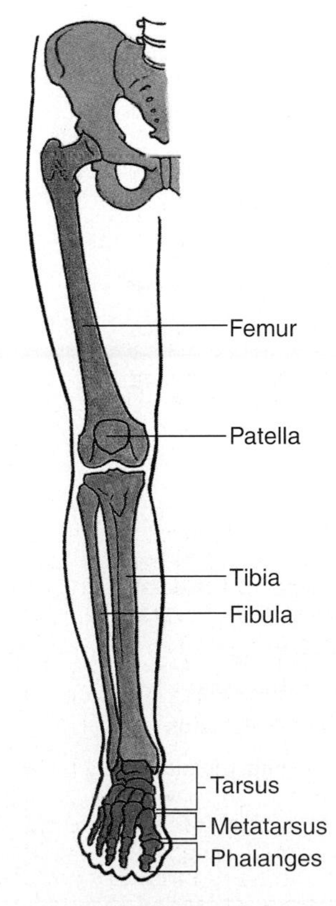

Lower limb bones

Bones

The skeleton of the lower limb is composed of the pelvic girdle (formed by the hip bones) and the skeleton of the free limb (Fig. 6.1). The *pelvic girdle*, together with the sacrum and coccyx, forms the bony pelvis.

HIP BONE

The hip bone joins the sacrum to the femur and forms the bony connection between the trunk and lower limb. Each hip bone consists of three bones: ilium, ischium, and pubis. Until puberty these bones are separated by a triradiate cartilage. The bones begin to fuse at 15–17 years of age, and little or no trace of their lines of fusion is visible in adults. The *ilium* is the superior, larger part of the hip bone and forms the superior part of the *acetabulum*, the deep cuplike cavity (socket) on the lateral aspect of the hip bone for articulation with the head of the femur. The *ischium* forms the posteroinferior part of the acetabulum and hip bone. The *pubis* forms the anterior part of the acetabulum and the anteromedial part of the hip bone.

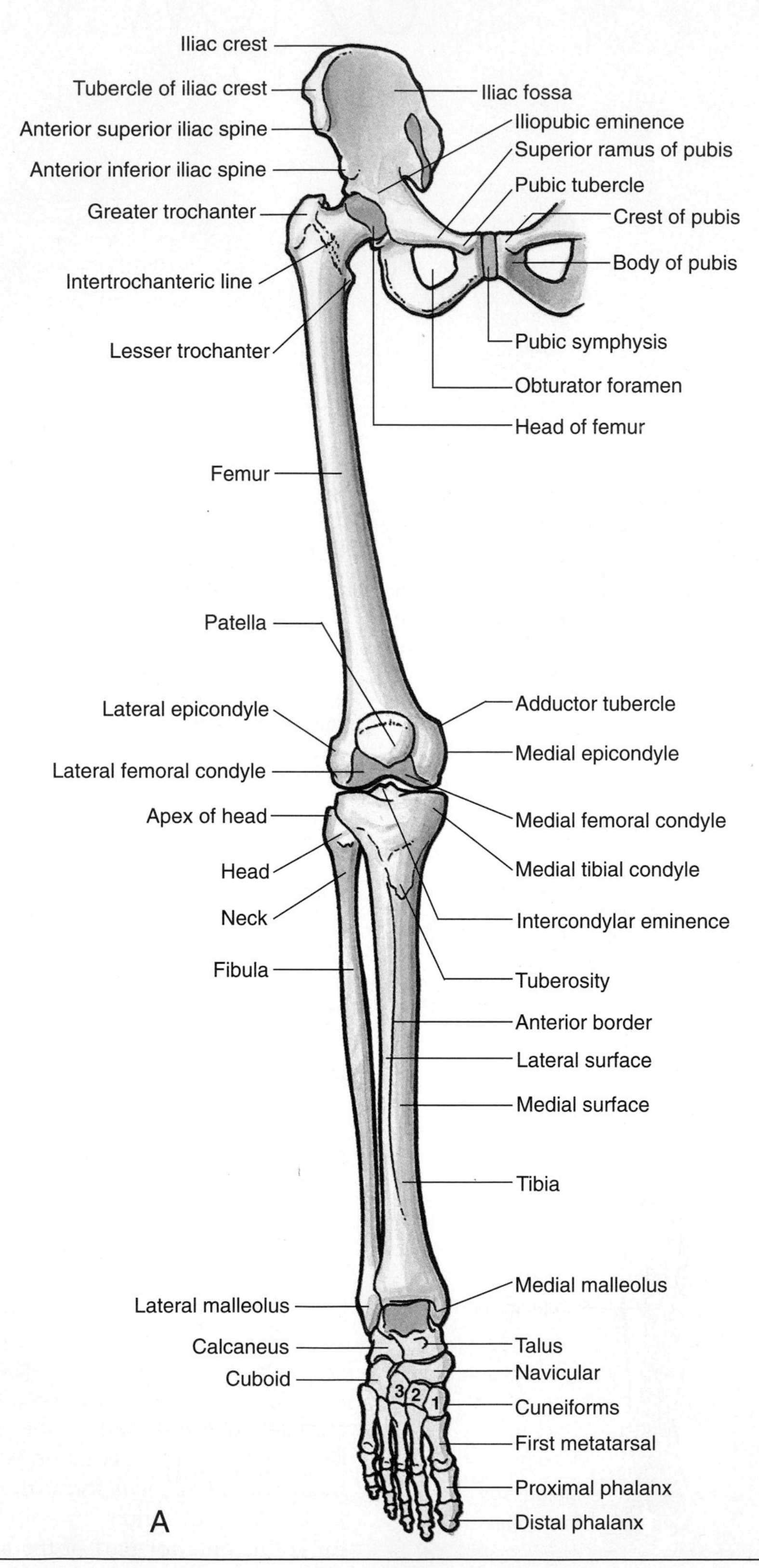

Figure 6.1. Bones of lower limb. **A.** Anterior view. **B.** Posterior view.

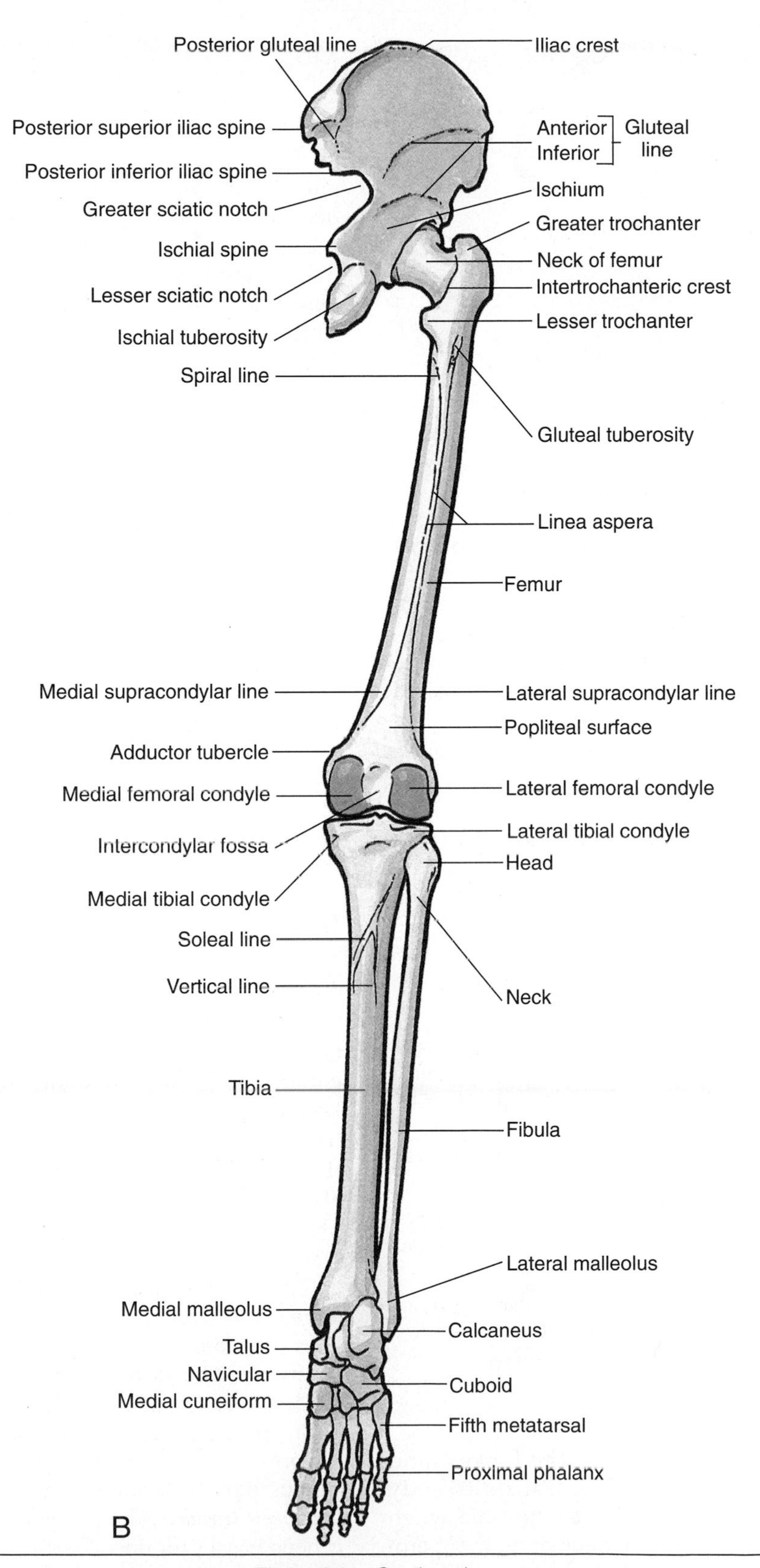

Figure 6.1. *Continued.*

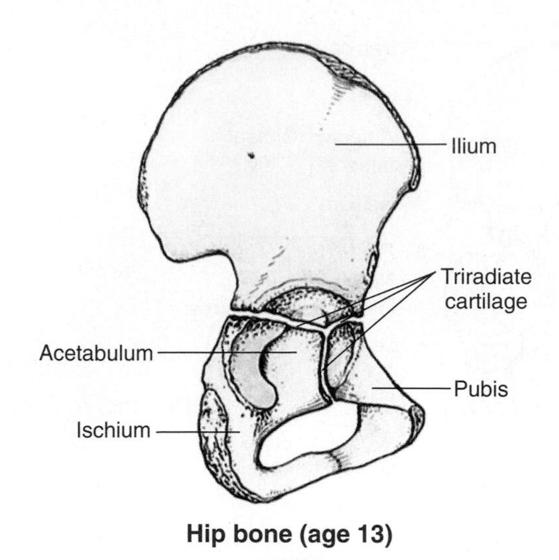

Hip bone (age 13)

To place the hip bone in the anatomical position, move it until the anterior superior iliac spine of the ilium and the pubic tubercle are in the same coronal plane. Observe that the

- Ischial spine and the superior end of the pubic symphysis are in the same horizontal plane
- Internal aspect of the body of the pubis faces almost directly superiorly
- Acetabulum faces inferolaterally
- Obturator foramen lies inferomedial to the acetabulum

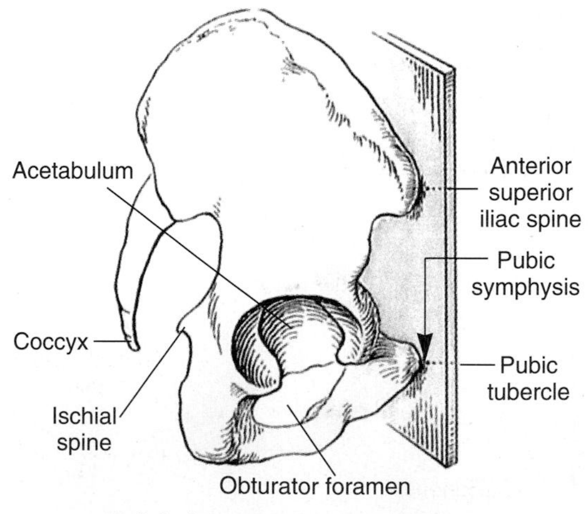

Pelvis in anatomical position

FEMUR

The femur (thigh bone), the longest and heaviest bone in the body, transmits body weight from the hip bone to the tibia when standing. The *head of the femur* projects superomedially and slightly anteriorly when articulating with the acetabulum. The superior end of the femur consists of a head, neck, and two trochanters (greater and lesser). The head and neck are at an angle (115–140°) to the long axis of the body (shaft) of the femur; the angle varies with age and sex. When this angle is decreased the condition is known as *coxa vara;* when it is increased, it is *coxa valga*. Although this architecture allows greater mobility of the femur at the hip joint, it imposes considerable strain on the neck of the femur. The *body of the femur* is curved, with the convexity anteriorly. The distal end of the femur ends in two spirally curved *condyles* (medial and lateral).

TIBIA AND FIBULA

The large weightbearing *tibia* articulates with the condyles of the femur superiorly and the talus inferiorly. The nutrient foramen of the tibia, the largest in the skeleton, is located on the posterior surface of the superior third of the bone. The nutrient canal runs a long inferior course in the bone before it opens into the medullary (marrow) cavity. The slender *fibula* lies posterolateral to the tibia and is mainly for the attachment of muscles and has little or no function in weightbearing. The bodies (shafts) of the tibia and fibula are connected by an *interosseous membrane.*

TARSUS, METATARSUS, AND PHALANGES

Tarsus

The tarsus consists of seven bones (Fig. 6.2): talus, calcaneus, cuboid, navicular, and three cuneiforms. Only one bone, the talus, articulates with the leg bones. The *talus* (ankle bone) has a body, neck, and head. The talus rests on the anterior two-thirds of the calcaneus and also articulates with the tibia, fibula, and navicular. The superior surface of the talus bears the weight of the body transmitted from the tibia.

The *calcaneus* (calcaneum, heel bone) is the largest and strongest bone in the foot. It articulates with the talus superiorly and the cuboid anteriorly. The *sustentaculum tali* is a shelflike process that projects from the superior border of the medial surface of the calcaneus and helps support the talus. The lateral surface of the calcaneus has an oblique ridge called the *fibular (peroneal) trochlea.* The posterior part of the calcaneus has a prominence—the tuberosity of the calcaneus (tuber calcanei)—which has medial,

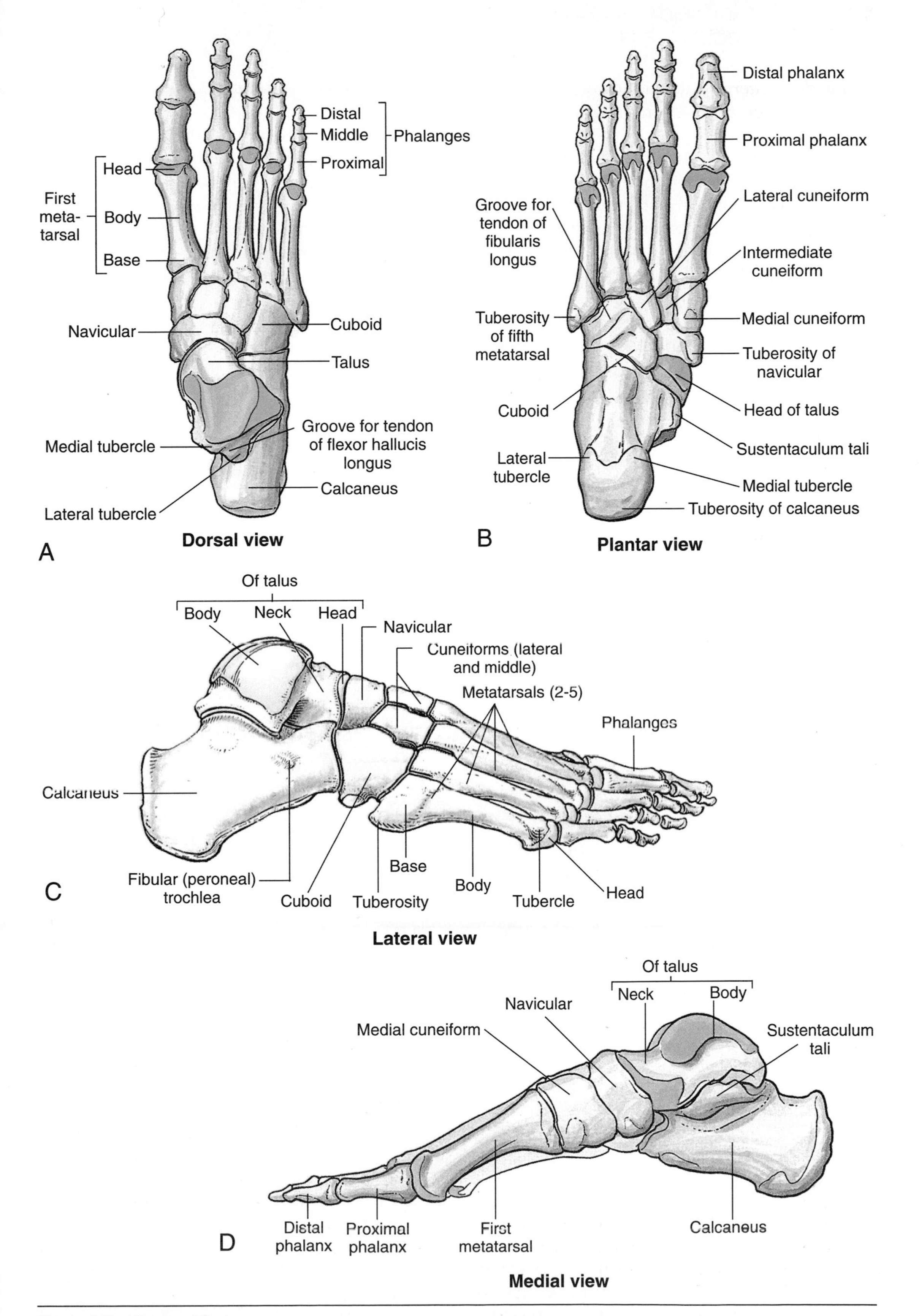

Figure 6.2. Bones of foot.

lateral, and anterior tubercles (processes). Only the medial tubercle rests on the ground during standing.

The *navicular* is located between the head of the talus and the cuneiforms. The *cuboid* is the most lateral bone in the distal row of the tarsus. Anterior to the tuberosity of the cuboid, on the lateral and inferior surfaces of the bone, there is a groove in the cuboid. There are three *cuneiforms*: medial (first), intermediate (second), and lateral (third). Each cuneiform articulates with the navicular posteriorly and the base of its appropriate metatarsal anteriorly. In addition, the lateral cuneiform articulates with the cuboid.

Metatarsus

The metatarsus consists of five metatarsals that are numbered from the medial side of the foot. Each bone consists of a base (proximally), a body, and a head (distally). The bases of the metatarsals articulate with the cuneiform and cuboid bones, and the heads articulate with the proximal phalanges. On the plantar surface of the head of the first metatarsal are prominent medial and lateral *sesamoid bones*. The base of the fifth metatarsal has a large tuberosity that projects over the lateral margin of the cuboid.

Phalanges

There are 14 phalanges: the first digit has two phalanges (proximal and distal); the other four digits have three each (proximal, middle, and distal). Each phalanx consists of a base (proximally), a body, and a head (distally). The phalanges of the first digit are short, broad, and strong.

Surface Anatomy of Bones of Lower Limb

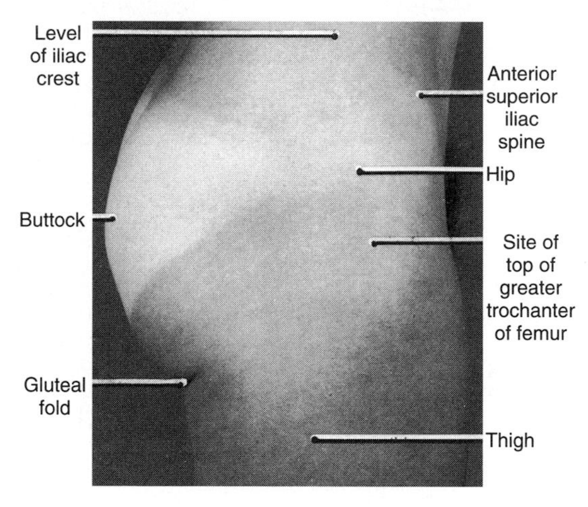

When you put your hand on your hip, it rests on the *iliac crest*, the superior margin of the ilium. This crest is easily palpated because it extends along the inferior margin of the side of the trunk. Posteriorly its highest point is at the level of the disc between L4 and L5 vertebrae. Clinically, this level is commonly used as a surface marking when performing lumbar punctures for obtaining cerebrospinal fluid (p. 213). The iliac crest ends anteriorly at the rounded *anterior superior iliac spine*, which is easy to palpate, especially in thin persons, because it is subcutaneous and often visible. The thick *tubercle of the iliac crest* is about 5 cm posterior to the anterior superior iliac spine and marks the widest point of the crest. Also palpate the *pubic tubercle* anteriorly, about 2.5 cm from the median plane. It is an important bony landmark in the diagnosis and surgical repair of inguinal hernias (p. 92).

The iliac crest ends posteriorly in a sharp *posterior superior iliac spine* (p. 148) that is difficult to palpate in most people, but its position is easy to locate because it lies at the bottom of a *skin dimple*, about 4 cm lateral to the median plane. The dimple exists because the skin and underlying fasciae attach to the spine. A line connecting the dimples lies at the level of the S2 spinous process and passes though the center of the *sacroiliac joints*. The dimples are useful landmarks when palpating the area of the sacroiliac joints in search of swelling or local tenderness. These dimples also indicate the termination of the iliac crests from which bone marrow and pieces of bone for grafts can be obtained. The sacroiliac joint is not palpable because of the overhang of the ilium and the presence of ligaments (see Chapter 4).

The *ischial tuberosity* is easily palpated. The thick gluteus maximus muscle and fat cover it when

Surface Anatomy of Bones of Lower Limb. ***Continued***

the thigh is extended but not when the thigh is flexed (e.g., when sitting). The *gluteal fold*, a prominent skin fold formed by fat, coincides with the inferior border of the gluteus maximus. The *gluteal sulcus*, the skin crease inferior to the gluteal fold, separates the buttock from the thigh.

The posterior edge of the *greater trochanter of the femur* is easily palpable on the lateral side of the thigh. Because it lies close to the skin, the greater trochanter causes discomfort when lying on your side on a hard surface. In the anatomical position, a line joining the tips of the greater trochanters normally passes through the center of the femoral heads and the pubic tubercles. The *body of the femur* (shaft) is not usually palpable because it is covered with large muscles.

The *condyles of the femur* are subcutaneous and easily palpable when the knee is flexed or extended. At the center of each condyle is a prominent *epicondyle* that is easily palpable. The patellar surface of the femur is where the *patella* (kneecap) slides during flexion and extension of the leg at the knee joint. The lateral and medial margins of the patellar surface can be palpated when the leg is flexed. The *adductor tubercle*, a small prominence of bone, may be felt at the superior part of the medial femoral condyle.

The anteromedial surface of the tibia is subcutaneous, smooth, and flat. The skin covering it is freely movable. The prominence at the ankle, the *medial malleolus*, is also subcutaneous, and its inferior end is blunt. The *tibial tuberosity*, a rounded elevation, is palpable about 5 cm distal to the apex of the patella.

The *head of the fibula* can be easily palpated at the level of the superior part of the tibial tuberosity because the knoblike head is subcutaneous at the posterolateral aspect of the knee. The *neck of the fibula* can be palpated just distal to the head. Only the distal part of the body of the fibula is subcutaneous. Palpate your *lateral malleolus*, noting that it is subcutaneous and that its inferior end is sharp. Observe that the tip of the lateral malleolus extends farther distally (1–2 cm) and more posteriorly than does the tip of the medial malleolus.

The *head of the talus* is palpable anteromedial to the proximal part of the lateral malleolus when the foot is inverted and anterior to the medial malleolus when the foot is everted. Eversion of the foot makes the head more prominent as it moves out from the navicular. The head occupies the space between the sustentaculum tali and the tuberosity of the navicular. When the foot is plantarflexed, the superior surface of the body of the talus can be palpated on the anterior aspect of the ankle, anterior to the inferior end of the tibia.

The posterior, medial, and lateral surfaces of the *calcaneus* can be easily palpated. The weightbearing *medial tubercle of the calcaneus* on the plantar surface of the foot is broad and large but is not usually easily palpable unless it is associated with a heel spur (bony outgrowth). The *sustentaculum tali* can usually be felt as a small prominence approximately a finger breadth distal to the tip of the medial malleolus. The *fibular (peroneal) trochlea* (tubercle) may be detectable as a small tubercle on the lateral aspect of the calcaneus, anteroinferior to the tip of the lateral malleolus.

The *tuberosity of the navicular* (navicular tubercle) is easily seen and palpated on the medial aspect of the foot, inferoanterior to the tip of the medial malleolus. The *cuboid and cuneiforms* are difficult to identify individually by palpation. The cuboid can be felt on the lateral aspect of the foot, posterior to the base of the fifth metatarsal. The medial cuneiform can be palpated between the tuberosity of the navicular and the base of the first metatarsal.

The *head of the first metatarsal* forms a prominence on the medial aspect of the foot. The medial and lateral *sesamoids* inferior to the head of this metatarsal can be felt to slide when the first digit (great toe) is moved passively. The *tuberosity of the fifth metatarsal* forms a prominent landmark on the lateral aspect of the foot and can easily be palpated at the midpoint of the lateral border of the foot. The *bodies of the metatarsals and phalanges* can be felt on the dorsum of the foot between the extensor tendons. Usually palpation of the bony prominences on the plantar surface of the foot is difficult because of the thick fascia and pads of fat.

Fractures of the hip bone often occur in serious vehicular accidents. Anteroposterior compression of these bones fractures the pubic rami. Lateral compression of the pelvis may fracture the acetabulum, as may "falls on the feet," which forcefully drive the femora superiorly into the acetabula.

Fractures of the femoral neck are fairly common, especially in older persons with osteoporosis (reduction in the quantity of bone). These fractures often interrupt the blood supply to the femoral head, which may lead to degeneration of bone from ischemia.

Fracture of the tibia through the nutrient canal usually damages the nutrient artery and predisposes to nonunion of the bone fragments. Because the body of the tibia is unprotected anteromedially throughout its course and is relatively slender at the junction of its inferior and middle thirds, not surprisingly the tibia is the most common long bone to be fractured. It is also the most frequent site of a compound (open) fracture.

Fracture of the fibula commonly occurs 2–6 cm proximal to the distal end of the lateral malleolus and is often associated with fracture-dislocations of the ankle joint. When a person slips, forcing the foot into an excessively inverted position, the ankle ligaments often tear, forcing the talus to tilt against the lateral malleolus, which shears it off. *The fibula is a common source of bone for grafting.* Even after a long piece of the fibula has been removed, walking, running, and jumping can be normal.

Fractures of the neck of the talus may occur during severe dorsiflexion of the ankle (e.g., when a person is pressing extremely hard on the brake pedal of a car during a head-on collision). In some injuries the body of the talus dislocates posteriorly.

Fractures of the calcaneus occur in persons who fall on their heels (e.g., from a ladder). Usually the bone breaks into several fragments that disrupt the subtalar joint, the articulation between the talus and calcaneus (p. 277).

Fractures of the metatarsals and phalanges usually occur when a heavy object falls on the foot or the foot is run over.

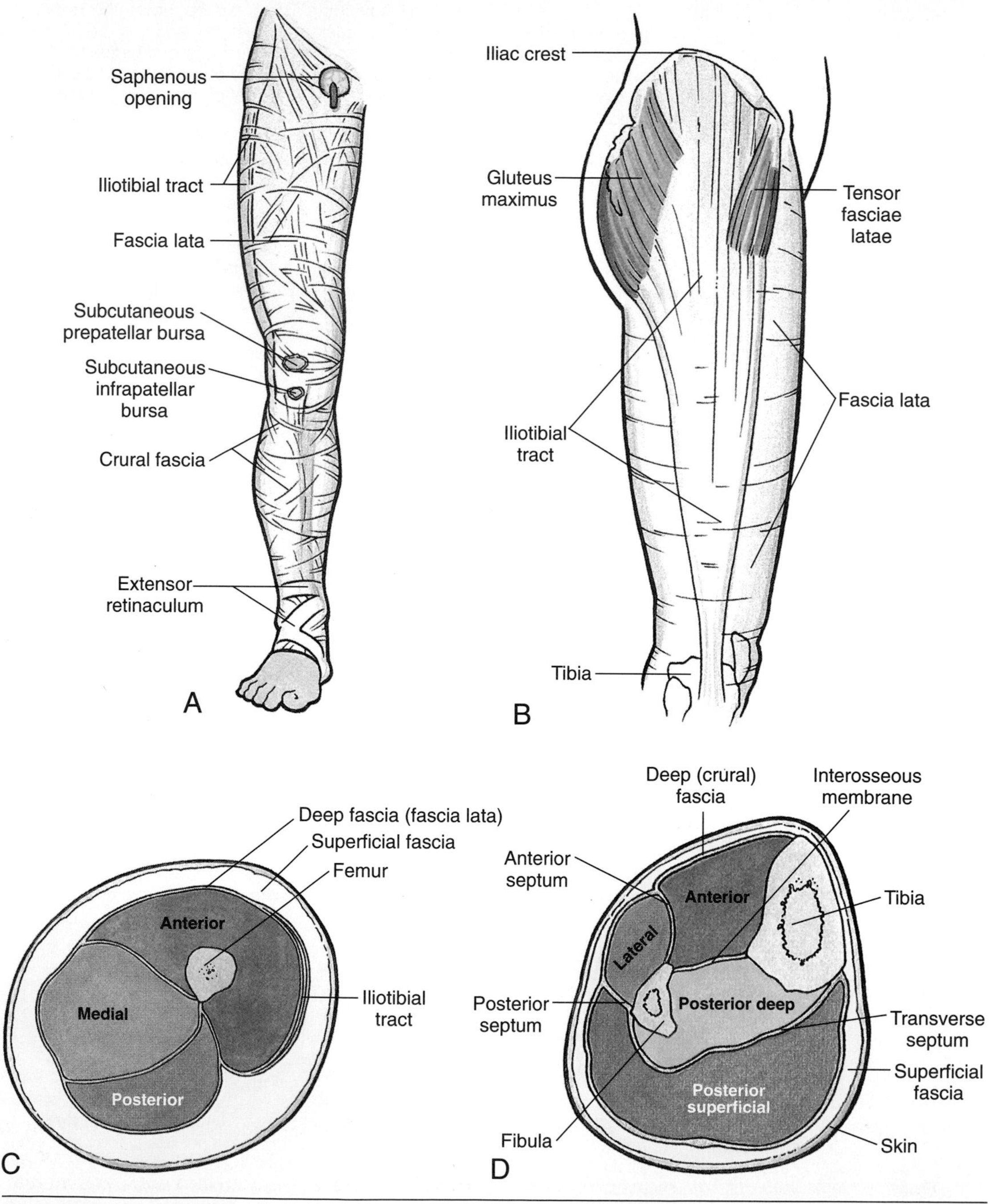

Figure 6.3. Fasciae of lower limb. **A.** Anterior view of deep fascia. **B.** Lateral view of hip and thigh, especially to show iliotibial tract of fascia lata. **C.** Transverse section of thigh showing its fascial compartments. **D.** Transverse section of leg showing its fascial compartments.

Fasciae

The fasciae of the lower limb consist of superficial and deep layers (Fig. 6.3).

The *superficial fascia* lies deep to the skin and consists of loose connective tissue that contains a variable amount of fat, cutaneous nerves, superficial veins (great and small saphenous veins and their tributaries), lymphatic vessels, and lymph nodes. The superficial fascia of the hip and thigh is continuous with that of the inferior part of the anterolateral abdominal wall and the buttock. At the knee the superficial fascia loses its fat and blends with the deep fascia, but fat is present in the superficial fascia of the leg.

The *deep fascia* is a dense layer of connective tissue between the superficial fascia and the muscles. The deep fascia invests the lower limb like an elastic stocking and is called fascia lata in the thigh and crural fascia in the leg.

The *fascia lata* is attached

- Superiorly to the inguinal ligament, pubic arch, body of the pubis, and pubic tubercle
- Laterally and posteriorly to the iliac crest
- Posteriorly to the sacrum, coccyx, sacrotuberous ligament, and ischial tuberosity

Distally the fascia lata is attached to the exposed bony parts around the knee and is continuous with the crural fascia. The fascia lata is very strong laterally where it extends from the iliac tubercle to the tibia. This part, the *iliotibial tract*, receives tendinous reinforcements from the tensor fasciae latae and gluteus maximus muscles. The distal end of the straplike iliotibial tract is attached to the lateral condyle of the tibia.

The *saphenous opening* is a deficiency (gap) in the fascia lata inferior to the inguinal ligament and inferolateral to the pubic tubercle. The sievelike *cribriform fascia*, derived from the subcutaneous connective tissue, fills the saphenous opening. The great saphenous vein passes through the saphenous opening and cribriform fascia to join the femoral vein. Some efferent lymphatic vessels from the superficial inguinal lymph nodes also pass through this opening and fascia to enter the deep inguinal lymph nodes.

The thigh muscles are separated into three groups (anterior, medial, and posterior) by three fascial *intermuscular septa* (Fig. 6.3C) that arise from the deep aspect of the fascia lata and are attached to the linea aspera of the femur. The lateral intermuscular septum is strong; the other two septa are relatively weak.

The *crural fascia* [deep fascia of leg (L. crus, leg)] is continuous with the fascia lata. The crural fascia is attached to the anterior and medial borders of the tibia, where it is continuous with the periosteum. The fascia is very thick in the proximal part of the anterior aspect of the leg, where it forms part of the proximal attachments of the underlying muscles. Although thin in the distal part of the leg, the crural fascia is thicker where it forms the superior and inferior *extensor retinacula*. From the deep surface of the crural fascia, anterior and posterior *intermuscular septa* pass deeply to attach to the corresponding margins of the fibula. These septa along with the interosseous membrane divide the leg into compartments (Fig. 6.3D):

- Anterior (extensor) compartment
- Lateral [fibular (peroneal)] compartment
- Posterior (flexor) compartment, which is subdivided into superficial and deep compartments

Nerves and Vessels

CUTANEOUS NERVES

Several cutaneous nerves in the superficial fascia supply the skin of the lower limb (see Fig. 6.4, Table 6.3). These nerves, except for some proximal ones, are branches of the lumbar and sacral plexuses (pp. 135 and 152).

SUPERFICIAL VEINS

There are two major veins in the superficial fascia, the great (long) and small (short) *saphenous veins* (Fig. 6.5). Veins from the digits join to form a *dorsal venous arch* on the dorsum of the foot. Veins leave this arch and converge medially to form the great saphenous vein and laterally to form the small saphenous vein. The superficial veins of the sole unite to form a plantar

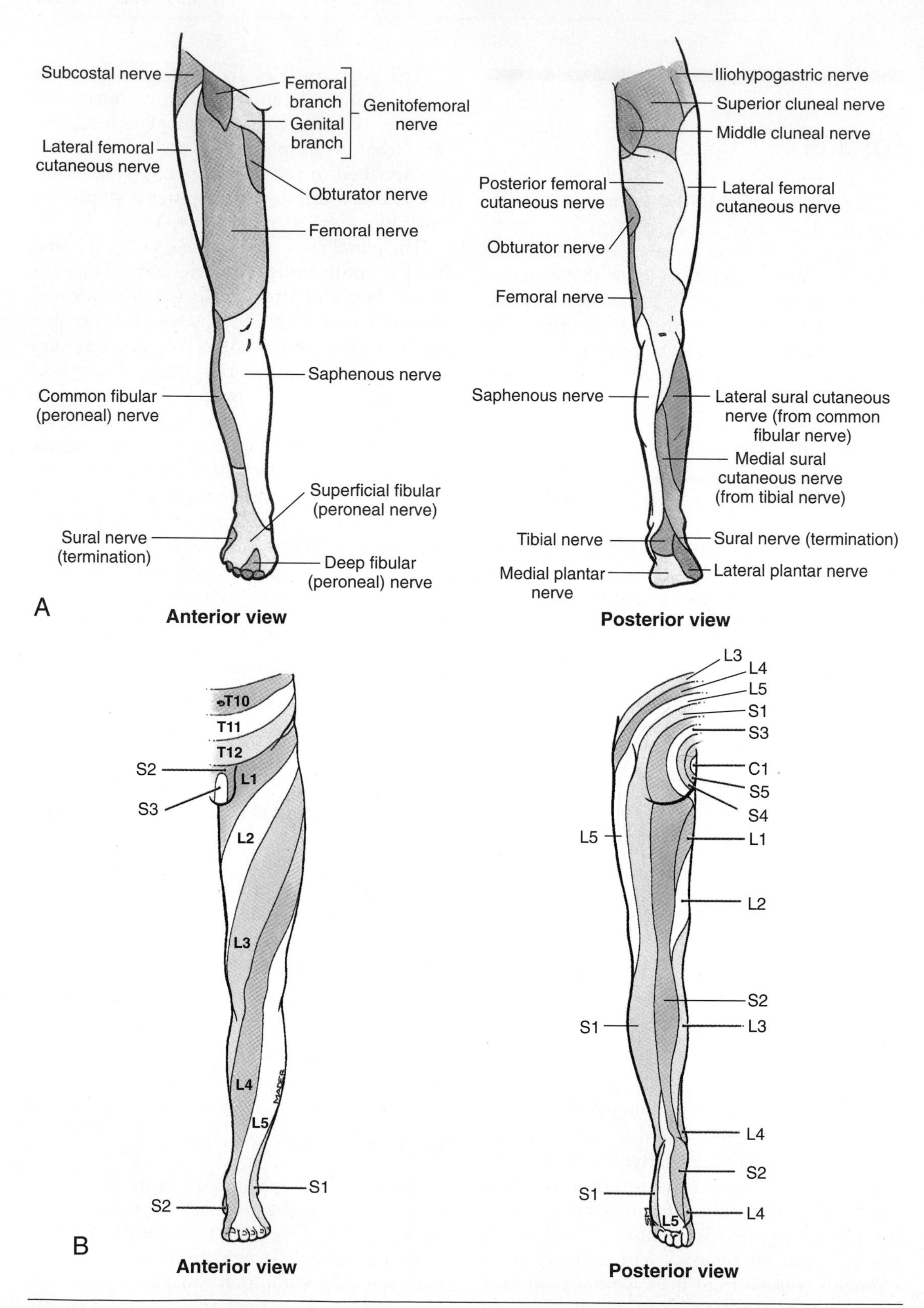

Figure 6.4. Cutaneous innervation of lower limb. **A.** Anterior and posterior views showing distribution of cutaneous nerves, usually containing fibers from more than one spinal nerve. **B.** Similar views of lower limb showing dermatomes (areas of distribution of each spinal nerve to skin).

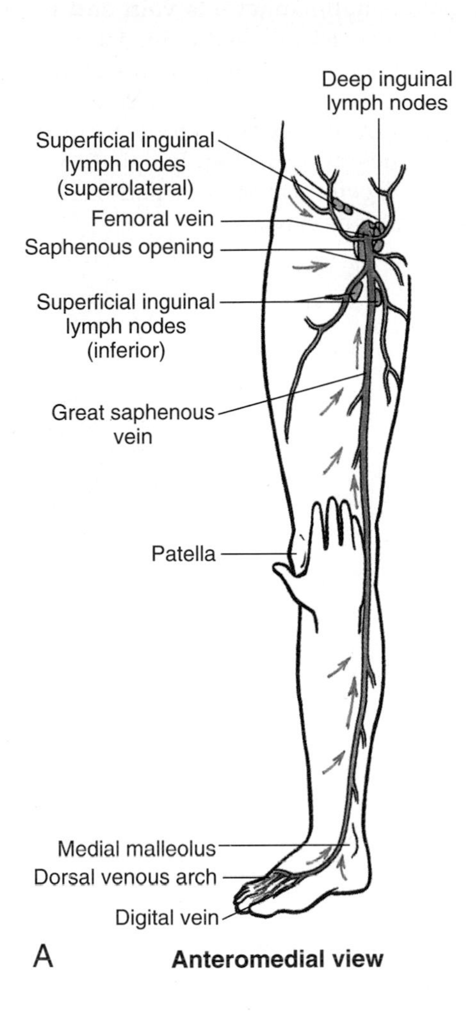

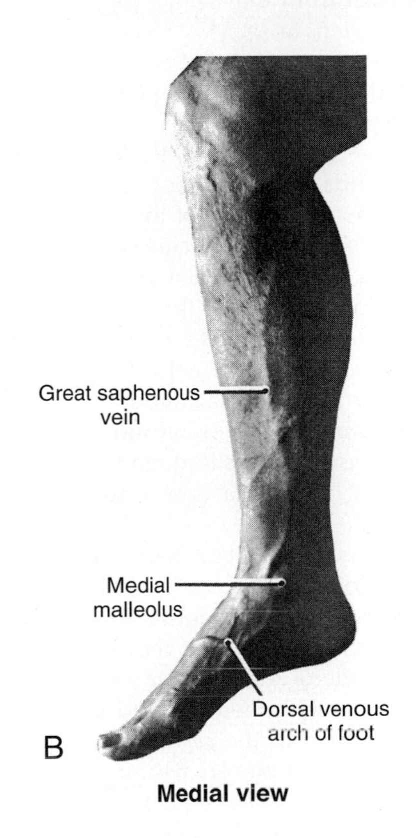

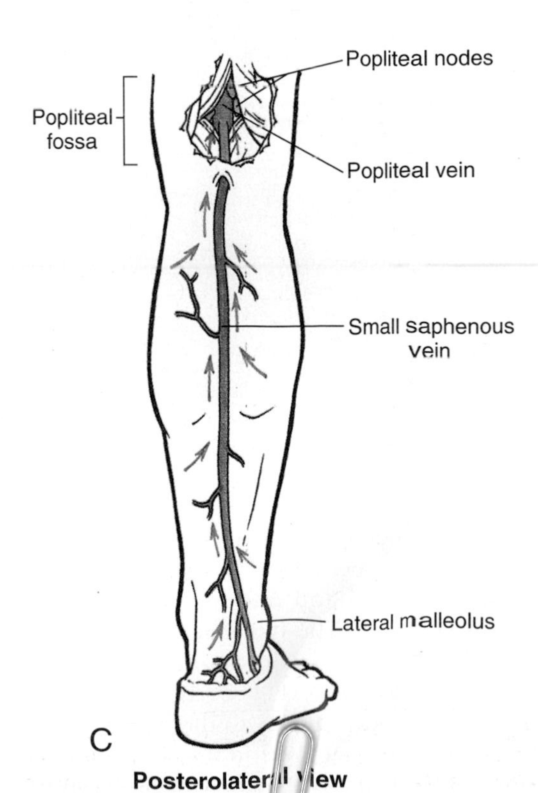

Figure 6.5. Superficial venous and lymphatic drainage of lower limb. **A.** Great saphenous vein showing its relationship to the medial malleolus and knee. *Green arrows*, superficial lymphatic drainage to inguinal lymph nodes. **B.** Photograph showing course of great saphenous vein. **C.** Short saphenous vein and superficial lymphatic drainage (*green arrows*) to popliteal lymph nodes.

venous arch from which the great and small saphenous veins arise. Most blood returns from the foot via deep veins that communicate with the superficial veins through perforating (communicating) veins.

The *great saphenous vein*, the longest in the body, ascends from the foot to the groin. It begins at the medial end of the dorsal venous arch of the foot and passes anterior to the medial malleolus. The vein then passes obliquely across the inferior third of the tibia and ascends to the medial aspect of the knee, where it lies superficial to the medial epicondyle, about 10 cm posterior to the medial border of the patella. From here, it ascends superolaterally in the superficial fascia, passes through the saphenous opening in the fascia lata, and enters the femoral vein.

The *small saphenous vein* begins at the lateral end of the dorsal venous arch and passes posterior to the lateral malleolus with the sural nerve. The small saphenous vein ascends along the lateral side of the calcaneal tendon (tendo calcaneus or Achilles tendon) and the posterior aspect of the leg to the *popliteal fossa*, the diamond-shaped area posterior to the knee. In the popliteal fossa the small saphenous vein perforates the deep popliteal fascia and usually ends in the popliteal vein.

When the valves of the perforating veins are incompetent (dilated so that their cusps do not close), contractions of the calf muscles, which normally propel the blood superiorly, cause a reverse flow of blood. As a result, the superficial veins become enlarged and tortuous, forming *varicose veins*, often on the posterior aspect of the leg.

Vein grafts obtained from the great saphenous vein are used to bypass obstructions in blood vessels (e.g., an atheromatous occlusion of a coronary artery or one of its branches). When a portion of the vein is used as a bypass, it is reversed so that the valves do not obstruct blood flow.

LYMPHATIC DRAINAGE

There are deep and superficial lymphatic vessels. Most superficial vessels accompany the saphenous veins and their tributaries in the superficial fascia. The lymphatic vessels along the great saphenous vein end in the *superficial inguinal lymph nodes* (Fig. 6.5*A*). Most lymph from these nodes passes to the external iliac lymph nodes; some goes to the deep inguinal lymph nodes. The lymphatic vessels accompanying the small saphenous vein end in the *popliteal lymph nodes* located in the fat posterior to the knee. The deep lymphatic vessels accompanying the deep blood vessels also end in the popliteal lymph nodes. Lymph from these nodes ascends through deep lymphatic vessels to the *deep inguinal lymph nodes* and then to the external iliac lymph nodes.

Lymph nodes enlarge when diseased. *Minor sepsis* (presence of pathogenic microorganisms or their toxins in blood or other tissues) and abrasions of the lower limb produce slight enlargement of the inguinal lymph nodes in otherwise healthy people. Malignancies of the external genitalia and perineal abscesses also result in enlargement of these nodes.

Organization of Thigh Muscles

The thigh muscles are organized into three compartments by intermuscular septa that pass between the muscles from the fascia lata to the femur. The compartments are named anterior, medial, and posterior on the basis of their location, actions, and nerve supply (see Fig. 6.11).

ANTERIOR THIGH MUSCLES

The anterior thigh muscles are located in the anterior compartment. For attachments, nerve supply, and main actions of these muscles see Table 6.1 and Figures 6.6 and 6.7. The anterior thigh muscles are the

- Iliopsoas: two separate muscles at its origin, the psoas major and iliacus, that arise in the abdomen and enter the thigh deep to the inguinal ligament to insert on the lesser trochanter of the femur
- Tensor fasciae latae: a fusiform, straplike muscle that lies on the lateral side of the thigh and is enclosed between two layers of fascia lata
- Pectineus: a flat quadrangular muscle that adducts the thigh
- Sartorius: the most superficial muscle in the anterior thigh that acts across two joints, the hip and knee, and throughout much of its course covers the femoral artery
- Quadriceps femoris: the extensor of the leg

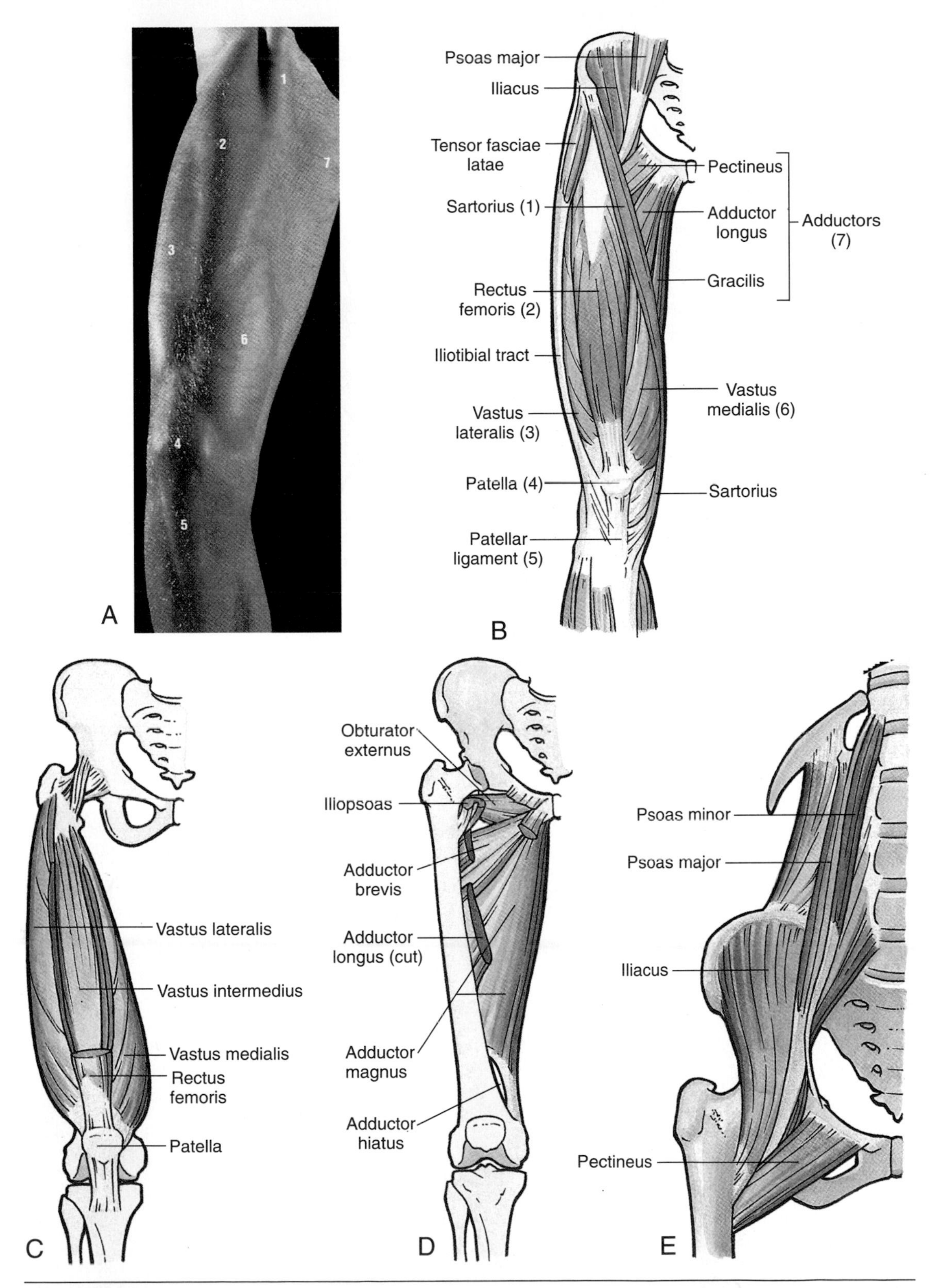

Figure 6.6. **A.** Surface anatomy of thigh and proximal part of leg. **B.** Muscles in anterior part of thigh. *Numbers* in **A** refer to structures labeled in **B. C.** Three vastus muscles; most of rectus femoris has been removed to show vastus intermedius. **D.** Deep dissection of medial compartment of thigh to show adductor magnus. **E.** Iliopsoas (psoas major and iliacus) and pectineus.

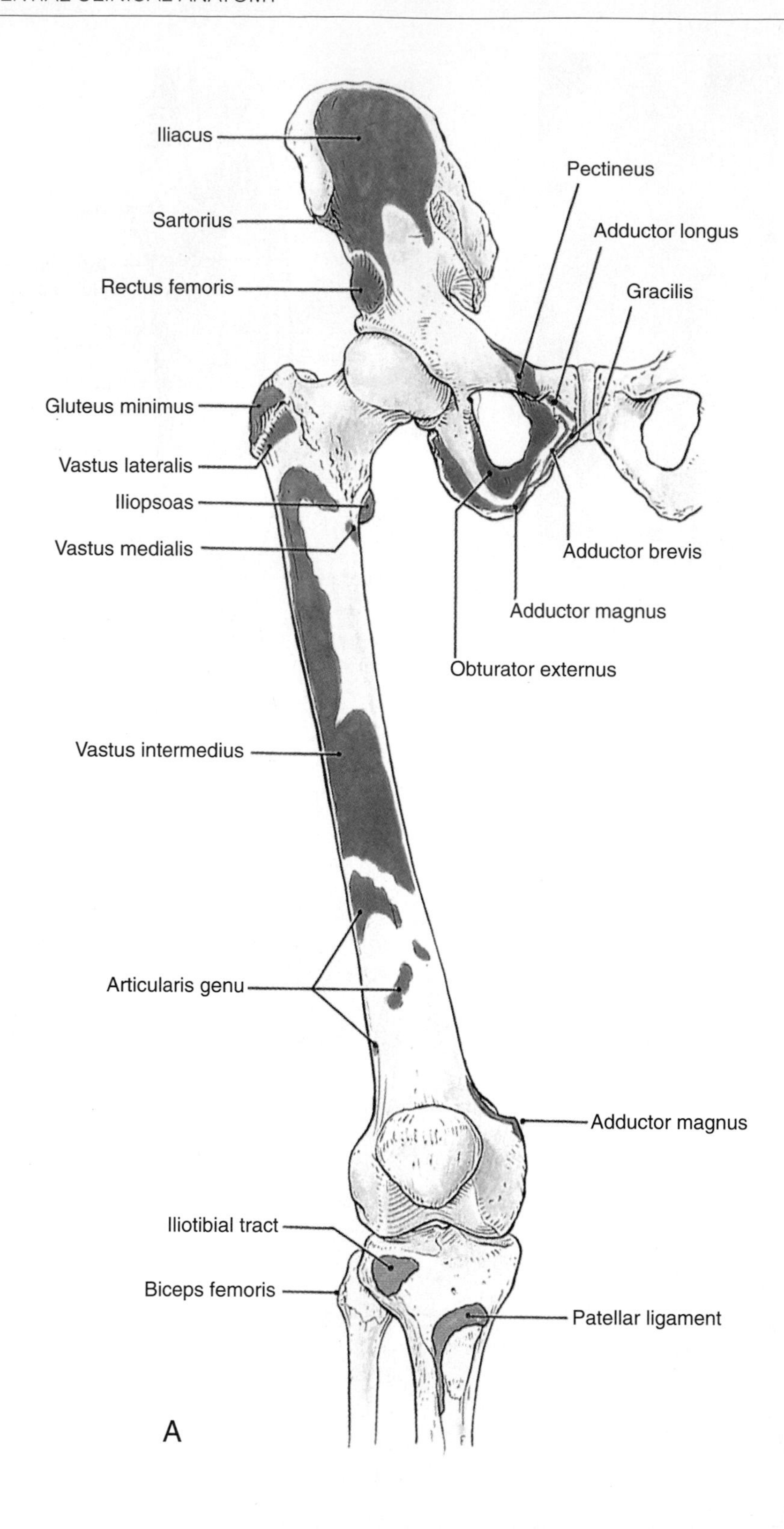

Figure 6.7. Bones of lower limb showing attachments of muscles. **A.** Anterior view. **B.** Posterior view.

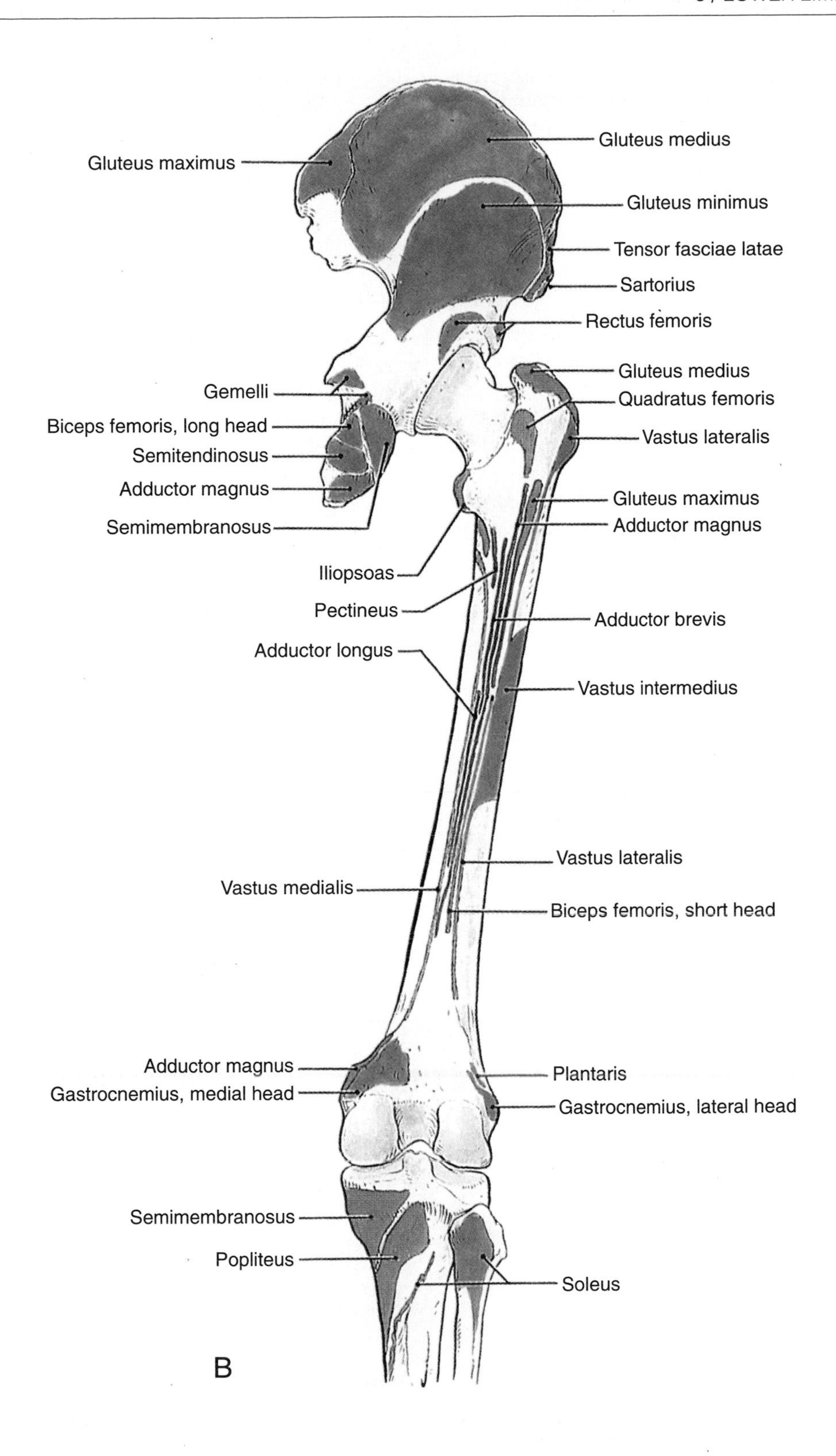

Figure 6.7. *Continued.*

at the knee joint that covers the anterior, medial, and lateral aspects of the femur and is divided into (*a*) rectus femoris located on the anterior aspect of the thigh, (*b*) vastus lateralis lying on the lateral side of the thigh, (*c*) vastus medialis covering the medial side of the thigh, and (*d*) vastus intermedius located deep to the rectus femoris and between the vastus medialis and vastus lateralis

A small flat muscle, the *articularis genu,* may blend with the vastus intermedius. It attaches superiorly to the inferior part of the anterior aspect of the femur and inferiorly to the synovial capsule of the knee joint and the wall of the suprapatellar bursa. The articularis genu pulls the synovial capsule superiorly during extension of the leg so it will not be caught between the patella and femur.

The quadriceps femoris is tested with the patient lying on the back with the knee partly flexed. The patient is asked to extend the knee against resistance. If the quadriceps is functioning normally, it can be seen and felt easily; if it is paralyzed, the person cannot extend the leg. A person with *paralysis of the quadriceps femoris* often presses on the distal end of the thigh during walking to prevent flexion of the knee joint.

A common knee problem for runners is *chondromalacia patellae* ("runner's knee"). The soreness and aching around or deep to the patella is due to *quadriceps imbalance.* This condition may be caused by a blow to the patella or from extreme flexion of the knee joint (e.g., during squatting and kneeling). Such overstressing of the knee can occur in any running sport such as jogging, basketball, or soccer.

A direct blow on the patella may fracture it in two or more fragments. Transverse *patellar fractures* may result from a blow to the knee or from sudden contraction of the quadriceps (e.g., when one slips and attempts to prevent a backward fall). The proximal fragment is pulled superiorly with the quadriceps tendon, and the distal fragment remains with the patellar ligament. If the patella is removed, more force (about 30%) may be required by the quadriceps to extend the leg completely.

The tendons of the four parts of the quadriceps femoris unite to form the *quadriceps tendon.* This broad band attaches to the patella. The *patellar ligament* attaches the patella to the *tibial tuberosity* and is the continuation of the quadriceps tendon. The apex of the triangular patella, directed inferiorly, indicates the level of the knee joint when the patellar ligament is taut. The *patella,* a sesamoid bone, lies anterior to the distal end of the femur; here it articulates posteriorly with the condyles of the femur. The patella increases the power of the already strong quadriceps femoris by holding the quadriceps tendon away from the distal end of the femur. This improves the tendon's angle to the tibial tuberosity and increases its leverage.

MEDIAL THIGH MUSCLES

The medial thigh muscles (adductor group) are located in the medial compartment of the thigh. The adductor group of muscles consists of the

- Adductor longus: the most anterior adductor muscle
- Adductor brevis: deep to adductor longus and anterior to the adductor magnus
- Adductor magnus: largest adductor muscle that is composed of adductor and hamstring parts
- Gracilis: a long, straplike muscle lying along the medial side of the thigh and knee
- Obturator externus: a deeply placed fan-shaped muscle in the superomedial part of the thigh

An opening in the aponeurotic attachment of the adductor magnus, the *adductor hiatus,* leads into the popliteal fossa posterior to the knee (Figs. 6.5*C* and 6.6*D*). All adductor muscles, except the pectineus and part of the adductor magnus, are supplied by the *obturator nerve* (L2–L4). The pectineus is supplied by the femoral nerve (L2–L4), and the "hamstring part" of the adductor magnus is supplied by the sciatic nerve (L4). For attachments, nerve supply, and main actions of the medial thigh muscles, see Figures 6.6 and 6.7 and Table 6.2.

A "pulled groin" or groin injury means that there is a strain, stretching, and probably some tearing of the superior attachments of one or more of the medial thigh muscles. The injury, usually involving the muscles in the adductor group, occurs in sports that require quick starts (e.g., 50-meter dash, basketball, football, baseball, and soccer).

Femoral Triangle

The femoral triangle is a triangular fascial space in the superomedial third of the thigh (Fig. 6.8). It appears as a depression inferior to the inguinal ligament when the thigh is flexed, abducted, and laterally rotated.

Table 6.1
Anterior Thigh Muscles

Muscle	Proximal Attachment	Distal Attachment	Innervation[a]	Main Actions
Iliopsoas				
Psoas major	Sides of T12–L5 vertebrae and discs between them; transverse processes of all lumbar vertebrae	Lesser trochanter of femur	Ventral rami of lumbar nerves (**L1**, **L2**, and L3)	Acting jointly in flexing thigh at hip joint and in stabilizing this joint[b]
Iliacus	Iliac crest, iliac fossa, ala of sacrum, and anterior sacroiliac ligaments	Tendon of psoas major, lesser trochanter, and femur distal to it	Femoral nerve (**L2** and L3)	
Tensor fasciae latae	Anterior superior iliac spine and anterior part of iliac crest	Iliotibial tract that attaches to lateral condyle of tibia	Superior gluteal (L4 and L5)	Abducts, medially rotates, and flexes thigh; helps to keep knee extended; steadies trunk on thigh
Sartorius	Anterior superior iliac spine and superior part of notch inferior to to it	Superior part of medial surface of tibia	Femoral nerve (L2 and L3)	Flexes, abducts, and laterally rotates thigh at hip joint; flexes leg at knee joint[c]
Quadiceps femoris				
Rectus femoris	Anterior inferior iliac spine and ilium superior to acetabulum	Base of patella and by patellar ligament to tibial tuberosity[d]	Femoral nerve (L2, **L3**, and **L4**)	Extends leg at knee joint; rectus femoris also steadies hip joint and helps iliosoas to flex thigh
Vastus lateralis	Greater trochanter and lateral lip of linea aspera of femur			
Vastus medialis	Intertrochanteric line and medial lip of linea aspera of femur			
Vastus intermedius	Anterior and lateral surfaces of body of femur			

[a] Numbers indicate spinal cord segmental innervation of nerves [e.g., **L1**, **L2**, and L3 indicate that nerves supplying psoas major are derived from first three lumbar segments of the spinal cord; boldface type (**L1**, **L2**) indicates main segmental innervation]. Damage to one or more of these spinal cord segments or to motor nerve roots arising from them results in paralysis of the muscles concerned.

[b] Psoas major is also a postural muscle that helps control deviation of trunk and is active during standing.

[c] Four actions of sartorius (L. sartor, tailor) produce the once common crosslegged sitting postion used by tailors—hence the name.

[d] See Figure 6.7 for attachments of muscles.

The femoral triangle is bounded

- Superiorly by inguinal ligament
- Medially by adductor longus
- Laterally by sartorius

The *base* of the femoral triangle is formed by the inguinal ligament. The *apex* is the point where the medial border of the sartorius crosses the medial border of the adductor longus. The muscular *floor* is formed from lateral to medial by the iliopsoas, pectineus, and adductor longus. The *roof* is formed by fascia lata and cribriform fascia.

From lateral to medial, the main contents of the femoral triangle are the

- Femoral nerve and its branches
- Femoral artery and its branches

Table 6.2
Medial Thigh Muscles

Muscle	Proximal Attachment	Distal Attachment[a]	Innervation[b]	Main Actions
Pectineus	Superior ramus of pubis	Pectineal line of femur, just inferior to lesser trochanter	Femoral nerve (**L2**, and L3); may receive a branch from obturator nerve	Adducts and flexes thigh; assists with medial rotation of thigh
Adductor longus	Body of pubis inferior to pubic crest	Middle third of linea aspera of femur	Obturator nerve, anterior branch (L2, **L3**, and L4)	Adducts thigh
Adductor brevis	Body and inferior ramus of pubis	Pectineal line and proximal part of linea aspera of femur	Obturator nerve (L2, **L3**, and L4)	Adducts thigh and to some extent flexes it
Adductor magnus	Inferior ramus of pubis, ramus of ischium (adductor part), and ischial tuberosity (hamstring part)	Gluteal tuberosity, linea aspera, medial supracondylar line (adductor part), and adductor tubercle of femur (hamstring part)	*Adductor part*: obturator nerve (L2, **L3**, and **L4**) *Hamstring part*: tibial part of sciatic nerve (**L4**)	Adducts thigh; its adductor part also flexes thigh, and its hamstring part extends it
Gracilis	Body and inferior ramus of pubis	Superior part of medial surface of tibia	Obturator nerve (**L2** and L3)	Adducts thigh, flexes leg, and helps rotate it medially
Obturator externus	Margins of obturator foramen and obturator membrane	Trochanteric fossa of femur	Obturator nerve (L3 and **L4**)	Laterally rotates thigh; steadies head of femur in acetabulum

Collectively, the first five muscles listed are the adductors of the thigh, but their actions are more complex (e.g., they act as flexors of the hip joint during flexion of the knee joint and are active during walking).

[a] See Figure 6.7 for muscle attachments.

[b] See Table 6.1 for explanation of segmental innervation.

- Femoral vein and its tributaries, including the great saphenous vein
- Femoral sheath

Femoral Nerve. The femoral nerve (L2–L4)—the largest branch of the lumbar plexus—forms in the abdomen within the psoas major and descends posterolaterally through the pelvis to the midpoint of the inguinal ligament. It then passes deep to the inguinal ligament lateral to the femoral vessels. After entering the femoral triangle, the femoral nerve divides into several terminal branches that supply the anterior thigh muscles (Tables 6.1 and 6.3). It also sends articular branches to the hip and knee joints and gives several branches to the skin on the anteromedial side of the lower limb.

The *saphenous nerve*, a cutaneous branch of the femoral nerve (Fig. 6.8), descends through the femoral triangle, lateral to the femoral sheath (p. 238). The saphenous nerve accompanies the femoral artery in the adductor canal and becomes superficial by passing between the sartorius and gracilis muscles. It runs anteroinferiorly to supply the skin and fascia on the anterior and medial aspects of the knee, leg, and foot.

Femoral Artery. The femoral artery—the chief artery to the lower limb—is the continuation of the external iliac artery (Fig. 6.8*A*, Table 6.4). The femoral artery enters the femoral triangle deep to the midpoint of the inguinal ligament and lateral to the femoral vein. It is posterior to the fascia lata and descends on the psoas major, pectineus, and adductor longus muscles in the floor of the femoral triangle. The femoral artery bisects the femoral triangle, and at its apex, it runs deep to the sartorius within the *adductor canal* and then passes through the adductor hiatus to become the popliteal artery.

The *deep femoral artery* (profunda femoris artery)—the chief artery to the thigh—is the largest branch of the femoral artery. It arises from the lateral side of the femoral artery with-

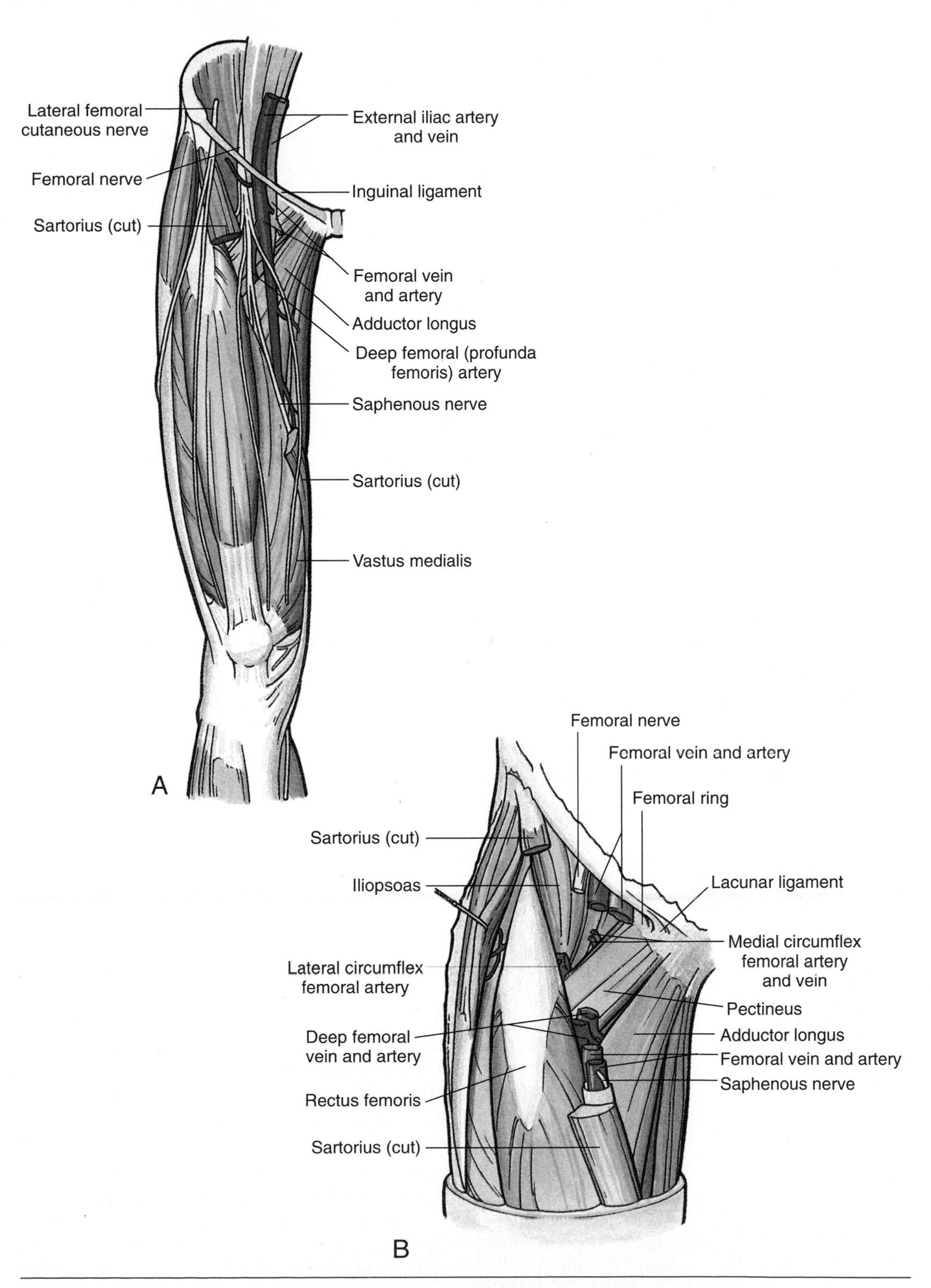

Figure 6.8. **A.** Dissection of anterior part of thigh showing muscles, vessels, and nerves. **B.** Deep dissection showing floor of femoral triangle and its relationship to medial and lateral circumflex femoral vessels.

Table 6.3
Nerves of Thigh

Nerve	Origin	Course	Distribution
Genitofemoral	Lumbar plexus (L1 and L2)	Descends on anterior surface of psoas major and divides into genital and femoral branches	Femoral branch supplies skin over femoral triangle; genital branch supplies scrotum or labia majora
Ilioinguinal	Lumbar plexus (L1)	Passes through inguinal canal and divides into femoral and scrotal or labial branches	Femoral branch supplies skin over femoral triangle
Posterior femoral cutaneous	Sacral plexus (S2–S3)	Passes through greater sciatic foramen inferior to piriformis m., runs deep to gluteus maximus, and emerges from its inferior border	In addition to buttock, it supplies skin over posterior aspect of thigh
Lateral femoral cutaneous	Lumbar plexus (L2 and L3)	Passes deep to inguinal ligament, 2–3 cm medial to anterior superior iliac spine	Supplies skin on anterior and lateral aspects of thigh
Medial and intermediate femoral cutaneous	Femoral n.	Arise in femoral triangle and pierce fascia lata of thigh	Skin on medial and anterior aspect of thigh
Femoral	Lumbar plexus (L2–L4)	Passes deep to midpoint of inguinal ligament, lateral to femoral vessels, and divides into muscular and cutaneous branches	Supplies anterior thigh muscles, hip and knee joints, and skin on anteromedial side of thigh
Sciatic	Sacral plexus (L4–S3)	Enters gluteal region through greater sciatic foramen, inferior to piriformis m., descends along posterior aspect of thigh, and divides proximal to knee into tibial and common fibular peroneal nn.	Innervates hamstrings by its tibial division, except for short head of biceps femoris, which is innervated by its common fibular division; provides articular branches to hip and knee joints
Obturator	Lumbar plexus (L3–L4)	Enters thigh through obturator foramen and divides; its anterior branch descends between adductor longus and adductor brevis; its posterior branch descends between adductor brevis and adductor magnus	Anterior branch supplies adductor longus, adductor brevis, gracilis, and pectineus; posterior branch supplies obturator externus and adductor magnus

in the femoral triangle, about 4 cm inferior to the inguinal ligament. It runs lateral to the femoral artery and then passes posterior to it and the femoral vein. The deep femoral artery leaves the femoral triangle between the pectineus and adductor longus and descends posterior to the latter muscle, giving off perforating arteries that supply the adductor magnus and hamstrings (p. 247).

The *circumflex femoral arteries* are usually branches of the deep femoral artery, but they may arise directly from the femoral artery. They encircle the thigh, anastomose with each other and other arteries, and supply the thigh muscles and the proximal end of the femur. The *medial circumflex femoral* supplies most of the blood to the head and neck of the femur. It passes deeply between the iliopsoas and pectineus to reach the posterior part of the thigh. The *lateral circumflex femoral* passes laterally, deep to the sartorius and rectus femoris, and between the branches of the femoral nerve. Here it divides into branches that supply the head of the femur and muscles on the lateral side of the thigh.

The *obturator artery* assists the deep femoral artery to supply the adductor muscles of the thigh. Arising either from the internal iliac artery or as "an accessory obturator" from the inferior epigastric artery, it passes through the obturator foramen, enters the thigh, and divides into anterior and posterior branches.

Table 6.4
Arterial Supply to Thigh

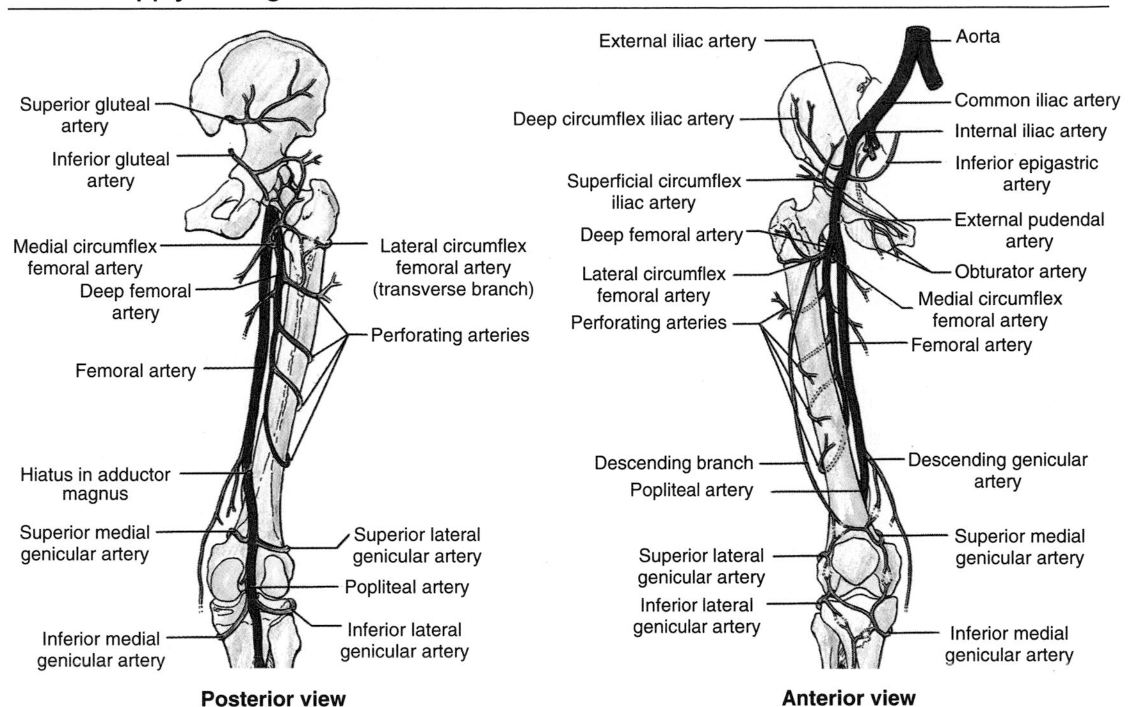

Artery	Origin	Course	Distribution
Femoral	Continuation of external iliac a. distal to inguinal ligament	Descends through femoral triangle, enters adductor canal, and ends at adductor hiatus	Anterior and anteromedial surfaces of thigh
Deep femoral (profunda femoris)	Femoral a. about 4 cm distal to inguinal ligament	Passes inferiorly, deep to adductor longus	Perforating branches pass through adductor magnus m. to posterior compartment of thigh
Lateral circumflex femoral	Deep femoral a.; may arise from femoral a.	Passes laterally deep to sartorius and rectus femoris and divides into three branches	Ascending branch supplies anterior part of gluteal region; transverse branch winds around femur; descending branch descends to knee and joins genicular anastomoses
Medial circumflex femoral	Deep femoral a. or may arise from femoral a.	Passes medially and posteriorly between pectineus and iliopsoas, enters gluteal region, and divides into two branches	Supplies most blood to head and neck of femur; transverse branch takes part in cruciate anastomosis of thigh (p. 245); ascending branch joins inferior gluteal a.
Obturator	Internal iliac a.	Passes through obturator foramen, enters medial compartment of thigh, and divides into anterior and posterior branches	Anterior branch supplies obturator externus, pectineus, adductors of thigh, and gracilis; posterior branch supplies muscles attached to ischial tuberosity

The posterior branch gives off an *acetabular branch* that supplies the head of the femur, but the main supply of the head is from the medial circumflex femoral artery.

Femoral Vein. The femoral vein enters the thigh medial to the femoral artery and passes over the pectineus. The femoral vein ends posterior to the inguinal ligament, where it becomes the external iliac vein (Fig. 6.8*A*). In the inferior part of the femoral triangle, the femoral vein lies deep to the femoral artery and receives the deep femoral vein (profunda femoris vein), the great saphenous vein, and other tributaries.

Femoral Sheath. The femoral sheath is a funnel-shaped, fascial tube that encloses proximal parts of the femoral vessels and the femoral canal. *The femoral sheath does not enclose the femoral nerve.* The femoral sheath, an inferior prolongation of the transversalis fascia in the abdomen (p. 81), ends about 4 cm inferior to the inguinal ligament by becoming continuous with the loose connective tissue covering of the femoral vessels. The medial wall of the femoral sheath is pierced by the great saphenous vein and lymphatic vessels (Fig. 6.5*A*). The femoral sheath allows the femoral artery and vein to glide deep to the inguinal ligament during movements of the hip joint. The femoral sheath is subdivided by two vertical septa into three compartments: a lateral compartment for the femoral artery, an intermediate compartment for the femoral vein, and a medial compartment (space) that is the femoral canal.

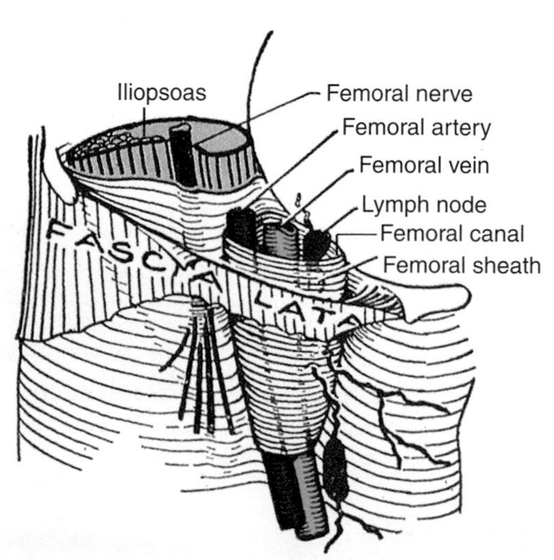

Fascia of superior thigh

The femoral artery can be compressed by pressing posteriorly at the midinguinal point. The *pulse of the femoral artery* is usually palpable just inferior to the inguinal ligament, at a point halfway between the anterior superior iliac spine and the pubic tubercle. Normally the pulse is strong, but if the lumina of the common or external iliac arteries are partially occluded, the pulse may be diminished. The femoral artery is easily exposed and cannulated [e.g., for cardioangiography (radiography of the heart and great vessels after introduction of contrast material)]. For *left cardiac angiography*, a long slender catheter is inserted into the femoral artery and passed superiorly to the openings of the coronary arteries (p. 63).

In about 20% of people an enlarged pubic branch of the inferior epigastric artery takes the place of the obturator artery or forms an *accessory obturator artery*. This artery runs close to or across the femoral ring to reach the obturator foramen. Here, it is closely related to the free margin of the lacunar ligament and the neck of a femoral hernia. Consequently, this artery could be involved in a *strangulated femoral hernia* (one that is so tightly constricted that the blood supply to the hernia is compromised).

The femoral vein is not usually palpable, but its position can be located by feeling the pulsations of the femoral artery, which lies just lateral to it. In thin people the femoral vein is surprisingly close to the surface and may be mistaken for the great saphenous vein.

To secure blood samples and take pressure recordings from the right chambers of the heart and/or from the pulmonary artery, or for *right cardiac angiography*, a long slender catheter is inserted into the femoral vein as it ascends through the femoral triangle. By using fluoroscopic control, the catheter is passed through the external and common iliac veins and inferior vena cava into the right atrium of the heart (p. 58).

The femoral ring (Fig. 6.8*B*) is a weak area in the anterior abdominal wall that is the usual site of a *femoral hernia*, a protrusion of abdominal viscera (often small intestine) through the femoral ring into the femoral canal. The hernial sac compresses the contents of the femoral canal and distends its wall. Initially the hernia is small because it is contained within the femoral canal, but it can enlarge by passing inferiorly through the saphenous opening into the loose connective tissue of the thigh. *Strangulation of a femoral hernia* interferes with the blood supply to the herniated intestine, and this vascular impairment may result in death of the tissues.

The *femoral canal* is the short, conical medial compartment of the femoral sheath that lies between the medial edge of the femoral sheath and the femoral vein. This space allows the femoral vein to expand when there is increased venous return from the lower limb. The femoral canal contains loose connective tissue, fat, a few lymphatic vessels, and sometimes a deep inguinal lymph node. The canal is widest at its abdominal end, the femoral ring, and extends distally to the level of the proximal edge of the saphenous opening.

The *femoral ring* is the small superior end or abdominal opening of the femoral canal (Fig. 6.8*B*). The ring, about 1 cm wide, is closed by extraperitoneal fatty tissue called the *femoral septum*. This septum is pierced by the lymphatic vessels connecting the inguinal and external iliac lymph nodes. The boundaries of the femoral ring are

- Laterally, the partition between the femoral ring and femoral vein
- Posteriorly, the superior ramus of the pubis covered by the pectineus and its fascia
- Medially, the lacunar ligament and conjoint tendon
- Anteriorly, the medial part of the inguinal ligament

Adductor Canal

The adductor (subsartorial) canal, about 15 cm long, is a narrow fascial tunnel in the thigh (Fig. 6.8). Located deep to the middle third of the sartorius, it provides an intermuscular passage through which the femoral vessels pass to reach the popliteal fossa and become popliteal vessels. The adductor canal begins where the sartorius crosses over the adductor longus and ends at the *adductor hiatus* in the tendon of the adductor magnus (Fig. 6.6*D*). The contents of the adductor canal are the

- Femoral artery and vein
- Saphenous nerve
- Nerve to the vastus medialis

Gluteal Region

The gluteal region (buttocks) lies posterior to the pelvis between the iliac crest and the inferior border of the gluteus maximus. The *gluteal sulcus*, inferior to the *gluteal fold*, indicates the inferior border of the gluteus maximus.

GLUTEAL LIGAMENTS

The parts of the bony pelvis (hip bones, sacrum, and coccyx) are bound together by dense ligaments (Fig. 4.2). The *sacrotuberous and sacrospinous ligaments* convert the sciatic notches in the hip bones into greater and lesser sciatic foramina. The *greater sciatic foramen* is the passageway for structures entering or leaving the pelvis (e.g., sciatic nerve), whereas the *lesser sciatic foramen* is a passageway for structures entering or leaving the perineum (e.g., pudendal nerve).

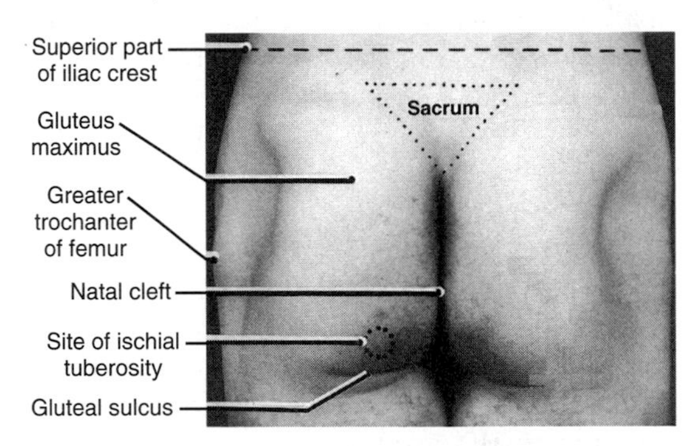

Gluteal region

GLUTEAL MUSCLES

The gluteal muscles (Fig. 6.9) consist of

- Large glutei (maximus, medius, and minimus) that are mainly extensors and abductors of the thigh at the hip joint
- A deeper group of smaller muscles (piriformis, obturator internus, gemelli, and quadratus femoris) that are mainly lateral (external) rotators of the thigh at the hip joint

For the attachments, nerve supply, and main actions of the gluteal muscles, see Figures 6.7 and 6.9 and Table 6.5. Three *gluteal bursae* separate the gluteus maximus from underlying structures. Bursae are membranous sacs containing a capillary layer of synovial fluid; they are located in the superficial fascia between the skin and bone. The purpose of bursae is to reduce friction.

- The trochanteric bursa separates the gluteus maximus from the lateral side of the greater trochanter of the femur
- The gluteofemoral bursa separates the gluteus maximus from the superior part of the proximal attachment of the vastus lateralis
- The ischial bursa separates the gluteus maximus from the ischial tuberosity

GLUTEAL NERVES

The skin of the gluteal region is richly inner-

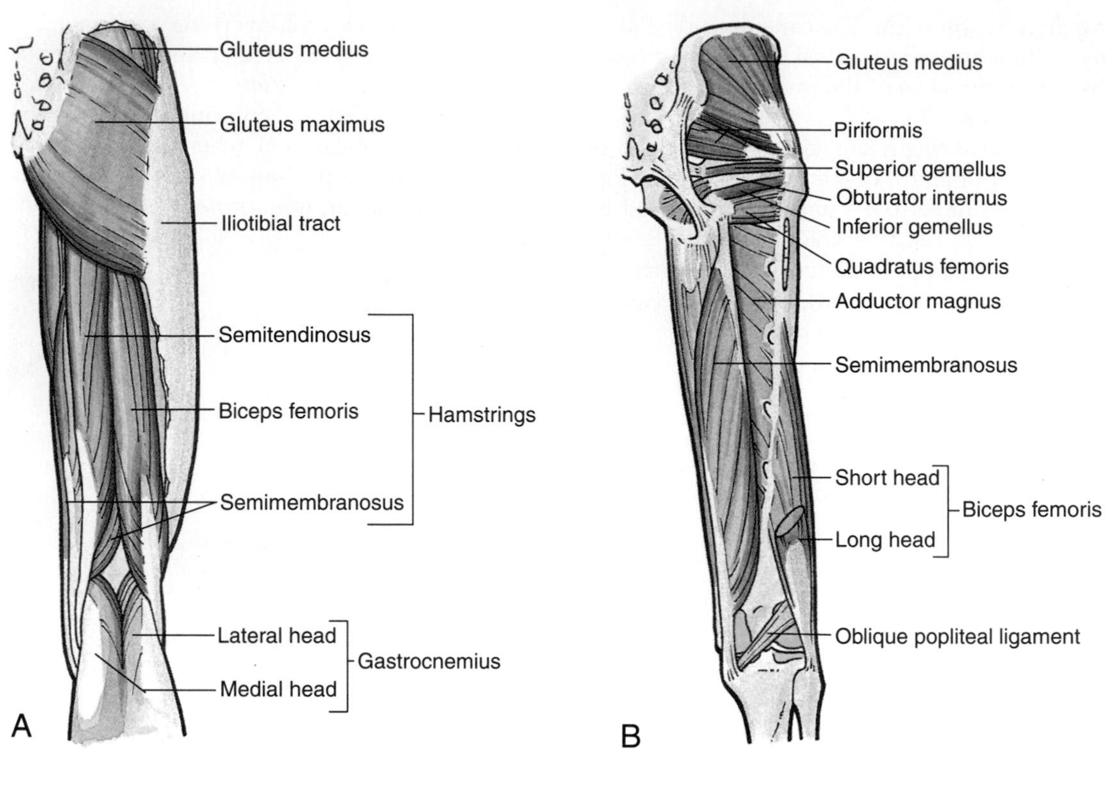

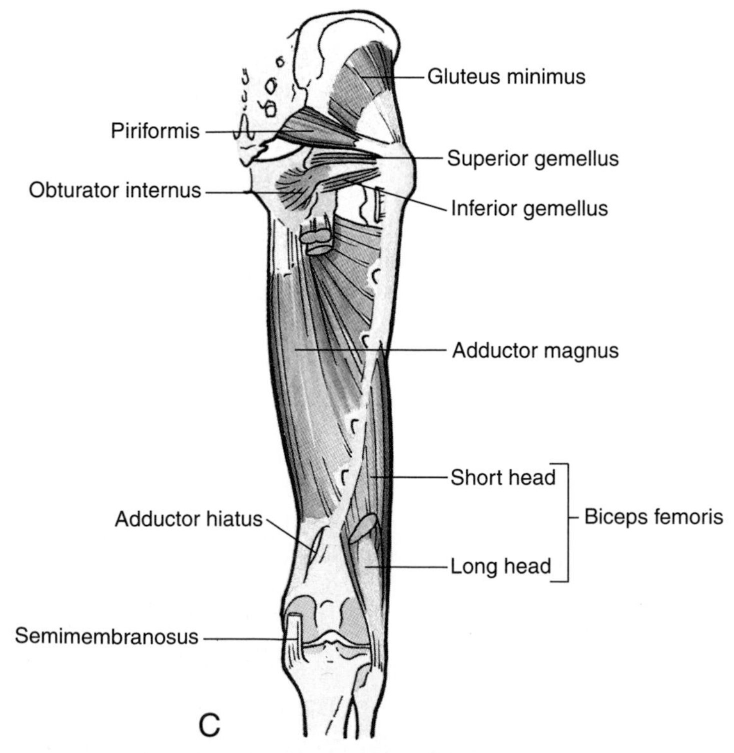

Figure 6.9. Muscles of gluteal region and posterior aspect of thigh. **A.** Superficial dissection of gluteus maximus and hamstrings. **B.** Deeper dissection showing gluteus medius, lateral rotators of hip, semimembranosus, and short head of biceps femoris. **C.** Still deeper dissection showing gluteus minimus and adductor magnus.

Table 6.5
Muscles of Gluteal Region

Muscle	Proximal Attachment	Distal Attachment[a]	Innervation[b]	Main Actions
Gluteus maximus	Ilium posterior to posterior gluteal line (Fig. 6.1*B*), dorsal surface of sacrum and coccyx, and sacrotuberous ligament	Most fibers end in iliotibial tract that inserts into lateral condyle of tibia; some fibers insert on gluteal tuberosity of femur	Inferior gluteal nerve (L5, **S1**, and **S2**)	Extends thigh and assists in its lateral rotation; steadies thigh and assists in raising trunk from flexed position
Gluteus medius	External surface of ilium between anterior and posterior gluteal lines	Lateral surface of greater trochanter of femur	Superior gluteal nerve (**L5** and S1)	Abduct and medially rotate thigh; steady pelvis on leg when opposite leg is raised
Gluteus minimus	External surface of ilium between anterior and inferior gluteal lines	Anterior surface of greater trochanter of femur		
Piriformis	Anterior surface of sacrum and sacrotuberous ligament	Superior border of greater trochanter of femur	Branches of ventral rami of S1 and S2	Laterally rotate extended thigh and abduct flexed thigh; steady femoral head in acetabulum
Obturator internus	Pelvic surface of obturator membrane and surrounding bones	Medial surface of greater trochanter of femur[c]	Nerve to obturator internus (L5 and **S1**)	
Gemelli, superior and inferior	Superior, ischial spine; inferior, ischial tuberosity		Superior gemellus: same nerve supply as obturator internus inferior gemellus: same nerve supply as quadratus femoris	
Quadratus femoris	Lateral border of ischial tuberosity	Quadrate tubercle on intertrochanteric crest of femur and inferior to it	Nerve to quadratus femoris (L5 and S1)	Laterally rotates thigh[d]; steadies femoral head in acetabulum

[a] See Figures 6.7 and 6.9 for muscle attachments.

[b] See Table 6.1 for explanation of segmental innervation.

[c] Gemelli muscles blend with tendon of obturator internus muscle as it attaches to greater trochanter of femur.

[d] There are six lateral rotators of the thigh: piriformis, obturator internus, gemelli (superior and inferior), quadratus femoris, and obturator externus. These muscles also stabilize the hip joint.

vated by superficial gluteal nerves, the *cluneal nerves* (Fig. 6.4*B*). The cluneal nerves (L. clunes, buttocks) supply the skin over the iliac crest, between the posterior superior iliac spines, and over the iliac tubercles. These nerves are vulnerable to injury when bone is taken from the ilium for grafting.

The *deep gluteal nerves* are the sciatic, posterior femoral cutaneous, superior gluteal and inferior gluteal nerves, nerve to the quadratus femoris, pudendal nerve, and nerve to the obturator internus (Table 6.6). All these nerves are branches of the sacral plexus (p. 152) and leave the pelvis through the greater sciatic foramen. Except for the superior gluteal nerve, they all emerge inferior to the piriformis (Fig. 6.10).

GLUTEAL ARTERIES

The gluteal arteries are *branches of the internal iliac arteries* (Fig. 6.10, Table 6.7). The major branches include the

- Superior gluteal artery

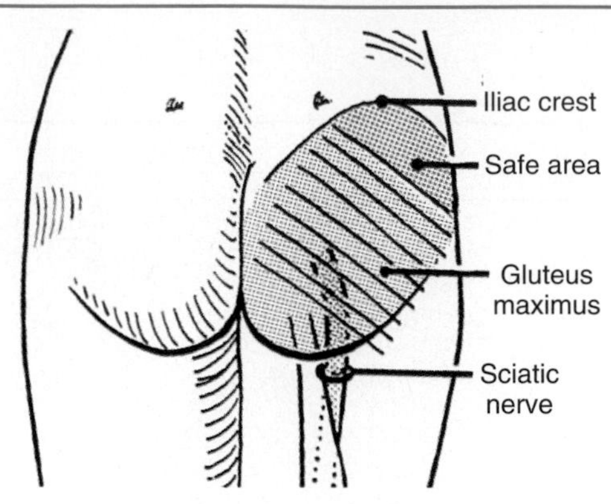

Intragluteal injection

Diffuse deep pain in the gluteal and lateral thigh regions may be caused by *trochanteric bursitis*, which is characterized by tenderness over the greater trochanter of the femur. The ischial bursa may become inflamed with excessive friction, producing a friction bursitis known as *ischial bursitis*.

If the gluteus medius and minimus muscles are weak or paralyzed, their supportive and steadying effect on the pelvis when walking is lost. Hence, when the foot is raised on the normal side, the pelvis falls on that side. Similarly, the person walks with a waddling gait known as a gluteal gait or a *gluteus medius limp*, which is characterized by a falling of the pelvis toward the unaffected side at each step. The supportive action of the glutei medius and minimus when the opposite foot is raised depends on normal

- Muscles and innervation
- Articulation of components of hip joint
- Femoral neck (i.e., one that is intact and normally angulated)

When any of the above is abnormal (e.g., paralysis of the gluteus medius and/or gluteus minimus, congenital dislocation of the hip joint, or nonunion of a fracture of the femoral neck or hip bone), the supporting mechanism fails, and the pelvis sinks when an attempt is made to stand on the affected limb. This is called Trendelenburg's sign.

Injury to the gluteal nerves may occur in wounds of the buttock (e.g., gunshot or stab wounds). With respect to the sciatic nerve, the buttock has a side of safety (its lateral side) and a side of danger (its medial side). Wounds or surgery on the medial side may injure the sciatic nerve and its branches to the hamstrings in the posterior aspect of the thigh (Fig. 6.10). Paralysis of these muscles results in impairment of extension of the thigh and flexion of the leg.

The gluteal region is a common injection site because the gluteal muscles are thick and large; consequently they provide a large surface area for absorption of drugs. Injections can be made safely only into the superolateral part of the buttock. Because of the danger of injuring gluteal nerves, drugs are sometimes injected into the anterolateral muscles of the thigh.

- Inferior gluteal artery
- Internal pudendal artery

The superior and inferior gluteal arteries leave the pelvis through the greater sciatic foramen and pass superior and posterior to the piriformis, respectively. The internal pudendal arteries pass through the gluteal region on their way to the perineum, but they do not supply any structures in the buttocks.

GLUTEAL VEINS

The gluteal veins draining the gluteal region are tributaries of the *internal iliac veins*. The superior and inferior gluteal veins accompany the corresponding arteries through the greater sciatic foramen, superior and inferior to the piriformis muscle, respectively. They communicate with tributaries of the femoral vein, thereby providing an alternate route for the return of blood from the lower limb if the femoral vein is occluded or has to be ligated. The *internal pudendal veins* accompany the internal pudendal arteries and join to form a single vein that enters the internal iliac vein. These veins drain blood from the external genitalia and perineal region.

Posterior Thigh Muscles

The muscles in the posterior aspect of the thigh are the *hamstrings*:

- Semitendinosus

Table 6.6
Nerves of Gluteal Region

Nerve	Origin	Course	Distribution[a]
Cluneal (superior, middle, and inferior)	*Superior*: dorsal rami of L1–L3 nn. *Middle*: dorsal rami of S1–S3 nn. *Inferior*: posterior femoral cutaneous n.	Superior and middle nn. exit through posterior sacral foramina and enter gluteal region Inferior nn. curve around inferior border of gluteal maximus	Skin of buttock or gluteal region as far as greater trochanter
Sciatic	Sacral plexus (L4–S3)	Leaves pelvis through greater sciatic foramen, inferior to piriformis, and enters gluteal region	Supplies no muscles in gluteal region
Posterior femoral cutaneous	Sacral plexus (S1–S3)	Leaves pelvis through greater sciatic foramen inferior to piriformis, runs deep to gluteus maximus, and emerges from its inferior border	Skin of buttock through inferior cluneal branches; skin over posterior aspect of thigh and calf
Superior gluteal	Ventral rami of L4–S1 nn.	Leaves pelvis through greater sciatic foramen, superior to piriformis and runs between gluteus medius and minimus	Gluteus medius, minimus, and tensor fasciae latae
Inferior gluteal	Ventral rami of L5–S2 nn.	Leaves pelvis through greater sciatic foramen, inferior to piriformis, and divides into several branches	Gluteus maximus
Nerve to quadratus femoris	Ventral rami of L4, L5, and S1 nn.	Leaves pelvis through greater sciatic foramen deep to sciatic n.	Hip joint, inferior gemellus, and quadratus femoris
Pudendal	Ventral rami of S2–S4 nn.	Enters gluteal region through greater sciatic foramen inferior to piriformis; descends posterior to sacrospinous ligament; enters perineum through lesser sciatic foramen	Supplies most innervation to the perineum; supplies no structures in gluteal region
Nerve to obturator internus	Ventral rami of L5, S1, and S2 nn.	Enters gluteal region through greater sciatic foramen inferior to piriformis; descends posterior to ischial spine; enters lesser sciatic foramen and passes to obturator internus	Superior gemellus and obturator internus

[a] See Figure 6.4 for cutaneous innervation of lower limb.

- Semimembranosus
- Biceps femoris

The hamstrings span the hip and knee joints; hence they are *extensors of the thigh* at the hip joint and *flexors of the leg* at the knee joint. Both actions cannot be performed fully at the same time. For the attachments, nerve supply, and actions of these muscles see Figures 6.9–6.11 and Table 6.8.

The *sciatic nerve* descends from the gluteal region into the posterior aspect of the thigh, where it lies on the adductor magnus and is crossed posteriorly by the long head of the biceps femoris. Although the sciatic nerve may separate into the tibial and common fibular (peroneal) nerves in the gluteal region or the thigh, the separation usually occurs in the inferior third of the thigh. The sciatic nerve supplies no structure in the gluteal region. The sci-

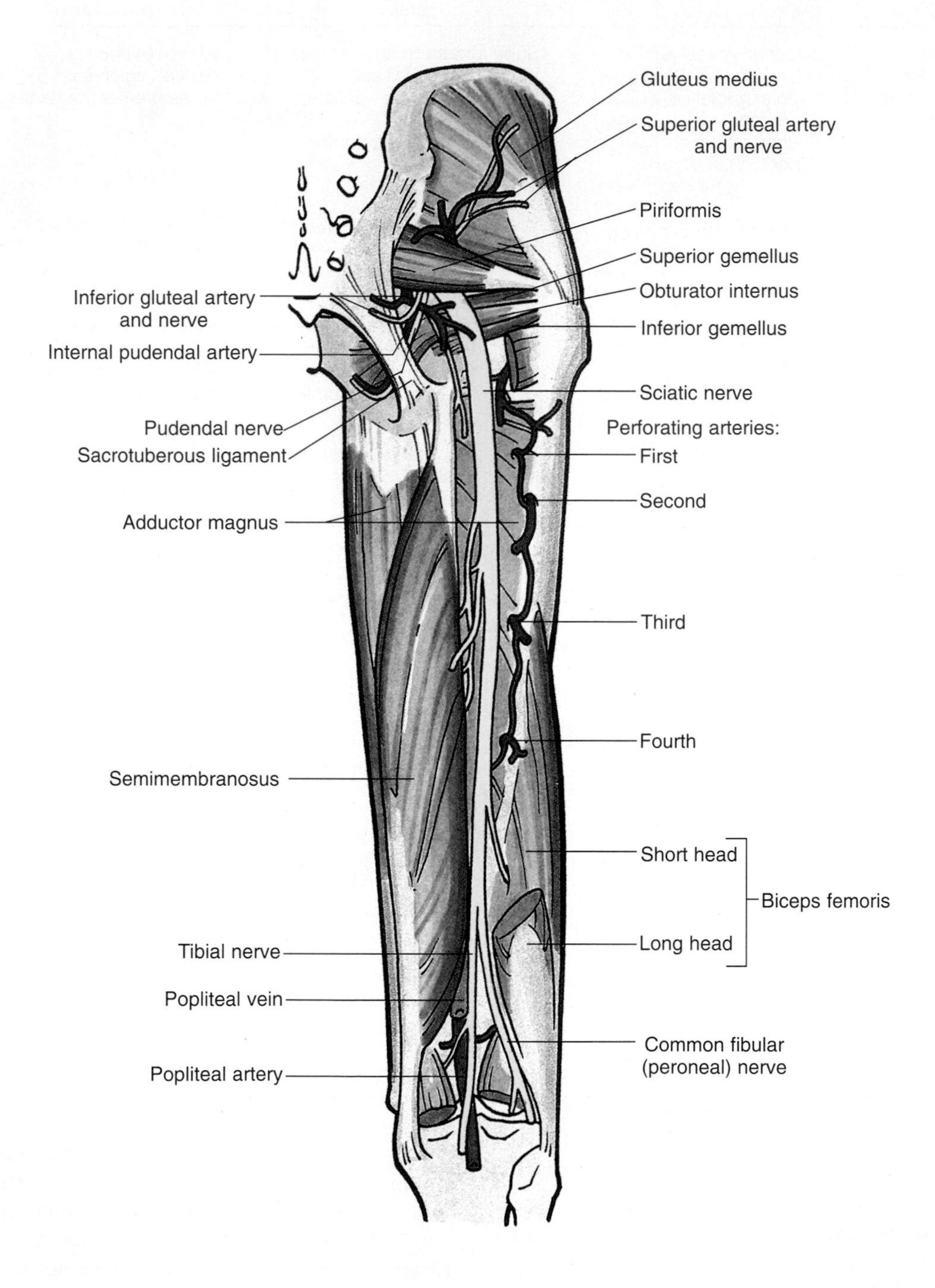

Figure 6.10. Dissection showing relation of vessels and nerves to muscles of gluteal region and posterior thigh.

Table 6.7
Arteries of Gluteal Region

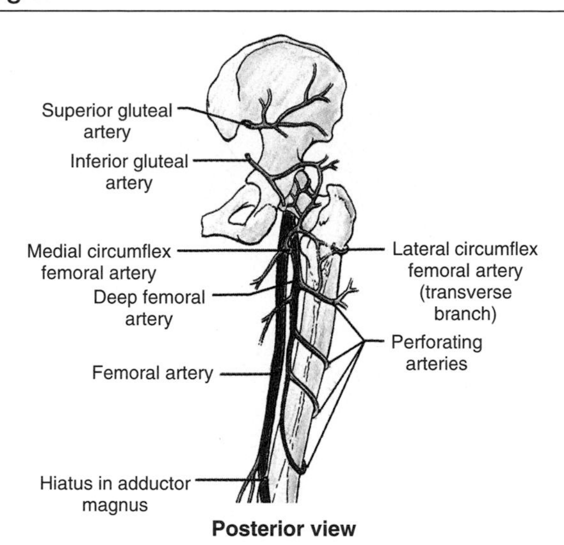

Posterior view

Artery[a]	Course	Distribution
Superior gluteal	Enters gluteal region through greater sciatic foramen, superior to piriformis, and divides into superficial and deep branches; anastomoses with inferior gluteal and medial circumflex femoral arteries	*Superficial branch*: gluteus maximus *Deep branch*: runs between gluteus medius and minimus and supplies them and tensor fasciae latae
Inferior gluteal	Enters gluteal region through greater sciatic foramen, inferior to piriformis, and descends on medial side of sciatic n.; anastomoses with superior gluteal artery and participates in cruciate anastomosis of thigh, involving first perforating artery of deep femoral and medial and lateral circumflex femoral arteries	Supplies gluteus maximus, obturator internus, quadratus femoris, and superior parts of hamstrings
Internal pudendal	Enters gluteal region through greater sciatic foramen and descends posterior to ischial spine; enters perineum through lesser sciatic foramen	Supplies external genitalia and muscles in the pelvic region; does not supply gluteal region

[a] All these arteries arise from internal iliac artery (Table 6.4, anterior view).

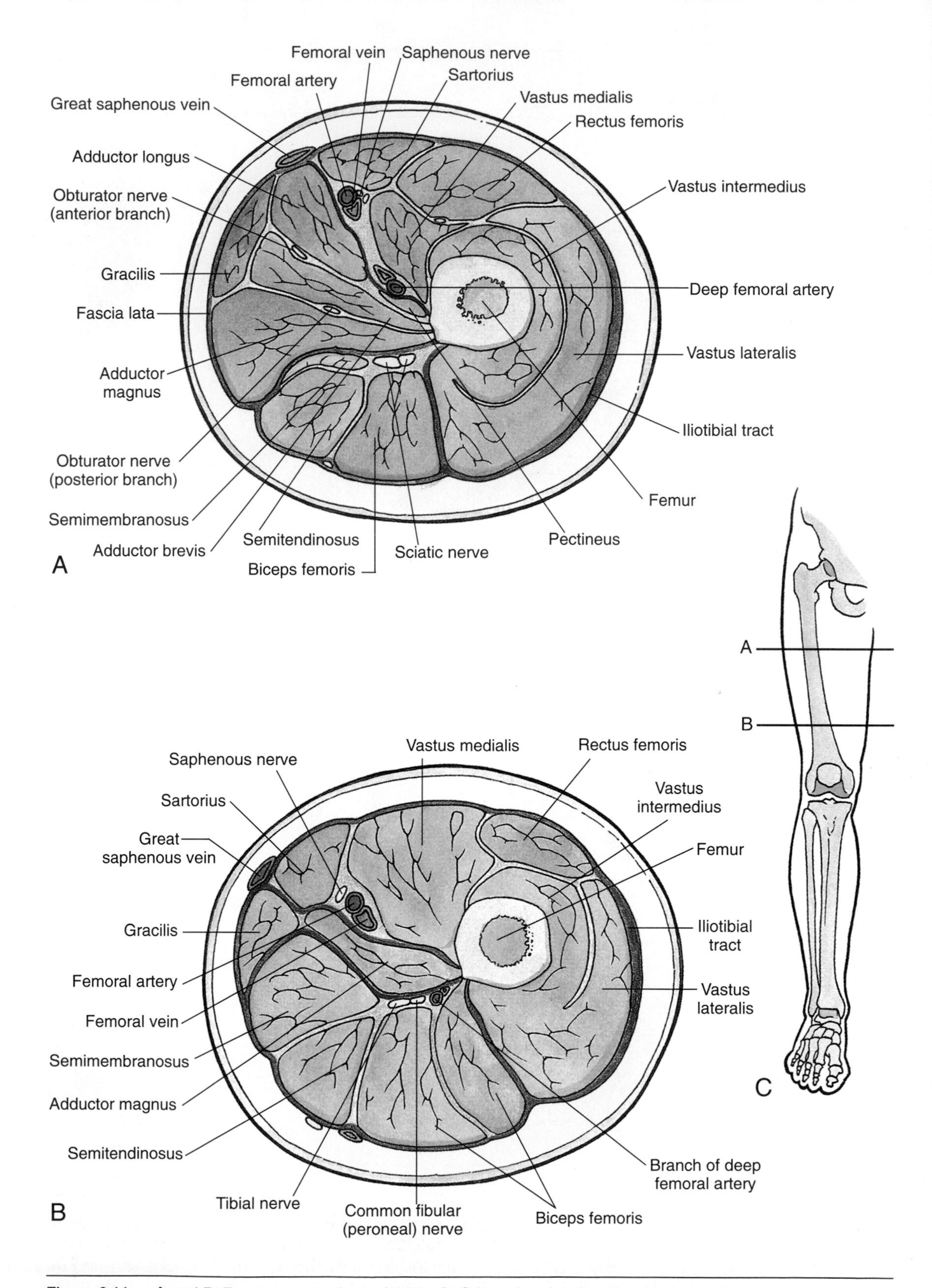

Figure 6.11. **A** and **B.** Transverse sections of thigh. **C.** Orientation drawing showing level of sections.

Table 6.8
Posterior Thigh Muscles

Muscle[a]	Proximal Attachment[b]	Distal Attachment	Innervation[c]	Main Actions
Semitendinosus	Ischial tuberosity	Medial surface of superior part of tibia	Tibial division of sciatic n. (**L5**, **S1**, and S2)	Extend thigh; flex leg and rotate it medially; when thigh and leg are flexed, they can extend trunk
Semimembranosus	Ischial tuberosity	Posterior part of medial condyle of tibia	Tibial division of sciatic n. (**L5**, **S1**, and S2)	Extend thigh; flex leg and rotate it medially; when thigh and leg are flexed, they can extend trunk
Biceps femoris	*Long head*: ischial tuberosity *Short head*: linea aspera and lateral supracondylar line of femur	Lateral side of head of fibula; tendon is split at this site by fibular collateral ligament of knee	Long head: tibial division of sciatic n. (L5, **S1**, and S2) Short head: common fibular (peroneal) division of sciatic n. (L5, **S1**, and S2)	Flexes leg and rotates it laterally; extends thigh (e.g., when starting to walk)

[a] Collectively these three muscles are known as hamstrings.
[b] See Figure 6.7 for muscle attachments.
[c] See Table 6.1 for explanation of segmental innervation.

atic nerve supplies branches to the hip joint and muscular branches to the hamstrings.

The *deep femoral artery*, the largest branch of the femoral artery, passes deeply into the thigh and lies posterior to the femoral artery and vein on the medial side of the thigh (Fig. 6.11*A*). The deep femoral artery ends in the inferior one-third of the thigh as the *fourth perforating artery*, which pierces the adductor magnus and supplies the hamstrings. Usually three *perforating arteries* arise from the posterior aspect of the deep femoral artery (Fig. 6.10). They supply the hamstrings and anastomose with each other in the vastus lateralis.

Pulled hamstrings are common in persons who run very hard (e.g., in quick-start sports such as baseball and soccer). The violent muscular exertion required to accelerate tears part of the attachments of the hamstrings to the ischial tuberosity. Hamstring injuries often result from inadequate warming up before competition.

Popliteal Fossa

The diamond-shaped popliteal fossa is on the posterior aspect of the knee (Fig. 6.12). The popliteal fossa is formed

- Superolaterally by the biceps femoris
- Superomedially by the semimembranosus and semitendinosus
- Inferolaterally and inferomedially by the lateral and medial heads of gastrocnemius, respectively
- Posteriorly (roof) by skin and fascia
- Anteriorly (floor) by the popliteal surface of the femur, the oblique popliteal ligament, and the fascia over the popliteus (see Fig. 6.16, *C* and *D*)

The main contents of the popliteal fossa are the

- Popliteal vessels (popliteal artery and vein)
- Tibial and common fibular (peroneal) nerves
- Small saphenous vein
- Popliteal lymph nodes and lymphatic vessels

The *deep popliteal fascia* is a strong sheet of deep fascia that forms a protective covering for the neurovascular structures passing from the thigh through the popliteal fossa to the leg.

The *popliteal artery* is the continuation of the femoral artery and begins when this artery passes through the adductor hiatus (Fig. 6.6*D*, Table 6.9). The popliteal artery passes inferolaterally through the popliteal fossa and ends at the inferior border of the popliteus by dividing

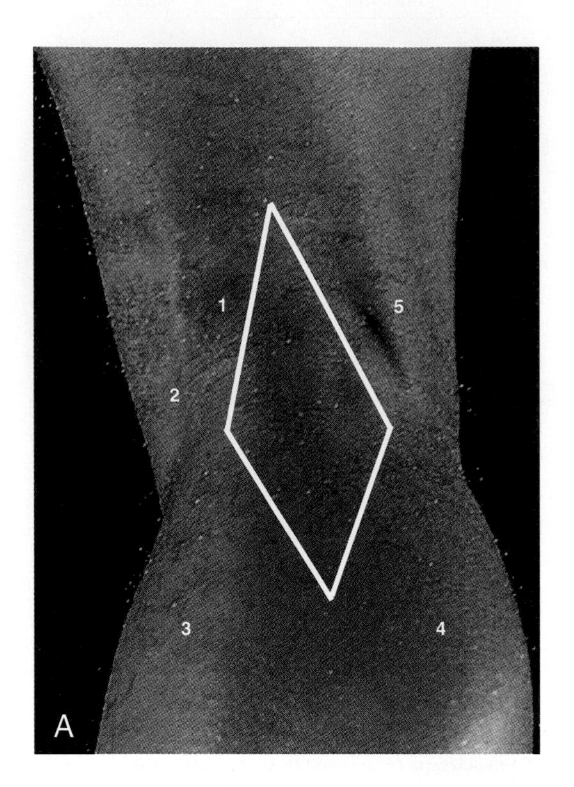

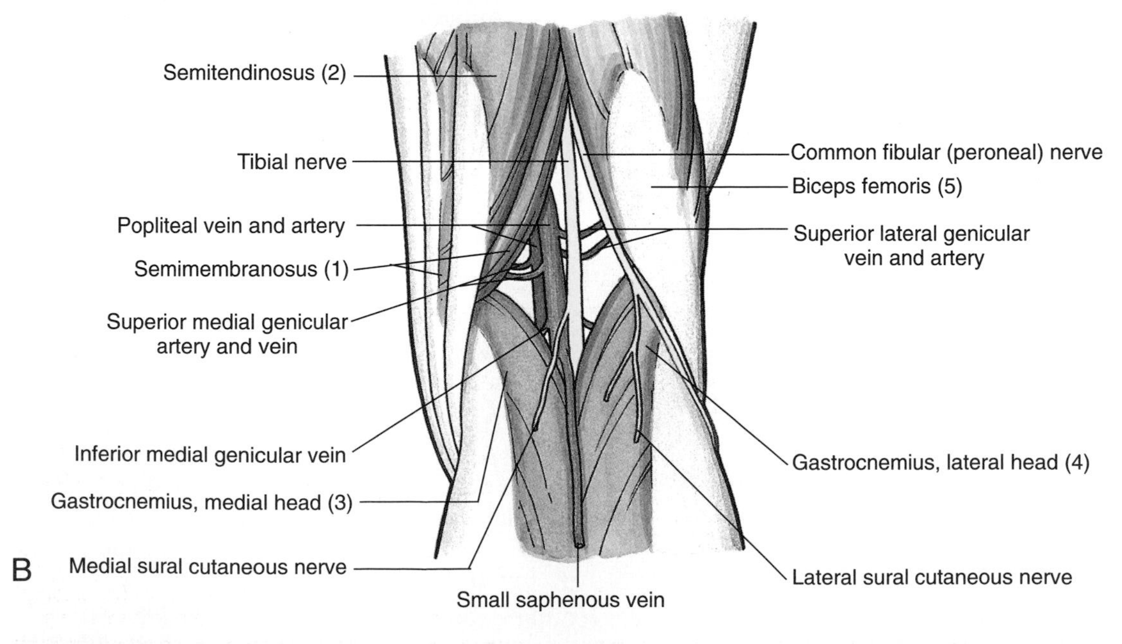

Figure 6.12. Right popliteal fossa. **A.** Surface anatomy. *Numbers* refer to structures in **B.** Diamond-shaped popliteal fossa is outlined. **B.** Dissection of popliteal fossa showing its boundaries and contents.

Table 6.9
Arterial Supply to Leg

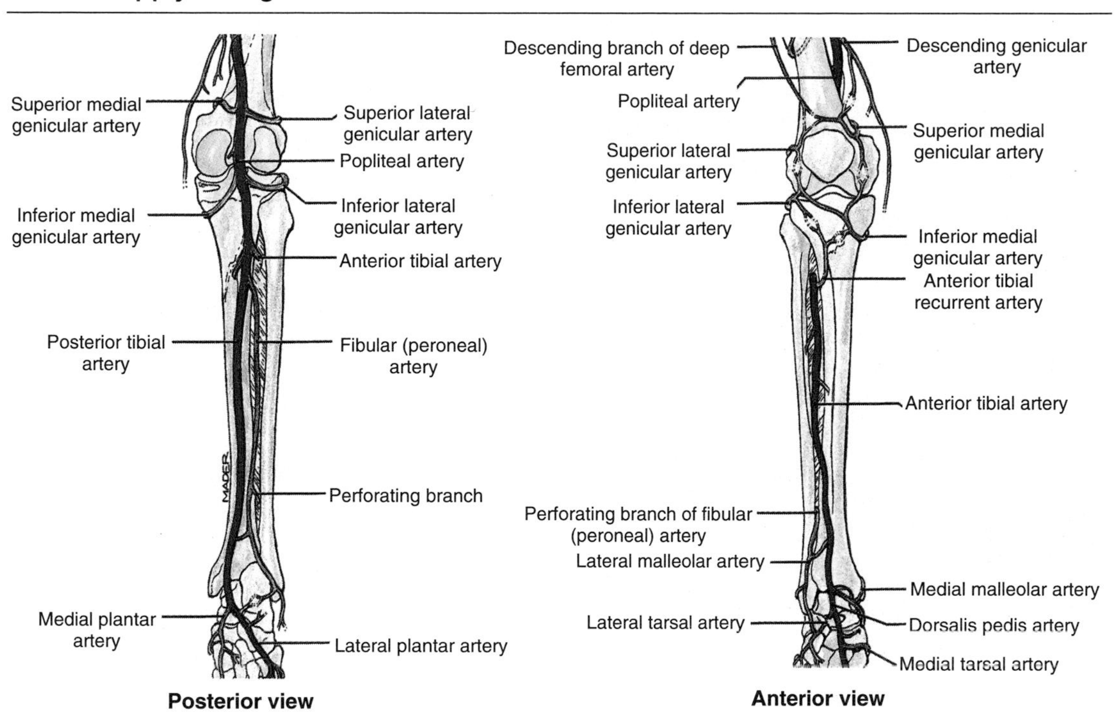

Artery	Origin	Course	Distribution
Popliteal	Continuation of femoral a. at adductor hiatus in adductor magnus	Passes through popliteal fossa to leg; ends at lower border of popliteus m. by dividing into anterior and posterior tibial aa.	Superior, middle, and inferior genicular aa. to both lateral and medial aspects of knee
Anterior tibial	Popliteal a.	Passes into anterior compartment through gap in superior part of interosseous membrane and descends this membrane between tibialis anterior and extensor digitorum longus	Anterior compartment of leg
Dorsalis pedis	Continuation of anterior tibial a. distal to inferior extensor retinaculum	Descends anteromedially to first interosseous space and divides into plantar and arcuate aa.	Muscles on dorsum of foot; pierces first dorsal interosseous m. to contribute to formation of plantar arch
Posterior tibial	Popliteal a.	Passes through posterior compartment of leg and terminates distal to flexor retinaculum by dividing into medial and lateral plantar aa.	Posterior and lateral compartments of leg; circumflex fibular branch joins anastomoses around knee; nutrient artery passes to tibia
Fibular (peroneal)	Posterior tibial a.	Descends in posterior compartment adjacent to posterior intermuscular septum	Posterior compartment of leg: perforating branches supply lateral compartment of leg

into the anterior and posterior tibial arteries. The popliteal artery, the deepest structure in the fossa, runs close to the articular capsule of the knee joint. Five genicular branches of the popliteal artery supply the articular capsule and ligaments of the knee joint. The *genicular arteries* are the lateral superior, medial superior, middle, lateral inferior, and medial inferior genicular arteries. They join to form the *genicular anastomosis*, a network of arterial vessels around the knee (L. genu, knee). The muscular branches of the popliteal artery supply the hamstring, gastrocnemius, soleus, and plantaris muscles. The superior muscular branches of the popliteal artery have clinically important anastomoses with the terminal part of the deep femoral and gluteal arteries.

Because the deep popliteal fascia is strong and limits expansion, pain from an abscess or tumor in the popliteal fossa is usually severe. *Popliteal abscesses* tend to spread superiorly and inferiorly because of the toughness of the popliteal fascia.

Because the popliteal artery is deep, it may be difficult to feel the *popliteal pulse*. Palpation of this pulse is commonly performed by placing the patient in the prone position with the leg flexed to relax the popliteal fascia and hamstrings. When the femoral artery is obstructed, an obvious sign is weakening or loss of the popliteal pulse.

A *popliteal aneurysm* (dilation of popliteal artery) can cause swelling and pain in the popliteal fossa. If the femoral artery has to be ligated, blood can bypass the occlusion via the genicular anastomoses and reach the popliteal artery distal to the ligation.

Because of its protected position in the popliteal fossa the *tibial nerve* is not commonly injured. It may, however, be injured by deep lacerations. Severance of the tibial nerve results in paralysis of the flexor muscles in the leg and the intrinsic muscles in the sole of the foot. Persons with tibial nerve injury cannot plantarflex their ankle or flex their toes. There is also a loss of sensation in the sole of the foot.

The *popliteal vein* is formed at the distal border of the popliteus muscle, or it may be represented by several small veins. Throughout its course, the popliteal vein lies superficial to and in the same fibrous sheath as the popliteal artery. The popliteal vein ends at the adductor hiatus where it becomes the femoral vein.

The *sciatic nerve* usually ends at the superior angle of the popliteal fossa by dividing into the tibial and common fibular (peroneal) nerves (Figs. 6.10–6.12, Table 6.10). The *tibial nerve*—the medial, larger terminal branch of the sciatic nerve—is the most superficial of the three main central components of the popliteal fossa (i.e., nerve, vein, and artery). The *common fibular (peroneal) nerve*—the lateral, smaller terminal branch of the sciatic nerve—begins at the superior angle of the popliteal fossa and follows the medial border of the biceps femoris muscle and tendon along the superolateral boundary of the popliteal fossa. It leaves the fossa by passing superficial to the lateral head of the gastrocnemius and then passes over the posterior aspect of the head of the fibula before winding around the neck of this bone (Fig. 6.13*B*).

Leg

The leg is divided into three compartments: anterior, lateral, and posterior (Fig. 6.3*D*).

ANTERIOR COMPARTMENT

The anterior compartment of the leg is the extensor compartment anterior to the interosseous membrane, between the lateral surface of the tibia and the anterior intermuscular septum. The four muscles of the anterior compartment are (Figs. 6.13 and 6.14)

- Tibialis anterior
- Extensor hallucis longus
- Extensor digitorum longus
- Fibularis (peroneus) tertius

These muscles are mainly dorsiflexors of the ankle joint and extensors of the toes. For their attachments, nerve supply, and main actions see Table 6.11.

The *superior extensor retinaculum* is a strong, broad band of deep fascia, passing from the fibula to the tibia, proximal to the malleoli. It binds down the tendons of muscles in the anterior compartment, preventing them from bowstringing anteriorly during dorsiflexion of the ankle joint.

The *inferior extensor retinaculum*, a Y-shaped band of deep fascia, attaches laterally to the anterosuperior surface of the calcaneus. It forms a strong loop around the tendons of the fibularis (peroneus) tertius and the extensor digitorum longus muscles.

The *deep fibular (peroneal) nerve* (Fig. 6.13*B*, Table 6.10), the nerve of the anterior compartment, is one of two terminal branches of the

Table 6.10
Nerves of Leg

Nerve	Origin	Course	Distribution
Saphenous	Femoral n.	Descends with femoral vessels through femoral triangle and adductor canal and then descends with great saphenous v.	Supplies skin on medial side of leg and foot
Sural	Usually arises from both tibial and common fibular (peroneal) nn.	Descends between heads of gastrocnemius and becomes superficial at the middle of the leg; descends with small saphenous v. and passes inferior to the lateral malleolus to the lateral side of foot	Supplies skin on posterior and lateral aspects of leg and lateral side of foot
Tibial	Sciatic n.	Descends through popliteal fossa and lies on popliteus; runs inferiorly on the tibialis posterior with the posterior tibial vessels; terminates beneath the flexor retinaculum by dividing into the medial and lateral plantar nn.	Supplies posterior muscles of leg and knee joint
Common fibular (peroneal)	Sciatic n.	Separates from tibial n. at apex of popliteal fossa and follows medial border of biceps femoris and its tendon; passes over posterior aspect of head of fibula and then winds around neck of fibula, deep to fibularis longus, where it divides into deep and superficial fibular nn.	Supplies skin on lateral part of the posterior aspect of leg via its branch, lateral sural cutaneous n; also supplies knee joint via its articular branch
Superficial fibular (peroneal)	Common fibular (peroneal) n.	Arises between fibularis longus and neck of fibula and descends in lateral compartment of the leg; pierces deep fascia at distal third of leg to become subcutaneous	Supplies fibularis longus and brevis and skin on distal third of anterior surface of leg and dorsum of foot
Deep fibular (peroneal)	Common fibular (peroneal) n.	Arises between fibularis longus and neck of fibula; passes through extensor digitorum longus and descends on interosseous membrane; crosses distal end of tibia and enters dorsum of foot	Supplies anterior muscles of leg and skin of first interdigital cleft; sends articular branches to joints it crosses

common fibular (peroneal) nerve. It accompanies the anterior tibial artery between the extensor hallucis longus and tibialis anterior muscles.

The common fibular nerve is the most commonly injured nerve in the lower limb, mainly because it winds superficially around the neck of the fibula. This nerve may be severed during fracture of the neck of the fibula or severely stretched when the knee joint is injured. In addition, it is susceptible to pressure exerted on the head and neck of the fibula by a tightly applied plaster cast or by a restraining strap on the operating table.

Severance of the common fibular nerve results in paralysis of all the dorsiflexor muscles of the ankle and eversion muscles of the foot (Tables 6.10 and 6.11). The loss of eversion of the foot and dorsiflexion of the ankle causes the foot to hang down, a condition known as *foot-drop*. The patient has a high stepping gait in which the foot is raised higher than is necessary so the toes do not hit the ground. In addition, the foot is brought down suddenly, producing a distinctive "clop." There is also a variable loss of sensation on the anterolateral aspect of the leg and the dorsum of the foot.

The *anterior compartment syndrome* (shin splints) is a painful condition of the anterior compartment of the leg that follows vigorous and/or lengthy exercise. The anterior tibial muscles swell from sudden overuse. The swelling reduces blood flow to the muscles, and they become painful and tender

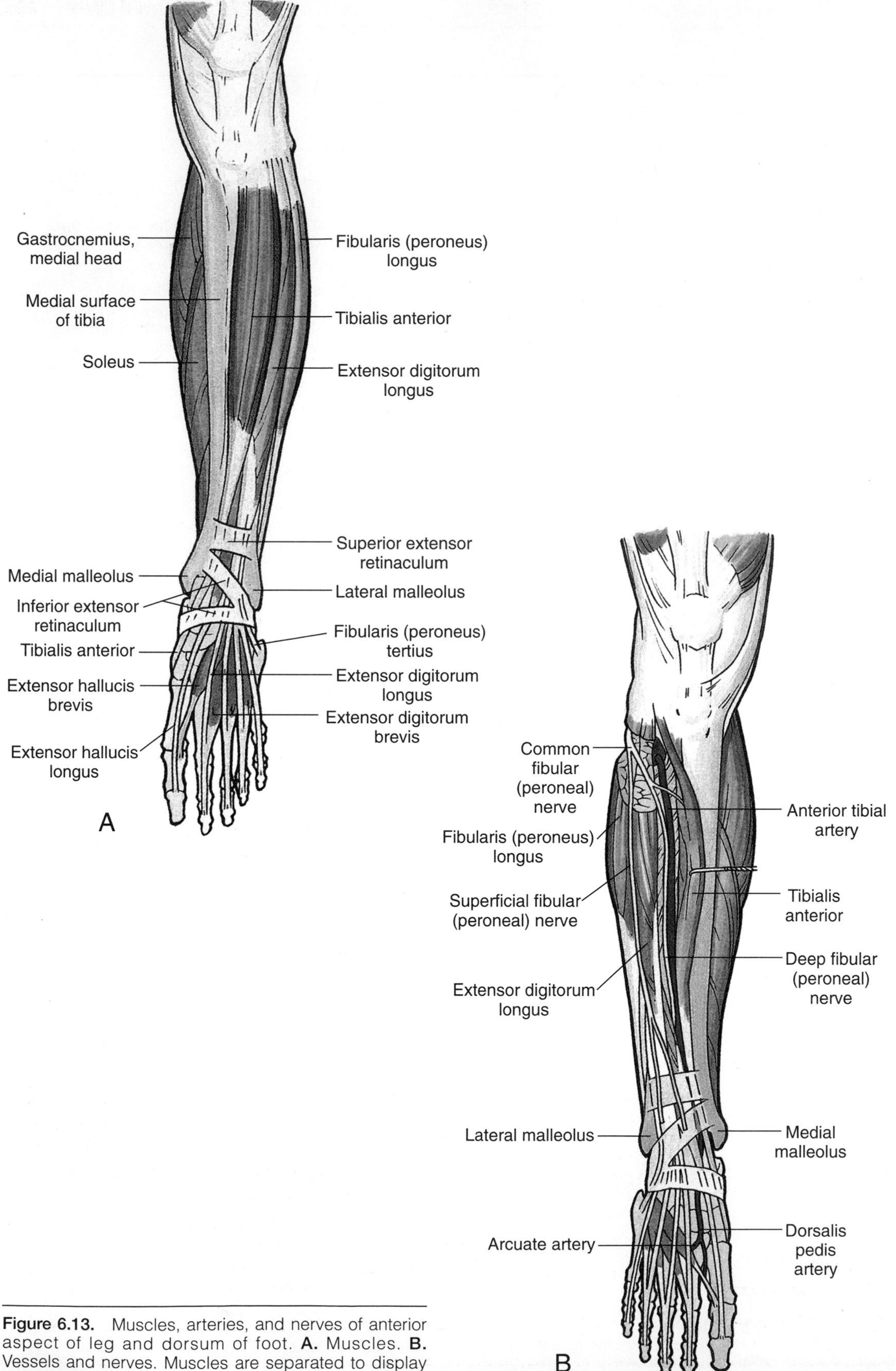

Figure 6.13. Muscles, arteries, and nerves of anterior aspect of leg and dorsum of foot. **A.** Muscles. **B.** Vessels and nerves. Muscles are separated to display these structures.

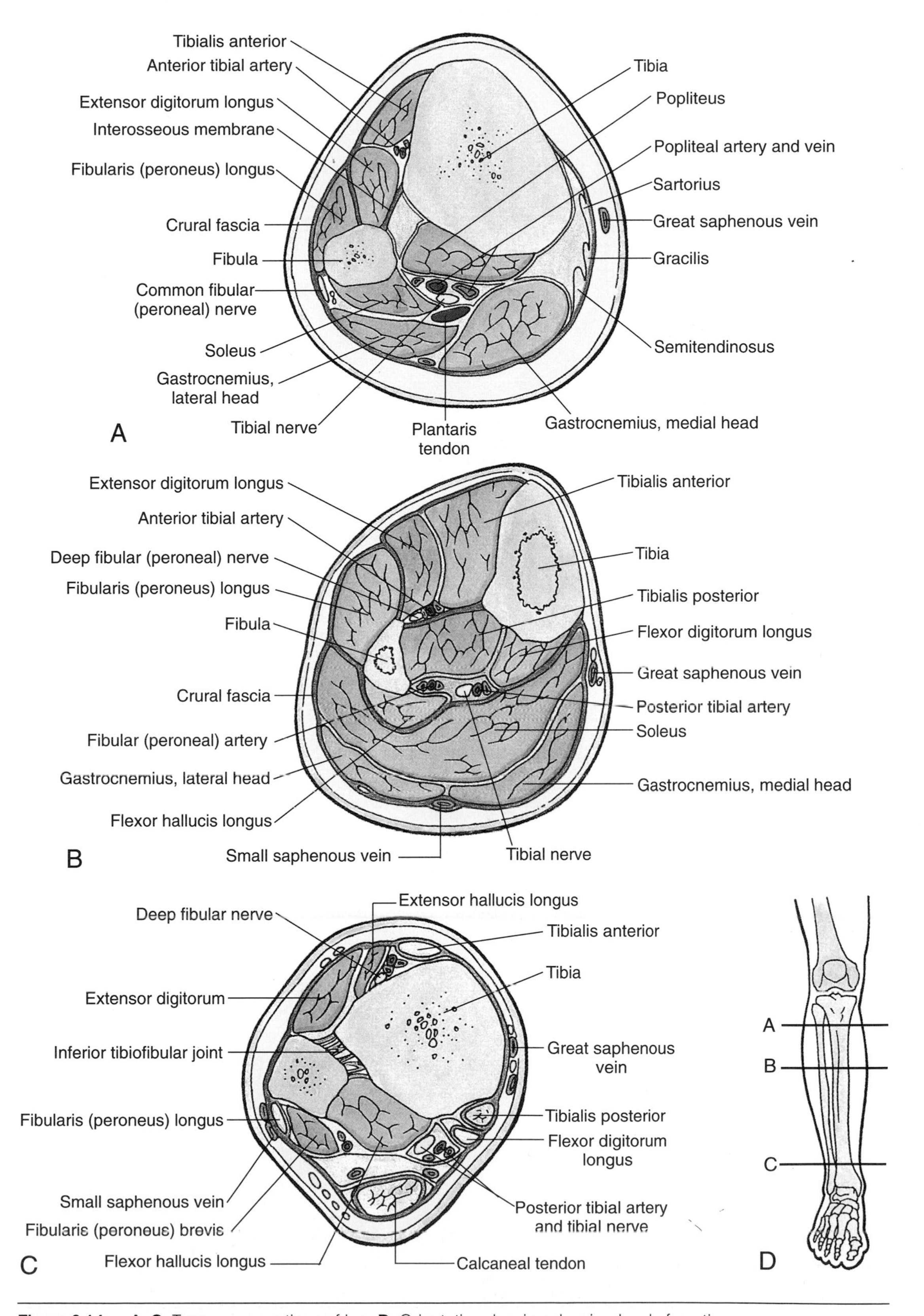

Figure 6.14. **A–C.** Transverse sections of leg. **D.** Orientation drawing showing level of sections.

Table 6.11
Muscles in Anterior and Lateral Compartments of Leg

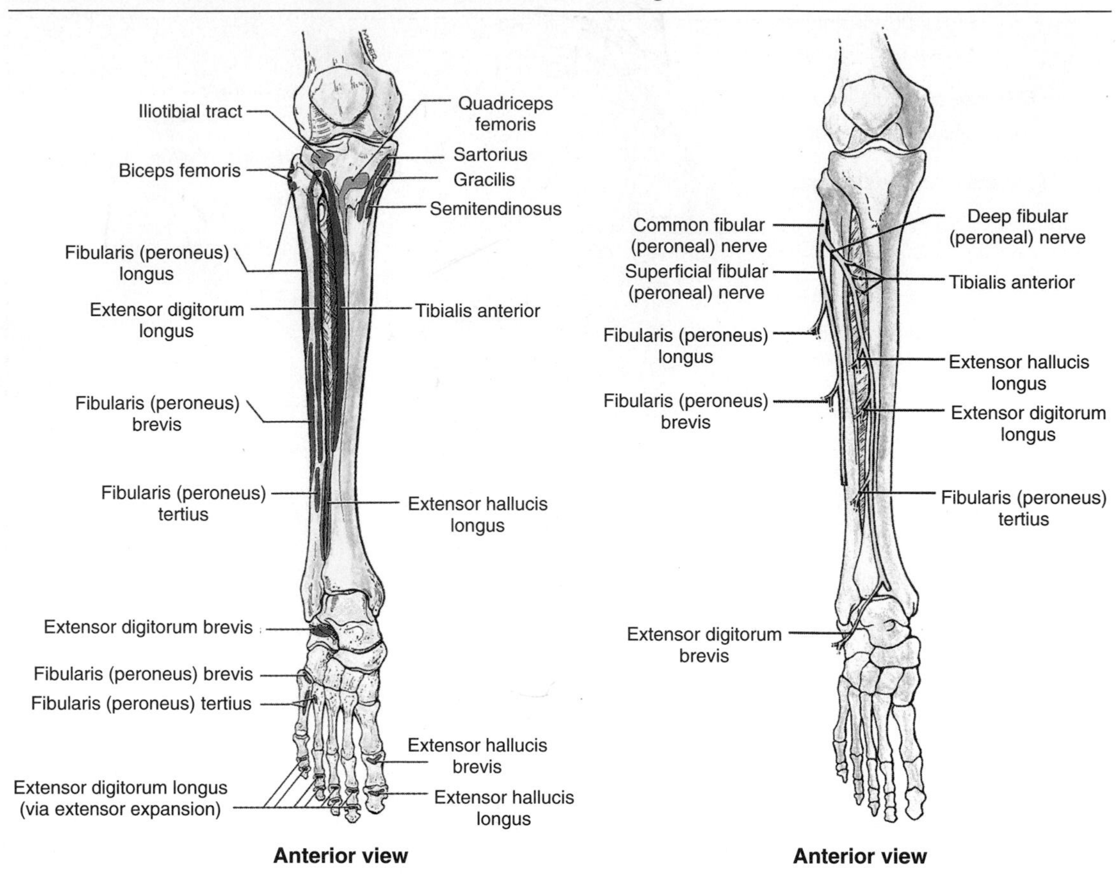

Muscle	Proximal Attachment	Distal Attachment	Innervation[a]	Main Actions
Anterior compartment				
Tibialis anterior	Lateral condyle and superior half of lateral surface of tibia	Medial and inferior surfaces of medial cuneiform and base of first metatarsal	Deep fibular (peroneal) n. (**L4** and L5)	Dorsiflexes ankle and inverts foot
Extensor hallucis longus	Middle part of anterior surface of fibula and interosseous membrane	Dorsal aspect of base of distal phalanx of great toe (hallux)	Deep fibular (peroneal) n. (L5 and S1)	Extends great toe and dorsiflexes ankle
Extensor digitorum longus	Lateral condyle of tibia and superior three-fourths of anterior surface of interosseous membrane	Middle and distal phalanges of lateral four digits	Deep fibular (peroneal) n. (L5 and S1)	Extends lateral four digits and dorsiflexes ankle
Fibularis (peroneus) tertius	Inferior third of anterior surface of fibula and interosseous membrane	Dorsum of base of fifth metatarsal	Deep fibular (peroneal) n. (L5 and S1)	Dorsiflexes ankle and aids in eversion of foot
Lateral compartment				
Fibularis (peroneus) longus	Head and superior two-thirds of lateral surface of fibula	Base of first metatarsal and medial cuneiform	Superficial fibular (peroneal) nerve (**L5**, **S1**, and **S2**)	Evert foot and weakly plantarflex ankle
Fibularis (peroneus) brevis	Inferior two-thirds of lateral surface of fibula	Dorsal surface of tuberosity on lateral side of base of fifth metatarsal	Superficial fibular (peroneal) nerve (**L5**, **S1**, and **S2**)	Evert foot and weakly plantarflex ankle

[a] See Table 6.1 for explanation of segmental innervation.

Surface Anatomy of Anterolateral Surface of Leg

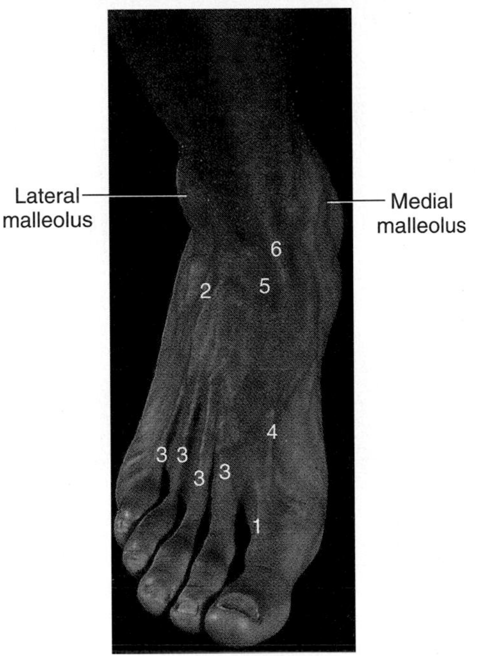

The *tibialis anterior* lies superficially and is easily palpable just lateral to the anterior border of the tibia. As the foot is dorsiflexed, the large tendon of the tibialis anterior (*6*) can be seen and felt as it runs distally and slightly medially over the anterior surface of the ankle joint to the medial side of the foot. If the first digit is extended, the tendon of the *extensor hallucis longus* (*4*) can be palpated just lateral to the tendon of the tibialis anterior. Also observe the tendon of the extensor hallucis brevis (*1*). As the other toes are dorsiflexed, the tendons of the extensor digitorum longus (*3*) can be palpated lateral to the extensor hallucis longus and followed to the four lateral digits (*2–5*). The fibularis (peroneus) tertius (*2*) is part of the extensor digitorum longus. The *fibularis longus* is subcutaneous throughout its course. The tendons of this muscle and the fibularis brevis are palpable when the foot is everted as they pass around the posterior aspect of the lateral malleolus.

The *anterior tibial artery* (Table 6.9) supplies structures in the anterior compartment. The smaller terminal branch of the popliteal artery, the anterior tibial artery, begins at the inferior border of the popliteus and passes anteriorly over the interosseous membrane to descend on its anterior surface. It ends at the ankle joint, midway between the malleoli, where it becomes the *dorsalis pedis artery* (dorsal pedal artery).

LATERAL COMPARTMENT

The lateral compartment of the leg is bounded by the lateral surface of the fibula, the anterior and posterior intermuscular septa, and the crural fascia (Figs. 6.14 and 6.15). The lateral compartment contains the fibularis (peroneus) longus and brevis muscles. See Table 6.11 for their attachments, nerve supply, and main actions.

The *superficial fibular (peroneal) nerve* (Table 6.10), a branch of the common fibular nerve, is the nerve of the lateral compartment of the leg. It supplies the skin on the distal part of the anterior surface of the leg and nearly all the dorsum of the foot and digits (Fig. 6.4). There is no artery in the lateral compartment of the leg; the muscles are supplied by perforating branches of the fibular (peroneal) artery.

POSTERIOR COMPARTMENT

The muscles in the posterior compartment are divided into superficial and deep groups by the transverse septum (Fig. 6.3*D*). The tibial nerve and posterior tibial vessels supply both divisions of the posterior compartment and run between the superficial and deep groups of muscle, just deep to the transverse intermuscular septum.

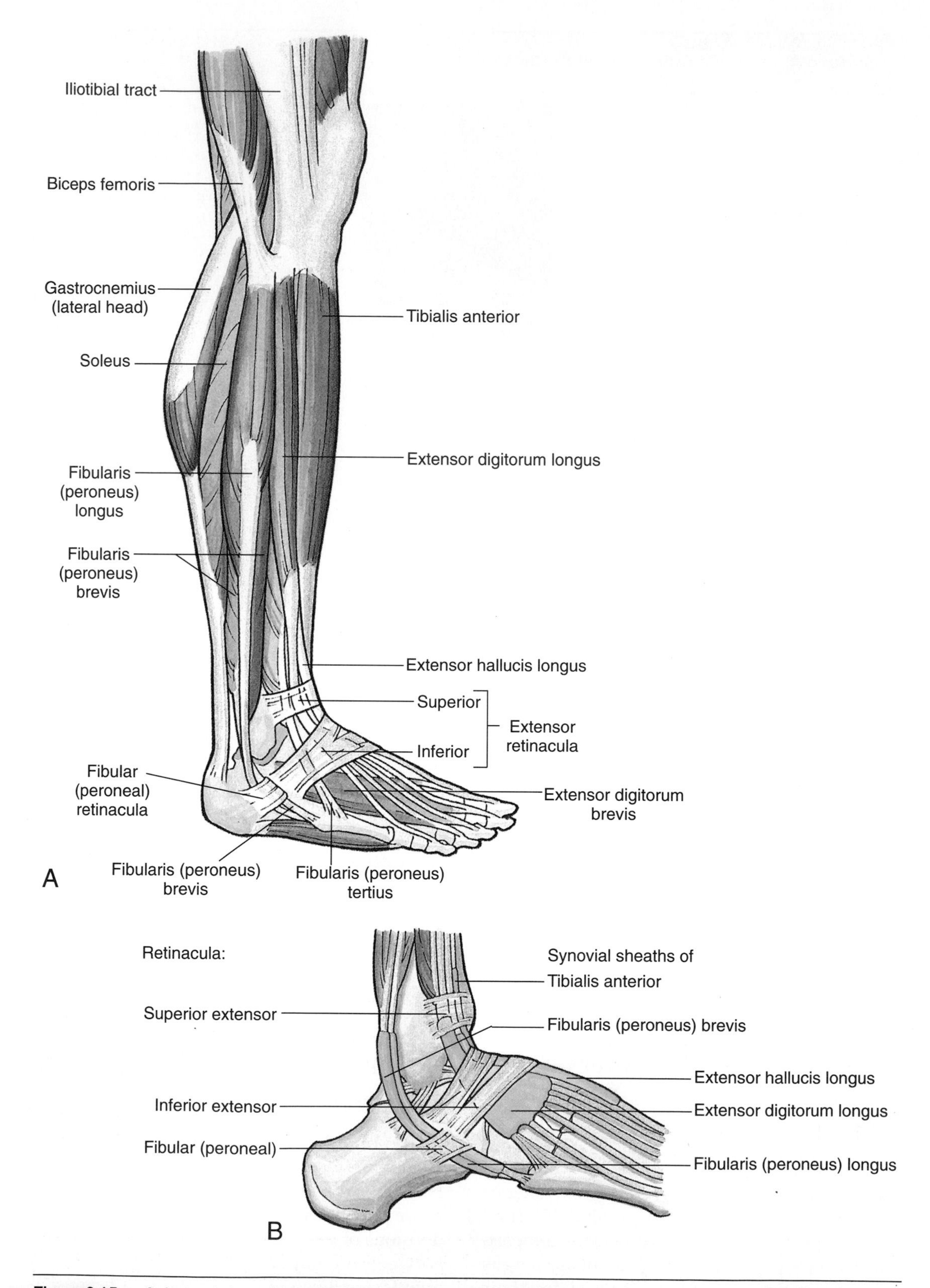

Figure 6.15. **A.** Lateral view of muscles of leg and foot. Also shown are extensor and fibular (peroneal) retinacula. **B.** Lateral view of distal part of leg and proximal foot showing retinacula and synovial sheaths of tendons at ankle.

Superficial Muscle Group

Three muscles comprise the superficial group in the posterior compartment of the leg: gastrocnemius, soleus, and plantaris (Figs. 6.14 and 6.16). For attachments, nerve supply, and main actions of these muscles see Table 6.12. Together the two-headed gastrocnemius and soleus form the *triceps surae*. This tripartite muscle has a common tendon of insertion into the calcaneus called the calcaneal tendon. This muscle plantarflexes the ankle joint.

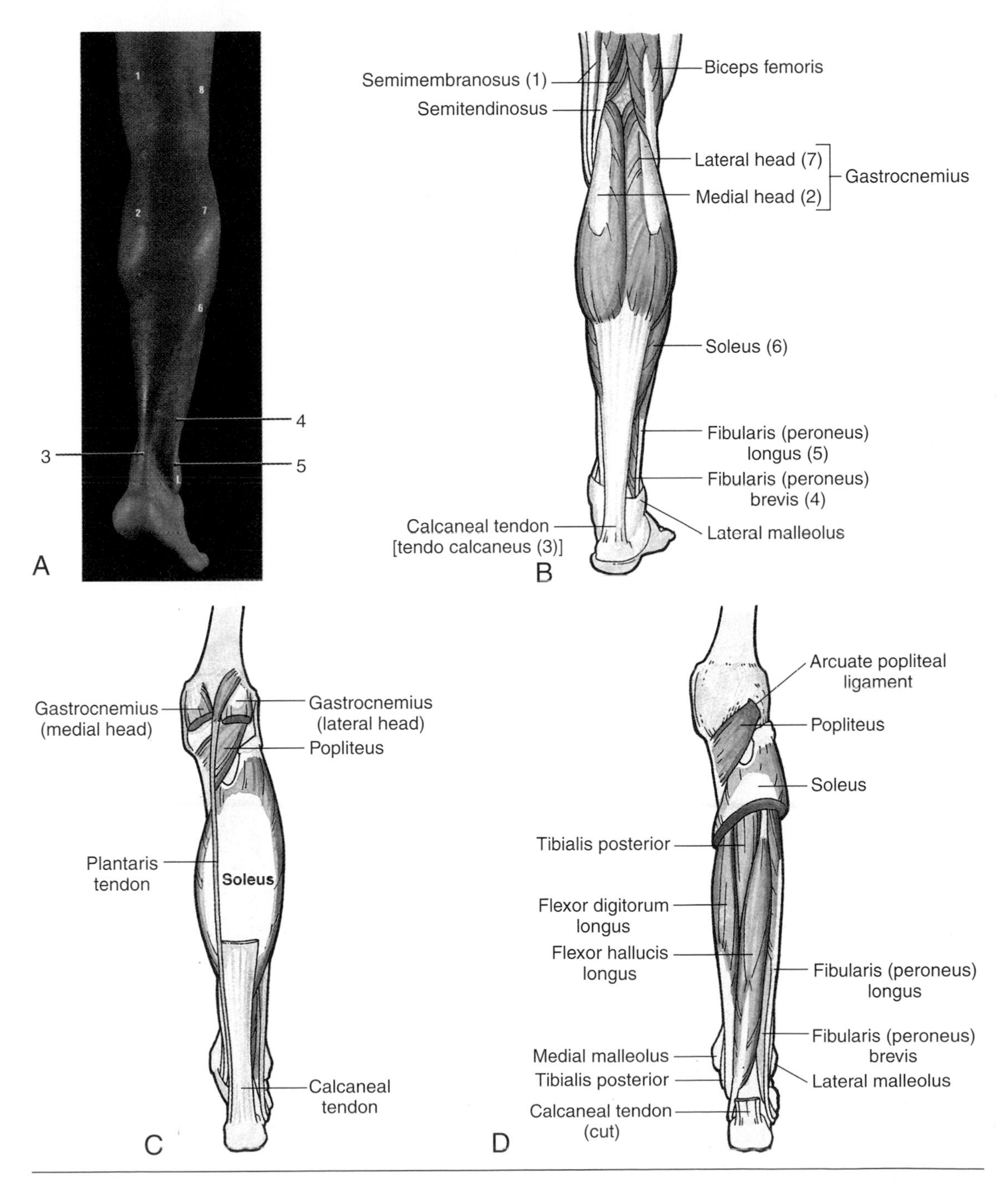

Figure 6.16. Muscles of posterior compartment of leg. **A.** Surface anatomy of lower limb. *Numbers* refer to muscles labeled in **B.** *L*, lateral malleolus. **B.** Superficial dissection of leg muscles. **C.** Most of gastrocnemius has been removed to show soleus, popliteus, and plantaris. **D.** Gastrocnemius and most of soleus have been removed to show deep muscles in posterior compartment.

Table 6.12
Muscles in Posterior Compartment of Leg

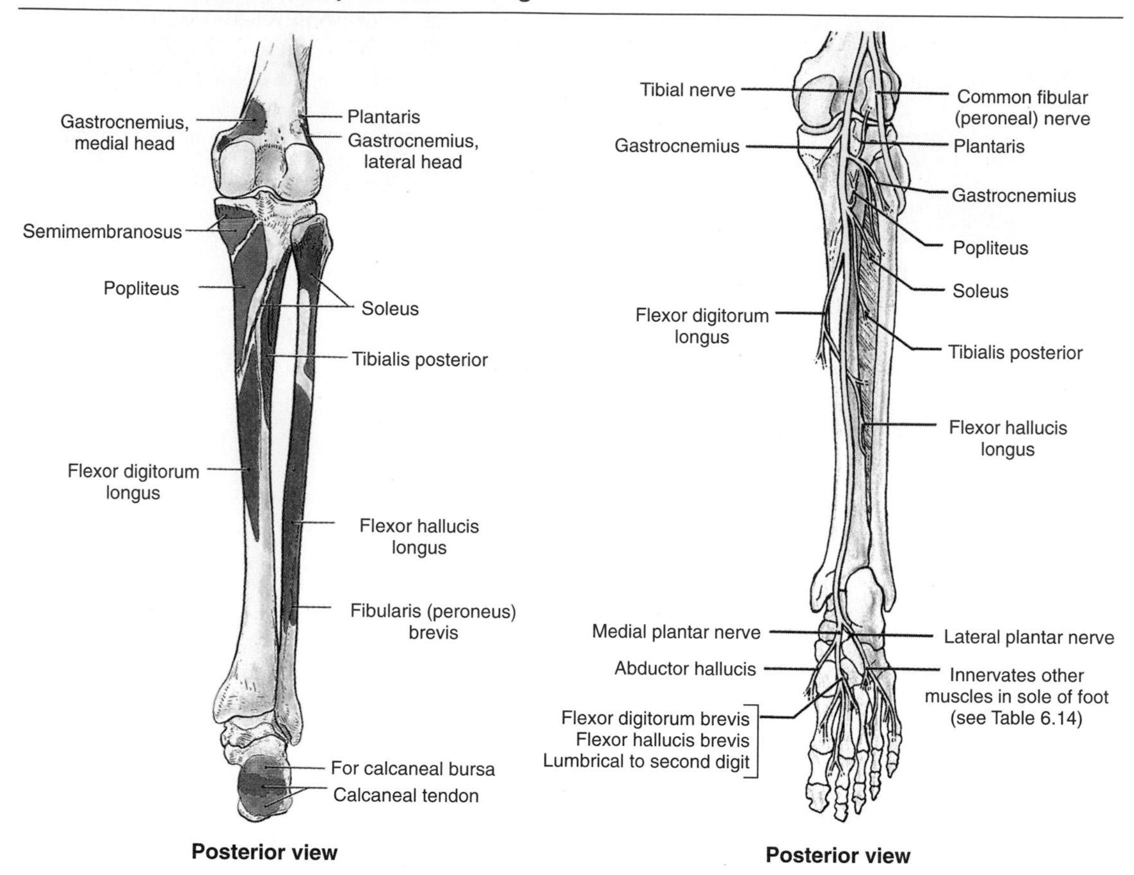

Muscle	Proximal Attachment	Distal Attachment	Innervation[a]	Main Actions
Superficial muscles				
Gastrocnemius	*Lateral head*: lateral aspect of lateral condyle of femur *Medial head*: popliteal surface of femur, superior to medial condyle	Posterior surface of calcaneus via calcaneal tendon (tendo calcaneus)	Tibial nerve (S1 and **S2**)	Plantarflexes ankle, raises heel during walking, and flexes leg at knee joint
Soleus	Posterior aspect of head of fibula, superior fourth of posterior surface of fibula, soleal line and medial border of tibia			Plantarflexes ankle and steadies leg on foot
Plantaris	Inferior end of lateral supracondylar line of femur and oblique popliteal ligament			Weakly assists gastrocnemius in plantarflexing ankle and flexing knee
Deep muscles				
Popliteus	Lateral surface of lateral condyle of femur and lateral meniscus	Posterior surface of tibia, superior to soleal line	Tibial nerve (**L4**, **L5**, and S1)	Weakly flexes knee and unlocks it

Table 6.12. *Continued*

Muscle	Proximal Attachment	Distal Attachment	Innervation[a]	Main Actions
Flexor hallucis longus	Inferior two-thirds of posterior surface of fibula and inferior part of interosseous membrane	Base of distal phalanx of of great toe (hallux)	Tibial nerve (**S2** and S3)	Flexes great toe at all joints and plantarflexes ankle; supports medial longitudinal arch of foot
Flexor digitorum longus	Medial part of posterior surface of tibia inferior to soleal line, and by a broad tendon to fibula	Bases of distal phalanges of lateral four digits	Tibial nerve (**S2** and S3)	Flexes lateral four digits and plantarflexes ankle; supports longitudinal arches of foot
Tibialis posterior	Interosseous membrane, posterior surface of tibia inferior to soleal line, and posterior surface of fibula	Tuberosity of navicular, cuneiform, and cuboid and bases of second, third, and fourth metatarsals	Tibial nerve (L4 and L5)	Plantarflexes ankle and inverts foot

[a] See Table 6.1 for explanation of segmental intervation.

Inflammation, strain, and rupture of the calcaneal tendon (tendo calcaneus) are fairly common. These painful injuries often occur during running and quick-start sports (e.g., squash). *Rupture of the calcaneal* tendon usually results in abrupt pain in the posterior aspect of the leg. The person cannot use the limb, and a lump in the calf appears owing to shortening of the triceps surae muscle. After rupture of the calcaneal tendon, the foot can be dorsiflexed to a greater extent than normal, and the patient cannot plantarflex the foot.

The *ankle reflex* [(S1) ankle jerk, Achilles tendon reflex] is induced by striking the calcaneal tendon with a reflex hammer. If the S1 nerve root is cut or compressed, the ankle reflex is virtually absent. *Tennis leg* is a painful calf injury resulting from partial tearing of the medial belly of the gastrocnemius at or near its musculotendinous junction. It is caused by overstretching the muscle by concomitant full extension of the knee and dorsiflexion of the ankle joint. *Calcaneal bursitis*, inflammation and swelling of the calcaneal bursa (Fig. 6.17), is fairly common in long distance runners and is caused by excessive friction on the bursa as the tendon continuously slides over it.

When a person is standing, the venous return of the leg depends largely on muscular activity of the triceps surae muscle (often called the *calf pump*). Contraction of the calf muscles pumps blood superiorly in the deep veins (p. 22).

The *gastrocnemius*, the most superficial muscle in the posterior compartment, forms most of the prominence of the calf. Because its fibers are mainly vertical, contractions of the gastrocnemius produce rapid movements during running and jumping. Although the gastrocnemius acts on both the knee and ankle joints, it cannot exert its full power on both joints at the same time. The *soleus* lies deep to the gastrocnemius and is very powerful. The *plantaris* is usually small and may be absent.

Deep Muscle Group

Four muscles comprise the deep group in the posterior compartment of the leg (Figs. 6.14 and 6.16):

- Popliteus
- Flexor hallucis longus
- Flexor digitorum longus
- Tibialis posterior

The *popliteus* acts on the knee joint, whereas the others act on the ankle and foot joints. See Table 6.12 for attachments, nerve supply, and main actions of these muscles.

The *flexor hallucis longus* is the powerful "push-off" muscle during walking, running, and jumping. It provides much of the spring to the step. The *flexor digitorum longus* is smaller than the flexor hallucis longus, even though it moves four digits. It passes diagonally into the sole of the foot and divides into four tendons,

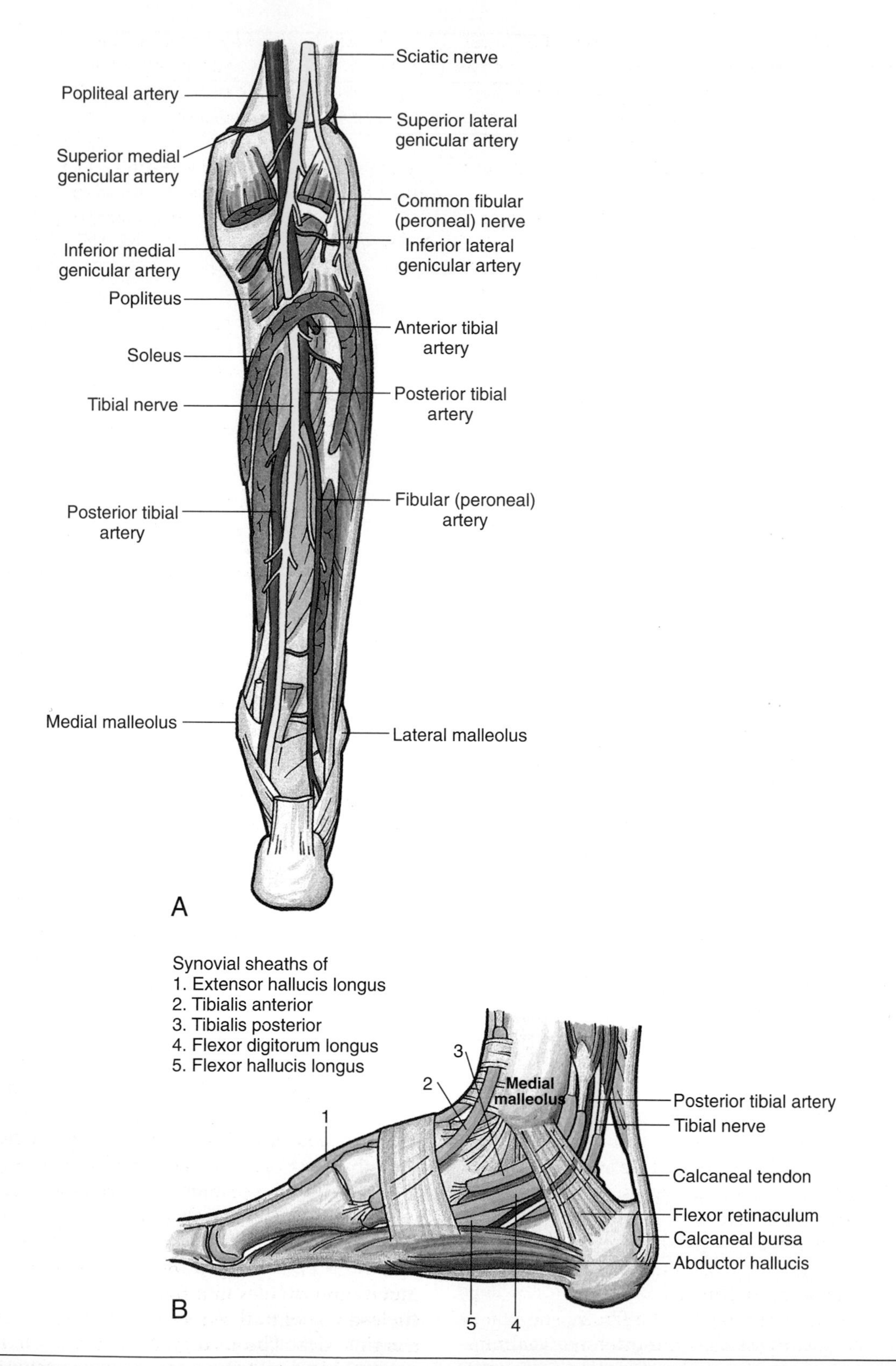

Figure 6.17. **A.** Deep dissection of posterior aspect of leg showing nerves and arteries. Soleus is largely cut away. **B.** Medial view of distal part of leg, ankle, and foot showing retinacula and synovial sheaths of tendons at ankle.

which pass to the distal phalanges of the lateral four digits. The *tibialis posterior,* the deepest muscle in the posterior compartment, lies between the flexor digitorum longus and the flexor hallucis longus in the same plane as the tibia and fibula. Its tendon can be seen and felt posterior to the medial malleolus.

The *tibial nerve* (Fig. 6.17, Table 6.10) supplies all muscles in the posterior compartment of the leg. It descends in the median plane of the calf, deep to the soleus (Fig. 6.14*B*). Posteroinferior to the malleolus, the tibial nerve divides into the medial and lateral plantar nerves. A cutaneous branch of the tibial nerve, the *medial sural cutaneous nerve,* usually unites with the communicating branch of the common fibular (peroneal) nerve to form the *sural nerve.* This nerve supplies the skin of the lateral and posterior part of the inferior third of the leg and the lateral side of the foot. Articular branches of the tibial nerve supply the knee joint, and medial calcaneal branches supply the skin of the heel.

The *posterior tibial artery* (Table 6.9) provides the main blood supply to the foot. It is the larger terminal branch of the popliteal artery that begins at the distal border of the popliteus and passes deep to the origin of the soleus. After giving off the fibular (peroneal) artery, its largest branch, the posterior tibial artery passes inferomedially on the posterior surface of the tibialis posterior. During its descent, it is accompanied by the tibial nerve and veins. At the ankle the posterior tibial artery runs posterior to the medial malleolus, from which it is separated by the tendons of the tibialis posterior and flexor digitorum longus muscles. Inferior to the medial malleolus, it runs between the tendons of the flexor hallucis longus and flexor digitorum longus. Deep to flexor retinaculum and the origin of the abductor hallucis, the posterior tibial artery divides into medial and lateral plantar arteries.

The *fibular (peroneal) artery* begins inferior to the distal border of the popliteus and the tendinous arch of the soleus. It descends obliquely toward the fibula and passes along its medial side, usually within the flexor hallucis longus. The fibular artery gives muscular branches to the popliteus and other muscles in the posterior and lateral compartments of the leg. It also supplies a nutrient artery to the fibula. The fibular artery usually pierces the interosseous membrane and passes to the dorsum of the foot, where it anastomoses with the arcuate artery.

The *circumflex fibular artery* arises from the posterior tibial artery at the knee and passes laterally over the neck of the fibula to the anastomoses around the knee. The *nutrient artery of the tibia,* the largest nutrient artery in the body, arises from the posterior tibial artery near its origin. Other branches of the posterior tibial artery are the *calcaneal arteries,* which supply the heel. A malleolar branch joins the network of vessels on the medial malleolus.

The pulse of the posterior tibial artery is not always easy to palpate; it is absent in about 15% of people. The posterior tibial pulse can usually be palpated between the posterior surface of the medial malleolus and the medial border of the calcaneal tendon. It is usually easier to palpate when the foot is relaxed and not bearing weight. Palpation of the pulse of the posterior tibial artery is essential for examining patients with occlusive peripheral artery disease. *Intermittent claudication* is characterized by leg cramps that develop during walking and disappear soon after rest. This painful condition results from ischemia of the leg muscles caused by narrowing or occlusion of the leg arteries.

Foot

The foot, the distal part of the lower limb, supports the weight of the body and is important in locomotion.

DEEP FASCIA

The deep fascia of the foot is thin on the dorsum of the foot where it is continuous with the inferior extensor retinaculum. Over the lateral and posterior aspects of the foot, the deep fascia is continuous with the *plantar fascia* (deep fascia of the sole). The central part of the plantar fascia is thick and forms the strong *plantar aponeurosis* (Fig. 6.18*A*). It consists of a thick central part and weaker and thinner medial and lateral portions. This aponeurosis, consisting of longitudinally arranged bands of dense fibrous connective tissue, helps to support the longitudinal arches of the foot and to hold the parts of the foot together. It arises posteriorly from the calcaneus and divides into five bands that split to enclose the digital tendons that attach to the margins of the fibrous digital sheaths and to the sesamoid bones of the great toe. From the margins of the central part of the aponeurosis, verti-

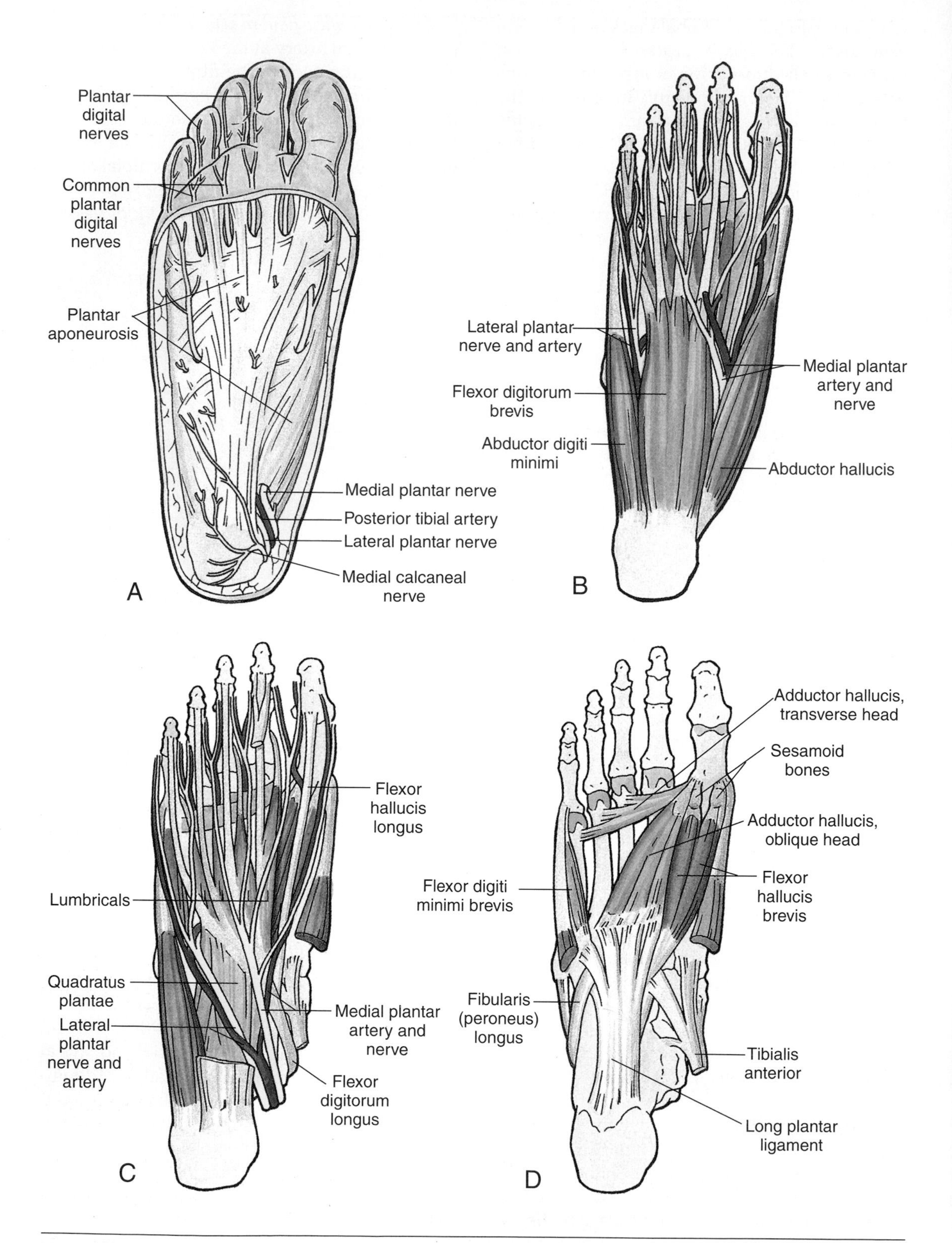

Figure 6.18. Dissections of sole of foot. **A.** Superficial dissection primarily to show plantar aponeurosis. **B.** First layer of plantar muscles, digital nerves, and arteries. **C.** Second layer of plantar muscles. **D.** Third layer of plantar muscles.

Table 6.13
Muscles in Sole of Foot

Muscle	Proximal Attachment	Distal Attachment	Innervation	Main Actions
First layer				
Abductor hallucis	Medial tubercle of tuberosity of calcaneus, flexor retinaculum, and plantar aponeurosis	Medial side of base of proximal phalanx of first digit	Medial plantar nerve (S2 and **S3**)	Abducts and flexes first digit (great toe, hallux)
Flexor digitorum brevis	Medial tubercle of tuberosity of calcaneus, plantar aponeurosis, and intermuscular septa	Both sides of middle phalanges of lateral four digits		Flexes lateral four digits
Abductor digiti minimi	Medial and lateral tubercles of tuberosity of calcaneus, plantar aponeurosis, and intermuscular septa	Lateral side of base of proximal phalanx of fifth digit	Lateral plantar nerve (S2 and **S3**)	Abducts and flexes fifth digit
Second layer				
Quadratus plantae	Medial surface and lateral margin of plantar surface of calcaneus	Posterolateral marginof tendon of flexor digitorum longus	Lateral plantar nerve (S2 and **S3**)	Assists flexor digitorum longus in flexing lateral four digits
Lumbricals	Tendons of flexor digitorum longus	Medial aspect of expansion over lateral four digits	*Medial one:* medial plantar nerve (S2 and **S3**) *Lateral three:* lateral plantar nerve (S2 and **S3**)	Flex proximal phalanges and extend middle and distal phalanges of lateral four digits
Third layer				
Flexor hallucis brevis	Plantar surfaces of cuboid and lateral cuneiforms	Both sides of base of proximal phalanx of first digit	Medial plantar nerve (S2 and **S3**)	Flexes proximal phalanx of first digit
Adductor hallucis	*Oblique head:* bases of metatarsals 2–4 *Transverse head:* plantar ligaments of metatarsophalangeal joints	Tendons of both heads attach to lateral side of base of proximal phalanx of first digit	Deep branch of lateral plantar nerve (S2 and **S3**)	Adducts first digit; assists in maintaining transverse arch of foot
Flexor digiti minimi brevis	Base of fifth metatarsal	Base of proximal phalanx of fifth digit	Superficial branch of lateral plantar nerve (S2 and **S3**)	Flexes proximal phalanx of fifth digit, thereby assisting with its flexion
Fourth layer				
Plantar interossei (three muscles)	Bases and medial sides of metatarsals 3–5	Medial sides of bases of proximal phalanges of third to fifth digits	Lateral plantar nerve (S2 and **S3**)	Adduct digits (2–4) and flex metatarsophalangeal joints
Dorsal interossei (four muscles)	Adjacent sides of metatarsals 1–5	*First:* medial side of proximal phalanx of second digit *Second to fourth:* lateral sides of second to fourth digits		Abduct digits (2–4) and flex metatarsophalangeal joints

cal septa extend deeply to form three compartments of the sole of the foot (Fig. 6.18): a *medial compartment* (abductor hallucis, flexor hallucis brevis, and medial plantar nerve and vessels), a *lateral compartment* (abductor and flexor digiti minimi brevis), and a *central compartment* (flexor digitorum brevis, flexor digitorum longus, quadratus plantae, lumbricals, proximal part of tendon flexor hallucis longus, and lateral plantar nerve and vessels). The muscles, nerves, and vessels in the sole may be described according to these compartments, but the muscles are usually dissected and described by layers.

MUSCLES

There are four muscular layers in the sole of the foot (Fig. 6.18, Table 6.13). These muscles help maintain the arches of the foot and enable humans to stand on uneven ground. The muscles are of little importance individually because the fine control of the individual toes is not important to most people. There are two *neurovascular planes in the foot*: a superficial plane between the first and second muscular layers and a deep plane between the third and fourth muscular layers.

There are two closely connected muscles on the dorsum of the foot, the *extensor digitorum brevis* and *extensor hallucis brevis* (Fig. 6.15). These broad thin muscles form a fleshy mass on the lateral part of the dorsum of the foot, anterior to the lateral malleolus. The extensor hallucis brevis muscle is part of the extensor digitorum brevis.

> Functionally the extensor digitorum brevis and extensor hallucis brevis muscles are relatively unimportant. Probably the only clinical reason for knowing about their presence is that contusion and tearing of the muscle fibers result in a *hematoma*, which produces a swelling anteromedial to the lateral malleolus. Most people who have not seen this swelling presume they have a badly sprained ankle.

NERVES

The medial and lateral plantar nerves, branches of the tibial nerve, supply the intrinsic muscles of the foot, except for the extensor digitorum brevis, which is supplied by the deep fibular (peroneal) nerve (Fig. 6.18, Table 6.14). The cutaneous innervation of the foot is supplied by

- Medial and lateral plantar nerves, sole of the foot
- Tibial nerve (calcaneal branches), heel
- Sural nerve, lateral margin of the foot and fifth digit
- Saphenous nerve, medial side of the foot as far as the head of the first metatarsal
- Superficial and deep fibular nerves, dorsum of the foot

ARTERIES

The *dorsalis pedis artery* is the direct continuation of the anterior tibial artery distal to the ankle joint. It provides a secondary blood supply to the foot, augmenting that provided by the posterior tibial artery (p. 266). The dorsalis pedis begins midway between the malleoli and runs anteromedially, deep to the inferior extensor retinaculum between the extensor hallucis longus and the extensor digitorum longus tendons on the dorsum of the foot. It passes to the first interosseous space where it divides into a deep plantar artery that passes to the sole of the foot and a first dorsal interosseous artery. The *deep plantar artery* passes deeply through the first interosseous space to join the lateral plantar artery in the formation of the deep plantar arch. The *arcuate artery* runs laterally across the bases of the metatarsal bones, deep to the extensor tendons, where it gives off the second, third, and fourth dorsal metatarsal arteries. These vessels run to the clefts of the toes where each of them divides into two dorsal digital arteries for the sides of adjoining toes.

> Palpation of the *dorsalis pedis pulse* is usually easy because the artery is subcutaneous. The dorsalis pedis pulse can usually be felt on the dorsum of the foot, where the artery passes over the navicular and cuneiform bones just lateral to the extensor hallucis longus tendon (p. 255). It may also be felt distal to this at the proximal end of the first interosseous space. A diminished or absent dorsalis pedis pulse suggests *arterial insufficiency* caused by vascular disease.

The arteries in the sole of the foot are derived from the posterior tibial artery (Fig. 6.18, *A–C*). It divides deep to the abductor hallucis muscle to form the medial and lateral *plantar arteries*, which run parallel to the similarly named nerves. The *plantar arch* begins opposite the base of the fifth metatarsal bone as the continuation of the lateral plantar artery (p. 266). It is completed medially by union with the

Table 6.14
Nerves of Foot

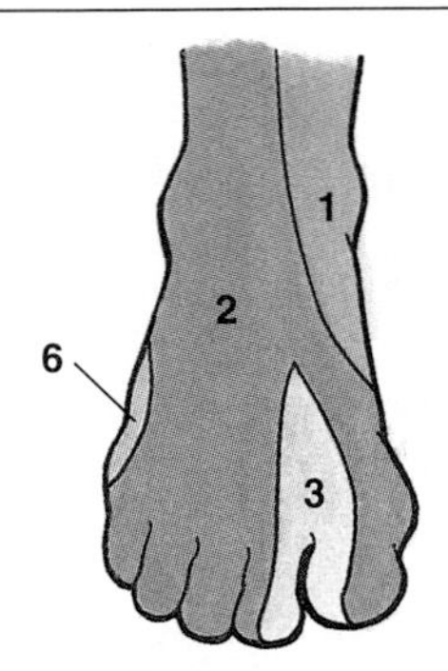

Dorsal surface

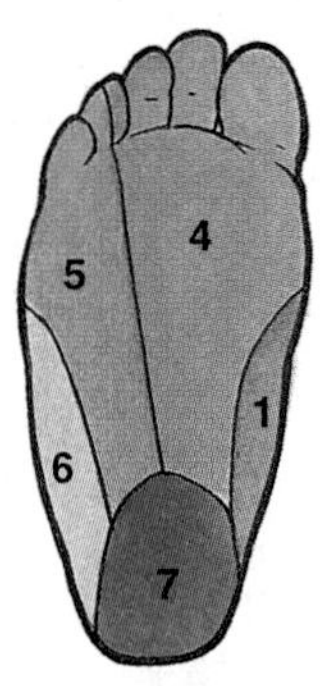

Plantar surface

Nerve[a]	Origin	Course	Distribution
Saphenous (*1*)	Femoral n.	Arises in femoral triangle and descends through thigh and leg; accompanies great saphenous v. anterior to medial malleolus and ends on medial side of foot	Supplies skin on medial side of the foot as far anteriorly as head of first metatarsal
Superficial fibular (*2*)	Common fibular (peroneal) n.	Pierces deep fascia in distal third of leg to become cutaneous and send branches to foot and digits	Supplies skin on dorsum of foot and all digits, except lateral side of fifth and adjoining sides of the first and second digits
Deep fibular (*3*)	Common fibular (peroneal) n.	Passes deep to extensor retinaculum to enter dorsum of foot	Supplies extensor digitorum brevis and skin on contiguous sides of first and second digits
Medial plantar (*4*)	Larger terminal branch of the tibial n.	Passes distally in foot between abductor hallucis and flexor digitorum brevis and divides into muscular and cutaneous branches	Supplies skin of medial side of sole of foot and sides of first three digits; also supplies abductor hallucis, flexor digitorum brevis, flexor hallucis brevis, and first lumbrical n.
Lateral plantar (*5*)	Smaller terminal branch of the tibial n.	Passes laterally in foot between quadratus plantae and flexor digitorum brevis mm. and divides into superficial and deep branches	Supplies quadratus plantae, abductor digiti minimi, and flexor digiti minimi brevis; deep branch supplies plantar and dorsal interossei, lateral three lumbricals, and adductor hallucis; supplies skin on sole lateral to a line splitting fourth digit
Sural (*6*)	Usually arises from both tibial and common fibular (peroneal) nn.	Passes inferior to the lateral malleolus to lateral side of foot	Lateral aspect of foot
Calcaneal branches (*7*)	Tibial and sural nn.	Pass from distal part of the posterior aspect of leg to skin on heel	Skin of heel

[a] Numbers in parentheses refer to drawings.

deep plantar artery, a branch of the dorsalis pedis artery. As it crosses the foot, the plantar arch gives off four *plantar metatarsal arteries,* three perforating arteries, and branches to the tarsal joints and the muscles in the sole of the foot. These arteries join with the superficial branches of the medial and lateral plantar arteries to form the *plantar digital arteries.*

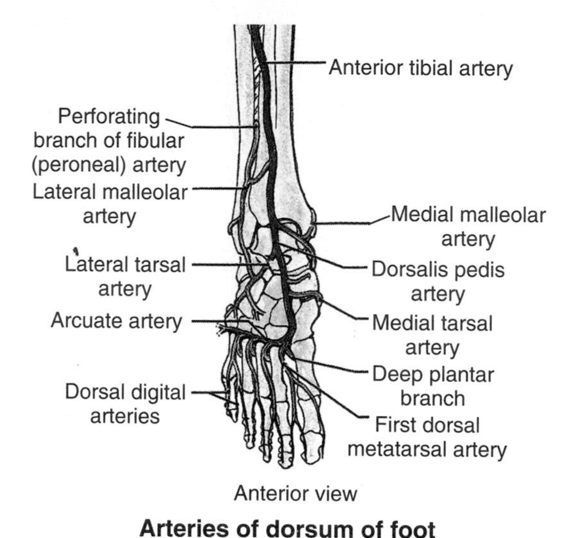

Arteries of dorsum of foot

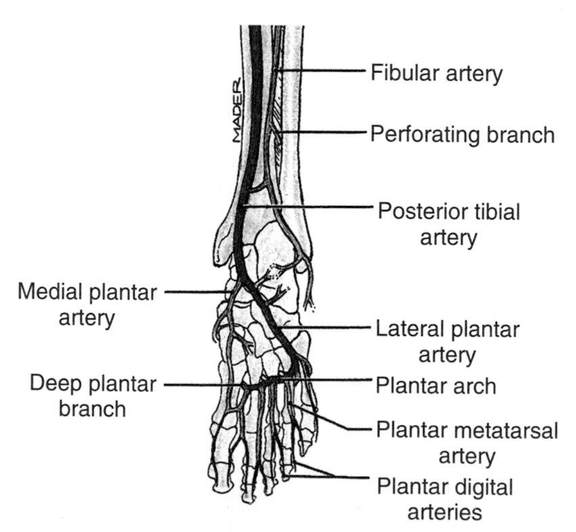

Arteries of sole of foot

Wounds of the foot involving the plantar arterial arch result in severe bleeding. Ligature of the arch is difficult because of its depth and the structures that are related to it.

Joints

HIP JOINT

The hip is a multiaxial ball and socket type of synovial joint.

Articular Surfaces. The head of the femur articulates with the acetabulum of the hip bone. The depth of the acetabulum is increased by the fibrocartilaginous *acetabular labrum* (L. labrum, lip), which attaches to the bony rim of the acetabulum and the transverse acetabular ligament (*A*, roof; *P*, posterior rim of acetabulum; *F*, fovea; *T*, "teardrop" appearance caused by the superimposition of structures at the inferior margin of the acetabulum; *G*, greater trochanter; *I*, intertrochanteric crest; *L*, lesser trochanter).

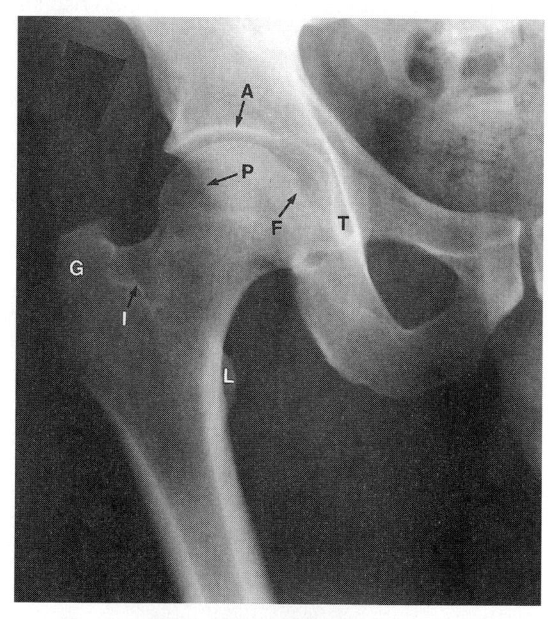

Radiograph of hip joint

Articular Capsule. The strong *fibrous capsule* attaches proximally to the acetabulum and the transverse acetabular ligament. It attaches distally to the neck of the femur as follows:

- Anteriorly to the *intertrochanteric line* and the root of the greater trochanter
- Posteriorly to the neck proximal to the *intertrochanteric crest*

Most capsular fibers take a spiral course from the hip bone to the intertrochanteric line, but some deep fibers pass circularly around the neck, forming an *orbicular zone*. These fibers form a collar around the neck that constricts the capsule and helps to hold the femoral head in the acetabulum. Some deep longitudinal fibers of the capsule form *retinacula*, which are reflected superiorly along the neck as longitudinal bands that blend with the periosteum. The retinacula contain blood vessels that supply the head and neck of the femur (p. 268).

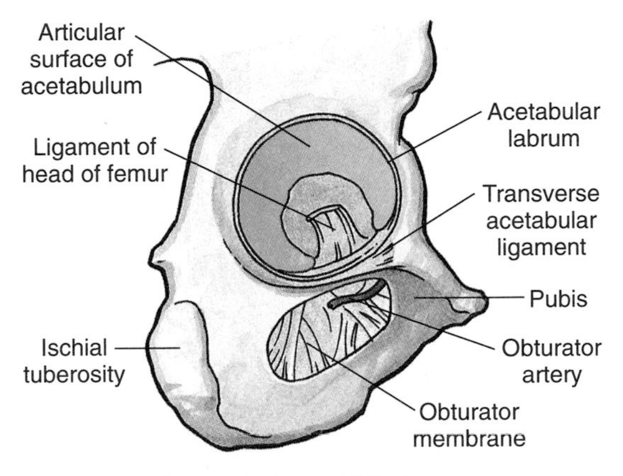

Lateral view of hip bone

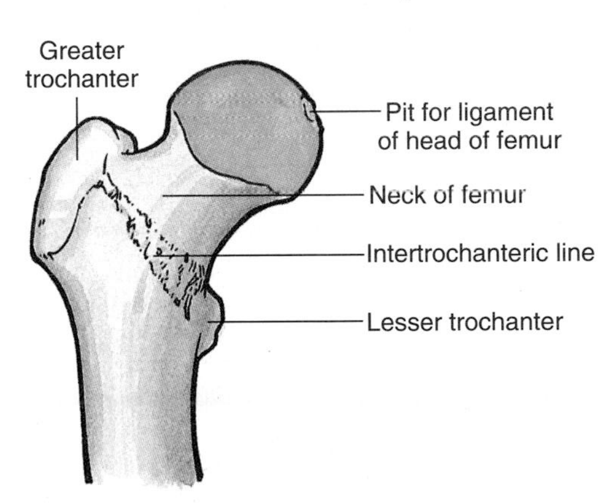

Anterior view of femur (proximal end)

The *synovial membrane* lines the fibrous capsule and also covers the

- Neck of the femur between the attachment of the fibrous capsule and the edge of the articular cartilage of the head
- Nonarticular area of the acetabulum and forms a covering for the ligament of the head of the femur

Ligaments. The fibrous capsule is reinforced anteriorly by a strong, Y-shaped ligament, the *iliofemoral ligament*, that attaches to the anterior inferior iliac spine and the acetabular rim proximally and to the intertrochanteric line distally. This ligament prevents overextension of the hip joint during standing by screwing the head of the femur into the acetabulum.

The fibrous capsule is reinforced inferiorly and anteriorly by the *pubofemoral ligament* that attaches to the pubic part of the acetabular rim and the iliopubic eminence; it blends with the medial part of the iliofemoral ligament and tightens during extension and abduction of the hip joint, preventing overabduction of the joint.

The fibrous capsule is reinforced posteriorly by the *ischiofemoral ligament* that arises from the ischial portion of the acetabular rim and spirals superolaterally to the neck of the femur, medial to the base of the greater trochanter; this ligament tends to screw the femoral head medially into the acetabulum during extension of the hip joint, thereby preventing hyperextension.

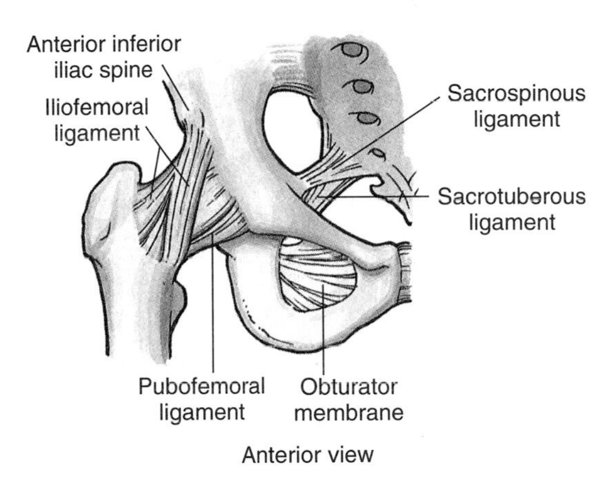

Anterior view

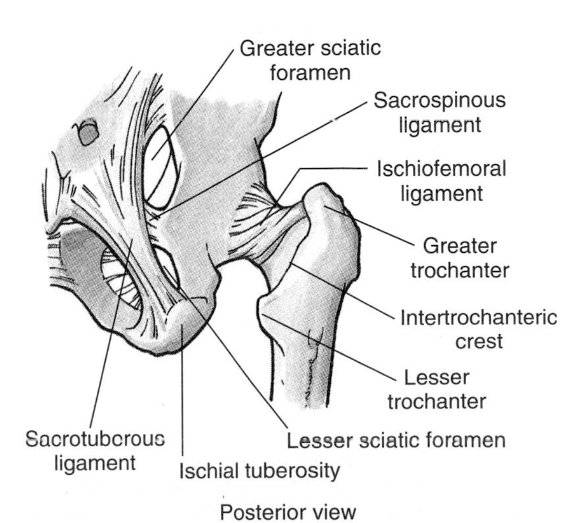

Posterior view

Ligaments of hip joint

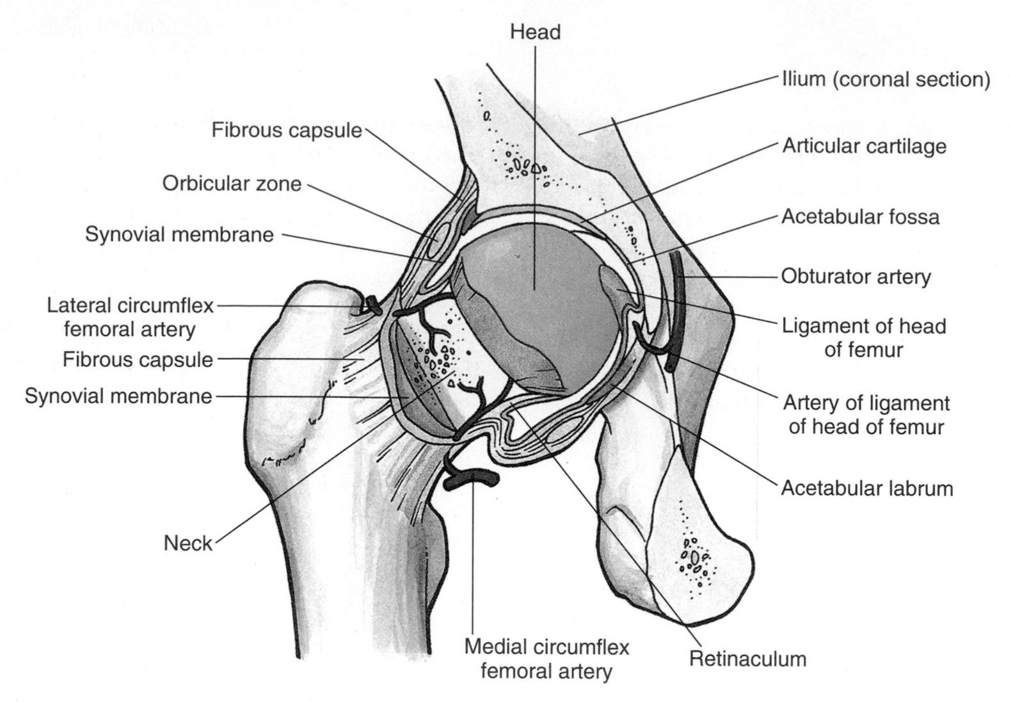

Blood supply of femoral head and neck

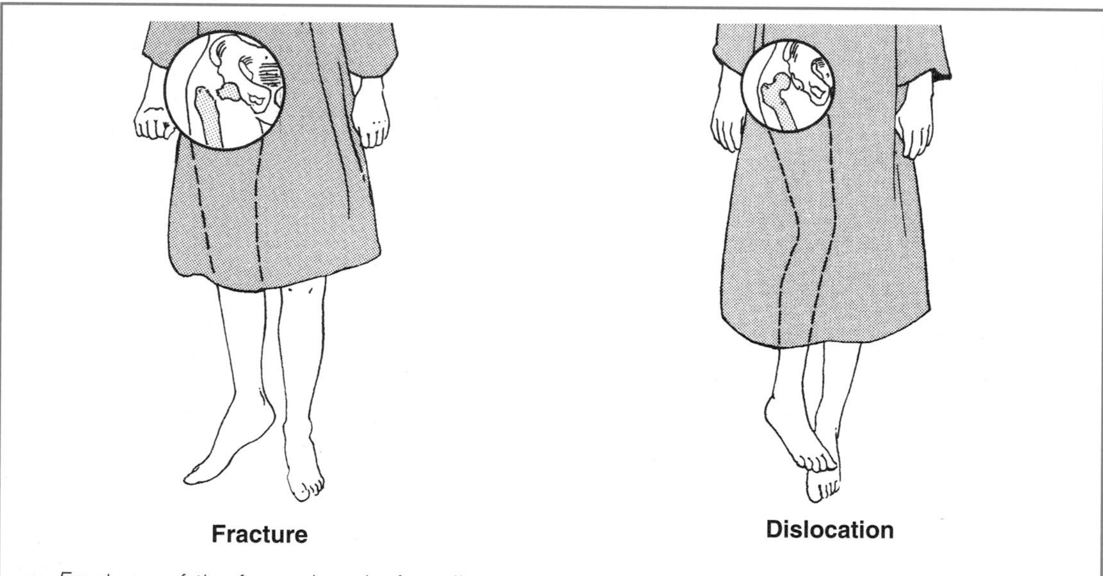

Fracture

Dislocation

Fractures of the femoral neck often disrupt blood supply to the head of the femur. The medial circumflex femoral artery is clinically important because it supplies most of the blood to the head and neck of the femur. It is often torn when the femoral neck is fractured or the hip joint is dislocated. In some cases the blood supplied via the artery in the ligament of the head may be the only blood received by the proximal fragment of the femoral head. If the blood vessels are ruptured, the fragment of bone may receive no blood and undergo *aseptic necrosis*.

Congenital dislocation of a hip joint is common, occurring in about 1.5 per 1000 live births, and is bilateral in about half of those cases. Inability to abduct the thigh is characteristic of congenital dislocation. In addition, the affected limb appears shorter because the dislocated femoral head is more superior than on the normal side. *Acquired dislocation* of a hip joint is uncommon because this articulation is so strong and stable. Nevertheless, dislocation may occur during an automobile accident when the hip is flexed, adducted, and medially rotated. The fibrous capsule ruptures inferiorly and posteriorly, allowing the head to pass through the tear in the capsule and over the posterior margin of the acetabulum. Often the acetabular margin fractures, producing a *fracture dislocation* of the hip joint. When the femoral head dislocates, it usually carries the acetabular bone fragment and the acetabular labrum with it.

The *ligament of the head of the femur* is weak and appears to be of little importance in strengthening the hip joint. Its wide end attaches to the margins of the acetabular notch and the transverse acetabular ligament; its narrow end attaches to the fovea (pit) in the head of the femur. Usually it contains a small artery to the head of the femur.

Movements. Hip movements are flexion-extension, abduction-adduction, medial-lateral rotation, and circumduction.

Blood Supply. Branches of the medial and lateral *circumflex femoral arteries* and the artery to the head of the femur, a branch of the obturator artery.

Nerve Supply. Femoral nerve (branch to rectus femoris), obturator nerve (anterior division), sciatic nerve (nerve to quadratus femoris), and superior gluteal nerve.

KNEE JOINT

The knee is a hinge type of synovial joint that permits some rotation when flexed.

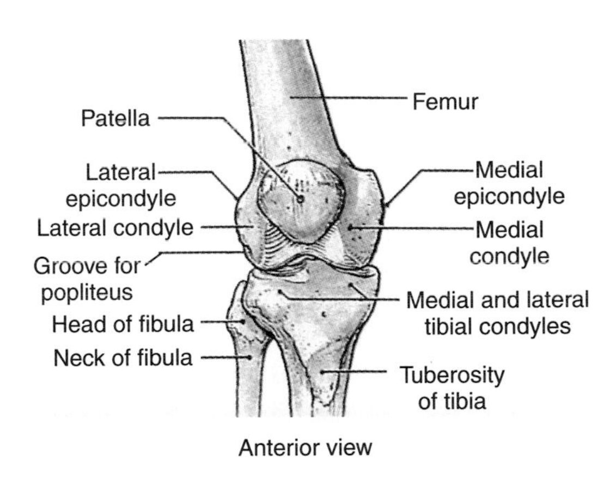

Anterior view

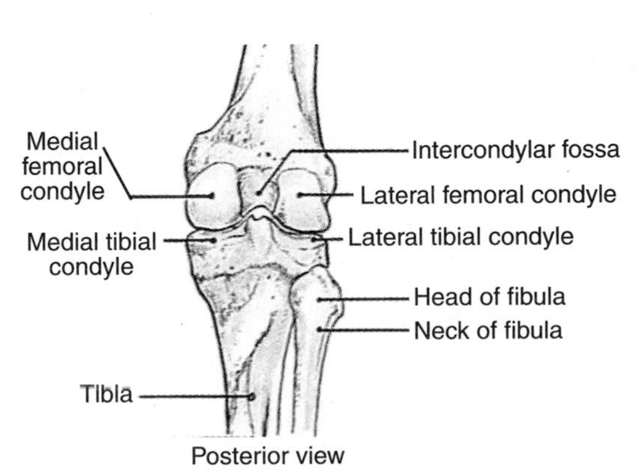

Posterior view

Bones of knee joint

Articular Surfaces. The knee joint consists of three articulations:

- Lateral and medial articulations between the femoral and tibial condyles
- Intermediate articulation between the patella and femur

The knee joint is relatively weak mechanically because of the configurations of its articular surfaces. For strength it relies on ligaments that bind the femur to the tibia.

Articular Capsule. The strong *fibrous capsule* attaches to the femur superiorly, just proximal to the articular margins of the condyles and also to the intercondylar fossa (notch) posteriorly. It is deficient on the lateral condyle to allow the tendon of the popliteus to pass out of the joint. Inferiorly the fibrous capsule of the knee joint attaches to the articular margin of the tibia, except where the tendon of the popliteus crosses the bone.

The extensive *synovial membrane* lines the internal aspect of the fibrous capsule and attaches to the periphery of the patella and the edges of the menisci. The synovial membrane reflects from the posterior aspect of the joint onto the cruciate ligaments. The reflection of the membrane between the tibia and the patella covers the infrapatellar fatpad. The synovial membrane covering the fatpad and cruciate ligaments separates them from the joint cavity. The median *infrapatellar synovial fold* extends posteriorly from the fatpad to the intercondylar fossa of the femur. *Alar folds* project from the median fold to the lateral edges of the patella.

The knee joint cavity extends superior to the patella as the *suprapatellar bursa,* which lies deep to the articularis genu and vastus intermedius.

Ligaments. The fibrous capsule is strengthened by five intrinsic ligaments: patellar ligament, fibular collateral ligament, tibial collateral ligament, oblique popliteal ligament, and arcuate popliteal ligament.

The round *fibular collateral ligament* extends inferiorly from the lateral epicondyle of the femur to the lateral surface of the head of the fibula. The tendon of the popliteus passes deep to the fibular collateral ligament, separating it from the lateral meniscus. The tendon of the biceps femoris is also split by this ligament.

The *tibial collateral ligament* is a strong flat band that extends from the medial epicondyle of the femur to the medial condyle and superi-

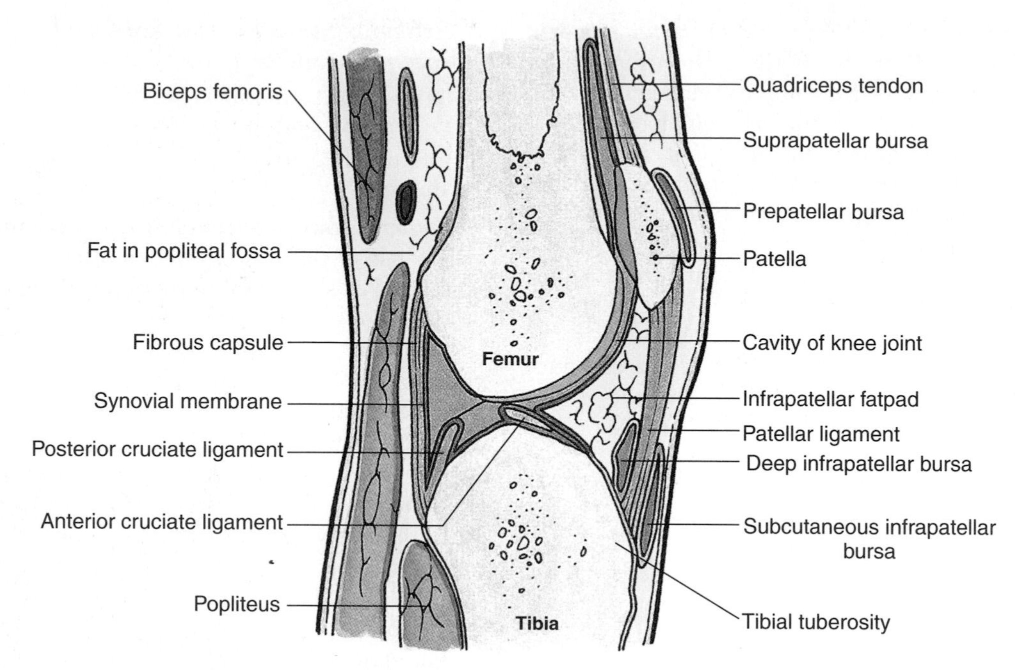

Sagittal section of knee joint

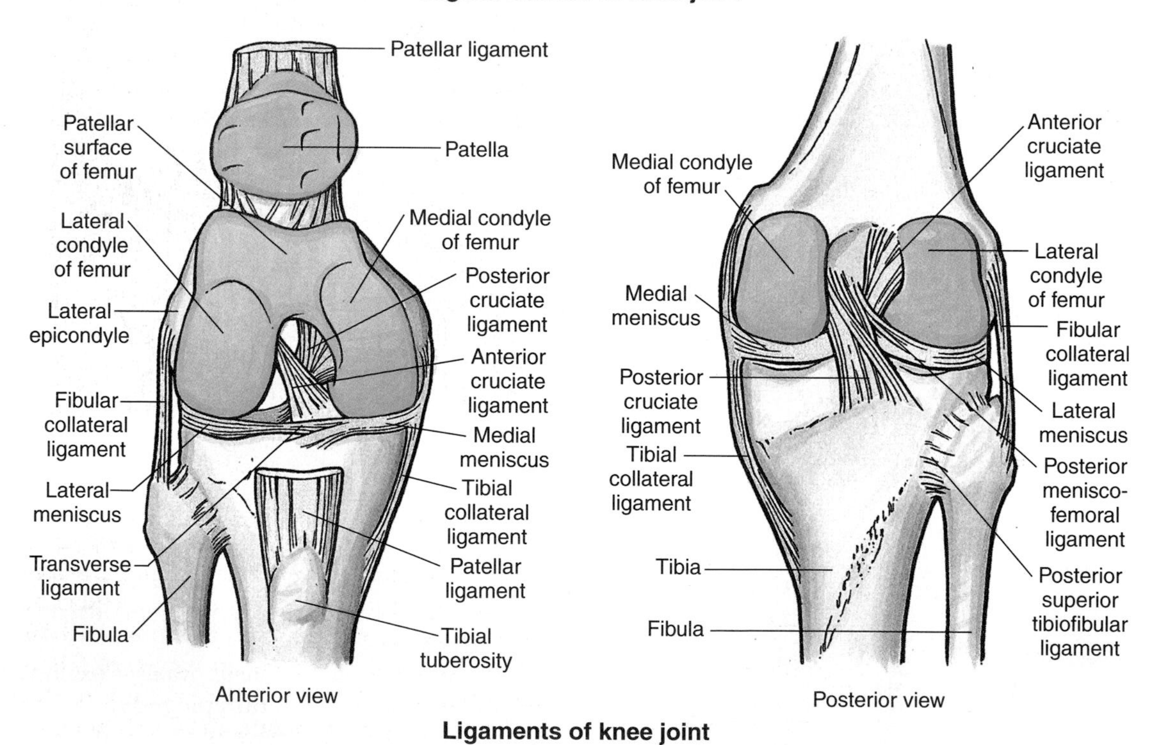

Ligaments of knee joint

or part of the medial surface of the tibia. The deeper fibers of the tibial collateral ligament are firmly attached to the medial meniscus.

The *oblique popliteal ligament* (Fig. 6.9B) is an expansion of the tendon of the semimembranosus that strengthens the fibrous capsule posteriorly. It arises posterior to the medial condyle of the tibia and passes superolaterally to attach to the central part of the posterior aspect of the fibrous capsule.

The *arcuate popliteal ligament* (Fig. 6.16D) strengthens the fibrous capsule posteriorly. It arises from the posterior aspect of the head of the fibula, passes superomedially over the tendon of the popliteus, and spreads over the posterior surface of the knee joint.

The *cruciate ligaments* join the femur and tibia within the articular capsule of the joint but outside the synovial cavity. The cruciate ligaments cross each other obliquely like the letter X.

The *anterior cruciate ligament*, the weaker of the two cruciate ligaments, arises from the anterior part of the intercondylar area of the tibia, just posterior to the attachment of the medial meniscus. It extends superiorly, posteriorly, and laterally to attach to the posterior part of the medial side of the lateral condyle of the femur. The ligament is slack when the knee is flexed and taut when it is fully extended, preventing posterior displacement of the femur on the tibia and hyperextension of the knee joint. When the joint is flexed at a right angle, the tibia cannot be pulled anteriorly because it is held by the anterior cruciate ligament.

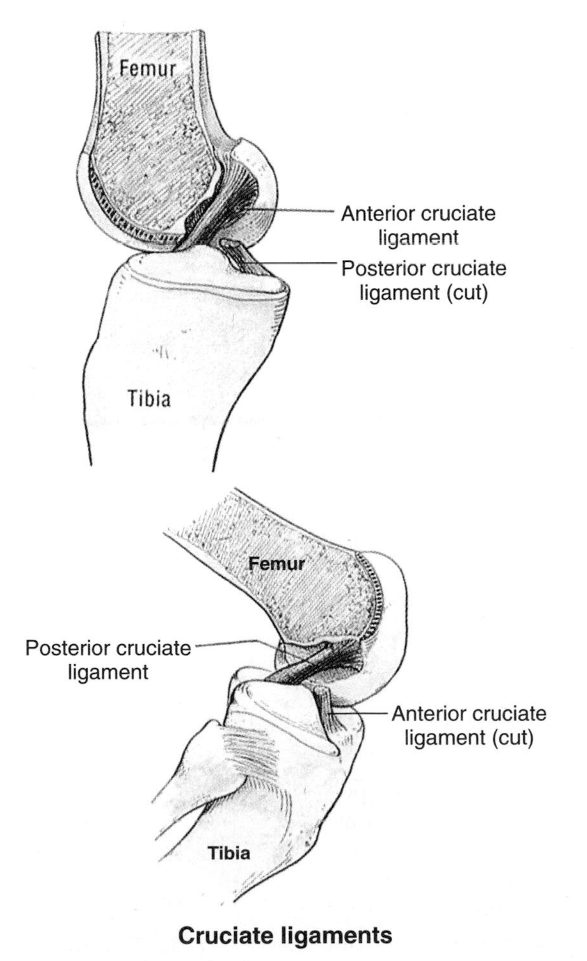

Cruciate ligaments

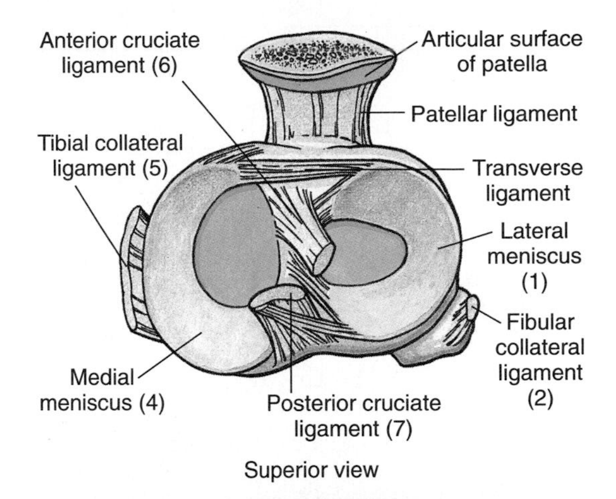

Superior view

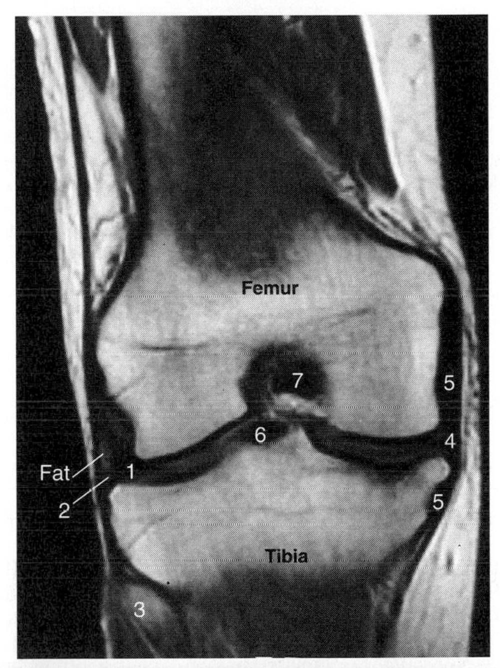

Coronal view (magnetic resonance image)

Knee joint

The *posterior cruciate ligament*, the stronger of the two cruciate ligaments, arises from the posterior part of the intercondylar area of the tibia. It passes superiorly and anteriorly on the medial side of the anterior cruciate ligament to attach to the anterior part of the lateral surface of the medial condyle of the femur. The ligament tightens during flexion of the knee joint, preventing anterior displacement of the femur on the tibia or posterior displacement of the tibia on the femur. It also helps to prevent hyperflexion of the knee joint. In the weightbearing flexed knee, the posterior cruciate ligament stabilizes the femur (e.g., when walking downhill).

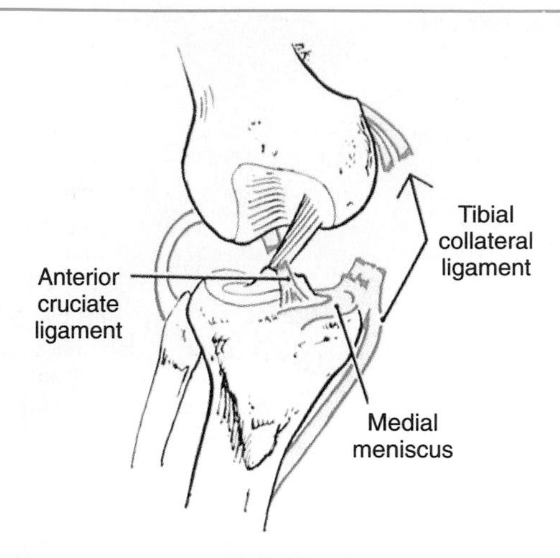

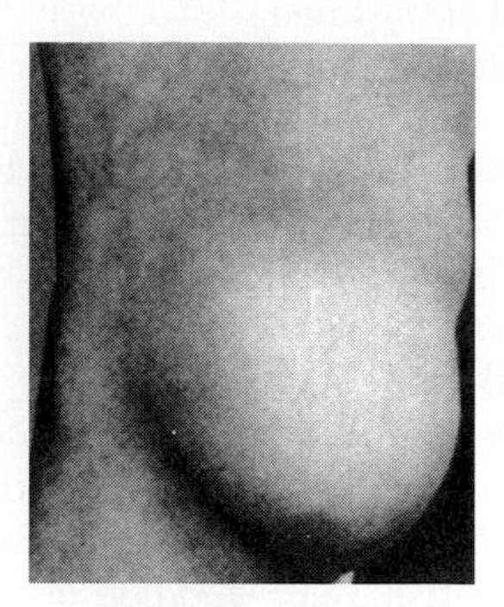

Prepatellar bursitis

Knee joint injuries are common because the knee is a major weightbearing joint and its stability depends almost entirely on its associated ligaments and muscles. The tibial and fibular collateral ligaments normally prevent disruption of the sides of the knee joint. They are tightly stretched when the leg is extended and prevent rotation of the tibia laterally or the femur medially. Because the collateral ligaments are slack during flexion of the leg, they permit some rotation of the tibia on the femur in this position. The fibular collateral ligament is not commonly torn because it is very strong. The firm attachment of the tibial collateral ligament to the medial meniscus is of considerable clinical significance because *tearing of the tibial collateral ligament frequently results in concomitant tearing of the medial meniscus.* The damage is frequently caused by a blow to the lateral side of the knee. Injury to the medial meniscus results from a twisting strain that is applied to the knee joint when it is flexed. Because the meniscus is firmly adherent to the tibial collateral ligament, twisting strains of this ligament may tear and/or detach the medial meniscus from the fibrous capsule. This injury is common in athletes who twist their flexed knees while running (e.g., in football and volleyball).

The anterior cruciate ligament may be torn when the tibial collateral ligament ruptures after the knee is hit hard from the lateral side while the foot is fixed (e.g., in the ground). First the tibial collateral ligament ruptures, opening the joint on the medial side. This may tear the medial meniscus and the anterior cruciate ligament. The posterior cruciate ligament may be injured when the superior part of the tibia is struck with the knee flexed. This kind of injury may occur when a car passenger's leg is driven against the dashboard. If the knee joint is severely hyperflexed, the posterior cruciate ligament may be torn.

Injections may be made into the synovial cavity of the knee joint for diagnostic and/or therapeutic purposes. They are generally made into the lateral side of the knee joint. When the knee joint is inflamed, the amount of synovial fluid may increase. Aspiration of this fluid may be necessary to relieve pressure in the joint or to obtain a sample of it for diagnostic studies.

The suprapatellar (quadriceps) bursa communicates freely with the synovial cavity of the knee joint. Hence, stab or puncture wounds superior to the patella may infect the knee joint through the suprapatellar bursa. This bursa may also be involved in fractures of the distal end of the femur.

Prepatellar bursitis is a friction bursitis caused by friction between the skin and the patella. If the inflammation is chronic, the bursa becomes distended with fluid and forms a swelling anterior to the knee.

Subcutaneous infrapatellar bursitis results from excessive friction between the skin and the tibial tuberosity. The swelling occurs over the proximal end of the tibia. *Deep infrapatellar bursitis* results in a swelling between the patellar ligament and the tibia, superior to the tibial tuberosity. Enlargement of this bursa obliterates the dimples on each side of the patellar ligament when the leg is extended.

The *menisci of the knee joint* are crescentic plates of fibrocartilage on the articular surface of the tibia that deepen the articulating surface and act as shock absorbers. Wedge-shaped in transverse section, the menisci are firmly attached at their ends to the intercondylar area of the tibia. Their external margins are attached to the fibrous capsule of the knee joint. The *coronary ligaments* are capsular fibers that attach the margins of the menisci to the tibial condyles. A slender fibrous band, the *transverse ligament of the knee*, joins the anterior edges of the two menisci, allowing them to move together during movements of the femur on the tibia.

The *medial meniscus* is broader posteriorly than anteriorly. Its anterior end (horn) attaches to the anterior intercondylar area of the tibia, anterior to the attachment of the anterior cruciate ligament. Its posterior end (horn) attaches to the posterior intercondylar area, anterior to the attachment of the posterior cruciate ligament. The medial meniscus is firmly adherent to the deep surface of the tibial collateral ligament.

The *lateral meniscus* is nearly circular and is smaller and more freely movable than the medial meniscus. The tendon of the popliteus separates the lateral meniscus from the fibular collateral ligament. A strong tendinous slip, the *posterior meniscofemoral ligament*, joins the lateral meniscus to the posterior cruciate ligament and the medial femoral condyle.

Movements. The movements are mainly flexion and extension; some rotation occurs when the knee is flexed. When the leg is fully extended, the knee "locks" because of medial rotation of the femur on the tibia. This makes the lower limb a solid column and more adapted for weightbearing. To "unlock" the knee the popliteus contracts, rotating the femur laterally so that flexion of the knee can occur.

Blood Supply. The arteries are branches of the vessels that form the *genicular anastomoses* around the knee (Table 6.9). The middle genicular artery penetrates the fibrous capsule and supplies the cruciate ligaments, synovial membrane, and peripheral margins of the menisci.

Nerve Supply. The nerves are branches of the obturator, femoral, tibial, and common fibular nerves.

Bursae around Knee. There are many bursae around the knee (Table 6.15) because most tendons around the knee run parallel to the bones (p. 270) and pull lengthwise across the joint. Some bursae communicate with the synovial cavity of the knee joint: suprapatellar (quadriceps) bursa, popliteus bursa, anserine bursa, and gastrocnemius bursa.

TIBIOFIBULAR JOINTS

The tibia and fibula articulate at their proximal and distal ends at tibiofibular joints. Movement at the proximal tibiofibular joint is impossible without movement at the distal one.

Proximal (Superior) Tibiofibular Joint

The proximal (superior) tibiofibular joint is a plane type of synovial joint.

Articular Surfaces. The flat facet on the head of the fibula articulates with a similar facet located posterolaterally on the lateral condyle of the tibia.

Articular Capsule and Ligaments. The fibrous capsule surrounds the joint and is attached to the margins of the articular surfaces of the fibula and tibia. It is strengthened by anterior and posterior ligaments. The synovial membrane lines the fibrous capsule. A pouch of synovial membrane, the *popliteus bursa*, passes deep to the tendon of the popliteus. This bursa may communicate with the synovial cavity of the knee joint.

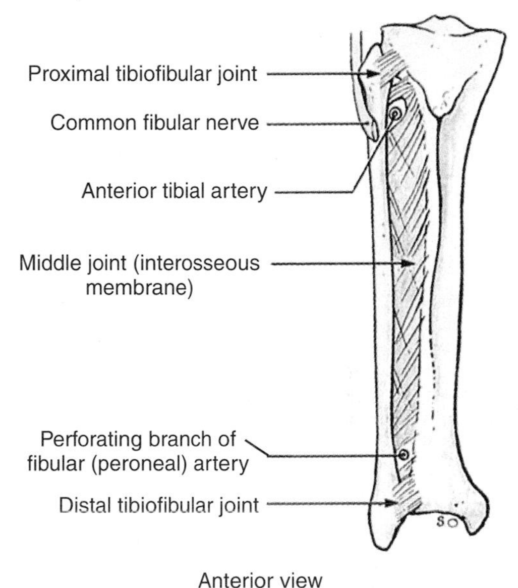

Tibiofibular articulations

Movements. Slight movement occurs during dorsiflexion and plantarflexion of the foot.

Blood Supply. The arteries are from the inferior lateral genicular and anterior tibial recurrent arteries.

Nerve Supply. The nerves are from the common fibular (peroneal) nerve and the nerve to the popliteus.

Distal (Inferior) Tibiofibular Joint

The distal (inferior) tibiofibular joint is a fibrous joint (syndesmosis).

Articular Surfaces. The rough, triangular articular area on the medial surface of the inferior end of the fibula articulates with a facet on the inferior end of the tibia. A small projection of the synovial capsule of the ankle joint extends superiorly into the inferior part of the distal tibiofibular joint.

Ligaments. A strong *interosseous ligament*, continuous superiorly with the interosseous membrane, forms the principal connection between the tibia and fibula. The joint is also strengthened anteriorly and posteriorly by the

Table 6.15
Bursae around Knee

Bursae	Locations	Comments
Suprapatellar	Between femur and tendon of quadriceps femoris	Held in position by articularis genu muscle; communicates freely with synovial cavity of knee joint
Popliteus	Between tendon of popliteus and lateral condyle of tibia	Opens into synovial cavity of knee joint, inferior to lateral meniscus
Anserine	Separates tendons of sartorius, gracilis, and semitendinosus from tibia and tibial collateral ligament	Area where tendons of these muscles attach to tibia resembles the foot of a goose (L, pes, foot; L. anser, goose)
Gastrocnemius	Lies deep to proximal attachment of tendon of medial head of gastrocnemius	This bursa is an extension of synovial cavity of knee joint
Semimembranosus	Located between medial head of gastrocnemius and semimembranosus tendon	Related to the distal attachment of semimembranosus
Subcutaneous prepatellar	Lies between skin and anterior surface of patella	Allows free movement of skin over patella during movements of leg
Subcutaneous infrapatellar	Located between skin and tibial tuberosity	Helps knee to withstand pressure when kneeling
Deep infrapatellar	Lies between patellar ligament and anterior surface of tibia	Separated from knee joint by infrapatellar fatpad

strong anterior and posterior *inferior tibiofibular ligaments*. The inferior deep part of the posterior inferior tibiofibular ligament is called the *transverse tibiofibular ligament*.

Movements. Slight movement of the joint occurs to accommodate the talus during dorsiflexion of the foot.

Blood Supply. The arteries are from the perforating branch of the fibular artery and from medial malleolar branches of the anterior and posterior tibial arteries.

Nerve Supply. The nerves are from the deep fibular (peroneal), tibial, and saphenous nerves.

The posterior tibiofibular ligament is much stronger than the anterior tibiofibular ligament. In severe ankle injuries, the anterior tibiofibular ligament may avulse the posteroinferior part of the tibia. In these cases the fracture enters the ankle joint. If, in addition, the medial and lateral malleoli are fractured, the injury is referred to as a "trimalleolar fracture" (i.e., a fracture of both malleoli and the posterior part of the inferior border of the tibia).

ANKLE JOINT

The ankle is a hinge type of synovial joint.

Articular Surfaces. The inferior ends of the tibia and fibula form a mortise (deep socket) into which the *talus* fits. The medial surface of the lateral malleolus articulates with the lateral surface of the talus. The tibia articulates with the talus in two places:

- Its inferior surface forms the roof of the mortise
- Its medial malleolus articulates with the medial surface of the talus

The malleoli grip the talus tightly as it rocks anteriorly and posteriorly during movements of the joint. The ankle joint is very stable during dorsiflexion because in this position the superior articulating surface of the talus (trochlea) fills the mortise formed by the malleoli. The grip of the malleoli on the trochlea is strongest during dorsiflexion of the foot because this movement forces the anterior part of the trochlea posteriorly, spreading the tibia and fibula slightly apart. This

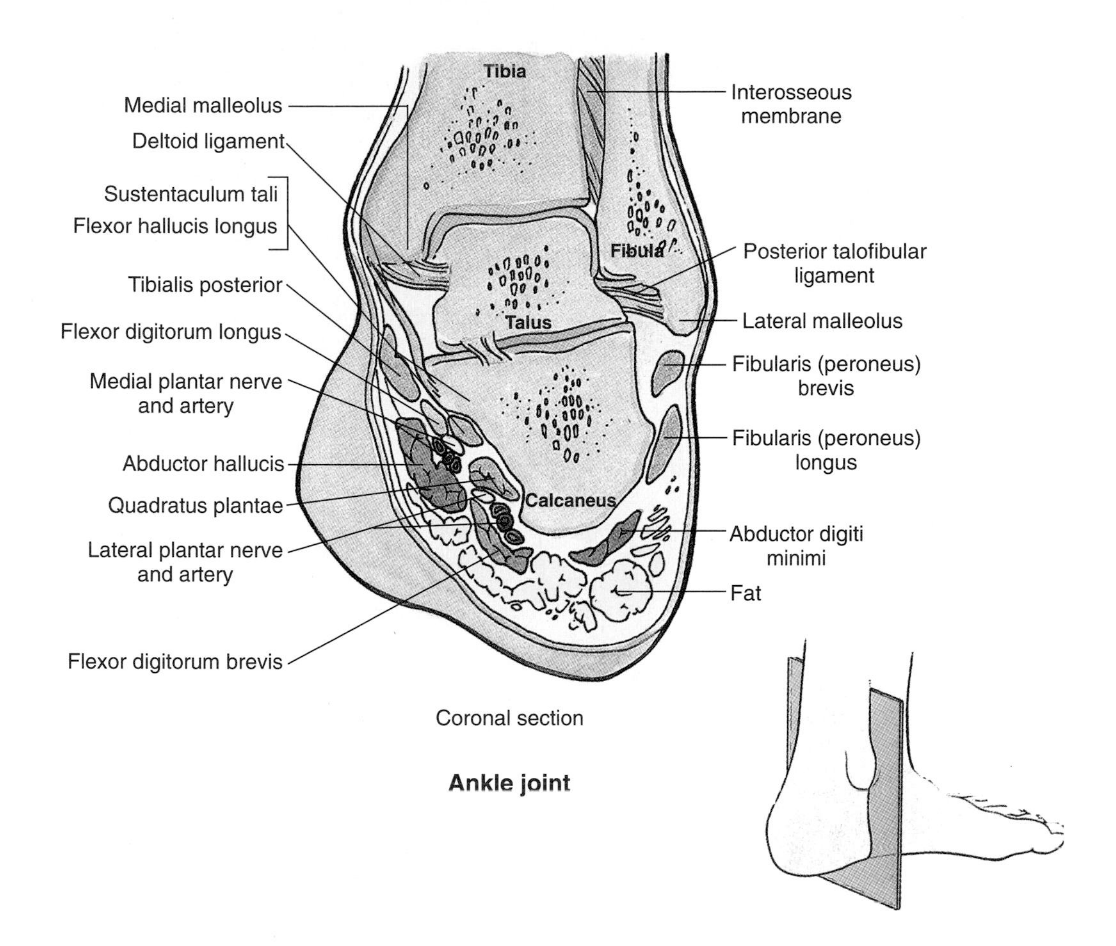

Ankle joint

spreading is limited by the strong interosseous ligament and by the anterior and posterior inferior tibiofibular ligaments that unite the leg bones. The ankle joint is relatively unstable during plantarflexion because the superior articulating surface of the talus is narrower posteriorly and therefore does not fill the mortise.

Articular Capsule. The *fibrous capsule* is thin anteriorly and posteriorly but is supported on each side by strong collateral ligaments. It is attached superiorly to the borders of the articular surfaces of the tibia and the malleoli and inferiorly to the talus.

Ligaments. The fibrous capsule is reinforced medially by the strong medial (deltoid) ligament that attaches proximally to the medial malleolus and fans out to attach distally to the talus, calcaneus, and navicular (tibionavicular, anterior and posterior tibiotalar, and tibiocalcaneal ligaments).

The fibrous capsule is reinforced laterally by the lateral ligament consisting of three parts:

- The weak anterior talofibular ligament is a flat band that extends anteromedially from the lateral malleolus to the neck of the talus
- The posterior talofibular ligament is a thick and fairly strong band that runs horizontally medially and slightly posteriorly from the malleolar fossa to the lateral tubercle of the talus
- The calcaneofibular ligament is a round cord that passes posteroinferiorly from the tip of the lateral malleolus to the lateral surface of the calcaneus

The *synovial membrane* is loose and extends superiorly between the tibia and fibula.

Movements. The movements are dorsiflexion and plantarflexion. When the foot is plantarflexed, some rotation, abduction, and adduction of the ankle joint are possible.

Blood Supply. The arteries are derived from the malleolar branches of the fibular and anterior and posterior tibial arteries.

Nerve Supply. The nerves are derived from the tibial nerve and the deep fibular nerve, a division of the common fibular nerve.

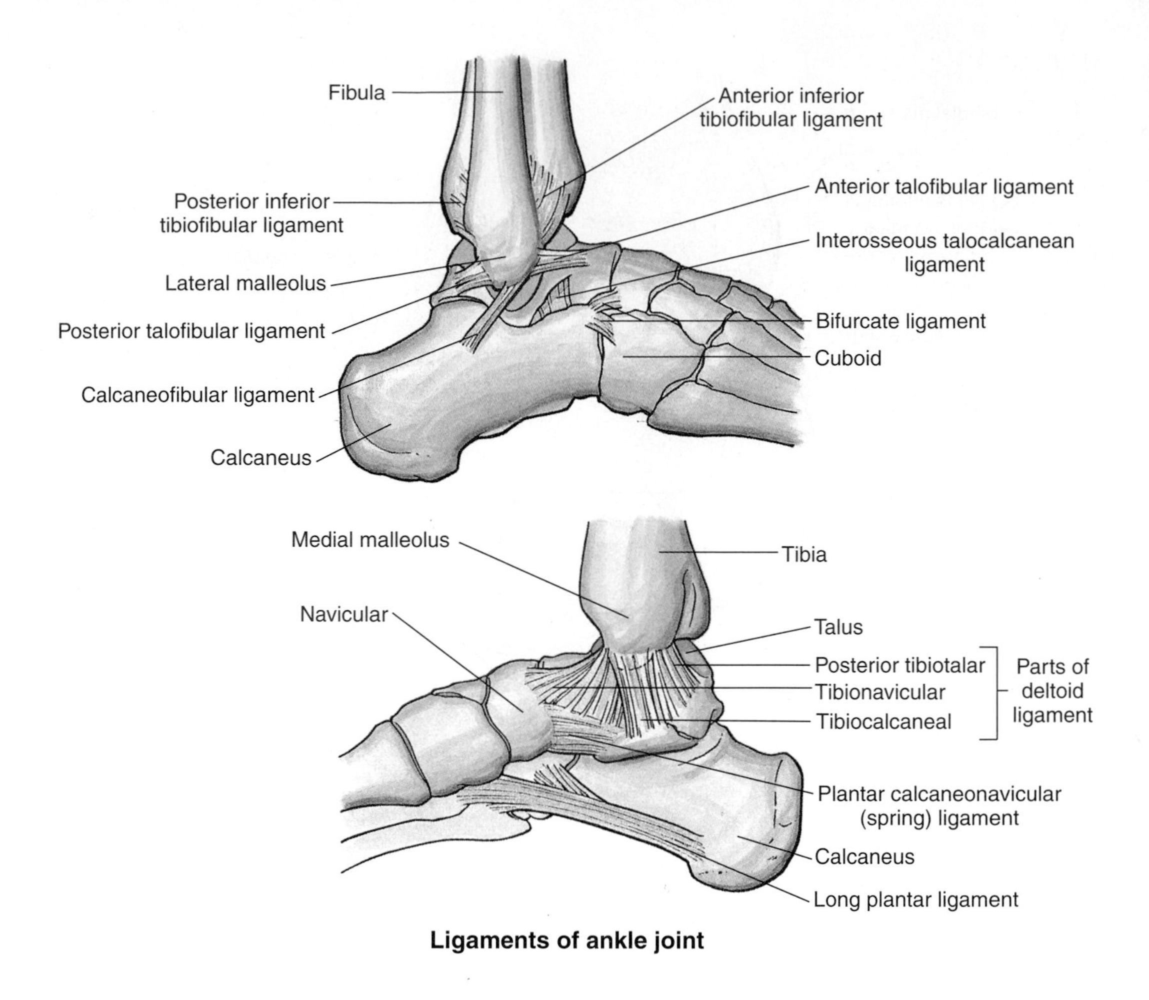

Ligaments of ankle joint

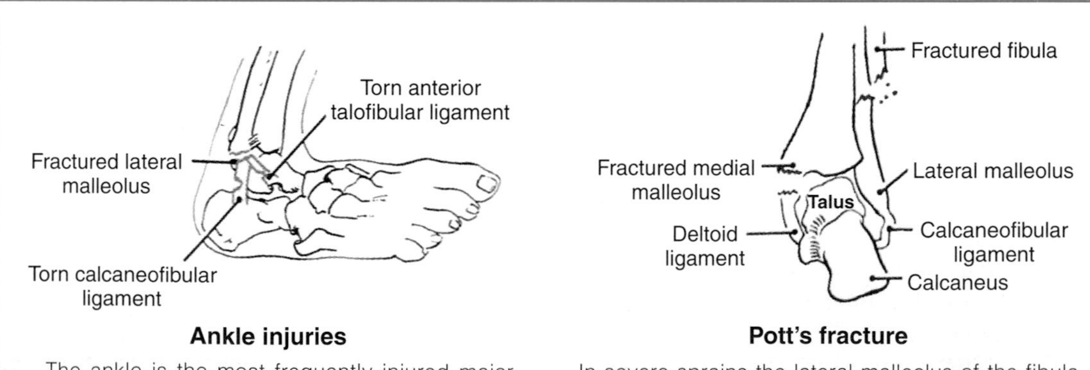

Ankle injuries

The ankle is the most frequently injured major joint in the body. Its lateral ligament is often injured because it is less strong than the medial ligament. A *sprained ankle* results from twisting of the weightbearing foot and is nearly always an inversion injury (i.e., the foot is forcefully inverted). In severe ankle sprains many fibers of the lateral ligament are torn, either partially or completely, resulting in instability of the ankle joint. The two most frequently torn parts of the lateral ligament are the calcaneofibular and anterior talofibular ligaments.

Pott's fracture

In severe sprains the lateral malleolus of the fibula is usually fractured.

The common *Pott's fracture* (fracture-dislocation of the ankle) occurs when the foot is forcibly everted. This pulls on the extremely strong medial (deltoid) ligament, often tearing off the medial malleolus. The talus then moves laterally, shearing off the lateral malleolus or, more commonly, breaking the fibula superior to the inferior tibiofibular joint. If the tibia is carried anteriorly, the posterior margin of the distal end of the tibia is also sheared off by the talus.

FOOT JOINTS

The many joints of the foot involve the tarsals, metatarsals, and phalanges (Table 6.16). The important intertarsal joints are the subtalar and transverse tarsal joints. The *transverse tarsal joint* consists of the talocalcaneonavicular and calcaneocuboid joints. The *subtalar (talocalcaneal) joint* occurs where the talus rests on the calcaneus. The other joints in the foot are relatively small and are so tightly joined by ligaments that only slight movement occurs between them. All the foot bones are united by dorsal and plantar ligaments.

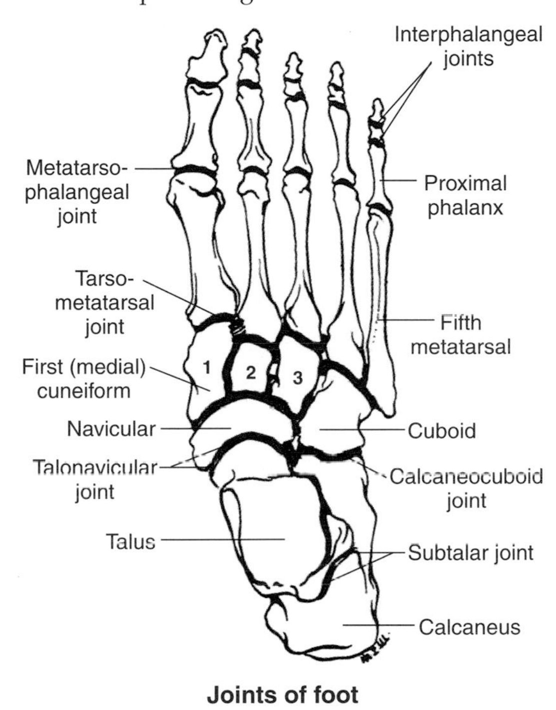

Joints of foot

The *plantar calcaneonavicular ligament* (spring ligament) extends from the sustentaculum tali to the posteroinferior surface of the navicular. This ligament is important in maintaining the longitudinal arch of the foot.

The *long plantar ligament* passes from the plantar surface of the calcaneus to the groove on the cuboid. Some of its fibers extend to the bases of the metatarsals, thereby forming a tunnel for the tendon of the fibularis (peroneus) longus. The long plantar ligament is important in maintaining the arches of the foot. The *plantar calcaneocuboid ligament* (short plantar ligament) is deep to the long plantar ligament. It extends from the anterior aspect of the inferior surface of the calcaneus to the inferior surface of the cuboid.

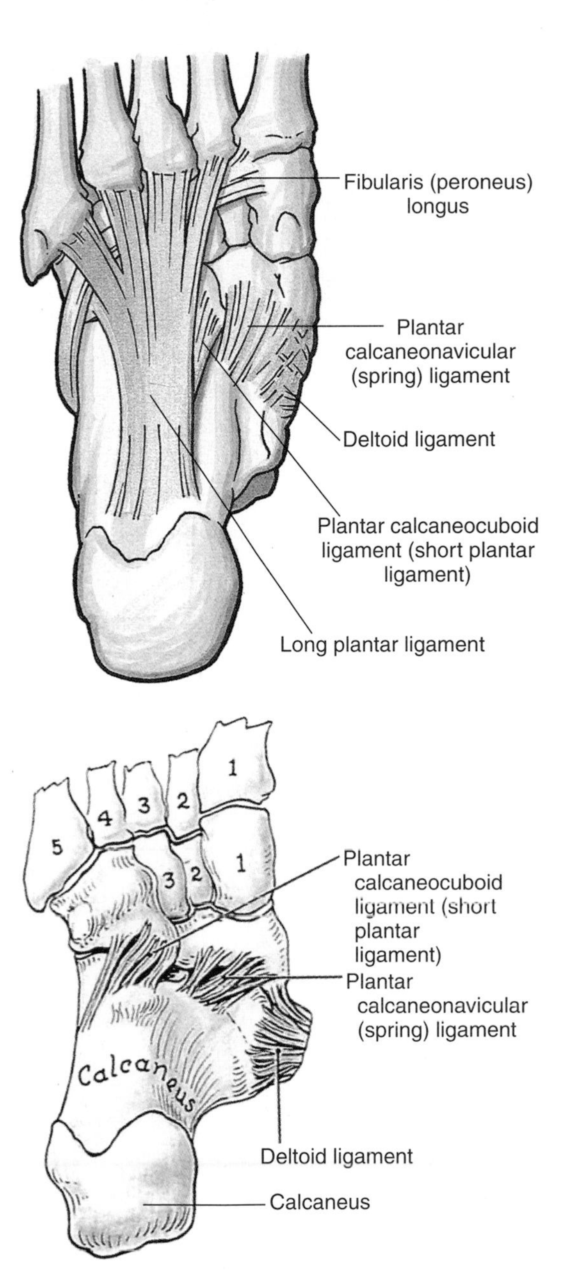

Plantar ligaments

ARCHES OF FOOT

The bones are arranged in longitudinal and transverse arches of the foot. They are designed as shock absorbers for supporting the weight of the body and for propelling it during movement. The resilient arches of the foot make it adaptable to surface and weight changes.

The weight of the body is transmitted to the talus from the tibia and fibula. Then it is transmitted posteroinferiorly to the calcaneus and anteroinferiorly to the heads of the second to

Table 6.16
Joints of Foot

Joint	Type	Articular Surface	Articular Capsule	Ligaments	Movements	Blood Supply	Nerve Suppy
Subtalar	Plane type of synovial joint	Inferior surface of body of talus articulates with superior surface of calcaneus	Fibrous capsule is attached to margins of articular surfaces	Medial, lateral, and posterior talocalcaneal ligaments support capsule; interosseous talocalcaneal ligament binds bones together	Inversion and eversion of foot	Posterior tibial and fibular aa.	Plantar aspect, medial or lateral plantar nn.; dorsal aspect deep fibular n.
Talocalcaneonavicular	Synovial joint; talonavicular part is ball and socket type	Head of talus articulates with calcaneus and navicular bones	Fibrous capsule incompletely encloses joint	Plantar calcaneonavicular ("spring") ligament supports head of talus	Gliding and rotatory movements are possible	Anterior tibial a. via lateral tarsal a.	
Calcaneocuboid	Plane type of synovial joint	Anterior end of calcaneus articulates with posterior surface of cuboid	Fibrous capsule encloses joint	Dorsal calcaneocuboid ligament, plantar calcaneocuboid ligament, and long plantar ligament support fibrous capsule	Inversion and eversion of foot	Anterior tibial a. via lateral tarsal a.	
Tarsometatarsal	Plane type of synovial joint	Anterior tarsal bones articulate with bases of metatarsal bones	Fibrous capsule encloses joint	Dorsal, plantar, and interosseous ligaments	Gliding or sliding	Lateral tarsal a., a branch of dorsalis pedis a.	Deep fibular. medial and lateral plantar, and sural nn.
Intermetatarsal	Plane type of synovial joint	Bases of metatarsal bones articulate with each other	Fibrous capsule encloses each joint	Dorsal, plantar, and interosseous ligaments bind bones together	Little individual movement of bones possible	Lateral metatarsal a., a branch of dorsalis pedis a.	Digital nn.
Metatarsophalangeal	Condyloid type of synovial joint	Heads of metatarsal bones articulate with bases of proximal phalanges	Fibrous capsule encloses each joint	Collateral ligaments support capsule on each side; plantar ligament supports plantar part of capsule	Flexion, extension, and some abduction, adduction, and circumduction	Lateral tarsal a., a branch of dorsalis pedis a.	
Interphalangeal	Hinge type of synovial joint	Head of one phalanx articulates with base of one distal to it	Fibrous capsule encloses each joint	Collateral and plantar ligaments support joints	Flexion and extension	Digital branches of plantar arch	

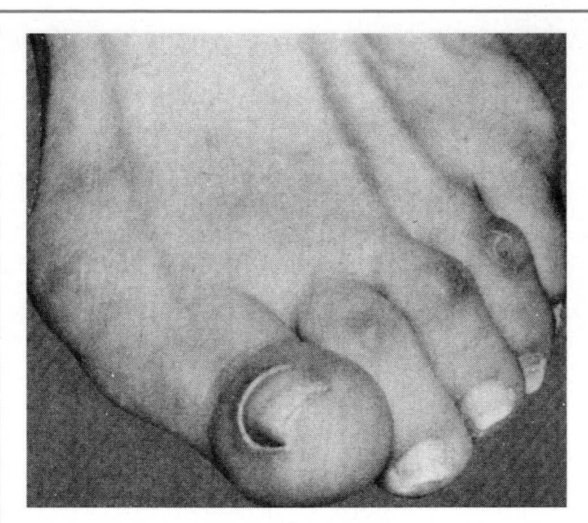

Hallux valgus

The first metatarsophalangeal joint may become enlarged and deformed (*hallux valgus*) with permanent lateral displacement of the first digit (hallux). Patients cannot move their first digit away from their second digit because the sesamoids under the head of the first metatarsal are usually displaced and lie in the space between the heads of the first and second metatarsals.

Hammer toe is a common deformity in which the proximal phalanx is permanently dorsiflexed at the metatarsophalangeal joint and the middle phalanx is plantarflexed at the interphalangeal joint. The distal phalanx is also flexed or extended, giving the digit (usually the second) a hammerlike appearance. This deformity may result from weakness of the lumbricals and interossei that flex the metatarsophalangeal joints and extend the interphalangeal joints.

fifth metatarsals and the sesamoids of the first digit. Between these weightbearing points are the relatively elastic arches of the foot that become slightly flattened by body weight during standing, but they normally resume their curvature (recoil) when body weight is removed (e.g., during sitting).

The *longitudinal arch of the foot* is composed of medial and lateral parts. Functionally, both parts act as a unit with the transverse arch, spreading the weight in all directions. The medial longitudinal arch is higher and more important. The *medial longitudinal arch* is composed of the calcaneus, talus, navicular, three cuneiforms, and three metatarsal bones. The *lateral longitudinal arch* is much flatter than the medial part of the arch and rests on the ground during standing. It is composed of the calcaneus, cuboid, and lateral two metatarsals.

The *transverse arch of the foot* runs from side to side. It is formed by the cuboid, cuneiforms, and bases of the metatarsals. The medial and lateral parts of the longitudinal arch serve as pillars for the transverse arch. The tendon of the fibularis (peroneus) longus, crossing the sole of the foot obliquely, helps maintain the curvature of the transverse arch (p. 277).

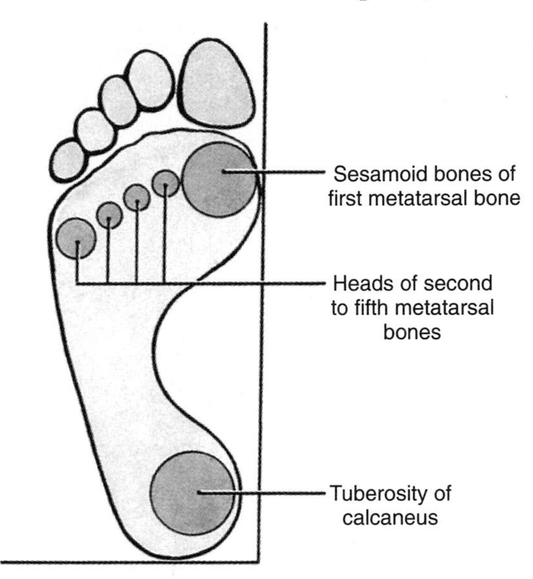

Weightbearing areas

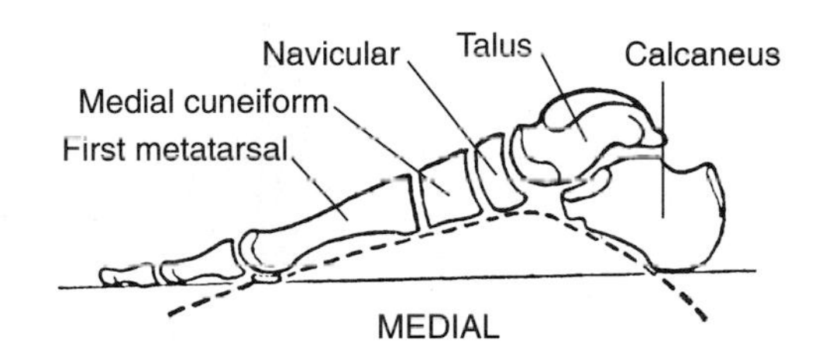

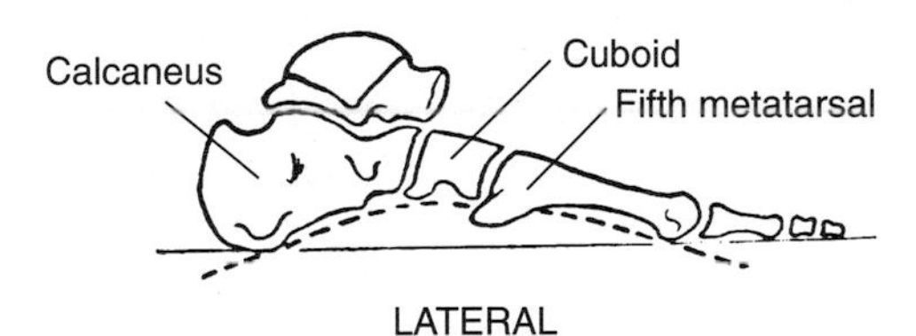

Longitudinal arches of foot

The integrity of the bony arches of the foot is maintained by the

- Shape of the interlocking bones
- Strength of the plantar ligaments, especially the plantar calcaneonavicular (spring) ligament and the long and short plantar ligaments
- Plantar aponeurosis
- Action of muscles through the bracing action of their tendons

Of these factors, the plantar ligaments and the plantar aponeurosis bear the greatest stress and are most important in maintaining the arches.

Flatfeet in adolescents and adults are caused by "fallen arches," usually the medial parts of the longitudinal arches. During standing the plantar ligaments and plantar aponeurosis stretch somewhat under the body weight. If these ligaments become abnormally stretched during long periods of standing, the plantar calcaneonavicular (spring) ligament can no longer support the head of the talus. As a result, some flattening of the medial part of the longitudinal arch occurs, and there is concomitant lateral deviation of the forefoot. In the common type of flatfoot, the foot resumes its arched form when the weight is removed from it. *Flatfeet are common in older persons*, particularly if they undertake much unaccustomed standing or gain weight rapidly. This results from added stress on the muscles and increased strain on the ligaments supporting the arches.

7/ UPPER LIMB

The upper limb is characterized by considerable mobility and is adapted for grasping and manipulating. It consists of five parts:

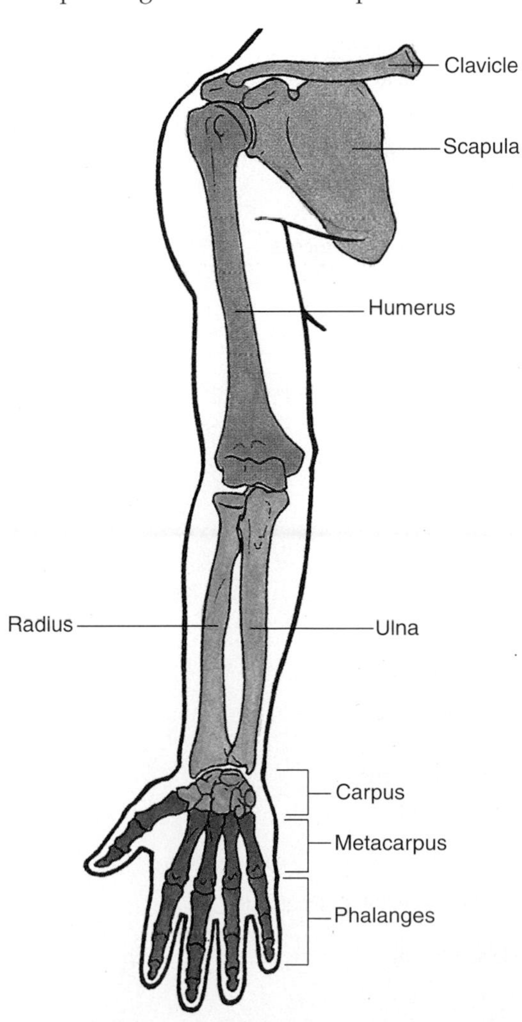

Upper limb

- Shoulder (*blue*), the junction of arm and trunk, containing the clavicle (collar bone) and scapula (shoulder blade)
- Arm (*purple*) containing the humerus, which connects the shoulder and elbow
- Forearm (*pink*) containing the ulna and radius, which connect the elbow and wrist
- Wrist (*orange*) containing the carpus, which connects the forearm and hand
- Hand (*red*) containing the metacarpus and phalanges

Bones

The skeleton of the upper limb is composed of the pectoral girdle and the skeleton of the limb. The *pectoral girdle* (shoulder girdle), formed by the clavicles and scapulae and joined anteriorly to the manubrium of the sternum, connects the upper limbs to the axial skeleton. Although very mobile, the pectoral girdle is supported and stabilized by muscles that are connected to the ribs, sternum, and vertebrae.

CLAVICLE

The clavicle connects the upper limb and trunk. Its medial end articulates with the manubrium of the sternum at the *sternoclavicular joint*. Its lateral end articulates with the acromion of the scapula at the *acromioclavicular joint*. The medial two-thirds of the body (shaft) of the clavicle is convex anteriorly, whereas the lateral third is flattened and concave anteriorly.

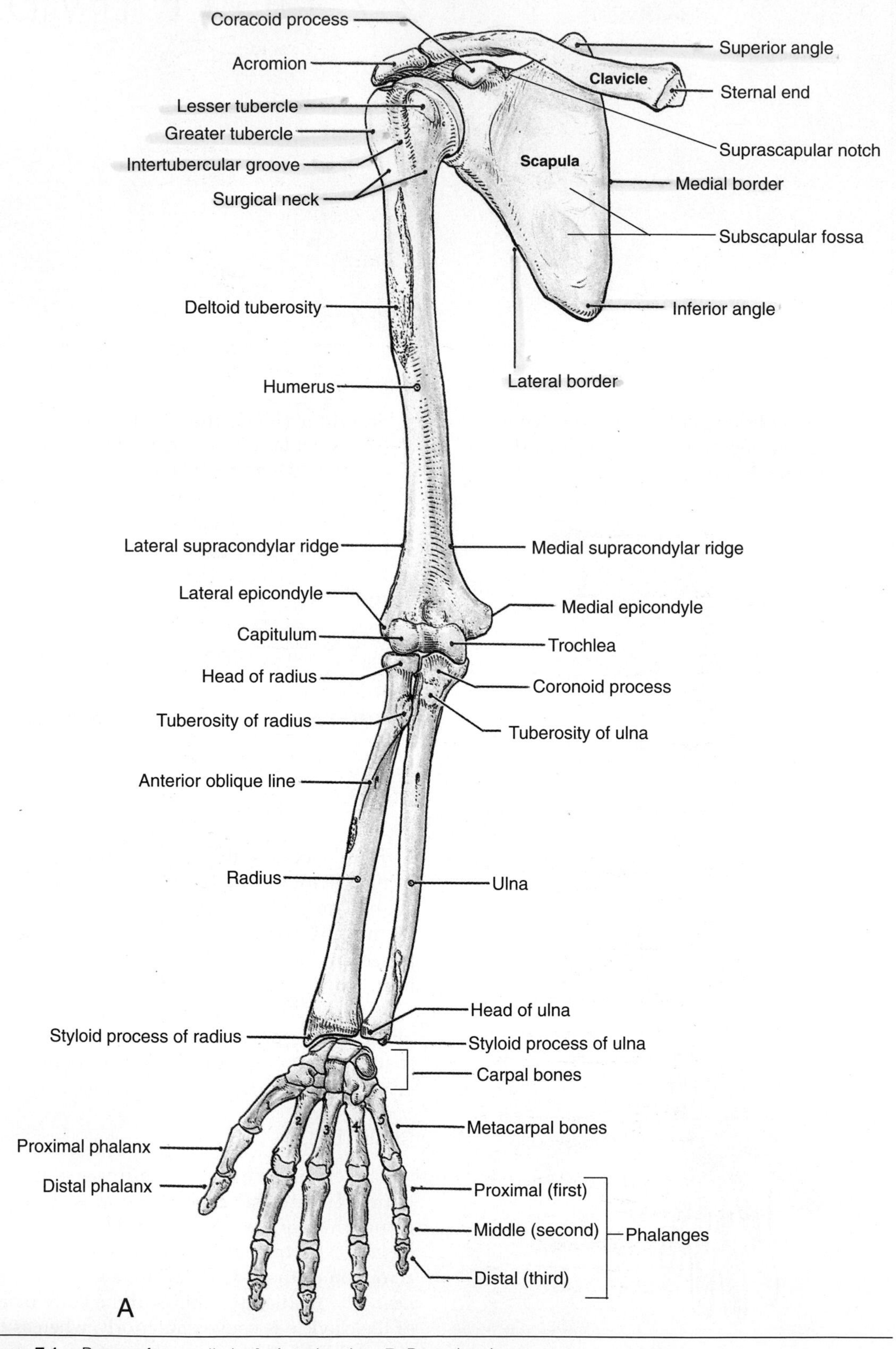

Figure 7.1. Bones of upper limb. **A.** Anterior view. **B.** Posterior view.

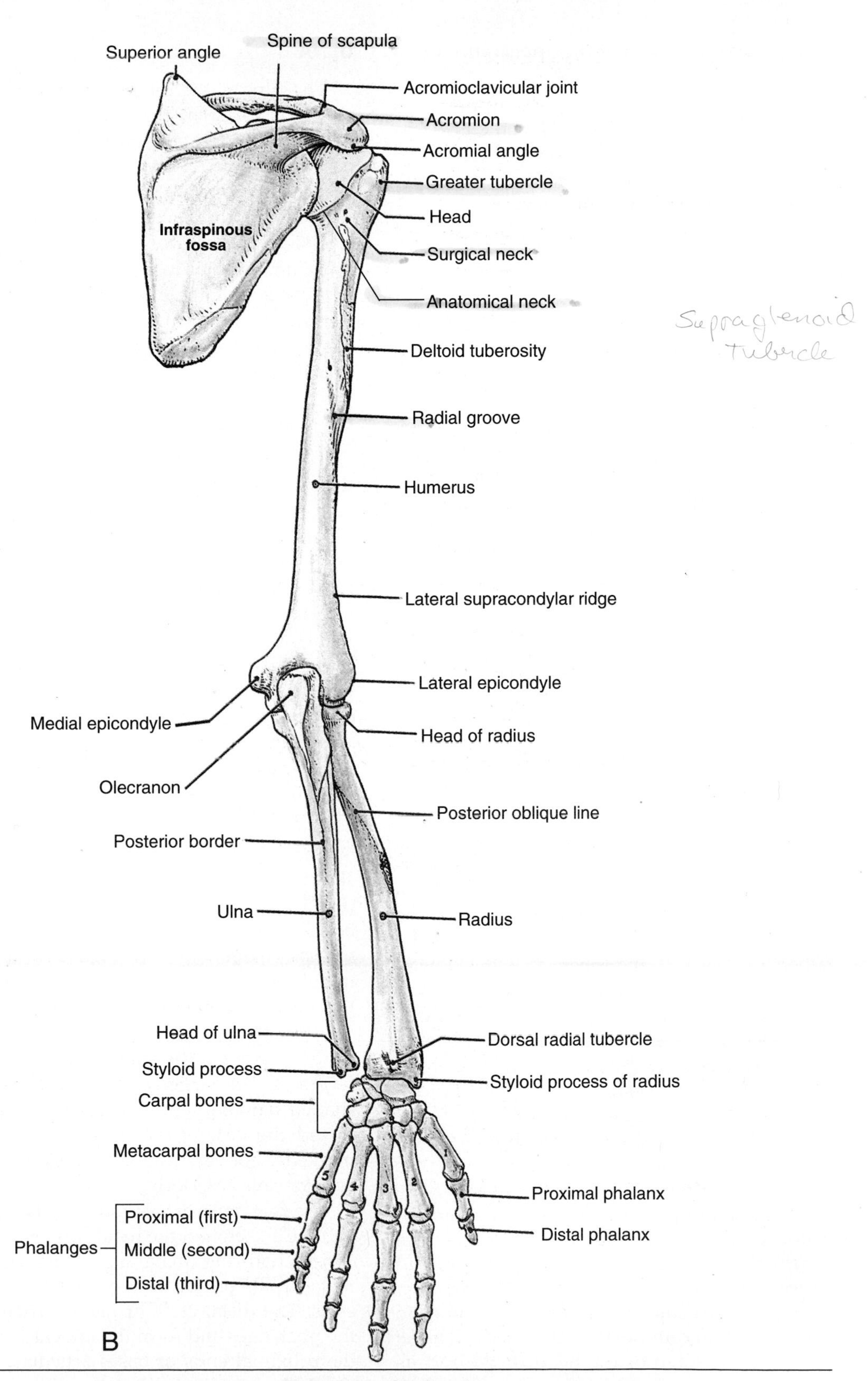

Figure 7.1. *Continued.*

These curvatures give it the appearance of an elongated capital S. The clavicle

- Serves as a strut to keep the upper limb away from the thorax so that the arm has maximum freedom of movement
- Transmits shocks (impacts) from the upper limb to the axial skeleton

SCAPULA

The scapula lies on the posterolateral aspect of the thorax, overlying the 2nd to 7th ribs (p. 11). The triangular body of the scapula is thin and translucent. Its concave anterior or costal surface has a large *subscapular fossa,* and its convex posterior or dorsal surface is separated by the spine of the scapula into a smaller *supraspinous fossa* and a larger *infraspinous fossa.* The spine continues laterally as the *acromion* that forms the point of the shoulder and articulates with the clavicle. Superolaterally, the lateral surface of the scapula forms the *glenoid cavity* (fossa), which articulates with the head of the humerus at the shoulder joint (scapulohumeral joint). The beaklike *coracoid process* is superior to the glenoid cavity and projects anterolaterally.

HUMERUS

The humerus (arm bone) articulates with the scapula at the shoulder joint and the radius and ulna at the elbow joint (Fig. 7.1). The ball-shaped head of the humerus articulates with the glenoid cavity of the scapula. The *intertubercular groove* separates the lesser tubercle from the greater tubercle. Just distal to the head, the *anatomical neck* separates the head from the tubercles. Distal to the tubercles is the *surgical neck,* which is where the humerus narrows to become the body (shaft). The body has two prominent features, the *deltoid tuberosity* laterally and the *radial groove* posteriorly. The sharp medial and lateral *supracondylar ridges* end distally in prominent medial and lateral epicondyles. The distal end of the humerus has two articular surfaces, a lateral *capitulum* (little head) for articulation with the head of the radius and a medial *trochlea* (pulley) for articulation with the ulna. Superior to the trochlea anteriorly is the *coronoid fossa* for the coronoid process of the ulna and posteriorly the *olecranon fossa* for the olecranon of this bone. Superior to the capitulum anteriorly is the shallow *radial fossa* for the edge of the head of the radius when the forearm is flexed.

ULNA

The ulna is the longer and more medial of the two forearm bones. Its proximal end includes the *olecranon* posteriorly and the *coronoid process* anteriorly. The anterior surface of the olecranon has the *trochlear notch* that receives the trochlea of the humerus. On the lateral side of the coronoid process is the *radial notch,* and inferior to this process is the *tuberosity of the ulna.* Initially the body (shaft) is thick, but it becomes narrow at its distal end, which has a large rounded head and a small conical *styloid process.*

RADIUS

The radius is the shorter and more lateral of the two forearm bones. Its proximal end consists of a disclike head, a short neck, and tuberosity (Fig. 7.1*A*). Proximally the smooth head is concave for articulation with the capitulum of the humerus. The neck is the constriction distal to the head. The *tuberosity of the radius,* just distal to the neck, separates the proximal end of the radius from the body (shaft). The distal end of the radius has an *ulnar notch* medially, a *styloid process* laterally, and a *dorsal tubercle* dorsally.

CARPUS, METACARPUS, AND PHALANGES

The eight carpals forming the *carpus* and the skeleton of the wrist are arranged in two rows (Fig. 7.2). From lateral to medial, the three large bones in the proximal row are the boat-shaped *scaphoid,* the moon-shaped *lunate,* and the pyramidal *triquetrum;* the small, pea-shaped *pisiform* lies on the palmar surface of the triquetrum. The distal row, from lateral to medial, consists of the *trapezium,* the somewhat wedge-shaped *trapezoid,* the large *capitate,* and the wedge-shaped hamate, which has a hooklike process called the *hook of the hamate.* The carpus is markedly convex from side to side posteriorly and concave anteriorly.

The five metacarpals forming the *metacarpus* and the skeleton of the hand proper connect the carpus with the phalanges in the digits. Each metacarpal consists of a body (shaft) and two ends. The distal ends or heads articulate with the phalanges and form the knuckles of the fist; the proximal ends or bases articulate with the carpal bones. Each digit has three *phalanges*

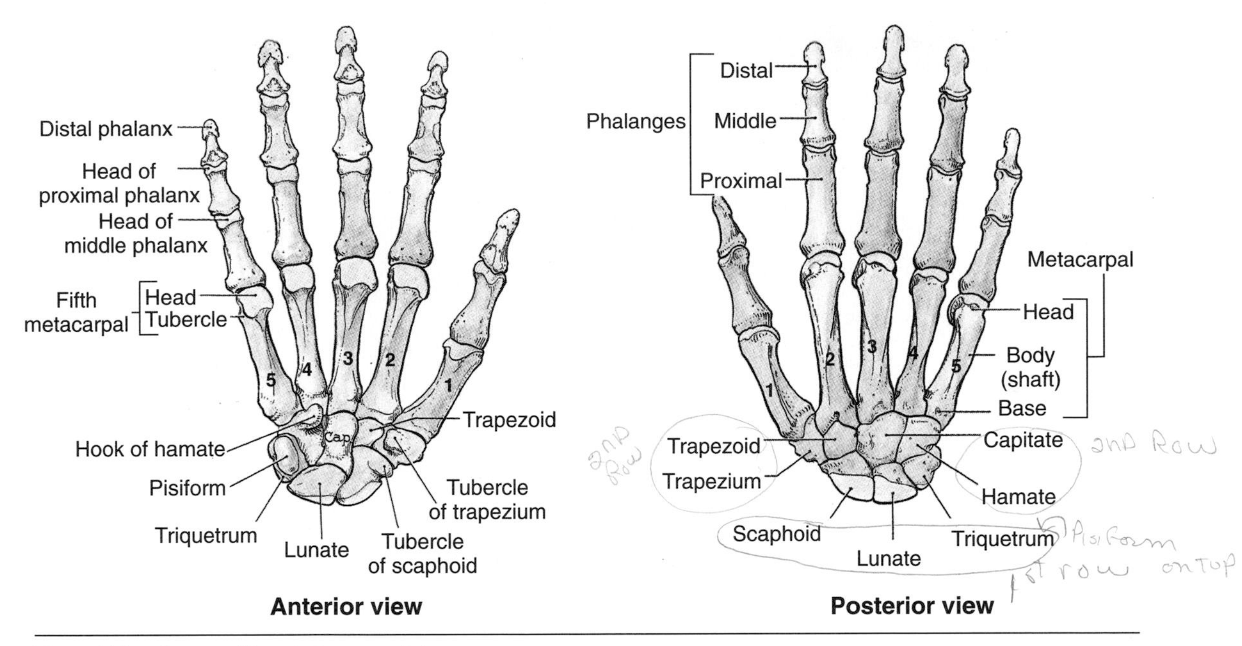

Figure 7.2. Bones of hand. *Cap*, capitate.

Curvatures of the clavicle help to increase its resilience when shocks are transmitted from the limb to the trunk (e.g., when a person falls on an outstretched limb). However, *fractures of the clavicle* are relatively common. A fracture results when the force of a fall on the shoulder is greater than the strength of the clavicle. The weakest part of the clavicle is the junction of its medial two-thirds and its lateral third. After a fracture, the lateral fragment of the clavicle drops from the weight of the upper limb.

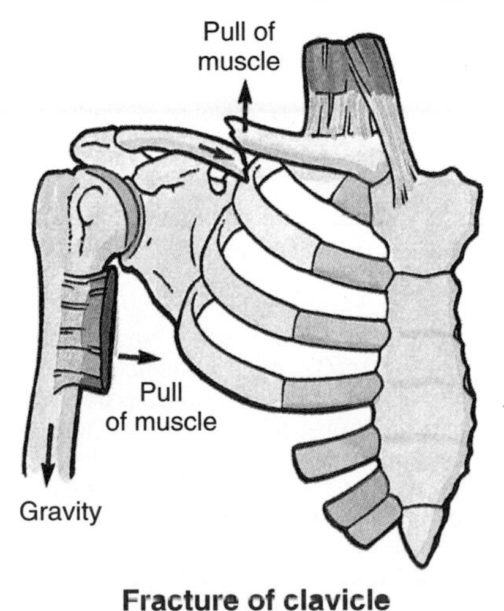

Fracture of clavicle

Fractures of the surgical neck of the humerus are common in elderly persons and usually result from falls on the elbow when the arm is abducted. Because nerves are in contact with the humerus (e.g., surgical neck-axillary nerve, radial groove-radial nerve, and medial epicondyle-ulnar nerve), the axillary, radial, median, and ulnar nerves may be injured in humeral fractures.

When a person falls on an outstretched hand with the forearm pronated, the main force moves through the carpus to the distal end of the radius and then proximally to the humerus, scapula, and clavicle. During such falls, fractures may occur in any of these bones, but the radius tends to break proximal to the wrist joint producing a *Colles' fracture.* The distal fragment of bone is often comminuted (broken into pieces), and the fragments are displaced posteriorly and superiorly, producing shortening of the radius. Displacement of the distal part of the radius may break off the ulnar styloid process.

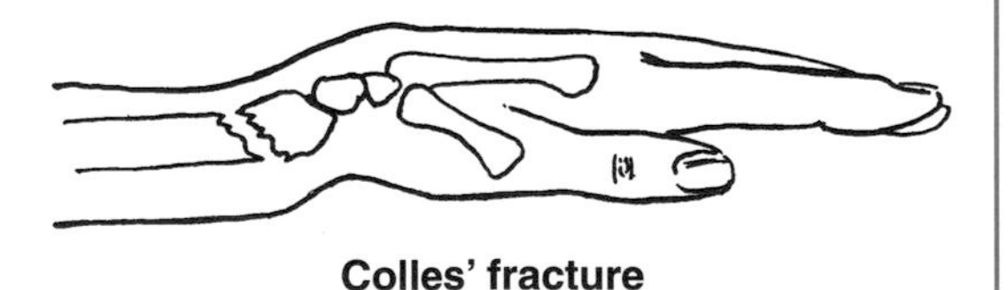

Colles' fracture

Fractures of the hand are common, and disability can result if normal relationships of the bones are not restored. *Fracture of the scaphoid* is also common, especially when the person falls on the palm with the hand abducted. Because the scaphoid is in the floor of the *anatomical snuff box* (p. 321), a clinical sign of fracture of this bone is tenderness in the snuff box.

Surface Anatomy of Upper Limb Bones

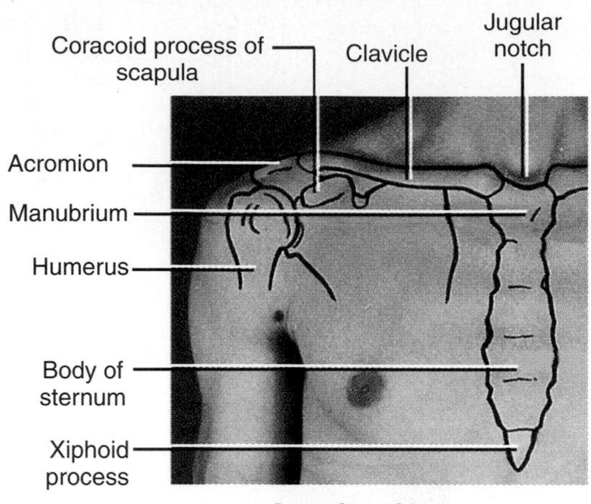

Anterior view

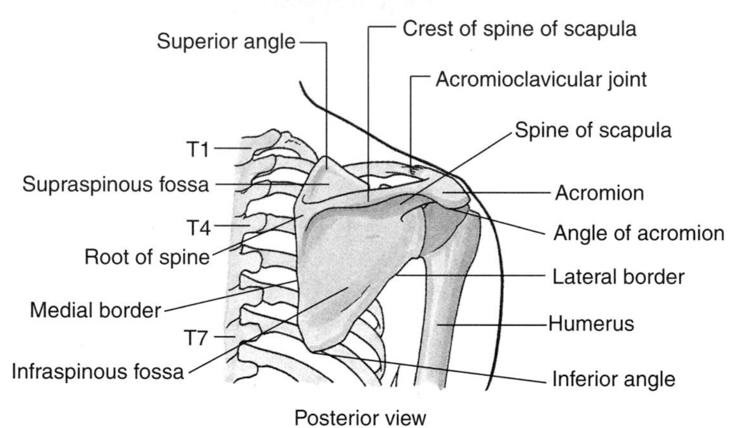

Posterior view

Bony landmarks

The clavicle is subcutaneous and can be easily palpated throughout its length. Its medial end projects superior to the manubrium. Between the medial elevations of the clavicles is the *jugular notch* (suprasternal notch). The lateral end of the clavicle can be palpated 2–3 cm medial to the lateral border of the acromion, particularly when the upper limb swings anteriorly and posteriorly. The point of the shoulder is the lateral part of the *acromion of the scapula*, which is easily felt and may be visible. The lateral and posterior borders of the acromion meet to form the *angle of the acromion*. This angle is the point from which the length of the upper limb is measured. Inferior to the acromion is the rounded curve of the shoulder formed by the deltoid muscle. The *crest of the scapular spine* is subcutaneous throughout and is easily felt.

When the upper limb is in the anatomical position,

- The superior angle of the scapula lies at the level of T2 vertebra
- The root of the scapular spine, its medial end, is opposite the spinous process of T3 vertebra
- The inferior angle of the scapula lies at the level of T7 vertebra near the inferior border of the 7th rib

The *coracoid process* of the scapula can be palpated deeply at the lateral side of the *deltopectoral triangle* (see Fig. 7.6*A*). The medial border of the scapula is palpable as it crosses the 2nd to 7th ribs, but the lateral border is not easily palpated because it is covered by thick muscles.

The *head of the radius* can be palpated and felt to rotate in the depression on the posterolateral aspect of the extended elbow joint, just distal to the lateral epicondyle of the humerus. The *radial styloid process* can be easily palpated on the lateral side of the wrist; it is about 1 cm more distal than the ulnar styloid process. The relationship of the radial and ulnar styloid processes is important in the diagnosis of certain injuries (e.g., wrist fractures). Proximal to the radial styloid process, the anterior, lateral, and posterior surfaces of the radius are palpable for several centimeters.

Surface Anatomy of Upper Limb Bones

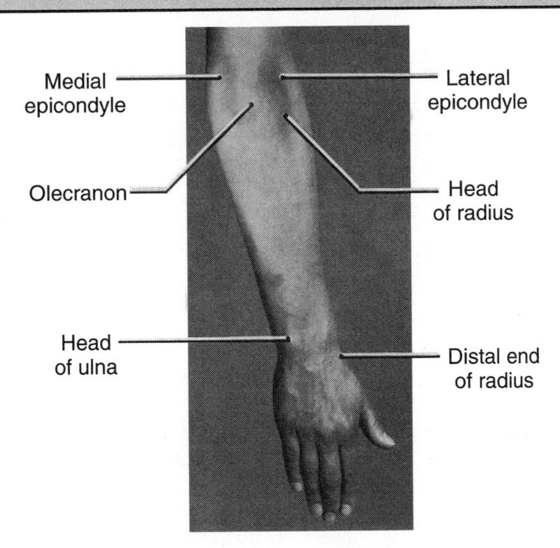

The *olecranon* and the posterior border of the ulna can be easily palpated. The *ulnar nerve* feels like a thick cord where it passes posterior to the medial epicondyle of the humerus. The *head of the ulna* forms a large rounded prominence that can be easily seen and felt on the medial part of the dorsal aspect of the wrist. The *ulnar styloid process* may be felt slightly distal to the head when the hand is supinated.

Some carpal bones can be easily palpated. The *pisiform* can be felt on the anterior aspect of the medial border of the wrist and can be moved from side to side when the hand is relaxed. The *hook of the hamate* can be palpated on deep pressure over the medial side of the palm, about 2 cm distal and lateral to the pisiform. The *tubercles of the scaphoid and trapezium* can be palpated at the proximal end of the thenar eminence (ball of thumb) when the hand is extended at the wrist joint. The metacarpals, although covered by the long extensor tendons of the digits, can be palpated on the dorsum of the hand. The heads of these bones form the knuckles of the fist. The dorsal aspects of the phalanges can be easily palpated. The knuckles of the fingers are formed by the heads of the proximal and middle phalanges.

except the first one, the thumb, which has only two. Each phalanx has a base proximally, a head distally, and a body between the base and head.

Fasciae

The fascia of the pectoral region is attached to the clavicle and sternum. The *pectoral fascia* invests the pectoralis major muscle and is continuous inferiorly with the fascia of the abdominal wall. The pectoral fascia leaves the lateral border of the pectoralis major and becomes the *axillary fascia* (Fig. 7.3*A*), which forms the floor of the axilla (armpit). A fascial layer extends from the axillary fascia as the *clavipectoral fascia*. This fascia encloses the pectoralis minor and subclavious muscles and then attaches to the clavicle. Because the inferior part of the clavipectoral fascia supports the axillary fascia, this part of the clavipectoral fascia is referred to as the *suspensory ligament of the axilla*.

A sheath of deep fascia—the *brachial fascia*—encloses the arm (Fig. 7.3, *B* and *C*) and is continuous superiorly with the pectoral and axillary fasciae. The brachial fascia is attached inferiorly to the epicondyles of the humerus and to the olecranon of the ulna. It also is continuous with the deep fascia of the forearm. Two *intermuscular septa* extend from the brachial fascia and attach to the medial and lateral supracondylar ridges of the humerus. The medial and lateral intermuscular septa divide the arm into anterior (flexor) and posterior (extensor) *fascial compartments*, each of which contains muscles, nerves, and blood vessels.

The deep fascia of the forearm—the *antebrachial fascia*—is a tough fibrous membrane that surrounds the muscles of the forearm. The

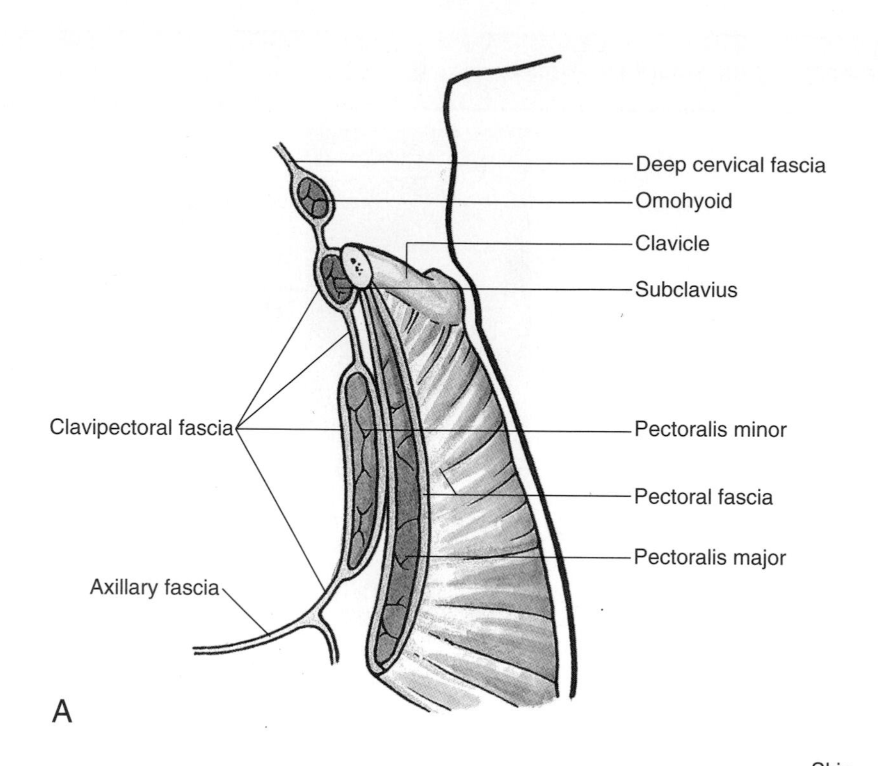

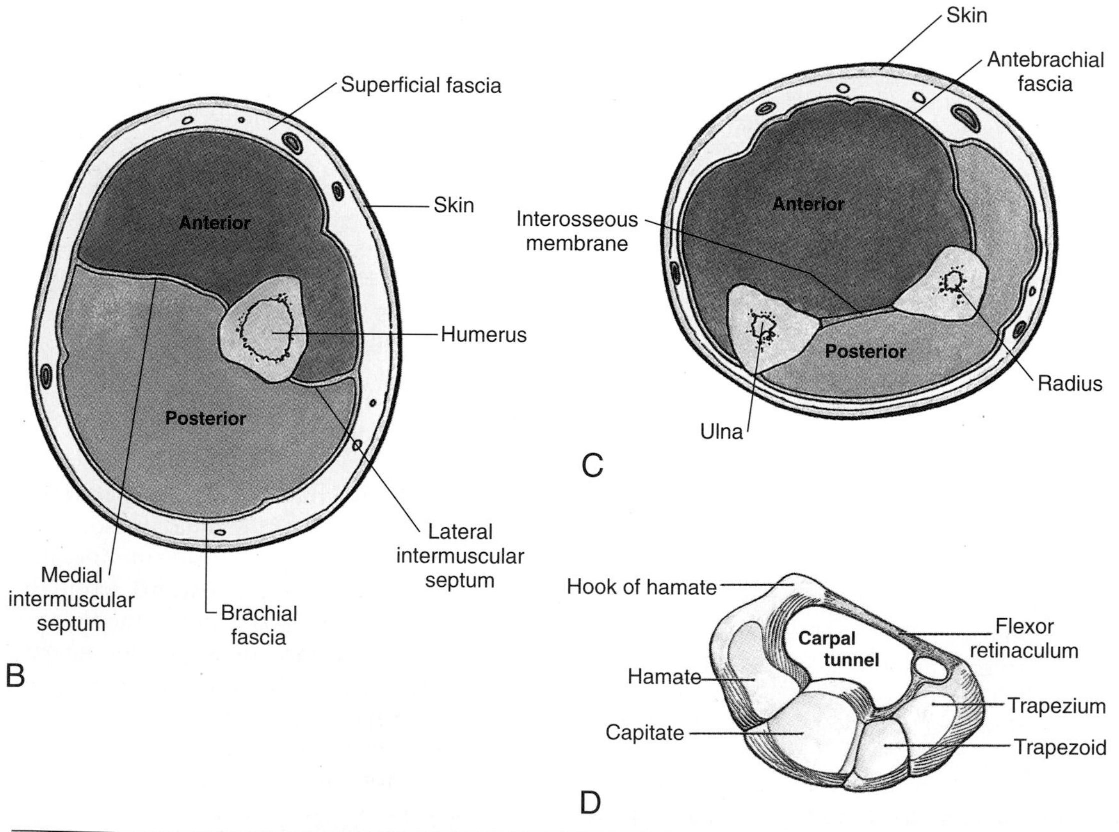

Figure 7.3. Fasciae of upper limb. **A.** Sagittal section showing fascia of pectoral region. **B.** Transverse section of arm showing its fascial compartments. **C.** Transverse section of forearm showing its fascial compartments. **D.** Transverse section through distal row of carpal bones and flexor retinaculum to show carpal tunnel.

antebrachial fascia is continuous with the brachial fascia of the arm and with the deep fascia of the hand. The antebrachial fascia is attached posteriorly to the olecranon and the subcutaneous border of the ulna. The forearm is organized into anterior and posterior compartments separated by an *interosseous membrane* connecting the radius and ulna. The antebrachial fascia thickens posteriorly to form a transverse band at the wrist, the *extensor retinaculum*, that retains the extensor tendons in position. The deep fascia also thickens anteriorly at the wrist to form the *flexor retinaculum*, a fibrous band that converts the anterior concavity of the carpus into a *carpal tunnel* through which the flexor tendons and median nerve pass (Fig. 7.3*D*). The deep fascia of the hand is continuous through the retinacula with the antebrachial fascia. The central part of the palmar fascia, the *palmar aponeurosis*, is very thick and tendinous (p. 325).

Cutaneous Nerves

Most cutaneous nerves of the upper limb (Fig. 7.4) are derived from the *brachial plexus* (see Fig. 7.7), and the nerves to the shoulder are derived from the ventral rami of the first four cervical nerves that form the *cervical plexus* (see Chapter 9).

Superficial Veins

The main superficial veins are the cephalic and basilic veins that originate from the dorsal venous arch in the hand. The *cephalic vein* (Fig. 7.5) ascends in the superficial fascia along the lateral border of the wrist and along the anterolateral surface of the forearm and arm. Superiorly the vein passes between the deltoid and pectoralis major muscles and enters the deltopectoral triangle (p. 305), where it joins the axillary vein. The *basilic vein* runs in the superficial fascia on the medial side of the forearm and the inferior part of the arm. It then passes deeply and runs superiorly into the axilla, where it joins deep brachial veins to form the axillary vein. The *median cubital vein* is the communication between the basilic and cephalic veins in the anterior part of the elbow region (cubital fossa). The superficial veins communicate with the deep veins via *perforating veins*.

Because of the prominence and accessibility of the superficial veins, they are commonly used for *venipuncture* (withdrawal of blood samples, introduction of fluids and transfusions, and passage of cardiac catheters). Considerable variation occurs in the connection of the basilic and cephalic veins in the cubital fossa. If the median cubital vein is very large, most blood from the cephalic vein enters the basilic vein. In these cases the superior part of the cephalic vein may be diminished or absent.

Lymphatic Drainage

Superficial lymphatic vessels arise from lymphatic plexuses in the digits, palm, and dorsum of the hand and ascend with the superficial veins of the upper limb (Fig. 7.5). Some lymphatic vessels accompanying the basilic vein pass through cubital lymph nodes, located superior to the medial epicondyle and medial to the basilic vein. The efferent vessels from the cubital nodes ascend in the arm and terminate in axillary lymph nodes. Most lymphatic vessels accompanying the cephalic vein cross the arm and enter the axillary nodes; however, some vessels continue alongside the cephalic vein and enter the deltopectoral (infraclavicular) lymph nodes. *Deep lymphatic vessels* accompany the radial, ulnar, interosseous, and brachial arteries and terminate in the axillary nodes.

Pectoral Muscles

Four pectoral muscles move the pectoral girdle and attach to the thoracic wall (pectoralis major, pectoralis minor, subclavius, and serratus anterior).

The *pectoralis major* covers the superior part of the thorax (Fig. 7.6*A*). Its inferolateral border forms the *anterior axillary fold* and most of the anterior wall of the axilla (p. 293). The pectoralis major and deltoid diverge slightly from

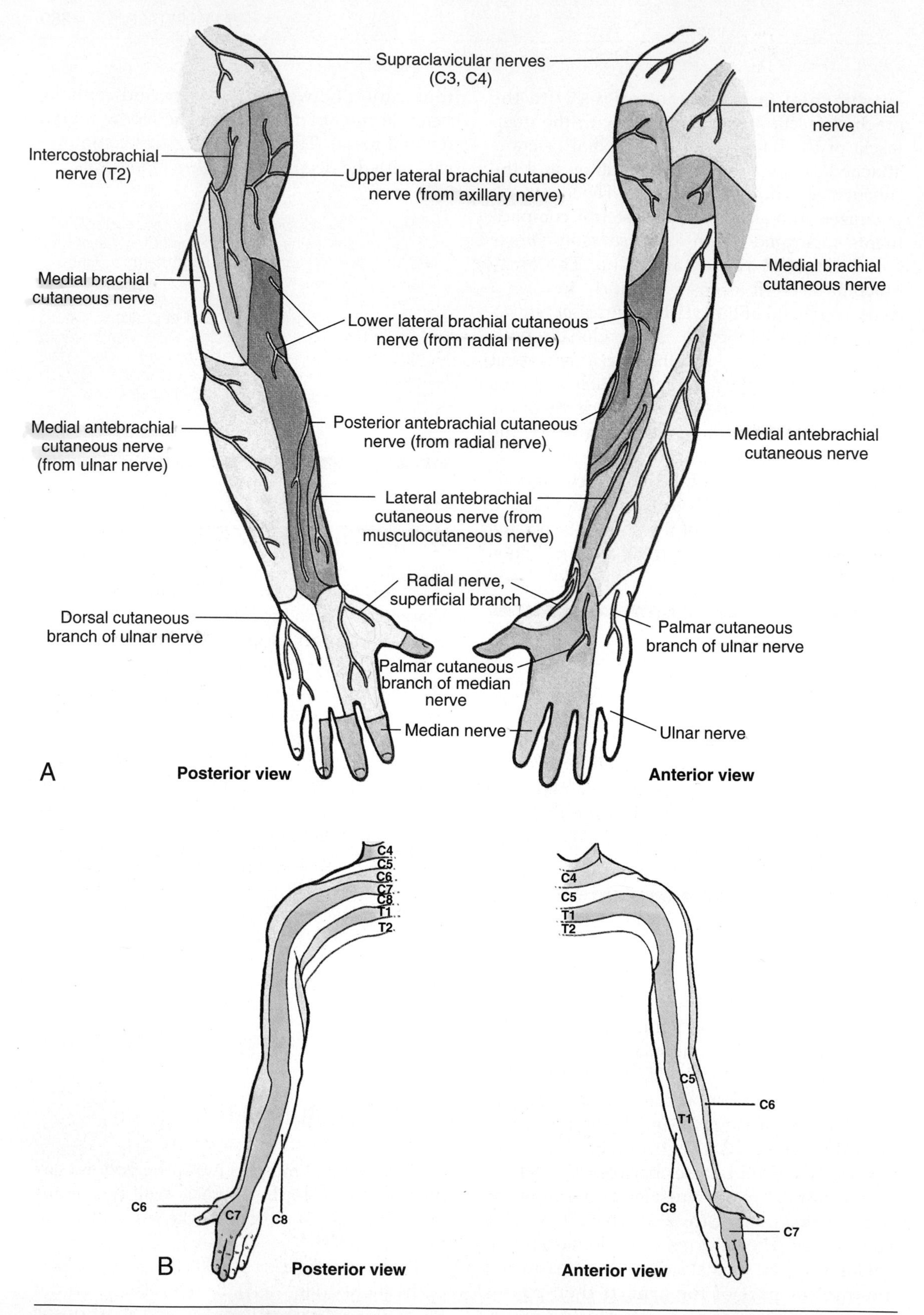

Figure 7.4. Cutaneous innervation of upper limb. **A.** Posterior and anterior views showing distribution of cutaneous nerves, usually containing fibers from more than one spinal nerve. **B.** Similar views of upper limb showing dermatomes (areas of distribution of each spinal nerve to skin).

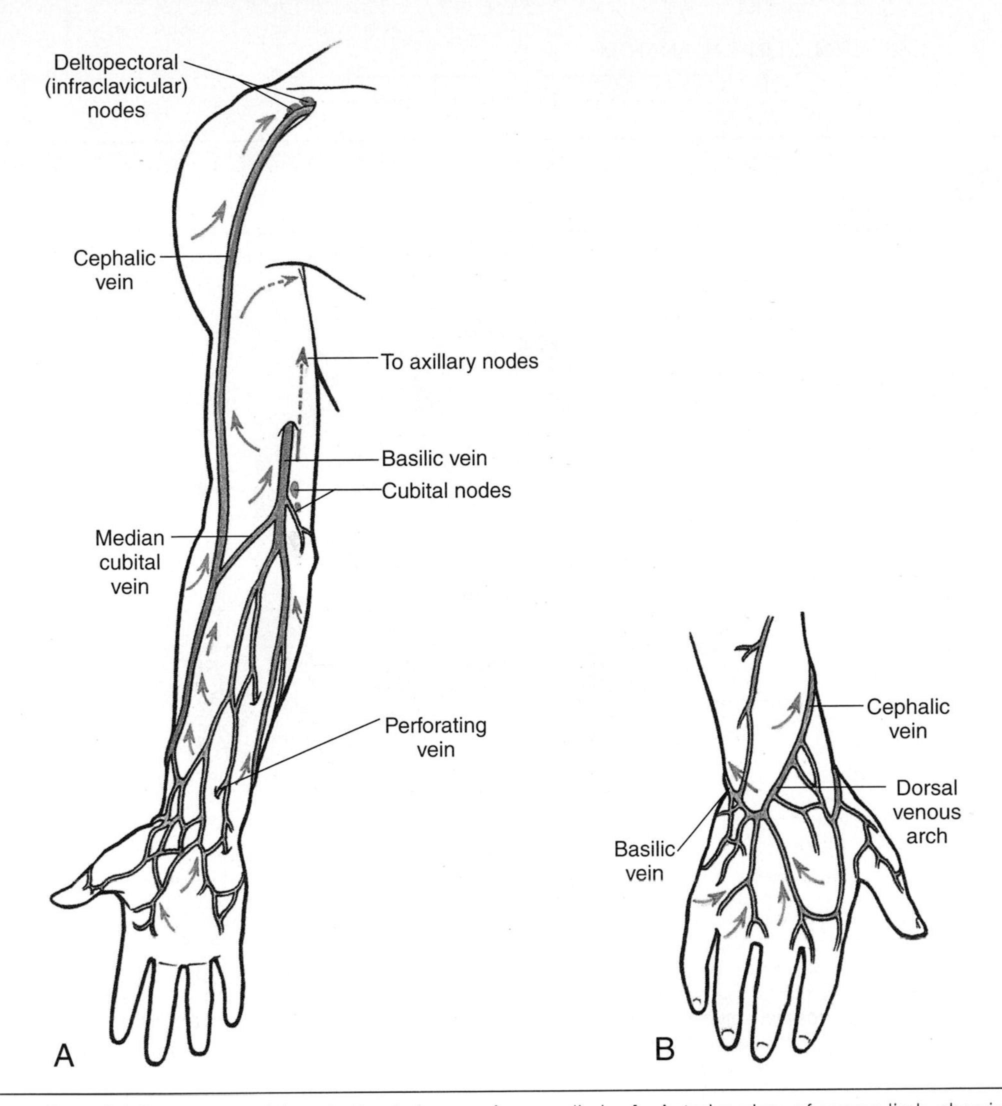

Figure 7.5. Superficial venous and lymphatic drainage of upper limb. **A.** Anterior view of upper limb showing cephalic and basilic veins and their tributaries. *Green arrows*, superficial lymphatic drainage to lymph nodes. **B.** Dorsal view of hand showing dorsal venous arch.

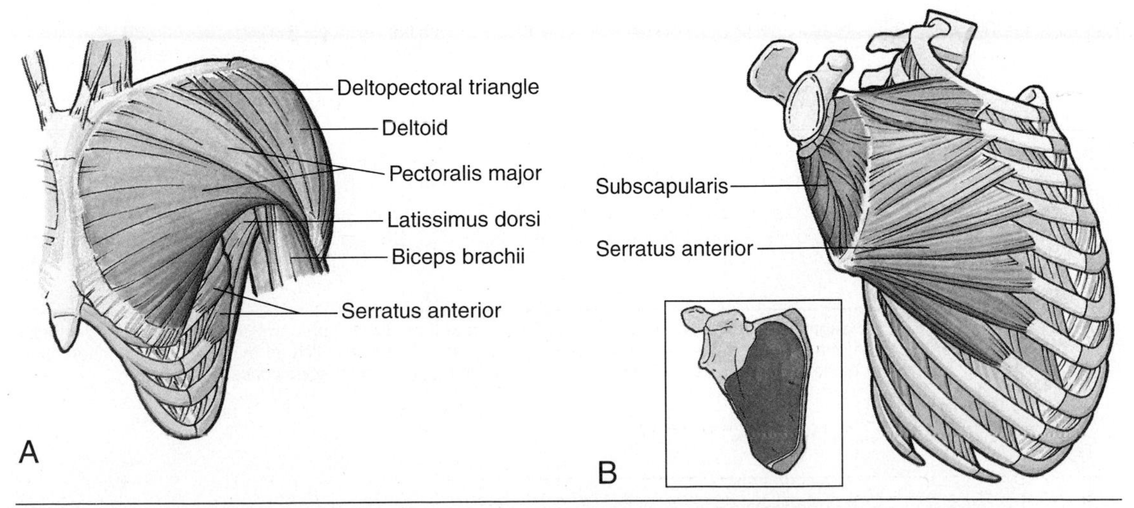

Figure 7.6. **A.** Muscles of pectoral region and axilla. **B.** Serratus anterior and subscapularis muscles. *Inset*, scapular attachments of subscapularis (*red*) and serratus anterior (*blue*).

Table 7.1
Pectoral Muscles

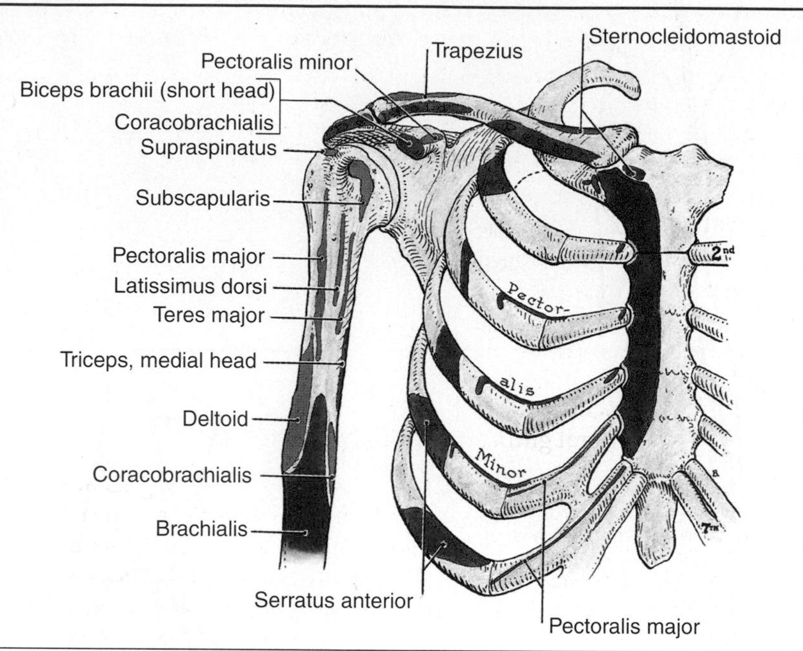

Muscle	Proximal Attachment	Distal Attachment	Innervation[a]	Main Actions
Pectoralis major	*Clavicular head:* anterior surface of medial half of clavicle *Sternocostal head:* anterior surface of sternum, superior six costal cartilages, and aponeurosis of external oblique muscle	Lateral lip of intertubercular groove of humerus	Lateral and medial pectoral nerves; clavicular head (C5 and **C6**), sternocostal head (**C7**, **C8**, and T1)	Adducts and medially rotates humerus Draws scapula anteriorly and inferiorly *Acting alone:* clavicular head flexes humerus and sternocostal head extends it
Pectoralis minor	3rd to 5th ribs near their costal cartilages	Medial border and superior surface of coracoid process of scapula	Medial pectoral nerve (C8 and T1)	Stabilizes scapula by drawing it inferiorly and anteriorly against thoracic wall
Subclavius	Junction of 1st rib and its costal cartilage	Inferior surface of middle third of clavicle	Nerve to subclavius (**C5** and C6)	Anchors and depresses clavicle
Serratus anterior	External surfaces of lateral parts of 1st to 8th ribs	Anterior surface of medial border of scapula	Long thoracic nerve (C5, **C6**, and **C7**)	Protracts scapula and holds it against thoracic wall; rotates scapula

[a] Numbers indicate spinal cord segmental innervation (e.g., C5 and C6 indicate that nerves supplying clavicular head of pectoralis major muscle are derived from fifth and sixth cervical segments of spinal cord). Boldface indicates main segmental innervation. Damage to these segments, or to motor nerve roots arising from them, results in paralysis of muscles concerned.

each other superiorly and, along with the clavicle, form the deltopectoral triangle. The *pectoralis minor* lies in the anterior wall of the axilla, where it is largely covered by the larger pectoralis major. The small, round *subclavius* lies inferior to the clavicle and affords some protection to the subclavian vessels when the clavicle fractures. The *serratus anterior* overlies the lateral portion of the thorax (Fig. 7.6*B*). It was given its name because of the saw-toothed appearance (L. serratus, a saw) of its fleshy slips (digitations). See Table 7.1 for attachments, nerve supply, and actions of the pectoral muscles.

The following muscles connect the upper limb to the vertebral column: latissimus dorsi, trapezius, levator scapulae, and rhomboids (p. 302). Only the latissimus dorsi attaches to the humerus; the others attach to the pectoral girdle.

When the serratus anterior is paralyzed from injury of the long thoracic nerve, the medial border of the scapula stands out, giving it the appearance of a wing when the person presses anteriorly (e.g., against a wall). When the arm is raised, the scapula pulls away from the thoracic wall—consequently the condition *winged scapula*. In addition, the arm cannot be abducted farther than the horizontal position because the serratus anterior cannot rotate the glenoid cavity superiorly (face upward) to allow complete abduction of the arm above the horizontal. This rotation results from the combined actions of the trapezius and serratus anterior muscles.

Axilla

The axilla is the pyramidal area at the junction of the arm and thorax. The shape and size of the axilla vary, depending on the position of the arm.

- Apex of the axilla lies between the 1st rib, the clavicle, and the superior edge of the subscapularis muscle; arteries, veins, lymphatics, and nerves pass through the apex to reach the arm
- Base of the axilla (armpit) is formed by the concave skin and fascia extending from the arm to the thoracic wall
- Anterior wall of the axilla is formed mainly by the pectoralis major and minor
- Posterior wall of the axilla is formed chiefly by the scapula, subscapularis, teres major, and latissimus dorsi
- Medial wall of the axilla is formed by the 1st to 4th ribs, intercostal muscles, and the overlying serratus anterior
- Lateral wall of the axilla is the intertubercular groove in the humerus (Fig. 7.1*A*)

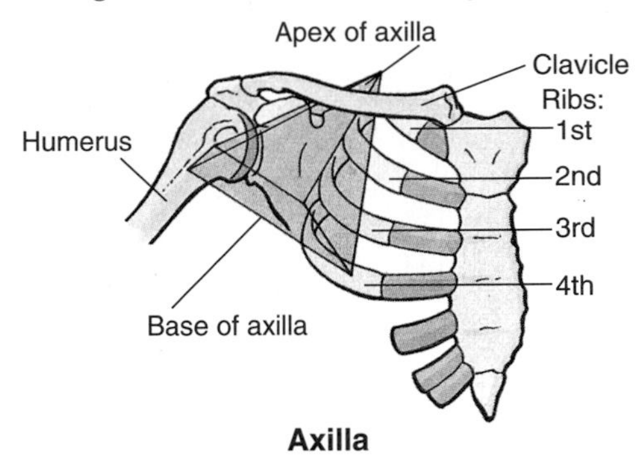

Axilla

AXILLARY ARTERY

The axillary artery begins at the lateral border of the 1st rib as the continuation of the subclavian artery and ends at the inferior border of the teres major. Here it passes posterior to the pectoralis minor and continues into the arm as the *brachial artery* (Fig. 7.7, Table 7.2). The axillary artery is divided into three parts by the pectoralis minor (the number of the part also indicates its number of branches).

- The first part, between the lateral border of the 1st rib and the superior border of the pectoralis minor, is enclosed in the *axillary sheath* along with the axillary vein and cords of the brachial plexus and has one small branch, the superior thoracic artery
- The second part lies posterior to the pectoralis minor and has two branches, the thoracoacromial and lateral thoracic arteries
- The third part extends from the inferior border of the pectoralis minor to the inferior border of the teres major and has three branches, the subscapular, anterior circumflex humeral, and posterior circumflex humeral arteries

AXILLARY VEIN

The axillary vein lies on the medial side of the axillary artery (Fig. 7.7, *A* and *D*). The axillary vein begins as the continuation of the basilic vein at the inferior border of the teres major and ends at the lateral border of the 1st rib where it becomes the subclavian vein. The axillary vein receives tributaries that correspond to the branches of the axillary artery, and at the inferior margin of the subscapularis, it receives the pair of brachial veins that accompany the brachial artery (venae comitantes).

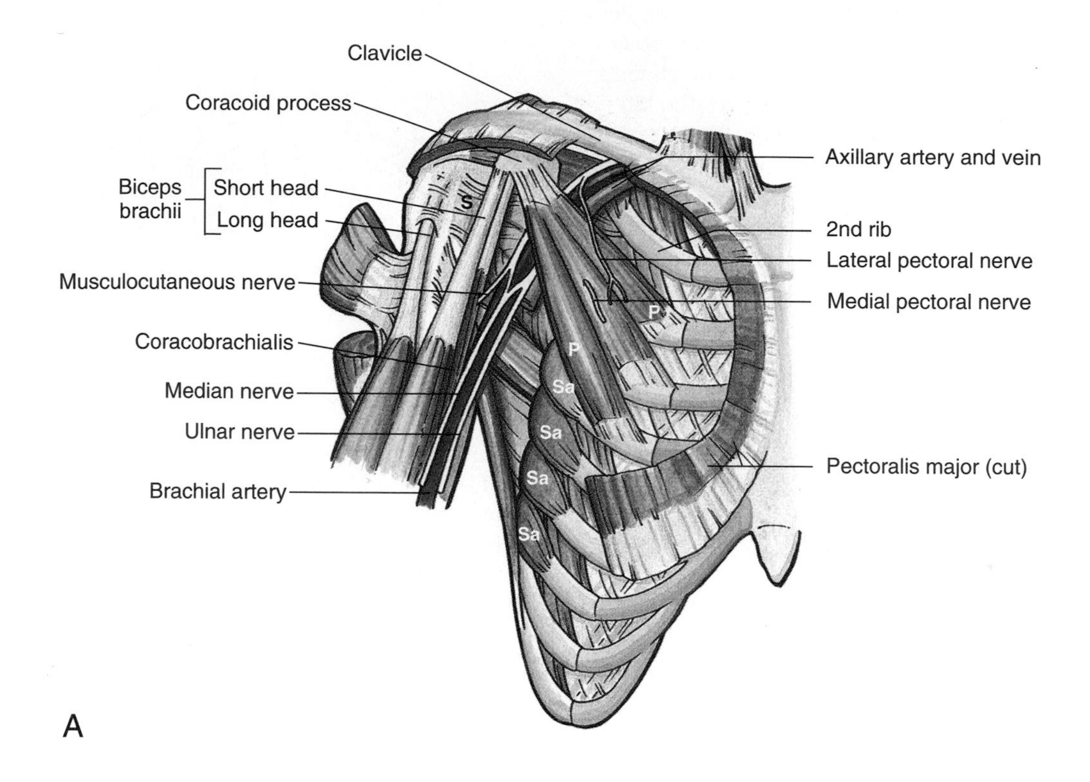

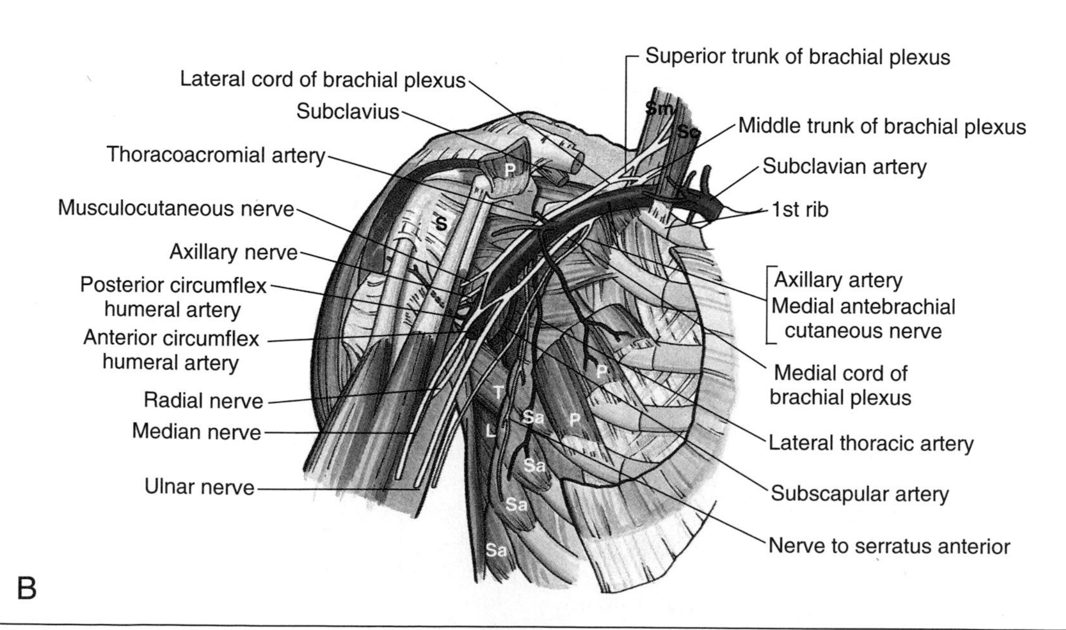

Figure 7.7. **A.** Anterior structures of axilla. **B.** Posterior and medial walls of axilla, showing axillary artery and brachial plexus. **C.** Posterior wall of axilla demonstrating posterior cord of brachial plexus and its branches. **D.** Transverse section of shoulder and axilla. *P*, pectoralis minor; *S*, subscapularis; *Sa*, serratus anterior; *Sc*, scalenus anterior; *Sm*, scalenus medius; *T*, teres major; *L*, latissimus dorsi.

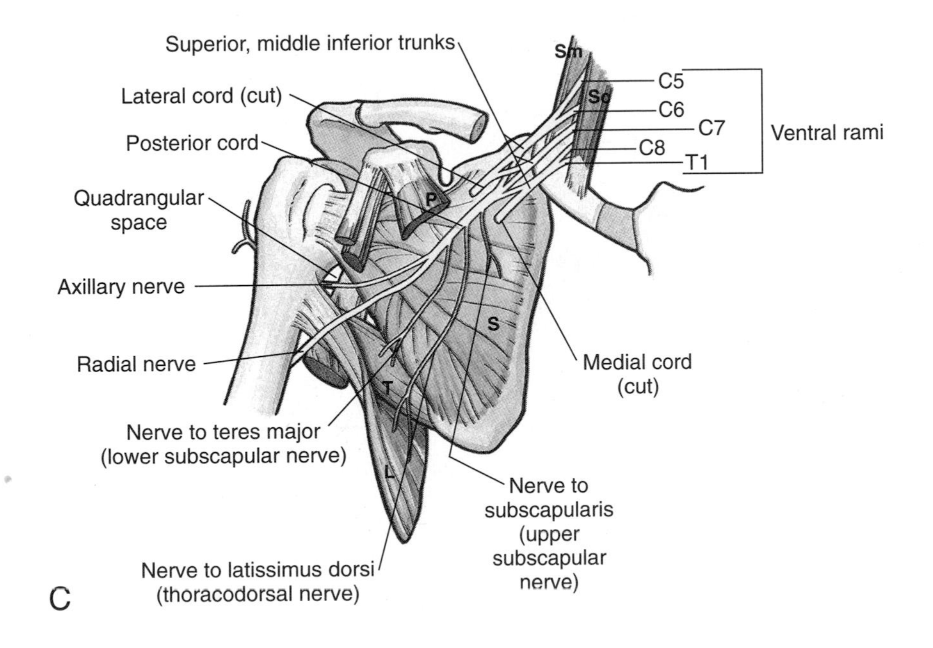

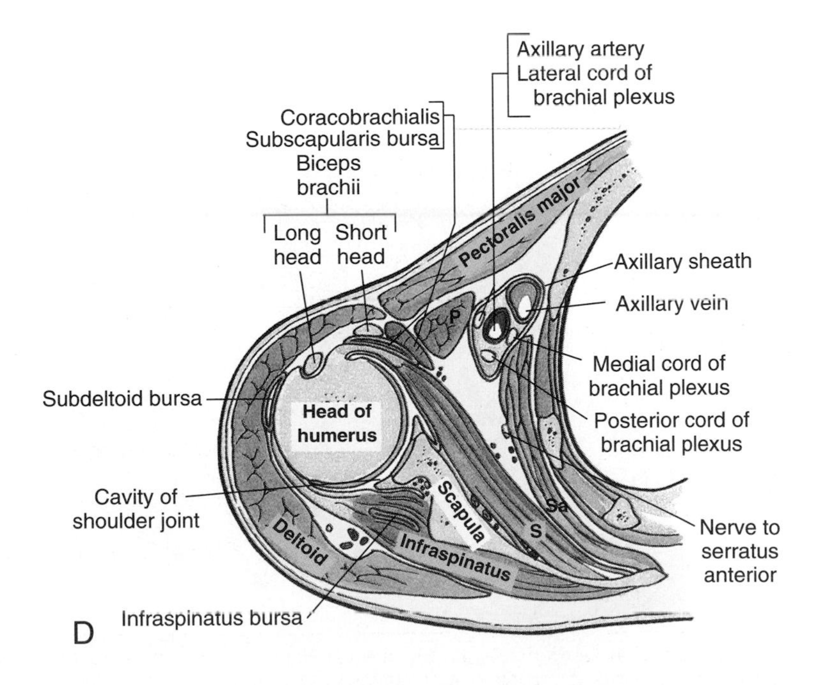

Figure 7.7. *Continued.*

Table 7.2
Branches of Subclavian, Axillary, and Brachial Arteries

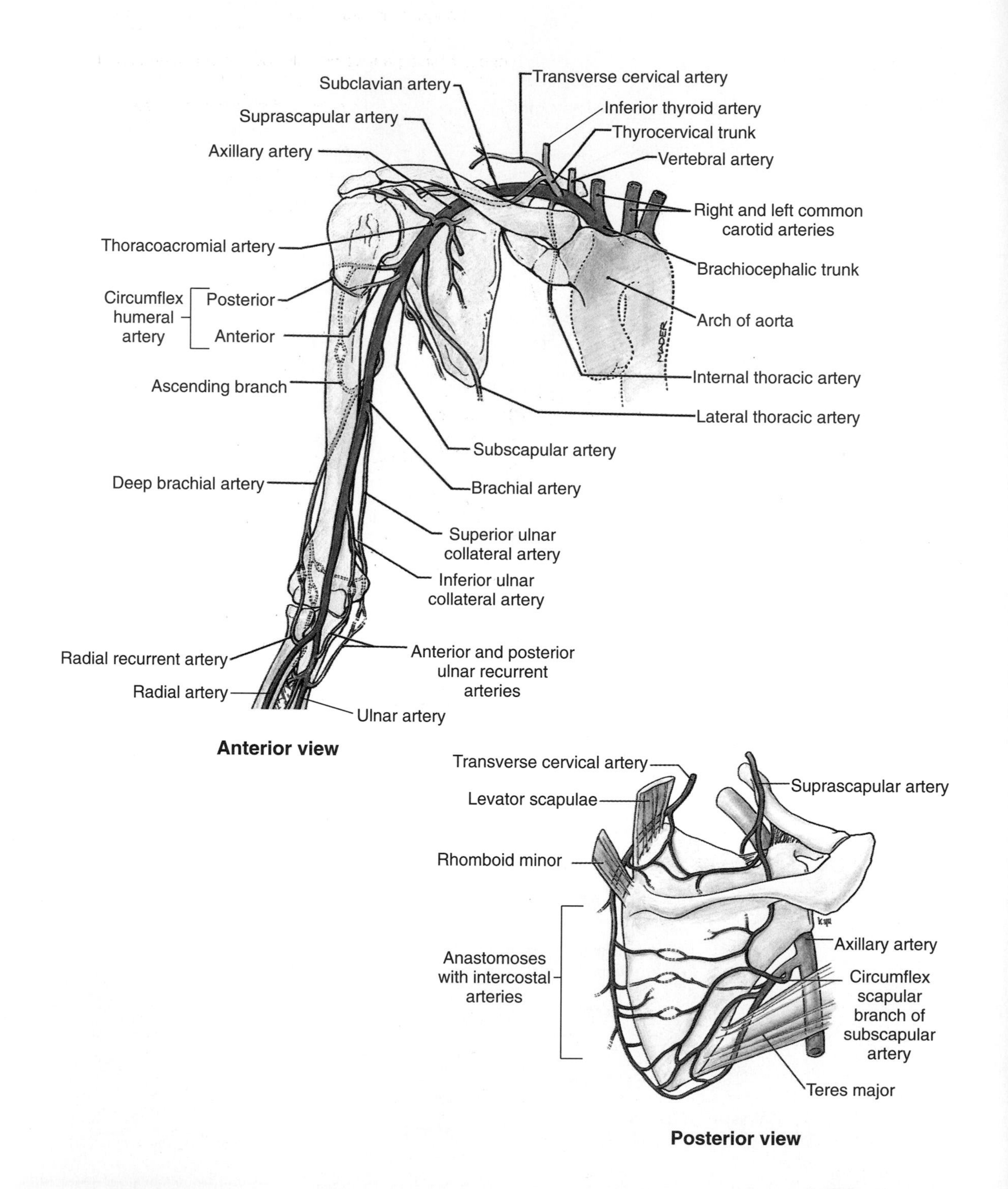

Table 7.2. ***Continued***

Artery	Origin	Course
Vertebral	Superior aspect of first part of subclavian a.	Ascends through transverse foramina of cervical vertebrae, except for C7, and enters skull through the foramen magnum
Internal thoracic	Inferior surface of subclavian a.	Descends, inclining anteromedially, posterior to sternal end of clavicle and first costal cartilage, and enters thorax
Thyrocervical trunk	Anterior aspect of first part of subclavian a.	Ascends as a short, wide, trunk and gives rise to three branches: inferior thyroid, suprascapular, and transverse cervical arteries
Suprascapular	Thyrocervical trunk	Passes inferolaterally over anterior scalene muscle and phrenic nerve, crosses subclavian a. and brachial plexus, and runs laterally posterior and parallel to clavicle; it then passes to posterior aspect of scapula and supplies supraspinatus and infraspinatus muscles
Superior thoracic	Only branch of first part of axillary a.	Runs anteromedially along superior border of pectoralis minor and then passes between it and pectoralis major to thoracic wall
Thoracoacromial	Second part of axillary a., deep to pectoralis minor	Curls around superomedial border of pectoralis minor, pierces clavipectoral fascia, and divides into four branches
Lateral thoracic	Second part of axillary a.	Descends along axillary border of pectoralis minor and follows it onto thoracic wall
Subscapular	Third part of axillary a.	Descends along lateral border of subscapularis and axillary border of scapula to its inferior angle, where it passes onto thoracic wall
Circumflex scapular a.	Subscapular a.	Curves around axillary border of scapula and enters infraspinous fossa
Thoracodorsal	Subscapular a.	Continues course of subscapular a. and accompanies thoracodorsal n.
Anterior and posterior circumflex humeral	Third part of axillary a.	These arteries anastomose to form a circle around surgical neck of humerus; larger posterior circumflex humeral a. passes through quadrangular space with axillary n.
Deep brachial	Brachial a. near its origin	Accompanies radial n. through groove in humerus and takes part in anastomosis around elbow joint
Ulnar collateral (superior and inferior)	Superior ulnar collateral a. arises from brachial a. near middle of arm; Inferior ulnar collateral a. arises from brachial a. just superior to elbow	Superior ulnar collateral a. accompanies ulnar n. to posterior aspect of elbow; inferior ulnar collateral a. divides into anterior and posterior branches; both ulnar collateral aa. take part in anastomosis around elbow joint.

In lacerations of the axilla, compression of the axillary artery may be necessary to stop the bleeding. To do so, the third part of the artery is pressed against the superior part of the humerus. Wounds in the axilla often involve the axillary vein because of its large size and exposed position. A wound in the superior part of the axillary vein is particularly dangerous not only because of profuse hemorrhage but also because of the risk of air entering the vessel and producing air bubbles (air emboli).

Extensive *arterial anastomoses* surround the scapula (Table 7.2). Several arteries—dorsal scapular, suprascapular, and subscapular—join to form networks on the anterior and posterior surfaces of the scapula. The clinical importance of the collateral (accessory) circulation that is possible because of these anastomoses becomes apparent when the main arterial pathway to the upper limb is disrupted. For example, if the axillary artery is obstructed between the thyrocervical trunk and subscapular artery, the direction of blood flow in the subscapular artery is reversed, enabling blood to reach the distal part of the axillary artery. Slow occlusion of an artery (e.g., resulting from disease or trauma) often enables sufficient collateral circulation to develop and prevents ischemia. Sudden occlusion usually does not allow sufficient time for a good collateral circulation to develop.

Table 7.3
Branches of Brachial Plexus

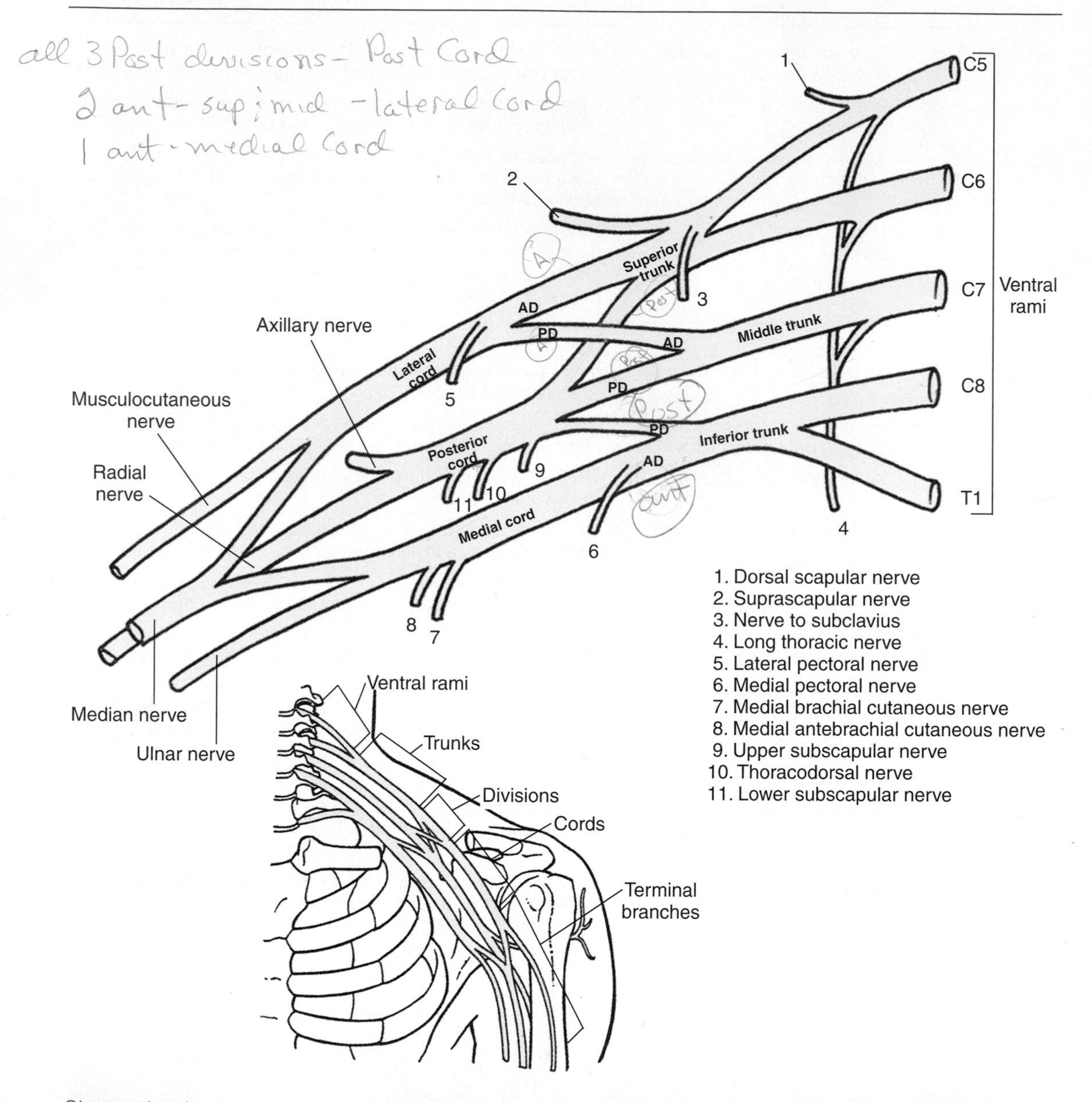

Observe that three nerves (musculocutaneous, median, and ulnar) are arranged like limbs of a capital M. Also observe anterior divisions (*AD*) and posterior divisions (*PD*).

Nerve	Origin	Course	Distribution
Supraclavicular branches			
Dorsal scapular	Ventral ramus of C5 with a frequent contribution from C4	Pierces scalenus medius, descends deep to levator scapulae, and enters deep surface of rhomboids	Innervates rhomboids and occasionally supplies levator scapulae
Long thoracic	Ventral rami of C5–C7	Descends posterior to C8 and T1 rami and passes distally on external surface of serratus anterior	Innervates serratus anterior

Table 7.3. *Continued*

Nerve to subclavius	Superior trunk receiving fibers from C5 and C6 and often C4	Descends posterior to clavicle and anterior to brachial plexus and subclavian artery	Innervates subclavius and sternoclavicular joint
Suprascapular	Superior trunk receiving fibers from C5 and C6 and often C4	Passes laterally across posterior triangle of neck, through scapular notch under superior transverse scapular ligament	Innervates supraspinatus, infraspinatus, and shoulder joint
Infraclavicular branches			
Lateral pectoral	Lateral cord receiving fibers from C5–C7	Pierces clavipectoral fascia to reach deep surface of pectoral muscles	Primarily supplies pectoralis major but sends a loop to medial pectoral n. that innervates pectoralis minor
Musculocutaneous	Lateral cord receiving fibers from C5–C7	Enters deep surface of coracobrachialis and descends between biceps brachii and brachialis	Innervates coracobrachialis, biceps brachii, and brachialis; continues as lateral antebrachial cutaneous n.
Median	Lateral root is a continuation of lateral cord, receiving fibers from C6 and C7; medial root is a continuation of medial cord receiving fibers from C8 and T1	Lateral root joins medial root to form median n. lateral to axillary artery	Innervates flexor mm. in forearm (except flexor carpi ulnaris, ulnar half of flexor digitorum profundus, and five hand mm.)
Medial pectoral	Medial cord receiving fibers from C8 and T1	Passes between axillary a. and v. and enters deep surface of pectoralis minor	Innervates the pectoralis minor and part of pectoralis major
Medial brachial cutaneous	Medial cord receiving fibers from C8 and T1	Runs along the medial side of axillary v. and communicates with intercostobrachial n.	Supplies skin on medial side of arm
Medial antebrachial cutaneous	Medial cord receiving fibers from C8 and T1	Runs between axillary a. and v.	Supplies skin over medial side of forearm
Ulnar	A terminal branch of medial cord receiving fibers from C8 and T1 and often C7	Passes down medial aspect of arm and runs posterior to medial epicondyle to enter forearm	Innervates one and one-half flexor muscles in forearm, most small muscles in hand, and skin of hand medial to a line bisecting fourth digit (ring finger)
Upper subscapular	Branch of posterior cord receiving fibers from C5 and C6	Passes posteriorly and enters subscapularis	Innervates superior portion of subscapularis
Thoracodorsal	Branch of posterior cord receiving fibers from C6–C8	Arises between upper and lower subscapular nn. and runs inferolaterally to latissimus dorsi	Innervates latissimus dorsi
Lower subscapular	Branch of posterior cord receiving fibers from C5 and C6	Passes inferolaterally, deep to subscapular a. and v., to subscapularis and teres major	Innervates inferior portion of subscapularis and teres major
Axillary	Terminal branch of posterior cord receiving fibers from C5 and C6	Passes to posterior aspect of arm through quadrangular space[a] in company with posterior circumflex humeral a. and then winds around surgical neck of humerus; gives rise to lateral brachial cutaneous n.	Innervates teres minor and deltoid, shoulder joint, and skin over inferior part of deltoid
Radial	Terminal branch of posterior cord receiving fibers from C5–C8 and T1	Descends posterior to axillary a.; enters radial groove with deep brachial a. to pass between long and medial heads of triceps	Innervates triceps brachii, anconeus, brachioradialis, and extensor muscles of forearm; supplies skin on posterior aspect of arm and forearm via posterior cutaneous nerves of arm and forearm

[a] Quadrangular space is bounded superiorly by subscapularis and teres minor, inferiorly by teres major, and medially by long head of triceps, and laterally by humerus.

AXILLARY LYMPH NODES

There are many lymph nodes in the fibrofatty connective tissue of the axilla. The axillary lymph nodes are arranged in five principal groups: four (pectoral, subscapular, lateral, and central) lie inferior and deep to the pectoralis minor, and one (apical) lies superior to it.

The lateral group of axillary lymph nodes is the first to be involved in *lymphangitis* (inflammation of lymphatic vessels, e.g., hand infection). Lymphangitis is characterized by the development of red, warm, tender streaks in the skin. Infections in the pectoral region, including the superior part of the abdomen, can also produce enlargement of the axillary lymph nodes. These nodes are also the most common site of metastases from cancer of the breast (p. 38).

BRACHIAL PLEXUS

Most nerves in the upper limb arise from the brachial plexus, a large network of nerves that originates in the neck and extends into the axilla. The brachial plexus is formed by the union of the ventral rami of C5–C8 nerves and the greater part of T1 ventral ramus (Fig. 7.7, Table 7.3). Small contributions from the ventral rami of C4 or T2 may be added to these five rami. As the rami enter the neck, they unite to form three trunks:

- C5 and C6 unite to form the superior trunk
- C7 continues as the middle trunk
- C8 and T1 unite to form the inferior trunk

Each of the three trunks of the brachial plexus divides into anterior and posterior divisions, posterior to the clavicle. The anterior divisions supply anterior (flexor) parts of the upper limb and posterior divisions supply posterior (extensor) parts:

- Posterior divisions of all three trunks unite to form the posterior cord
- Anterior divisions of the superior and middle trunks unite to form the lateral cord
- Anterior division of the inferior trunk continues as the medial cord

The cords of the plexus bear the relationship to the second part of the axillary artery that is indicated by their names (e.g., lateral cord is lateral to axillary artery).

The *lateral cord*, carrying fibers primarily from C5 to C7, has three branches (Table 7.3):

- One side branch, the lateral pectoral nerve
- Two terminal branches, the musculocutaneous nerve and the lateral root of the median nerve

The *lateral pectoral nerve* supplies the pectoralis major. The larger *musculocutaneous nerve* pierces the coracobrachialis and supplies it and the biceps brachii and brachialis. It terminates as the lateral antebrachial cutaneous nerve. The lateral root of the median nerve unites with the medial root of the median to form the *median nerve*, which supplies primarily flexor muscles in the forearm, the skin of part of the hand, and five muscles of the hand.

The *medial cord*, carrying fibers from C8 and T1, has five branches:

- Three side branches, the medial pectoral, medial brachial cutaneous, and medial antebrachial cutaneous nerves
- Two terminal branches, the ulnar nerve and medial root of the median nerve

The *medial pectoral nerve* passes through the pectoralis minor, supplying it and the pectoralis major. The *medial brachial cutaneous nerve* supplies skin on the medial side of the arm and the superior part of the forearm. The *medial antebrachial cutaneous nerve* supplies skin on the medial surface of the forearm.

The *ulnar nerve* passes through the arm to the forearm, where it supplies one and one-half muscles in the forearm [flexor carpi ulnaris and ulnar portion of flexor digitorum profundus (FDP)], most small muscles in the hand, and skin on the medial side of the hand. The medial root of the median nerve unites with the lateral root to form the *median nerve*, the distribution of which has been described.

The *posterior cord*, carrying fibers from C5 to T1, has five branches:

- Three side branches, the upper subscapular, thoracodorsal, and lower subscapular nerves
- Two terminal branches, the axillary and radial nerves

The *upper subscapular nerve* supplies the subscapularis; the *thoracodorsal nerve* supplies the latissimus dorsi, and the *lower subscapular nerve* supplies the teres major. The *axillary nerve* supplies the deltoid and teres major and skin over the inferior half of the deltoid. The *radial nerve*, the largest branch of the brachial plexus, supplies the extensor muscles of the upper limb and skin on the posterior aspect of the arm and forearm.

The brachial plexus is divided into supraclavicular and infraclavicular parts by the clavicle (Fig. 7.4*A*, Table 7.3). *Supraclavicular branches* arise from the ventral rami and trunks of the brachial plexus and are approachable through the neck, and *infraclavicular branches* arise from the cords of the plexus and are approachable through the axilla.

AXILLARY SHEATH

The axillary artery and vein and cords of the brachial plexus are enclosed in a thin fascial sheath. Traced superiorly into the neck, the axillary sheath is continuous with the prevertebral layer of cervical fascia anterior to the subclavian artery (Fig. 7.7*D*).

Back and Shoulder Regions

This section describes *extrinsic back muscles*—the superficial and intermediate groups that attach the upper limb to the axial skeleton. Other back muscles are described in Chapter 5.

Shoulder muscles are divided into three groups (Fig. 7.8, Table 7.4):

- Superficial extrinsic muscles (trapezius and latissimus dorsi)
- Deep extrinsic muscles (levator scapulae, rhomboids, and serratus anterior)

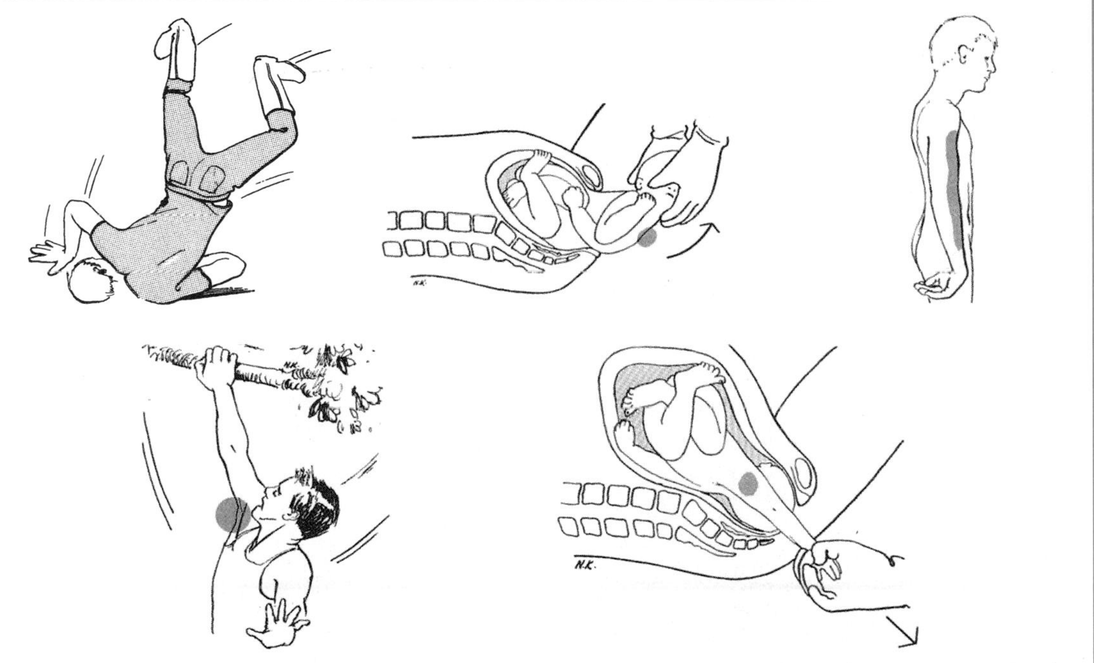

Disease, stretching, and wounds in the posterior triangle of the neck or axilla can produce *brachial plexus injuries*. Signs and symptoms depend on which part of the brachial plexus is involved. Injuries to the brachial plexus result in loss of muscular movement (paralysis) and often a loss of cutaneous sensation (anesthesia). Testing the patient's ability to perform movements can assess the degree of paralysis. In complete paralysis no movement is detectable. In incomplete paralysis not all muscles are paralyzed; hence the patient can move, but movements are weak compared with those on the normal side. Determining the ability of the person to feel pain (e.g., by a pinprick) can test the degree of anesthesia (loss of sensation).

Injuries to superior parts of the plexus usually result from excessive separation of the neck and shoulder. These brachial plexus injuries can happen to a person who is thrown from a motorcycle or a horse and lands on the shoulder in a way that widely separates the neck and shoulder. These injuries can also happen to a newborn when excessive stretching of the neck occurs during delivery. Injury to the superior trunk of the brachial plexus is apparent by the characteristic position of the limb ("waiter's tip position"), in which the limb hangs by the side in medial rotation.

Injuries to inferior parts of the plexus are uncommon, but they may occur when the upper limb is suddenly pulled superiorly when a person grasps something to break a fall or when a newborn's upper limb is pulled too hard during delivery. These events injure the inferior trunk of the brachial plexus (C8 and T1) and may pull the dorsal and ventral roots of the spinal nerves from the spinal cord. The short muscles of the hand are affected, and a *clawhand* results (p. 333).

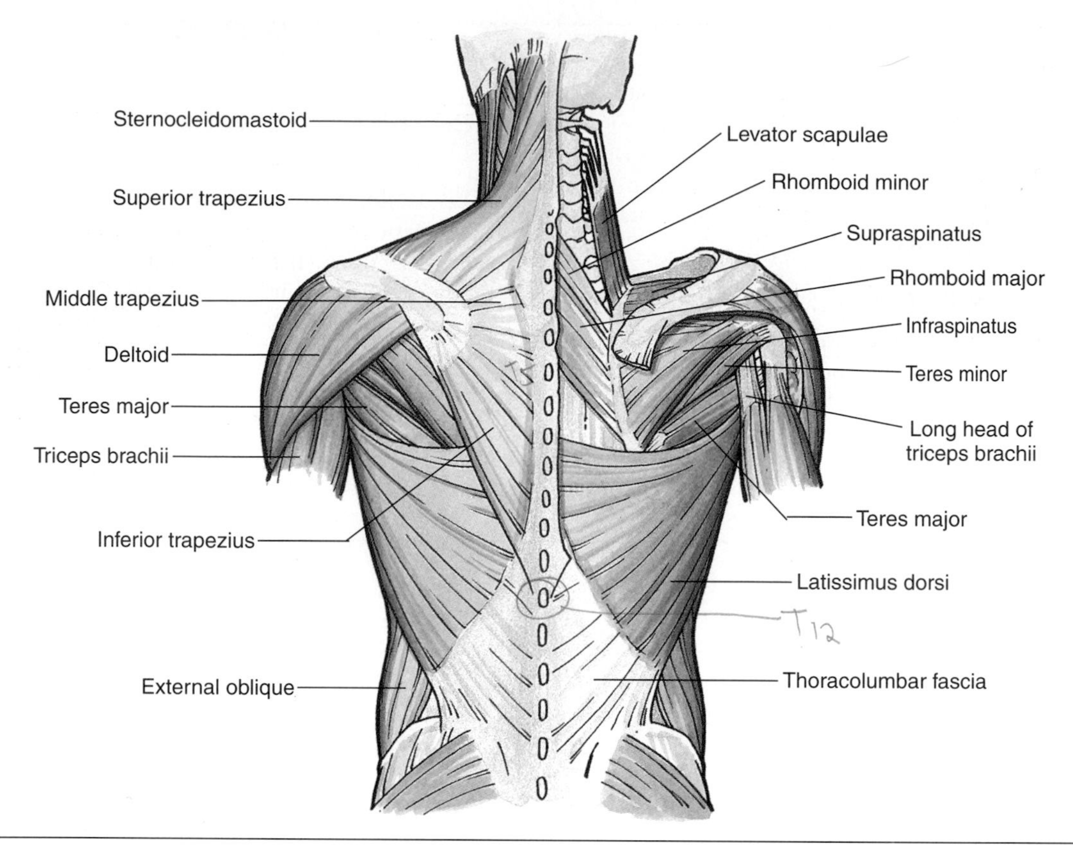

Figure 7.8. Superficial muscles of back and shoulder regions. *Left side*, superficial dissection; *right side*, deeper dissection.

- Intrinsic muscles (deltoid, teres major, and rotator cuff muscles)

SUPERFICIAL EXTRINSIC MUSCLES

The superficial extrinsic muscles are the trapezius and latissimus dorsi. The large, triangular trapezius covers the posterior aspect of the neck and superior half of the trunk. The *trapezius* attaches the pectoral girdle to the skull and vertebral column and assists in suspending the upper limb. The large, fan-shaped *latissimus dorsi* covers the inferior part of the back from T6 vertebra to the iliac crest. It passes between the trunk and the humerus and acts on the shoulder joint and indirectly on the pectoral girdle.

DEEP EXTRINSIC MUSCLES

The deep extrinsic muscles are the levator scapulae, rhomboids, and serratus anterior. The superior third of the straplike *levator scapulae* lies deep to the sternocleidomastoid; the inferior third is deep to the trapezius. The *rhomboid major and minor* lie deep to the trapezius and form parallel bands that pass inferolaterally from the vertebrae to the medial border of the scapula. The rhomboid major is about two times wider than the rhomboid minor.

INTRINSIC MUSCLES

The intrinsic muscles are the short *scapular muscles* (deltoid, teres major, supraspinatus, infraspinatus, subscapularis, and teres minor) that pass from the scapula to the humerus and act on the shoulder joint.

The *deltoid* forms the contour of the shoulder and, as its name indicates, is shaped like an inverted Greek delta (∇). The deltoid is divided into anterior, middle, and posterior parts and can act in part or as a whole. The *teres major* forms a raised oval area on the inferolateral third of the dorsum of the scapula. The teres major and the tendon of the latissimus dorsi form the *posterior axillary fold*.

Four scapular muscles (supraspinatus, infraspinatus, teres minor, and subscapularis) are called *rotator cuff muscles* (Fig. 7.4). All except the supraspinatus are rotators of the humerus;

Table 7.4
Superficial Back, Scapular, and Arm Muscles

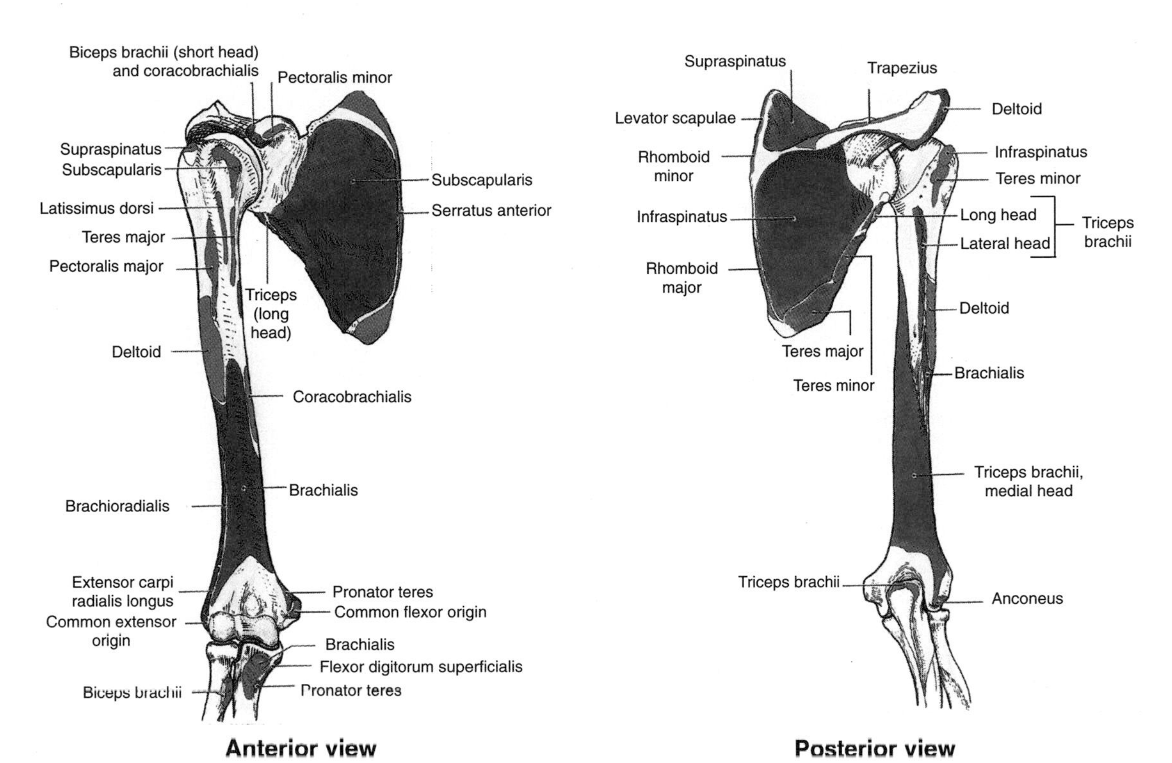

Muscle	Proximal Attachment	Distal Attachment	Innervation[a]	Main Actions
Trapezius	Medial third of superior nuchal line; external occipital protuberance, ligamentum nuchae, and spinous processes of C7–T12 vertebrae	Lateral third of clavicle, acromion, and spine of scapula	Spinal root of accessory n. (CN XI) and cervical nn. (C3 and C4)	Elevates, retracts, and rotates scapula; superior fibers elevate, middle fibers retract, and inferior fibers depress scapula; superior and inferior fibers act together in superior rotation of scapula
Latissimus dorsi	Spinous processes of inferior six thoracic vertebrae, thoracolumbar fascia, iliac crest, and inferior three or four ribs	Floor of intertubercular groove of humerus	Thoracodorsal n. (**C6**, **C7**, and C8)	Extends, adducts, and medially rotates humerus; raises body toward arms during climbing
Levator scapulae	Posterior tubercles of transverse processes of C1–C4 vertebrae	Superior part of medial border of scapula	Dorsal scapular (C5) and cervical (C3 and C4) nn.	Elevates scapula and tilts its glenoid cavity inferiorly by rotating scapula
Rhomboid minor and major	*Minor:* ligamentum nuchae and spinous processes of C7 and T1 vertebrae *Major:* spinous processes of T2–T5 vertebrae	Medial border of scapula from level of spine to inferior angle	Dorsal scapular n. (C4 and **C5**) rotate	Retracts scapula and rotates it to depress glenoid cavity; fixes scapula to thoracic wall
Deltoid	Lateral third of clavicle, acromion, and spine of scapula	Deltoid tuberosity of humerus	Axillary n. (**C5** and C6)	*Anterior part:* flexes and medially rotates arm *Middle part:* abducts arm *Posterior part:* extends and laterally rotates arm
Supraspinatus[b]	Supraspinous fossa of scapula	Superior facet on greater tubercle of humerus	Suprascapular n. (C4, **C5**, and C6)	Helps deltoid to abduct arm and acts with rotator cuff muscles[b]

Table 7.4. *Continued*

Infraspinatus[b]	Infraspinous fossa of scapula	Middle facet on greater tubercle of humerus	Suprascapular n. (**C5** and C6)	Laterally rotate arm; help to hold humeral head in glenoid cavity of scapula
Teres minor[b]	Superior part of lateral border of scapula	Inferior facet on greater tubercle of humerus	Axillary n. (**C5** and C6)	
Teres major	Dorsal surface of inferior angle of scapula	Medial lip of intertubercular groove of humerus	Lower subscapular n. (**C6** and C7)	Adducts and medially rotates arm
Subscapularis[b]	Subscapular fossa	Lesser tubercle of humerus	Upper and lower subscapular nn. (C5, **C6**, and C7)	Medially rotates arm and adducts it; helps to hold humeral head in glenoid cavity
Biceps brachii	*Short head:* tip of coracoid process of scapula *Long head*: supraglenoid tubercie of scapula	Tuberosity of radius and fascia of forearm via bicipital aponeurosis	Musculocutaneous n. (C5 and **C6**)	Supinates forearm and when it is supine, flexes forearm
Brachialis	Distal half of anterior surface of humerus	Coronoid process and tuberosity of ulna		Flexes forearm in all positions
Coracobrachialis	Tip of coracoid process of scapula	Middle third of medial surface of humerus	Musculocutaneous n. (C5, **C6**, and C7)	Helps to flex and adduct arm
Triceps brachii	*Long head:* infraglenoid tubercle of scapula *Lateral head:* posterior surface of humerus, superior to radial groove *Medial head:* posterior surface of humerus, inferior to radial groove	Proximal end of olecranon of ulna and fascia of forearm	Radial n. (C6, **C7**, and **C8**)	Extends forearm; it is chief extensor of forearm; long head steadies head of abducted humerus
Anconeus	Lateral epicondyle of humerus	Lateral surface of olecranon and superior part of posterior surface of ulna	Radial n. (C7, C8, and T1)	Assists triceps in extending forearm; stabilizes elbow joint; abducts ulna during pronation

[a] See Table 7.1 for explanation of nomenclature.

[b] Collectively, the supraspinatus, infraspinatus, teres minor, and subscapularis muscles are referred to as the rotator cuff muscles. Their prime function during all movements of shoulder joint is to hold head of humerus in glenoid cavity of scapula.

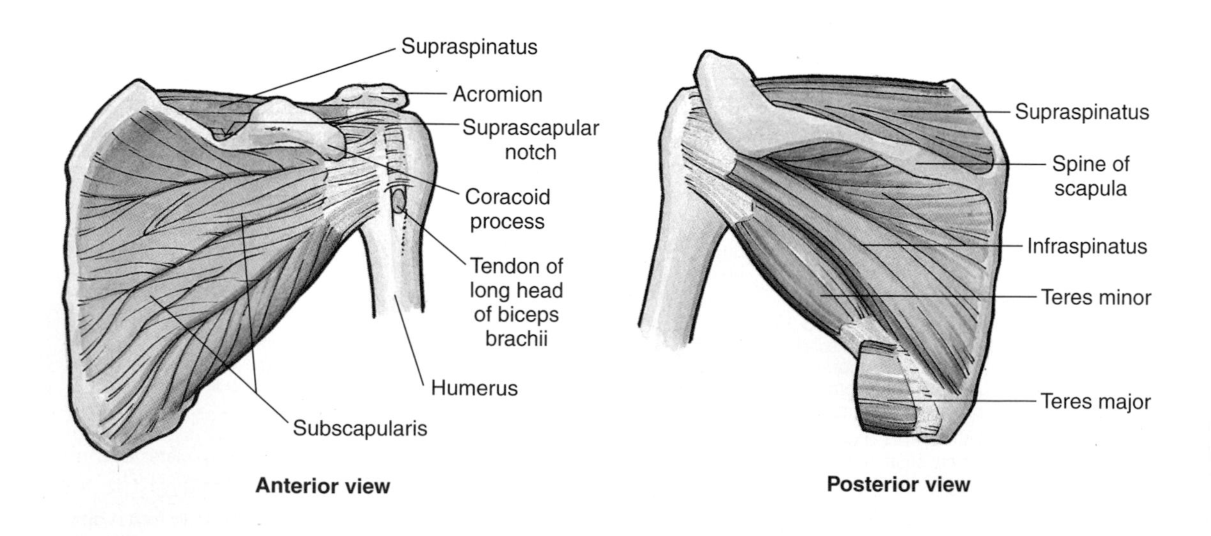

Figure 7.9. Rotator cuff muscles.

Surface Anatomy of Pectoral Region and Back

Trapezius
Posterior triangle of neck
Deltoid
Deltopectoral triangle
Clavicular head of pectoralis major
Anterior axillary fold
Sternocostal head of pectoralis major
Posterior axillary fold
Serratus anterior
Clavicle

The *clavicle* passes subcutaneously almost horizontally across the root of the neck and is palpable throughout its length. The *deltopectoral triangle* marks the interval between the deltoid and pectoralis major. The two heads of the pectoralis major are obvious in muscular people, and its inferolateral border forms the *anterior axillary fold*. The digitations of the serratus anterior appear inferolateral to the pectoralis major.

The *acromion* and crest of the spine of the scapula are subcutaneous and easily palpable. The acromion forms the point of the shoulder. The inferior angle of the scapula lies opposite the 7th rib and can be palpated. The superolateral border of the trapezius becomes obvious when the shoulder is shrugged. The latissimus dorsi is easily palpated. Its inferolateral border and the teres major form the *posterior axillary fold*.

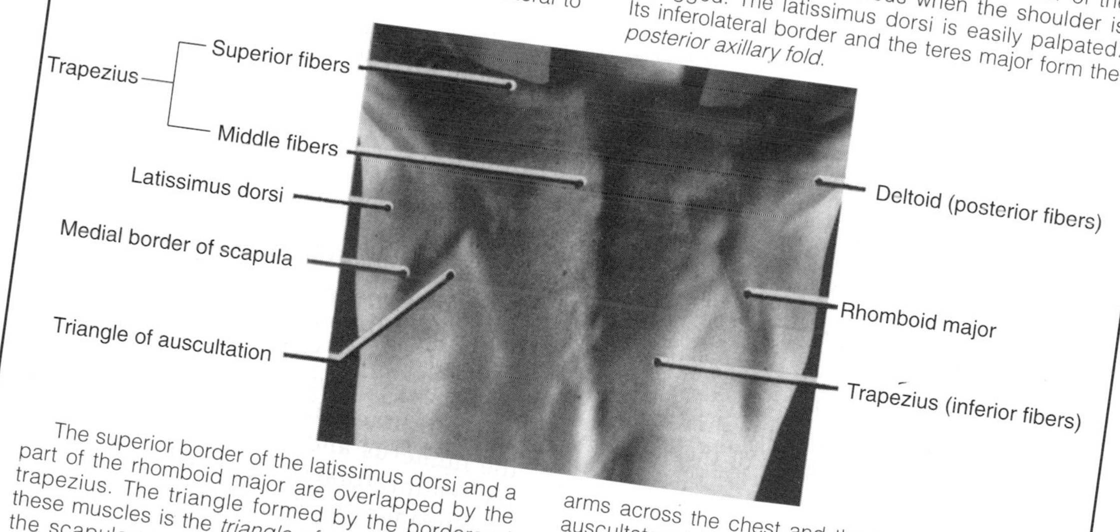

The superior border of the latissimus dorsi and a part of the rhomboid major are overlapped by the trapezius. The triangle formed by the borders of these muscles is the *triangle of auscultation*. When the scapulae are drawn anteriorly by folding the arms across the chest and the trunk is flexed, the auscultatory triangles enlarge, and the sixth intercostal spaces become subcutaneous; consequently, respiratory sounds are clearly audible with a stethoscope.

the supraspinatus is an abductor of the humerus. The tendons of these muscles blend with the articular capsule of the shoulder joint to form a *musculotendinous rotator cuff* for the shoulder joint. This cuff protects the joint and gives it stability by holding the head of the humerus in the glenoid cavity of the scapula. *Bursae around the shoulder*—between the tendons of the rotator cuff muscles and the fibrous capsule of the shoulder joint—reduce friction on the tendons passing over bones or other areas of resistance.

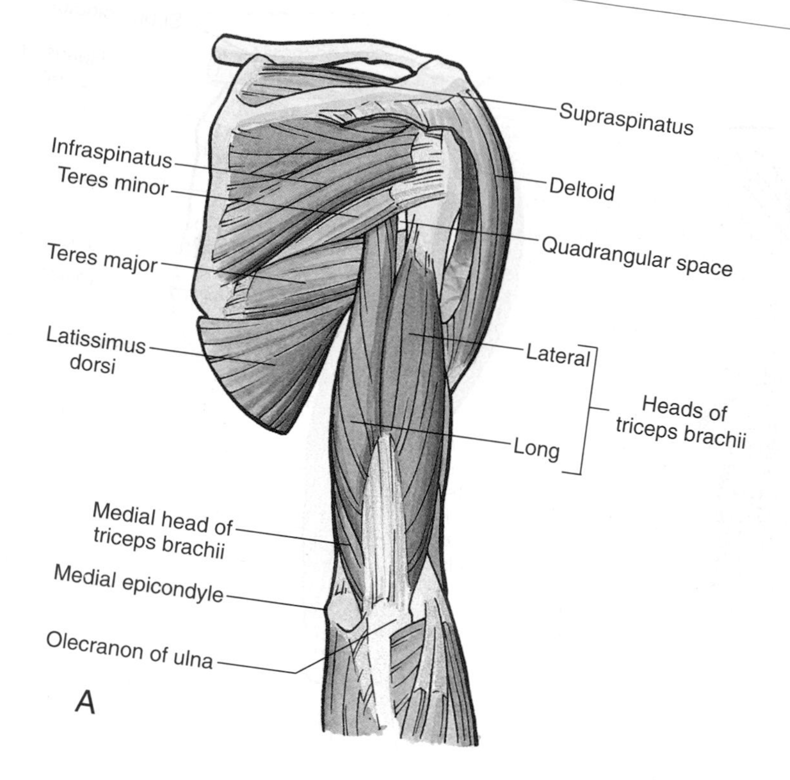

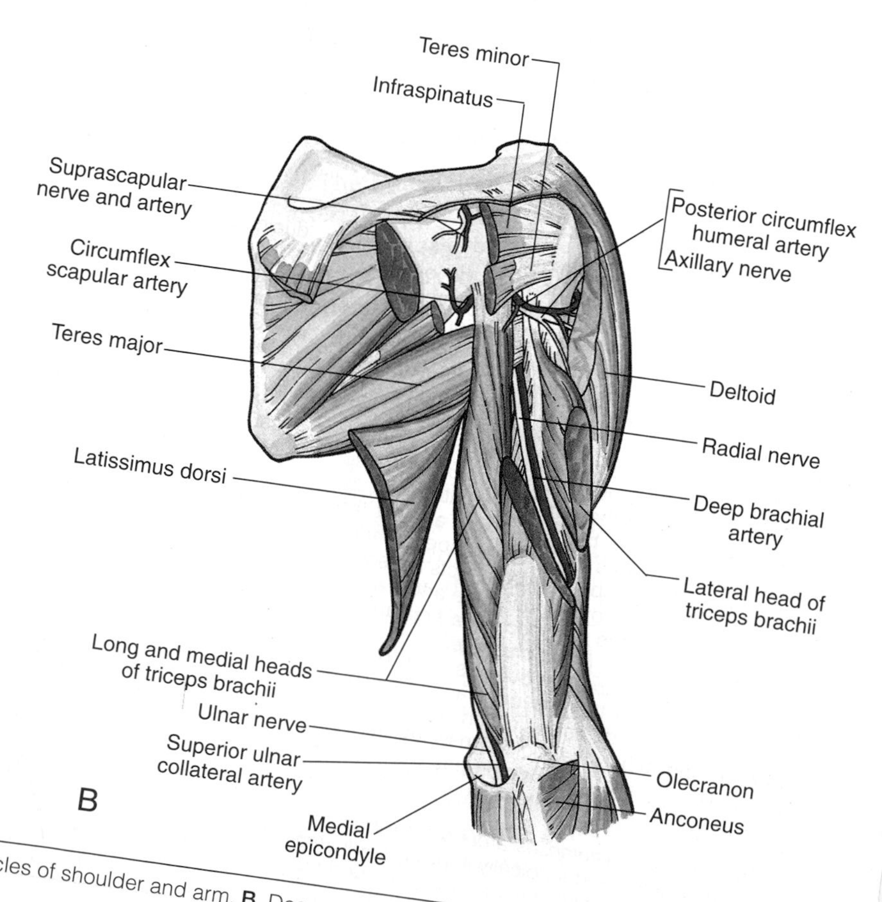

Figure 7.11. **A.** Posterior muscles of shoulder and arm. **B.** Deeper dissection showing vessels and nerves.

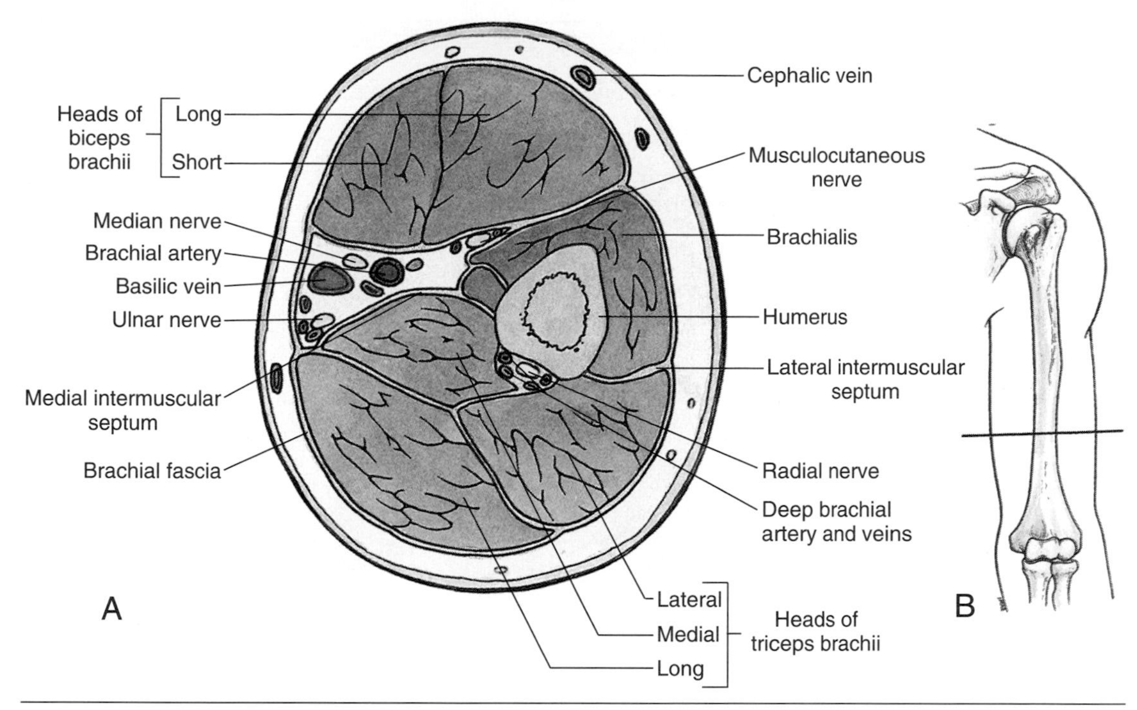

Figure 7.12. **A.** Transverse section of arm. **B.** Orientation drawing.

NERVES OF ARM

Four main nerves pass through the arm: median, ulnar, musculocutaneous, and radial (Figs. 7.10–7.12, Table 7.3). The first two give no branches to the arm. After arising from the brachial plexus, the median and ulnar nerves descend on the medial side of the arm and enter the forearm.

The *musculocutaneous nerve* supplies the muscles of the anterior (flexor) compartment of the arm. This nerve begins opposite the inferior border of the pectoralis minor, pierces the coracobrachialis, and continues distally between the biceps brachii and brachialis. It supplies all three of these muscles. In the interval between the biceps and brachialis, the musculocutaneous nerve becomes the *lateral antebrachial cutaneous nerve* and supplies the skin on the lateral aspect of the forearm (Fig. 7.4).

The *radial nerve* supplies the muscles of the posterior (extensor) compartment of the arm. This nerve enters the arm posterior to the brachial artery, medial to the humerus, and anterior to the long head of the triceps. It passes inferolaterally with the deep brachial artery around the body of the humerus in the *radial groove*. When it reaches the lateral bor-

Injury to the musculocutaneous nerve in the axilla results in paralysis of the coracobrachialis, biceps, and brachialis muscles. As a result, flexion of the elbow joint and supination of the forearm are greatly weakened. There may also be loss of sensation on the lateral surface of the forearm supplied by the lateral antebrachial cutaneous nerve.

Injury to the radial nerve proximal to the origin of the triceps results in paralysis of the triceps, brachioradialis, supinator, and extensor muscles of the wrist and digits. Loss of sensation in areas of skin supplied by this nerve also occurs. When the nerve is injured in the radial groove, the triceps is not completely paralyzed, but there is paralysis of muscles in the posterior compartment of the forearm that are supplied by more distal branches of the nerve. The characteristic clinical sign of radial nerve injury is *wrist-drop* (inability to extend the wrist and digits).

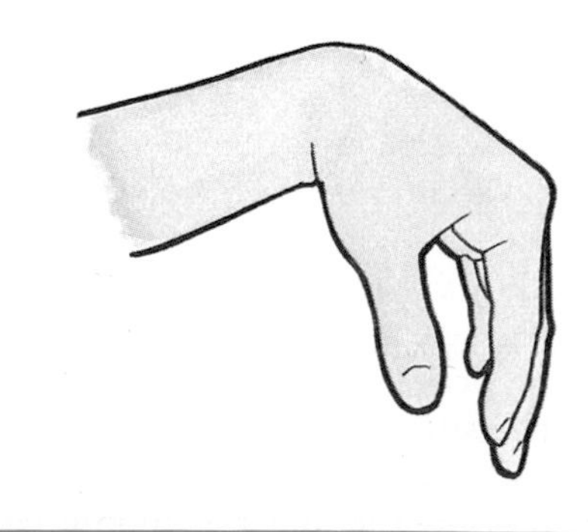

Surface Anatomy of Arm

The attachment of the deltoid can be easily palpated on the lateral surface of the arm. The three heads of the triceps (long head, lateral head, medial head) form a bulge on the posterior aspect of the arm and are identifiable when the forearm is extended against resistance. The olecranon, to which the triceps is attached distally, is easily palpated.

The biceps forms a bulge on the anterior aspect of the arm; its belly becomes more prominent when the elbow is flexed and supinated against resistance. Its tendon can be observed and palpated in the cubital fossa. The proximal part of the *bicipital aponeurosis* can be palpated where it passes obliquely over the brachial artery and median nerve. Medial and lateral *bicipital grooves* separate the bulges formed by the biceps and triceps and indicate the location of the medial and lateral intermuscular septa (Fig. 7.12). The cephalic vein runs superiorly in the lateral bicipital groove, and the basilic vein ascends in the medial bicipital groove.

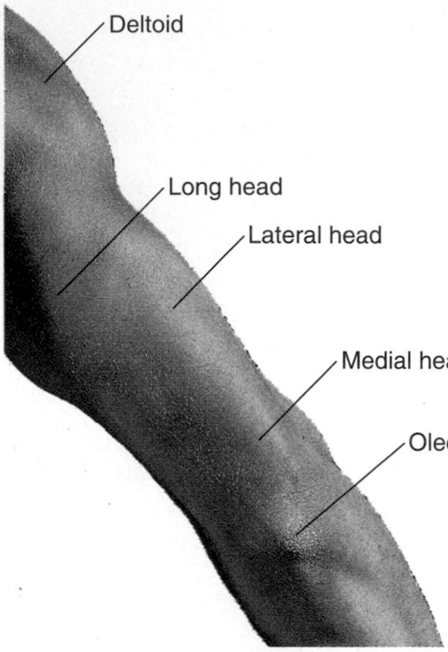

Posterior aspect of arm

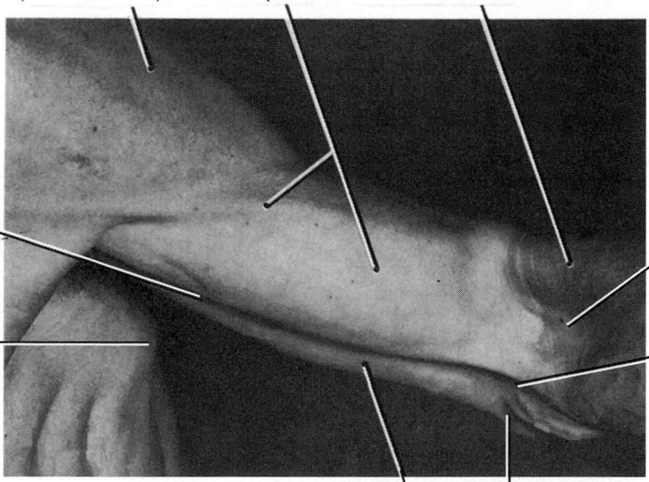

der of this bone, the radial nerve pierces the lateral intermuscular septum and continues inferiorly between the brachialis and brachioradialis to the level of the lateral epicondyle of the humerus. Here it divides into deep and superficial branches (see Fig. 7.14). The deep branch of the radial nerve is entirely muscular and articular in its distribution; the superficial branch of the radial nerve supplies sensory fibers to the dorsum of the hand and digits.

CUBITAL FOSSA

The cubital fossa is the hollow area on the anterior surface of the elbow (Figs. 7.13 and 7.14). The *boundaries of the cubital fossa* are superiorly an imaginary line connecting the medial and lateral epicondyles, medially the pronator teres, and lat-

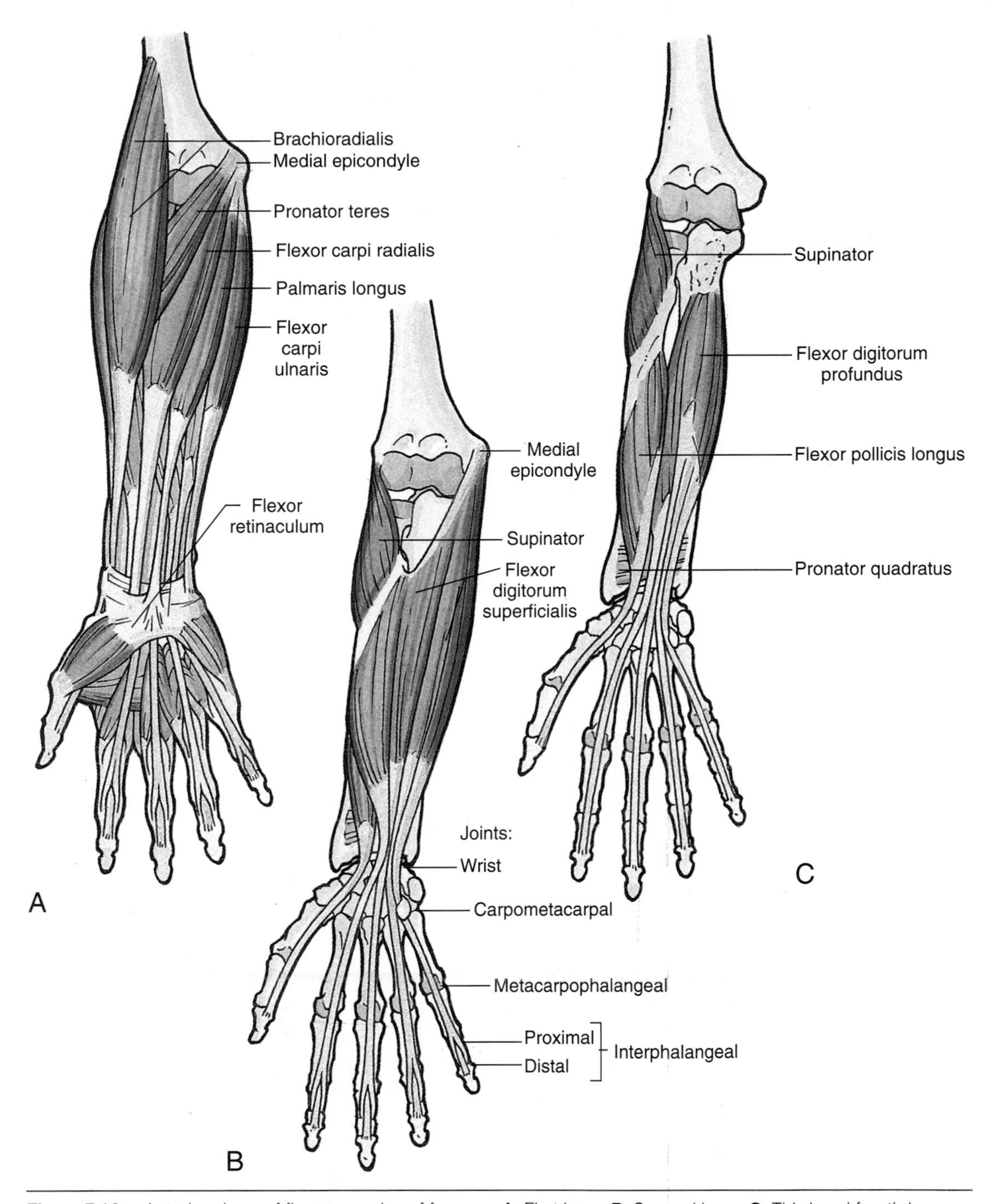

Figure 7.13. Anterior views of flexor muscles of forearm. **A.** First layer. **B.** Second layer. **C.** Third and fourth layers.

erally the brachioradialis. The floor of the cubital fossa is formed by the brachialis and supinator muscles of the arm and forearm, respectively. The roof of the cubital fossa is formed by deep fascia that blends with the bicipital aponeurosis, superficial fascia, and skin.

The contents of the cubital fossa are

- Brachial artery and its terminal branches, the radial and ulnar arteries
- Brachial and median cubital veins
- Median, radial, and medial and lateral antebrachial cutaneous nerves (Fig. 7.14*B*)

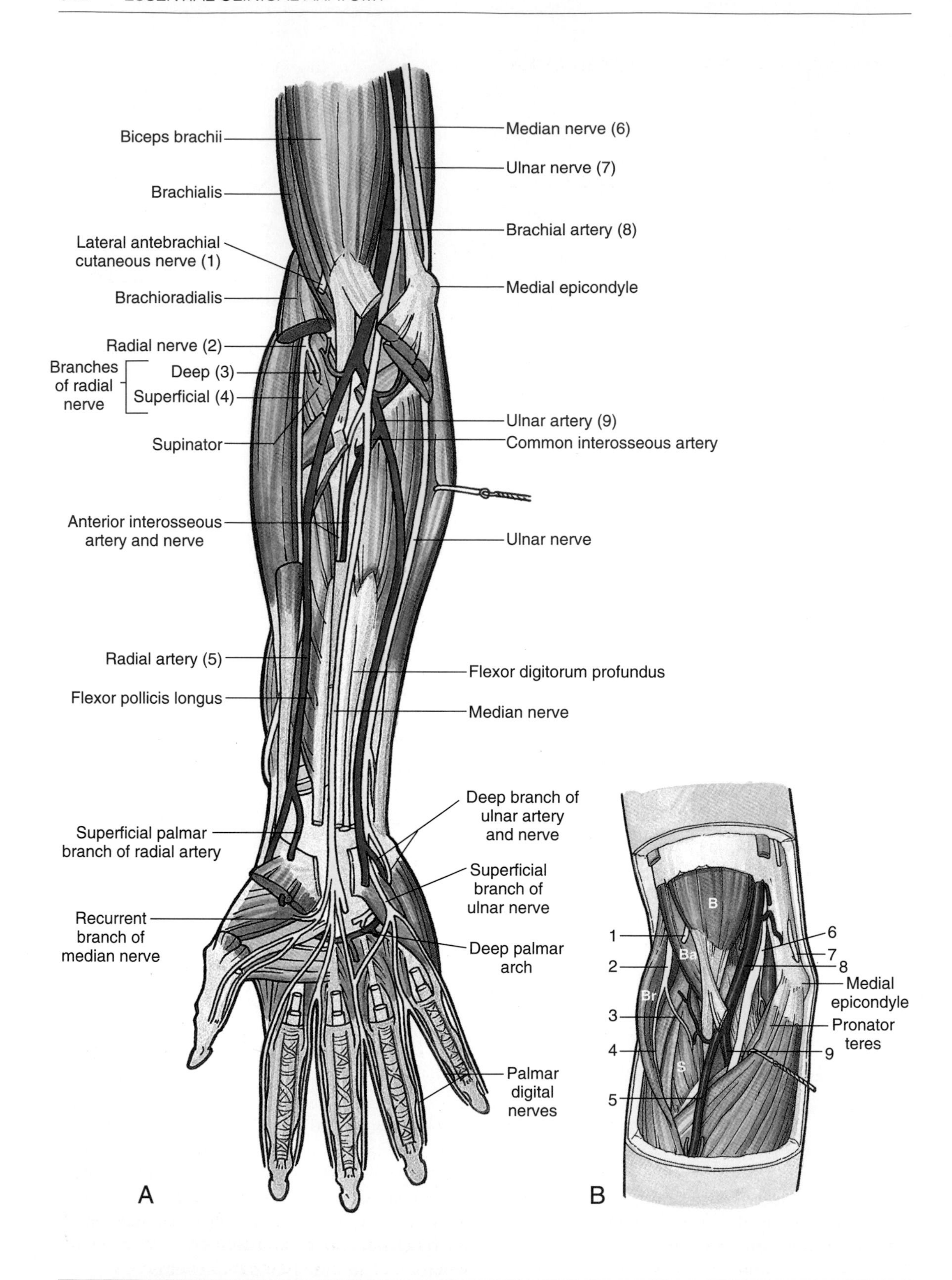

Figure 7.14. **A.** Dissection of anterior aspect of upper limb, primarily to show vessels and nerves. **B.** Contents of cubital fossa. *B*, biceps brachii; *Ba*, brachialis; *Br*, brachioradialis; *S*, supinator.

Forearm

The forearm extends from the elbow to the wrist and contains two bones, the radius and ulna. It contains many muscles, the tendons of which pass mostly to the hand.

MUSCLES OF FOREARM

Muscles of the forearm act on the joints of the elbow, wrist, and digits. In the proximal part of the forearm, the muscles form fleshy masses inferior to the medial and lateral epicondyles (Fig. 7.13). The tendons of these muscles pass through the distal part of the forearm and continue into the hand. The *flexor-pronator muscle group* arises by a common flexor tendon from the medial epicondyle, the common flexor attachment. The *extensor-supinator muscle group* arises by a common extensor tendon from the lateral epicondyle, the common extensor attachment.

Flexor Muscles

The tendons of most flexor muscles of the forearm are located on the anterior surface of the wrist and are held in place by the *flexor retinaculum*, a thickening of the antebrachial fascia. The flexor muscles are divided into two groups (Fig. 7.13, Table 7.5):

- A superficial group (pronator teres, flexor carpi radialis, palmaris longus, flexor carpi ulnaris, and flexor digitorum superficialis)
- A deep group [flexor digitorum profundis (FDP), flexor pollicis longus, and pronator quadratus]

The flexor digitorum superficialis is sometimes regarded as forming an intermediate group. All muscles in the anterior part of the forearm, except the brachioradialis, are supplied by the median and ulnar nerves. Functionally the brachioradialis is a flexor of the forearm, but it is supplied by the radial nerve. Hence, this muscle is a major exception to the rule that the radial nerve supplies only extensor muscles.

The long flexors of the digits [flexor digitorum superficialis (FDS) and FDP] also flex the metacarpophalangeal and wrist joints. The FDP flexes the digits in slow action, but this action is reinforced by the FDS when speed and flexion against resistance are required. When the wrist, metacarpophalangeal, and interphalangeal joints are flexed, the flexor muscles shorten, and their action is consequently weaker. Tendons of the long flexor muscles pass through the distal part of the forearm and continue into the medial four digits. FDS flexes the middle phalanges, and FDP flexes the middle and distal phalanges of these digits.

Extensor Muscles

The extensor muscles of the forearm are in the posterior part of the forearm (Figs. 7.15–7.17, Table 7.6). These extensor muscles can be organized into three functional groups:

- Muscles that extend and abduct or adduct the hand at the wrist joint (extensor carpi radialis longus, extensor carpi radialis brevis, and extensor carpi ulnaris)
- Muscles that extend the medial four digits (extensor digitorum, extensor indicis, and extensor digiti minimi)
- Muscles that extend or abduct the first digit or thumb (abductor pollicis longus, extensor pollicis brevis, and extensor pollicis longus)

As the tendons of the extensor muscles pass over the dorsum of the wrist they are surrounded by *synovial sheaths* that reduce friction between the extensor tendons and the associated bones. The extensor tendons are held in place by the *extensor retinaculum*, which prevents bowstringing of the tendons when the hand is hyperextended at the wrist joint.

The extensor muscles of the forearm are also divided into superficial and deep groups. Four of the *superficial extensors* (extensor carpi radialis brevis, extensor digitorum, extensor digiti minimi, and extensor carpi ulnaris) are attached by a *common extensor tendon* to the lateral epicondyle. The four flat tendons of the extensor digitorum muscles pass deep to the extensor retinaculum to the medial four digits. The common tendons of the index and little fingers are joined on their medial sides near the knuckles by the respective tendons of the extensor indicis and extensor digiti minimi muscles. The extensor indicis tendon enters the hand in the same tunnel as the tendons of the extensor digitorum. The tendon of the extensor digiti minimi has its own tunnel. Usually three oblique bands unite the four tendons proximal to the knuckles, restricting independent actions of the fingers so that no one digit can remain flexed when the other ones pass into full extension.

On the distal ends of the metacarpals and on the phalanges, the extensor tendons flatten to

Table 7.5
Muscles on Anterior Surface of Forearm

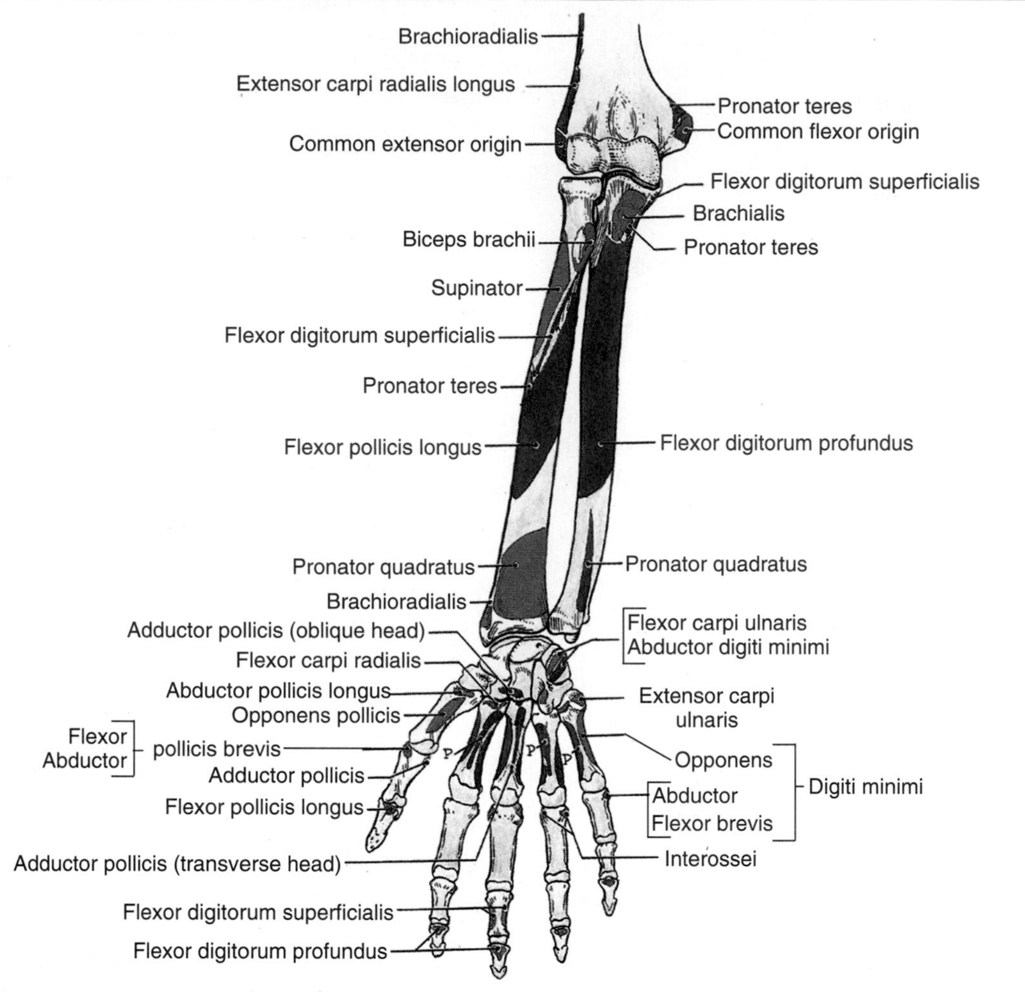

Muscle	Proximal Attachment	Distal Attachment	Innervation[a]	Main Actions
Pronator teres	Medial epicondyle of humerus and coronoid process of ulna	Middle of lateral surface of radius	Median n. (C6 and **C7**)	Pronates forearrm and flexes it
Flexor carpi radialis	Medial epicondyle of humerus	Base of second metacarpal bone	Median n. (C6 and **C7**)	Flexes hand and abducts it
Palmaris longus	Medial epicondyle of humerus	Distal half of flexor retinaculum and palmar aponeurosis	Median n. (C7 and C8)	Flexes hand and tightens palmar aponeurosis
Flexor carpi ulnaris	*Humeral head:* medial epicondyle of humerus *Ulnar head:* olecranon and posterior border of ulna	Pisiform bone, hook of hamate bone, and fifth metacarpal bone	Ulnar n. (C7 and **C8**)	Flexes hand and adducts it
Flexor digitorum superficialis	*Humeroulnar head:* medial epicondyle of humerus, ulnar collateral ligament, and coronoid process of ulna *Radial head:* superior half of anterior border of radius	Bodies of middle phalanges of medial four digits	Median n. (C7, **C8**, and T1)	Flexes middle phalanges of medial four digits; acting more strongly, it flexes proximal phalanges and hand

Table 7.5. *Continued*

Flexor digitorum profundus	Proximal three-fourths of medial and anterior surfaces of ulna and interosseous membrane	Bases ot distal phalanges of medial four digits	Medial part: ulnar n. (**C8** and T1) Lateral part: median n. (**C8** and T1)	Flexes distal phalanges of medial four digits; assists with flexion of hand
Flexor pollicis longus	Anterior surface of radius and adjacent interosseous membrane	Base of distal phalanx of thumb	Anterior interosseous n. from median (**C8** and T1)	Flexes phalanges of first digit (thumb)
Pronator quadratus	Distal fourth of anterior surface of ulna	Distal fourth of anterior surface of radius		Pronates forearm; deep fibers bind radius and ulna together

[a] See Table 7.1 for explanation of nomenclature.

form *extensor expansions*. Each expansion is wrapped around the dorsum and sides of a head of the metacarpal and proximal phalanx. The visorlike hood over the head of the metacarpal is anchored on each side to the *palmar ligament* (p. 341). The hood serves to retain the extensor tendon in the middle of the digit. The dorsal expansion divides into a median band that passes to the base of the middle phalanx and two lateral bands that pass to the base of the distal phalanx. The interosseous and lumbrical muscles of the hand are attached to the sides of the extensor expansion.

The *deep extensors of the forearm* (abductor pollicis longus, extensor pollicis brevis, and extensor pollicis longus) act on the thumb, and the extensor indicis helps to extend the index finger (Table 7.6). The three muscles acting on the thumb are deep to the superficial extensors and crop out along a furrow on the lateral part of the forearm that divides the extensors. Because of this characteristic, these muscles are referred to as *outcropping (thumb) muscles*.

Elbow tendinitis (golfer's or tennis elbow) is a painful musculoskeletal condition that may follow repetitive use of the superficial extensor muscles of the forearm. Pain is experienced over the lateral epicondyle. Repeated forceful flexion and extension at the wrist strain the common tendon attachment and may produce inflammation of the epicondyle (*epicondylitis*). The pain is usually caused by inflammation of the common extensor attachment of the muscles.

Sudden severe tension on a long extensor tendon may avulse part of its attachment to the phalanx. The most common result of the injury is *mallet finger* (also known as cricket or baseball finger). This deformity results from the distal interphalangeal joint suddenly being forced into extreme flexion (hyperflexion). This avulses the attachment of the tendon into the base of the distal phalanx. As a result, the patient cannot extend the distal interphalangeal joint.

NERVES OF FOREARM

The nerves of the forearm are the median, ulnar, and radial (Figs. 7.14–7.17). Their origins, courses, and distributions are described in Tables 7.3 and 7.7.

The *median nerve* is the principal nerve of the anterior compartment. It leaves the cubital fossa by passing between the heads of the pronator teres. The median nerve then passes deep to the FDS and continues distally between this muscle and the FDP.

The *ulnar nerve* enters the forearm by passing between the heads of the flexor carpi ulnaris. It then passes inferiorly between the flexor carpi ulnaris and FDP. The ulnar nerve becomes superficial at the wrist and supplies skin on the medial side of the hand.

The *radial nerve* appears in the cubital region between the brachialis and brachoradialis muscles. Soon after it enters the forearm, the radial nerve divides into deep and superficial branches. The deep branch arises anterior to the lateral epicondyle of the humerus and then pierces the supinator. The *deep branch of the radial nerve* winds around the lateral aspect of the neck of the radius and enters the posterior compartment of the forearm. The *posterior interosseous nerve* is the continuation of the deep branch of the radial nerve. The superficial branch of the radial nerve is a cutaneous and articular nerve that descends in the forearm under cover of the brachioradialis. It emerges in the distal part of the forearm and is distributed to skin on the dorsum of the hand and to many joints in the hand.

ARTERIES OF FOREARM

The main arteries of the forearm are the ulnar and radial arteries. The brachial artery

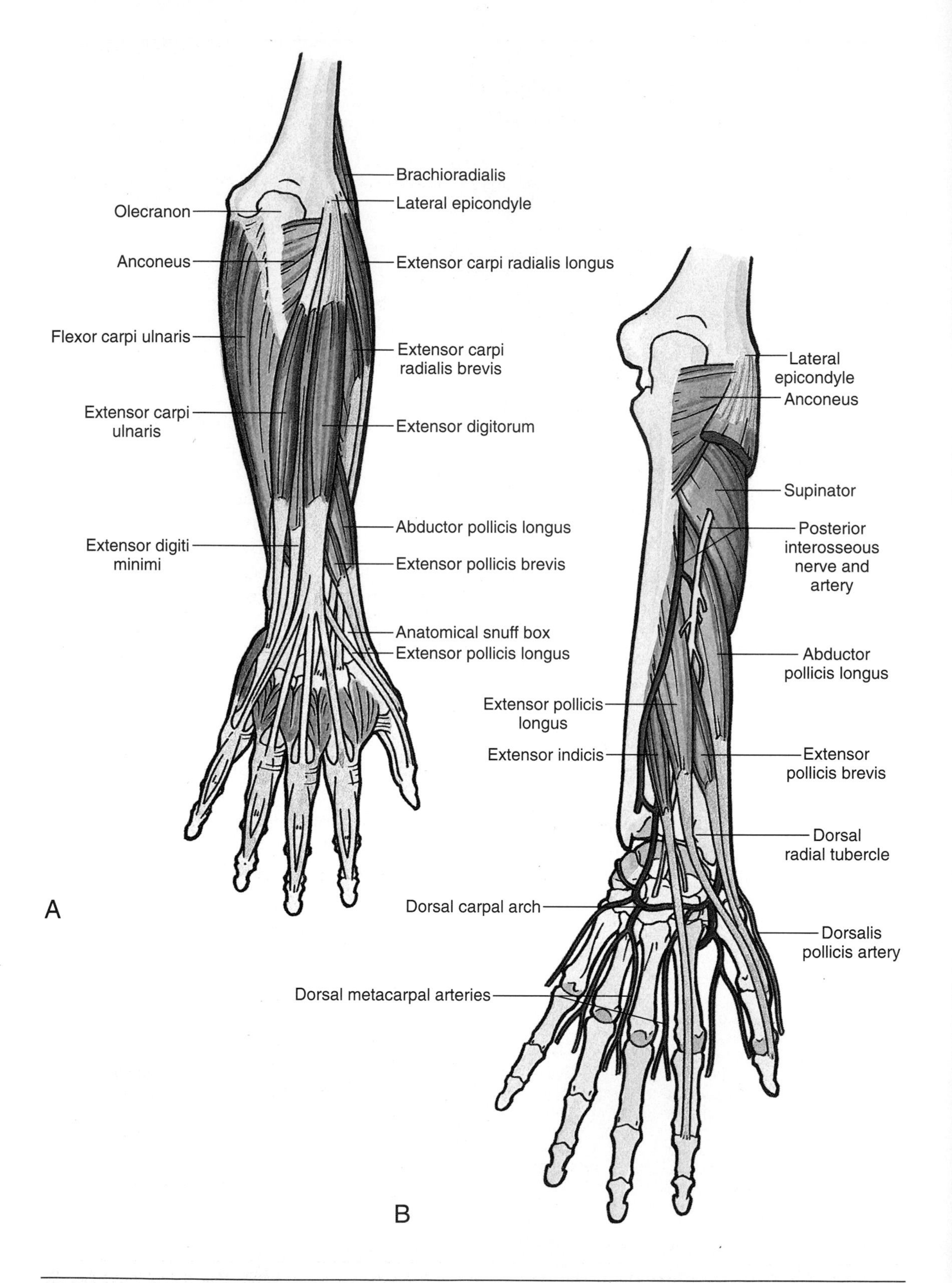

Figure 7.15. **A.** Superficial dissection of extensor muscles of forearm. **B.** Deeper dissection of supinator and outcropping muscles showing arteries and nerves.

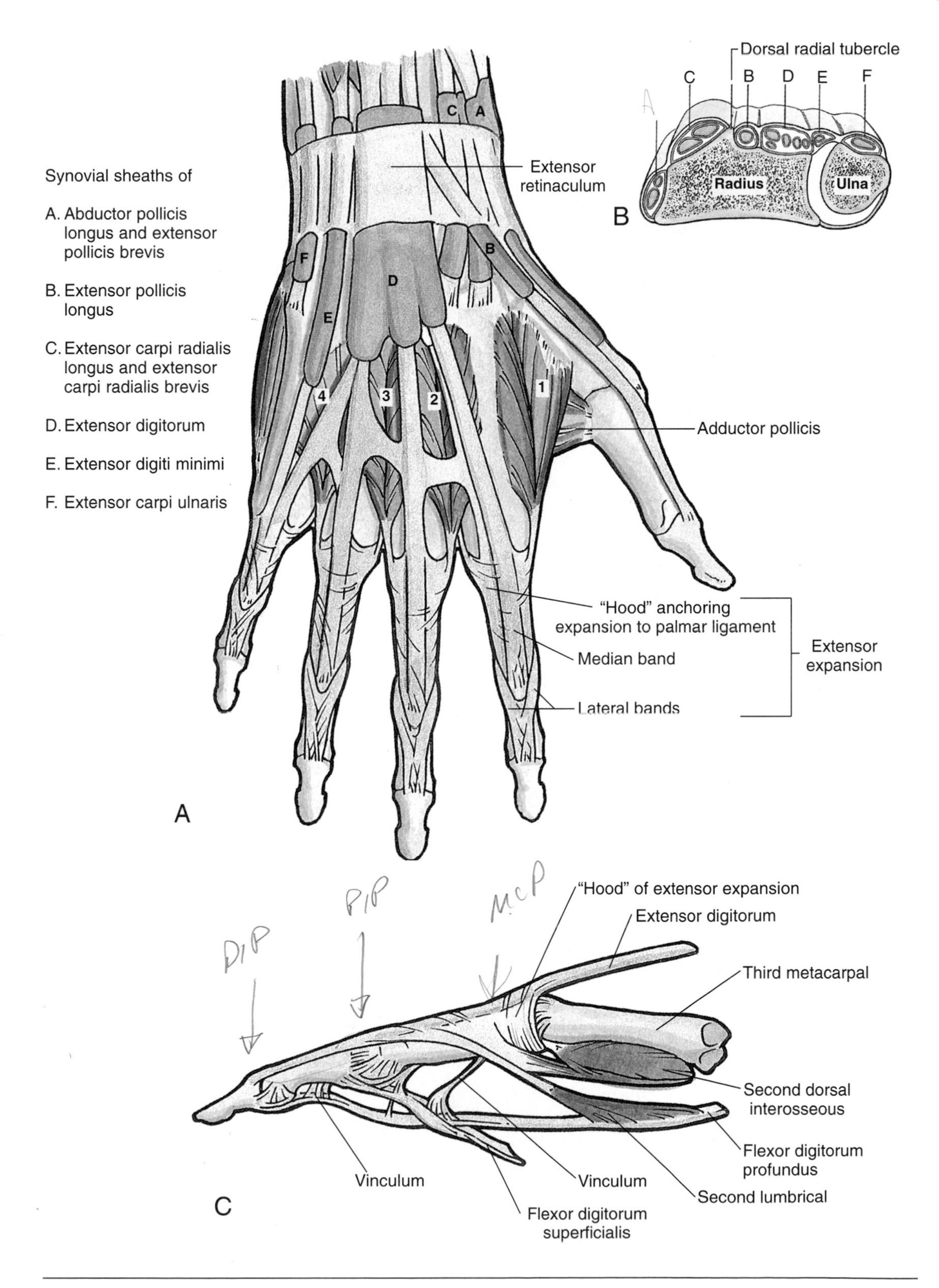

Figure 7.16. A. Synovial sheaths of extensor tendons and digital extensor expansions; *1–4*, dorsal interossei. **B.** Transverse section through distal radius and ulna showing tendons in their synovial sheaths. **C.** Lateral view of extensor expansion.

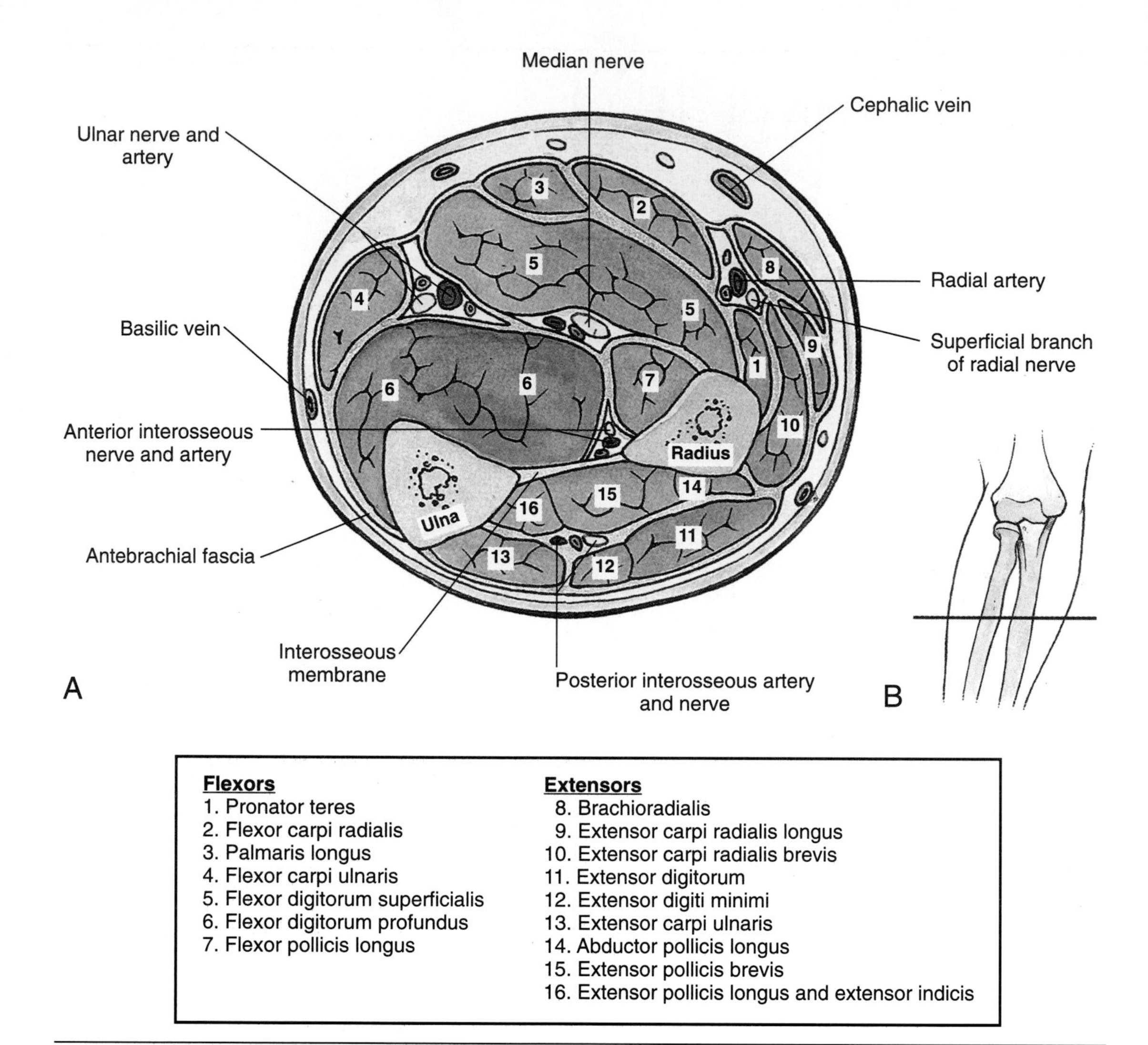

Figure 7.17. **A.** Transverse section through midforearm. **B.** Orientation drawing.

ends opposite the neck of the radius in the inferior part of the cubital fossa. Here it divides into its two terminal branches, the ulnar and radial arteries. Each of these arteries has several branches (Fig. 7.14, Table 7.8). The *ulnar artery,* the larger of the two terminal branches of the brachial artery, begins in the cubital fossa and descends through the anterior compartment of the forearm and enters the palm of the hand. The *radial artery,* the smaller of the two terminal branches of the brachial artery, also begins in the cubital fossa and passes inferolaterally deep to the brachioradialis. In the distal part of the forearm, the radial artery lies on the anterior surface of the radius and is covered only by skin and fascia. The radial artery leaves the forearm by winding around the lateral aspect of the wrist to reach the hand. In the hand the ulnar and radial arteries form anastomotic connections known as the superficial and deep palmar arches.

Wrist and Hand

Because of the importance of manual dexterity in occupational and recreational activities, a good understanding of the structure and function of the hand is essential for all who are involved in maintaining or restoring its activities: free motion, power grip, precision handling, and pinching.

Table 7.6
Muscles on Posterior Surface of Forearm

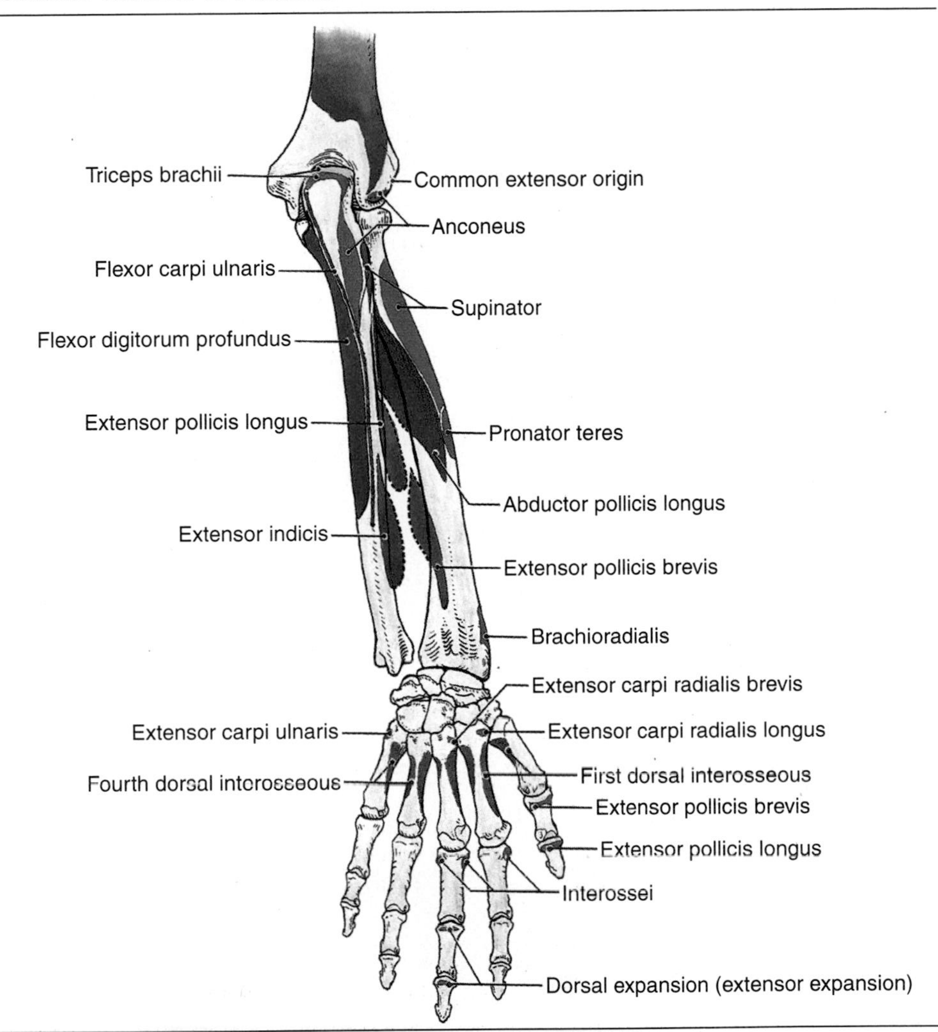

Muscle	Proximal Attachment	Distal Attachment	Innervation[a]	Main Actions
Brachioradialis	Proximal two-thirds of lateral supracondylar ridge of humerus	Lateral surface of distal end of radius	Radial n. (C5, **C6**, and C7)	Flexes forearm
Extensor carpi radialis longus	Lateral epicondyle of humerus	Base of second metacarpal bone	Radial n. (C6 and C7)	Extend and abduct hand at wrist joint
Extensor carpi radialis brevis		Base of third metacarpal bone	Deep branch of radial n. (**C7** and C8)	
Extensor digitorum		Extensor expansions of medial four digits	Posterior interosseous n. (**C7** and C8), a branch of the radial n.	Extends medial four digits at metacarpophalangeal joints; extends hand at wrist joint
Extensor digiti minimi		Extensor expansion of fifth digit		Extends fifth digit at metacarpophalangeal and interphalangeal joints
Extensor carpi ulnaris	Lateral epicondyle of humerus and posterior border of ulna	Base of fifth metacarpal bone		Extends and adducts hand at wrist joint

Table 7.6. *Continued*

Anconeus	Lateral epicondyle humerus	Lateral surface of olecranon and superior part of posterior surface of ulna	Radial n. (C7, C8, and T1)	Assists triceps in extending elbow joint; stabilizes elbow joint; abducts ulna during pronation
Supinator	Lateral epicondyle of humerus, radial collateral and anular ligaments, supinator fossa, and crest of ulna	Lateral, posterior, and anterior surfaces of proximal third of radius	Deep branch of radial n. (C5 and **C6**)	Supinates forearm, i.e., rotates radius to turn palm anteriorly
Abductor pollicis longus	Posterior surfaces of ulna, radius, and interosseous membrane	Base of first metacarpal bone	Posterior interosseous n. (C7 and **C8**)	Abducts thumb and extends it at carpometacarpal joint
Extensor pollicis brevis	Posterior surface of radius and interosseous membrane	Base of proximal phalanx of thumb		Extends proximal phalanx of thumb at carpometacarpal joint
Extensor pollicis longus	Posterior surface of middle third of ulna and interosseous membrane	Base of distal phalanx of thumb		Extends distal phalanx of thumb at metacarpophalangeal and interphalangeal joints
Extensor indicis	Posterior surface of ulna and interosseous membrane	Extensor expansion of second digit		Extends second digit and helps to extend hand

[a] See Table 7.1 for explanation of nomenclature.

Surface Anatomy of Wrist and Hand

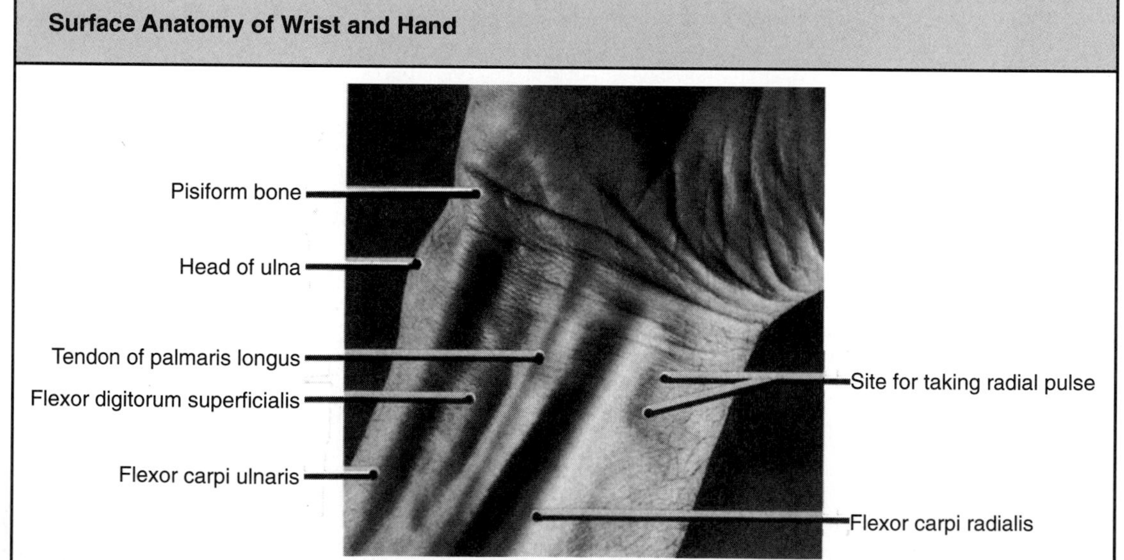

A common place for measuring the pulse rate is where the radial artery lies on the anterior surface of the distal end of the radius, lateral to the tendon of the flexor carpi radialis. Here the artery can be compressed against the distal end of the radius, where it lies between the tendons of the flexor carpi radialis and abductor pollicis longus. The *radial pulse*, like other palpable pulses, is a peripheral reflection of cardiac action.

The tendons of flexor carpi radialis and palmaris longus can be observed by flexing the closed fist against resistance. The median nerve lies deep to the palmaris longus tendon. The flexor carpi ulnaris tendon can also be traced to the pisiform. The ulnar artery and nerve pass into the palm lateral to the pisiform. The *ulnar pulse* is usually difficult to palpate. The tendons of the FDS can be palpated as the digits are alternately flexed and extended.

Surface Anatomy of Wrist and Hand

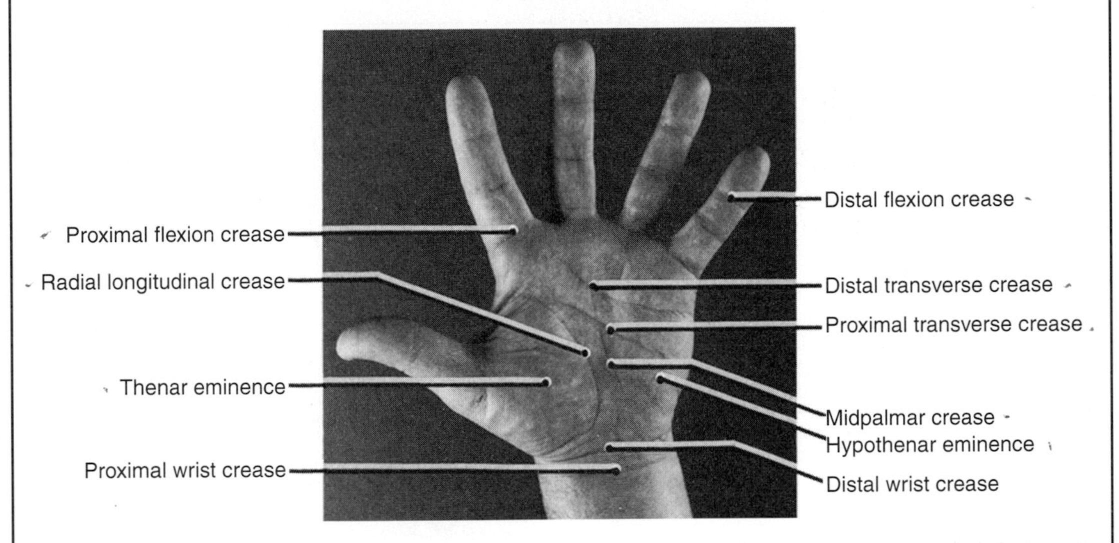

The skin on the palm is thick and richly supplied with sweat glands, but it contains no hair or sebaceous glands. The skin has longitudinal and transverse *flexion creases* where the skin is firmly bound to the deep fascia. The transverse creases indicate where folding of the skin occurs during flexion of the hand. The longitudinal creases deepen when the thumb is opposed (p. 326). The distal wrist crease indicates the proximal border of the flexor retinaculum.

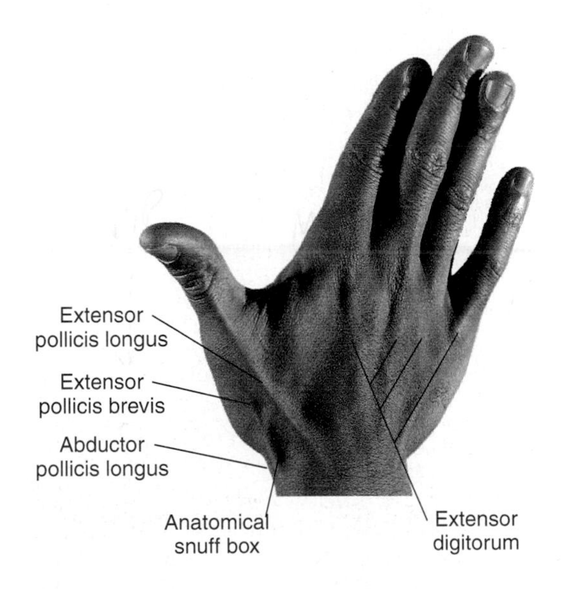

If the dorsum of the hand is examined with the wrist and digits extended against resistance, the extensor tendons stand out. The tendons of the extensor digitorum are not visible far beyond the knuckles because they flatten here to form the extensor expansions of the digits. Deep to the tendons, the metacarpals can be palpated. The tendons of the abductor pollicis longus and extensor pollicis brevis muscles indicate the anterior boundary of the *anatomical snuff box*, and the tendon of the extensor pollicis longus indicates its posterior boundary. The radial artery passes through the snuff box where its pulsations may be felt. The scaphoid and trapezium lie in the floor of the snuff box.

Table 7.7
Nerves of Forearm and Hand

Nerve	Origin	Course	Distribution
Median	By two roots from lateral (C6 and C7) and medial (C8 and T1) cords of brachial plexus	Enters cubital fossa medial to brachial a., passes between heads of pronator teres, descends between flexor digitorum superficialis and flexor digitorum profundus, and passes close to flexor retinaculum as it passes through carpal tunnel to reach hand	
Recurrent branch of median	Arises from median as soon as it passes distal to flexor retinaculum	Loops around distal border of flexor retinaculum to reach thenar muscles	
Lateral branch of median (1)	Lateral division of median n. as it enters palm of hand	Runs laterally supplying first lumbrical and cutaneous branches to anterior surface of thumb and lateral side of index finger	
Medial branch of median (1)	Medial division of the median n. as it enters the palm of the hand	Runs medially to second lumbrical muscle and sends cutaneous branches to adjacent sides of index and middle fingers and adjacent sides of middle and ring fingers	
Palmar cutaneous branch of median (2)	Median n. just proximal to flexor retinaculum	Passes between tendons of palmaris longis and flexor carpi radialis and runs superficial to flexor retinaculum	
Anterior interosseous	Median n. in distal part of cubital fossa	Passes inferiorly on interosseous membrane to supply flexor digitorum profundus, flexor pollicis longus, and pronator quadratus	
Ulnar (3)	Medial cord of brachial plexus (C8 and T1), but it often receives fibers from ventral ramus of C7	Passes posterior to medial epicondyle of humerus and enters forearm between heads of flexor carpi ulnaris; descends through forearm between flexor carpi ulnaris and flexor digitorum profundus; becomes superficial in distal part of forearm and passes superficial to flexor retinaculum	

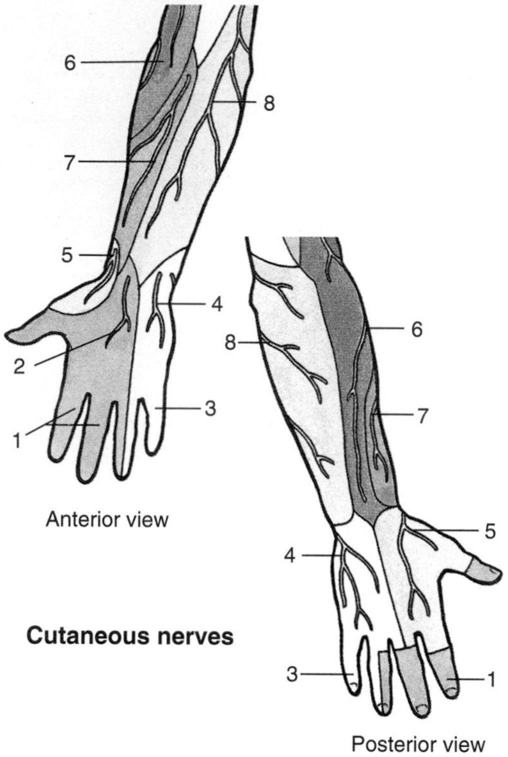

Cutaneous nerves

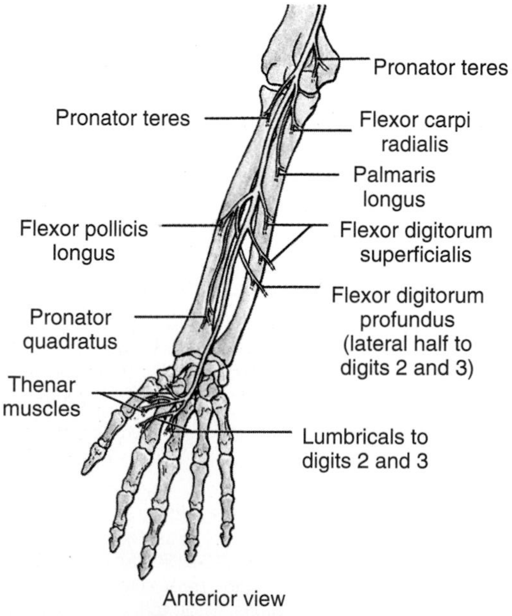

Median nerve

Table 7.7. *Continued*

Superficial branch of ulnar n.	Arises from ulnar n. at wrist as it passes between pisiform and hamate bones	Passes to palmaris brevis and to skin of medial one and one-half digits
Deep branch of ulnar n.	Arises from ulnar n. as described above	Supplies muscles of hypothenar eminence and then curves around inferior edge of hamate to pass deeply in palm to supply muscles shown in diagram
Palmar cutaneous branch of ulnar n. (4)	Ulnar n. near middle of forearm	Descends on ulnar a. and perforates deep fascia in the distal third of forearm
Radial	Posterior cord of brachial plexus (C5–C8 and T1)	Passes into cubital fossa and descends between brachialis and brachioradialis; at level of lateral epicondyle of humerus, it divides into superficial and deep branches
Superficial branch of radial n. (5)	Continuation of radial n. after deep branch is given off	Passes distally, anterior to pronator teres and deep to brachio-radialis; pierces deep fascia at wrist and passes onto dorsum of hand
Deep branch of radial n.	Arises from radial n. just distal to elbow	Winds around neck of radius in supinator; enters posterior compartment to supply muscles shown in diagram
Posterior interosseus	Terminal branch of deep branch of radial m.	Passes deep to extensor pollicis longus and lies on interosseous membrane
Posterior antebrachial cutaneous n. (6) -	Arises in arm from radial n.	Perforates lateral head of triceps and descends along lateral side of arm and posterior aspect of forearm to wrist
Lateral antebrachial cutaneous n. (7)	Continuation of musculocu-taneous n.	Descends along lateral border of forearm to wrist
Medial antebrachial cutaneous n. (8)	Medial cord of brachial plexus, receiving fibers from C8 and T1	Runs down arm on medial side of brachial a.; pierces deep fascia in cubital fossa and runs along medial aspect of forearm

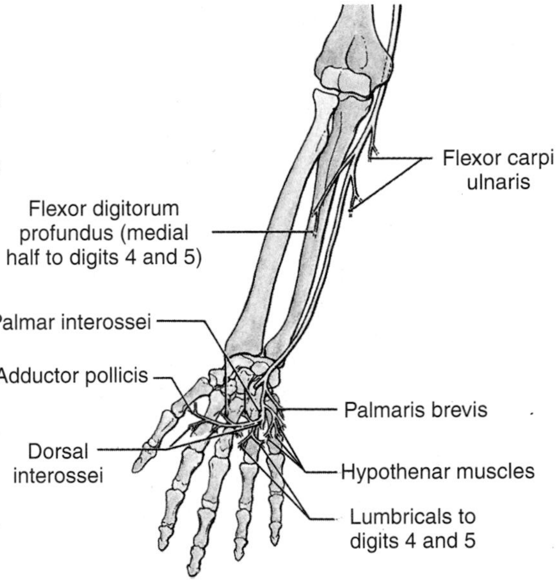

Anterior view

Ulnar nerve

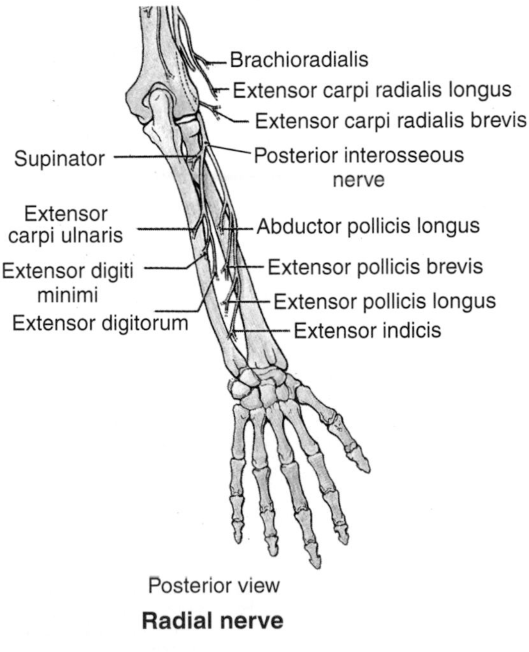

Posterior view

Radial nerve

[a] Numbers in parentheses refer to cutaneous nerves in diagrams.

Table 7.8
Arteries of Forearm

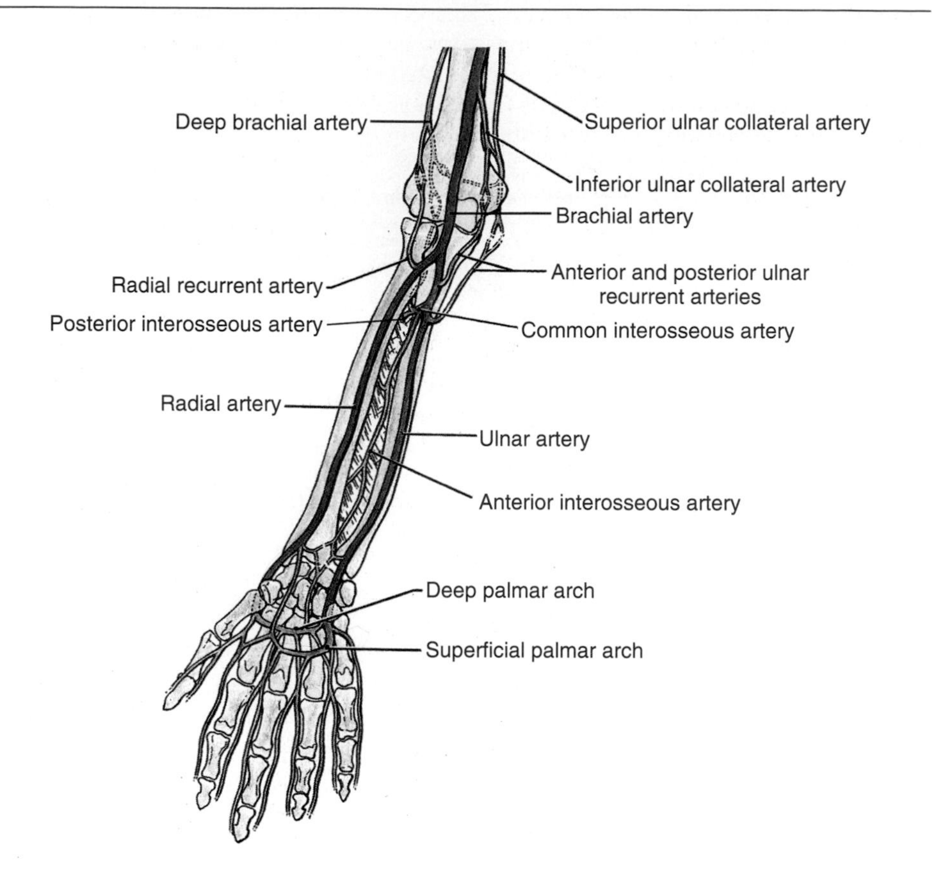

Artery	Origin	Course
Radial	Smaller terminal division of brachial in cubital fossa	Runs inferolaterally under cover of brachioradialis and distally lies lateral to flexor carpi radialis tendon; winds around lateral aspect of radius and crosses floor of anatomical snuff box to pierce fascia; ends by forming deep palmar arch with deep branch of ulnar a.
Ulnar	Larger terminal branch of brachial in cubital fossa	Passes inferomedially and then directly inferiorly, deep to pronator teres, palmaris longus, and flexor digitorum superficialis to reach medial side of forearm; passes superficial to flexor retinaculum at wrist and gives a deep palmar branch to deep arch and continues as superficial palmar arch
Radial recurrent	Lateral side of radial, just distal to its origin	Ascends on supinator and then passes between brachioradialis and brachialis
Anterior and posterior ulnar recurrent	Ulnar, just distal to elbow joint	Anterior ulnar recurrent a. passes superiorly and posterior ulnar collateral a. passes posteriorly to anastomose with ulnar collateral and interosseous recurrent aa.
Common interosseous	Ulnar, just distal to bifurcation of brachial	After a short course, terminates by dividing into anterior and posterior interosseous aa.
Anterior and posterior interosseous	Common interosseous a.	Pass to anterior and posterior sides of interosseous membrane

FASCIA OF PALM

Fascia of the palm is continuous proximally with the antebrachial fascia and with the fascia of the dorsum of the hand. The palmar fascia is thin over the thenar and hypothenar eminences but is thick centrally where it forms the fibrous *palmar aponeurosis* and in the digits where it forms the fibrous digital sheaths (Fig. 7.18). The palmar aponeurosis, a strong well-defined part of the deep fascia of the hand, covers the soft tissues and overlies the long flexor tendons. The proximal end of the palmar aponeurosis is continuous with the flexor retinaculum and the palmaris longus tendon. The distal end of the aponeurosis divides into four longitudinal digital bands that attach to the bases of the proximal phalanges and fuse with the fibrous digital sheaths.

A fibrous medial septum extends deeply from the medial border of the palmar aponeurosis to the fifth metacarpal. Medial to this septum is the medial or *hypothenar compartment* containing the hypothenar muscles (*green*). Similarly, a fibrous lateral septum extends deeply from the lateral border of the palmar aponeurosis to the first metacarpal. Lateral to this septum is the lateral or *thenar compartment* containing the thenar muscles (*blue*). Between the hypothenar and thenar compartments is the *central compartment* (*orange*) containing the flexor tendons and their sheaths, the lumbricals, and the digital vessels and nerves. The deepest muscular plane of the palm is the *adductor compartment* containing the adductor pollicis.

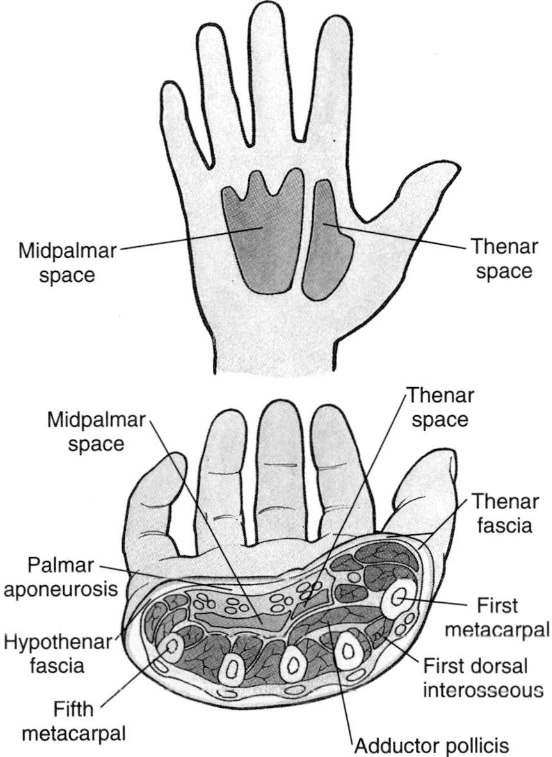

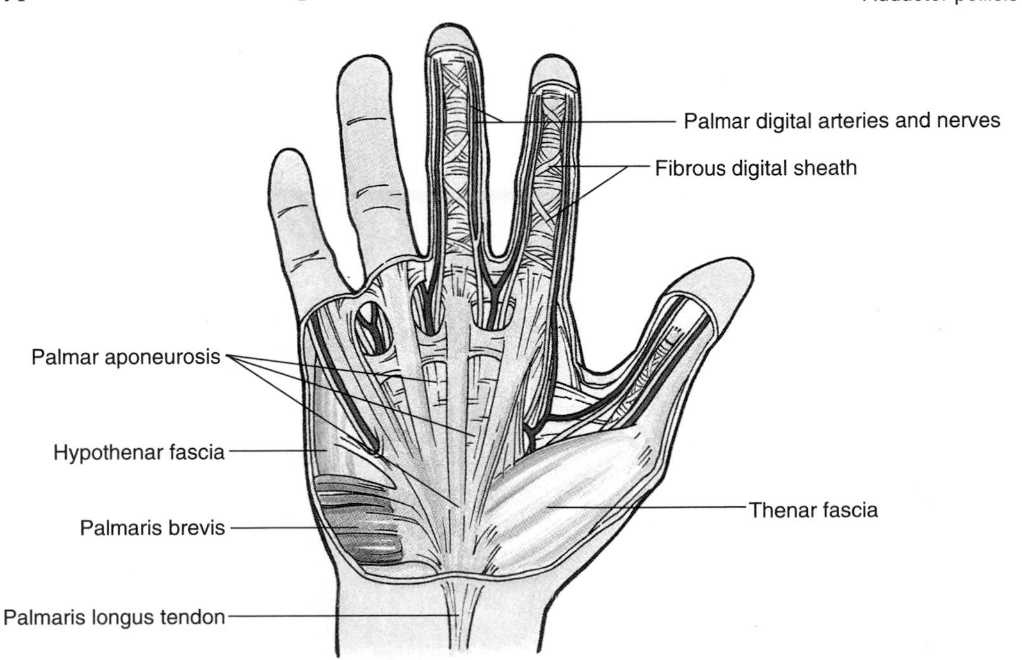

Figure 7.18. Dissection of hand showing deep fascia, fibrous digital sheaths, and digital nerves and vessels.

Between the flexor tendons and the fascia covering the deep palmar muscles are two potential spaces, the thenar and midpalmar spaces. The spaces are bounded by fibrous septa passing from the edges of the palmar aponeurosis to the metacarpal bones. Between the two spaces there is an especially strong fibrous septum that is attached to the third metacarpal.

The science of studying dermal ridge patterns of the palm—*dermatoglyphics*—is a valuable extension of the conventional physical examination of patients with certain congenital anomalies and genetic diseases. For example, persons with trisomy 21 (Down syndrome) often have only one transverse palmar crease (*simian crease*); however, about 1% of the general population has this crease with no other clinical features of the syndrome.

Dupuytren's contracture, a common hand abnormality, is a progressive fibrosis (increase in fibrous tissue) of the palmar aponeurosis, resulting in shortening and thickening of the digital bands. These shortened bands pull the digits (especially the fourth and fifth) into varying degrees of fixed flexion. The digits cannot straighten without corrective surgery on the palmar aponeurosis.

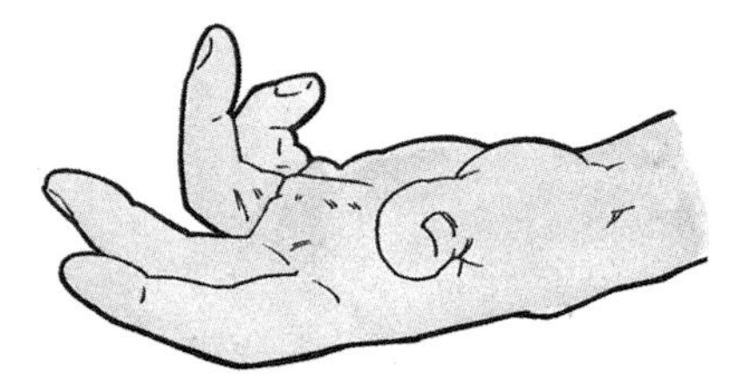

Dupuytren's contracture

The potential fascial spaces of the palm may become infected (e.g., after a puncture wound or because of secondary involvement from a long-neglected tendon sheath infection). These spaces determine the extent and direction of the spread of pus resulting from these infections. Because of the widespread use of antibiotics, infection rarely spreads from the fascial compartments of the hand, but an untreated infection can spread proximally from them through the carpal tunnel into the forearm.

MUSCLES OF HAND

The intrinsic muscles of the hand are in four groups (Figs. 7.19 and 7.20, Table 7.9):

- Thenar muscles in thenar compartment
- Adductor pollicis in adductor compartment
- Hypothenar muscles in hypothenar compartment
- Short muscles of the hand (lumbricals in the central compartment and the interossei between the metacarpals)

The *thenar muscles* (abductor pollicis brevis, flexor pollicis brevis, and opponens pollicis) are chiefly responsible for *opposition of the thumb.* This complex movement involves extension initially and then abduction, flexion, medial rotation, and usually adduction. The reinforcing action of the *adductor pollicis* and flexor pollicis longus increases the pressure that the opposed thumb can exert on the fingertips.

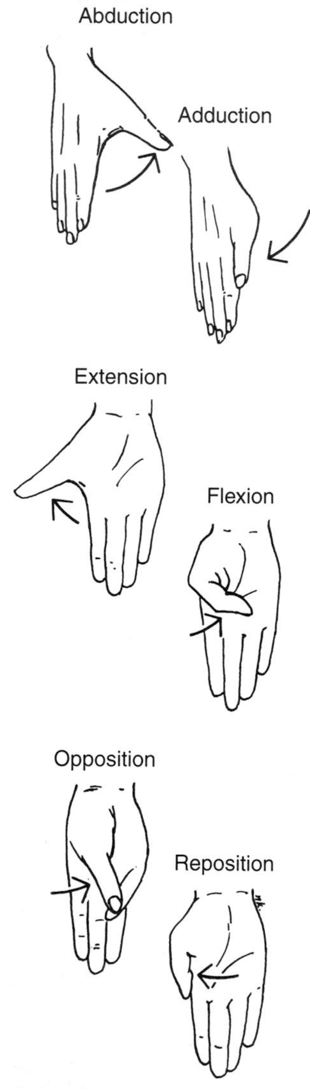

Movements of thumb

The deeply placed adductor pollicis muscle has two heads of origin that are separated by the radial artery as it enters the palm to form the deep palmar arch (Fig. 7.20).

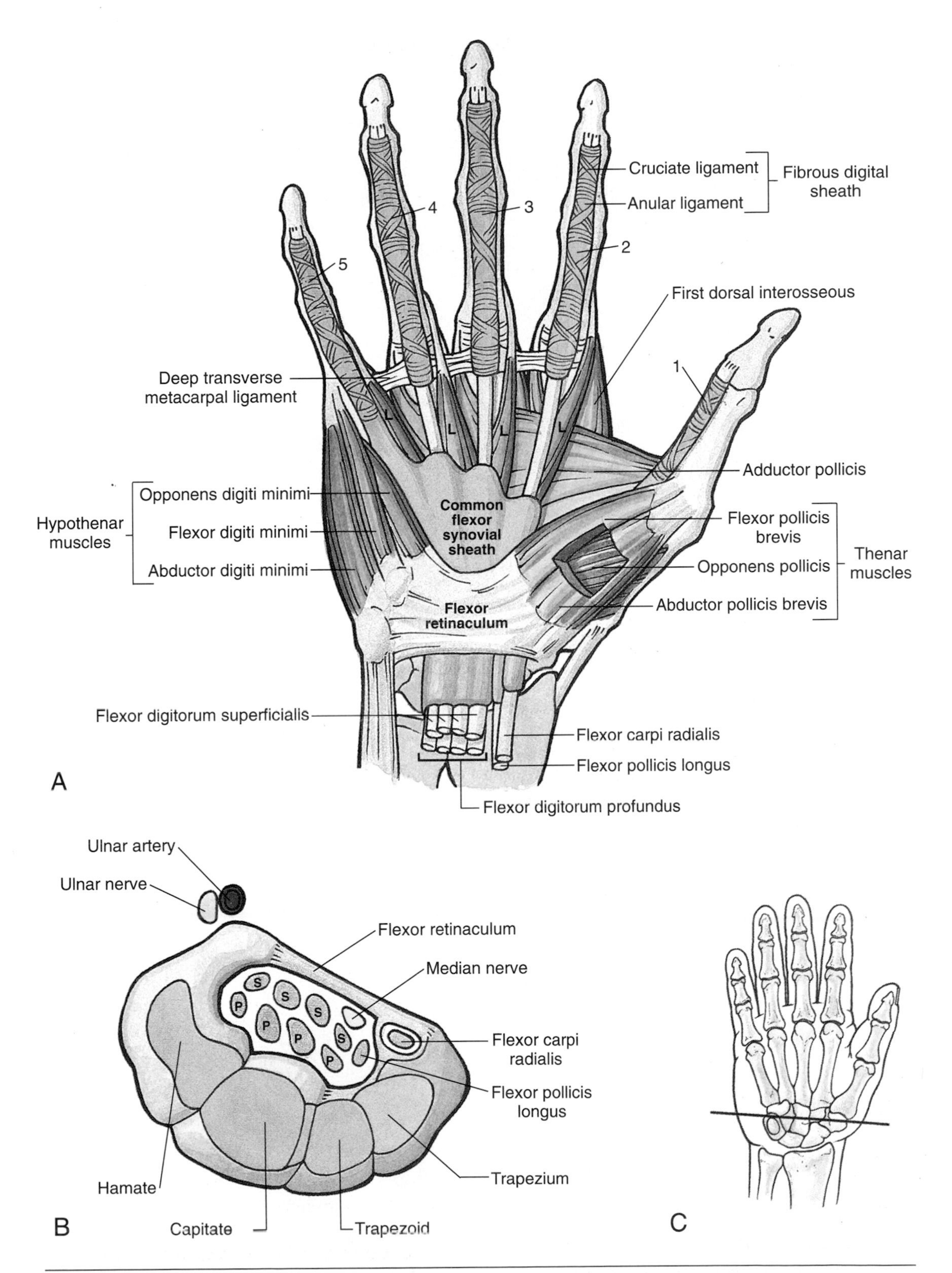

Figure 7.19. **A.** Dissection of hand showing digital synovial sheaths 1–5 (*blue*) of long flexor tendons and muscles. *L*, lumbricals. **B.** Transverse section of wrist showing carpal tunnel and its contents. *P*, flexor digitorum profundus; *S*, flexor digitorum superficialis. **C.** Orientation drawing for **B.**

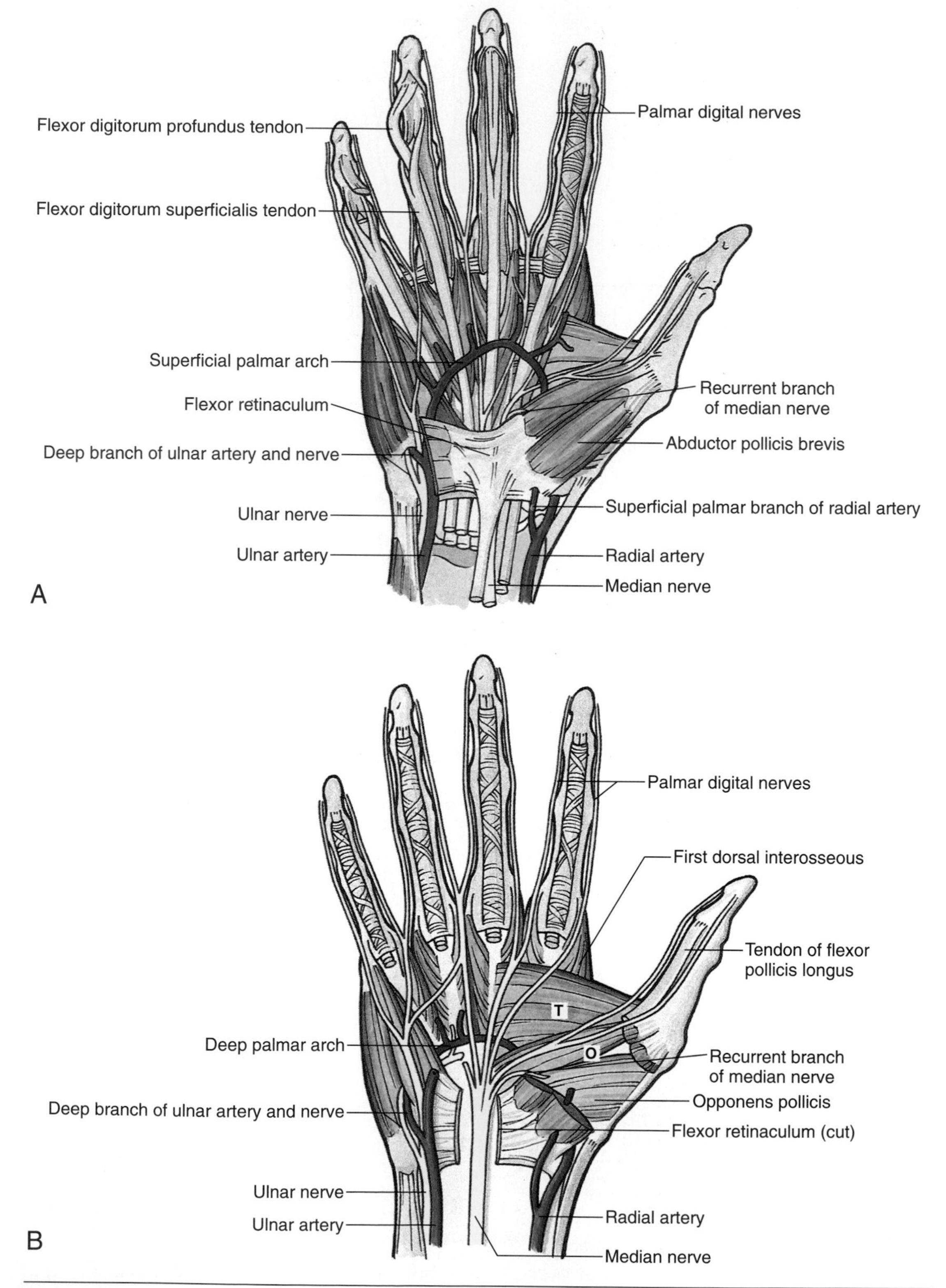

Figure 7.20. **A.** Superficial dissection of palm of hand, showing superficial palmar arch and distribution of median and ulnar nerves. **B.** Deeper dissection showing deep palmar arch and deep branch of ulnar nerve. *T*, transverse head of adductor pollicis; *O*, oblique head of adductor pollicis.

Table 7.9
Muscles of Hand

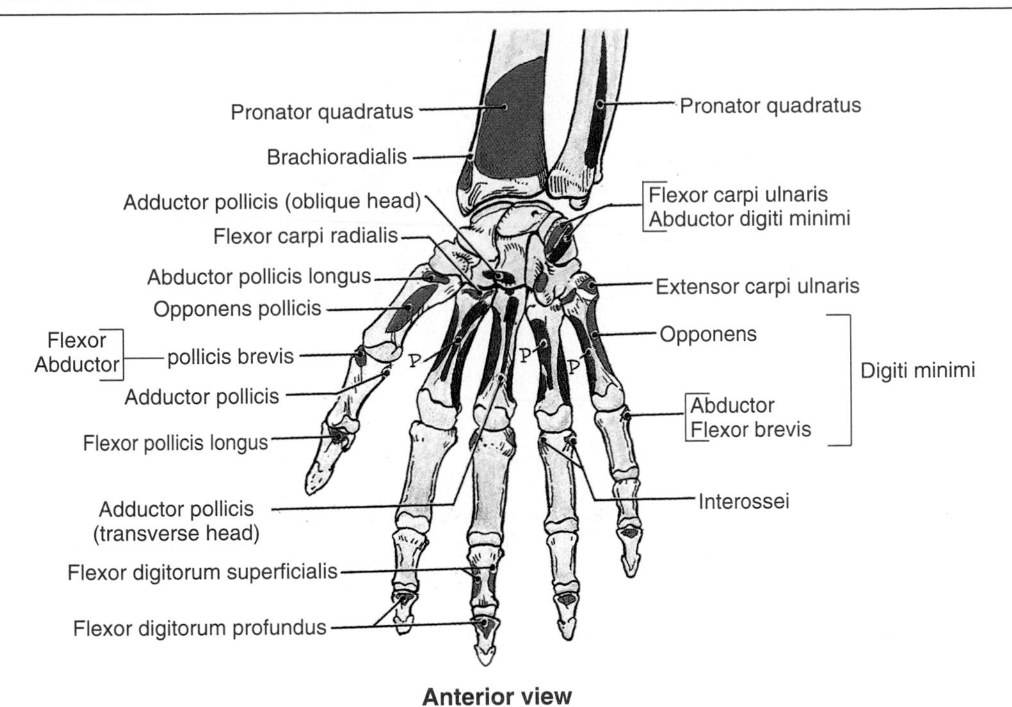

Anterior view

Brachioradialis
Extensor carpi radialis brevis
Extensor carpi ulnaris
Extensor carpi radialis longus
Fourth dorsal interosseous
First dorsal interosseous
Extensor pollicis brevis
Extensor pollicis longus
Interossei
Dorsal expansion
(extensor expansion)

Posterior view

P, proximal attachments of palmar interossei.

Muscle	Proximal Attachment	Distal Attachment	Innervation[a]	Main Actions
Abductor pollicis brevis	Flexor retinaculum and tubercles of scaphoid and trapezium bones	Lateral side of base of proximal phalanx of thumb	Recurrent branch of median n. (**C8** and T1)	Abducts thumb and helps oppose it
Flexor pollicis brevis	Flexor retinaculum and tubercle of trapezium bone	Lateral side of base of proximal phalanx of thumb	Recurrent branch of median n. (**C8** and T1)	Flexes thumb
Opponens pollicis	Flexor retinaculum and tubercle of trapezium bone	Lateral side of first metacarpal bone	Recurrent branch of median n. (**C8** and T1)	Opposes thumb toward center of palm and rotates it medially

Table 7.9. *Continued*

Adductor pollicis	*Oblique head:* bases of second and third metacarpals, capitate, and adjacent carpal bones *Transverse head:* anterior surface of body of third metacarpal bone	Medial side of base of proximal phalanx of thumb	Deep branch of ulnar n. (C8 and **T1**)	Adducts thumb toward middle digit
Abductor digiti minimi	Pisiform bone	Medial side of base of proximal phalanx of digit 5	Deep branch of ulnar n. (C8 and **T1**)	Abducts digit 5
Flexor digiti minimi brevis	Hook of hamate bone and flexor retinaculum	Medial side of base of proximal phalanx of digit 5	Deep branch of ulnar n. (C8 and **T1**)	Flexes proximal phalanx of digit 5
Opponens digiti minimi	Hook of hamate bone and flexor retinaculum	Medial border of fifth metacarpal bone	Deep branch of ulnar n. (C8 and **T1**)	Draws fifth metacarpal bone anteriorly and rotates it, bringing digit 5 into opposition with thumb
Lumbricals 1 and 2	Lateral two tendons of flexor digitorum profundus	Lateral sides of extensor expansions of digits 2–5	*Lumbricals 1 and 2:* median n. (C8 and **T1**)	Flex digits at metacarpophalangeal joints and extend interphalangeal joints
Lumbricals 3 and 4	Medial three tendons of flexor digitorum profundus	Lateral sides of extensor expansions of digits 2–5	*Lumbricals 3 and 4:* deep branch of ulnar n. (C8 and **T1**)	Flex digits at metacarpophalangeal joints and extend interphalangeal joints
Dorsal interossei 1–4	Adjacent sides of two metacarpal bones	Extensor expansions and bases of proximal phalanges of digits 2–4	Deep branch of ulnar n. (C8 and **T1**)	Adduct digits and assist lumbricals
Palmar interossei 1–3	Palmar surfaces of second, fourth, and fifth metacarpal bones	Extensor expansions of digits and bases of proximal phalanges of digits 2, 4, and 5	Deep branch of ulnar n. (C8 and **T1**)	Adduct digits and assist lumbricals

[a] See Table 7.1 for explanation of nomenclature.

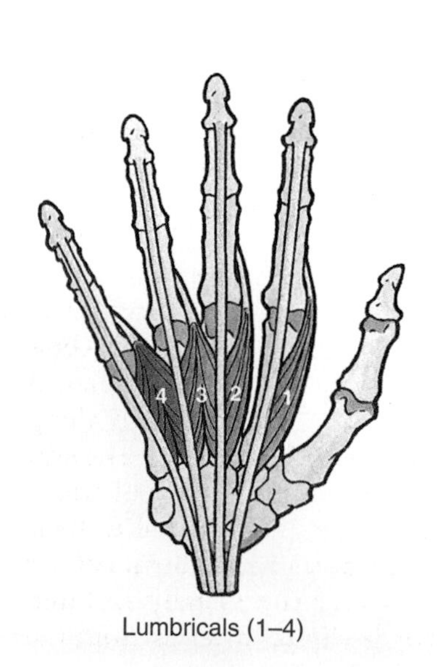

Lumbricals (1–4)

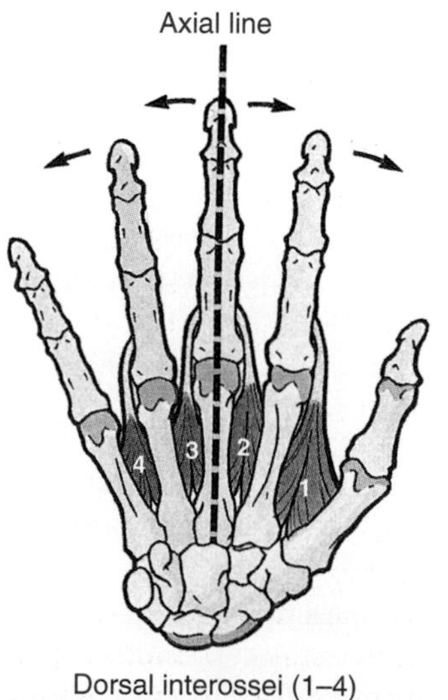

Dorsal interossei (1–4)

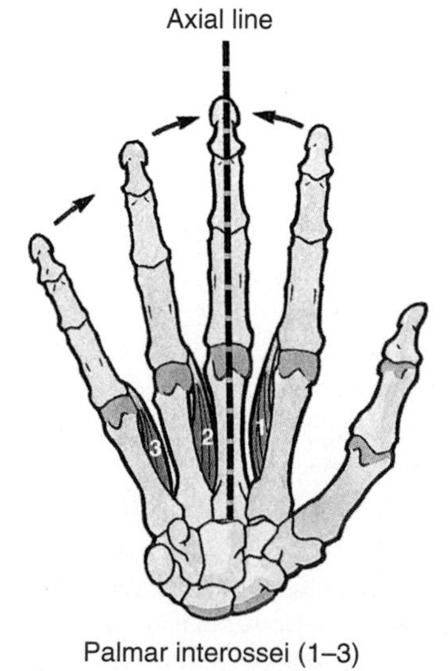

Palmar interossei (1–3)

Lumbricals and interossei

The distal attachment of the heads of adductor pollicis is the medial aspect of the base of the proximal phalanx. The tendon of insertion contains a sesamoid bone.

The *hypothenar muscles* (abductor digiti minimi, flexor digiti minimi brevis, and opponens digiti minimi) produce the *hypothenar eminence* and move the fifth digit.

The *short muscles of the hand* include the lumbricals and interossei. The *lumbricals* act on the medial four digits; the interossei act on all five digits (Table 7.9). The *interossei*, between the metacarpals, are arranged in two layers: three palmar and four dorsal muscles. The dorsal interossei abduct the digits, and the palmar interossei adduct them. A mnemonic device is to make an acronym of Dorsal ABduct (DAB) and Palmar ADuct (PAD).

FLEXOR TENDONS OF EXTRINSIC HAND MUSCLES

Tendons of the FDS and FDP enter the *common flexor synovial sheath* deep to the flexor retinaculum (Fig. 7.19). They enter the central compartment of the hand and then fan out to enter their respective *digital synovial sheaths*. These sheaths enable the tendons to slide freely over each other during movements of the digits. Near the base of the proximal phalanx, the tendon of the FDS splits and surrounds the tendon of the FDP. The halves of the FDS tendon are attached to the margins of the middle phalanx. The tendon of the FDP, after passing through the split in the FDS tendon, passes distally to attach to the base of the distal phalanx (Fig. 7.16C).

The *fibrous digital sheaths* are strong coverings of the flexor tendons and their synovial sheaths. The fibrous digital sheaths extend from the heads of the metacarpals to the bases of the distal phalanges and prevent the tendons from pulling away from the digits. The fibrous digital sheaths combine with the bones to form osseofibrous tunnels through which the tendons pass to reach the digits. The fibrous digital sheaths have thick and thin areas; the thick areas form the anular and cruciate ligaments (pulleys).

The long flexor tendons are supplied by small blood vessels that pass within synovial folds (*vincula*) from the periosteum of the phalanges. The tendon of the flexor pollicis longus passes deep to the flexor retinaculum to the thumb within its own synovial sheath. At the head of the metacarpal, the tendon runs between two *sesamoid bones*, one in the combined tendon of the flexor pollicis brevis and abductor pollicis brevis and the other in the tendon of the adductor pollicis.

Injuries such as a puncture from a rusty nail can cause infection of the synovial sheaths. When inflammation of the tendon and synovial sheath occurs (*tenosynovitis*), the digit swells, and movement becomes painful. Because the tendons of the second, third, and fourth digits nearly always have separate synovial sheaths, the infection is usually confined to the infected digit. In neglected infections, however, the proximal ends of these sheaths may rupture, allowing the infection to spread to the midpalmar space. Because the synovial sheaths of the thumb and fifth digit are often continuous with the common flexor synovial sheath, tenosynovitis in these digits may spread to the common sheath. Just how far an infection spreads from the digits depends on variations in their connections with the common flexor synovial sheath.

NERVES OF HAND

The nerves of the hand are the median, ulnar, and radial (Figs. 7.19 and 7.20, Table 7.7). The *median nerve* passes deep to the flexor retinaculum and through the *carpal tunnel*, where it lies superficial to the long flexor tendons. The *ulnar nerve* is bound by fascia to the anterior surface of the flexor retinaculum. The *radial nerve* supplies no hand muscles, but its superficial branch supplies skin mainly on the dorsum of the hand.

ARTERIES OF HAND

The arteries of the hand, the radial and ulnar and their branches, provide all the blood to the hand (Fig. 7.20, Table 7.10). The *superficial palmar arch*, formed mainly by the ulnar artery, gives rise to three common palmar digital arteries that anastomose with the palmar metacarpal arteries from the deep palmar arch. Each common *palmar digital artery* divides into a pair of proper palmar digital arteries that run along the sides of the second to fourth digits. The *deep palmar arch*, formed mainly by the radial artery, lies across the metacarpals just distal to their bases. The deep arch gives rise to three palmar metacarpal arteries that run distally and join the common palmar digital arteries from the superficial palmar arch.

Table 7.10
Arteries of Hand

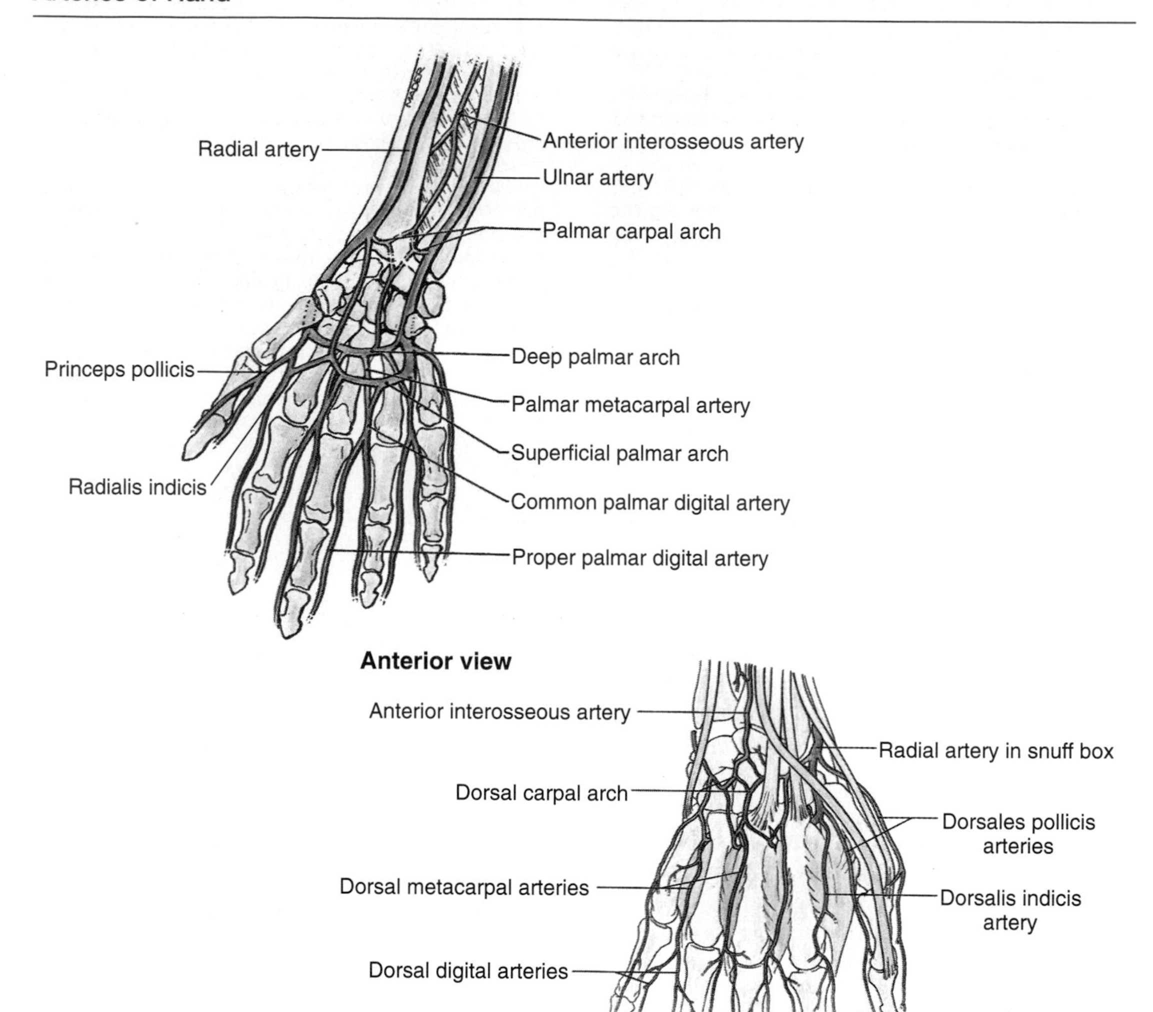

Artery	Origin	Course
Superficial palmar arch	Direct continuation of ulnar a.; arch is completed on lateral side by superficial branch of radial a. or another of its branches	Curves laterally deep to palmar aponeurosis and superficial to long flexor tendons; curve of arch lies across palm at level of distal border of extended thumb
Deep palmar arch	Direct continuation of radial a.; arch is completed on medial side by deep branch of ulnar a.	Curves medially deep to long flexor tendons and is in contact with bases of metacarpals
Common palmar digitals	Superficial palmar arch	Pass distally on lumbricals to webbings of digits
Proper palmar digitals	Common palmar digital aa.	Run along sides of digits 2–5
Princeps pollicis	Radial a. as it turns into palm	Descends on palmar aspect of first metacarpal and divides at the base of proximal phalanx into two branches that run along sides of thumb
Radialis indicis	Radial a. but may arise from princips pollicis a.	Passes along lateral side of index finger to its distal end
Dorsal carpal arch	Radial and ulnar aa.	Arches within fascia on dorsum of hand

Carpal tunnel syndrome results from any lesion (e.g., the inflammation of synovial sheaths) that significantly reduces the size of the carpal tunnel and compresses the median nerve. Because this nerve has two terminal branches that supply skin of the hand, tingling (*paresthesia*), absence of tactile sensation (*anesthesia*), or diminished sensation (*hypoesthesia*) may occur in the lateral three and one-half digits. Progressive loss of coordination and strength in the thumb (because of weakness of abductor pollicis brevis and opponens pollicis) may occur if the cause of the compression is not alleviated. Patients with carpal tunnel syndrome have difficulty performing fine movements of the thumb. To relieve symptoms of the syndrome, partial or complete surgical division of the flexor retinaculum—*carpal tunnel release*—may be necessary.

In attempted suicides by wrist slashing, the median nerve is commonly injured just proximal to the flexor retinaculum. This results in paralysis of the thenar muscles and the first two lumbricals. Hence opposition of the thumb is not possible, and fine control movements of the second and third digits are impaired. There would also be loss of sensation over the thumb and adjacent two and one-half digits. *Median nerve injury* in the elbow region results in loss of flexion of the proximal and distal interphalangeal joints of the second and third digits. The ability to flex the metacarpophalangeal joints of these digits is also affected because digital branches of the median nerve supply the first and second lumbricals. *Pronator syndrome* is caused by entrapment and compression of the median nerve in the elbow region. The nerve may be compressed between the heads of the pronator teres from trauma, muscular hypertrophy, or fibrous bands. Symptoms often follow activities that involve repeated elbow movements. Patients are first seen clinically with pain and tenderness in the proximal aspect of the anterior forearm.

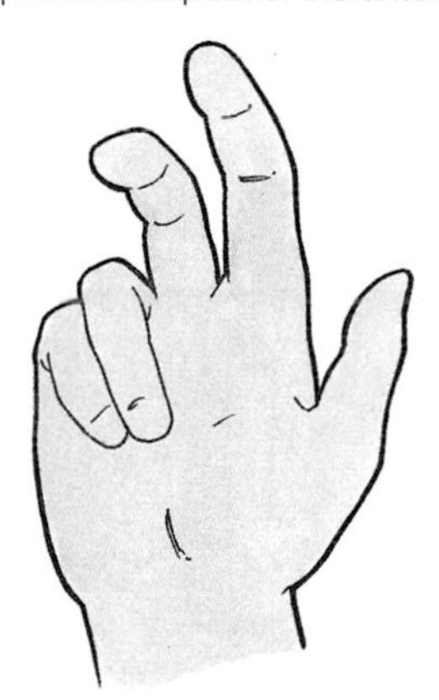

Median nerve injury

The *recurrent branch of the median nerve* that supplies the thenar muscles lies superficially and may be severed by relatively minor lacerations involving the thenar eminence. Severance of this nerve paralyzes the thenar muscles, and the thumb loses much of its usefulness. During carpal tunnel release the incision is made toward the medial side of the wrist to avoid possible injury to the recurrent branch of the median nerve.

Most nerve injuries in the upper limb affect opposition of the thumb. Undoubtedly injuries to the nerves supplying the intrinsic muscles of the hand, especially the median nerve, have the most severe effects on this complex movement. If the median nerve is severed in the forearm or at the wrist, the thumb cannot be opposed; however, the abductor pollicis longus and adductor pollicis (supplied by the posterior interosseous and ulnar nerves, respectively) may imitate opposition.

Ulnar nerve injury commonly occurs where the nerve passes posterior to the medial epicondyle of the humerus. Often the injury occurs when the elbow hits a hard surface and the epicondyle is fractured. Ulnar nerve injury may result in extensive motor and sensory loss to the hand with accompanying impaired power of adduction. Upon flexing the wrist joint, the hand is drawn to the lateral (radial) side by the flexor carpi radialis. After ulnar nerve injury, patients are likely to have difficulty making a fist because of paralysis of most intrinsic hand muscles. In addition they cannot flex their fourth and fifth digits at the distal interphalangeal joints when they try to straighten their fingers. This results in a characteristic appearance known as *clawhand*.

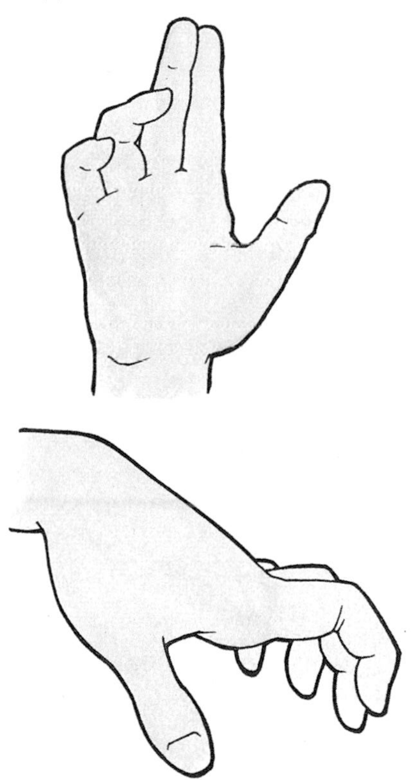

Ulnar nerve injury (clawhand)

Compression of the ulnar nerve may occur at the wrist near where it passes between the hook of the hamate and the pisiform. The depression between these bones is converted by the pisohamate ligament into an osseofibrous tunnel (Guyon's canal, tunnel of Guyon). Compression of the ulnar nerve in this canal may result in hypoesthesia in the medial one and one-half digits and weakness of the intrinsic muscles of the hand.

Although the radial nerve supplies no muscles in the hand, *radial nerve injury* in the arm or forearm can produce serious disability of the hand. The characteristic handicap is paralysis of the extensor muscles of the forearm and, consequently, inability to extend the wrist. The hand is flexed at the wrist and lies flaccid, a condition known as *wrist-drop* (p. 309). The digits are also flexed at the metacarpophalangeal joints. Severance of the deep branch of the radial nerve in deep wounds of the forearm results in an inability to extend the thumb and the metacarpophalangeal joints of the other digits. No loss of sensation occurs because the deep branch of the radial nerve is entirely muscular and articular in distribution. However, severance of the superficial branch of the radial nerve results in loss of sensation on the posterior surface of the forearm, hand, and proximal parts of the lateral three and one-half digits.

Because of the many arteries in the hand, bleeding is usually profuse when the palmar arches are lacerated. In lacerations of the palmar arterial arches, it may be useless to ligate only one forearm artery because these vessels usually have numerous communications in the forearm and hand. To obtain a bloodless surgical operating field in the hand for treating complicated injuries, it may be necessary to compress the brachial artery and its branches proximal to the elbow (e.g., using a pneumatic tourniquet). This procedure prevents blood from reaching the arteries of the forearm and hand through the anastomoses around the elbow.

VEINS OF HAND

The superficial and deep palmar arterial arches are accompanied by superficial and deep *palmar venous arches*, respectively. The dorsal digital veins drain into three dorsal metacarpal veins that unite to form a *dorsal venous network*. Superficial to the metacarpus, this network is prolonged proximally on the lateral side as the *cephalic vein*. On the medial side of the hand, the dorsal venous network is continuous proximally with the *basilic vein* (Fig. 7.5).

Joints of Upper Limb

JOINTS OF PECTORAL GIRDLE

The pectoral (shoulder) girdle involves the sternoclavicular, acromioclavicular, and shoulder joints. The mobility of the scapula is essential for the freedom of movement of the upper limb. When testing the range of motion of the pectoral girdle both scapulothoracic (movement of scapula on thoracic wall) and shoulder (glenohumeral) movements must be considered. When elevating the arm the movement occurs in a 2:1 ratio; for every 3° of elevation about 2° occur at the shoulder joint and 1° at the scapulothoracic articulation [see Fig. 6.31 in Agur (1991) listed under "Suggested Readings"].

STERNOCLAVICULAR JOINT

The sternoclavicular joint is a saddle type of synovial joint divided into two compartments by an articular disc.

Articulation. The sternal end of the clavicle articulates with the manubrium of the sternum and the first costal cartilage.

Articular Capsule. The fibrous capsule surrounds the joint and is attached to the margins of the articular surfaces. The synovial membrane lines the articular capsule and both surfaces of the articular disc.

Ligaments. Anterior and posterior sternoclavicular ligaments reinforce the capsule anteriorly and posteriorly. The costoclavicular ligament anchors the inferior surface of the medial end of the clavicle to the 1st rib and its costal cartilage.

Movements. The sternoclavicular joint moves in many directions: anteriorly, posteriorly, superiorly, and inferiorly.

Blood Supply. The sternoclavicular joint is supplied by internal thoracic and suprascapular arteries (Table 7.2).

Nerve Supply. Branches of the medial supraclavicular nerve and nerve to the subclavius supply the sternoclavicular joint (Table 7.3).

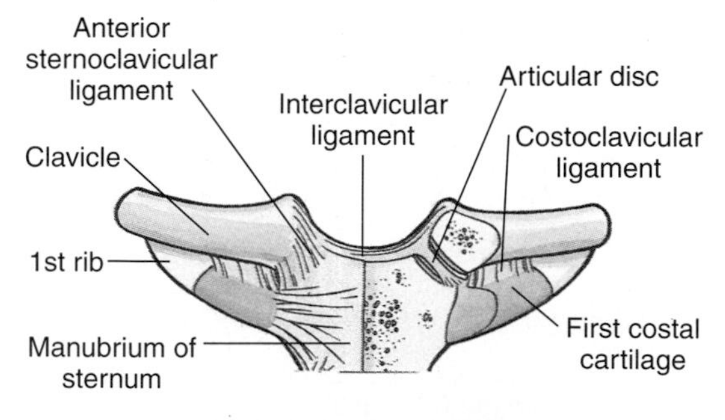

Sternoclavicular joint

The rarity of dislocation of the sternoclavicular joint attests to its strength. When a blow is received to the acromion of the scapula, or when the force of a fall on the outstretched hand transmits to the pectoral girdle, the force of the blow is usually transmitted along the long axis of the clavicle. The clavicle may break near the junction of its middle and lateral thirds, but it is uncommon for the sternoclavicular joint to dislocate.

ACROMIOCLAVICULAR JOINT

The acromioclavicular joint is a plane type of synovial joint.

Articulation. The lateral end of the clavicle articulates with the acromion of the scapula. The articular surfaces are separated by an incomplete articular disc.

Articular Capsule. The fibrous capsule is attached to the margins of the articular surfaces. A synovial membrane lines the fibrous capsule.

Ligaments. The acromioclavicular ligament strengthens the joint superiorly, and the coracoclavicular ligament anchors the clavicle to the coracoid process of the scapula.

Movements. The acromion of the scapula rotates on the clavicle. These movements are associated with scapulothoracic movements.

Blood Supply. The acromioclavicular joint is supplied by the suprascapular and thoracoacromial arteries (Table 7.2).

Nerve Supply. Supraclavicular, lateral pectoral, and axillary nerves supply the acromioclavicular joint (Table 7.3).

In contact sports such as football, soccer, and hockey, it is not uncommon for *dislocation of the acromioclavicular joint* to result from a hard fall on the shoulder or when a hockey player is driven into the boards. This injury, often called a "shoulder separation," is serious when both the acromioclavicular and coracoclavicular ligaments are torn. When the coracoclavicular ligament ruptures, the shoulder falls away from the clavicle because of the weight of the upper limb. If the fibrous capsule of the joint also ruptures, the acromion passes inferior to the acromial end of the clavicle. Dislocation of the acromioclavicular joint makes the acromion more prominent.

SHOULDER JOINT

The shoulder is a ball and socket type of synovial joint.

Articulation. The head of the humerus articulates with the glenoid cavity of the scapula, which is deepened by the fibrocartilaginous glenoid labrum.

Articular Capsule. The fibrous capsule surrounds the joint and is attached medially to the margin of the glenoid cavity of the scapula and laterally to the anatomical neck of the humerus. The synovial membrane that lines the fibrous

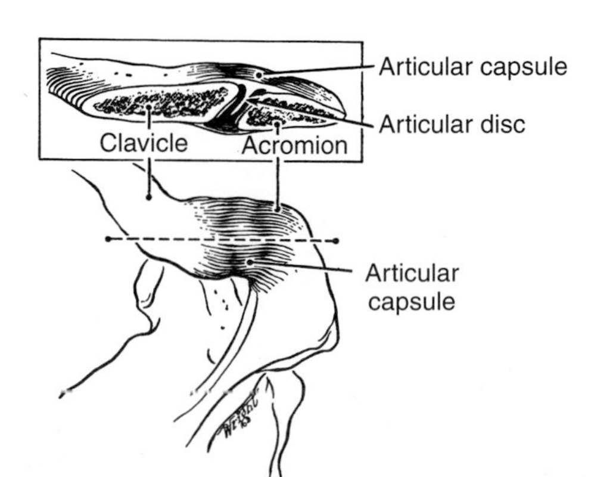

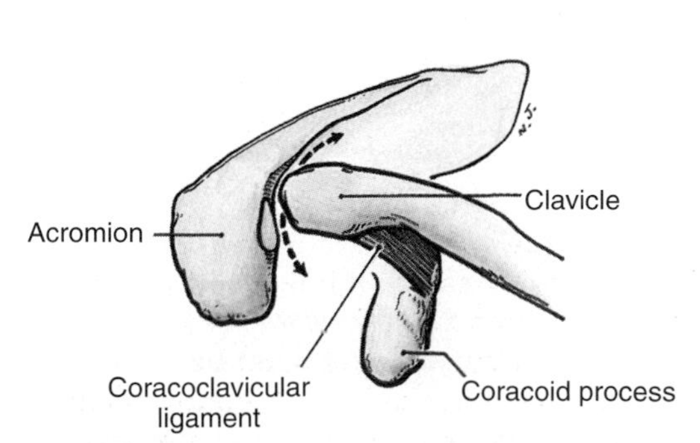

Acromioclavicular joint

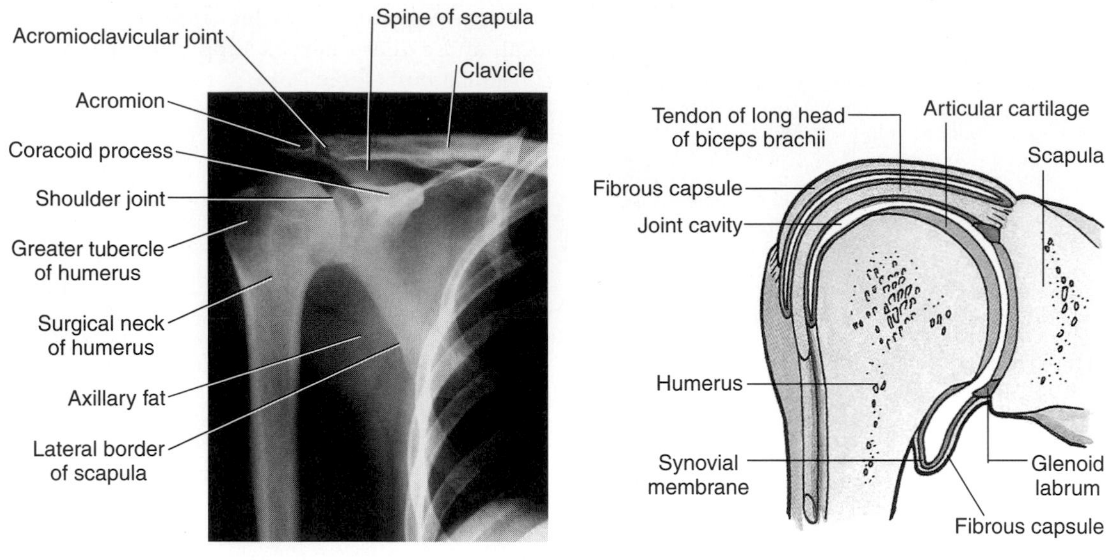

Coronal section

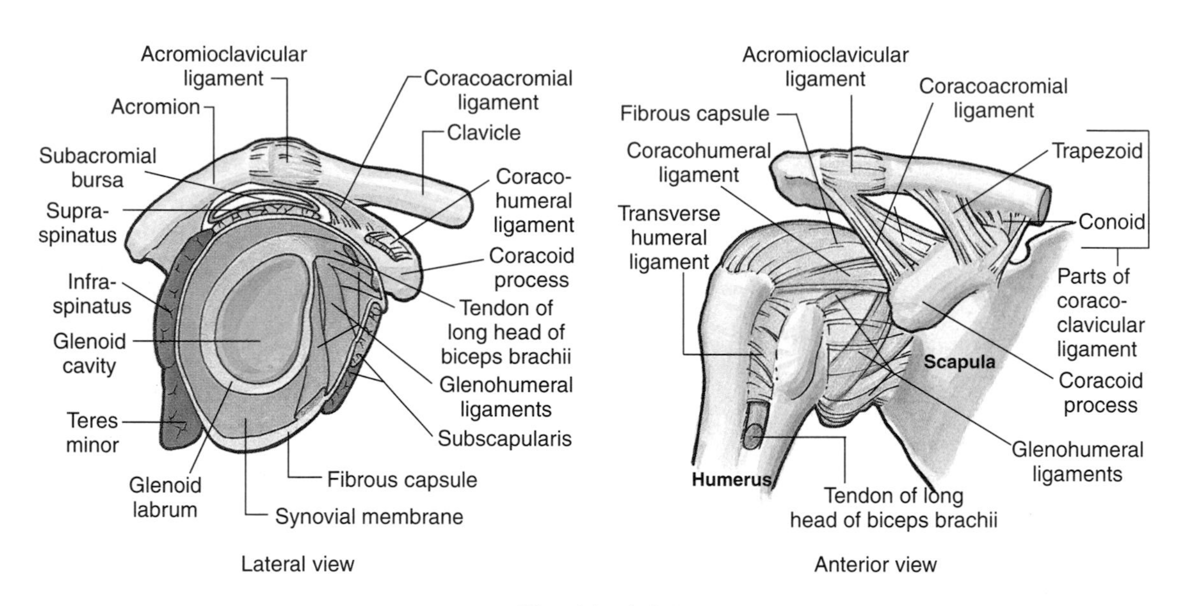

Lateral view

Anterior view

Shoulder joint

capsule forms a sheath for the tendon of the biceps brachii.

Ligaments. The glenohumeral ligaments strengthen the anterior aspect of the capsule, and the coracohumeral ligament strengthens the capsule superiorly. The capsule is also thickened by the transverse humeral ligament.

Movements. The shoulder movements are flexion-extension, abduction-adduction, rotation, and circumduction.

Blood Supply. The shoulder is supplied by the anterior and posterior circumflex humeral and suprascapular arteries (Table 7.2).

Nerve Supply. Suprascapular, axillary, and lateral pectoral nerves supply the shoulder (Table 7.3).

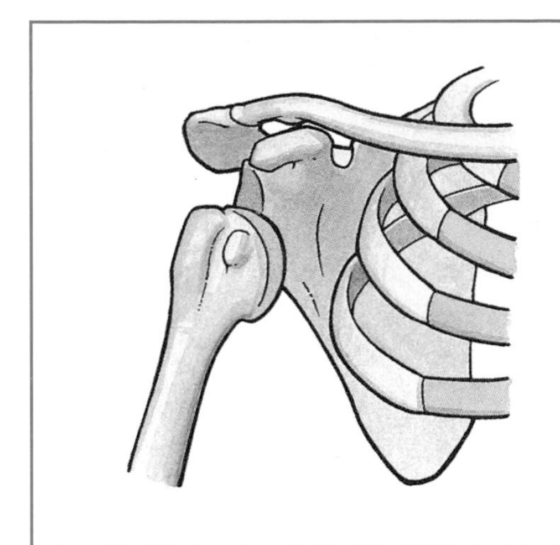

Because of its freedom of movement and instability, the shoulder joint is commonly dislocated from direct or indirect injury. *Anterior dislocation of the shoulder joint* occurs most often in young adults, particularly athletes. It is usually caused by excessive extension and lateral rotation of the humerus. The head of the humerus is driven anteriorly, and the fibrous capsule and glenoid labrum are usually stripped from the anterior aspect of the glenoid cavity. A hard blow to the humerus when the shoulder joint is fully abducted tilts the head of the humerus inferiorly onto the inferior weak part of the articular capsule. This may tear the capsule and dislocate the shoulder so that the humeral head comes to lie inferior to the glenoid cavity. The strong flexor and abductor muscles of the shoulder joint usually pull the humeral head anterosuperiorly into a subcoracoid position. Unable to use the arm, the patient commonly supports it with the other hand.

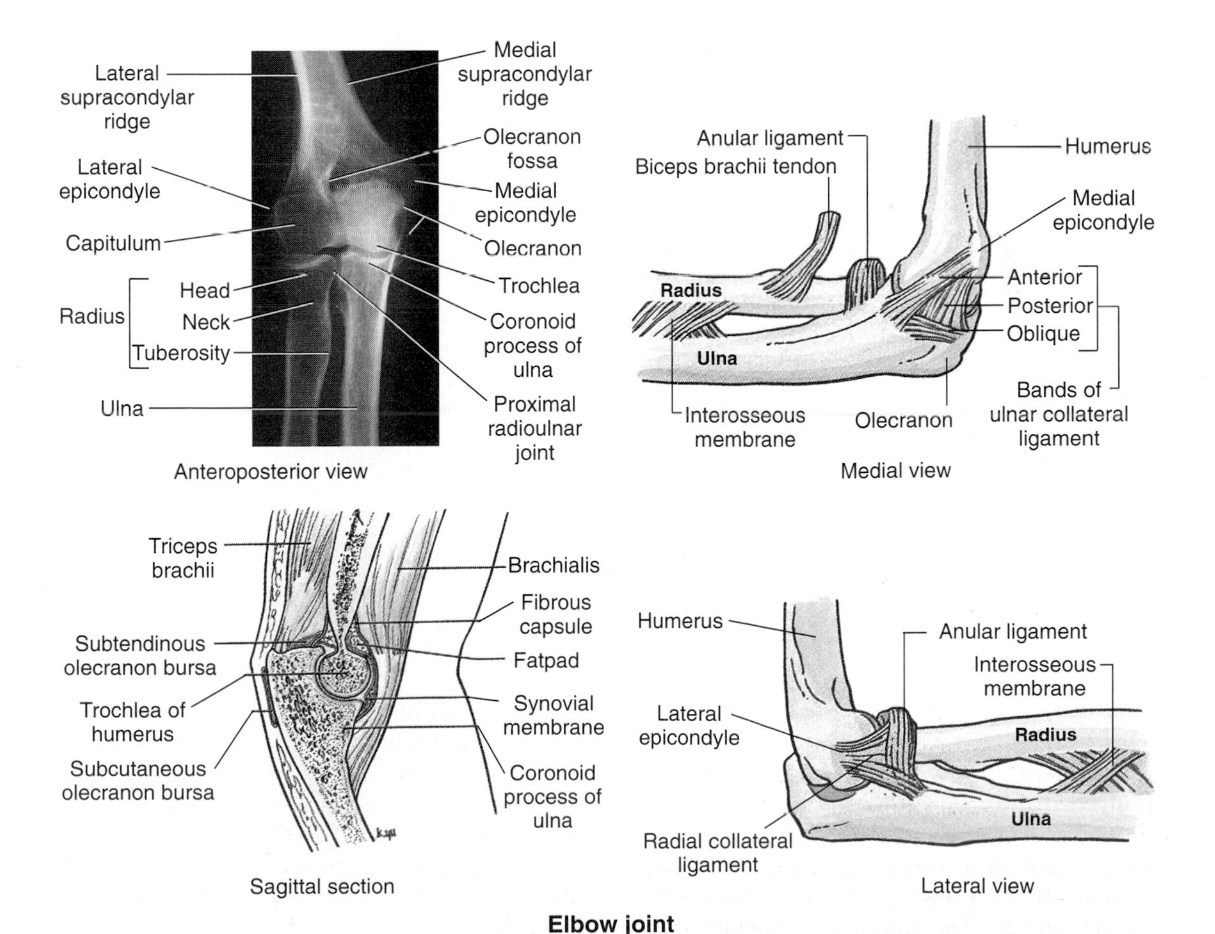

Elbow joint

ELBOW JOINT

The elbow is a hinge type of synovial joint.

Articulation. The trochlea and capitulum of the humerus articulate with the trochlear notch of the ulna and the head of the radius, respectively.

Articular Capsule. The fibrous capsule surrounds the joint. The synovial membrane lines the capsule and is continuous inferiorly with the synovial membrane of the proximal radioulnar joint. The articular capsule is weak anteriorly and posteriorly but is strengthened on each side by ligaments.

Ligaments. The fanlike radial collateral ligament extends from the lateral epicondyle to the anular ligament of the radius. The ulnar collateral ligament passes from the medial epicondyle to the coronoid process and olecranon of the ulna.

Movements. The elbow movements are flexion and extension.

Blood Supply. Arteries are derived from the anastomosis around the elbow joint.

Nerve Supply. The elbow is supplied mainly by the musculocutaneous and radial nerves.

Avulsion of the medial epicondyle in children can result from a fall that causes severe abduction of the extended elbow, an abnormal movement of this articulation. The resulting traction on the ulnar collateral ligament pulls the medial epicondyle distally. The anatomical basis of avulsion of the medial epicondyle is that the epiphysis for the medial epicondyle may not fuse with the distal end of the humerus until up to age 20. Usually fusion is complete radiographically at age 14 in females and age 16 in males. A *traction injury of the ulnar nerve* is a frequent complication of the abduction type of avulsion of the medial epicondyle. The anatomical basis for this stretching of the ulnar nerve is that it passes posterior to the medial epicondyle before entering the forearm.

Posterior dislocation of the elbow joint may occur when children fall on their hands with their elbows flexed. The distal end of the humerus is driven through the weak anterior portion of the fibrous capsule as the radius and ulna dislocate posteriorly.

PROXIMAL RADIOULNAR JOINT

The proximal radioulnar joint is a pivot type of synovial joint.

Articulation. The head of the radius articulates with the radial notch of the ulna.

Articular Capsule. The capsule encloses the joint and is continuous with that of the elbow joint.

Ligaments. The anular ligament, which is attached to the radial notch of the ulna, surrounds the joint and forms a collar around the head of the radius.

Movements. During pronation and supination of the forearm, the head of the radius rotates within the ring formed by the anular ligament and the radial notch of the ulna.

Blood Supply. Anterior and posterior interosseous arteries supply the proximal radioulnar joint.

Nerve Supply. The proximal radioulnar joint is supplied mainly from the musculocutaneous, median, and radial nerves.

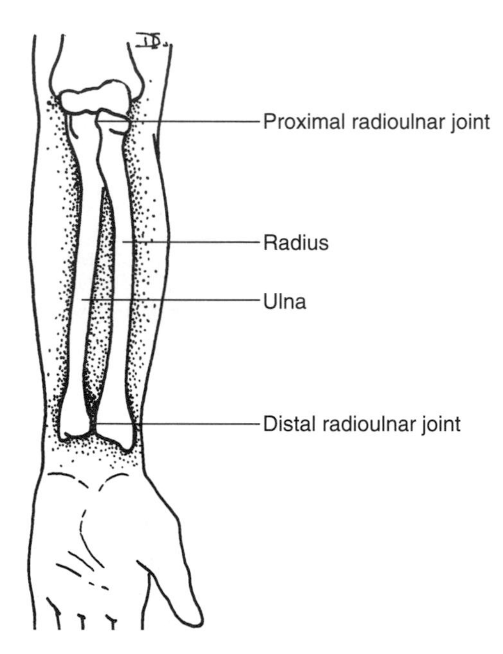

Supination

Preschool children are particularly vulnerable to subluxation (incomplete dislocation) of the head of the radius ("pulled elbow"). The history of these cases is typical. The child is suddenly lifted by the upper limb when the forearm is pronated (e.g., when lifting a child into a bus). The child cries out and refuses to use the limb, which is protected by holding it with the elbow flexed and the forearm pronated. The sudden pulling of the upper limb tears the attachment of the anular ligament, where it is loosely attached to the neck of the radius. The radial head then moves distally, partially out of the torn anular ligament. The proximal part of the ligament may become trapped between the head of the radius and the capitulum of the humerus.

DISTAL RADIOULNAR JOINT

The distal radioulnar joint is a pivot type of synovial joint.

Articulation. The head of the ulna articulates with the ulnar notch of the radius.

Articular Capsule. The fibrous capsule encloses the joint but is deficient superiorly. The synovial capsule extends superiorly between the radius and ulna to form the sacciform recess.

Ligaments. Weak anterior and posterior ligaments strengthen the capsule.

Movements. During pronation and supination of the forearm and hand, the head of the ulna rotates within the ulnar notch of the radius.

Blood Supply. Anterior and posterior interosseous arteries supply the distal radioulnar joint.

Nerve Supply. Anterior and posterior interosseous nerves supply the distal radioulnar joint.

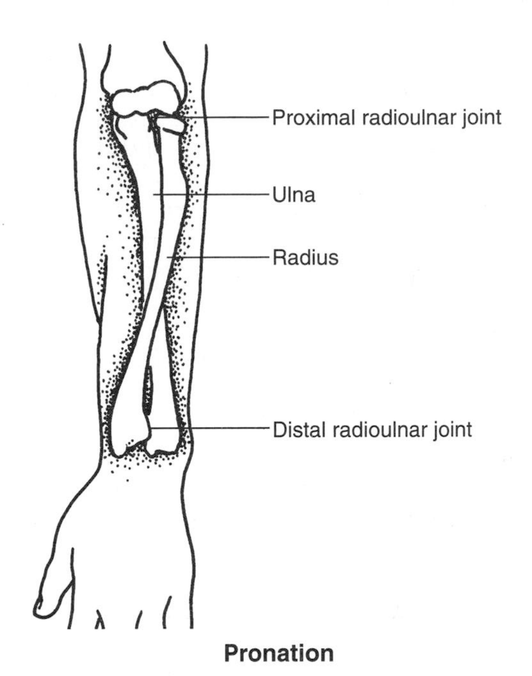

Pronation

Wrist fractures (e.g., Colles' fracture) involving the distal end of the radius are the most common type of fracture in persons older than 50 years. This injury commonly results when the person slips or trips and, in attempting to break the fall, lands on the outstretched hand with the forearm pronated. There is usually a complete transverse fracture of the distal 2–3 cm of the radius, and the fragment is displaced proximally, causing shortening of the radius. The fragment is usually tilted posteriorly, producing a characteristic hump described as the "dinner fork" deformity because of the resemblance of the broken wrist to an upside-down dinner fork (p. 285).

WRIST, INTERCARPAL, CARPOMETACARPAL, AND INTERMETACARPAL JOINTS

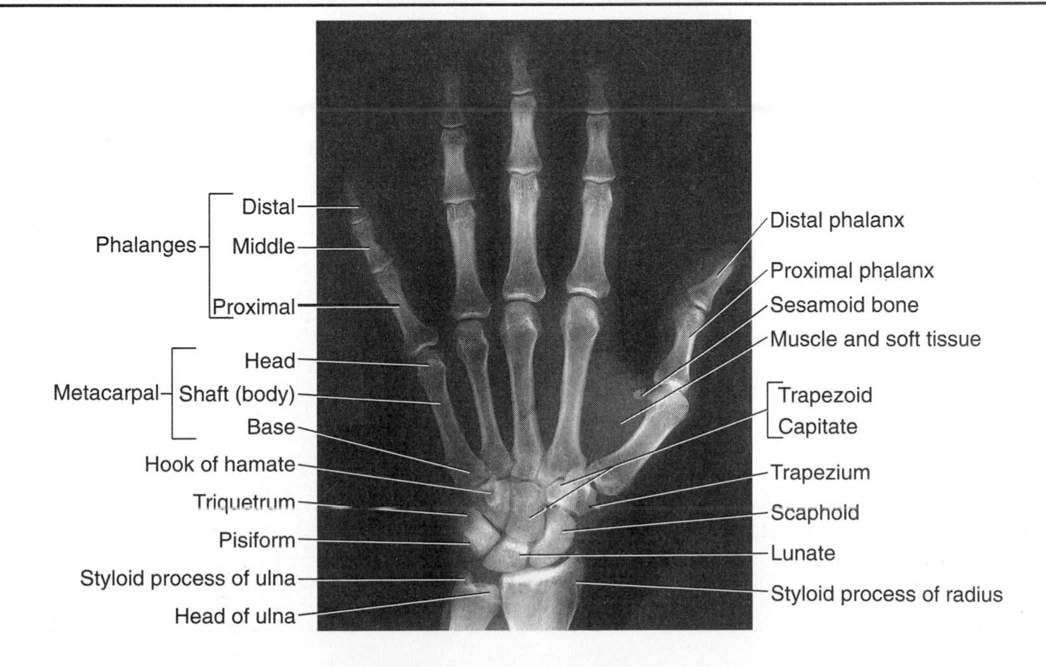

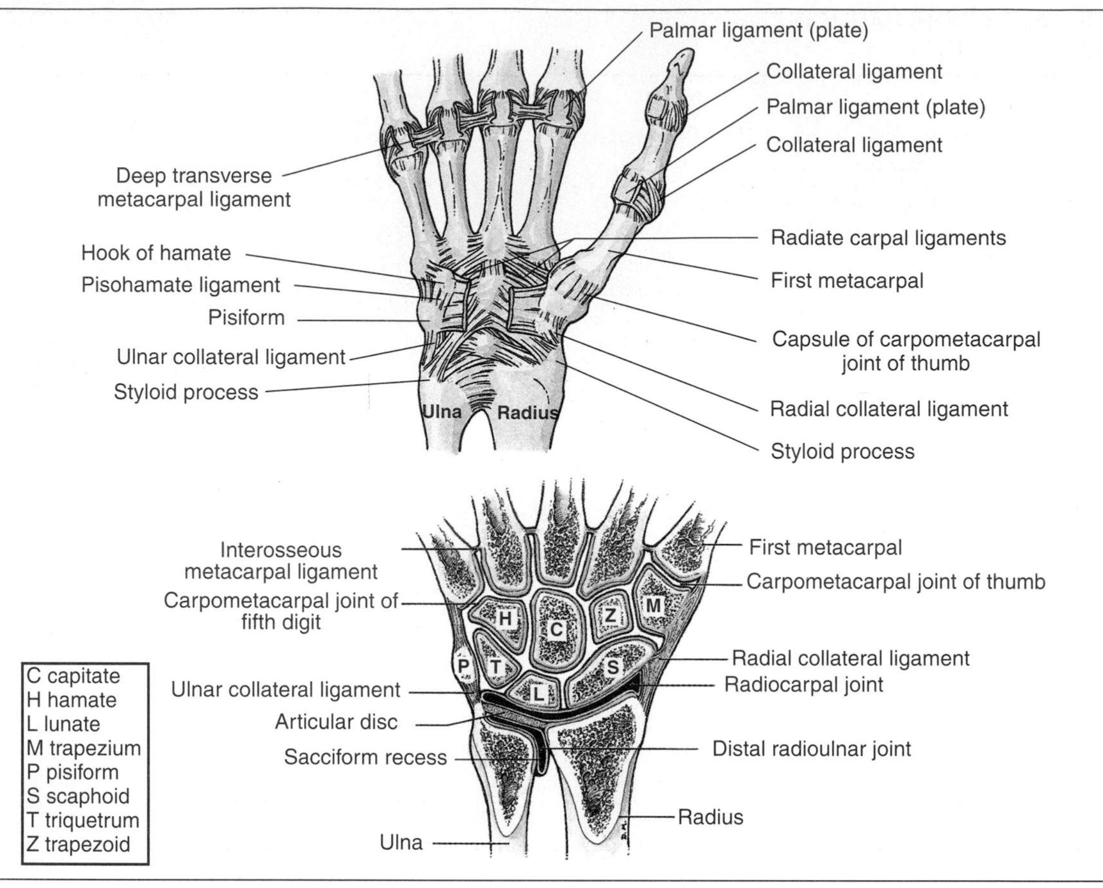

	Wrist Joint	Intercarpal Joints	Carpometacarpal and Intermetacarpal Joints
Type	Condyloid type of synovial joint	Plane type of synovial joint	Plane type of synovial joint, except for carpometacarpal joint of thumb, which is a saddle joint
Articulation	Distal end of radius and articular disc superiorly; articulate with scaphoid, lunate and triquetral bones inferiorly	Midcarpal joint is between proximal and distal rows of carpal bones; individual bones of proximal and distal rows also articulate with each other	Carpals and metacarpals articulate with each other as do metacarpals; carpometacarpal joint of thumb is between trapezium and base of first metacarpal
Articular capsule	Fibrous capsule surrounds joint and is attached to distal ends of radius and ulna and proximal row of carpal bones	Fibrous capsule surrounds the joint	Fibrous capsule surrounds the joint
Ligaments	Anterior and posterior ligaments strengthen the capsule; ulnar collateral ligament is attached to styloid process of ulna and triquetrum; radial collateral ligament is attached to styloid process of radius and scaphoid	Carpal bones are united by anterior, posterior, and interosseous ligaments	Bones are united by anterior, posterior, and interosseous ligaments
Movements	Flexion-extension, abduction-adduction, and circumduction	A small amount of gliding movement is possible; flexion and abduction of the hand occurs at midcarpal joint	Flexion-extension and abduction-adduction of carpometacarpal joint of first digit; almost no movement occurs at second and third digits; fourth digit is slightly mobile, and fifth digit is more mobile
Blood supply	Dorsal and palmar carpal arterial arches	Dorsal and palmar carpal arterial arches	Dorsal and palmar metacarpal arteries and deep carpal and deep palmar arterial arches
Nerve supply	All these joints are supplied by anterior interosseous branch of median nerve, posterior interosseous branch of radial nerve, and dorsal and deep branches of the ulnar nerve		

JOINTS OF DIGITS

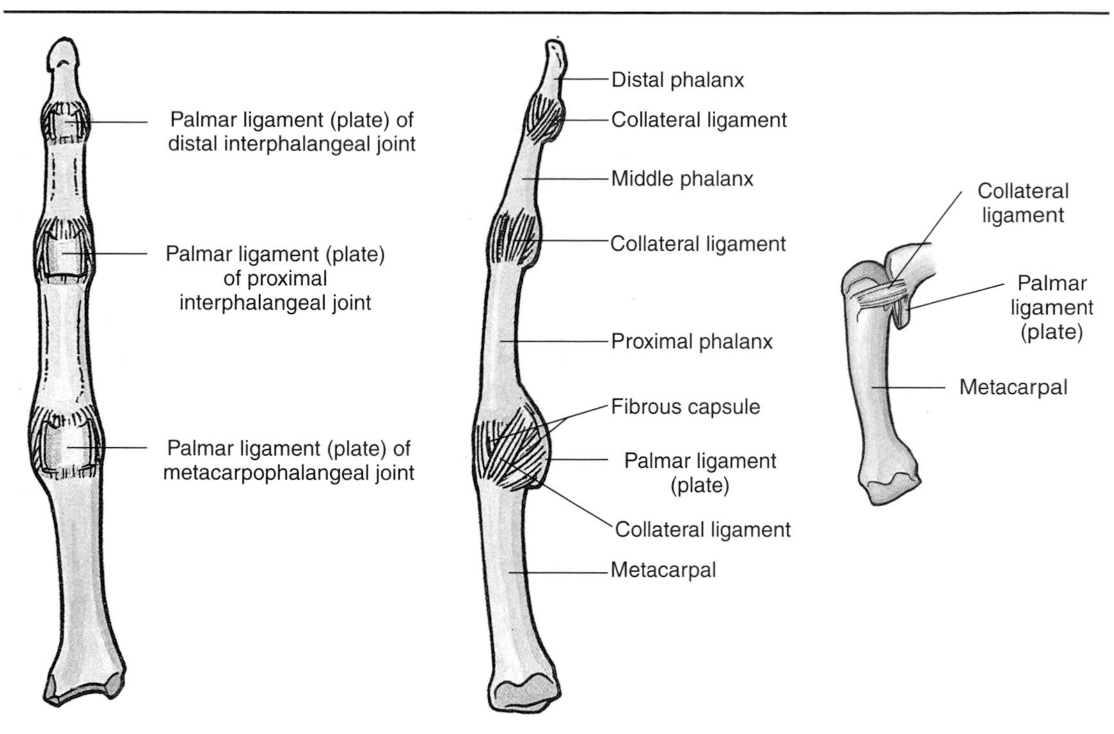

	Metacarpophalangeal Joints	Interphalangeal Joints
Type	Condyloid type of synovial joint	Hinge type of synovial joint
Articulation	Heads of metacarpals articulate with bases of proximal phalanges	Heads of phalanges articulate with bases of more distally located phalanges
Articular capsule	Fibrous capsule encloses each joint	Fibrous capsule encloses each joint
Ligaments	Strong palmar ligaments are attached to phalanges and metacarpals; deep transverse metacarpal ligaments unite second to fifth joints that hold heads of metacarpals together; collateral ligaments pass from heads of metacarpals to bases of phalanges	Ligaments are similar to those of metacarpophalangeal joints, except that they unite phalanges
Movements	Flexion-extension, abduction-adduction, and circumduction of second to fifth digits; flexion-extension of thumb occurs but abduction-adduction is limited	Flexion-extension
Blood supply	Deep digital arteries arise from superficial palmar arches	Digital arteries
Nerve supply	Digital nerves arise from ulnar and median nerves	Digital nerves arise from ulnar and median nerves

8 / HEAD

The head contains the skull, brain, cranial nerves, meninges, and special sense organs. It is also the site where food is ingested and air is inspired and expired.

Skull

The skull is the skeleton of the head. Its bones form the cranium and facial skeleton. The *cranium* encloses the brain and its meninges (coverings), proximal parts of the cranial nerves, and blood vessels. The *facial skeleton* contains the orbits (eye sockets) and nasal cavities and includes the maxilla and mandible (upper and lower jaws).

ANTERIOR ASPECT

Features of the anterior aspect of the skull are the frontal bone and zygomatic bones, orbits, nasal region, maxilla, and mandible (Fig. 8.1*A*). The *frontal bone* forms the skeleton of the forehead, articulating inferiorly with the nasal and zygomatic bones. The intersection of the frontal and two nasal bones is the nasion. The supraorbital margin of the frontal bone has a *supraorbital foramen* (or notch). Just superior to the supraorbital margin is a superciliary arch. Within the orbits are the superior and inferior *orbital fissures* and *optic canals*. Inferior to each orbit, there is an *infraorbital foramen* in the maxilla. The *zygomatic bones* form the prominences of the cheek. A small zygomaticofacial foramen pierces the lateral aspect of each bone. Inferior to the nasal bones is the oval piriform (nasal) aperture. Through this opening can be observed the bony nasal septum that divides the nasal cavity into right and left portions. On the lateral wall of each cavity are curved bony plates, the *conchae* (turbinates). The upper jaw is formed by the fused *maxillae;* their alveolar processes form the sockets and supporting bone for the *maxillary teeth.*

The alveolar processes of the *mandible* bear the mandibular teeth. Inferior to the second premolar teeth are mental foramina. The mental protuberance (chin) is a triangular elevation of bone inferior to the symphysis menti (mental symphysis), the region where the two halves of the fetal bone fused.

LATERAL ASPECT

The lateral aspect of the skull is formed by cranial and facial bones (Fig. 8.1*B*). The *temporal fossa* is bounded superiorly and posteriorly by temporal lines, anteriorly by the frontal and zygomatic bones, and inferiorly by the *zygomatic arch.* The superior border of this arch corresponds to the inferior limit of the cerebral hemisphere. The zygomatic arch is formed by the union of the temporal process of the zygomatic bone and the zygomatic process of the temporal bone. In the anterior part of the temporal fossa, about 4 cm superior to the midpoint of the zygomatic arch, is the *pterion.* It is indicated by the H-shaped formation of sutures that unite the frontal, parietal, sphenoid (greater wing), and temporal bones. The *external acoustic meatus* (ear canal) is the entrance to the tympanic membrane (eardrum). The *mastoid process* of the temporal bone lies posteroinferior to this meatus. Anterior to the mastoid process is the slen-

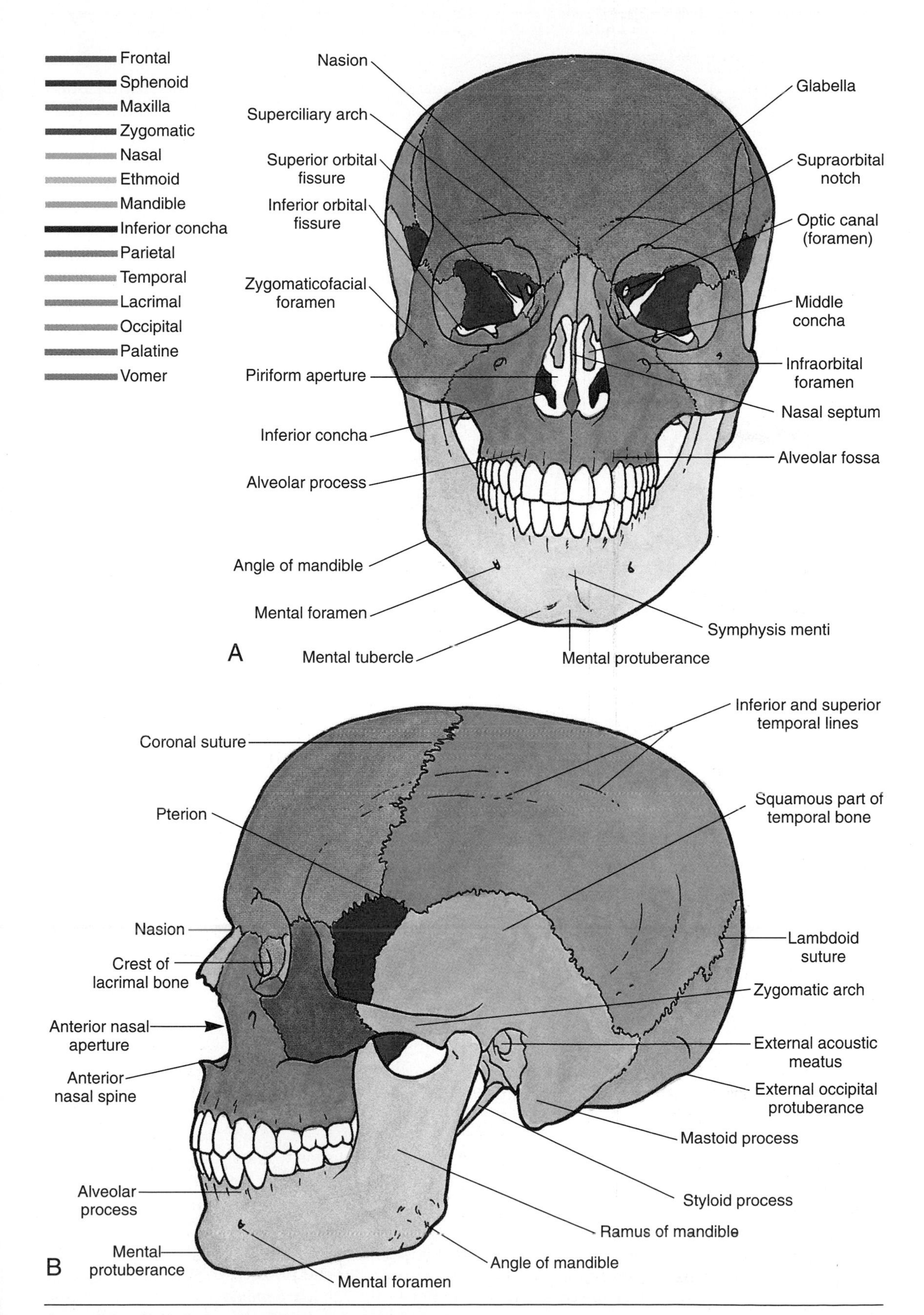

Figure 8.1. Skull. **A.** Anterior aspect. **B.** Lateral aspect.

der *styloid process* of the temporal bone. The *mandible* consists of two parts: a horizontal portion, the *body*, and a vertical portion, the *ramus*.

The *pterion* is an important clinical landmark because it overlies the anterior branches of the middle meningeal vessels that lie in the grooves on the internal aspect of the lateral wall of the calvaria (vault of skull, skull cap). A blow to the side of the head may fracture the thin bones that form the pterion, rupturing these vessels. The resulting collection of blood (*hematoma*) exerts pressure on the underlying cerebral cortex. Untreated middle meningeal artery hemorrhage may cause death in a few hours.

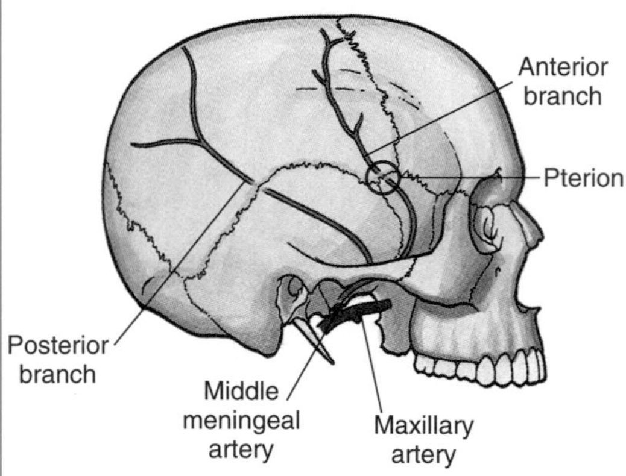

The convexity of the calvaria distributes and thereby minimizes the effects of a blow to it. However, hard blows to the head in thin areas of the cranium will likely produce fractures. *Linear skull fractures*, the most frequent type, usually occur at the point of impact, but fracture lines often radiate away from it in two or more directions. In a *contrecoup (counterblow) fracture*, no fracture occurs at the point of impact, but there is one on the opposite side of the skull.

POSTERIOR ASPECT

The posterior aspect of the skull (*occiput*) is formed by the occipital bone, portions of the parietal bones, and mastoid parts of the temporal bones (Fig. 8.2*A*). The *external occipital protuberance* is an easily palpable elevation in the median plane. The *superior nuchal line*, marking the superior limit of the neck, extends laterally from each side of this protuberance; the inferior nuchal line is less distinct.

SUPERIOR ASPECT

The superior aspect of the skull, usually somewhat oval in form, is broadened posterolaterally by the parietal eminences (Fig. 8.2*B*). The four bones that unite to form the *calvaria* are visible from this aspect: the frontal bone anteriorly, the right and left parietal bones laterally, and the occipital bone posteriorly. The *coronal suture* separates the frontal and parietal bones; the *sagittal suture* separates the parietal bones, and the *lambdoid suture* separates the parietal and temporal bones from the occipital bone. The *bregma* is the anthropological landmark formed by the intersection of the sagittal and coronal sutures. The *vertex*, the most superior point of the skull, is on the sagittal suture near its midpoint. The *lambda* indicates the intersection of the lambdoid and sagittal sutures; it can sometimes be felt as a depression.

INFERIOR ASPECT

The inferior aspect of the skull (base of skull) with the mandible removed shows the palatine processes of the maxillae and palatine bones, the sphenoid, vomer, temporal, and occipital bones (Fig. 8.3). The *hard palate* is formed by the palatine processes of the maxillae anteriorly and the horizontal plates of the palatine bones posteriorly. Posterior to the central incisor teeth is the *incisive fossa*. Posterolaterally are the greater and lesser palatine foramina. Superior to the posterior edge of the palate are the *choanae* (posterior nasal apertures). The *vomer*, a thin flat bone, makes a major contribution to the bony nasal septum. Wedged between the frontal, temporal, and occipital bones is the *sphenoid*, which consists of a body and three pairs of processes: greater wings, lesser wings, and pterygoid processes. The wings spread laterally from the body. The *pterygoid processes*, consisting of lateral and medial *pterygoid plates*, extend inferiorly on each side from the junction of the body and greater wings. The groove for the cartilaginous part of the *auditory tube* lies medial to the spine of the sphenoid. The *mandibular fossae* are depressions in the temporal bone that accommodate the condyles of the mandible.

The base of the skull is formed posteriorly by the *occipital bone*, which articulates with the sphenoid anteriorly. The four parts of the occipital bone are arranged around the large *foramen magnum*. On the lateral parts of the occipital bone are two large protuberances, the *occipital condyles*. The large opening between the occipital bone and the petrous part of the temporal bone is the *jugular foramen*. The *internal acoustic meatus* lies superolateral to the jugular foramen.

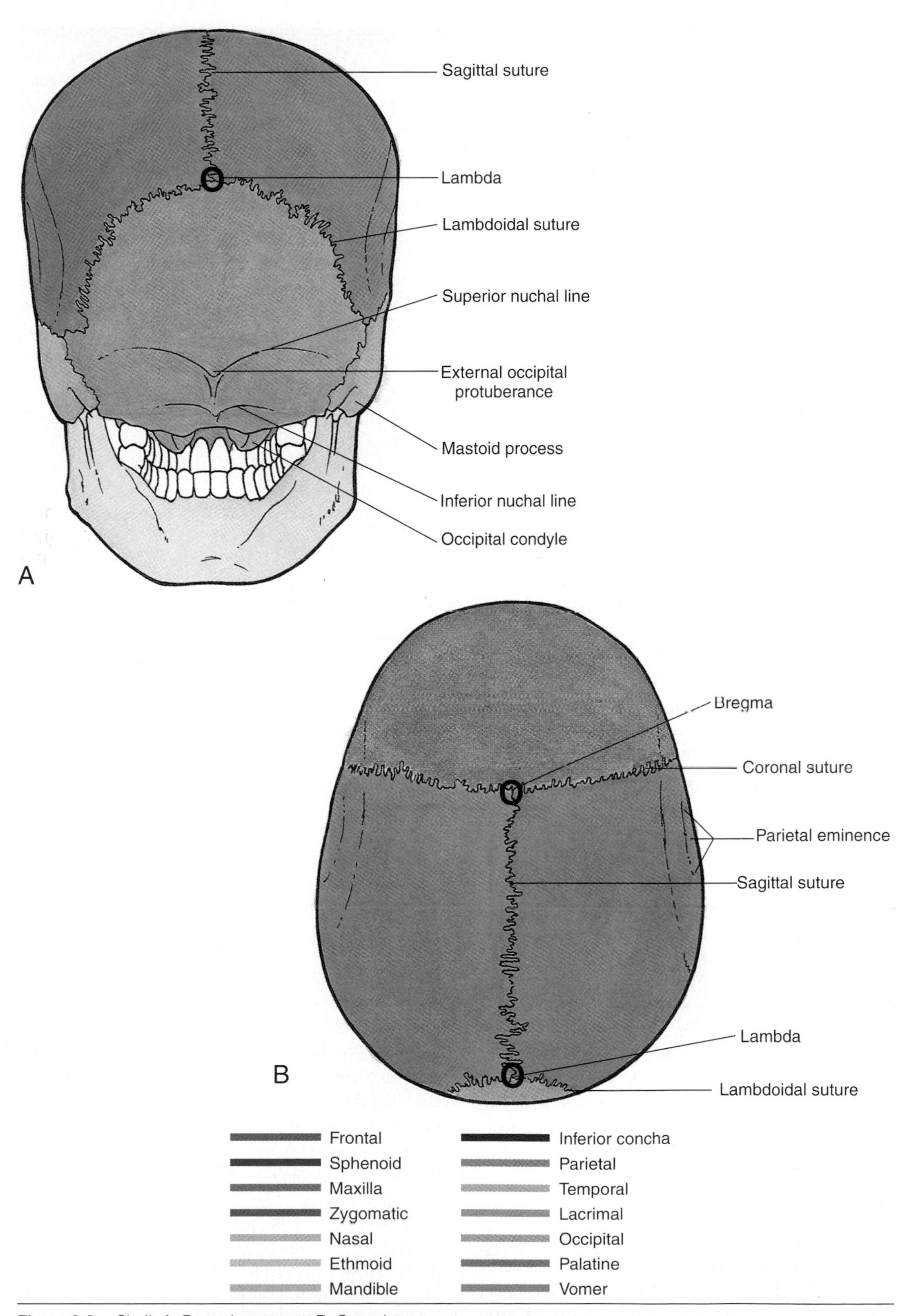

Figure 8.2. Skull. **A.** Posterior aspect. **B.** Superior aspect.

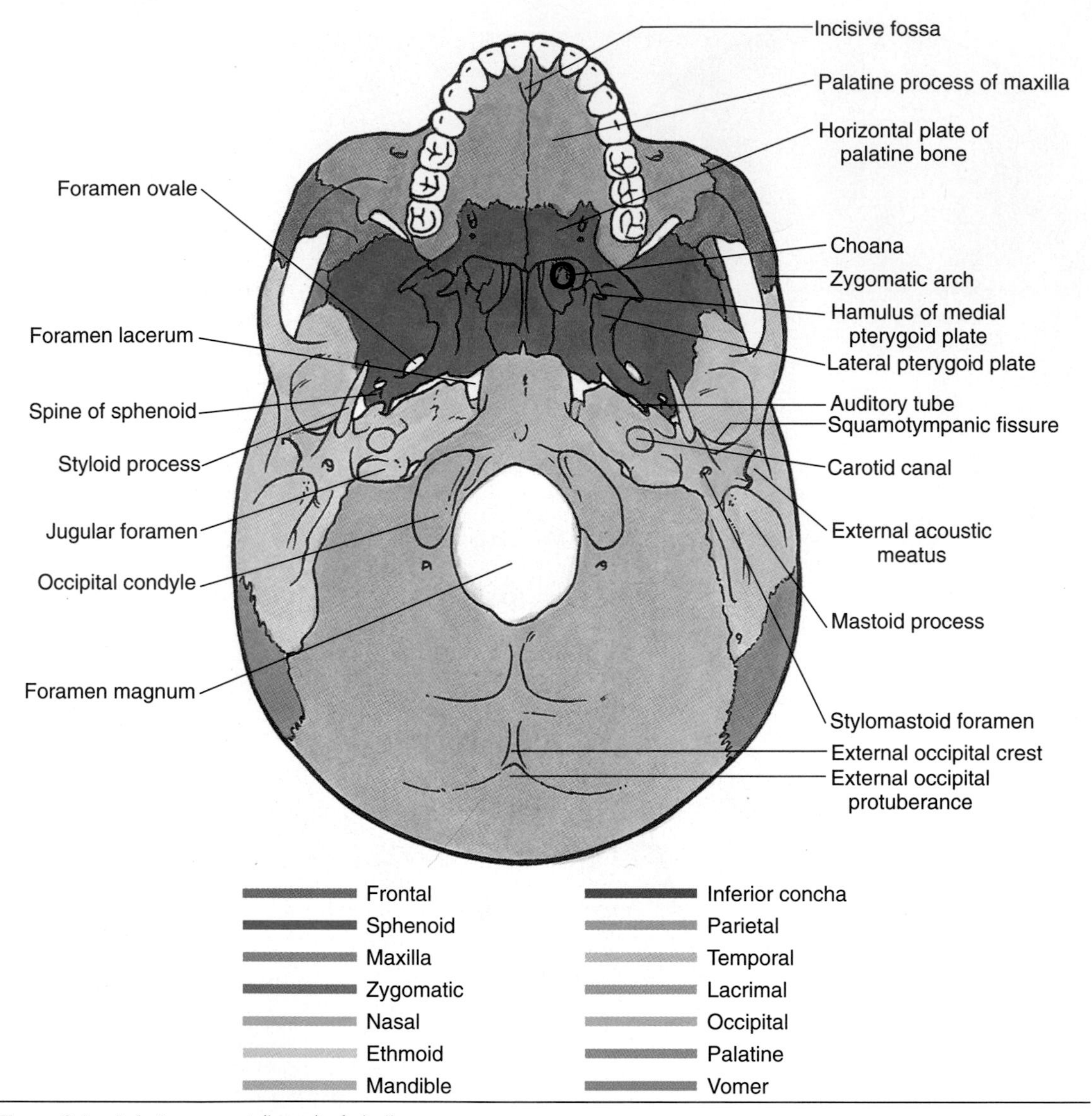

Figure 8.3. Inferior aspect (base) of skull.

The entrance to the *carotid canal* is located just anterior to the jugular foramen. The *mastoid process* is ridged because of the muscles that attach to it. The *stylomastoid foramen* lies posterior to the base of the styloid process.

INTERNAL ASPECT OF BASE

The internal aspect of the base of the skull has three *cranial fossae* (anterior, middle, and posterior) that form the bowl-shaped floor of the cranial cavity (Fig. 8.4*A*). The anterior cranial fossa is at the highest level, and the posterior cranial fossa is at the lowest level.

The *anterior cranial fossa* is formed by the frontal bone anteriorly, the ethmoid bone in the middle, and the body and lesser wings of the sphenoid posteriorly. The greater part of the anterior cranial fossa is formed by ridged *orbital plates* of the frontal bone (Fig. 8.4*B*), which support the frontal lobes of the brain and form the roofs of the orbits. The *frontal crest* is a median bony extension of the frontal bone, and the *crista galli* (cock's comb) is a median ridge of bone that projects superiorly from the ethmoid. On each side of the crista galli is the sievelike *cribriform plate* of the ethmoid.

The *middle cranial fossa* is composed of deep depressions on each side of the body of the sphenoid. The bones forming the fossa are the greater wings of the sphenoid, the squamous

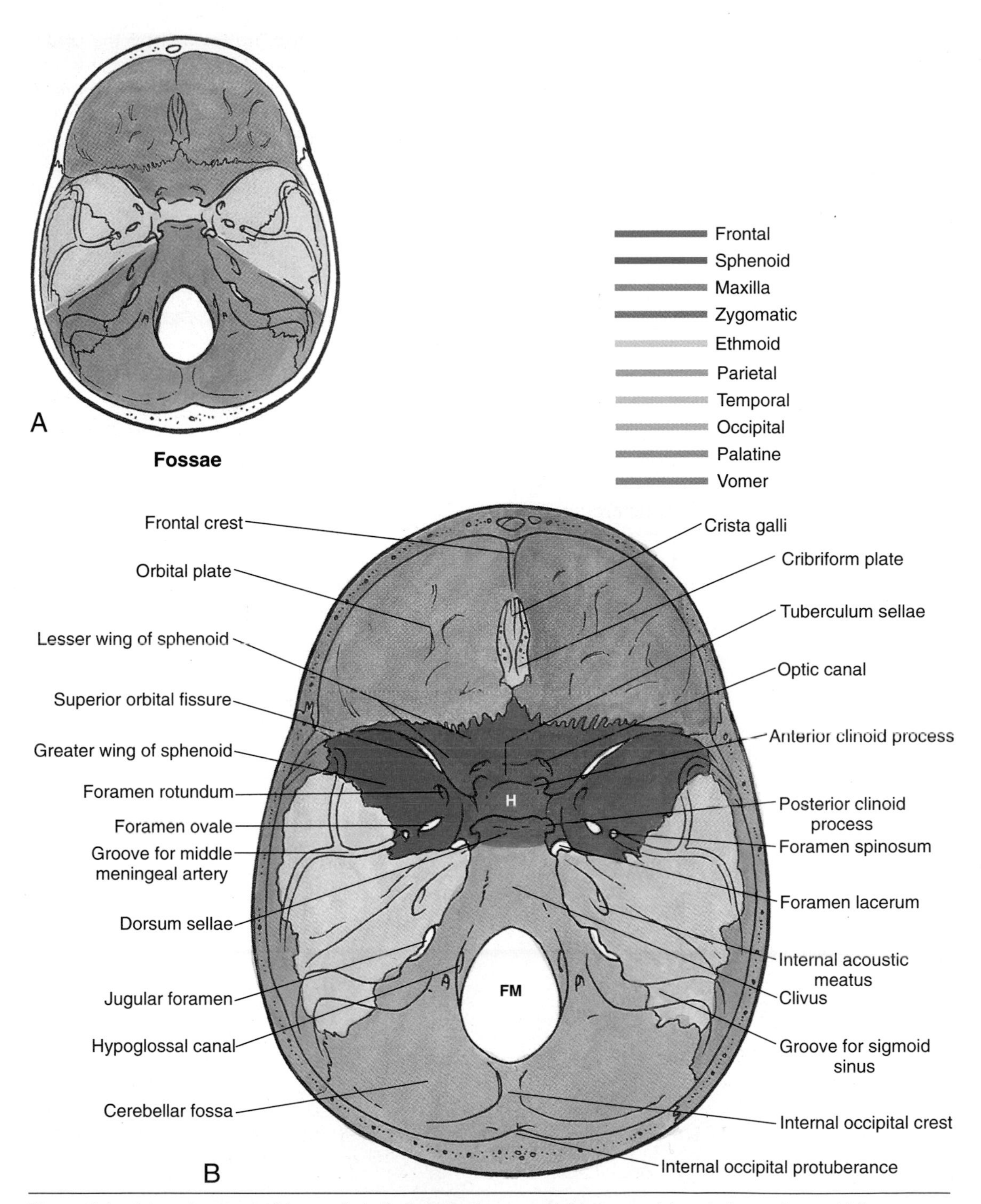

Figure 8.4. Internal aspect of base of skull. **A.** Note three cranial fossae: *pink*, anterior; *yellow*, middle; *orange*, posterior. **B.** Bones and foramina of fossae. *H*, hypophyseal fossa; *FM*, foramen magnum.

parts of the temporal bones laterally, and the petrous portions of the temporal bones posteriorly. This fossa is posteroinferior to the anterior cranial fossa and supports the temporal lobes of the brain. The boundary between the middle and posterior cranial fossae is the petrous ridge of the temporal bones laterally and the *dorsum sellae* of the sphenoid medially. The *sella turcica* (Turk's saddle), the saddlelike part of the sphenoid between the anterior and posterior *clinoid processes*, is composed of three parts:

- Tuberculum sellae, an olive-shaped swelling, anteriorly
- Hypophyseal fossa, a saddlelike depression for the hypophysis (pituitary gland), in the middle
- Dorsum sellae ("back of the saddle") posteriorly

The sharp posterior margins of the *lesser wings of the sphenoid* overhang the middle cranial fossa. These wings, joined by the sphenoidal jugum (L. jugum, yoke), end medially in two projections, the *anterior clinoid processes*. The *foramen lacerum* lies posterolateral to the hypophyseal fossa; the *optic canal* is at the base of the lesser wing, and the *superior orbital fissure* is between the greater and lesser wings. Observe the location of the foramen rotundum, foramen ovale, and foramen spinosum.

The *posterior cranial fossa* is formed largely by the occipital bone, but parts of the sphenoid and temporal bones contribute to it. The broad grooves in this fossa are formed by the transverse and sigmoid sinuses (p. 359). At the center of the posterior cranial fossa is the *foramen magnum*. Posterior to this large foramen, the fossa is partly divided by the *internal occipital crest* into two *cerebellar fossae*. The internal occipital crest ends in the internal occipital protuberance. The *jugular foramen* is at the base of the petrous ridge of the temporal bone, and the *hypoglossal canal* is superior to the anterolateral margin of the foramen magnum.

Face

The face is the anterior aspect of the head from the forehead to the chin and from ear to ear.

MUSCLES

Muscles of the face move the skin and change facial expressions to convey mood (Fig. 8.5, Table 8.1). Most muscles attach to

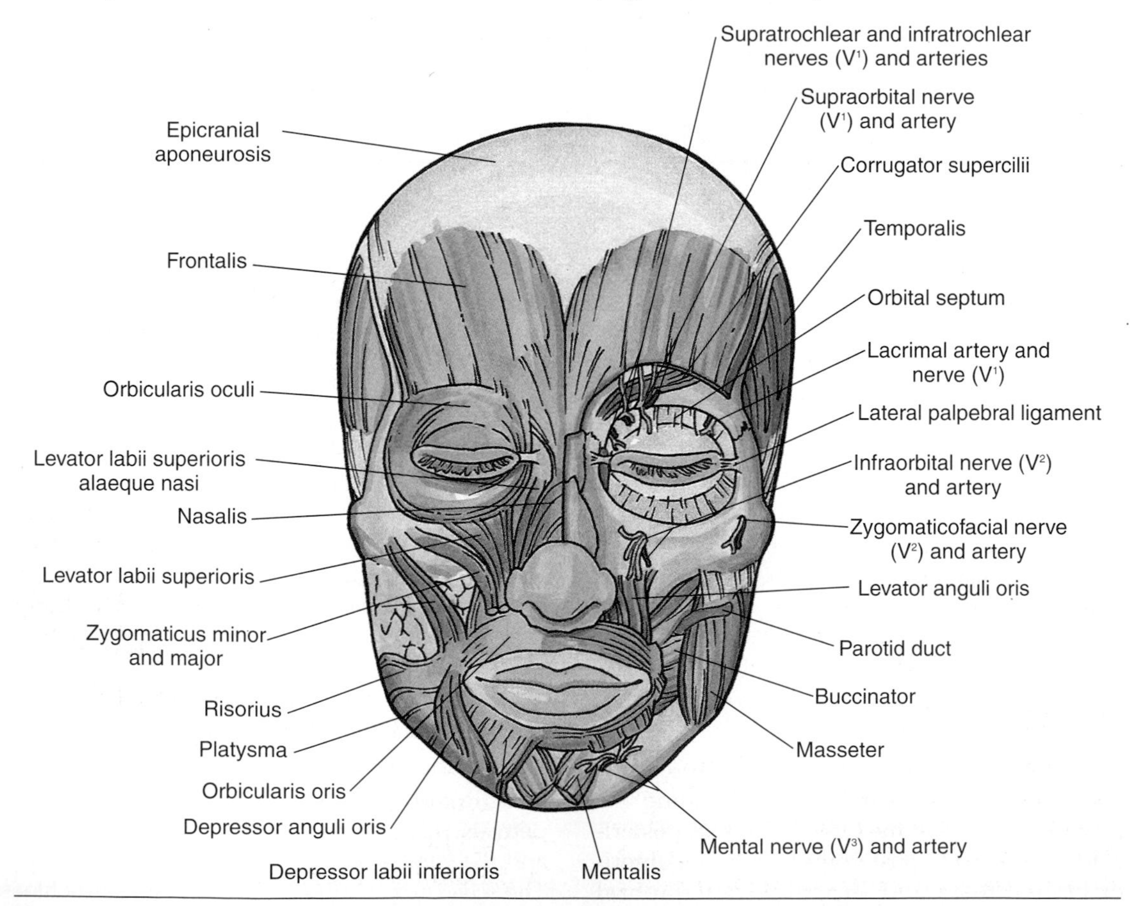

Figure 8.5. Muscles of face. Left side is a deeper dissection (see also Table 8.1).

Table 8.1
Main Muscles of Face[a]

Muscle	Origin	Insertion	Action(s)
Frontalis	Epicranial aponeurosis	Skin of forehead	Elevates eyebrows and forehead
Orbicularis oculi	Medial orbital margin, medial palpebral ligament, and lacrimal bone	Skin around margin of orbit; tarsal plate	Closes eyelids
Nasalis	Superior part of canine ridge of maxilla	Nasal cartilages	Draws ala (side) of nose toward nasal septum
Orbicularis oris	Some fibers arise near median plane of maxilla superiorly and mandible inferiorly; other fibers arise from deep surface of skin	Mucous membrane of lips	Compresses and protrudes lips (e.g., purses them during whistling and sucking)
Levator labii superioris	Frontal process of maxilla and infraorbital region	Skin of upper lip and alar cartilage of nose	Elevates lip, dilates nostril, and raises angle of mouth
Platysma	Superficial fascia of deltoid and pectoral regions	Mandible, skin of cheek, angle of mouth, and orbicularis oris	Depresses mandible and tenses skin of lower face and neck
Mentalis	Incisive fossa of mandible	Skin of chin	Elevates and protrudes lower lip
Buccinator	Mandible, pterygomandibular raphe, and alveolar processes of maxilla and mandible	Angle of mouth	Presses cheek against molar teeth, thereby aiding chewing; expels air from oral cavity as occurs when playing a wind instrument

[a] All these muscles are supplied by the facial nerve (CN VII).

bone or fascia and produce their effects by pulling the skin. The *muscles of facial expression* surround the orifices (mouth, eyes, nose, and ears) and act as sphincters and dilators that close and open the orifices. The *buccinator* (L. bucinator, trumpeter), active in smiling, also keeps the cheek taut, thereby preventing it from folding and being injured during chewing. The buccinator is also active during sucking, whistling, and blowing (e.g., playing a wind instrument). All muscles of facial expression receive their motor innervation from the *facial nerve* (CN VII).

NERVES

There is overlapping of the cutaneous nerves of the neck with those of the head. Cutaneous branches of the *cervical plexus* (Fig. 9.2) extend over the ear and posterior aspect of the neck (lesser occipital nerve) and over the parotid region of the face (great auricular nerve). However, the main sensory nerves of the face are derived from the *trigeminal nerve* (CN V) (Fig. 8.5, Table 8.2). Before leaving the skull, CN V divides into three primary divisions (see Fig. 8.10*A*): ophthalmic (CN V^1), maxillary (CN V^2), and mandibular (CN V^3) nerves; they are named according to their main areas of termination—regions of eye, maxilla, and mandible, respectively. The first two divisions (ophthalmic and maxillary) are wholly sensory; the mandibular division is largely sensory but contains fibers of the motor root of CN V.

The major cutaneous branches of the *ophthalmic nerve* (CN V^1) are the

- Lacrimal nerve
- Supraorbital nerve
- Supratrochlear nerve
- Infratrochlear nerve
- External nasal nerves

Table 8.2
Cutaneous Nerves of Face and Anterior Part of Scalp

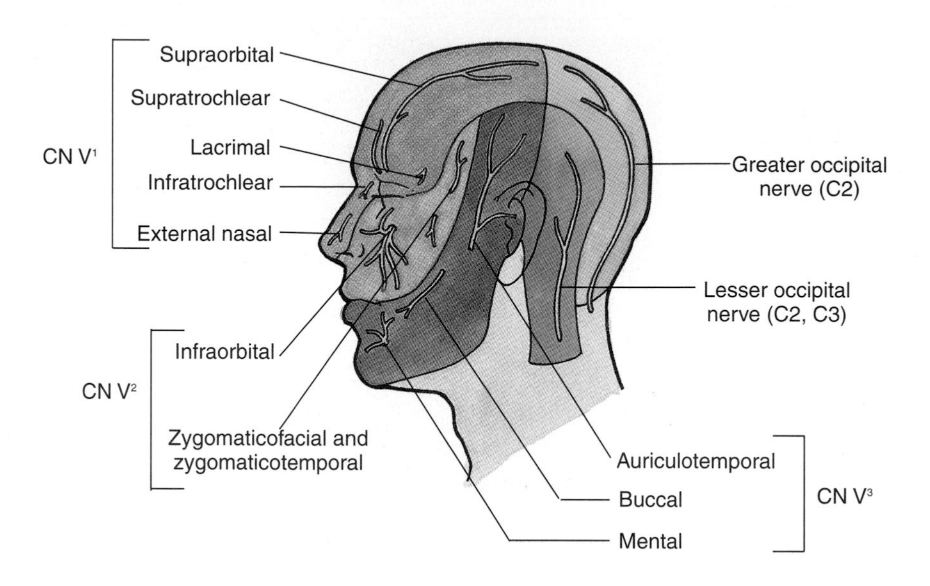

Nerve	Origin	Course	Distribution
Frontal	Ophthalmic n. (CN V^1)	Crosses orbit on superior aspect of levator palpebrae superioris m.; divides into supraorbital and supratrochlear branches	Skin of forehead, scalp, upper eyelid, and nose
Infratrochlear	Nasociliary n. (CN V^1)	Follows medial wall of orbit to upper eyelid	Skin of upper eyelid
External nasal	Anterior ethmoidal n. (CN V^1)	Runs in nasal cavity and emerges on face between nasal bone and nasal cartilage	Skin on dorsum of nose
Zygomatic	Maxillary n. (CN V^2)	Arises in floor of orbit, divides into zygomaticofacial and zygomatico-temporal nn., which traverse foramina of same name	Skin over zygomatic arch and anterior temporal region
Infraorbital	Terminal branch of maxillary n. (CN V^2)	Runs in floor of orbit and emerges at infraorbital foramen	Skin of cheek, lateral side of nose, and upper lip
Buccal	Mandibular n. (CN V^3)	From anterior division of CN V^3 in infratemporal fossa, it passes anteriorly to reach cheek	Skin of cheek
Auriculotemporal	Mandibular n. (CN V^3)	From posterior division of CN V^3, it passes between neck of mandible and external acoustic meatus to accompany superficial temporal artery	Skin anterior to ear and temporal region
Mental	Terminal branch of inferior alveolar n. (CN V^3)	Emerges from mandibular canal at mental foramen	Skin of chin and lower lip

The major cutaneous branches of the *maxillary nerve* (CN V^2) are the

- Infraorbital nerve
- Zygomaticotemporal nerve
- Zygomaticofacial nerve

The major cutaneous branches of the *mandibular nerve* (CN V^3) are the

- Auriculotemporal nerve
- Buccal nerve
- Mental nerve

The *motor nerves of the face* are the facial nerve (CN VII) to the muscles of facial expression and the mandibular nerve [third division of the trigeminal (CN V^3)] to the muscles of mastication (masseter, temporalis, medial, and lateral pterygoids) and to the mylohyoid, anterior belly of digastric, tensor veli palatini, and tensor tympani.

The *facial nerve* (CN VII) exits the skull via the stylomastoid foramen (Figs. 8.3 and 8.6*A*, Table 8.1). Its branches (temporal, zygomatic, buccal, mandibular, cervical, and posterior auricular nerves) supply the superficial muscle of the neck and chin (platysma), muscles of facial expression, muscle of the cheek (buccinator), muscles of the ear (auricular), and the scalp (occipitalis and frontalis).

VASCULATURE

Most arteries of the face are branches of the external carotid artery (Fig. 8.6*A*, Table 8.3). The *facial artery* provides the major arterial supply to the face. It arises from the external carotid artery and winds its way to the inferior border of the mandible, just anterior to the masseter. It then courses over the face to the medial canthus of the eye (the angle where the eye's upper and lower lids meet). The facial artery sends branches to the upper and lower lips (superior and inferior labial arteries), the side of the nose (lateral nasal artery), and terminates as the angular artery, which supplies the medial canthus.

The *superficial temporal artery* is the smaller terminal branch of the external carotid artery; the other branch is the maxillary artery. The superficial temporal artery emerges on the face between the temporomandibular joint (TMJ) and the ear and ends in the scalp by dividing into frontal and parietal branches (Fig. 8.6*A*). The *transverse facial artery* arises from the super-

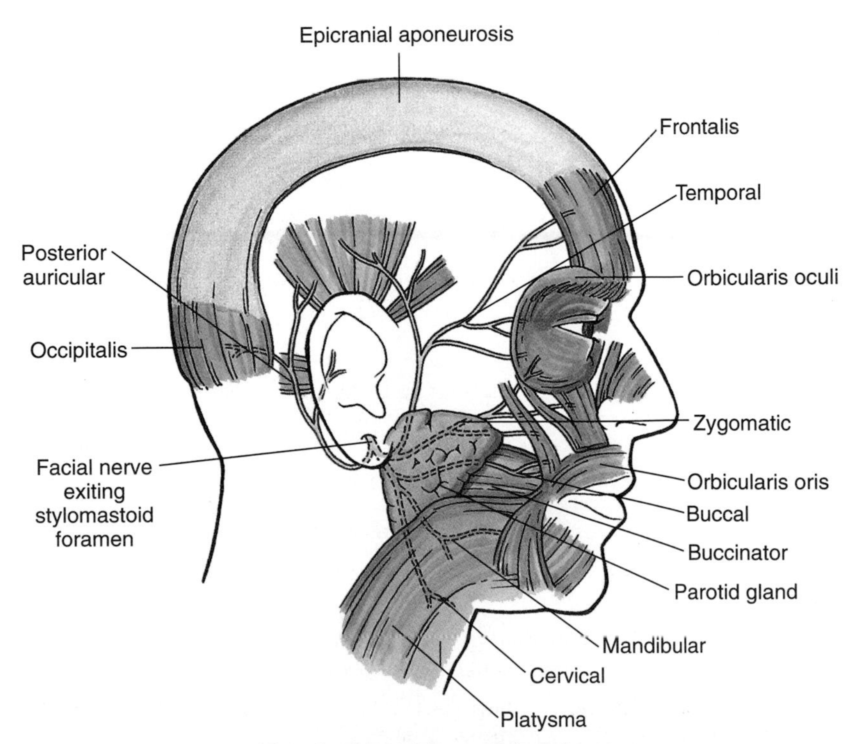

Terminal branches of facial nerve

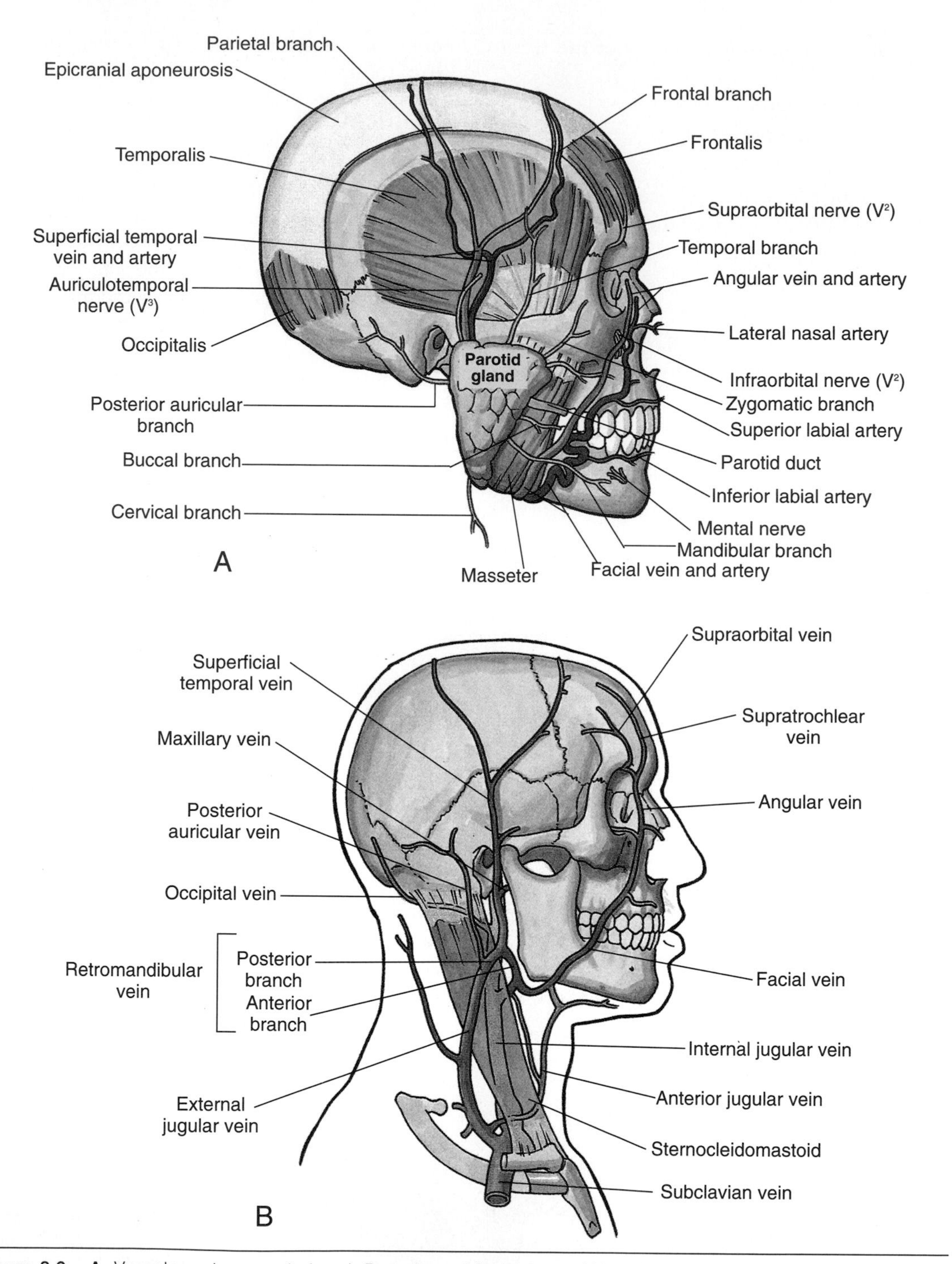

Figure 8.6. **A.** Vessels and nerves in head. Branches of facial nerve (CN VII) are temporal, zygomatic, posterior auricular, buccal, cervical, and mandibular. **B.** Venous drainage of head and neck.

ficial temporal artery within the parotid gland and crosses the face superficial to the masseter. It divides into numerous branches that supply the parotid gland and duct, the masseter, and skin of the face.

The *facial vein* (Fig. 8.6) provides the major venous drainage of the face. It begins at the medial canthus of the eye as the *angular vein* by the union of the supraorbital and supratrochlear veins. The facial vein then runs infero-

Because the face does not have a distinct layer of deep fascia and the superficial fascia is loose between the attachments of muscles, injury or infection causes marked swelling. Furthermore, *facial lacerations* tend to gape (part widely). Consequently the skin must be sutured carefully to prevent scarring. The looseness of the superficial fascia also enables tissue fluid and blood to accumulate after bruising of the face (e.g., a black eye).

Trigeminal neuralgia (tic douloureux) is a sensory disorder of the sensory division of CN V that is characterized by sudden attacks of excruciating facial pain. The cause is unknown, but it is thought that most patients have anomalous blood vessels that compress the nerve. The pain is often in the area of distribution of the mandibular nerve.

A lesion of the entire trigeminal nerve causes widespread anesthesia involving (*a*) the corresponding anterior half of the scalp, (*b*) the face, except for an area around the angle of the mandible, (*c*) the cornea and conjunctiva, and (*d*) the mucous membranes of the nose, mouth, and tongue (anterior two-thirds). Paralysis of the muscles of mastication also occurs. Sometimes the sensory root of CN V is cut to alleviate trigeminal pain; this is often done in the *trigeminal cave*, a dural recess that contains the roots of CN V and the trigeminal ganglion (see Fig. 8.10*A*).

For local anesthesia of the lower eyelid, side of the nose, upper lip, and superior part of the cheek, the infraorbital nerve, a branch of CN V^2, is infiltrated with an anesthetic agent (*infraorbital nerve block*) at the mouth of the infraorbital canal. A mental nerve block is performed by injecting anesthetic fluid around the mental nerve, a branch of CN V^3, as it emerges from the mental foramen.

Injury to the facial nerve (CN VII) or some of its branches produces paralysis of some or all facial muscles on the affected side. The most common cause of *facial paralysis* is inflammation of the facial nerve near the stylomastoid foramen (Bell's palsy). This produces edema, swelling, and compression of the nerve in the facial canal. Patients with Bell's palsy cannot close their lips and eyelids on the affected side. The cornea on the affected side is not lubricated with lacrimal fluid. Patients cannot whistle, blow a wind instrument, or chew effectively. The palsy weakens or paralyzes the buccinator, the cheek muscle that aids chewing and emptying the gutter between the teeth and cheek. Displacement of the mouth (drooping of its corner) is produced by contraction of unopposed contralateral facial muscles, resulting in food and saliva dribbling out of the side of the mouth.

Bell's palsy

posteriorly through the face, posterior to the facial artery. Inferior to the margin of the mandible, the facial vein is joined by the anterior branch of the retromandibular vein. The facial vein drains directly or indirectly into the internal jugular vein.

The *superficial temporal vein* drains the forehead and scalp and receives tributaries from the veins of the temple and face. Near the auricle, the superficial temporal vein enters the parotid gland. The *retromandibular vein*, formed by the union of the superficial temporal and maxillary veins, descends within the parotid gland, superficial to the external carotid artery and deep to the facial nerve. The retromandibular vein divides into an anterior branch that unites with the facial vein and a posterior branch that joins the posterior auricular vein to form the external jugular vein.

The *lymphatic vessels of the face* accompany other facial vessels (Fig. 8.7). Those from the lateral part of the face, including the eyelids, drain inferiorly to the *parotid lymph nodes*. Lymph from the deep parotid nodes drains into the deep cervical lymph nodes. Lymphatic vessels in the upper lip and in lateral parts of the lower lip drain into the *submandibular lymph nodes*, whereas lymphatic vessels in the chin and the central part of the lower lip drain into the *submental lymph nodes*.

The pulse of the facial artery can be taken where it winds around the inferior border of the mandible; the pulse of the superficial temporal artery can be taken as it passes anterior to the ear and crosses the zygomatic arch to supply the scalp.

Carcinomas of the lip usually involve the lower lip. Overexposure to sunshine over many years, as occurs with outdoor workers, is a common feature in these cases. Cancer cells from the central part of the lip, the floor of the mouth, and the tip of the tongue spread to the submental lymph nodes; cancer cells from the lateral part of the lip drain to the submandibular lymph nodes.

Table 8.3
Arterial Supply to Face and Anterior Part of Scalp

Artery	Origin	Course	Distribution
Facial	External carotid a.	Ascends deep to submandibular gland, winds around inferior border of mandible and enters face	Muscles of facial expression and face
Superior labial	Facial a. near angle of mouth	Runs medially in upper lip	Upper lip and ala (side) and septum of nose
Inferior labial	Facial a. near angle of mouth	Runs medially in lower lip	Lower lip and chin
Lateral nasal	Facial a. as it ascends alongside nose	Passes to ala of nose	Skin on ala and dorsum of nose
Angular	Terminal branch of facial a.	Passes to medial canthus (angle) of eye	Superior part of cheek and lower eyelid
Superficial temporal	Smaller terminal branch of external carotid a.	Ascends anterior to ear to temporal region and ends in scalp	Facial muscles and skin of frontal and temporal regions
Transverse facial	Superficial temporal a. within parotid gland	Crosses face superficial to masseter and inferior to zygomatic arch	Parotid gland and duct, muscles and skin of face
Mental	Terminal branch of inferior alveolar a.	Emerges from mental foramen and passes to chin	Facial muscles and skin of chin
Supraorbital	Terminal branch of ophthalmic a., a branch of internal carotid a.	Passes superiorly from supraorbital foramen	Muscles and skin of forehead and scalp
Supratrochlear	Terminal branch of ophthalmic a., a branch of internal carotid a.	Passes superiorly from supratrochlear notch	Muscles and skin of scalp

PAROTID GLAND

The parotid gland, the largest of the three paired salivary glands, is enclosed within a fascial *parotid sheath*. The area occupied by the gland, the *parotid bed*, is anteroinferior to the external acoustic meatus, where it is wedged between the ramus of the mandible and the mastoid process (Fig. 8.6*A*). The apex of the gland is posterior to the angle of the mandible, and its base is related to the zygomatic arch. The *parotid duct* passes horizontally from the anterior edge of the gland. At the anterior border of the masseter, the duct turns medially, pierces the buccinator, and enters the oral cavity opposite the second maxillary molar tooth. A small separate part of the gland (accessory parotid gland) may appear superior to the proximal part of the parotid duct.

The structures within the parotid gland, from superficial to deep, are the *facial nerve* and its branches, the *retromandibular vein*, and the *external carotid artery*. On the parotid sheath and within the gland are *parotid lymph nodes* (Fig. 8.7). These nodes also receive lymph from the forehead, lateral parts of the eyelids, temporal region, lateral surface of the auricle, anterior wall of the external acoustic meatus, and the middle ear. Lymph from the parotid nodes drains into the *cervical lymph nodes*.

The *auriculotemporal nerve*, a branch of CN V^3, is closely related to the parotid gland and passes superior to it with the superficial temporal vessels (Fig. 8.6*A*). The *great auricular nerve* (C2 and C3), a branch of the cervical plexus (Fig. 9.2*A*), passes external to the parotid gland. The parasympathetic compo-

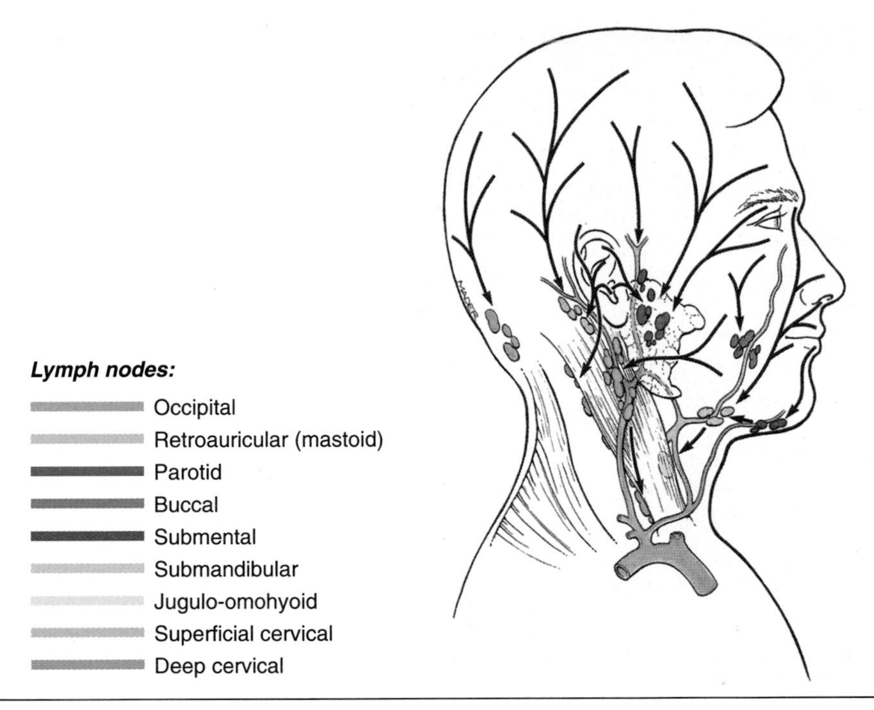

Figure 8.7. Lymphatic drainage of head and neck. *Arrows*, direction of lymph flow.

nent of the *glossopharyngeal nerve* (CN IX) supplies secretory fibers to the parotid gland, which are conveyed by the auriculotemporal nerve from the *otic ganglion* (p. 383). Stimulation of these fibers produces a thin, watery saliva. Sympathetic fibers are derived from the cervical ganglia through the *external carotid plexus of nerves* on the external carotid artery (p. 427). Sensory nerve fibers pass to the gland through the great auricular and auriculotemporal nerves.

Because CN VII and its branches pass through the parotid gland, they are in jeopardy during surgery of the parotid. An important step in *parotidectomy* (surgical excision of gland) is the identification, dissection, isolation, and preservation of the facial nerve.

The parotid gland may become infected through the bloodstream, as occurs in *mumps*, an acute communicable viral disease. Infection of the parotid gland causes inflammation (*parotiditis*) and swelling of the gland. Severe pain occurs because the parotid capsule limits swelling. Often the pain is worse during chewing because the enlarged gland is wrapped around the posterior border of the ramus of the mandible and is compressed against the mastoid process of the temporal bone when the mouth is opened. The mumps virus may also cause *inflammation of the parotid duct*, producing redness of the parotid papilla, the opening of the duct into the oral cavity. Because the pain produced by mumps may be confused with a toothache, redness of the papilla is often an early sign that the disease involves the gland and not a tooth. *Parotid gland disease* often causes pain in the auricle, external acoustic meatus, temple, and TMJ because the auriculotemporal nerve, from which the parotid gland receives sensory fibers, also supplies sensory fibers to the skin of the auricle and over the temporal fossa.

Scalp

The scalp covers the calvaria, extending from the superior nuchal lines on the occipital bone to the supraorbital margins of the frontal bone. Laterally it extends over the temporal fascia to the zygomatic arches. The scalp consists of five layers of tissue, the first three of which are connected intimately and move as a unit. Each letter of the word scalp serves as a memory key for its layers: **s**kin, **c**onnective tissue, **a**poneurosis epicranialis, **l**oose connective tissue, and **p**ericranium.

- The skin, thin except in the occipital region, contains many sweat and sebaceous glands and hair follicles; it has an abundant arterial supply and good venous and lymphatic drainage
- The connective tissue is a thick, richly vascularized, subcutaneous layer that is well supplied with nerves
- The aponeurosis epicranialis (epicranialis aponeurosis) is a strong tendinous sheet that covers the superior aspect of the calvaria; the aponeurosis is the membranous tendon of the fleshy bellies of the occipitalis and frontalis muscles (whereas the frontalis pulls the scalp anteriorly, wrinkles the forehead, and elevates the eyebrows, the occipitalis pulls the scalp posteriorly and wrinkles the skin on the posterior aspect of the neck)
- The loose connective tissue is somewhat like a sponge because it has many potential spaces that may distend with fluid that results from injury or infection; this layer allows free movement of the scalp proper (first three layers, skin, connective tissue, and epicranial aponeurosis)
- The pericranium, a dense layer of connective tissue, is the periosteum of the calvaria; it attaches firmly to the cranial bones, but the pericranium can be stripped fairly easily from the cranial bones of living persons, except where it is continuous with the fibrous tissue in the cranial sutures

Innervation of the scalp anterior to the auricles is through branches of all three divisions of CN V (Fig. 8.6*A*, Table 8.2). Posterior to the auricles, the nerve supply is from spinal cutaneous nerves (C2 and C3).

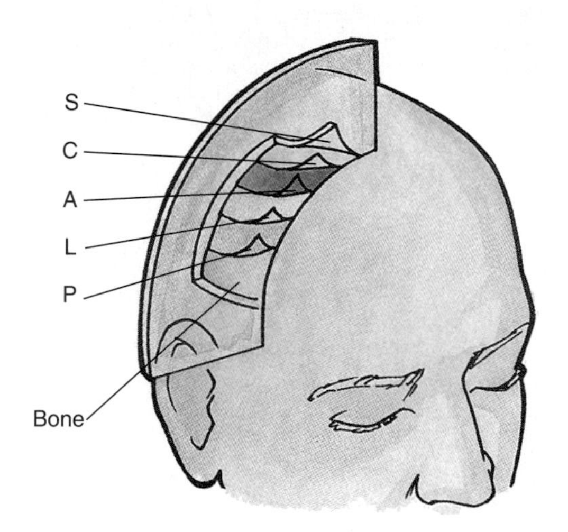

Layers of scalp

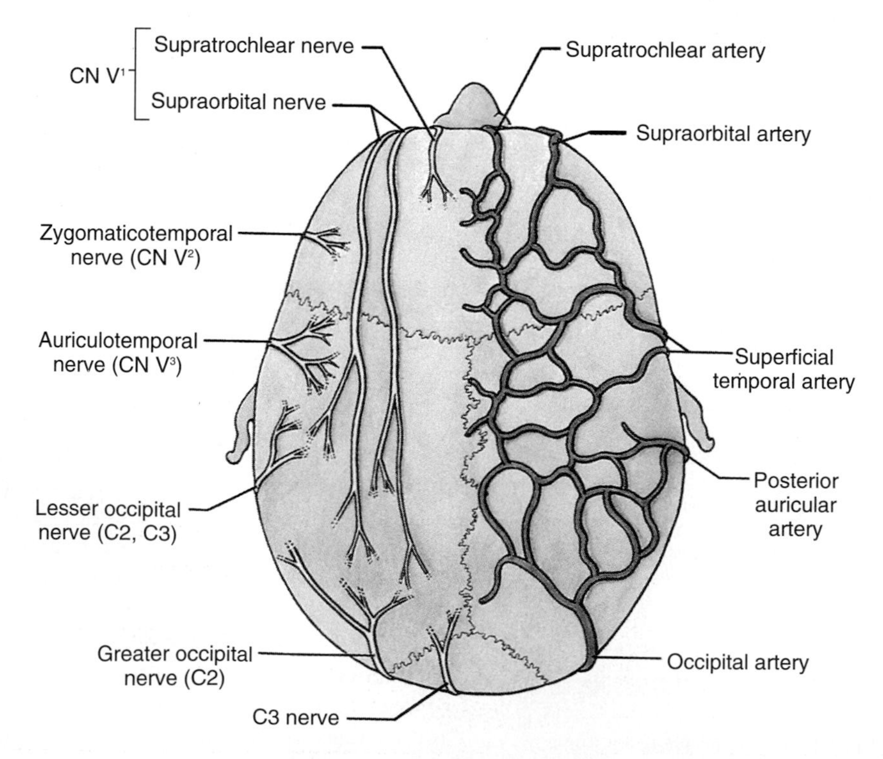

Arteries and nerves of scalp

Arterial supply of the scalp is from the external carotid arteries through the occipital, posterior auricular, and superficial temporal arteries and from the internal carotid arteries by way of the supratrochlear and supraorbital arteries (Table 8.3).

Venous drainage of the scalp is through the supraorbital and supratrochlear veins, which begin in the forehead and descend to unite at the medial canthus to form the facial vein (Fig. 8.6*B*). The superficial temporal veins and posterior auricular veins drain the scalp anterior and posterior to the auricles, respectively. The occipital veins drain the occipital region of the scalp.

Lymphatic drainage of the scalp is into the superficial ring of lymph nodes (submental, submandibular, parotid, retroauricular, and occipital) that is located at the junction of the head and neck (Fig. 8.7). Lymph from these nodes drains into the deep cervical lymph nodes along the internal jugular vein (Fig. 8.6*B*).

The first three layers of the scalp, the *scalp proper*, are often regarded clinically as a single layer because they remain together when a scalp flap is made during a *craniotomy* (surgical opening of cranium) and when the scalp is torn off during accidents. The loose connective tissue layer (fourth layer) is the dangerous area of the scalp because pus or blood spreads easily in it. Infection in this layer can also pass into the cranial cavity through *emissary veins* (Fig. 8.8) that pass through parietal foramina in the calvaria and infect intracranial structures (e.g., brain). An infection cannot pass into the neck because the occipitalis attaches to the occipital bone and mastoid parts of the temporal bones. A scalp infection cannot spread laterally beyond the zygomatic arches because the epicranial aponeurosis is continuous with the temporal fascia covering the temporalis that attaches to these arches (Fig. 8.6). An infection or fluid (e.g., pus or blood) can enter the eyelids and the root of the nose because the frontalis inserts into the skin and dense subcutaneous tissue and does not attach to the bone. Consequently, a black eye can result from an injury to the scalp or forehead. Most blood enters the upper eyelid, but some may enter the lower one.

Scalp lacerations are the most common type of head injury requiring surgical care. These wounds bleed profusely because the arteries enter around the periphery of the scalp and do not retract when lacerated because they are held open by the dense fibrous tissue in the second layer of the scalp. Hence, unconscious patients may bleed to death from scalp lacerations if bleeding is not controlled (e.g., by sutures).

Arteries of the scalp supply very little blood to the bones of the calvaria; these bones are supplied by the middle meningeal artery (p. 344). Hence, loss of the scalp does not produce necrosis (death) of the cranial bones.

Nerves and vessels of the scalp enter inferiorly and ascend through the second (connective tissue) layer of the scalp to the skin. Consequently, surgical flaps of the scalp are made so that they remain attached inferiorly to preserve the nerves and vessels and to promote good healing.

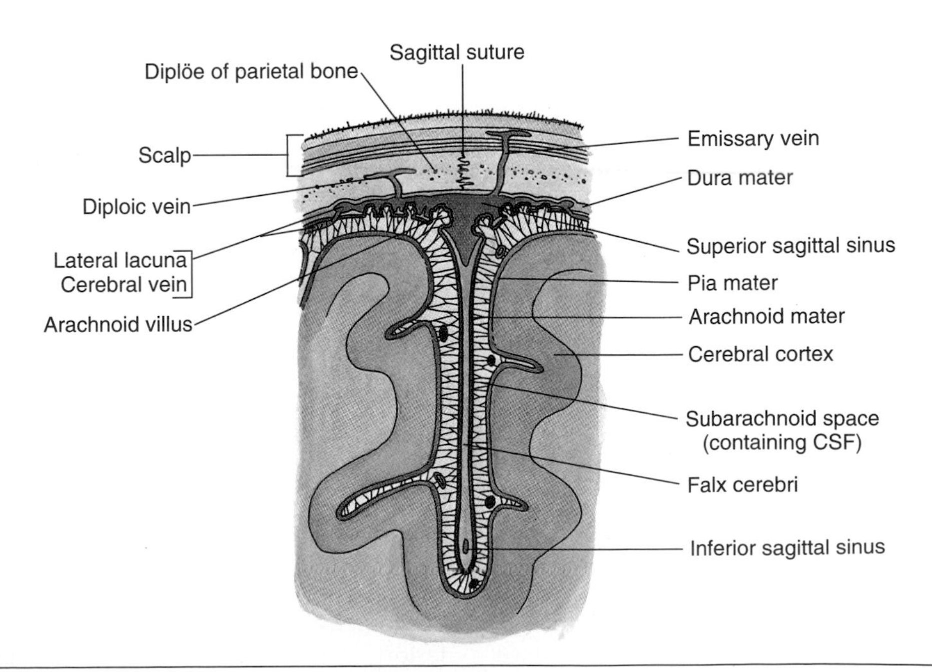

Figure 8.8. Coronal section of skull to show superior sagittal sinus and cranial meninges.

Cranial Meninges and Cerebrospinal Fluid

The *cranial meninges* (meningeal coverings of brain) consist of three layers (Fig. 8.8):

- Dura (mater), an external thick, tough layer
- Arachnoid (mater), an intermediate weblike layer
- Pia (mater), an internal delicate vascular layer

The cranial dura mater consists of two layers:

- The external endosteal (periosteal) layer is formed by the periosteum covering the internal surface of the calvaria
- The internal meningeal layer is a strong fibrous membrane that is continuous at the foramen magnum with the spinal dura mater covering the spinal cord

Cerebrospinal fluid (CSF) is a clear fluid similar to blood in constitution that is formed by the choroid plexuses of the ventricles (p. 365). CSF leaves the ventricular system and enters the subarachnoid space.

Dural septa (partitions) divide the cranial cavity into compartments and support parts of the brain (Figs. 8.8 and 8.9). The dural septa include the falx cerebri, tentorium cerebelli, falx cerebelli, and diaphragma sellae.

The *falx cerebri,* a sickle-shaped partition, lies in the longitudinal fissure between the cerebral hemispheres. It attaches in the median plane to the internal surface of the calvaria from the frontal crest and crista galli anteriorly to the internal occipital protuberance posteriorly.

The *tentorium cerebelli,* a wide crescentic fold, separates the occipital lobes of the cerebral hemispheres from the cerebellum. The falx cerebri attaches to the tentorium and holds it up. The tentorium is attached posteriorly to the occipital bone along the grooves for the transverse sinuses. Its concave anteromedial border is free, producing a gap, the *tentorial notch* (incisure), for the midbrain (Fig. 8.9*A*).

The *falx cerebelli,* a vertical fold in the posterior part of the posterior cranial fossa, separates the cerebellar hemispheres.

The *diaphragma sellae,* a small circular sheet of dura, forms the roof of the hypophyseal fossa. It covers the pituitary gland and has an aperture for passage of the infundibulum (pituitary stalk) and hypophyseal veins.

DURAL VENOUS SINUSES

Dural venous sinuses are endothelial-lined spaces between the endosteal and meningeal layers of the dura mater (Figs. 8.8 and 8.9). Blood from the brain drains into these sinuses and ultimately into the internal jugular veins.

The *superior sagittal sinus* lies in the convex border of the falx cerebri. It begins at the crista galli and ends near the internal occipital protuberance at the *confluence of sinuses*. The superior sagittal sinus receives the superior cerebral veins and communicates on each side through slitlike openings with the *lateral lacunae* (venous spaces in dura mater).

The *inferior sagittal sinus,* much smaller than the superior sagittal sinus, runs in the inferior concave border of the falx cerebri and ends in the straight sinus.

The *straight sinus* is formed by the union of the inferior sagittal sinus with the great cerebral vein. It runs inferoposteriorly along the line of attachment of the falx cerebri to the tentorium cerebelli, where it becomes continuous with one of the transverse sinuses, usually the left.

The *transverse sinuses* pass laterally from the confluence of sinuses and groove the occipital bones and the posteroinferior angles of the parietal bones. The transverse sinuses leave the tentorium cerebelli and become the sigmoid sinuses.

The *sigmoid sinuses* follow S-shaped courses in the posterior cranial fossa, forming deep grooves in the temporal and occipital bones (Fig. 8.4*B*). Each sigmoid sinus turns anteriorly and then continues as the internal jugular vein inferior to the jugular foramen.

The *occipital sinus* lies in the attached border of the falx cerebelli and ends superiorly in the confluence of sinuses. The occipital sinus communicates inferiorly with the internal vertebral venous plexus (p. 216).

The *cavernous sinuses* (Fig. 8.9*A*) are located on each side of the sella turcica and the body of the sphenoid. Each sinus extends from the superior orbital fissure anteriorly to the apex of the petrous part of the temporal bone posteriorly. Each cavernous sinus receives blood from the superior and inferior ophthalmic

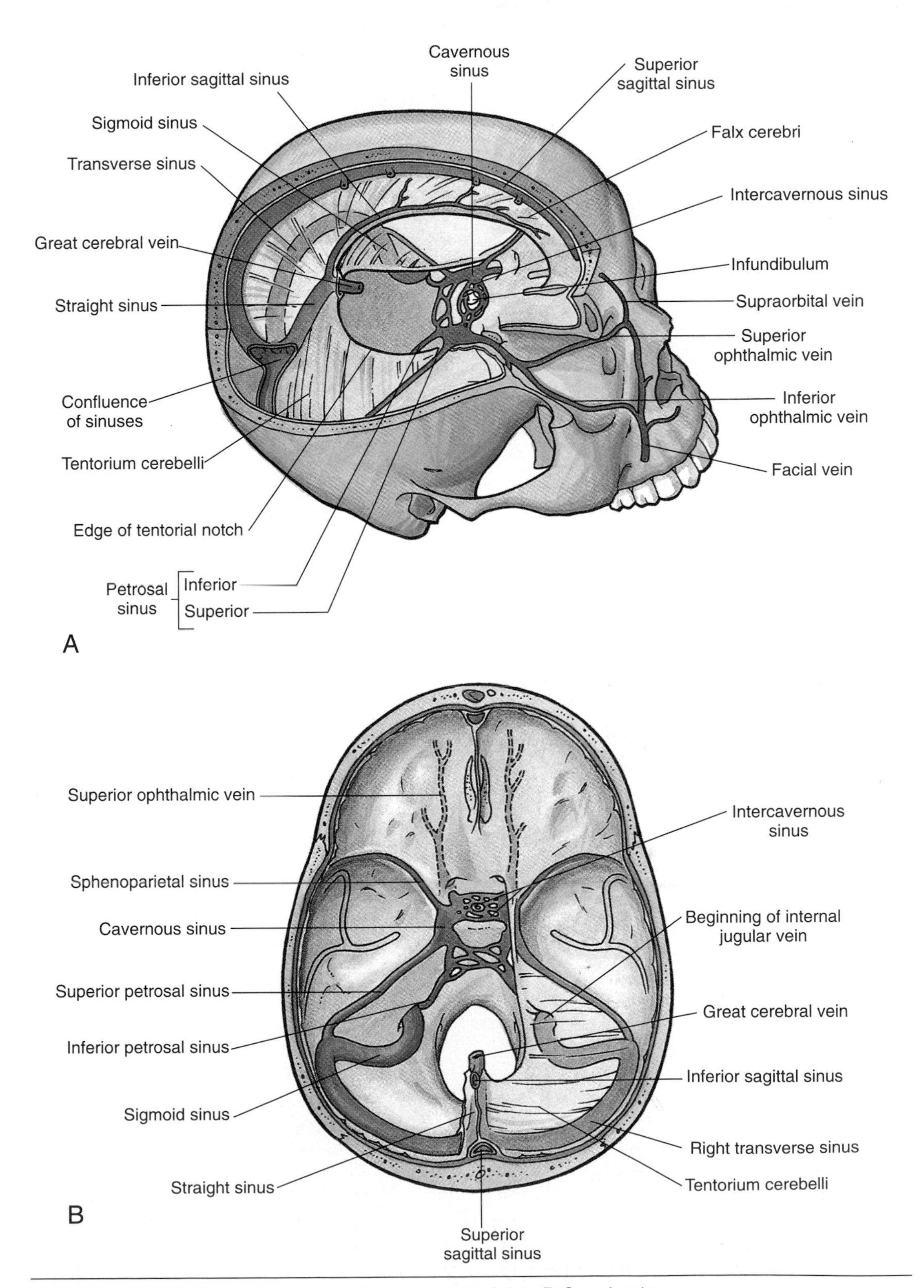

Figure 8.9. Dural folds and dural venous sinuses. **A.** Lateral view. **B.** Superior view.

veins, the superficial middle cerebral vein, and the sphenoparietal sinus. The cavernous sinuses communicate with each other through *intercavernous sinuses*. The cavernous sinuses drain posteriorly and inferiorly through the superior and inferior petrosal sinuses and the pterygoid plexuses. Inside each cavernous sinus is the *internal carotid artery* (Fig. 8.10) with its sympathetic plexus and abducent nerve (CN VI). From superior to inferior, the lateral wall of each cavernous sinus contains the

- Oculomotor nerve (CN III)
- Trochlear nerve (CN IV)
- CN V^1 and CN V^2 divisions of the trigeminal nerve

The *superior petrosal sinuses* run from the posterior ends of the cavernous sinuses to the transverse sinuses, at the site where these sinuses curve inferiorly to form the sigmoid sinuses. Each superior petrosal sinus lies in the margin of the tentorium cerebelli, which attaches to the superior margin of the petrous part of the temporal bone.

The *inferior petrosal sinuses* drain the cavernous sinuses into the internal jugular veins. Commencing at the posterior end of the cavernous sinus, each sinus runs inferiorly in a groove between the petrous part of the temporal bone and the basilar part of the occipital bone.

The *basilar sinus* connects the inferior petrosal sinuses and communicates inferiorly with the internal vertebral venous plexus.

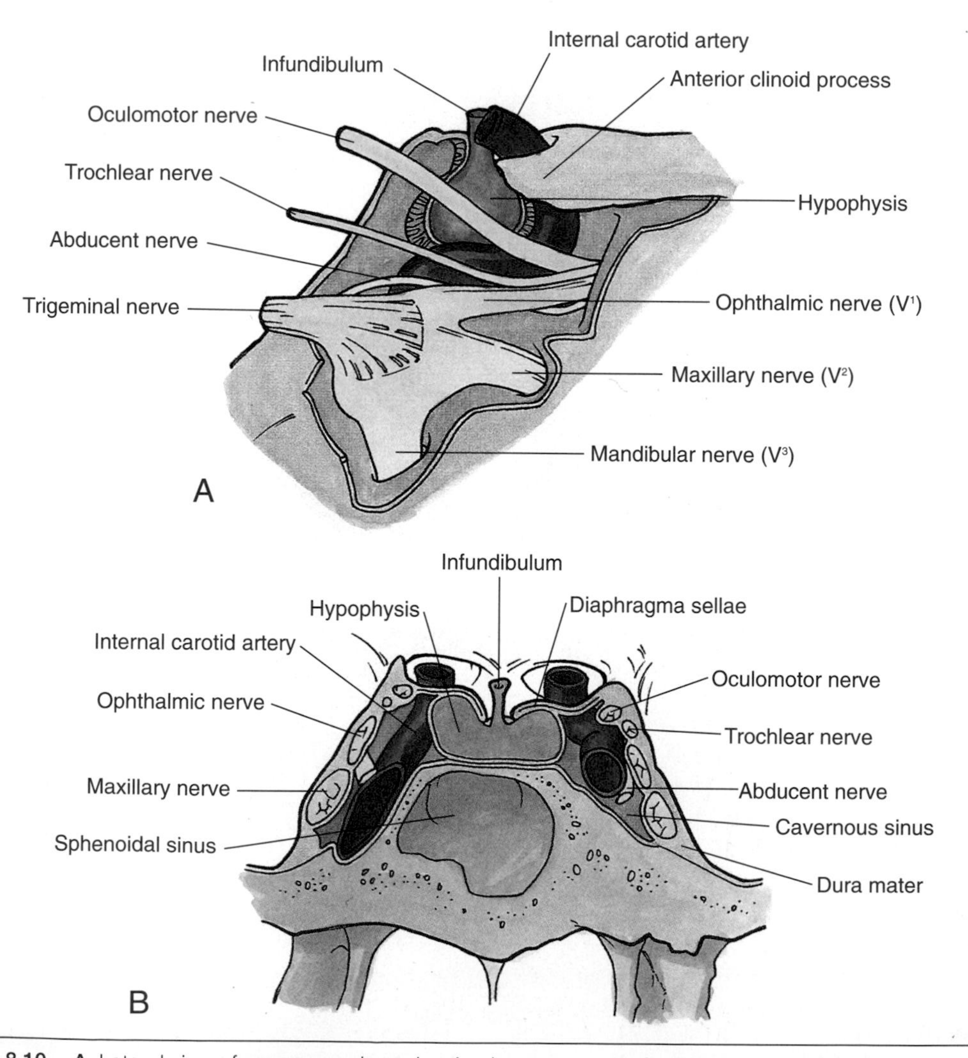

Figure 8.10. **A.** Lateral view of cavernous sinus showing its contents. **B.** Coronal section of cavernous sinuses.

Blood from the medial canthus of the eye, nose, and lips usually drains inferiorly into the facial vein (Fig. 8.6*B*). However, because the facial vein has no valves, blood may pass superiorly to the superior ophthalmic veins and enter the cavernous sinus (Fig. 8.9). In patients with *thrombophlebitis of the facial vein* [inflammation of the vein with secondary thrombus (clot) formation], pieces of a thrombus may produce *thrombophlebitis of the cavernous sinuses.* Infection of the facial veins spreading to the dural venous sinuses may be initiated by squeezing pustules on the side of the nose and upper lip.

The basilar and occipital sinuses communicate through the foramen magnum with the *internal vertebral venous plexuses.* Because these venous channels are valveless, compression of the thorax, abdomen, or pelvis—as occurs during heavy coughing and straining—may force venous blood from these regions into the vertebral venous system and from it into the dural venous sinuses. As a result, pus in abscesses and tumor cells in these regions may spread to the vertebrae and brain.

In *fractures of the base of the skull*, the internal carotid artery may tear within the cavernous sinus, producing an *arteriovenous fistula.* Arterial blood rushes into the cavernous sinus, enlarging it and forcing blood into the connecting veins, especially the superior ophthalmic veins. As a result, the eye protrudes (*exophthalmos*) and the conjunctiva becomes engorged (*chemosis*). The protruding eye pulsates in synchrony with the radial pulse, a phenomenon known as *pulsating exophthalmos.* Because CN III, IV, V^1, V^2, and VI lie in or close to the lateral wall of the cavernous sinus (Fig. 8.10), they may also be affected when injuries of this sinus occur.

VASCULATURE AND INNERVATION OF DURA MATER

Arteries of the dura mater supply more blood to the calvaria than they do to the dura. The largest of the meningeal arteries, the *middle meningeal artery*, is a branch of the maxillary artery (p. 380). The middle meningeal artery enters the cranial cavity through the *foramen spinosum*, runs laterally on the floor of the middle cranial fossa, and turns superoanteriorly on the greater wing of the sphenoid, where it divides into anterior and posterior branches. The anterior branch runs superiorly to the pterion and then curves posteriorly to ascend toward the vertex of the skull. The posterior branch runs posterosuperiorly and ramifies over the posterior aspect of the skull. *Veins of the dura mater* accompany the meningeal arteries and may also be torn in fractures of the calvaria.

The *nerve supply of the dura mater* is largely through the three divisions of CN V. Sensory branches are also received from the vagus nerve (CN X) and the superior three cervical nerves. The sensory endings are more numerous in the dura along each side of the superior sagittal sinus and in the tentorium cerebelli than they are in the floor of the cranium. Pain fibers are also numerous where arteries and veins pierce the dura.

The attachment of the endosteal layer of the dura mater to the floor of the cranium is firmer than it is to the calvaria. Consequently, a blow to the head can detach the dura from the calvaria without fracturing the bones, whereas a basal fracture usually tears the dura and arachnoid, resulting in leakage of CSF (e.g., into the neck).

Pituitary tumors may extend superiorly through the aperture in the diaphragma sellae and/or cause bulging of the diaphragm. Pressure from these tumors on the pituitary gland and optic chiasma may produce endocrine and/or visual symptoms.

MENINGEAL SPACES

Three meningeal spaces relate to the cranial meninges:

- Extradural (epidural) space is between the cranial bones and the endosteal layer of the dura (because the dura is attached to the bones, the extradural space is a potential space; it becomes real space if blood from torn meningeal vessels accumulates in it)
- Subdural space is a potential space that may develop in the deepest part of the dura after a head injury
- Subarachnoid space, between the arachnoid and pia mater, is filled with CSF

Brain

The brain is composed of the cerebrum, cerebellum, and brainstem (midbrain, pons, and medulla). The cranial cavity lodges the brain and meninges. The roof of the cranial cavity is formed by the calvaria, and its floor is formed by the base of the skull. The following brief discussion of the gross structure of the brain is presented to show how the brain relates to the cranium, cranial nerves, CSF, and meninges.

When the calvaria and dura mater are removed, folds (*gyri*) and grooves (*sulci and fissures*) of the cerebral cortex are visible through the delicate layers of arachnoid and pia mater. The sulci and fissures of the brain are distinctive landmarks that subdivide the cerebral hemispheres into smaller areas such as lobes (Fig. 8.11).

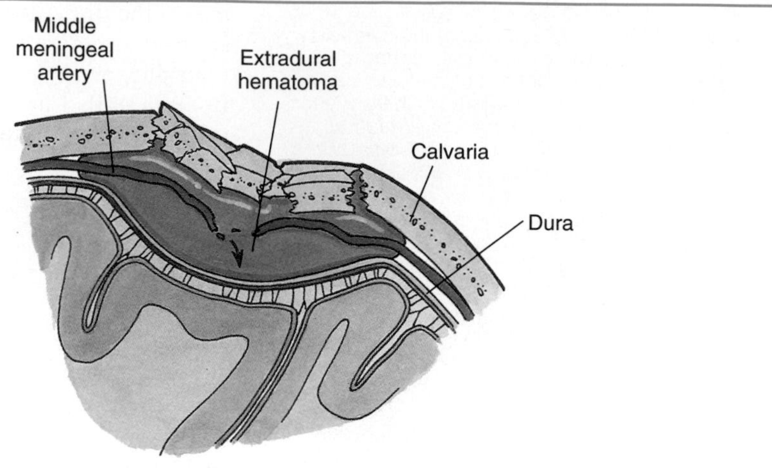

Extradural hematoma

Head injuries may be associated with various types of intracranial hemorrhage. *Extradural (epidural) hemorrhage* between the endosteal layer of dura and the calvaria may follow a blow to the head. Typically, a brief concussion results, followed by a lucid interval of some hours. This is succeeded by drowsiness and coma. Most bleeding is from the torn meningeal arteries and results in an *extradural (epidural) hematoma*, a slow, localized accumulation of blood. As the blood mass increases, compression of the brain occurs, necessitating evacuation of the blood and occlusion of the bleeding vessels.

Subdural hemorrhage may follow a blow to the head that jerks the brain inside the skull and injures it. Displacement of the brain is greatest in elderly people in whom some shrinkage of the brain has occurred. Subdural hemorrhage commonly results from tearing of a cerebral vein as it enters the superior sagittal sinus (Fig. 8.8). Subdural hemorrhage results in subdural hematoma. Although the dura and arachnoid are parts of a single membrane [see Haines (1991) listed under "Suggested Readings"], blood may collect in an abnormal subdural space that forms when trauma separates them.

Subarachnoid hemorrhage into the subarachnoid space usually follows the rupture of an *aneurysm* [dilatation of an intracranial blood vessel (p. 367)]. Subarachnoid hemorrhages are also associated with skull fractures and cerebral lacerations. This bleeding results in meningeal irritation, which produces severe headache, stiff neck, and often loss of consciousness.

Headaches may be dural in origin. The dura mater is sensitive to pain, especially around the superior sagittal sinus. A headache occurring after a *lumbar puncture* for removal of CSF (p. 213) may result from stimulation of sensory nerve endings in the dura. When CSF is removed the brain sags slightly, pulling on the dura and its vessels and nerves. For this reason, patients are asked to keep their heads down after a lumbar puncture to minimize or prevent headaches.

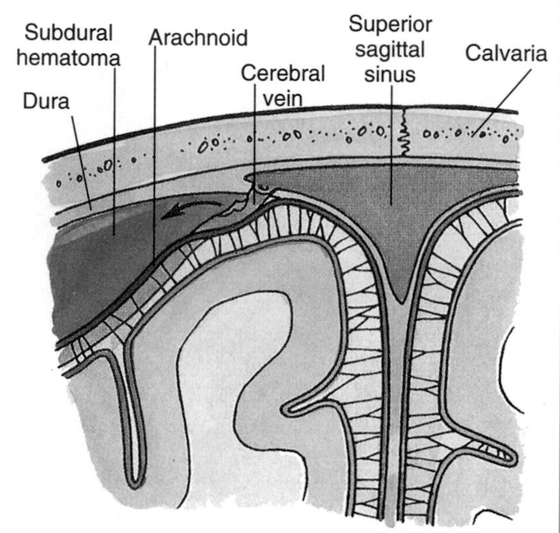

Subdural hematoma

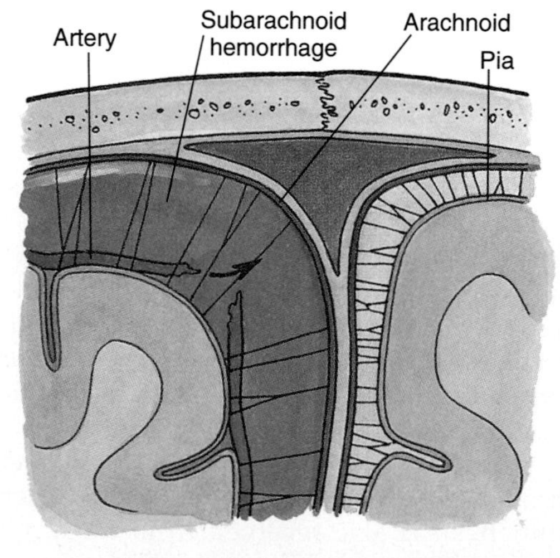

Subarachnoid hemorrhage

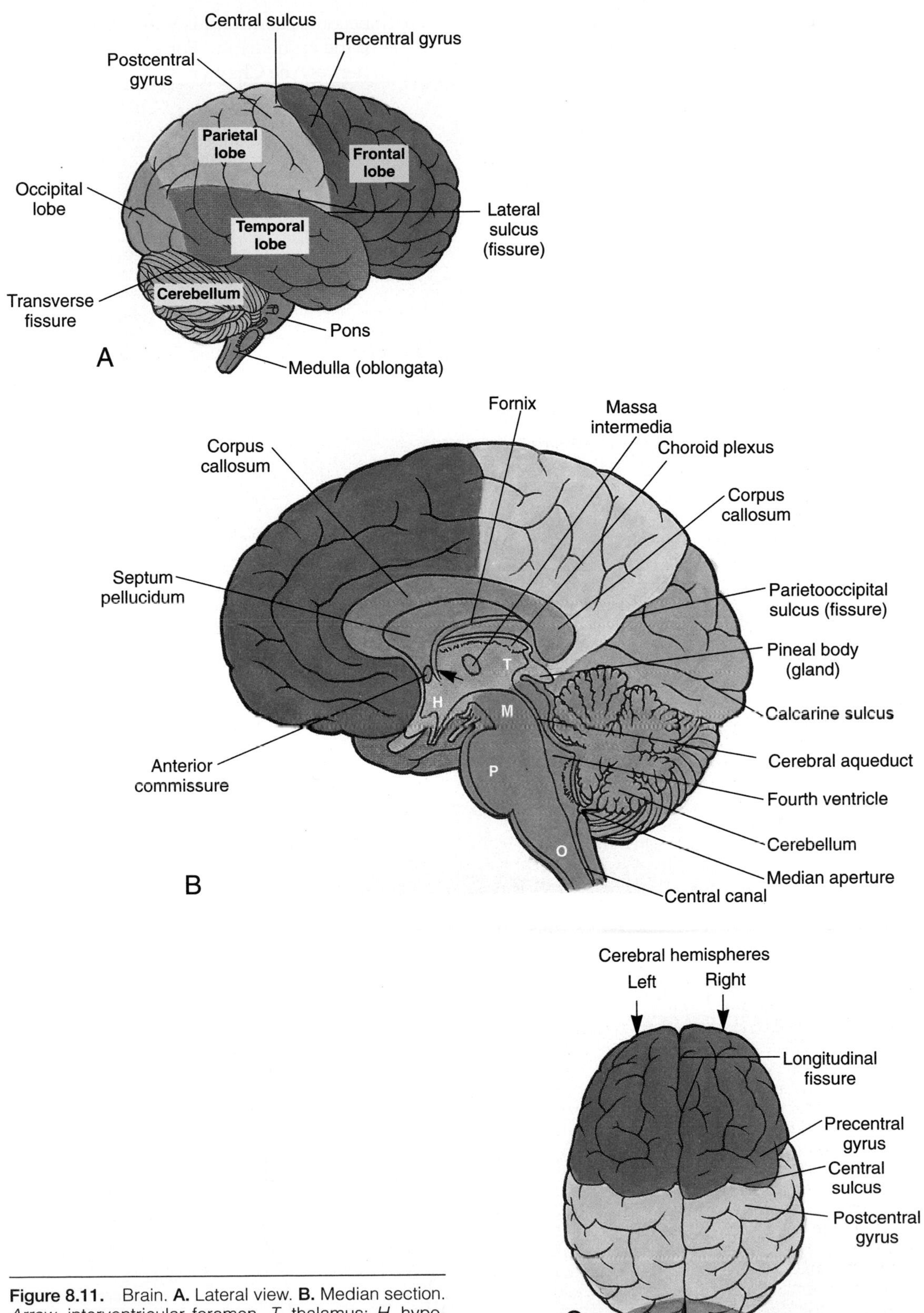

Figure 8.11. Brain. **A.** Lateral view. **B.** Median section. *Arrow*, interventricular foramen. *T*, thalamus; *H*, hypothalamus; *M*, midbrain; *P*, pons; *O*, medulla. **C.** Superior view.

- Cerebral hemispheres form the largest part of the brain; they occupy the anterior and middle cranial fossae and extend posteriorly over the tentorium cerebelli and cerebellum (the cavity in each hemisphere, a lateral ventricle, is part of the ventricular system)
- Diencephalon (the largest parts being the thalamus and hypothalamus) forms the central core of the brain and surrounds the third ventricle of the brain; the cavity between the right and left halves of the diencephalon forms the narrow third ventricle
- Midbrain, the rostral part of the brainstem, lies at the junction of the middle and posterior cranial fossae; the cavity of the midbrain forms a narrow canal, the *cerebral aqueduct*, that conducts CSF from the lateral and third ventricles to the fourth ventricle
- Pons, the middle part of the brainstem, lies in the anterior part of the posterior cranial fossa; the cavity in the pons forms the superior part of the fourth ventricle
- Medulla oblongata, the caudal part of the brainstem, lies in the posterior cranial fossa and is continuous with the spinal cord; the cavity of the medulla forms the inferior part of the fourth ventricle
- Cerebellum overlies the pons and medulla and lies beneath the tentorium cerebelli in the posterior cranial fossa

Cranial nerves arise from the brain. In general, cranial nerves are surrounded by a dural sheath as they leave the cranium; the dural sheath becomes continuous with the connective tissue of the epineurium. For a summary of the cranial nerves, see Chapter 10.

Cerebral concussion is an abrupt transient loss of consciousness immediately after a blow to the head. *Cerebral contusion* (bruising) results from trauma. The pia is stripped from the injured area of the brain and may be torn, allowing blood to enter the subarachnoid space. The bruising results either from the sudden impact of the moving brain against the stationary skull or the skull against the brain. A contusion may result in an extended loss of consciousness. *Cerebral lacerations* (tearing of neural tissue) are often associated with depressed skull fractures or gunshot wounds. Lacerations result in rupture of blood vessels and bleeding into the brain and subarachnoid space. As a result, there is an increase in intracranial pressure and cerebral compression. *Cerebral compression* may be produced by

- Intracranial collections of blood
- Obstruction of CSF circulation or absorption
- Intracranial tumors or abscesses
- Edema of brain (e.g., after a head injury)

VENTRICULAR SYSTEM

The ventricular system of the brain consists mainly of four ventricles (Figs. 8.11*B* and 8.12). The first and second ventricles, or *lateral ventricles*, are the largest parts of the system and occupy large parts of the cerebral hemispheres. Each lateral ventricle opens into the third ventricle

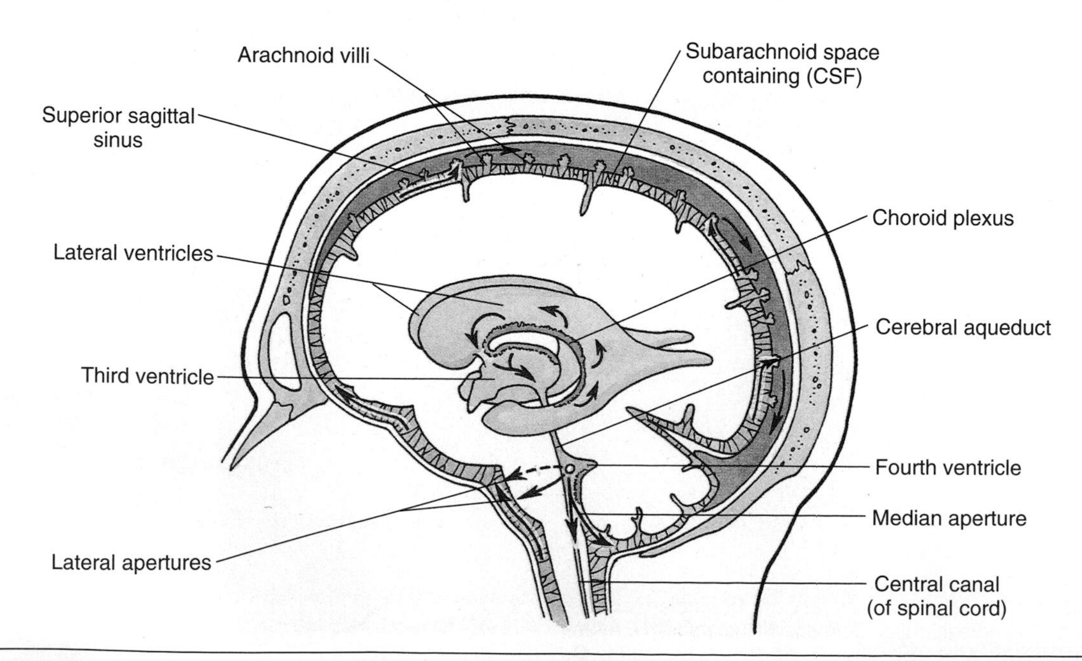

Figure 8.12. Ventricular system of brain. *Arrows*, direction of CSF flow.

through an interventricular foramen. The *third ventricle* is a slitlike cavity between the right and left halves of the diencephalon. It is continuous posteroinferiorly with the *cerebral aqueduct* that connects the third and fourth ventricles. The *fourth ventricle* in the posterior parts of the pons and medulla extends inferoposteriorly where it is continuous with the central canal of the inferior part of the medulla and throughout the spinal cord. CSF drains from the fourth ventricle through a single median and paired lateral apertures into the subarachnoid space. These apertures are the only means by which CSF enters the subarachnoid space. If they are blocked, the ventricles distend, producing compression of the cerebral hemispheres.

At certain places, mainly at the base of the brain, the arachnoid and pia mater are widely separated by large pools of CSF (cisterns). The main *subarachnoid cisterns* include the

- Cerebellomedullary cistern (cisterna magna), between the inferior surface of the cerebellum and the roof of the fourth ventricle
- Pontine cistern on the anterior surface of the pons and medulla
- Interpeduncular cistern between the cerebral peduncles of the midbrain
- Superior cistern between the posterior part of the corpus callosum and the superior surface of the cerebellum

The main source of CSF secretion is in the choroid plexuses (Fig. 8.12). *Choroid plexuses* are located in the roofs of the third and fourth ventricles and on the floors of the bodies and inferior horns of the lateral ventricles. Although choroid plexuses are the main source of CSF and the arachnoid villi are the main sites of CSF absorption, there are exchanges between blood plasma and CSF elsewhere (e.g., across the lining of the ventricles). CSF from the lateral and third ventricles passes through the *cerebral aqueduct* into the fourth ventricle. It leaves the fourth ventricle through its *median and lateral apertures* and passes into the subarachnoid space, where it collects in the cerebellomedullary and pontine cisterns. From these cisterns, some CSF passes inferiorly into the subarachnoid space around the spinal cord and posterosuperiorly over the cerebellum. However, most CSF flows into the interpeduncular and superior cisterns. CSF from

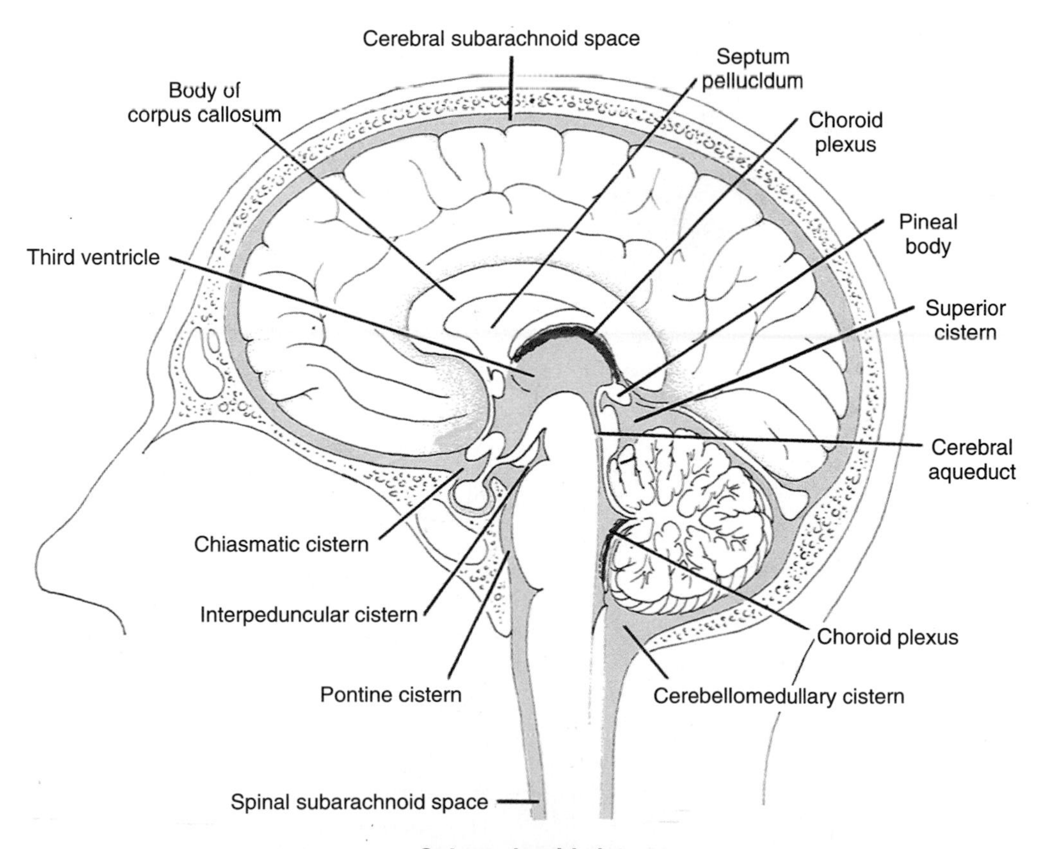

Subarachnoid cisterns

the various cisterns spreads superiorly through the sulci and fissures on the medial and superolateral surfaces of the cerebral hemispheres. CSF also passes into the extensions of the subarachnoid space around the cranial nerves, the most important of which are those surrounding the optic nerves (see Fig. 8.17).

The main site of CSF absorption into the venous system is through the tiny *arachnoid villi* [protrusions of arachnoid into the walls of dural venous sinuses, especially the superior sagittal sinus and lateral lacunae (Figs. 8.8 and 8.12)]. The arachnoid villi become hypertrophied with age, when they are called *arachnoid granulations.*

CSF may be obtained from the cerebellomedullary cistern, a procedure known as *cisternal puncture.* The subarachnoid space or the ventricular system may also be entered for measuring or monitoring CSF pressure, injecting antibiotics, or administering contrast media for radiography.

Overproduction of CSF, obstruction of its flow, or interference with its absorption results in an excess of CSF in the ventricles and enlargement of the head, a condition known as *hydrocephalus.* Excess CSF dilates the ventricles, thins the brain, and separates the cranial bones in infants.

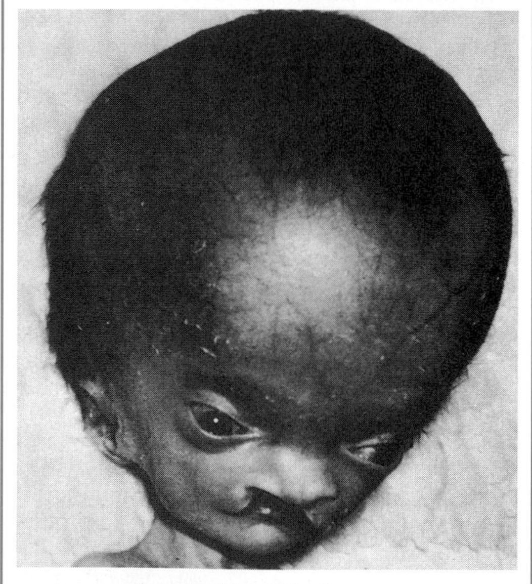

Hydrocephalus

Fractures in the floor of the middle cranial fossa may result in leakage of CSF from the ear (*CSF otorrhea*) if the meninges superior to the middle ear are torn and the tympanic membrane (eardrum) is ruptured. Fractures in the floor of the anterior cranial fossa may involve the cribriform plate of the ethmoid, resulting in leakage of CSF through the nose (*CSF rhinorrhea*). CSF otorrhea and CSF rhinorrhea present a risk of *meningitis* because an infection may spread to the meninges from the ear or nose.

BLOOD SUPPLY

The blood supply to the brain is from branches of the internal carotid and vertebral arteries (Fig. 8.13, Table 8.4). The *internal carotid arteries* arise in the neck from the common carotid arteries. The terminal branches of the internal carotid are the anterior and middle cerebral arteries. The *vertebral arteries* begin in the root of the neck as branches of the first part of the subclavian arteries and unite at the caudal border of the pons to form the *basilar artery.* The basilar artery, so-named because of its close relationship to the base of the skull, runs through the pontine cistern to the superior border of the pons, where it ends by dividing into the two *posterior cerebral arteries.*

In general, each cerebral artery supplies a surface and a pole of the brain as follows:

- The anterior cerebral artery supplies most of the medial and superior surfaces and the frontal pole (*green*)

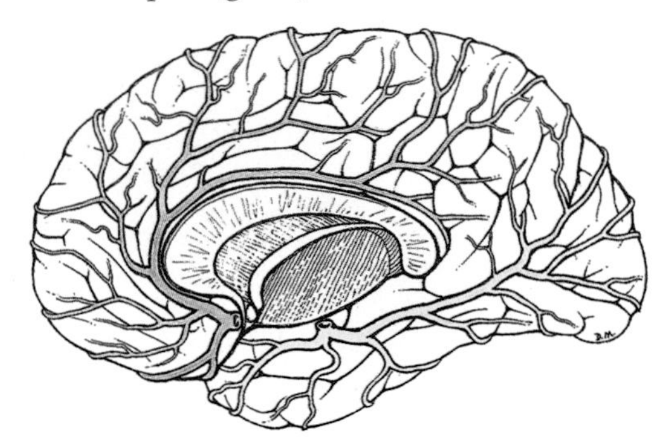

Medial view

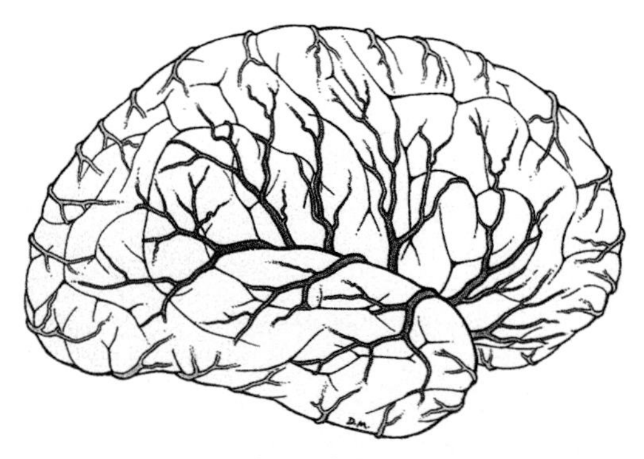

Lateral view

Blood supply to brain

- The middle cerebral artery supplies the lateral surface and temporal pole (*purple*)
- The posterior cerebral artery supplies the inferior surface and occipital pole (*yellow*)

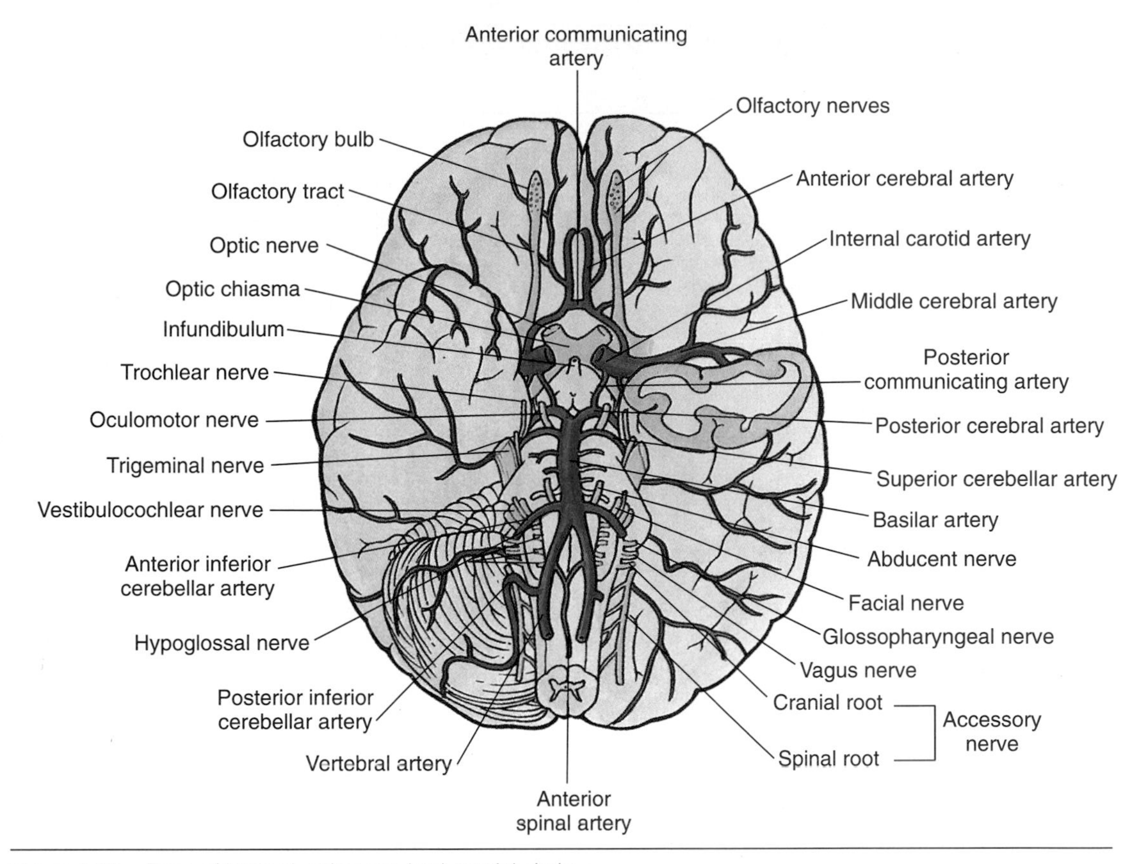

Figure 8.13. Base of brain showing cerebral arterial circle.

The *cerebral arterial circle* (of Willis) at the base of the brain is an important anastomosis between the four arteries (vertebrals and internal carotids) that supply the brain. It is formed by the posterior cerebral, posterior communicating, internal carotid, anterior cerebral, and anterior communicating arteries. Variations in the size of the vessels that form the circle are common.

The cerebral arterial circle is an important means of collateral circulation in the event one of the arteries forming the circle is obstructed. In elderly persons, the anastomoses are often inadequate when a large artery (e.g., the internal carotid) is suddenly occluded; a *vascular stroke* results. The most common causes of vascular strokes are spontaneous *cerebrovascular accidents* such as cerebral thrombosis, cerebral hemorrhage, cerebral embolism, and subarachnoid hemorrhage. *Hemorrhagic stroke* follows from rupture of an artery or an aneurysm. The most common type of aneurysm is a *berry aneurysm*, occurring in the vessels of or near the cerebral arterial circle and the medium-sized arteries at the base of the brain. In time, especially in persons with high blood pressure (hypertension), the weak part of the wall expands and may rupture, allowing blood to enter the subarachnoid space.

Orbit

The orbit is a pyramidal cavity in the facial skeleton with its base anterior and its apex posterior. The orbits contain the eyeballs, protecting them and their associated muscles, nerves, and vessels, together with most of the lacrimal apparatus. The bones forming the orbital cavity are lined with *periorbita* (periosteum). The periorbita forms a fascial sheath for the orbital contents. At the optic canal and superior orbital fissure, the periorbita is continuous with the endosteal layer of dura. It is also continuous over the orbital margins and through the inferior orbital fissure with the periosteum covering the external surface of the skull (pericranium). The orbit has four walls and an apex:

- The superior wall (roof) is formed mainly by the orbital plate of the frontal bone (Fig. 8.4*B*) that separates the orbital cavity from the anterior cranial fossa; near the apex the

Table 8.4
Arterial Supply to Brain

Artery	Origin	Distribution
Vertebral	Subclavian a.	Cranial meninges and cerebellum
Posterior inferior cerebellar	Vertebral a.	Posteroinferior aspect of cerebellum
Basilar	Formed by junction of vertebral aa.	Brainstem, cerebellum, and cerebrum
Pontine	Basilar a.	Numerous branches to brainstem
Anterior inferior cerebellar	Basilar a.	Inferior aspect of cerebellum
Superior cerebellar	Basilar a.	Superior aspect of cerebellum
Internal carotid	Common carotid a. at superior border of thyroid cartilage	Gives branches in cavernous sinus and provides primary supply to brain
Anterior cerebral	Internal carotid a.	Cerebral hemispheres, except for occipital lobes (p. 366)
Middle cerebral	Continuation of the internal carotid a. distal to anterior cerebral a.	Most of lateral surface of cerebral hemispheres
Posterior cerebral	Terminal branch of basilar a.	Inferior aspect of cerebral hemisphere and occipital lobe (p. 366)
Anterior communicating	Anterior cerebral a.	Cerebral arterial circle
Posterior communicating	Posterior cerebral a.	Cerebral arterial circle

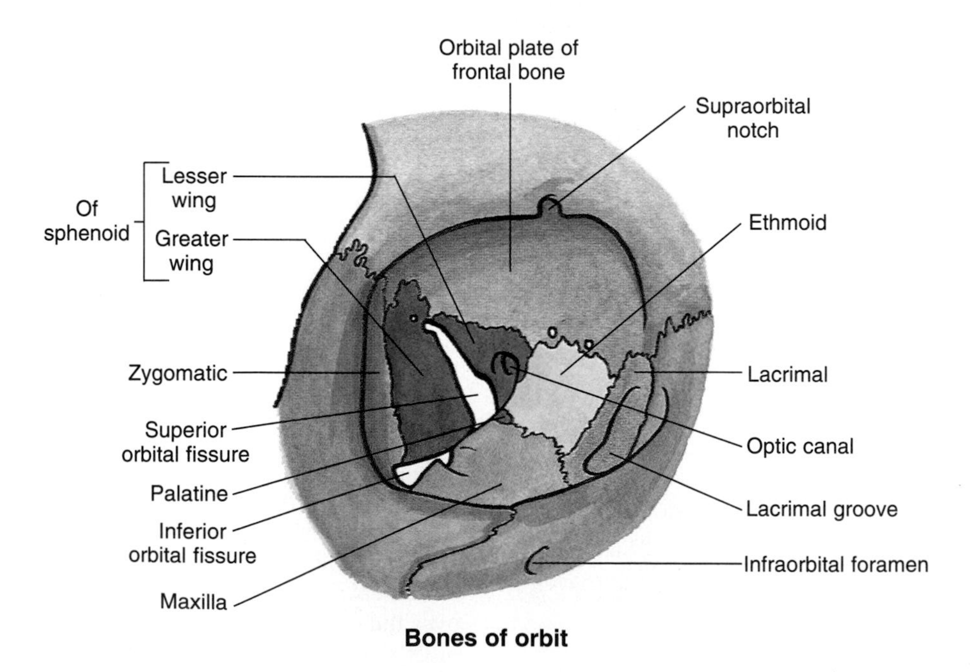

Bones of orbit

wall is formed by the lesser wing of the sphenoid

- The medial wall is formed by the ethmoid, along with contributions from the frontal, lacrimal, and sphenoid bones; anteriorly, the paper-thin medial wall is indented by the *lacrimal groove*, which accommodates the lacrimal sac and the proximal part of the nasolacrimal duct
- The inferior wall (floor) is formed mainly

by the maxilla and partly by the zygomatic and palatine bones; the thin floor is partly separated from the lateral wall of the orbit by the inferior orbital fissure

Because of the thinness of the medial and inferior walls, a blow to the eye may fracture the orbit. Fractures of the medial wall may involve the ethmoidal and sphenoidal sinuses, whereas fractures in the floor may involve the maxillary sinus (p. 401). Although the superior wall is stronger than the medial and inferior walls, it is thin enough to be translucent and may be readily penetrated. Thus a sharp object may pass through it into the frontal lobe of the brain.

- The lateral wall is formed by the frontal process of the zygomatic and the greater wing of the sphenoid; the lateral wall is thick, especially its posterior part that separates the orbit from the middle cranial fossa
- The apex is at the optic canal, just medial to the superior orbital fissure

EYELIDS AND LACRIMAL APPARATUS

The eyelids (palpebrae) protect the eyes from injury and excessive light and keep the cornea moist. These movable folds are covered externally by thin skin and internally by *palpebral conjunctiva* (Fig. 8.14). The palpebral con-

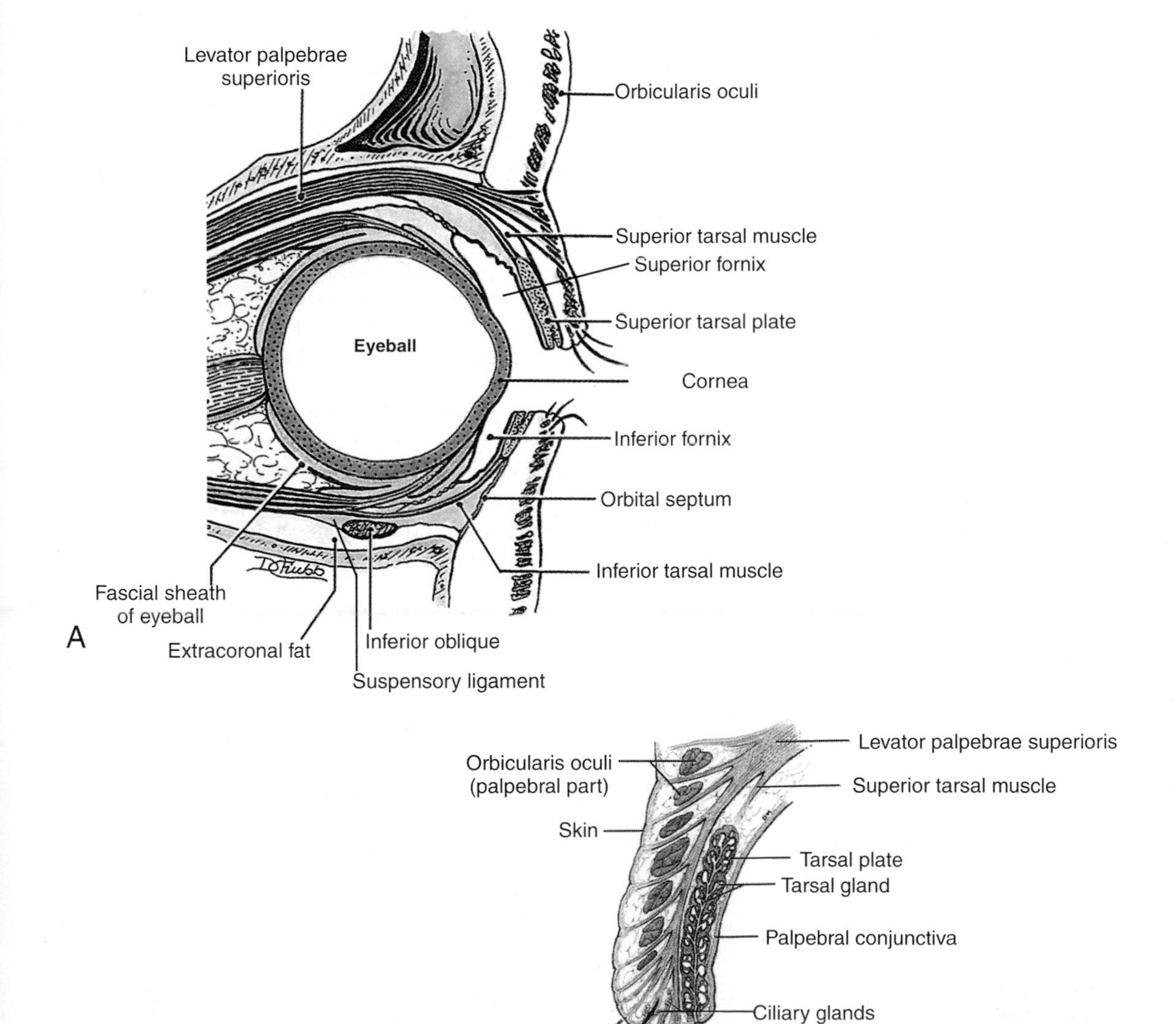

Figure 8.14. Sagittal sections of orbit (**A**) and eyelid (**B**).

junctiva is reflected onto the eyeball where it is continuous with the *bulbar conjunctiva*. The lines of reflection of the palpebral conjunctiva onto the eyeball form deep recesses, the superior and inferior *conjunctival fornices*. The eyelids are strengthened by dense bands of connective tissue, the superior and inferior *tarsal plates*. Fibers of the orbicularis oculi muscle are in the connective tissue between this plate and the skin of the eyelid. Embedded in the tarsal plates are *tarsal glands*, the secretion of which lubricates the edges of the eyelids and prevents them from sticking together when they close. The eyelashes, or cilia, are in the margins of the lids. The large sebaceous glands associated with the eyelashes are the *ciliary glands*.

Between the nose and the medial canthus (angle) of the eye is the *medial palpebral ligament* that connects the eyelids, including their muscles, to the medial margin of the orbit. A similar *lateral palpebral ligament* attaches the eyelids to the lateral margin of the orbit. The medial and lateral palpebral ligaments are connected by the tarsal plates. The *orbital septum* is a weak membrane that connects the tarsal plates to the margins of the orbit and then becomes continuous with the periosteum.

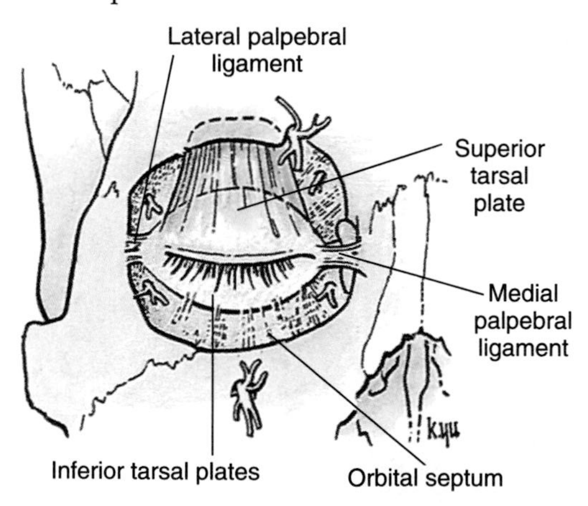

Skeleton of eyelids

A person with paralysis of the facial nerve (CN VII) cannot close the eyelids because of paralysis of the orbicularis oculi. Protective blinking of the eye is also lost; as a result, lacrimal fluid does not wash across the cornea to keep it moist (see Bell's palsy, p. 353).

If the ducts of the ciliary glands become obstructed or inflamed, a painful red swelling known as a *sty* develops on the eyelid. Cysts of the ciliary glands associated with the eyelashes (*chalazia*) may also form.

ORBITAL CONTENTS

The orbital contents are the eyeball, optic nerve, ocular muscles, fasciae, nerves, vessels, fat, and the lacrimal gland and sac. The *eyeball* has three layers (Fig. 8.15*A*):

- The external fibrous supporting layer consists of the sclera and cornea
- The middle vascular pigmented layer consists of the choroid, ciliary body, and iris
- The internal neural layer consists of the retina

External Fibrous Layer

The *sclera* is the posterior five-sixths, opaque part of the external coat. The anterior part of the sclera is visible through the bulbar conjunctiva as the "white of the eye" (p. 371). The *cornea* is the anterior one-sixth, transparent part of the external coat.

Middle Vascular Layer

The *choroid*, a dark brown membrane between the sclera and retina, forms the largest part of the middle coat and lines most of the sclera. It terminates anteriorly in the ciliary body. The choroid is firmly attached to the retina, but it can easily be stripped from the sclera.

The *ciliary body* connects the choroid with the circumference of the iris. The ciliary body has folds on its internal surface—*ciliary processes*—that secrete aqueous humor. This fluid fills the anterior and posterior chambers of the eyeball, which are fluid-filled spaces anterior and posterior to the iris.

The *iris*, anterior to the lens, is a contractile diaphragm with a pupil, the central aperture for transmitting light. When a person is awake, the size of the pupil varies continually to regulate the amount of light entering the eye. Two muscles control the size of the pupil: the sphincter pupillae closes the pupil, and the dilator pupillae opens it.

Internal Neural Layer or Retina

The internal neural layer or retina comprises two layers: a pigment cell layer and a neural layer. In the *fundus*, the posterior portion of the retina, there is a circular depressed spot—the *optic disc* (optic papilla)—where the optic nerve enters the eyeball. Because it contains nerve fibers and no photoreceptors, the optic disc is insensitive to light. Just lateral to this blind area is

Surface Anatomy of Eyelids, Eyeball, and Lacrimal Apparatus

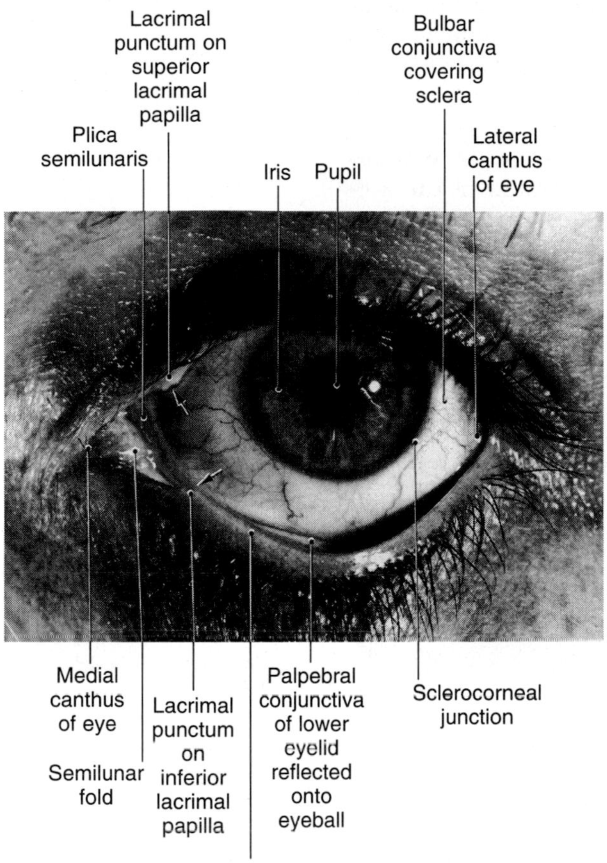

The place where the upper and lower eyelids meet is the canthus; each eye has medial and lateral canthi. In the medial canthus is a reddish area, the *lacrimal lake*, within which is the *lacrimal caruncle*, a small mound of moist, modified skin. Lateral to the caruncle is a semilunar fold of conjunctiva that slightly overlaps the eyeball. When the edge of the lower eyelid is everted, a small pit called the *lacrimal punctum* (*arrows*) is visible at its medial end on the summit of a small elevation called the *lacrimal papilla*. A similar punctum and papilla are on the upper eyelid.

The *lacrimal gland* in the superolateral part of the orbit produces lacrimal fluid (tears). The *lacrimal duct* opens into the superior fornix of the conjunctiva (Fig. 8.14*A*). When the cornea dries, the eyelids blink, carrying a film of fluid over the cornea. The lacrimal lake constitutes a small reservoir of lacrimal fluid. Each lacrimal punctum is the opening of a slender canal, the *lacrimal canaliculus*, which carries the fluid to the *lacrimal sac*. From here, the fluid passes through the *nasolacrimal duct* where it empties into the inferior meatus of the nose (p. 398).

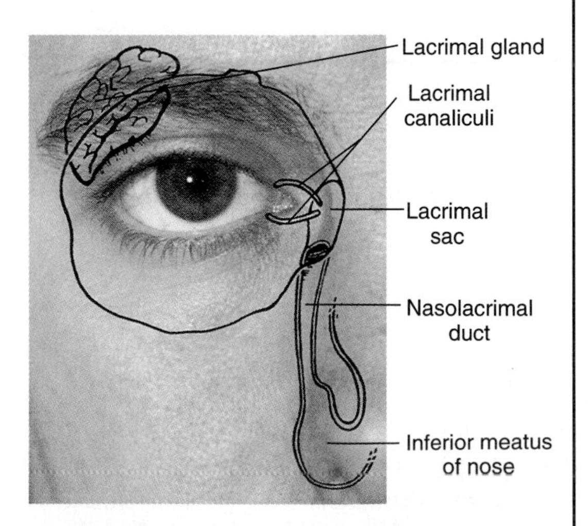

a yellow spot, the macula lutea; at its center there is a depression—the *fovea centralis*—the area of most acute vision. The retina is supplied by the central artery of the retina, a branch of the ophthalmic artery. A corresponding system of retinal veins unites to form the central vein of the retina.

The retina and optic nerve develop from an outgrowth of the embryonic forebrain, the *optic vesicle*, which carries meninges with it. Hence the optic nerve is invested with meninges containing an extension of the subarachnoid space. The central artery and vein of the retina run within the distal part of the optic nerve and cross the subarachnoid space around it.

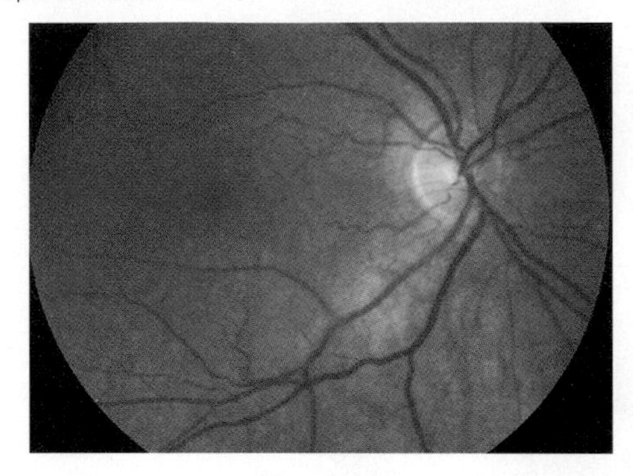

Fundus

Physicians view the fundus with an *ophthalmoscope*. The retinal arteries and veins radiate over the fundus from the optic disc. An increase in CSF pressure slows venous return from the retina, causing edema (fluid accumulation). Normally the optic disc is flat and does not form a papilla. *Edema of the retina* is obvious during ophthalmoscopy as swelling of the optic disc (papilledema). *Papilledema* results from increased intracranial pressure, which increases CSF pressure in the extension of the subarachnoid space around the optic nerve. Observe the pale, oval optic disc with retinal vessels radiating from its center in this view of the retina seen through an ophthalmoscope.

Refractive Media of Eye

On their way to the retina, light waves pass through the *refractive media of the eye*: cornea, aqueous humor, lens, and vitreous humor (Fig. 8.15*A*).

The *cornea* is the circular area of the anterior part of the external fibrous coat of the eyeball; it is largely responsible for refraction of the light that enters the eye. It is transparent, avascular, and sensitive to touch. The cornea is supplied by the ophthalmic nerve (CN V^1) and is nourished by aqueous humor, lacrimal fluid, and oxygen absorbed from the air.

The *aqueous humor* in the anterior and posterior chambers is produced by the ciliary processes. This clear watery solution provides nutrients for the avascular cornea and lens. After passing through the pupil from the posterior chamber into the anterior chamber, the aqueous humor drains into the venous sinus of the sclera, called the *sinus venosus sclerae* (canal of Schlemm).

The *lens* is a transparent biconvex structure enclosed in a lens capsule. The lens capsule is anchored to the ciliary body and retina by the suspensory ligament of the lens. Encircled by the ciliary processes, the lens is posterior to the iris and anterior to the vitreous humor. The curvatures of the surfaces of the lens, particularly the anterior surface, constantly vary to focus near or distant objects on the retina. The shape of the lens is changed by the ciliary muscle in the ciliary body (Fig. 8.15*B*).

The *vitreous humor* is the transparent gel that occupies the vitreous body in the posterior four-fifths of the eyeball between the lens and retina. In addition to transmitting light, the vitreous humor holds the retina in place and provides support for the lens.

When the sensory innervation of the cornea is damaged, the cornea can easily be injured by foreign particles that may produce *corneal ulcers*. Patients with scarred or opaque corneas often receive homologous *corneal transplants*. Corneal implants of nonreactive plastic material are also used.

As one gets older, the lenses become harder and more flattened. These changes gradually reduce the person's focusing power, a condition known as *presbyopia*. Some elderly people also experience a loss of transparency (clouding) of the lens from areas of opaqueness (*cataracts*). Cataract extraction is a common eye operation.

When drainage of aqueous humor is reduced significantly, pressure builds up in the chambers of the eye (glaucoma). Blindness can result from compression of the neural layer of the retina if aqueous humor production is not reduced to maintain normal intraocular pressure.

MUSCLES OF ORBIT

The muscles of the orbit are the levator palpebrae superioris, the four recti (superior, inferior, medial, and lateral), and the two oblique (superior and inferior). Their attachments, nerve supply, and actions are illustrated in Figure 8.16 and Table 8.5.

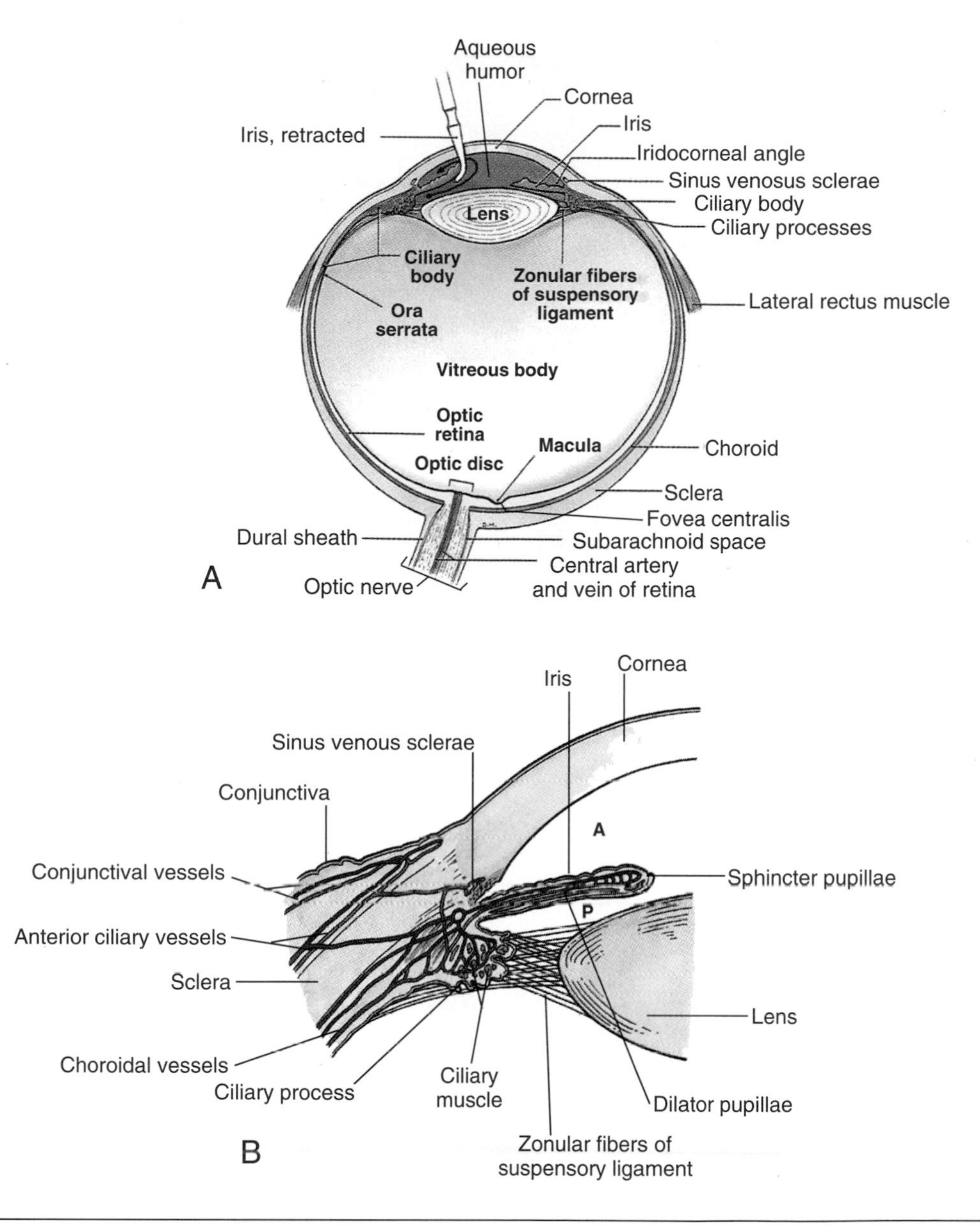

Figure 8.15. **A.** Horizontal section of eyeball showing its three coats. *Arrow*, flow of aqueous humor from posterior chamber to anterior chamber. **B.** Ciliary region of eyeball. *A*, anterior chamber; *P*, posterior chamber.

The four recti arise from a fibrous cuff, the *common tendinous ring*, that surrounds the optic canal and part of the superior orbital fissure (p. 376). Structures that enter the orbit through this canal and the adjacent part of the fissure lie at first in the cone of recti. The lateral and medial recti lie in the same horizontal plane, and the superior and inferior recti lie in the same vertical plane. All four recti attach to the sclera on the anterior half of the eyeball; their actions can be determined from this. The medial and lateral recti pull the eyeball medially and laterally, respectively. The superior rectus pulls the eyeball superiorly, and the inferior rectus pulls it inferiorly; however, neither muscle pulls directly parallel to the long axis of the eyeball. As a result both recti tend to pull the eyeball medially. This medial pull of the superior and inferior recti is normally overcome by the pull of the obliques. The inferior oblique directs the eyeball laterally and superiorly; therefore when it works with the superior rectus an upward movement of the eyeball occurs. Similarly, the superior

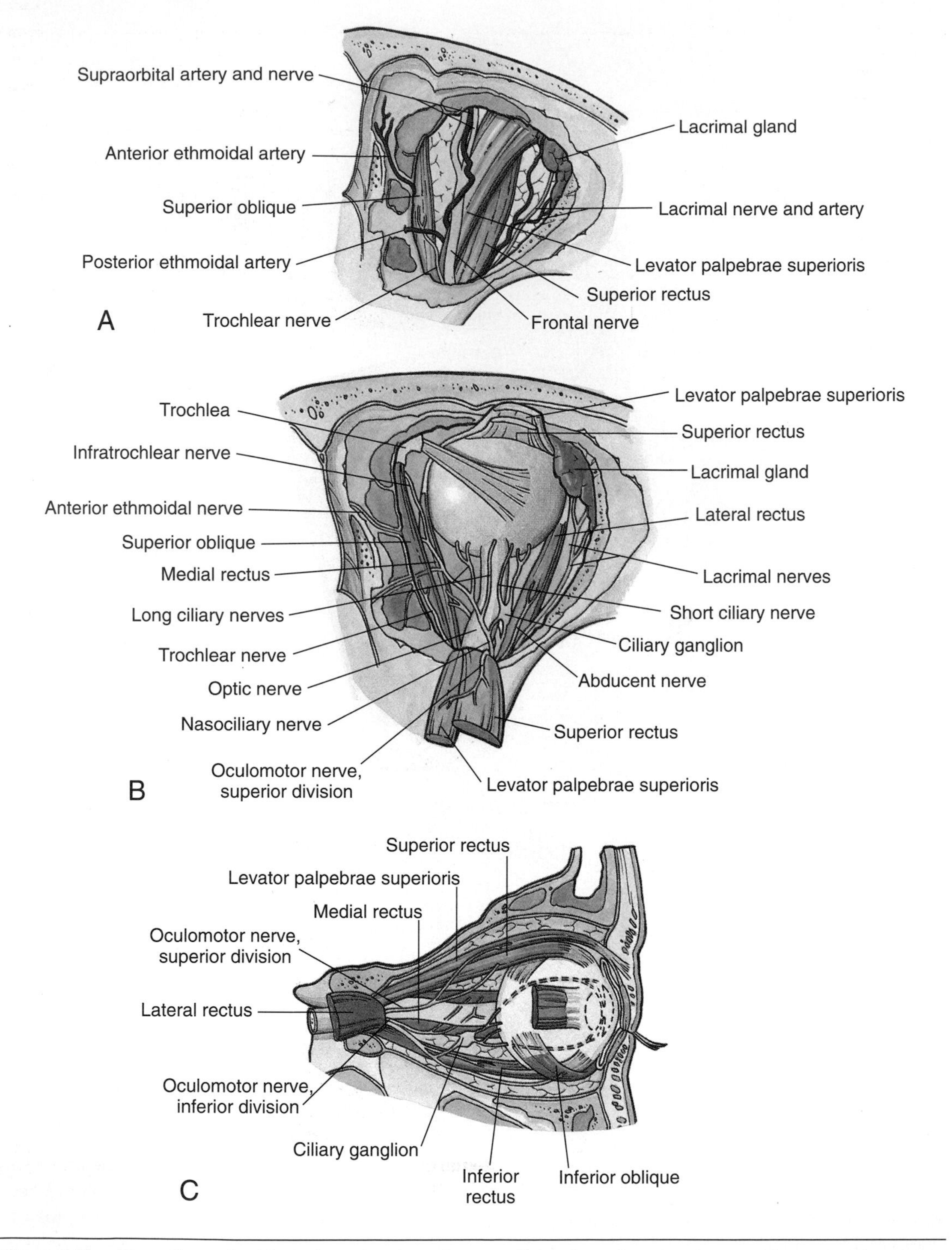

Figure 8.16. Dissections of orbit. **A.** Superior view of a superficial dissection showing muscles, nerves, and vessels. **B.** Similar view of a deep dissection. **C.** Lateral view.

oblique directs the eyeball inferiorly and laterally, and when it works with the inferior rectus an inferior movement is obtained. All eye muscles work together to move the eyes.

The fascial sheath (bulbar sheath, Tenon's capsule) envelops the eyeball from the optic nerve to the corneoscleral junction (Figs. 8.14*A* and 8.17). The *fascial sheath* is pierced by the

Table 8.5
Muscles of Orbit

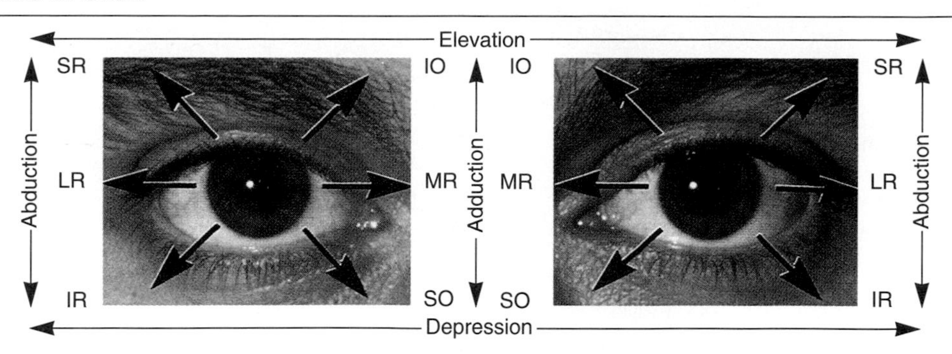

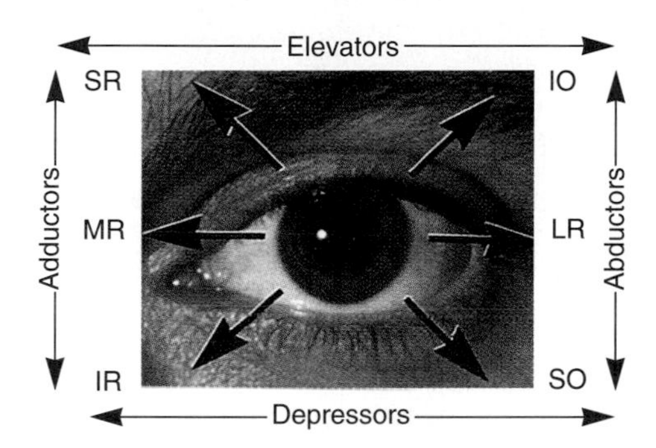

Individual anatomical actions of muscles as studied anatomically

It is essential to appreciate that all muscles are continuously involved in eye movements; thus, the individual actions are not usually tested clinically. *SR*, superior rectus (CN III); *LR*, lateral rectus (CN VI); *IR*, inferior rectus (CN III); *IO*, inferior oblique (CN III); *MR*, medial rectus (CN III); *SO*, superior oblique (CN IV).

Muscle	Origin	Insertion	Innervation	Action(s)
Levator palpebrae superioris	Lesser wing of sphenoid bone and superior and anterior to optic canal	Tarsal plate and skin of upper eyelid	Oculomotor n.; deep layer (superior tarsal m.) is supplied by sympathetic fibers	Elevates upper eyelid
Superior rectus	Common tendinous ring	Sclera just posterior to cornea	Oculomotor n.	Elevates, adducts, and rotates eyeball medially;
Inferior rectus	Common tendinous ring	Sclera just posterior to cornea	Oculomotor n.	depresses, adducts, and rotates eyeball medially
Lateral rectus	Common tendinous ring	Sclera just posterior to cornea	Abducent n.	Abducts eyeball
Medial rectus	Common tendinous ring	Sclera just posterior to cornea	Oculomotor n.	Adducts eyeball
Superior oblique	Body of sphenoid bone	Its tendon passes through a fibrous ring or trochlea and changes its direction and inserts into sclera deep to superior rectus m.	Trochlear n.	Abducts, depresses, and medially rotates eyeball
Inferior oblique	Anterior part of floor of orbit	Sclera deep to lateral rectus m.	Oculomotor n.	Abducts, elevates, and laterally rotates eyeball

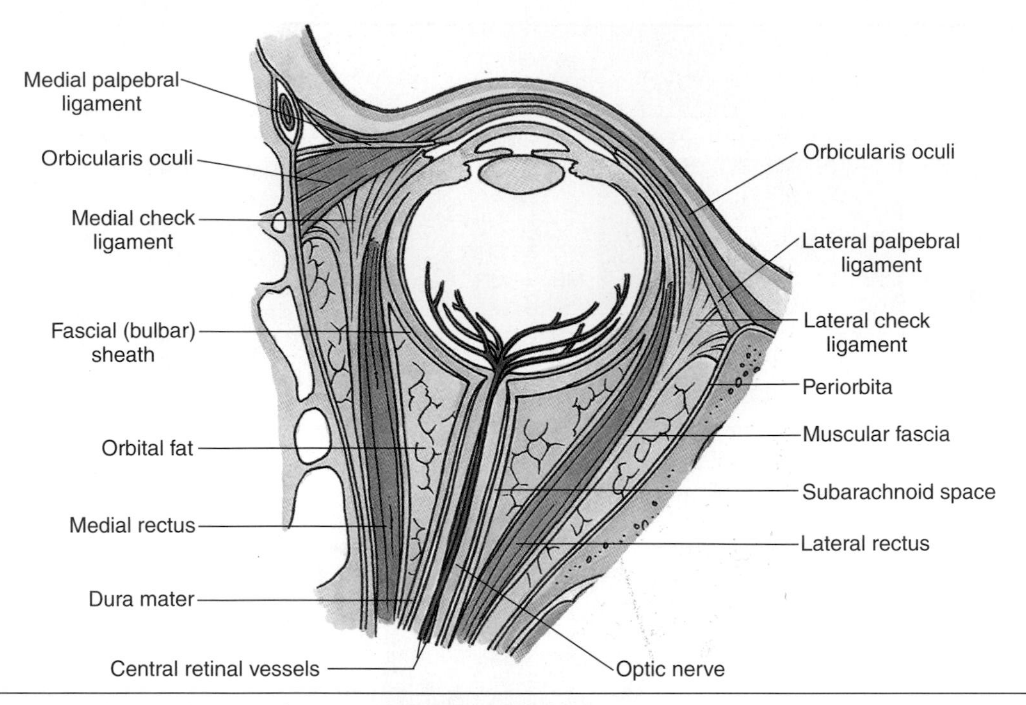

Figure 8.17. Horizontal section of right orbit, viewed superiorly, showing bulbar or fascial sheath of eyeball.

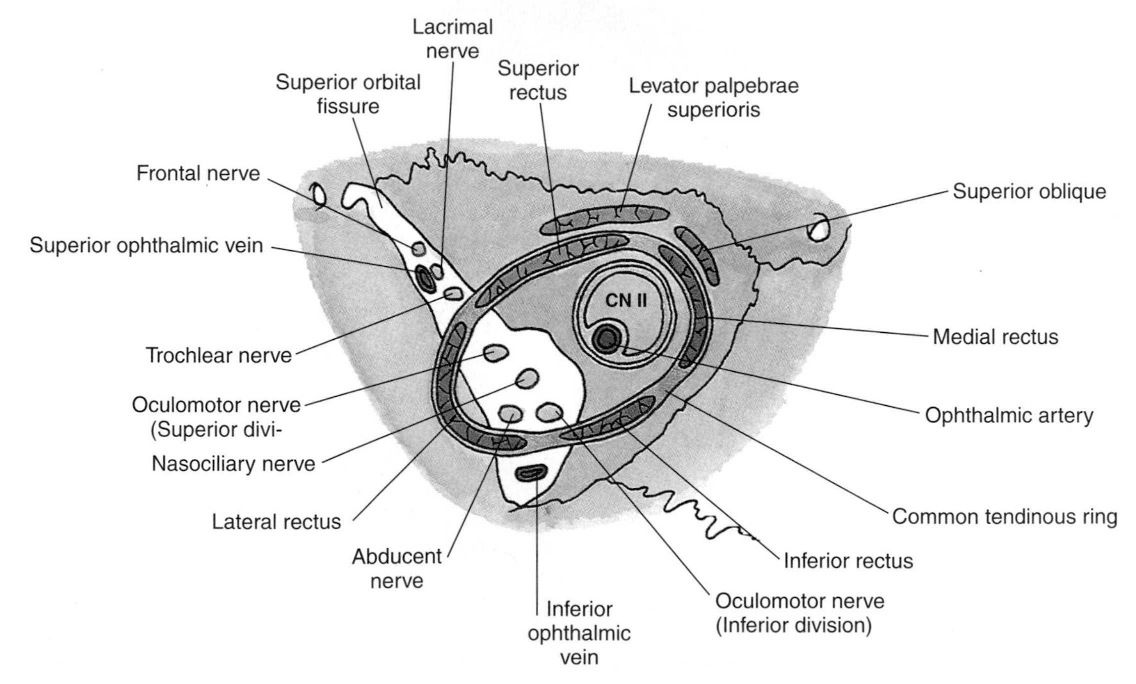

tendons of the extraocular muscles and is reflected onto each as a tubular sheath. There are triangular expansions from the sheaths of the medial and lateral recti called medial and lateral *check ligaments*, which are attached to the lacrimal and zygomatic bones, respectively. A blending of the check ligaments with the fasciae of the inferior rectus and inferior oblique muscles forms a hammocklike sling called the suspensory ligament of the eyeball.

NERVES OF ORBIT

In addition to the optic nerve (CN II), the nerves of the orbit include those that enter through the *superior orbital fissure* and supply the muscles of the eyeball (oculomotor, CN III; trochlear, CN IV; and abducent, CN VI).

- CN IV supplies superior oblique
- CN VI supplies lateral rectus
- CN III supplies levator palpebrae superioris, superior rectus, medial rectus, inferior rectus, and inferior oblique

In summary, all orbital muscles are supplied by CN III, except superior oblique (SO) and lateral rectus (LR), which are supplied by CN IV and CN VI, respectively. A memory key is LR6, SO4, all others CN III.

Several branches of CN V^1 supply structures in the orbit. The *lacrimal nerve* arises in the lateral wall of the cavernous sinus. It enters the orbit through the superior orbital fissure, passes to the lacrimal gland, and gives branches to the conjunctiva and skin of the upper eyelid (Fig. 8.16*A*). The *frontal nerve* also enters the orbit through the superior orbital fissure and divides into the supraorbital nerve and supratrochlear nerve, which supply the upper eyelid, scalp, and forehead.

The *nasociliary nerve*, the sensory nerve to the eye, also passes through the superior orbital fissure and supplies several branches to the orbit (Fig. 8.16*B*). The *short ciliary nerves* carry parasympathetic and sympathetic fibers to the ciliary body and iris. The *long ciliary nerves* transmit postganglionic sympathetic fibers to the dilator pupillae and afferent fibers from the iris and cornea. The *infratrochlear nerve* supplies the eyelids, conjunctiva, skin of the nose, and lacrimal sac. The anterior and posterior *ethmoidal nerves*, branches of the nasociliary nerve, supply the mucous membrane of the sphenoidal and ethmoidal sinuses and the nasal cavities (p. 401).

The *ciliary ganglion* is a very small group of nerve cell bodies between the optic nerve and lateral rectus, toward the posterior limit of the orbit (Fig. 8.16, *B* and *C*). The *short ciliary nerves*, branches of the ciliary ganglion, are distributed to the eyeball. They consist of postganglionic parasympathetic fibers originating in the ciliary ganglion, afferent fibers from the nasociliary nerve that pass through the ganglion, and postganglionic sympathetic fibers that also pass through it.

VASCULATURE OF ORBIT

Blood supply of the orbit is mainly from the *ophthalmic artery* (Table 8.6); the infraorbital artery also contributes to the supply of this region (Fig. 8.5). The central artery of the retina arises inferior to the optic nerve and runs within the dural sheath of this nerve until it approaches the eyeball. The *central artery* pierces the nerve and runs within it to emerge at the optic disc. Branches of the central artery spread over the internal surface of the retina (p. 372). Its terminal branches are end arteries that provide the only blood supply to the retina.

Venous drainage of the orbit is through the superior and inferior *ophthalmic veins* that pass through the superior orbital fissure and enter the cavernous sinus. The central vein of the retina usually enters the cavernous sinus directly, but it may join one ophthalmic vein.

Complete oculomotor nerve palsy affects most ocular muscles, the levator palpebrae superioris, and the sphincter pupillae. The upper eyelid droops (*ptosis*) and cannot be raised voluntarily because of the unopposed orbicularis oculi, supplied by CN VII. There is also a fully dilated, nonreactive pupil because of the unopposed dilator pupillae. The eyeball is fully abducted and depressed because of the unopposed lateral rectus and superior oblique, respectively.

Interruption of a cervical sympathetic trunk results in paralysis of the superior tarsal muscle supplied by sympathetic fibers, causing ptosis. This is part of the *Horner syndrome*, which also includes a constricted pupil, sinking of the eye, and redness, dryness, and increased temperature of the face on the affected side.

One or more extraocular muscles may be paralyzed by disease in the brainstem or by head injury; this results in *diplopia* (double vision). Paralysis of a muscle is apparent by limitation of eye movement in the field of action of the muscle and by the production of two images (diplopia) when one attempts to use the muscle. When the abducent nerve (CN VI) supplying only the lateral rectus is paralyzed, the patient cannot abduct the eyeball on the affected side. The eyeball is fully adducted by the unopposed pull of the medial rectus.

Because terminal branches of the central retinal artery are end arteries, their obstruction by an embolus results in instant and total blindness in the affected area of the eye. *Obstruction of the central retinal artery* is usually unilateral and occurs in old people. Because the central vein of the retina enters the cavernous sinus, thrombophlebitis (thrombus formation and inflammation) of this sinus or vein may result in the passage of thrombi to the central retinal vein and produce clotting in the small retinal veins. *Blockage of the central retinal vein* usually results in slow, painless loss of vision.

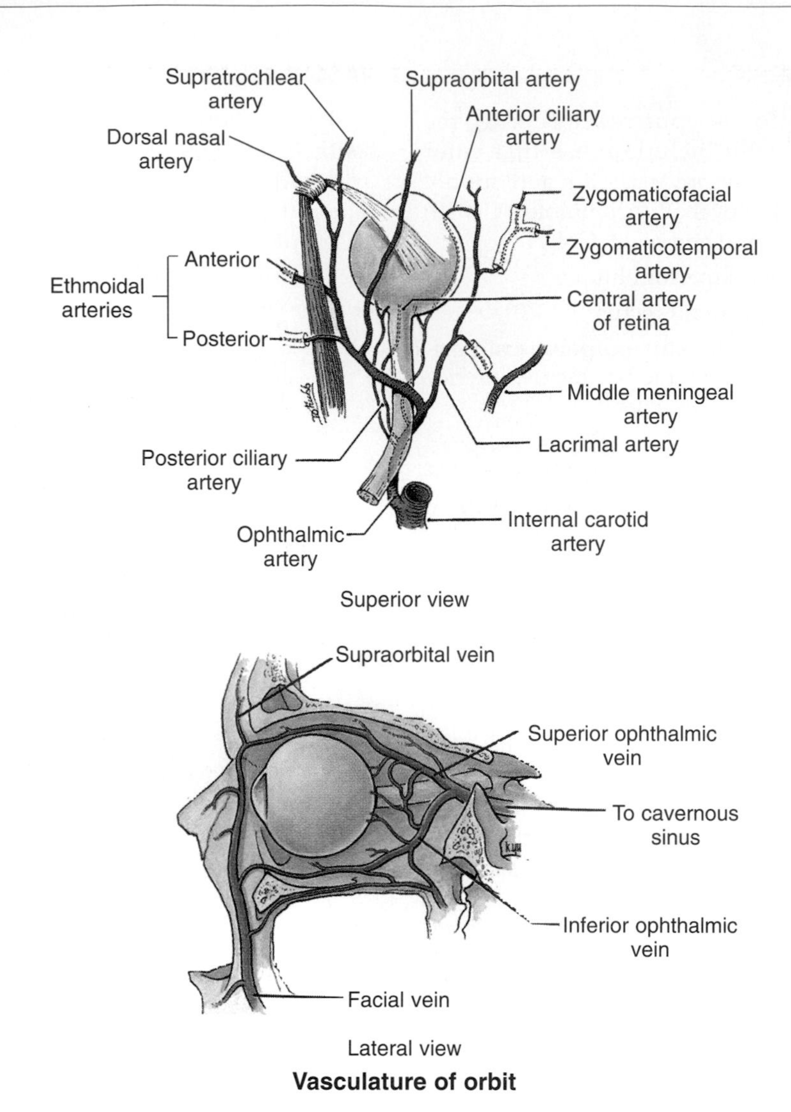

Vasculature of orbit

Temporal Region

The temporal region of the skull contains the temporal and infratemporal fossae, which are superior and inferior to the zygomatic arch, respectively (Figs. 8.1 and 8.18).

TEMPORAL FOSSA

The temporal fossa is bounded superiorly and posteriorly by the temporal lines and anteriorly by the frontal and zygomatic bones. The floor of the temporal fossa is formed by parts of the four bones that form the pterion (Fig. 8.1). *Temporal fascia* covers the temporalis muscle superior to the zygomatic arch. Inferiorly, the fascia splits into two layers that attach to the superior margin of the zygomatic arch.

INFRATEMPORAL FOSSA

The infratemporal fossa is deep and inferior to the zygomatic arch and posterior to the maxilla. The boundaries of the fossa are

- Laterally, ramus of mandible
- Medially, lateral pterygoid plate
- Anteriorly, maxilla
- Posteriorly, condylar process of mandible and styloid process of temporal bone
- Superiorly, inferior surface of greater wing of sphenoid
- Inferiorly, where medial pterygoid attaches to mandible near its angle

Table 8.6
Arterial Supply of Orbit

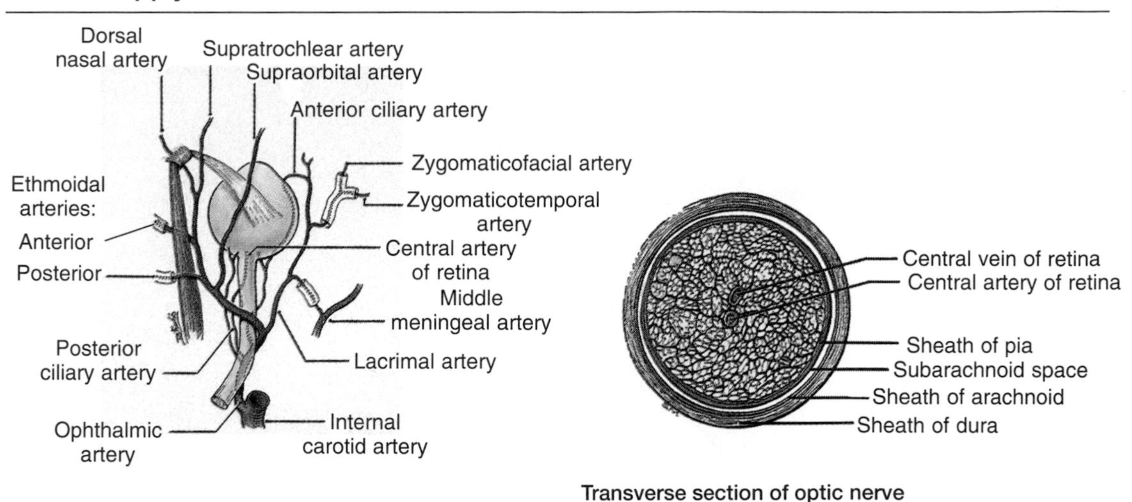

Transverse section of optic nerve

Artery	Origin	Course and Distribution
Ophthalmic	Internal carotid a.	Traverses optic foramen to reach orbital cavity
Central artery of retina	Ophthalmic a.	Runs in dural sheath of optic n. and pierces nerve near near eyeball.; appears at center of optic disc
Supraorbital	Ophthalmic a.	Passes superiorly and posteriorly from supraobital foramen to supply forehead and scalp
Supratrochlear	Ophthalmic a.	Passes from supraorbital margin to forehead and scalp
Lacrimal	Ophthalmic a.	Passes along superior border of lateral rectus m. to supply lacrimal gland, conjunctiva, and eyelids
Dorsal nasal	Ophthalmic a.	Courses along dorsal aspect of nose and supplies its surface
Long posterior ciliaries	Ophthalmic a.	Pierce sclera to supply ciliary body and iris
Posterior ethmoidal	Ophthalmic a.	Passes through posterior ethmoidal foramen to posterior ethmoidal cells
Anterior ethmoidal	Ophthalmic a.	Passes through anterior ethmoidal foramen to anterior cranial fossa; supplies anterior and middle ethmoidal cells, frontal sinus, nasal cavity, and skin on dorsum of nose
Infraorbital	Third part of maxillary a.	Passes along infraorbital groove and foramen to face
Anterior ciliary	Ophthalmic a.	Pierce sclera at the periphery of iris and form network in iris

The contents of the infratemporal fossa are the inferior part of the temporalis muscle, the medial and lateral pterygoid muscles, the maxillary artery, the pterygoid venous plexus, the mandibular, inferior alveolar, lingual, and buccal nerves, chorda tympani, and the otic ganglion.

The *medial pterygoid* lies on the medial aspect of the ramus of the mandible. Its two heads embrace the inferior head of the lateral pterygoid and unite (Fig. 8.19). The medial pterygoid passes inferoposteriorly and attaches to the medial surface of the mandible near its angle. The *lateral pterygoid*, which also has two heads of origin, passes posteriorly and attaches to the capsule of the TMJ. The attachments, nerve, supply, and actions of the pterygoids are described in Tables 8.7 and 8.8.

The *maxillary artery*, the larger of the two terminal branches of the external carotid artery, arises posterior to the neck of the mandible and passes anteriorly, deep to the

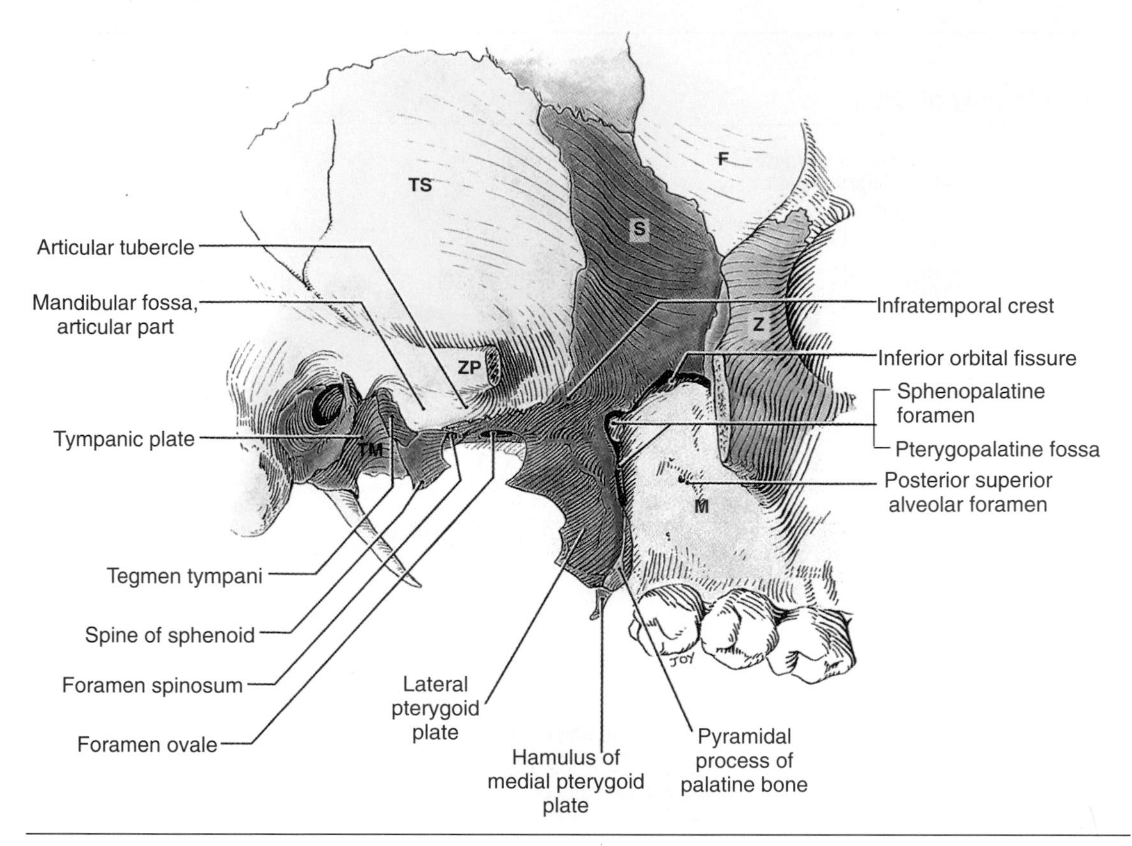

Figure 8.18. Lateral view of central part of skull showing infratemporal fossa. Mandible and most of zygomatic arch have been removed. *TS*, squamous part of temporal bone; *F*, frontal bone; *S*, sphenoid; *Z*, zygomatic bone; *ZP*, zygomatic process of temporal bone; *TM*, mastoid part of temporal bone; *M*, maxilla.

neck of the mandibular condyle. It passes superficial or deep to the lateral pterygoid and disappears in the infratemporal fossa. The maxillary artery is divided into three parts by the lateral pterygoid.

The branches of the first part of the maxillary artery supply chiefly the tympanic membrane, dura, and mandibular teeth:

- Deep auricular artery to external acoustic meatus
- Anterior tympanic artery to tympanic membrane
- Middle meningeal artery (clinically the most important branch) to dura and calvaria
- Accessory meningeal arteries to cranial cavity
- Inferior alveolar artery to mandible, gingivae (gums), and teeth

The branches of the second part of the maxillary artery supply the muscles of mastication:

- Masseteric artery
- Deep temporal artery
- Pterygoid artery
- Buccal artery

The branches of the third part of the maxillary artery arise just before and after the vessel enters the pterygopalatine fossa. These arteries supply the maxillary teeth, parts of the face and orbit, palate, and nasal cavity:

- Posterior superior alveolar artery
- Middle superior alveolar artery
- Infraorbital artery
- Descending palatine artery
- Artery of pterygoid canal
- Pharyngeal artery
- Sphenopalatine artery, termination of maxillary artery

The *pterygoid venous plexus* is partly between the temporalis and lateral pterygoid and partly

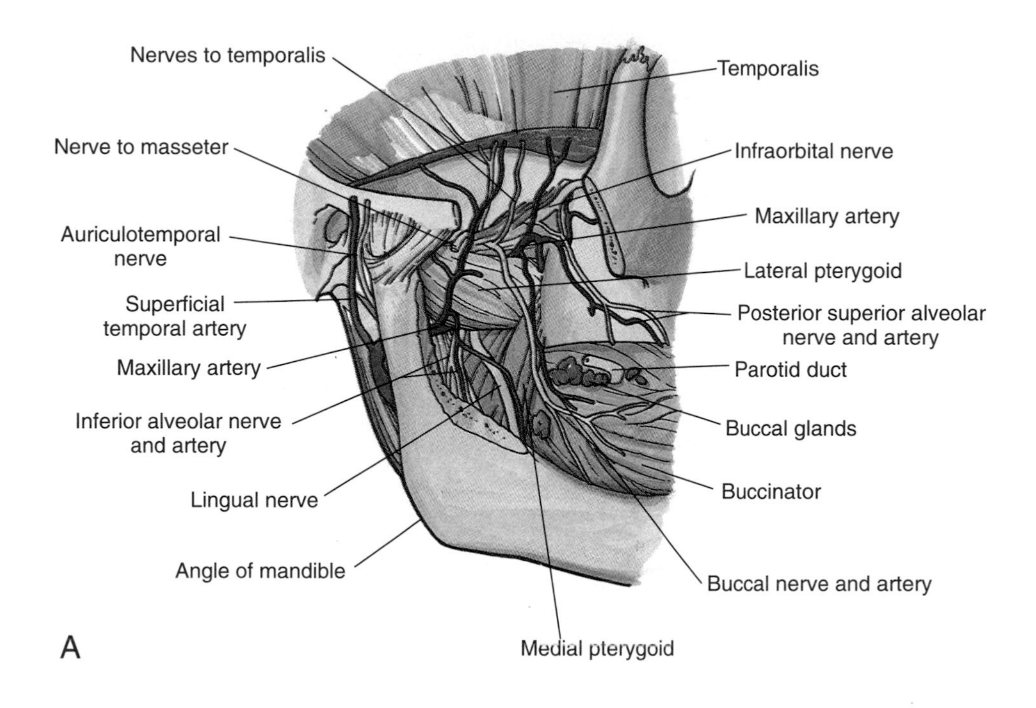

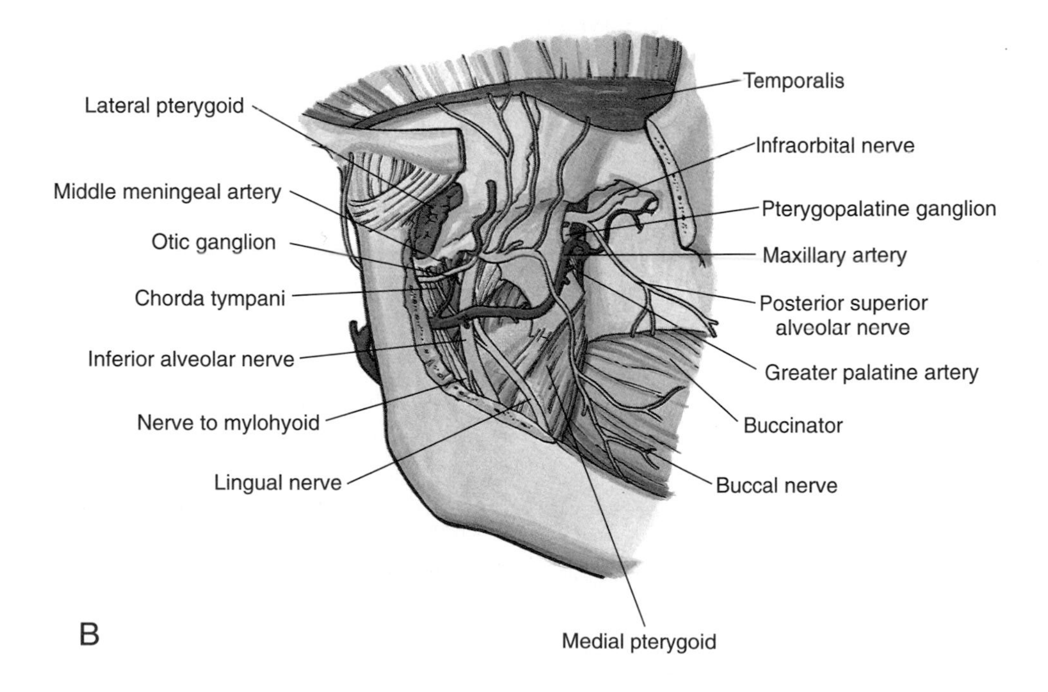

Figure 8.19. Dissections of infratemporal region. **A.** Superficial. **B.** Deep.

Table 8.7
Movements of Mandible at Tempomandibular Joint

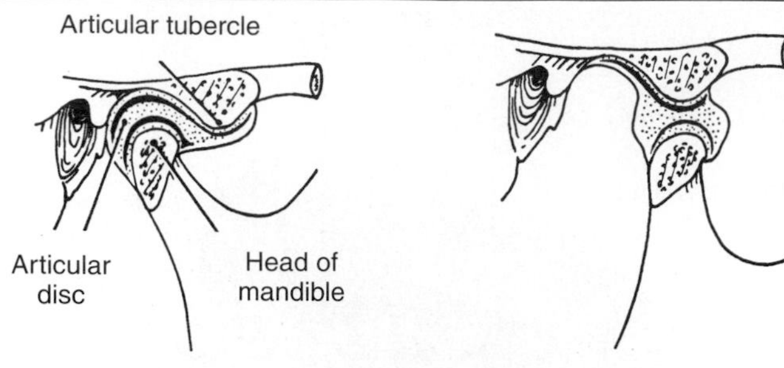

Elevation (Close Mouth)	Depression (Open Mouth)	Protrusion (Protrude Chin)	Retrusion (Retrude Chin)	Side-to-Side Movements (Grinding and Chewing)
Temporalis; masseter; medial pterygoid	Lateral pterygoid; suprahyoid and infrahyoid muscles	Masseter; lateral pterygoid; medial pterygoid	Temporalis masseter	Temporalis of same side Pterygoids of opposite side; masseter

Table 8.8
Muscles Acting on Temporomandibular Joint

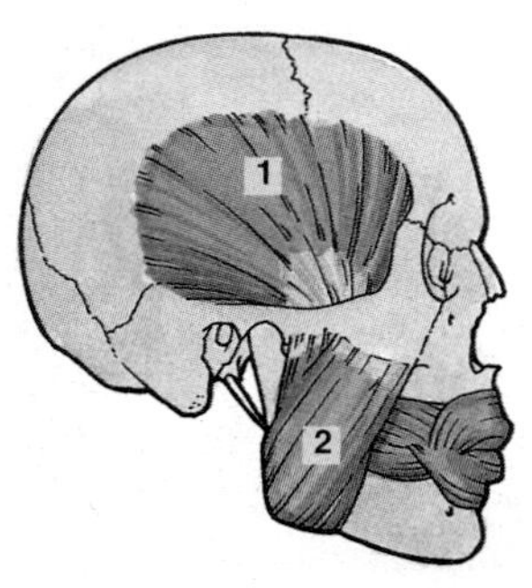

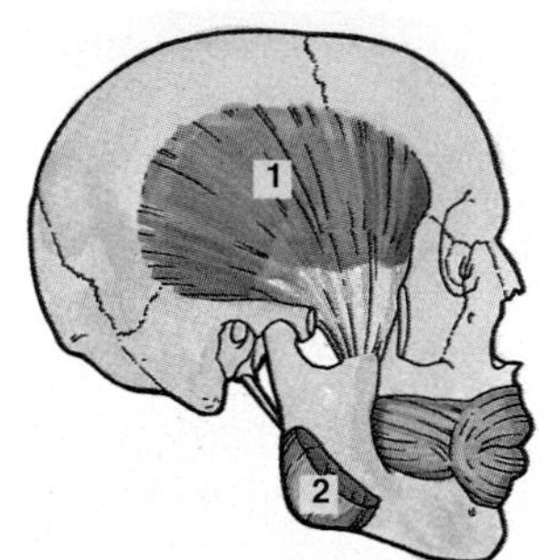

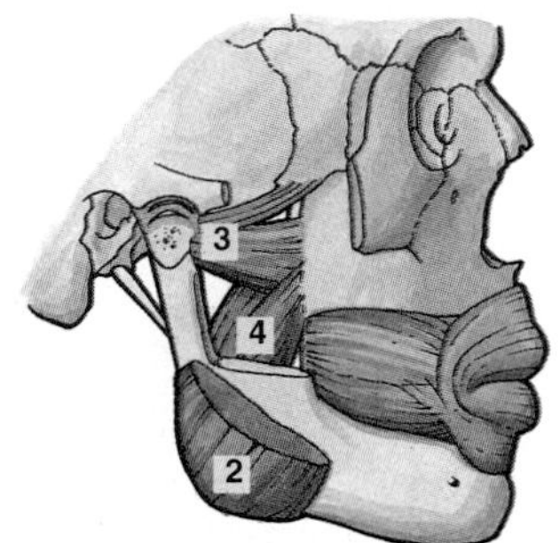

Muscle	Origin	Insertion	Innervation	Main Actions
Temporalis (1)	Floor of temporal fossa and deep surface of temporal fascia	Tip and medial surface of coronoid process and anterior border of ramus of mandible	Deep temporal branches of mandibular n. (CN V^3)	Elevates mandible, closing jaws; its posterior fibers retrude mandible after protrusion
Masseter (2)	Inferior border and medial surface of zygomatic arch	Lateral surface of ramus of mandible and its coronoid process	Mandibular n. via masseteric nerve that enters its deep surface	Elevates and protrudes mandible, thus closing jaws; deep fibers retrude it
Lateral pterygoid (3)	*Superior head:* infratemporal surface and infratemporal crest of greater wing of sphenoid bone *Inferior head:* lateral surface of lateral pterygoid plate	Neck of mandible, articular disc, and capsule of temporomandibular joint	Mandibular nerve (CN V^3) via lateral pterygoid nerve from anterior trunk, which enters its deep surface	Acting together, they protrude mandible and depress chin; acting alone and alternately, they produce side-to-side movements of mandible
Medial pterygoid (4)	*Deep head:* medial surface of lateral pterygoid plate and pyramidal process of palatine bone *Superficial head:* tuberosity of maxilla	Medial surface of ramus of mandible, inferior to mandibular foramen	Mandibular nerve (CN V^3) via medial pterygoid nerve	Helps to elevate mandible, closing jaws; acting together, they help to protrude mandible; acting alone, it protrudes side of jaw; acting alternately, they produce a grinding motion

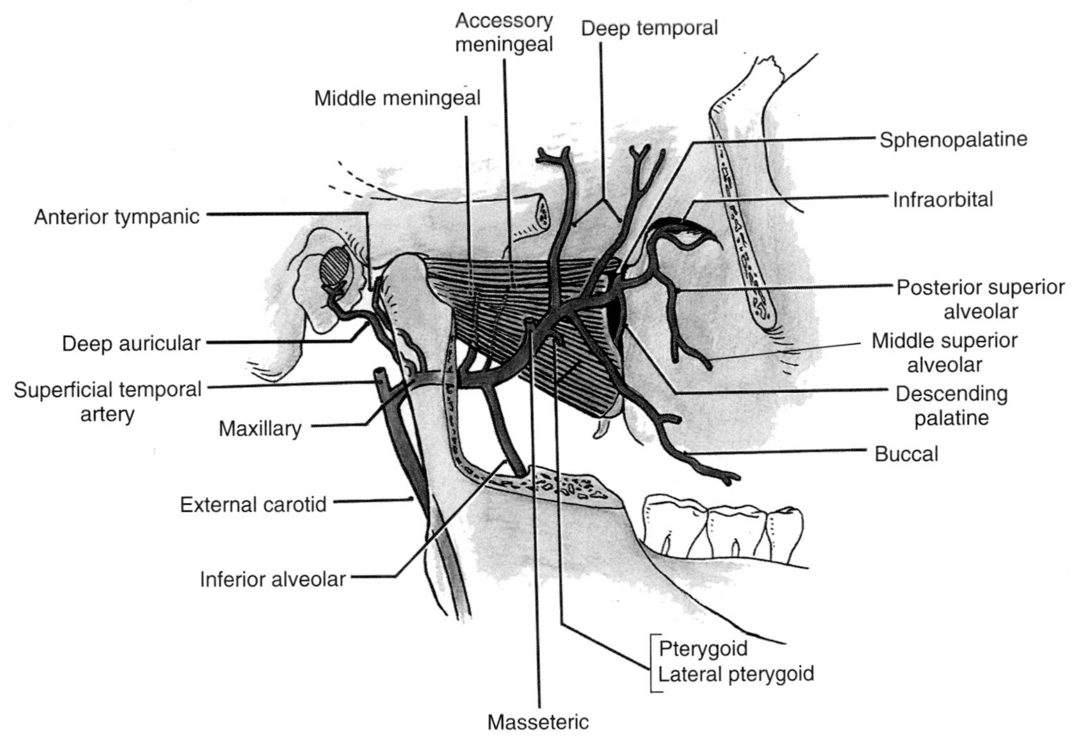

Branches of maxillary artery

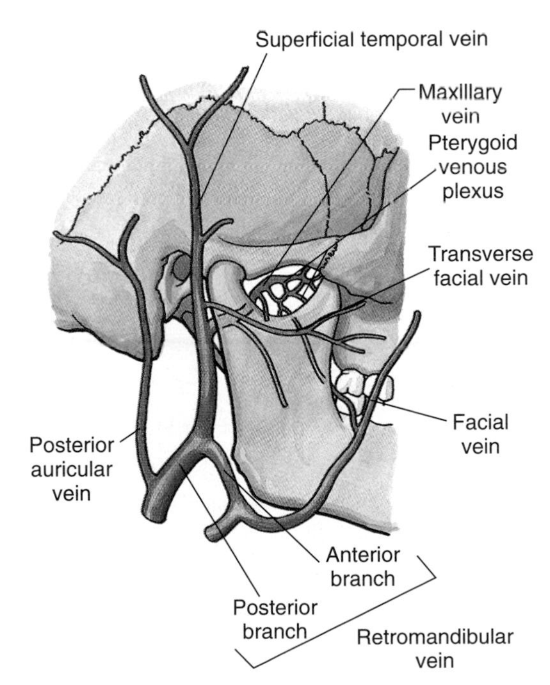

Venous drainage of infratemporal fossa

between the two pterygoids. The pterygoid venous plexus has connections with the facial vein through the cavernous sinus (Fig. 8.9).

The *mandibular nerve* (CN V^3) descends through the foramen ovale into the infratemporal fossa and divides into sensory and motor branches. The branches of CN V^3 are the auriculotemporal, inferior alveolar, lingual, and buccal nerves. Branches of the mandibular nerve also supply the four muscles of mastication (temporalis, masseter, and medial and lateral pterygoids) but not the buccinator, which is supplied by CN VII.

The *auriculotemporal nerve* (Fig. 8.19*A*) encircles the middle meningeal artery and breaks into numerous branches, the largest of which passes posteriorly, medial to the neck of the mandible, and supplies sensory fibers to the auricle and temporal region. This nerve also sends articular fibers to the TMJ and parasympathetic secretomotor fibers to the parotid gland.

The *otic ganglion* (parasympathetic) is in the infratemporal fossa (Fig. 8.19*B*), just inferior to the foramen ovale, medial to the mandibular nerve, and posterior to the medial pterygoid. Preganglionic parasympathetic fibers, derived mainly from the glossopharyngeal nerve (CN IX), synapse in the otic ganglion. Postganglionic parasympathetic fibers, which are secretory to the parotid gland, pass from the ganglion to this gland through the auriculotemporal nerve.

The *inferior alveolar nerve* passes through the mandibular canal and sends branches to all mandibular teeth on its side. One of its branches, the *mental nerve*, passes through the mental foramen and supplies the skin and mucous membrane of the lower lip, the skin of the chin, and the vestibular gingiva of the mandibular incisors.

The *lingual nerve* lies anterior to the inferior alveolar nerve. It is sensory to the anterior two-thirds of the tongue, the floor of the mouth, and the lingual gingivae. It enters the mouth between the medial pterygoid and the ramus of the mandible and passes anteriorly under cover of the oral mucosa, just inferior to the third molar tooth (see Fig. 8.21). The *chorda tympani*, a branch of CN VII carrying taste fibers from the anterior two-thirds of the tongue, joins the lingual nerve in the infratemporal fossa. The chorda tympani also carries secretomotor fibers for the submandibular and sublingual salivary glands.

In *mandibular nerve block* an anesthetic agent is applied to the mandibular nerve (CN V^3) where it enters the infratemporal fossa. The needle is passed through the mandibular notch of the ramus of the mandible into the infratemporal fossa (extraoral approach), and the anesthetic is injected. The following branches of CN V^3 are usually anesthetized: auriculotemporal, inferior alveolar, lingual, and buccal.

In *inferior alveolar nerve block* the inferior alveolar nerve, a branch of CN V^3, is anesthetized by injecting anesthetic around the mandibular foramen. When this nerve block is successful, all mandibular teeth are anesthetized to the median plane. The skin and mucous membrane of the lower lip, the labial alveolar mucosa and gingivae, and the skin of the chin are also anesthetized because they are supplied by the mental branch of this nerve.

Temporomandibular Joint

The TMJ is a modified hinge type of synovial joint (Table 8.7). The articular surfaces involved are the condyle of the mandible, the articular tubercle, and the mandibular fossa. An *articular disc* divides the joint cavity into superior and inferior compartments. The *articular capsule* of the TMJ is loose. The fibrous capsule attaches to the margins of the articular area on the temporal bone and around the neck of the mandible. The thick part of the capsule forms the *lateral (temporomandibular) ligament* that strengthens the TMJ laterally. There are two *synovial membranes*: one lines the fibrous capsule superior to the disc; the other lines the capsule inferior to the disc. Two ligaments, other than the lateral ligament, connect the mandible to the cranium, but neither adds much strength to the joint. The *stylomandibular ligament* runs from the styloid process to the angle of the mandible. The *sphenomandibular ligament* runs from the spine of the sphenoid to the lingula of the mandible.

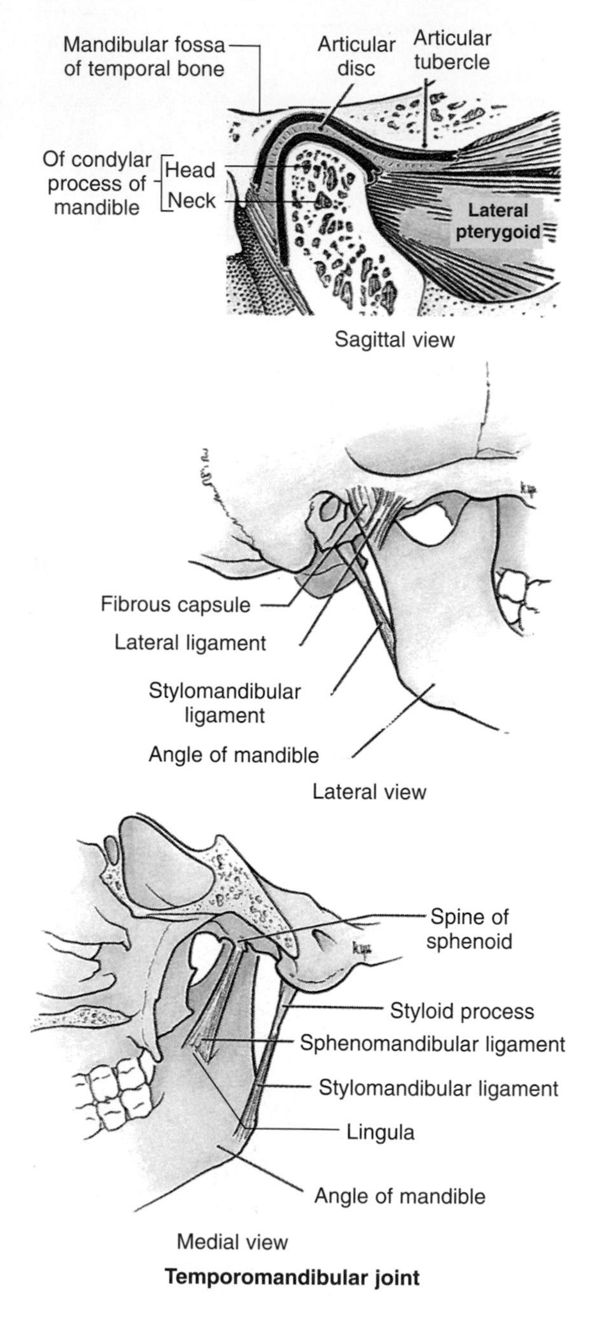

Temporomandibular joint

The *movements of the mandible* at the TMJs are depression, elevation (occlusion), protrusion (protraction), retrusion (retraction), and lateral movement (Table 8.7). When the mandible is depressed during opening of the mouth, the head of the mandible and articular disc move anteriorly on the articular surface until the head lies inferior to the articular tubercle. As this anterior gliding occurs, the head of the mandible rotates on the inferior surface of the articular disc, permitting simple chewing or grinding movements over a small range. During protrusion and retrusion of the mandible, the head and articular disc slide anteriorly and posteriorly on the articular surface of the temporal bone, with both sides moving together.

Movements of the TMJs result chiefly from the action of the muscles of mastication. The attachments, nerve supply, and actions of these muscles are described in Tables 8.7 and 8.8. The *temporalis*, a fan-shaped muscle, covers the temporal region. The *masseter*, a quadrangular muscle, covers the lateral aspect of the ramus and coronoid process of the mandible. The *lateral pterygoid* has two heads of origin; its apex points posteriorly. The *medial pterygoid*, a quadrilateral muscle, is deep to the ramus of the mandible; its two heads of origin embrace the inferior head of the lateral pterygoid.

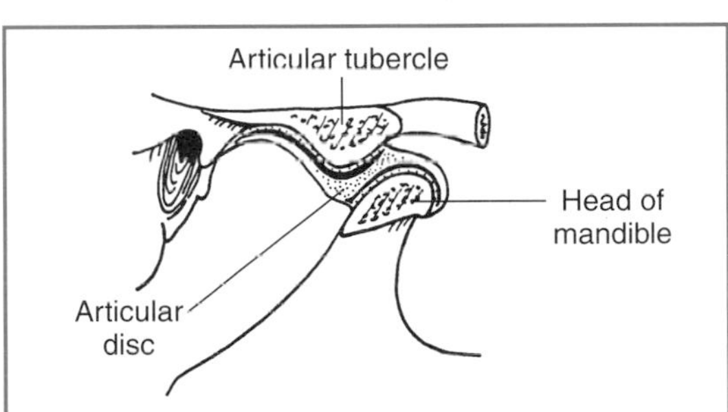

During yawning or taking a very large bite, excessive contraction of the lateral pterygoids may cause the heads of the mandible to dislocate (pass anterior to articular tubercles). In this position, the mandible remains wide open, and the person cannot close it. Most commonly the TMJ is dislocated by a blow to the chin when the mouth is open. TMJ dislocation may accompany fractures of the mandible. Because of the close relationship of the facial and auriculotemporal nerves to the TMJ, care must be taken during surgical procedures to preserve both the branches of the facial nerve overlying it and the articular branches of the auriculotemporal nerve that enter the posterior part of the joint.

Oral Cavity

The oral cavity (mouth) consists of two parts: the vestibule and mouth proper. The *vestibule* is the slitlike space between the lips and cheeks and the teeth and gingivae (gums). The vestibule communicates with the exterior through the orifice of the mouth. The *mouth proper* is limited laterally and anteriorly by the maxillary and mandibular alveolar arches housing the teeth. The roof of the oral cavity is formed by the palate. Posteriorly, the oral cavity communicates with the oropharynx (oral part of pharynx).

LIPS, CHEEKS, AND GINGIVAE

The *lips* are mobile, muscular folds surrounding the mouth that contain the orbicularis oris and the superior and inferior labial vessels and nerves (Figs. 8.5 and 8.6). They are covered externally by skin and internally by mucous membrane. The labial arteries, branches of the facial arteries, anastomose with each other to form an arterial ring. The pulse of these arteries can be palpated by grasping the lip lightly between the first two digits. The upper lip has a vertical groove called the *philtrum*. As the skin of the lip approaches the mouth, it changes color abruptly to red; this red margin of the lip is the *vermilion border*, a transitional zone between the skin and mucous membrane.

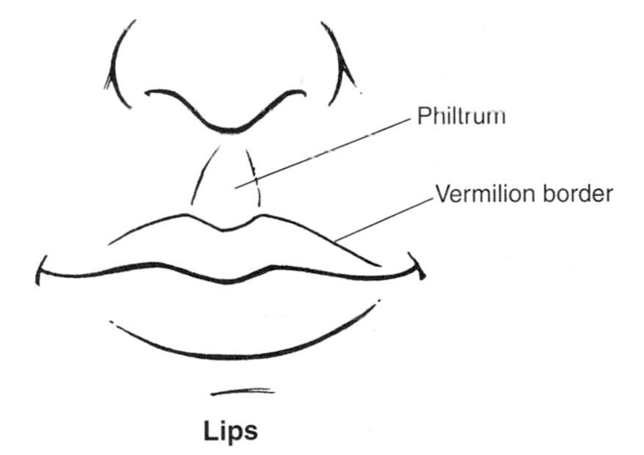

Lips

The *cheeks* have essentially the same structure as the lips, with which they are continuous. The principal muscle of the cheeks is the *buccinator*. The lips and cheeks function as an oral sphincter that pushes food from the vestibule into the mouth proper. The tongue and buccinator keep the food between the molar teeth during chewing. The buccinator compresses the cheek against the molar teeth and pushes the food onto the occlusal surfaces of the teeth. The *buccal glands* are small mucous glands between the mucous membrane and the buccinator (Fig. 8.19*A*).

The *gingivae* (gums) are composed of fibrous tissue covered with mucous membrane. They are firmly attached to the alveolar processes of the jaws and the necks of the teeth.

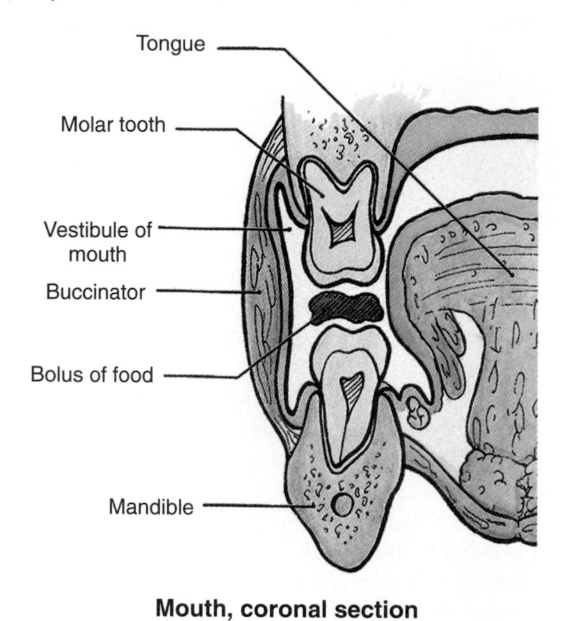

Mouth, coronal section

TEETH

There are 20 deciduous (primary) teeth. The first tooth usually erupts at age 6–8 months and the last by 20–24 months. Eruption of the permanent (secondary) teeth, normally 16 in each jaw (three molars, two premolars, one canine, and two incisors), is usually complete by age 18, except for the third molars ("wisdom teeth").

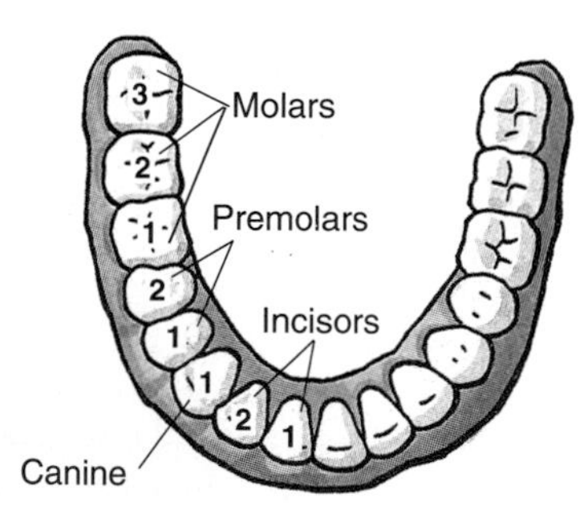

Adult mandibular teeth

A tooth is made up of a crown, neck, and root. The *crown* projects from the gingiva. Each type of tooth has a characteristic appearance. The *neck* is the part of the tooth between the crown and root. The *root* is fixed in the alveolus (tooth socket) by a fibrous periodontal membrane. Most of the tooth is composed of dentin that is covered by enamel over the crown and cementum over the root. The pulp cavity contains connective tissue, blood vessels, and nerves. The root canal transmits the nerves and vessels to and from the pulp cavity.

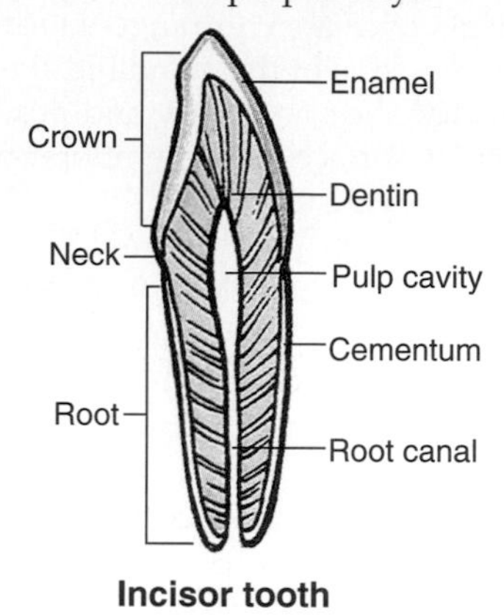

Incisor tooth

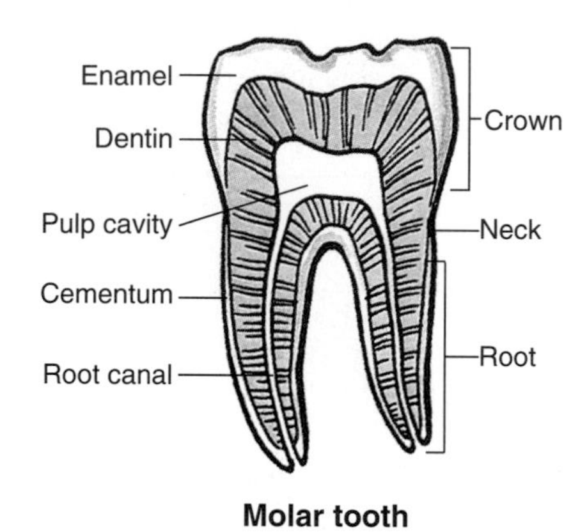

Molar tooth

Invasion of the pulp of the tooth by a carious lesion ("cavity") results in infection and irritation of the tissues in the pulp cavity. This condition causes an inflammatory process (*pulpitis*). Because the pulp cavity is a rigid space, the swollen pulpal tissues cause considerable pain (*toothache*).

Improper oral hygiene results in food deposits in tooth and gingival crevices, which may cause inflammation of the gingivae (*gingivitis*). If untreated, the disease spreads to other supporting structures (including the alveolar bone), producing *periodontitis.* Periodontitis results in inflammation of the gingivae and may result in absorption of alveolar bone and gingival recession. Gingival recession exposes the sensitive cementum of the teeth.

The superior and inferior *alveolar arteries,* branches of the maxillary, supply both the maxillary (upper) and mandibular (lower) teeth, respectively. *Veins* with the same names and

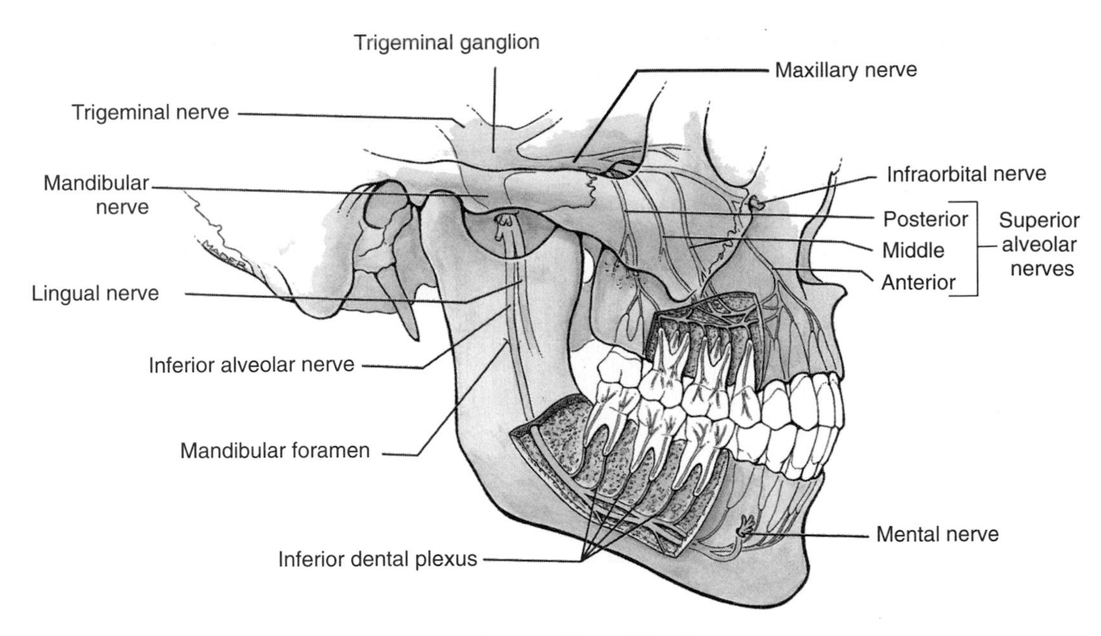

Innervation of teeth

distribution accompany the arteries. *Lymphatic vessels* from the teeth and gingivae pass mainly to the submandibular lymph nodes. The superior and inferior *alveolar nerves*, branches of CN V^2 and CN V^3, respectively, supply the maxillary and mandibular teeth.

PALATE

The palate forms the roof of the mouth and the floor of the nasal cavities. The palate consists of two regions: the hard palate (the anterior two-thirds or bony part) and the soft palate (the posterior one-third or fibromuscular part).

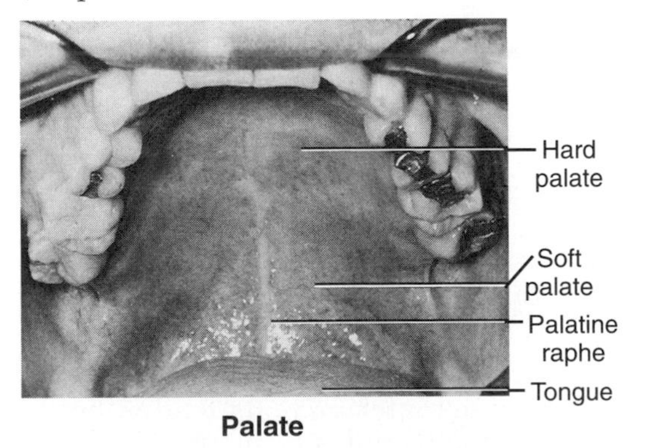

Palate

The *hard palate* is formed by the palatine processes of the maxillae and the horizontal plates of the palatine bones (Fig. 8.3). The *incisive foramen* is posterior to the central incisor teeth. The *incisive canal* transmits the nasopalatine nerve and the terminal branch of the sphenopalatine artery. Medial to the third molar tooth, the *greater palatine foramen* pierces the lateral border of the bony palate. The greater palatine vessels and nerve emerge from this foramen and run anteriorly on the palate (Fig. 8.20). The *lesser palatine foramina* transmit the lesser palatine nerves and vessels to the soft palate and adjacent structures.

The *soft palate* is the posterior fibromuscular part of the palate that is attached to the posterior edge of the hard palate. It extends posteroinferiorly as a curved free margin from which hangs a conical process, the *uvula*. When a person swallows, the soft palate moves posteriorly against the wall of the pharynx, thereby preventing regurgitation of food into the nasal cavity. Laterally the soft palate is continuous with the wall of the pharynx and is joined to the tongue and pharynx by the palatoglossal and palatopharyngeal arches, respectively. The *palatine tonsils*, often referred to as the "tonsils," are two masses of lymphoid tissue, one on each side of the oropharynx. Each is in a *tonsillar fossa*, bounded by the palatoglossal and palatopharyngeal arches and the tongue (Figs. 8.20 and 8.21*A*). The soft palate is strengthened by the *palatine aponeurosis*, formed by the expanded tendon of the tensor veli palatini. The aponeurosis, attached to the posterior margin of the hard palate, is thick anteriorly and thin posteriorly. The anterior part of the soft palate is formed mainly by the aponeurosis, whereas its posterior part is muscular.

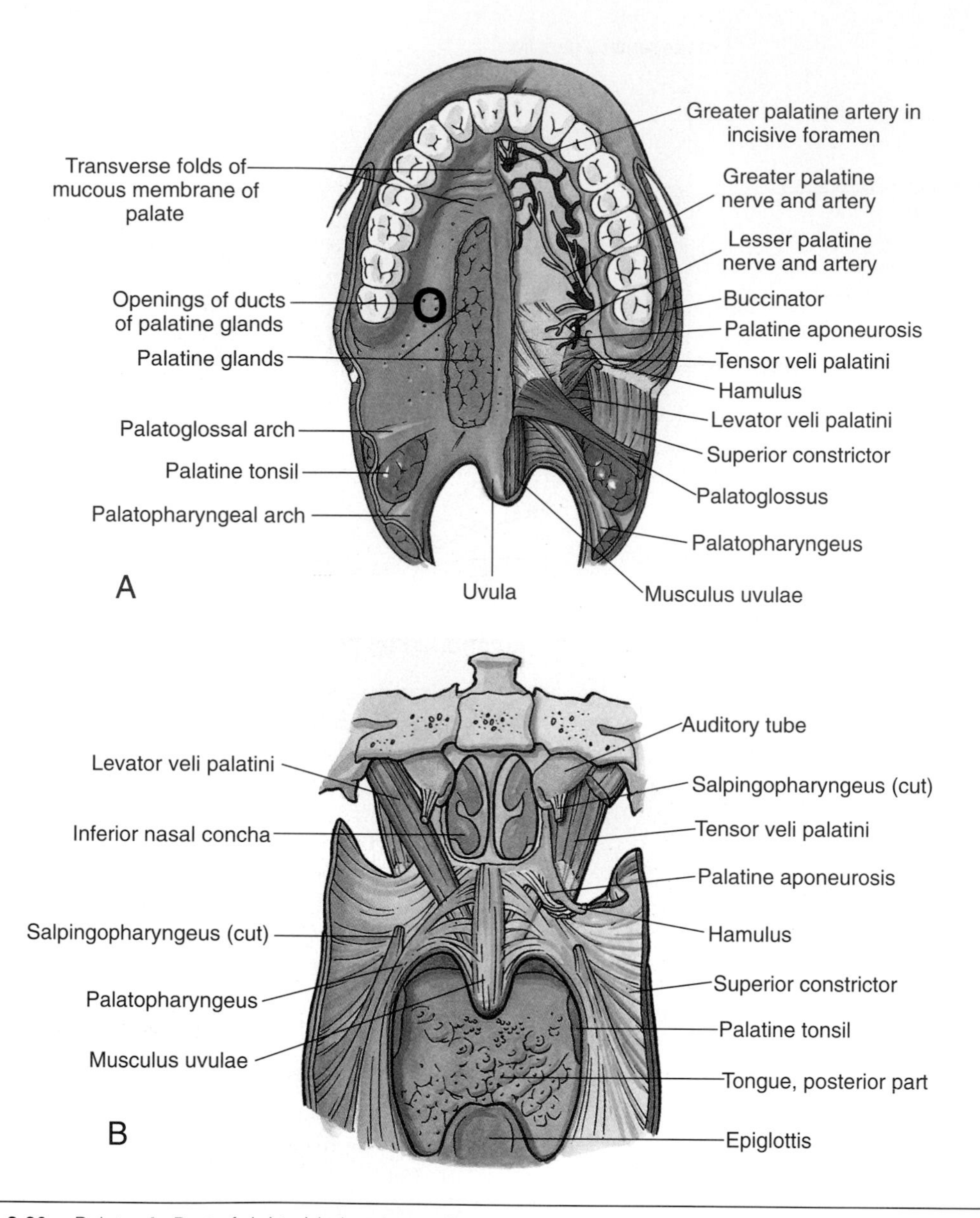

Figure 8.20. Palate. **A.** Part of right side has been dissected to show palatine glands. Left side has been dissected to show muscles of soft palate and palatine arteries and nerves. **B.** Posterior view of a dissection of soft palate showing muscles and their relationship to posterior part of tongue.

Muscles of Soft Palate

Muscles of the soft palate arise from the base of the cranium and descend to the palate. The soft palate may be raised so that it is in contact with the posterior wall of the pharynx. The soft palate can also be drawn inferiorly so that it is in contact with the posterior part of the tongue. For attachments, nerve supply, and actions of the muscles of the soft palate, see Table 8.9.

- Levator veli palatini is a cylindrical muscle that runs inferoanteriorly, spreading out in the soft palate where it attaches to the superior surface of the palatine aponeurosis
- Tensor veli palatini is a triangular muscle that passes inferiorly; its tendon hooks around the hamulus of the medial pterygoid plate before inserting into the palatine aponeurosis
- Palatoglossus, a slender slip of muscle covered with mucous membrane, forms the palatoglossal arch
- Palatopharyngeus, a thin flat muscle cov-

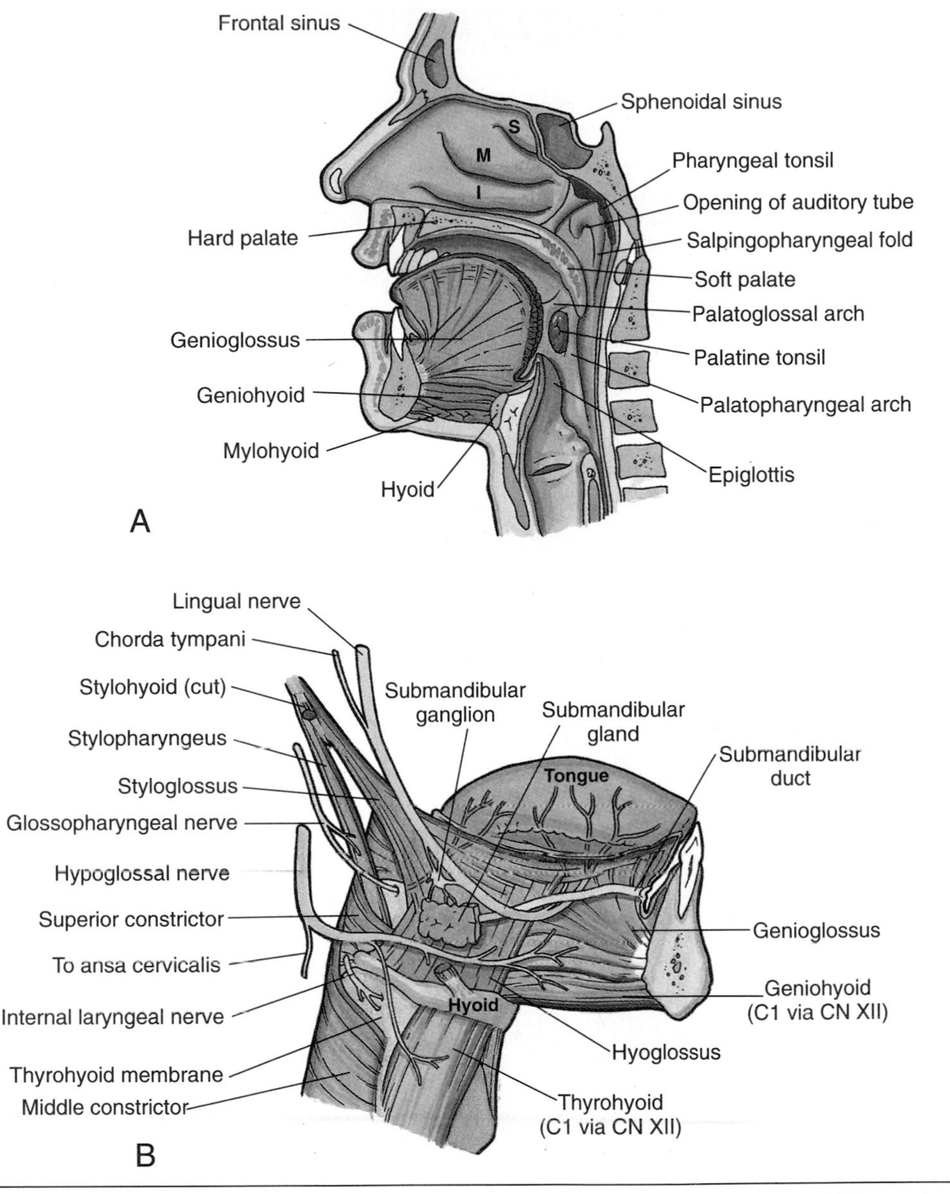

Figure 8.21. **A.** Sagittal section showing mouth, nose, and pharynx. *S*, superior concha; *M*, middle concha; *I*, inferior concha. **B.** Muscles and nerves of tongue.

ered with mucous membrane, forms the palatopharyngeal arch

- Musculus uvulae inserts into the mucosa of uvula

Vasculature and Innervation of Palate

The palate has a rich blood supply chiefly from the *greater palatine artery* on each side, a branch of the descending palatine artery (Fig. 8.20*A*). This artery passes through the greater palatine foramen and runs anteriorly and medially. The lesser palatine artery enters through the lesser palatine foramen and anastomoses with the ascending palatine artery, a branch of the facial artery. The *veins of the palate*, corresponding and accompanying the branches of the maxillary artery, are tributaries of the pterygoid plexus (p. 383).

The *sensory nerves of the palate* are branches of the pterygopalatine ganglion (Fig. 8.22). The greater palatine nerve supplies the gingivae,

Table 8.9
Muscles of Soft Palate

Muscle	Superior Attachment	Inferior Attachment	Innervation	Main Action(s)
Levator veli palatini	Cartilage of auditory tube and petrous part of temporal bone	Palatine aponeurosis	Pharyngeal branch of vagus n. via pharyngeal plexus	Elevates soft palate during swallowing and yawning
Tensor veli palatini	Scaphoid fossa of medial pterygoid plate, spine of sphenoid bone, and cartilage of auditory tube	Palatine aponeurosis	Medial pterygoid n. (a branch of mandibular n.) via otic ganglion	Tenses soft palate and opens mouth of auditory tube during swallowing and yawning
Palatoglossus	Palatine aponeurosis	Side of tongue	Cranial part of CN XI through pharyngeal branch of vagus n. (CN X) via pharyngeal plexus	Elevates posterior part of tongue and draws soft palate onto tongue
Palatopharyngeus	Hard palate and palatine aponeurosis	Lateral wall of pharynx	Cranial part of CN XI through pharyngeal branch of vagus n. (CN X) via pharyngeal plexus	Tenses soft palate and pulls walls of pharynx superiorly, anteriorly, and medially during swallowing
Musculus uvulae	Posterior nasal spine and palatine aponeurosis	Mucosa of uvula	Cranial part of CN XI through pharyngeal branch of vagus n. (CN X) via pharyngeal plexus	Shortens uvula and pulls it superiorly

mucous membrane, and glands of most of the hard palate. The nasopalatine nerve supplies the mucous membrane of the anterior part of the hard palate. The lesser palatine nerve supplies the soft palate. The palatine nerves accompany the arteries through the greater and lesser palatine foramina, respectively. Except for the tensor veli palatini supplied by CN V^2, all muscles of the soft palate are supplied through the pharyngeal plexus of nerves (p. 440).

The nasopalatine nerves can be anesthetized by injecting anesthetic into the mouth of the incisive canal. The needle is inserted posterior to the *incisive papilla*, a slight elevation of the mucosa that covers the incisive foramen. Both nasopalatine nerves can be anesthetized by the same injection where they emerge through the incisive foramen. The affected tissues are the palatal mucosa, the lingual gingivae and alveolar bone of the six anterior maxillary teeth, and the hard palate. Both right and left nerves are anesthetized by the same injection as they emerge from the incisive canal.

TONGUE

The tongue is a mobile muscular organ that varies greatly in shape. It is involved with mastication, taste, deglutition (swallowing), articulation (speech), and oral cleansing, but its main functions are squeezing food into the pharynx when swallowing and forming words during speaking. The tongue is mainly composed of muscles and is covered by mucous membrane.

The *dorsum of the tongue* is divided by a V-shaped sulcus terminalis into an anterior oral part (anterior two-thirds) and a posterior root, the pharyngeal part (posterior one-third). At the apex of the terminal sulcus is a small median depression, the *foramen cecum*, a remnant of the opening of the embryonic thyroglossal duct that was attached to the developing thyroid gland.

The *oral part* of the tongue is freely movable, but it is loosely attached to the floor of the mouth by the *lingual frenulum*. On each side of this fold, a deep lingual vein is visible through the thin mucous membrane (p. 393). The dorsal lingual veins drain the dorsum and sides of the tongue. There is a median groove on the dorsum of the oral part of the tongue. The mucous membrane on the oral part of the tongue is rough because of the presence of *lingual papillae*:

- Large flat-topped vallate papillae are anterior to the sulcus terminalis
- Foliate papillae are small lateral folds of the lingual mucosa
- The numerous long filiform papillae contain afferent nerve endings that are sensitive to touch

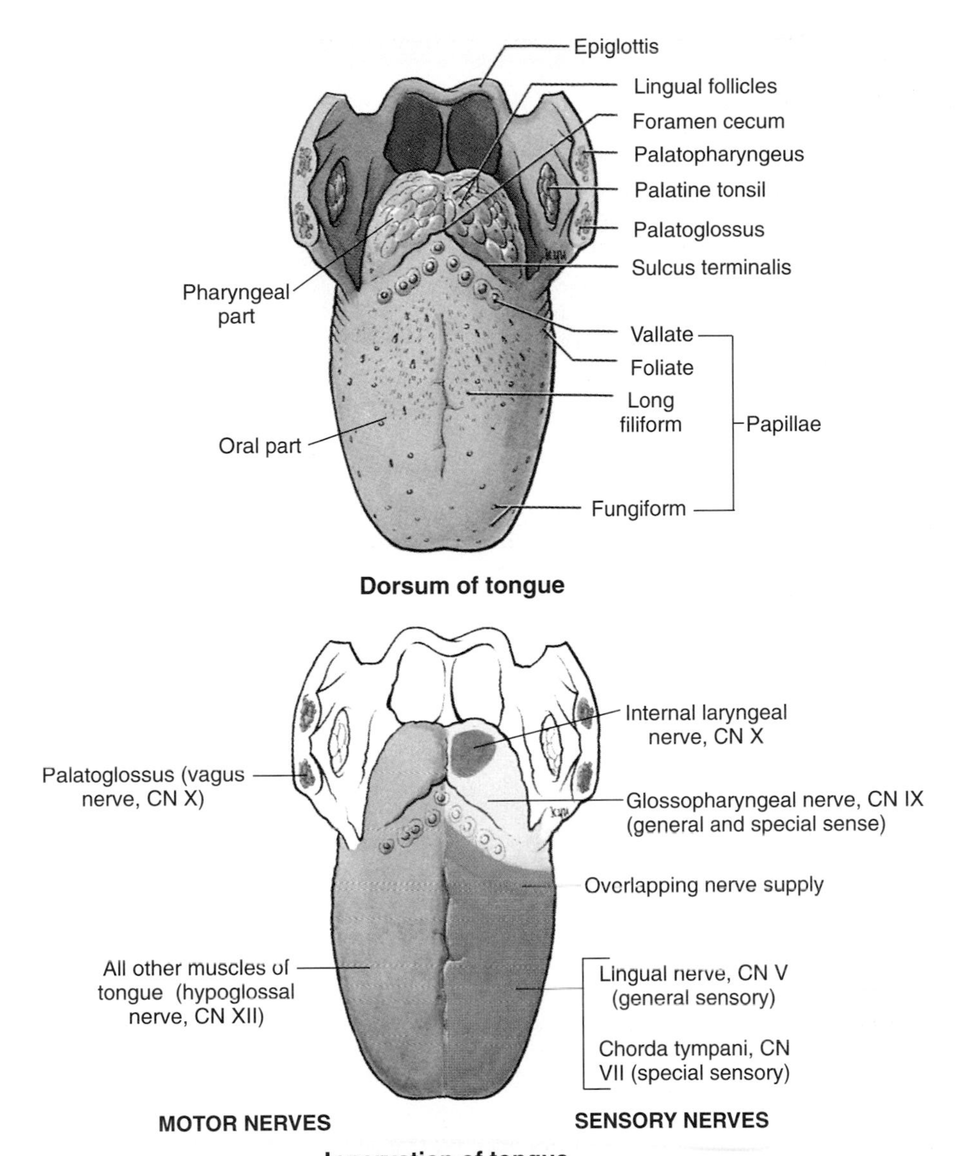

- Mushroom-shaped fungiform papillae appear as pink or red spots

The vallate, foliate, and most of the fungiform papillae contain taste receptors in the *taste buds*.

The *pharyngeal part* of the tongue is posterior to the sulcus terminalis and palatoglossal arches. Its mucous membrane has no papillae; however, the underlying nodules of lingual follicles give this part of the tongue a cobblestone appearance. The lymphatic nodules are collectively known as the *lingual tonsil.*

Muscles

Four intrinsic and four extrinsic muscles are in each half of the tongue (Fig. 8.21). The *intrinsic muscles* (superior and inferior longitudinal, transverse, and vertical) are confined to the tongue and are not attached to bone. The attachments, nerve supply, and actions of the *extrinsic muscles* (hyoglossus, genioglossus, styloglossus, and palatoglossus) are listed in Table 8.10. All muscles of the tongue, except the palatoglossus, are supplied by the *hypoglossal nerve* (CN XII).

Sensory Innervation

For general sensation, the mucosa of the anterior two-thirds of the tongue is supplied by the *lingual nerve* (Fig. 8.21*B*), a branch of CN V^3. For special sensation (taste), this part of the tongue, except the vallate papillae, is supplied

Table 8.10
Extrinsic Muscles of Tongue

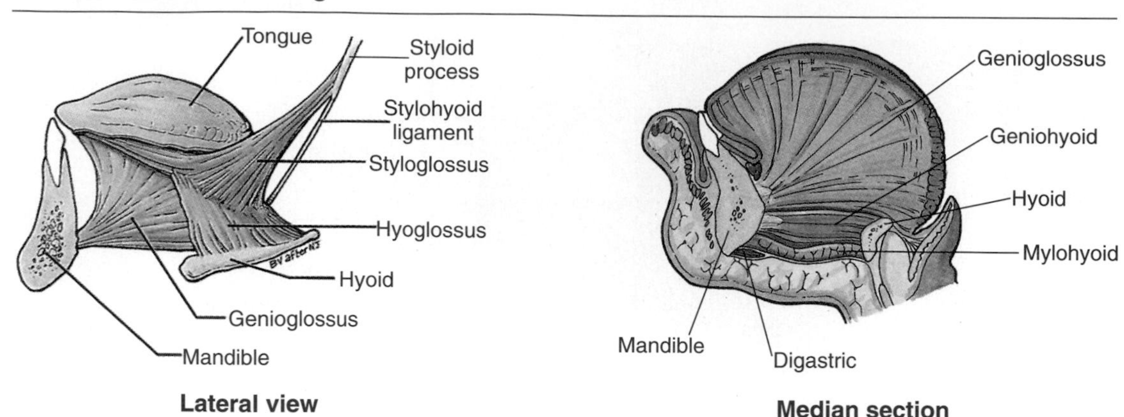

Muscle	Origin	Insertion	Innervation	Actions
Genioglossus	Superior part of mental spine of mandible	Dorsum of tongue and body of hyoid bone	Hypoglossal n. (CN XII)	Depresses tongue; its posterior part protrudes tongue
Hyoglossus	Body and greater horn of hyoid bone	Side and inferior aspect of tongue		Depresses and retracts tongue
Styloglossus	Styloid process and stylohyoid ligament			Retracts tongue and draws it up to create a trough for swallowing
Palatoglossus	Palatine aponeurosis of soft palate	Side of tongue	Cranial root of CN XI via pharyngeal branch of CN X and pharyngeal plexus	Elevates posterior part of tongue

through the *chorda tympani nerve,* a branch of CN VII. The chorda tympani joins the lingual nerve and runs anteriorly in its sheath. The mucous membrane of the posterior one-third of the tongue and the vallate papillae are supplied by the lingual branch of the *glossopharyngeal nerve* (CN IX), for general and special sensation (taste). Twigs of the *internal laryngeal nerve,* a branch of the vagus nerve (CN X), supply a small area of the tongue just anterior to the epiglottis. The sensory nerves also carry *parasympathetic secretomotor fibers* to the serous glands in the tongue. Parasympathetic fibers from the chorda tympani nerve travel with the lingual nerve to the submandibular and sublingual salivary glands. These nerve fibers synapse in the *submandibular ganglion* that hangs from the lingual nerve (Fig. 8.21*B*).

Vasculature

Arterial Supply. The arteries of the tongue are chiefly derived from the *lingual artery,* which arises from the external carotid. On entering the tongue, the lingual artery passes deep to the hyoglossus muscle. The main branches of the lingual artery are the

- Dorsal lingual arteries to the dorsum of the tongue
- Deep lingual artery to the tip of the tongue
- Sublingual artery to the sublingual gland and adjacent muscles

Venous Drainage. Venous drainage of the tongue is through the

- Dorsal lingual veins that accompany the lingual artery
- Deep lingual veins (ranine veins) that begin at the tip of the tongue and run posteriorly beside the lingual frenulum to join the sublingual vein

All these veins terminate, directly or indirectly, in the internal jugular vein.

Lymphatic Drainage. Lymphatic drainage from the tongue takes four routes (p. 394):

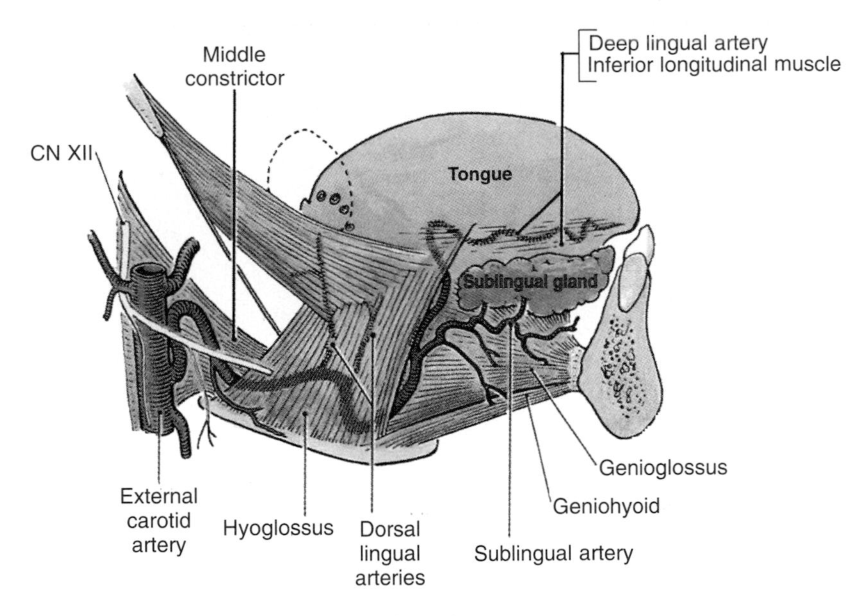

Arteries of tongue

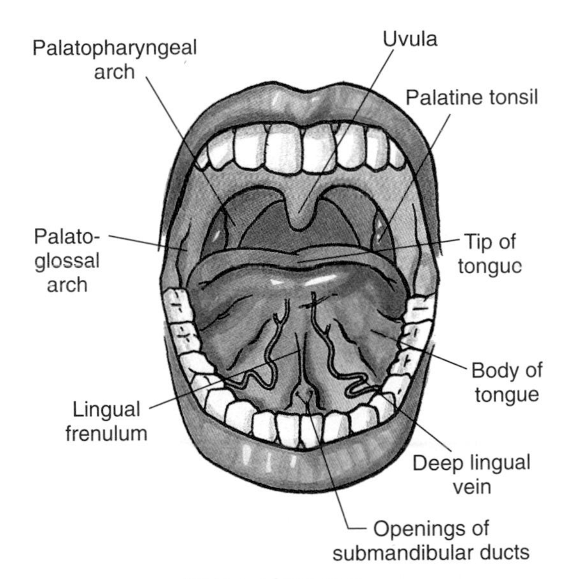

Veins of tongue

- Lymph from the posterior one-third of the tongue drains to the superior deep cervical lymph nodes on both sides
- Lymph from the medial part of the anterior two-thirds of the tongue drains directly to the inferior deep cervical lymph nodes
- Lymph from lateral parts of the anterior two-thirds of the tongue drains to the submandibular lymph nodes
- The tip of the tongue drains to the submental lymph nodes

One may touch the oral part of the tongue without feeling discomfort, but when the pharyngeal part is touched, one usually gags. CN IX and CN X are responsible for the muscular contraction of each side of the pharynx. Glossopharyngeal branches (CN IX) provide the afferent limb of the *gag reflex*.

When the genioglossus is paralyzed, the tongue has a tendency to fall posteriorly, obstructing the airway and presenting the risk of suffocation. Total relaxation of the genioglossus muscles occurs during general anesthesia; therefore the tongue of an anesthetized patient must be prevented from relapsing by inserting an airway.

Trauma, such as a fractured mandible, may injure the hypoglossal nerve. This results in paralysis and eventual atrophy of one side of the tongue. The tongue deviates to the paralyzed side during protrusion because of the action of the unaffected genioglossus muscle on the other side.

When quick absorption of a drug is desired [such as when nitroglycerin is used as a vasodilator in angina pectoris (chest pain)], the pill or spray is put under the tongue where it dissolves and enters the deep lingual veins in less than a minute.

Lymphatic drainage of the tongue is of particular importance because of the common occurrence of *lingual carcinoma*. Malignant tumors in the posterior one-third of the tongue metastasize to the superior deep cervical lymph nodes on both sides, whereas tumors in the anterior two-thirds usually do not metastasize to the inferior deep cervical lymph nodes until late in the disease. Because these nodes are closely related to the internal jugular vein, metastatic carcinoma from the tongue may be widely distributed through the submental and submandibular regions and along the internal jugular vein in the neck.

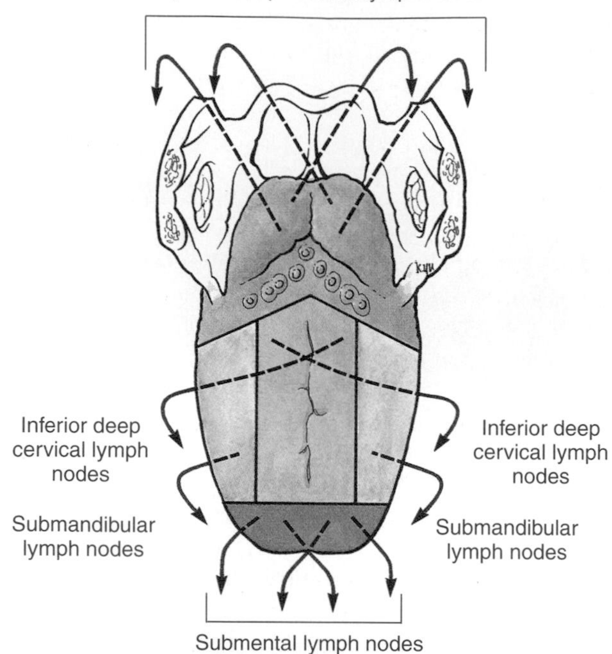

Lymphatic drainage of tongue

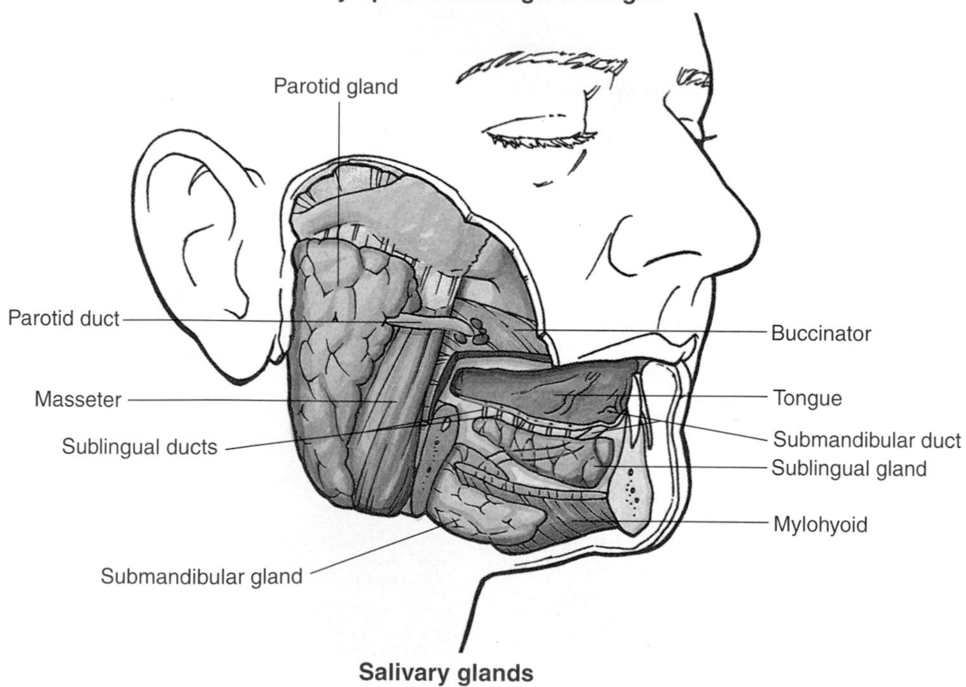

Salivary glands

SALIVARY GLANDS

The salivary glands are the parotid, submandibular, and sublingual glands.

The *parotid glands* are the largest of the three pairs of salivary glands; the *parotid duct*, piercing the buccinator, and the vasculature of the gland are described on page 354.

The *submandibular glands* lie along the body of the mandible, partly superior and partly inferior to the posterior half of the mandible and partly superficial and partly deep to the mylohyoid. The *submandibular duct* arises from the portion of the gland that lies between the mylohyoid and hyoglossus. The duct passes deep and then superficial to the lingual nerve. It opens by one to three orifices on a small sublingual papilla beside the lingual frenulum. Its orifice is readily visible, and often saliva can be seen trickling from it.

The *sublingual glands* are the smallest and most deeply situated. Each almond-shaped gland lies in the floor of the mouth between the mandible and the genioglossus. The glands from each side unite to form a horseshoe-shaped glandular mass around the lingual frenulum. Numerous small *sublingual ducts* open into the floor of the mouth.

Excision of a submandibular gland because of a *calculus* ("stone") in its duct or a tumor in the gland is not uncommon. The skin incision is made at least 2.5 cm inferior to the angle of the mandible to avoid the mandibular branch of the facial nerve (p. 351).

The parotid and submandibular salivary glands may be examined radiographically after the injection of a contrast medium into their ducts. This special type of radiograph (*sialogram*) demonstrates the salivary ducts and some secretory units. Because of the small size of the sublingual ducts of the sublingual glands, one cannot usually inject contrast medium into them.

Pterygopalatine Fossa

The pterygopalatine fossa is a small pyramidal space inferior to the apex of the orbit. The fossa lies between the pterygoid plates and the sphenoid posteriorly and the palatine bone medially. The maxilla lies anteriorly, and the fragile vertical plate of the palatine bone forms its medial wall. The incomplete *roof of the pterygopalatine fossa* is formed by the greater wing of the sphenoid bone. The *floor of the pterygopalatine fossa* is the pyramidal process of the palatine bone. Its superior, larger end opens into the *inferior orbital fissure*; its inferior end is closed except for the palatine foramina. Laterally the pterygopalatine fossa opens into the infratemporal fossa and can be partly inspected without separating the bones.

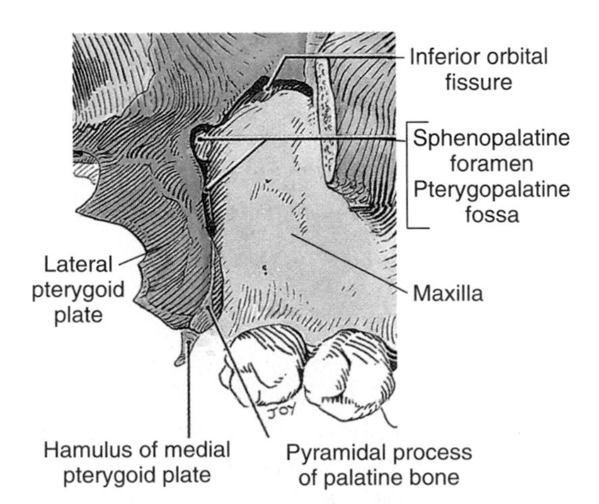

Pterygopalatine fossa

The pterygopalatine fossa communicates

- Laterally with the infratemporal fossa
- Medially with the nasal cavity through the sphenopalatine foramen
- Anteriorly with the orbit through the inferior orbital fissure
- Posterosuperiorly with the middle cranial fossa through the foramen rotundum and pterygoid canal

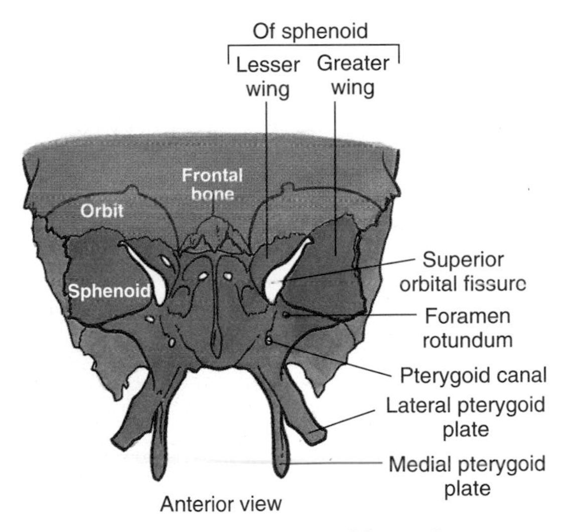

Opening of pterygoid canal

The contents of the pterygopalatine fossa (Fig. 8.22) are

- Terminal branches of maxillary artery
- Maxillary nerve (CN V^2)
- Nerve of pterygoid canal
- Pterygopalatine ganglion

The *maxillary nerve* (CN V^2) enters the pterygopalatine fossa through the foramen rotundum and runs anterolaterally in the posterior part of the fossa. Within the fossa, the maxillary nerve gives off the *zygomatic nerve*, which divides into zygomaticofacial and zygomati-

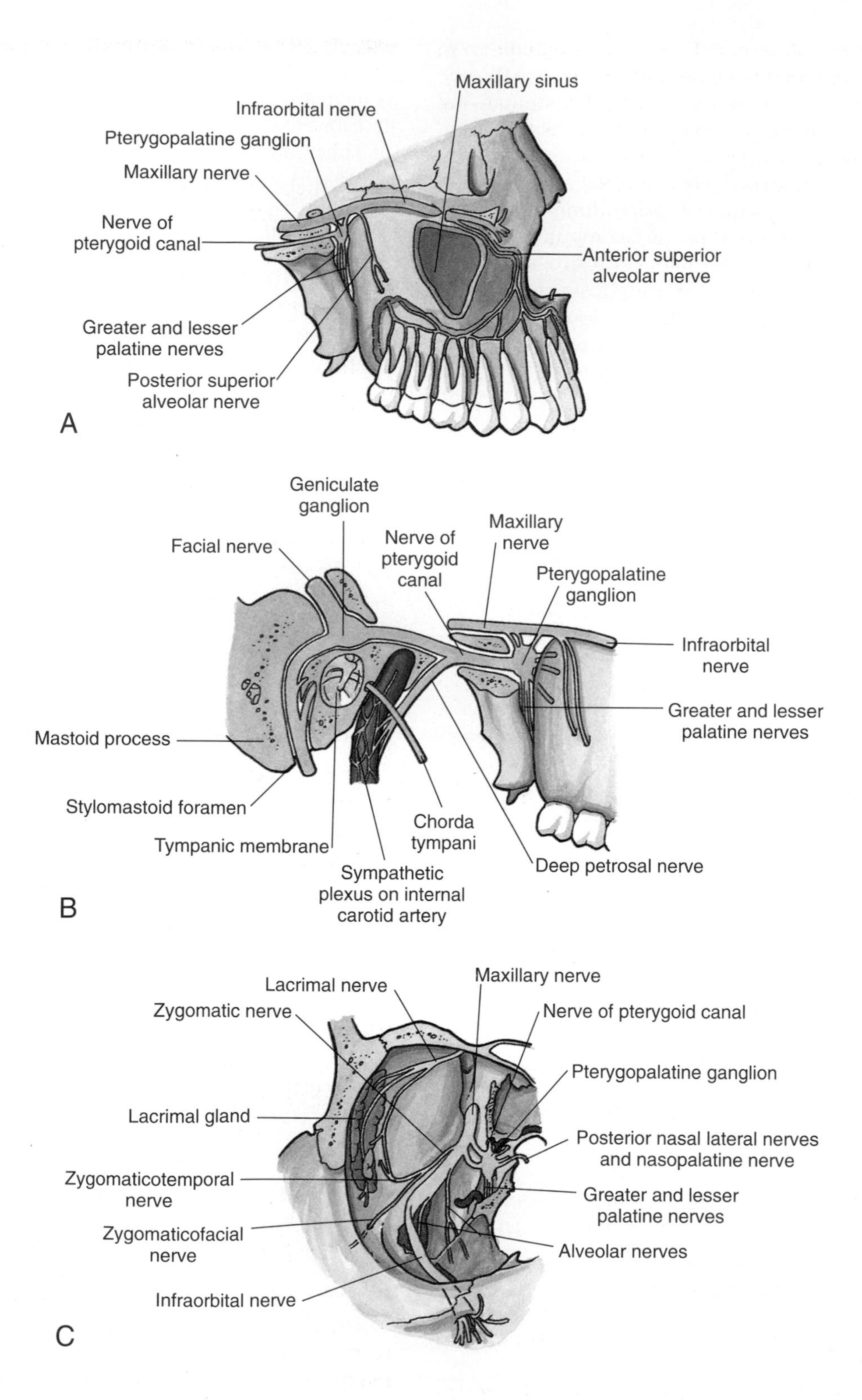

Figure 8.22. **A.** Nerves of pterygopalatine fossa. **B.** Autonomic fibers to pterygopalatine ganglion. **C.** Pterygopalatine fossa viewed through floor of orbit showing maxillary nerve (CN V^2) and its branches.

cotemporal nerves. These nerves emerge from the zygomatic bone through cranial foramina of the same name and supply the lateral region of the cheek and the temple. While in the pterygopalatine fossa, the maxillary nerve also gives off the two *pterygopalatine nerves* that suspend the parasympathetic *pterygopalatine ganglion* in the superior part of the pterygopalatine fossa.

The sensory fibers of the maxillary nerve pass through the pterygopalatine ganglion without synapsing and supply the nose, palate, tonsil, and gingivae. The maxillary nerve leaves the pterygopalatine fossa through the inferior orbital fissure, after which it is known as the *infraorbital nerve.* The parasympathetic fibers to the pterygopalatine ganglion come from the facial nerve via the *greater petrosal nerve.* This nerve joins the *deep petrosal nerve* as it passes through the foramen lacerum to form the *nerve of the pterygoid canal.* This nerve passes anteriorly through the pterygoid canal to the pterygopalatine fossa. The parasympathetic fibers of the greater petrosal nerve synapse in the pterygopalatine ganglion. The *deep petrosal nerve* is a sympathetic nerve from the internal carotid plexus. Its postganglionic fibers are from nerve cell bodies in the superior cervical sympathetic ganglion. The fibers do not synapse in the pterygopalatine ganglion but pass directly to join the branches of the maxillary nerve. The postsynaptic parasympathetic and sympathetic fibers pass to the lacrimal gland and the glands of the nasal cavity and upper pharynx. Fibers to the lacrimal gland reach it via the zygomaticotemporal branch of the zygomatic nerve (CN V^2).

The *pterygopalatine part of the maxillary artery,* its third part, passes through the pterygomaxillary fissure and enters the *pterygopalatine fossa,* where it lies anterior to the pterygopalatine ganglion. The artery breaks into branches that accompany all nerves in the fossa with the same names. The branches are the

- Posterior superior alveolar artery
- Descending palatine artery, which divides into the greater and lesser palatine arteries
- Artery of pterygoid canal
- Sphenopalatine artery, which divides into nasal branches to the nasal cavity, associated paranasal sinuses, and nasopalatine artery
- Infraorbital artery, which gives rise to the anterior superior alveolar artery and terminates as branches to the lower eyelid, nose, and upper lip

Nose

The nose contains the peripheral organ of smell. The functions of the nose and nasal cavities are

- Olfaction (smelling)
- Respiration
- Filtration of dust
- Humidification of inspired air
- Reception of secretions from the paranasal sinuses and nasolacrimal duct

The *external nose* varies considerably in size and shape, mainly because of differences in the nasal cartilages. The dorsum of the nose extends from its root at the face to its apex (tip). The inferior surface of the nose is pierced by two apertures, the *anterior nares* (nostrils), which are separated from each other by the nasal septum. The partly bony and cartilaginous *nasal septum* divides the chamber of the nose into two nasal cavities. The main components of the nasal septum are the

- Perpendicular plate of ethmoid
- Vomer
- Septal cartilage

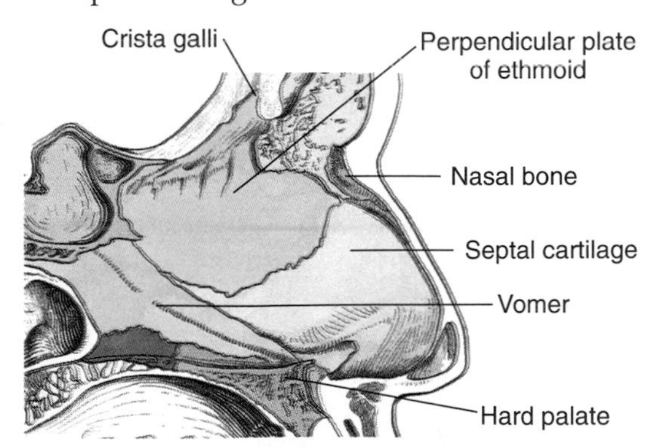

Nasal septum

The *perpendicular plate,* forming the superior part of the nasal septum, is thin and descends from the cribriform plate of the ethmoid. The *vomer,* a thin flat bone, forms the posteroinferior part of the nasal septum. It articulates with the perpendicular plate of the ethmoid and the septal cartilage.

The *bony part of the nose* consists of the

- Nasal bones

- Frontal processes of maxillae
- Nasal part of the frontal bone

The *cartilaginous part of the nose* consists of five main cartilages: two lateral cartilages, two alar cartilages, and a septal cartilage.

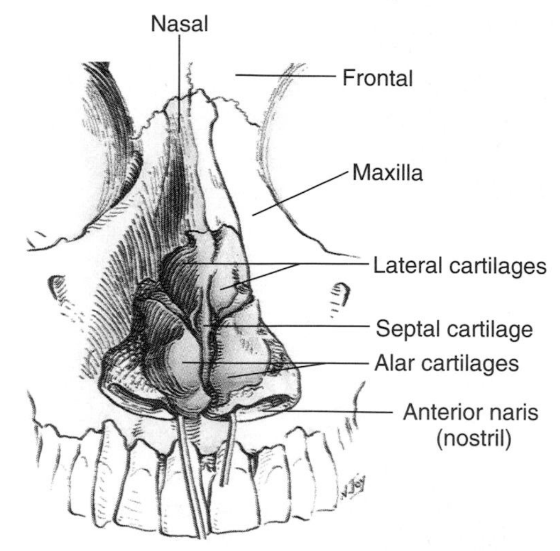

Nasal bones and cartilages

NASAL CAVITIES

The nasal cavities, entered through the anterior nares, open into the nasopharynx through the *choanae* (posterior nares) (Fig. 8.3). Mucosa lines the nasal cavities, except the *vestibule,* which is lined with skin (Fig. 8.23). The *nasal mucosa* is firmly bound to the periosteum and perichondrium of the supporting bones and cartilages of the nose. The mucosa is continuous with the lining of all the chambers with which the nasal cavities communicate: the nasopharynx posteriorly, the paranasal sinuses superiorly and laterally, and the lacrimal sac and conjunctiva superiorly. The inferior two-thirds of the nasal mucosa is the *respiratory area,* and the superior one-third is the *olfactory area.* Air passing over the respiratory area is warmed and moistened before it passes through the rest of the upper respiratory tract to the lungs. The olfactory area contains the peripheral organ of smell; sniffing draws air to the area.

Boundaries

- The roof of the nasal cavity is curved and narrow, except at the posterior end; it is divided into three parts (frontonasal, ethmoidal, and sphenoidal), which are named from the bones that form them
- The floor of the nasal cavity, wider than the roof, is formed by the palatine process of the maxilla and the horizontal plate of the palatine bone
- The medial wall of the nasal cavity is formed by the nasal septum
- The lateral wall of the nasal cavity is uneven because of three scroll-shaped elevations, the nasal conchae

The nasal conchae (superior, middle, and inferior) divide the nasal cavity into four passages (Fig. 8.23*A*): superior meatus, middle meatus, inferior meatus, and hiatus semilunaris.

The *superior meatus* is a narrow passage between the superior and middle nasal conchae into which the posterior ethmoidal sinuses open by one or more orifices.

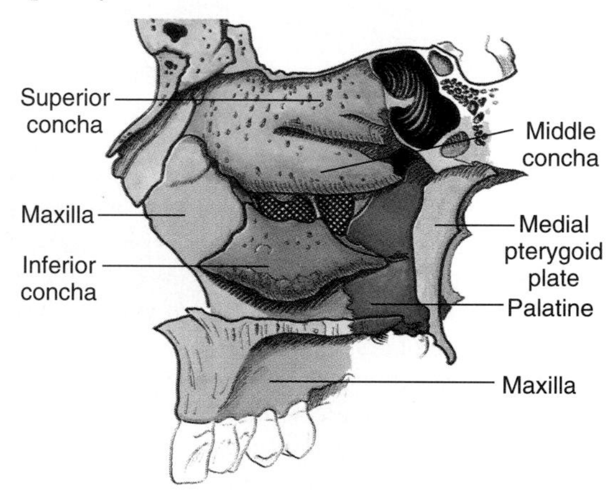

Lateral wall of nose

The *middle meatus* is longer and wider than the superior one. The anterosuperior part of this meatus leads into a funnel-shaped opening, the *infundibulum,* through which it communicates with the frontal sinus. The passage that leads inferiorly from each frontal sinus to the infundibulum is the *frontonasal duct.* The maxillary sinus also opens into the middle meatus.

The *inferior meatus* is a horizontal passage, inferolateral to the inferior nasal concha. The *nasolacrimal duct* from the lacrimal sac opens into the anterior part of this meatus.

The *hiatus semilunaris* is a semicircular groove into which the frontal sinus opens. The *ethmoidal bulla,* a rounded elevation located superior to the hiatus, is visible when the middle concha is removed. The bulla is formed by the *middle ethmoidal cells* that constitute the *ethmoidal sinuses.* Near the hiatus semilunaris are the openings of the anterior ethmoidal sinuses.

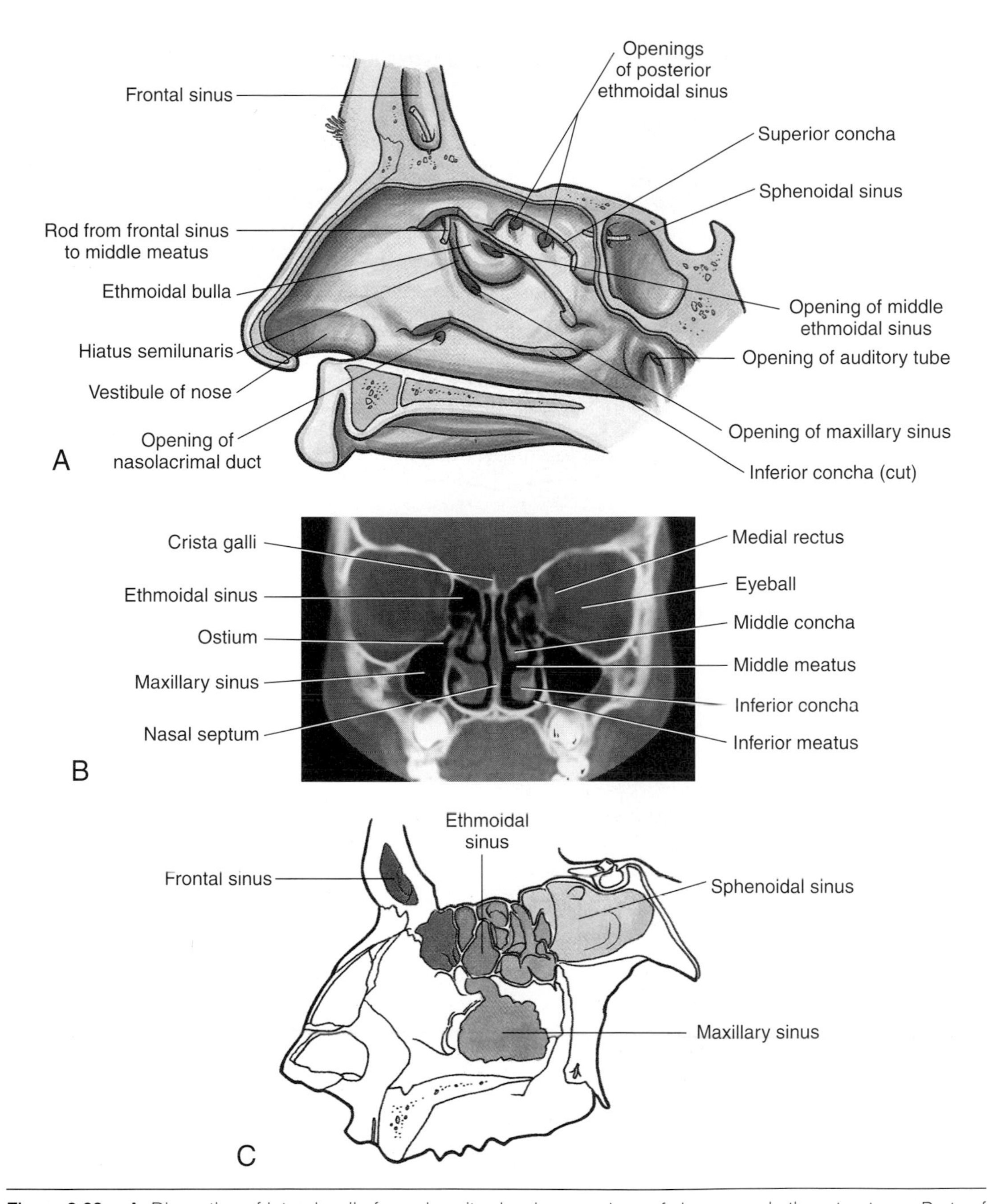

Figure 8.23. **A.** Dissection of lateral wall of nasal cavity showing openings of sinuses and other structures. Parts of conchae are cut away. **B.** Coronal computed tomographic scan showing sinuses. **C.** Medial view of a dissection showing paranasal sinuses.

Vasculature and Innervation

The *blood supply* of the medial and lateral walls of the nasal cavity (Fig. 8.24*A*) is from branches of the sphenopalatine artery, the anterior and posterior ethmoidal arteries, the greater palatine artery, and the superior labial artery and lateral nasal branches of the facial artery (Fig. 8.6*A*). A *plexus of veins* drains into the sphenopalatine, facial, and ophthalmic veins (Fig. 8.9*A*).

The *nerve supply* of the inferior two-thirds of the nasal mucosa is chiefly from the nasopala-

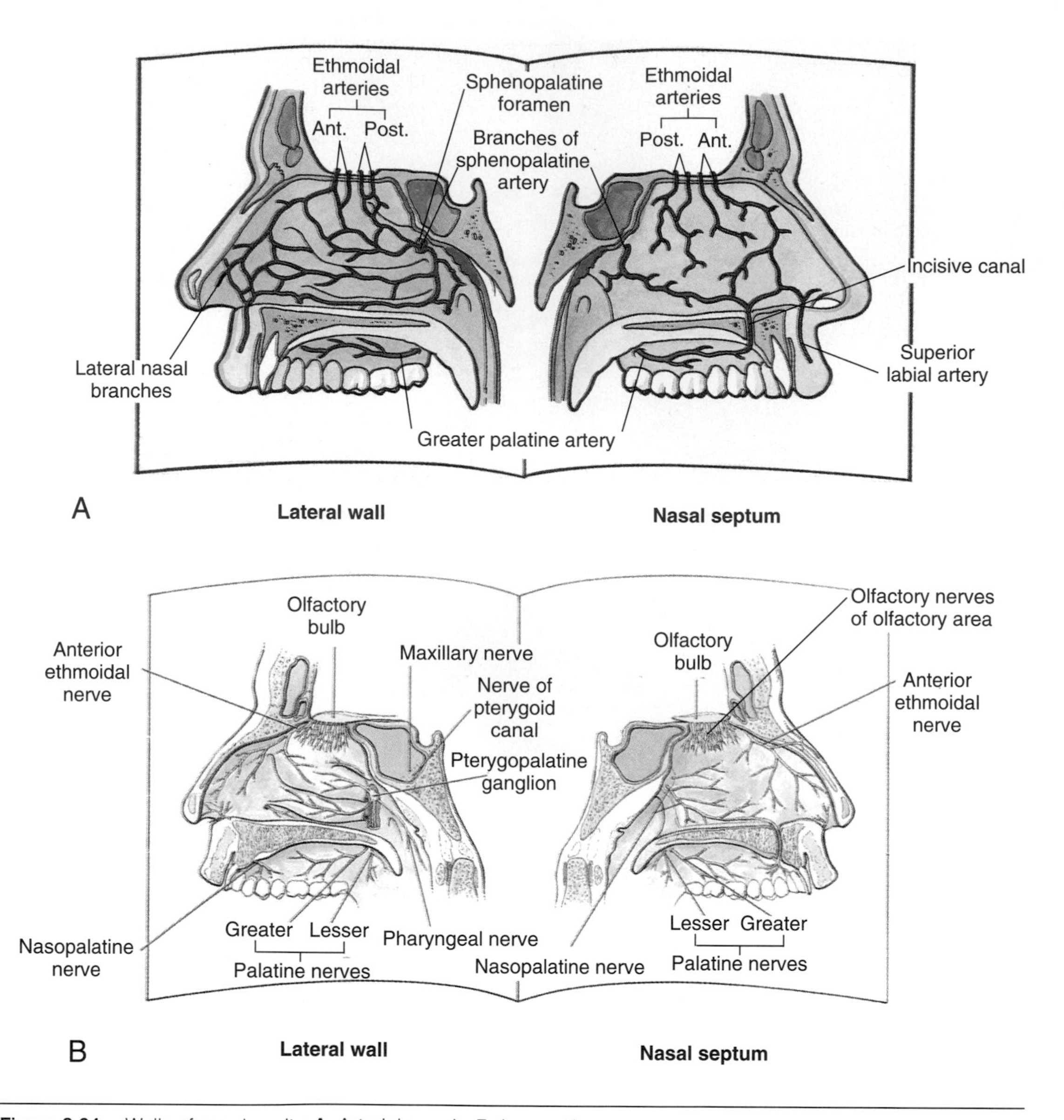

Figure 8.24. Walls of nasal cavity. **A.** Arterial supply. **B.** Innervation.

tine nerve, a branch of CN V^2 (Fig. 8.24*B*). Its anterior portion is supplied by the anterior ethmoidal nerve, a branch of the nasociliary nerve, derived from CN V^1. The lateral wall of the nasal cavity is supplied by nasal branches of the maxillary nerve (CN V^2), the greater palatine nerve, and the anterior ethmoidal nerve.

PARANASAL SINUSES

The paranasal sinuses are air-filled extensions of the respiratory part of the nasal cavity into the following cranial bones: frontal, ethmoid, sphenoid, and maxilla (Fig. 8.23). They are named according to the bones in which they are located.

The *frontal sinuses* are between the outer and inner tables of the frontal bone, posterior to the superciliary arches and the root of the nose. Each sinus drains through a *frontonasal duct* into the infundibulum, which opens into the hiatus semilunaris of the middle meatus (Figs. 8.21*A* and 8.23*A*). The frontal sinuses are innervated by branches of the *supraorbital nerves* (CN V^1).

Fractures of the nose are quite common because its bony parts are thin. If the injury results from a direct blow, the cribriform plate of the ethmoid (Fig. 8.4) may be fractured and the nasal septum may be displaced, deviating from the median plane. Sometimes the deviation is so severe that the nasal septum comes into contact with the lateral wall of the nasal cavity. Because this obstructs breathing, surgical repair may be necessary.

Although nasal discharges are commonly associated with upper respiratory tract infections, a nasal discharge after a head injury may be CSF. *CSF rhinorrhea* results from fracture of the cribriform plate, tearing of the cranial meninges, and leakage of CSF.

The nasal mucosa becomes swollen and inflamed (*rhinitis*) during upper respiratory infections and with some allergies (hayfever). Swelling of this mucous membrane occurs readily because of its vascularity. Infections of the nasal cavities may spread to the

- Anterior cranial fossa through the cribriform plate
- Nasopharynx and retropharyngeal soft tissues
- Middle ear through the auditory tube
- Paranasal sinuses
- Lacrimal apparatus and conjunctiva

Epistaxis (nosebleed) is relatively common because of the richness of the blood supply to the nasal mucosa. In most cases the cause is trauma, and the bleeding is located in the anterior third of the nose. Mild epistaxis often results from *nose picking*, which tears the veins in the vestibule of the nose. However, epistaxis is also associated with infections and *hypertension*. Spurting of blood from the nose results from the rupture of arteries.

The *ethmoidal sinuses* comprise several small cavities, ethmoidal cells, within the lateral mass of the ethmoid between the nasal cavity and orbit. The anterior *ethmoidal cells* may drain indirectly into the middle meatus through the infundibulum. The middle ethmoidal cells open directly into the middle meatus. The posterior ethmoidal cells open directly into the superior meatus. The ethmoidal sinuses are supplied by the anterior and posterior ethmoidal branches of the *nasociliary nerves* (CN V^1).

The *sphenoidal sinuses*, separated by a bony septum, are in the body of the sphenoid and may extend into its wings. Because of these sinuses, the body of the sphenoid is fragile. Only thin plates of bone separate the sinuses from several important structures: the optic nerves and optic chiasma, the pituitary gland, the internal carotid arteries, and the cavernous and intercavernous sinuses. The posterior ethmoidal nerve and posterior ethmoidal artery supply the sphenoidal sinuses (Fig. 8.16).

The *maxillary sinuses* are the largest of the paranasal sinuses. These pyramidal cavities occupy the entire bodies of the maxillae (Fig. 8.23, *B* and *C*). The apex of the maxillary sinus extends toward and often into the zygomatic bone. The base of the sinus forms the inferior part of the lateral wall of the nasal cavity. The roof of the sinus is formed by the floor of the orbit, and its narrow floor is formed by the alveolar part of the maxilla. The roots of the maxillary teeth, particularly the first two molars, often produce conical elevations in the floor of the sinus. Each sinus drains into the middle meatus of the nasal cavity through the hiatus semilunaris by an ostium (opening) in the superior part of its base (Fig. 8.23*B*). Because of the superior location of this opening, it is impossible for the sinus to drain when the head is erect until the sinus is full. *Innervation of the maxillary sinus* is from the anterior, middle, and posterior superior alveolar nerves (Fig. 8.22*A*), branches of the maxillary nerve (CN V^2). *Blood supply of the maxillary sinus* is mainly from the superior alveolar branches of the maxillary artery (Fig. 8.19), but branches of the greater palatine artery contribute blood to the floor of the maxillary sinus (Fig. 8.24).

If nasal drainage is blocked, infections of the ethmoidal cells may break through the fragile medial wall of the orbit. Severe infections may cause blindness because some posterior ethmoidal cells lie close to the optic canal. Spread of infection from these cells could also affect the dural sheath of the optic nerve and cause *optic neuritis*.

The maxillary sinuses are the most commonly involved in infection, probably because their apertures are located superior to the floor of the sinus, a poor location for natural drainage of the sinus. In addition, when the mucous membrane of the sinus is congested, the maxillary apertures may be obstructed. The proximity of the molar teeth to the floor of the maxillary sinus poses potentially serious problems (Fig. 8.23*B*). During removal of a maxillary molar tooth, fracture of a root may occur. If proper retrieval methods are not used, a piece of the root may be driven superiorly into the maxillary sinus. As a result, a communication may be created between the oral cavity and the maxillary sinus. The maxillary sinus can often be cannulated and drained by passing a cannula through the nostril and into the ostium of the sinus.

Ear

The ear is divided into external, middle, and internal parts (Fig. 8.25). The ear has two functions: balance and hearing. The *tympanic membrane* (eardrum) separates the external ear from the middle ear or tympanic cavity. The *auditory tube* joins the middle ear to the nasopharynx (Fig. 8.21*A*).

EXTERNAL EAR

The external ear is composed of the *auricle* that collects sounds and the external acoustic meatus that conducts them to the tympanic membrane (Fig. 8.25). The auricle, which has several named parts, consists of elastic cartilage covered with skin. The *external acoustic meatus* extends from the concha of the auricle to the tympanic membrane. The lateral third of this S-shaped passage is cartilaginous and is lined with the skin of the auricle, whereas its medial two-thirds is bony and is lined with thin skin that is continuous with the external layer of the tympanic membrane. The ceruminous and sebaceous glands produce cerumen (earwax).

The *tympanic membrane*, about 1 cm in diameter, is a thin, oval, semitransparent membrane at the medial end of the external acoustic meatus. It forms a partition between the external and middle parts of the ear. The tympanic membrane is covered with very thin skin externally and mucous membrane internally. The membrane shows a concavity toward the meatus with a central depression, the *umbo*. From the umbo, a bright area called the *cone of light* radiates anteroinferiorly. The tympanic membrane moves in response to air vibrations that pass to it through the external acoustic meatus. Movements of the membrane are transmitted by the *auditory ossicles* (malleus, incus, and stapes) through the middle ear to the internal ear. The external surface of the tympanic membrane is supplied mainly by the *auriculotemporal nerve*, a branch of the mandibular division of the trigeminal nerve (CN V^3). Some innervation is supplied by a small auricular branch of the vagus (CN X); this nerve contains some glossopharyngeal (CN IX) fibers and possibly some fibers of the facial nerve (CN VII). The internal surface of the tympanic membrane is supplied by CN IX.

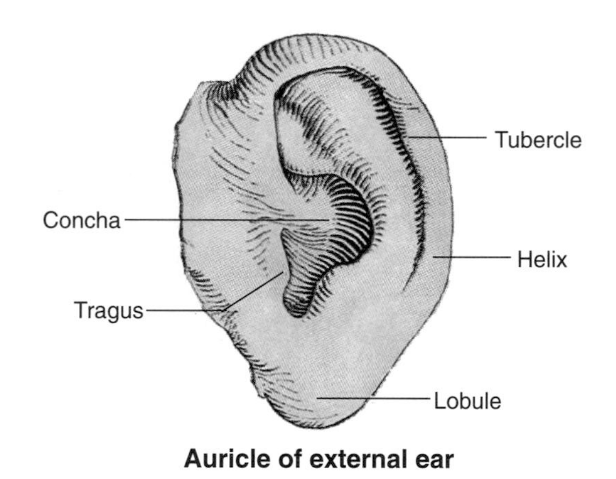

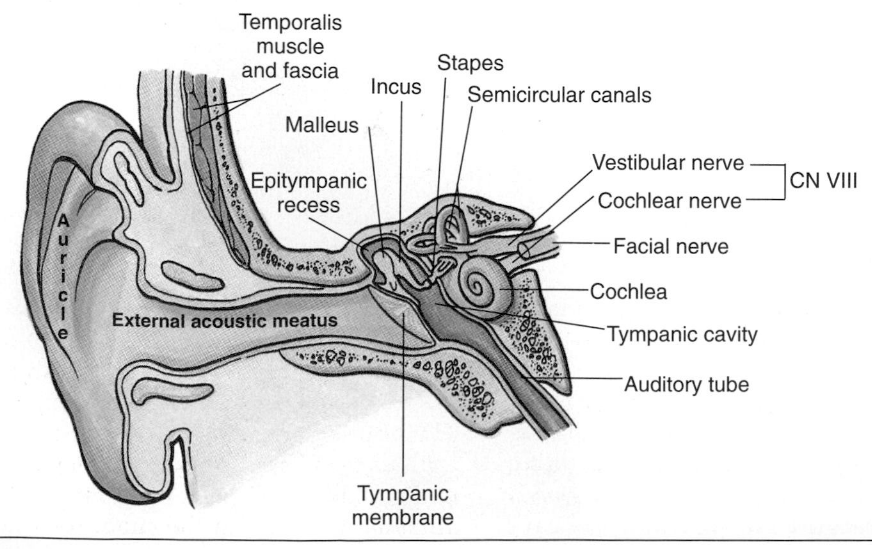

Figure 8.25. Coronal section of ear.

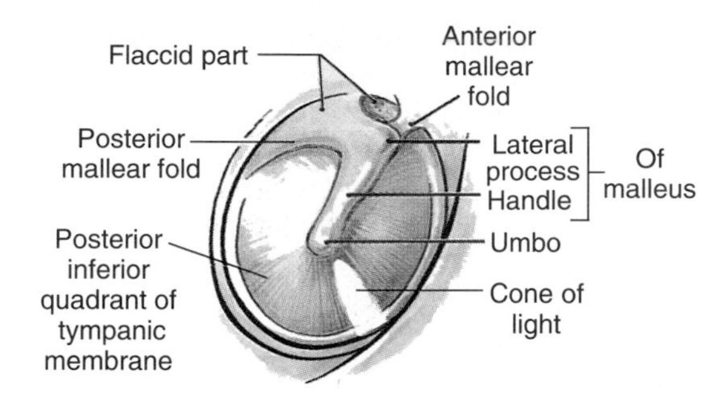

The tips of a *stethoscope* are angulated according to the anteromedial direction of the external acoustic meatus. Examination of the meatus in adults begins by grasping the helix of the auricle and pulling it posterosuperiorly. This movement reduces the curvature of the meatus, facilitating insertion of the *otic speculum*. It also provides a clue to tenderness, which can indicate inflammation of the meatus and auricle called *otitis externa* (e.g., swimmer's ear). The external acoustic meatus is relatively short in infants; therefore extra care must be taken to prevent damage to the tympanic membrane.

A reddened, bulging tympanic membrane is a sign of *otitis media*. Perforation of the tympanic membrane may occur with a middle ear infection and is one of several causes of middle ear deafness. Perforation may also result from foreign bodies in the external acoustic meatus, trauma, or excessive pressure (e.g., during scuba diving). Because the superior half of the tympanic membrane is much more vascular than the inferior half, incisions (e.g., to release pus) are made posteroinferiorly through the membrane. This site also avoids the chorda tympani nerve and auditory ossicles.

Severe bleeding or escape of CSF through a ruptured tympanic membrane and the external acoustic meatus (*CSF otorrhea*) may occur after a severe blow to the head. Fractures of the floor of the middle cranial fossa may tear the meninges and result in loss of CSF through a ruptured tympanic membrane into the external acoustic meatus.

MIDDLE EAR

The middle ear is in the petrous part of the temporal bone (Fig. 8.26). It includes the *tympanic cavity*, the space directly internal to the tympanic membrane, and the *epitympanic recess*, the space superior to the membrane. The middle ear is connected anteriorly with the nasopharynx by the auditory tube. Posterosuperiorly, the tympanic cavity connects with the mastoid cells through the *mastoid antrum*. The tympanic cavity is lined with mucous membrane that is continuous with that lining the auditory tube, mastoid cells, and mastoid antrum.

Contents of the middle ear are the

- Auditory ossicles (malleus, incus, and stapes)
- Stapedius and tensor tympani muscles
- Chorda tympani nerve branch of (CN VII)
- Tympanic plexus of nerves on promontory

Walls of Middle Ear (Tympanic Cavity)

The middle ear, shaped like a narrow box, has a roof, floor, and four walls (Fig. 8.26).

- The roof (tegmental wall) is formed by a thin plate of bone, the *tegmen tympani*, which separates the tympanic cavity from the dura on the floor of the middle cranial fossa
- The floor (jugular wall) is formed by a layer of bone that separates the tympanic cavity from the superior bulb of the internal jugular vein
- The lateral (membranous) wall is formed almost entirely by the tympanic membrane; superiorly it is formed by the lateral bony wall of the *epitympanic recess* (the handle of the malleus is incorporated in the tympanic membrane, and its head extends into the epitympanic recess)
- The medial or labyrinthine wall separates the tympanic cavity from the inner ear
- The anterior wall (carotid wall) separates the tympanic cavity from the carotid canal; superiorly it has the opening of the auditory tube and the canal for the tensor tympani
- The posterior wall (mastoid wall) is connected by the aditus to the mastoid antrum to the mastoid cells in the mastoid process; anteroinferiorly the mastoid antrum is related to the canal for the facial nerve

The *mastoid antrum* is a cavity in the mastoid process of the temporal bone (Fig. 8.26). The mastoid antrum is connected to the epitympanic recess of the tympanic cavity by the aditus to the mastoid antrum and is separated from the middle cranial fossa by a thin roof, the *tegmen tympani*. Its floor has several apertures through which the mastoid antrum communicates with the mastoid (air) cells in the mastoid process. The antrum and mastoid cells are lined by mucous membrane that is continuous with that of the middle ear. Anteroinferiorly the mastoid antrum is related to the canal for the facial nerve.

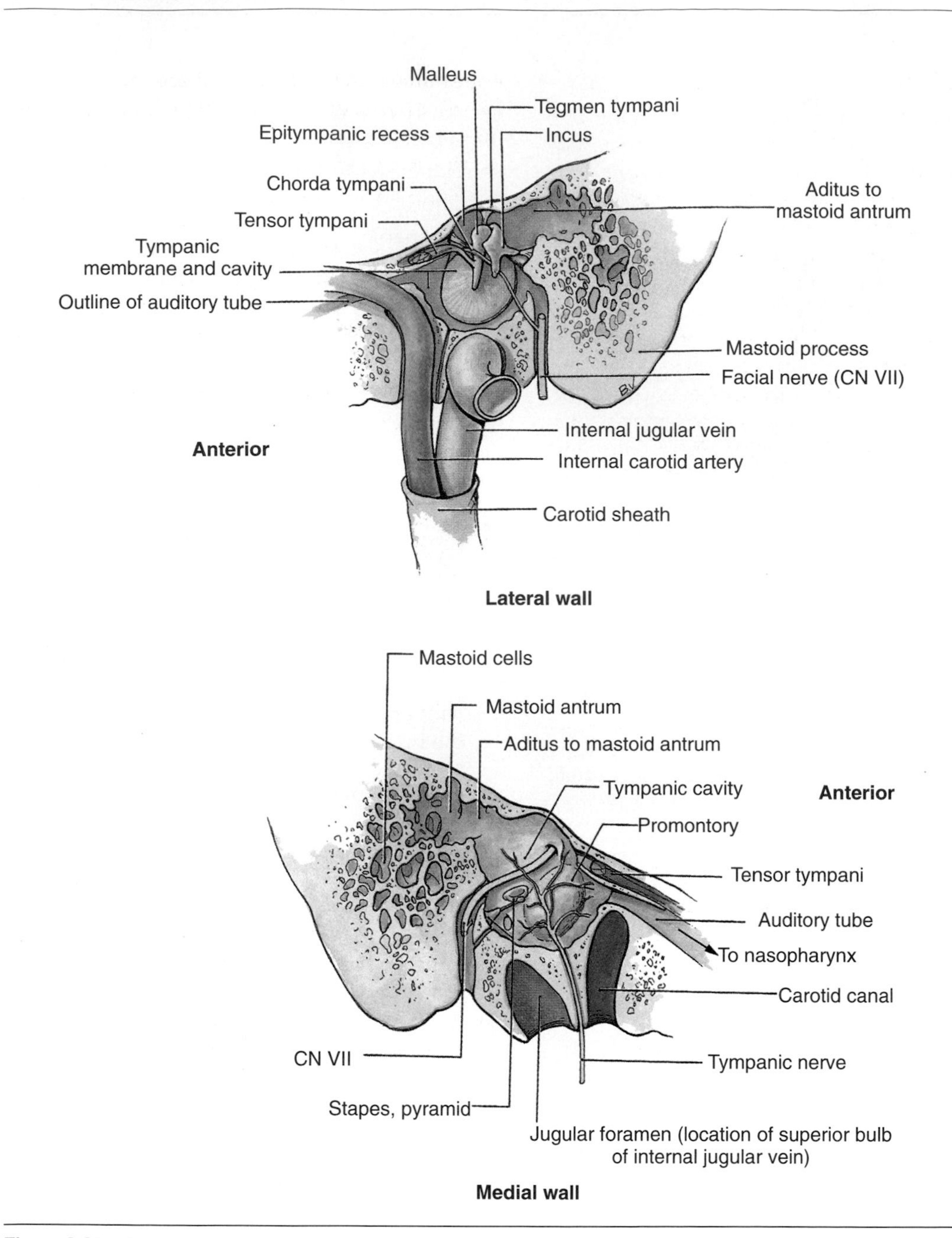

Figure 8.26. Dissections of middle ear showing contents and walls of tympanic cavity.

Infections of the mastoid antrum and cells (*mastoiditis*) result from a middle ear infection (*otitis media*). Infections may spread superiorly into the middle cranial fossa through the petrosquamous fissure in young children or may cause *osteomyelitis* (bone infection) of the tegmen tympani. Since the advent of antibiotics, mastoiditis is uncommon. During operations for mastoiditis, surgeons are conscious of the course of the facial nerve so that it will not be injured. One access to the tympanic cavity is through the mastoid antrum. In a child, only a thin plate of bone must be removed from the lateral wall of the mastoid antrum to expose the tympanic cavity. However, in adults, bone must be penetrated for 15 mm or more. Most *mastoidectomies* are endaural (i.e., performed through the posterior wall of the external acoustic meatus).

Auditory Tube

The auditory tube connects the tympanic cavity to the nasopharynx, where it opens posterior to the inferior meatus of the nasal cavity (Fig. 8.23). The posterior third of the tube is bony, and the remainder is cartilaginous (Fig. 8.26). The auditory tube is lined by mucous membrane that is continuous posteriorly with that of the tympanic cavity and anteriorly with that of the nasopharynx. The function of the auditory tube is to equalize pressure in the middle ear with the atmospheric pressure, thereby allowing free movement of the tympanic membrane. By allowing air to enter and leave the cavity, it balances the pressure on both sides of the membrane.

Arteries of the auditory tube are derived from the ascending pharyngeal artery, a branch of the external carotid artery, and the middle meningeal artery and artery of the pterygoid canal, branches of the maxillary artery. *Veins* drain into the pterygoid venous plexus (p. 383). *Nerves* arise from the *tympanic plexus*, which is formed by fibers of the facial (CN VII) and glossopharyngeal (CN IX) nerves. The auditory tube also receives fibers from the *pterygopalatine ganglion* (Fig. 8.24B).

Earache is a common symptom that has multiple causes, two of which are otitis externa and otitis media. Earache may also be referred pain from distant lesions (e.g., a dental abscess).

The auditory tube forms a route for infections to pass from the nasopharynx to the tympanic cavity. This tube is easily blocked by swelling of its mucous membrane, even by mild infections (e.g., a "head cold"), because the walls of its cartilaginous part are normally in apposition. When the auditory tube is occluded, residual air in the tympanic cavity is usually absorbed into the mucosal blood vessels. This results in lowering of pressure in the tympanic cavity, retraction of the tympanic membrane, and interference with its free movement. As a result, hearing is affected. Pressure changes resulting from air flights can be equalized by swallowing or chewing gum; these movements open the auditory tubes.

Auditory Ossicles

The auditory ossicles (malleus, incus, and stapes) form a chain of bones across the tympanic cavity from the tympanic membrane to the *oval window* (fenestra vestibuli) (Fig. 8.26). The malleus is attached to the tympanic membrane, and the stapes occupies the oval window. The incus is located between these two bones and articulates with them. The ossicles are covered with the mucous membrane that lines the tympanic cavity.

The rounded superior part, or head of the *malleus* (hammer), lies in the epitympanic recess. Its neck lies against the flaccid part of the tympanic membrane, and its handle is embedded in the tympanic membrane and moves with it. The head of the malleus articulates with the incus, and the tendon of the tensor tympani muscle inserts into its handle. The *chorda tympani* nerve crosses the medial surface of the neck of the malleus.

The large body of the *incus* (anvil) lies in the epitympanic recess where it articulates with the head of the malleus. The long process of the incus articulates with the stapes, and its short process is connected by a ligament to the posterior wall of the tympanic cavity.

The base of the *stapes* (stirrup), the smallest ossicle, fits into the oval window on the medial wall of the tympanic cavity. Its head, directed laterally, articulates with the incus.

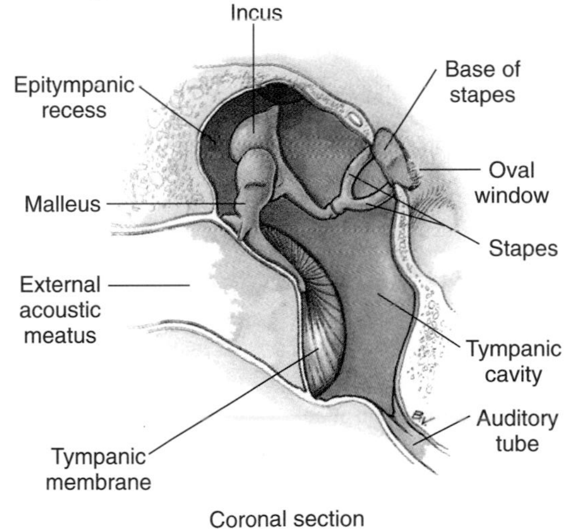

Middle ear

The malleus functions as a lever with the longer of its two arms attached to the tympanic membrane. The base of the stapes is considerably smaller than the tympanic membrane. As a result, the vibratory force of the stapes is about 10 times that of the tympanic membrane. Thus, the auditory ossicles increase the force but decrease the amplitude of the vibrations transmitted from the tympanic membrane.

Two muscles produce movements of the auditory ossicles that affect the tympanic membrane: tensor tympani and stapedius.

Tensor Tympani. This short muscle arises from the superior surface of the cartilaginous part of the auditory tube, the greater wing of the sphenoid, and the petrous part of temporal bone. It inserts into the handle of the malleus. The tensor tympani muscle is supplied by the mandibular nerve (CN V^3). The tensor tympani pulls the handle of the malleus medially, tensing the tympanic membrane and reducing the amplitude of its oscillations. This tends to prevent damage to the internal ear when one is exposed to loud sounds.

Stapedius. This tiny muscle is in the pyramidal eminence (pyramid). It arises from this eminence on the posterior wall of the tympanic cavity. Its tendon enters the tympanic cavity by traversing a pinpoint foramen in the apex of the pyramid and inserts on the neck of the stapes. The nerve to the stapedius arises from the facial nerve (CN VII). The stapedius pulls the stapes posteriorly and tilts its base in the oval window, thereby tightening the anular ligament and reducing the oscillatory range. It also prevents excessive movement of the stapes.

The tympanic muscles have a protective action in that they dampen large vibrations of the tympanic membrane resulting from loud noises. Paralysis of the stapedius muscle (e.g., resulting from a lesion of the facial nerve) is associated with excessive acuteness of hearing (*hyperacusia*). This condition results from uninhibited movements of the stapes.

INTERNAL EAR

The internal ear (*vestibulocochlear organ*) is involved with the reception of sound and the maintenance of balance. Buried in the petrous part of the temporal bone, the internal ear consists of the sacs and ducts of the *membranous labyrinth*. This membranous system contains endolymph and the end organs for hearing and balancing. The membranous labyrinth, surrounded by *perilymph*, is suspended in the bony labyrinth.

Bony Labyrinth

The bony labyrinth of the internal ear is composed of three parts: cochlea, vestibule, and semicircular canals. It occupies much of the lateral portion of the petrous part of the temporal bone (Figs. 8.25 and 8.27).

Cochlea. This shell-shaped part of the bony labyrinth contains the *cochlear duct*, the part of the internal ear that is concerned with hearing. The cochlea makes 2.5 turns around a bony core called the *modiolus*, in which there are canals for blood vessels and nerves. It is the large basal turn of the cochlea that produces the *promontory* on the medial wall of the tympanic cavity.

Vestibule. This small oval chamber (about 5 mm long) contains the *utricle* and *saccule*, parts of the balancing apparatus. The vestibule is continuous anteriorly with the bony cochlea, posteriorly with the semicircular canals, and

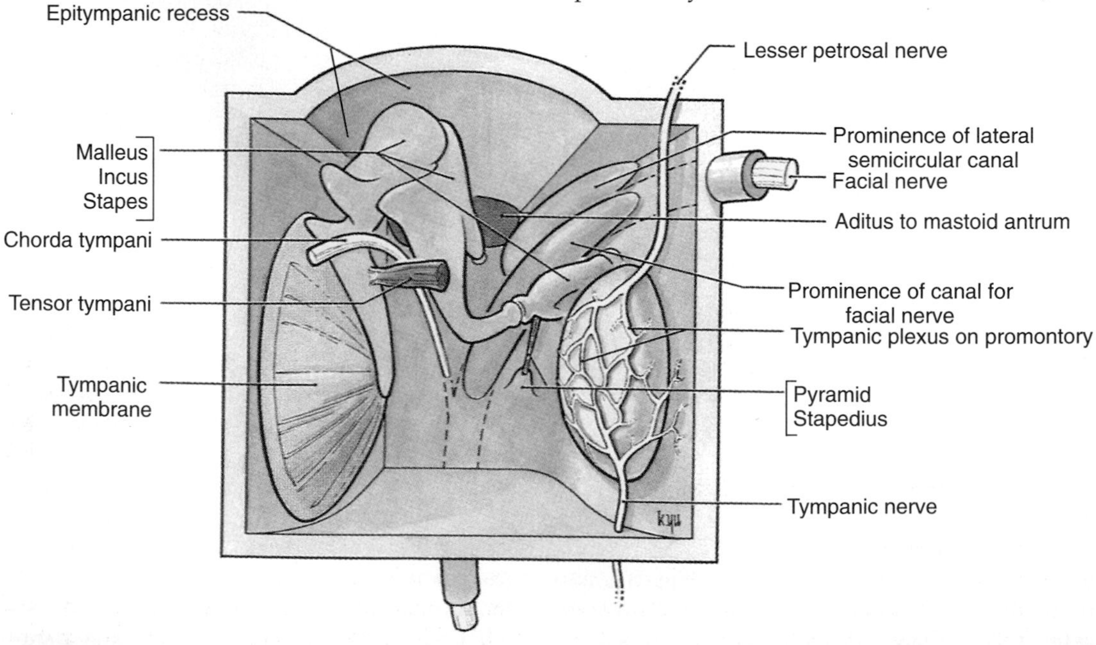

Walls of tympanic cavity

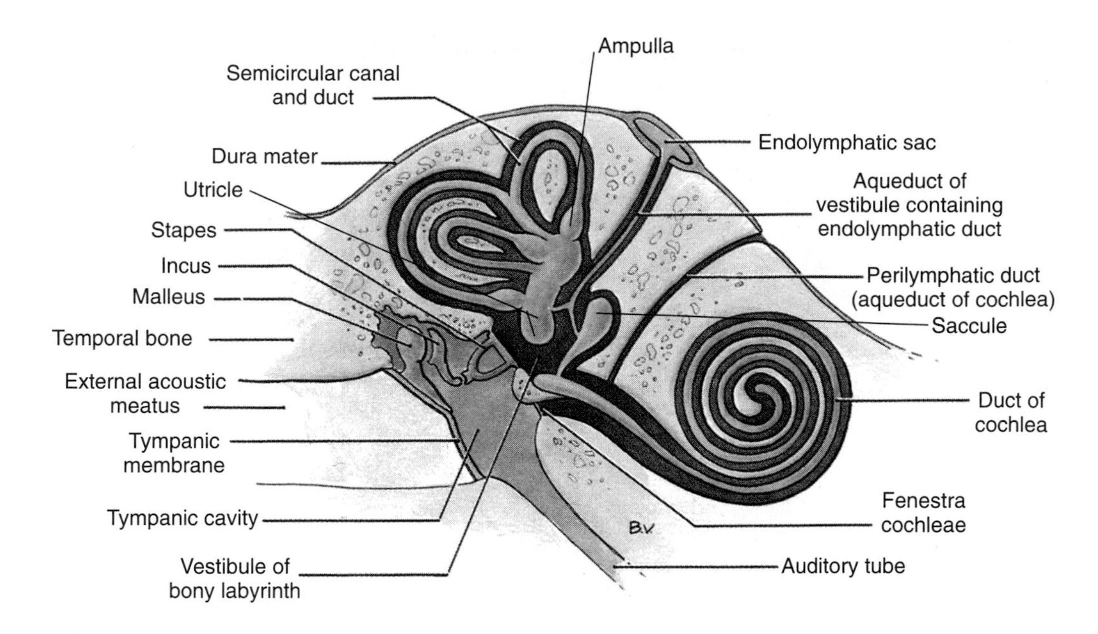

Figure 8.27. General scheme of ear showing relationship of middle and internal parts. Observe that membranous labyrinth is suspended in bony labyrinth.

with the posterior cranial fossa by the *aqueduct of the vestibule*. This aqueduct extends to the posterior surface of the petrous part of the temporal bone, where it opens posterolateral to the *internal acoustic meatus*. It contains the endolymphatic duct and two small blood vessels.

Semicircular Canals. These canals (anterior, posterior, and lateral) communicate with the vestibule of the bony labyrinth. The canals lie posterosuperior to the vestibule into which they open and are set at right angles to each other. They occupy three planes in space. Each semicircular canal forms about two-thirds of a circle and is about 1.5 mm in diameter, except at one end where there is a swelling called the *ampulla*. The canals have only five openings into the vestibule because the anterior and posterior canals have one stem common to both. Lodged within the canals are the *semicircular ducts*.

Membranous Labyrinth

The membranous labyrinth consists of a series of communicating sacs and ducts that are suspended in the bony labyrinth (Fig. 8.27). The membranous labyrinth contains *endolymph*, a watery fluid that differs in composition from the perilymph surrounding it in the bony labyrinth. The membranous labyrinth consists of three main parts:

- Utricle and saccule, two small communicating sacs in the vestibule of the bony labyrinth
- Three semicircular ducts in the semicircular canals
- Cochlear duct in the cochlea

The membranous labyrinth is suspended in the bony labyrinth. A spiral thickening of the periosteal lining of the cochlear canal, the *spiral ligament*, secures the cochlear duct to the cochlear canal. The various parts of the membranous labyrinth form a closed system of sacs and ducts that communicate with one another. The semicircular ducts open into the utricle through five openings, and the utricle communicates with the saccule through the utriculosaccular duct from which the *endolymphatic duct* arises. The saccule is continuous with the cochlear duct through a narrow communication known as the *ductus reuniens*.

The utricle and saccule both have specialized areas of sensory epithelium called maculae. The *macula utriculi* is in the floor of the utricle, parallel with the base of the skull, whereas the *macula sacculi* is vertically placed on the medial wall of the saccule. The hair cells in the maculae are innervated by fibers of the vestibular division of the vestibulocochlear nerve (CN VIII). The primary sensory neurons are in the *vestibular ganglion*, which is in the internal acoustic meatus. The *endolymphatic duct* emerges through the bone of the posterior cranial fossa and expands into a blind pouch called the *endolymphatic sac*. It is located under cover of the dura on the posterior surface of the petrous part of the temporal bone (Fig. 8.27). The endolymphatic sac is a storage reservoir for excess endolymph formed by the blood capillaries within the membranous labyrinth.

Each *semicircular duct* has an ampulla or expansion at one end containing a sensory area, the *crista ampullaris*. The cristae are sensors which record movements of the endolymph in the ampulla that result from rotation of the head in the plane of the duct. The hair cells of the cristae, like those of the maculae, are supplied by primary sensory neurons whose cell bodies are in the *vestibular ganglion*.

The *cochlear duct* is a spiral, blind tube firmly fixed to the internal and external walls of the cochlear canal by the spiral ligament (Fig. 8.27). This triangular duct lies between the osseous spiral lamina and the external wall of the cochlear canal. The roof of the cochlear duct is formed by the *vestibular membrane*, and its floor is formed by the *basilar membrane* and the external part of the osseous spiral lamina. The receptor of auditory stimuli is the *spiral organ* (of Corti), situated on the basilar membrane. It is overlaid by the gelatinous *tectorial membrane*. The spiral organ contains hair cells that respond to vibrations induced in the endolymph by sound waves.

Internal Acoustic Meatus

The internal acoustic meatus is a narrow canal that runs laterally for about 1 cm within the petrous part of the temporal bone (Fig. 8.4). The opening of the meatus is in the posteromedial part of this bone, in line with the external acoustic meatus. The internal acoustic meatus is closed laterally by a thin, perforated plate of bone that separates it from the internal ear. Through this plate pass the facial nerve (CN VII), branches of the vestibulocochlear nerve (CN VIII), and blood vessels (Fig. 8.25). The vestibulocochlear nerve divides near the lateral end of the internal acoustic meatus into a cochlear portion and a vestibular portion.

The maculae of the membranous labyrinth are primarily static organs for signaling the position of the head in space, but they also respond to quick tilting movements and to linear acceleration and deceleration. *Motion sickness* results mainly from fluctuating stimulation of the maculae.

Persistent exposure to excessively loud sounds causes degenerative changes in the spiral organ, resulting in *high tone deafness*. This type of hearing loss commonly occurs in workers who are exposed to loud noises and do not wear protective earmuffs (e.g., persons working for long periods around jet engines). Injury to the ear by an imbalance in pressure between ambient (surrounding) air and the air in the middle ear is called *otic barotrauma*. This type of injury occurs in fliers and divers.

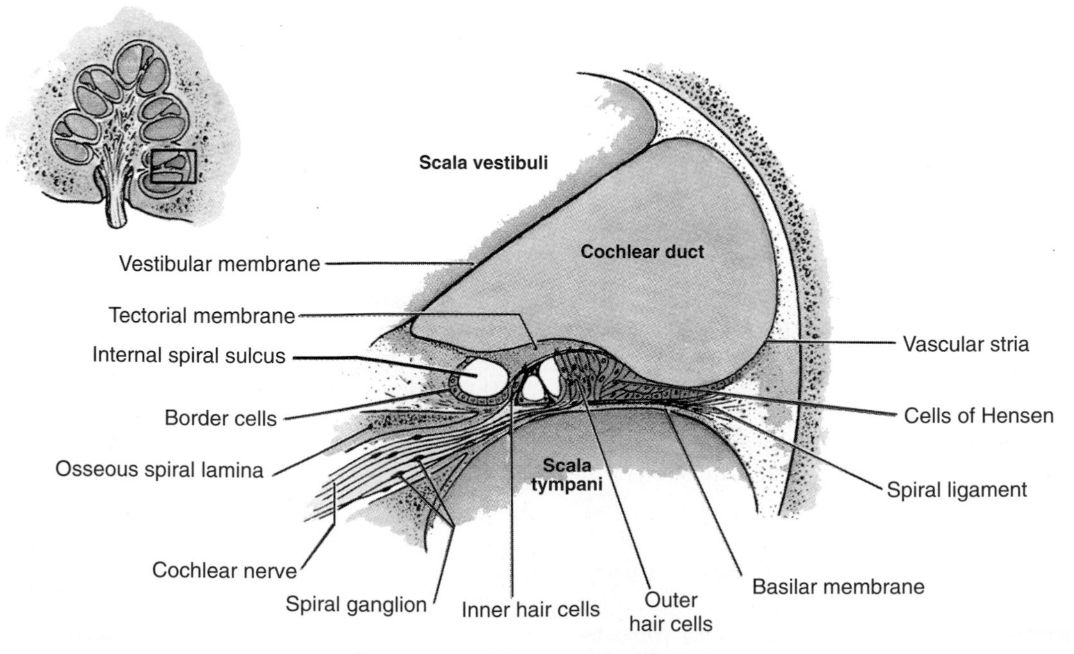

Section of cochlea and spiral organ

9 / NECK

The neck contains vessels, nerves, and other structures connecting the head, trunk, and limbs. It also contains important endocrine glands such as the thyroid gland. The skeleton of the neck is formed by the cervical vertebrae (Fig. 5.1C) and clavicles (p. 39).

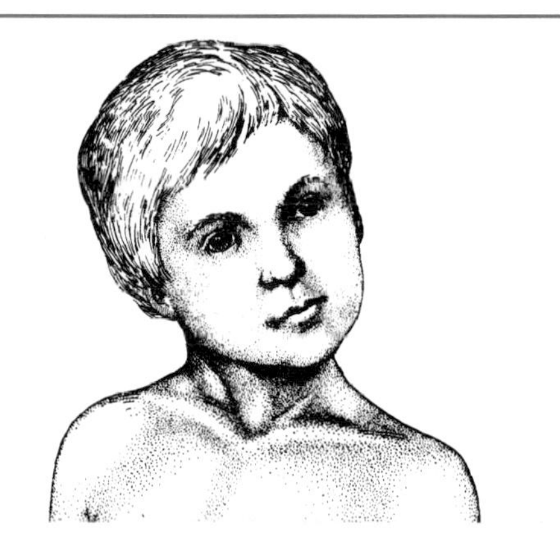

Occasionally the sternocleidomastoid (SCM) is injured by tearing its fibers when pulling an infant's head during a difficult birth. Later the lesion results in *torticollis* (wry neck), a flexion deformity of the neck. There is a tilt and rotation of the head to one side and restricted rotation to the other side. Stiffness of the neck results from fibrosis and shortening of the SCM. A fibrotic mass may be palpable in the muscle that may entrap a branch of the accessory nerve (CN XI), denervating part of the clavicular head of the SCM. Surgical release of a partially fibrotic SCM from its distal attachments may be necessary to enable the person to tilt and rotate the head normally. Torticollis may also follow stretching of CN XI during childbirth. This nerve supplies both the SCM and trapezius.

The platysma is paralyzed when the facial nerve (CN VII) is injured and/or when its cervical branch is severed. When the platysma is paralyzed, the skin of the neck forms slack folds.

Fasciae of Neck

The *superficial cervical fascia* (L. cervix, neck) is usually a thin layer of subcutaneous connective tissue that lies between the dermis of the skin and the deep cervical fascia. It contains the platysma, cutaneous nerves, blood and lymphatic vessels, and variable amounts of fat.

The *deep cervical fascia* consists of three fascial layers (Fig. 9.1): investing, pretracheal, and prevertebral. These fascial layers (sheaths) form natural lines of cleavage (planes) through which tissues may be separated during surgery, and they limit the spread of abscesses (collections of pus) resulting from infections. The cervical fascia also affords slipperiness that allows structures in the neck to move and pass over one another without difficulty, such as when swallowing and twisting the neck.

The *investing layer of cervical fascia* surrounds structures in the neck. It lies between the superficial fascia and muscles. Superiorly, the investing layer attaches to the

- Superior nuchal line of occipital bone
- Spinous processes of cervical vertebrae
- Mastoid processes of temporal bones
- Zygomatic arches
- Inferior border of mandible
- Hyoid bone

Inferiorly, the investing layer is attached to the

- Manubrium

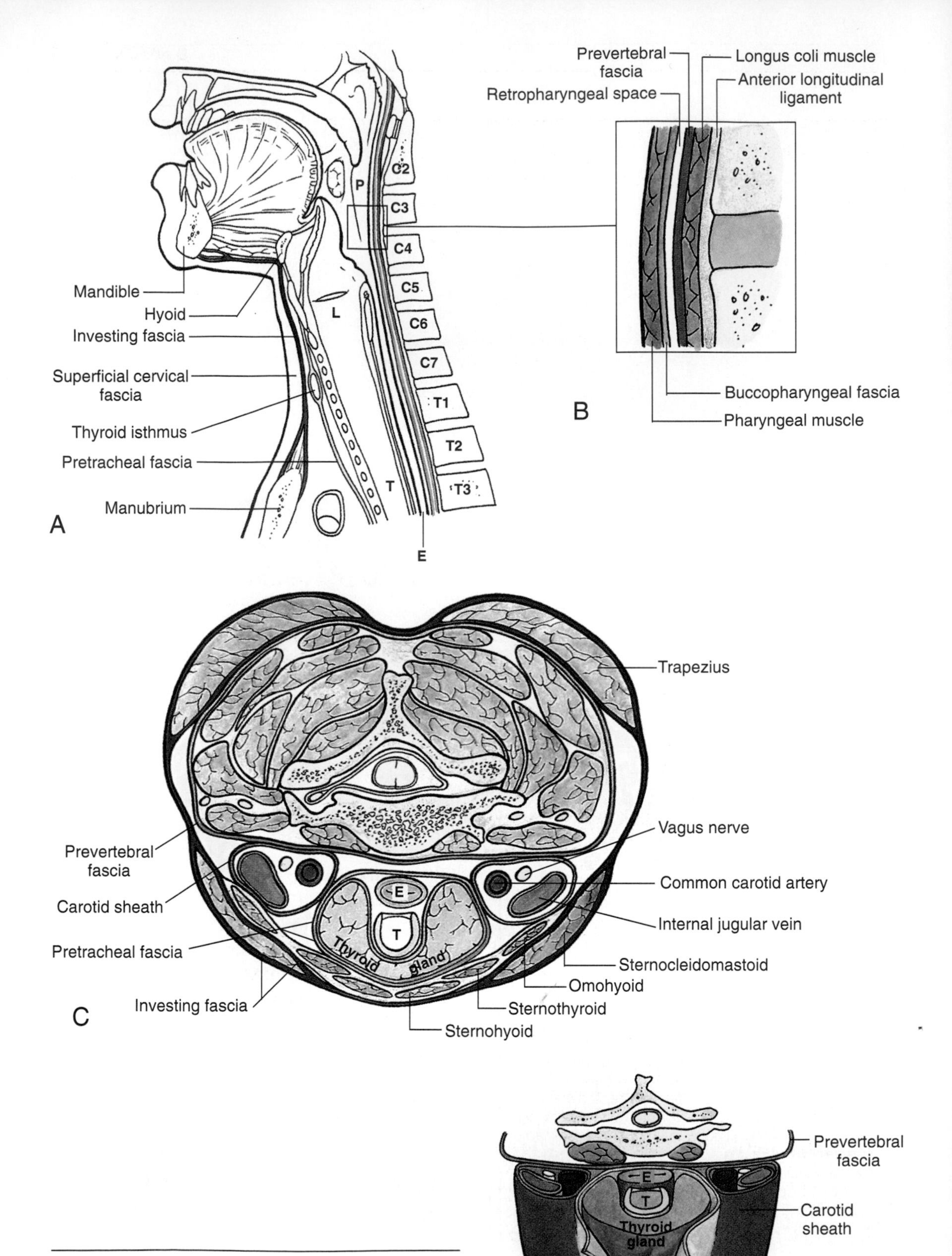

Figure 9.1. Fascia of neck: investing fascia; pretracheal fascia; prevertebral fascia, and carotid sheath. **A.** Median section of head and neck. *P*, pharynx; *L*, larynx; *T*, trachea; *E*, esophagus. **B.** Enlarged portion of **A** showing retropharyngeal space. **C.** Transverse section of neck through thyroid gland. **D.** Pretracheal fascia and its relationship to carotid sheath.

Surface Anatomy of Neck

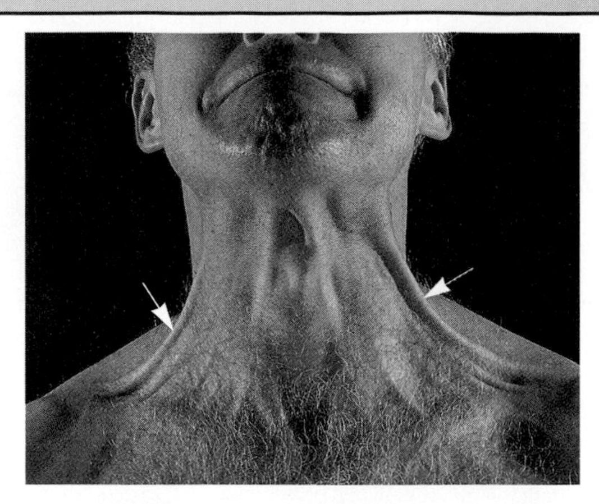

The skin of the neck is thin and pliable. The subcutaneous connective tissue, or superficial fascia, contains the *platysma*, a thin sheet of striated muscle that ascends to the face (*arrows*). Cutaneous nerves, fat, and blood and lymphatic vessels are also in this connective tissue.

The *sternocleidomastoid* (SCM) is the key landmark of the neck. This broad bulging muscle bisects the neck diagonally and stands out when the head is turned to the side. The SCM is crossed by the platysma and external jugular vein. The SCM arises by a tendinous head from the manubrium of the sternum and by a broader muscular head from the medial part of the clavicle (cleid means clavicle). The internal jugular vein passes in the interval (gap) between the sternal and clavicular heads. The two heads unite as the muscle extends superolaterally across the neck to attach to the mastoid process of the temporal bone. The *jugular notch* in the manubrium is easily palpated between the medial ends of the clavicles. Put a digit in your jugular notch, turn your head to the right, and move your digit to the left. Feel the narrow, tendinous sternal head of your left SCM. Move your digit laterally and palpate the broader clavicular head of the SCM. Lateral to this head is a large depression, the *supraclavicular fossa*.

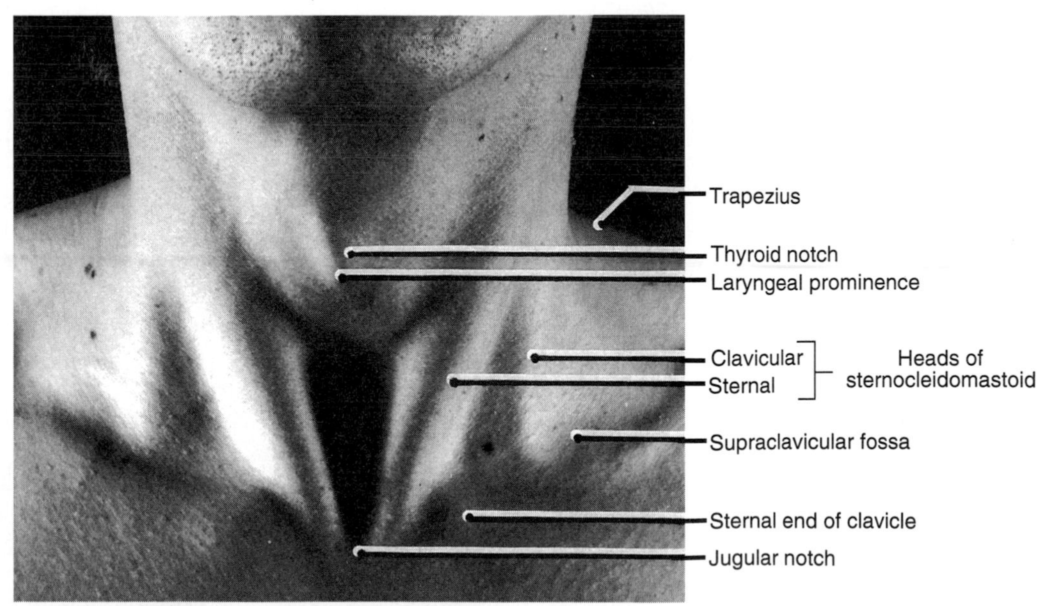

The *trapezius* extends over the posterior aspect of the neck and forms the sloping ridge of the neck. This muscle attaches the pectoral girdle (clavicle and scapula) to the skull and vertebral column; palpate it as you shrug your shoulder. The attachments, innervation, and actions of the SCM and trapezius are illustrated and described in Table 9.1.

The superior bony landmarks of the neck are the

- Inferior margin of mandible
- Mastoid process of temporal bone

Surface Anatomy of Neck ***Continued***

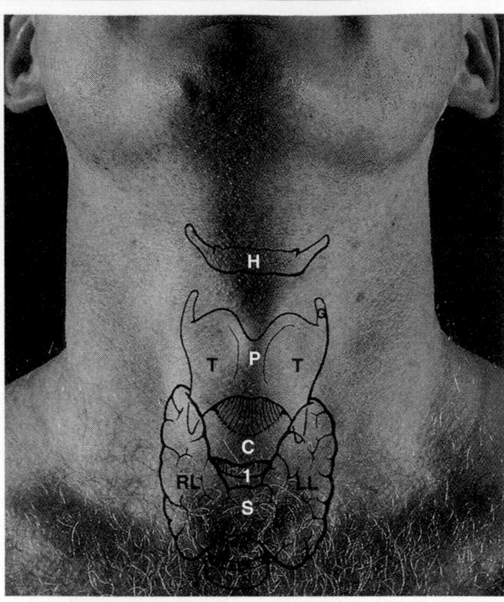

- External occipital protuberance of occipital bone

The inferior bony landmarks of the neck are the superior borders of the clavicles and the manubrium of sternum.

The *hyoid* (*H*) is a U-shaped bone that lies superior to the thyroid cartilage (*T*) at the level of C3 vertebra. The *laryngeal prominence* (*P*) is produced by the laminae of the *thyroid cartilage* that meet in the median plane. The laryngeal prominence ("Adam's apple") is palpable and frequently visible. The thyroid cartilage is at the level of C4 and C5 vertebrae. The superior (greater) horn of the cartilage (*G*) is palpable when the horn of the opposite side is steadied. The *cricoid cartilage* (*C*), another laryngeal cartilage, can be felt inferior to the laryngeal prominence. It lies at the level of C6 vertebra.

The *tracheal rings* are palpable in the inferior part of the neck. The second to fourth cartilaginous rings cannot be felt because the isthmus (*S*) connecting the right (*RL*) and left thyroid (*LL*) lobes covers them. The first tracheal ring (*1*) is just superior to the isthmus.

- Clavicles
- Acromions and spines of scapulae

Just superior to the manubrium, the investing layer of cervical fascia divides into two layers, one attached to the anterior and one to the posterior surface of the manubrium. A *suprasternal space* lies between these layers and encloses the sternal heads of the SCMs, the inferior ends of the anterior jugular veins, the jugular venous arch, fat, and a few lymph nodes.

The *pretracheal layer of cervical fascia*, limited to the anterior part of the neck, extends inferiorly from the thyroid and cricoid cartilages into the thorax, where it blends with the fibrous pericardium (p. 55). The pretracheal layer lies deep to the infrahyoid muscles and splits to enclose the thyroid, trachea, pharynx, and esophagus. It blends laterally with the carotid sheath.

The *prevertebral layer of cervical fascia* forms a tubular sheath for the vertebral column and the muscles associated with it. The prevertebral layer extends from the base of the skull to T3 vertebra, where it fuses with the anterior longitudinal ligament. The prevertebral layer extends laterally as the axillary sheath, which surrounds the axillary vessels and the brachial plexus (see Fig. 9.2).

The *carotid sheath* is a tubular, fascial condensation that extends from the base of the skull to the root of the neck (Fig. 9.1, *B* and *C*). This fascial sheath blends anteriorly with the investing and pretracheal layers and posteriorly with the prevertebral layer of cervical fascia. The carotid sheath contains the

- Common and internal carotid arteries
- Internal jugular vein
- Vagus nerve (CN X)

Table 9.1
Muscles in Lateral Aspect of Neck

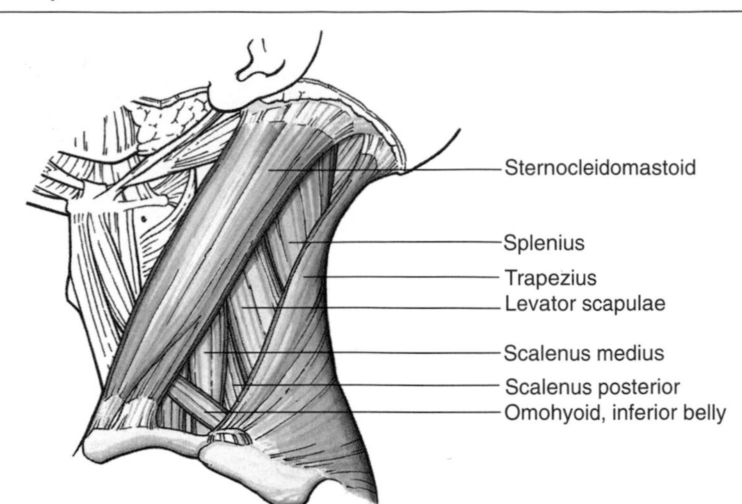

Muscle	Origin	Insertion	Innervation	Actions
Platysma	Inferior border of mandible, skin and subcutaneous tissues of lower face	Fascia covering superior parts of pectoralis major and deltoid muscles	Cervical branch of facial n. (CN VII)	Draws corners inferiorly and widens mouth as in expressions of sadness and fright; draws the skin of neck superiorly
Trapezius	Medial third of superior nuchal line, external occipital protuberance, ligamentum nuchae, spinous processes of C7–T12 vertebrae, and lumbar and sacral spinous processes	Lateral third of clavicle, acromion and spine of scapula	Spinal root of accessory n. (CN XI) and C3 and C4 nn.	Elevates, retracts, and and rotates scapula
Sternocleidomastoid	Lateral surface of mastoid process of temporal bone and lateral half of superior nuchal line	Sternal head: anterior surface of manubrium of sternum Clavicular head: superior surface of medial third of clavicle	Spinal root of accessory n. (CN XI) and C2 and C3 nn.	Tilts head to one side, i.e., laterally; flexes neck and rotates it so face is turned superiorly toward opposite side; acting together, the two muscles flex the neck
Splenius capitis	Inferior half of ligamentum nuchae and spinous processes of superior six thoracic vertebrae	Lateral aspect of mastoid process and lateral third of superior nuchal line	Dorsal rami of middle cervical spinal nn.	Laterally flexes and rotates head and neck to same side; acting bilaterally, they extend head and neck
Levator scapulae	Posterior tubercles of transverse processes of C1–C4 vertebrae	Superior part of medial border of scapula	Dorsal scapular n. (C5) and cervical spinal nn. (C3 and C4)	Elevates scapula and tilts its glenoid cavity inferiorly by rotating scapula
Scalenus posterior	Posterior tubercles of transverse processes of C4–C6 vertebrae	External border of 2nd rib	Ventral rami of cervical spinal nn. (C7 and C8)	Flexes neck laterally; elevates 2nd rib during forced inspiration
Scalenus medius	Posterior tubercles of transverse processes of C2–C7 vertebrae	Superior surface of 1st rib, posterior to groove for subclavian a.	Ventral rami of cervical spinal nn.	Flexes neck laterally; elevates 1st rib during forced inspiration

- Deep cervical lymph nodes
- Carotid sinus nerve
- Sympathetic fibers

The *retropharyngeal space* (Fig. 9.1*B*) is the largest and most important interfascial space in the neck. It is a potential space consisting of loose connective tissue between the prevertebral layer of cervical fascia and the *buccopharyngeal fascia* surrounding the pharynx superficially. Inferiorly, the buccopharyngeal fascia blends with the pretracheal layer of cervical fascia. The retropharyngeal space permits movement of the pharynx, esophagus, larynx, and trachea during swallowing. This space is closed superiorly by the base of the skull and on each side by the carotid sheath. It opens inferiorly into the superior mediastinum (p. 68).

The investing layer of cervical fascia helps to prevent the spread of abscesses. Pus from an abscess posterior to the prevertebral layer of cervical fascia may extend laterally in the neck and form a swelling posterior to the SCM. The pus may perforate the investing layer and enter the retropharyngeal space, producing a bulge in the pharynx (*retropharyngeal abscess*). This swelling may cause difficulty in swallowing and speaking.

Infections in the head or cervical vertebrae may also spread inferiorly, posterior to the esophagus and enter the posterior mediastinum (p. 68) or anterior to the trachea and enter the anterior mediastinum. Infections in the retropharyngeal space may also extend inferiorly into the superior mediastinum. Similarly, air from a ruptured trachea, bronchus, or esophagus can pass superiorly in the neck.

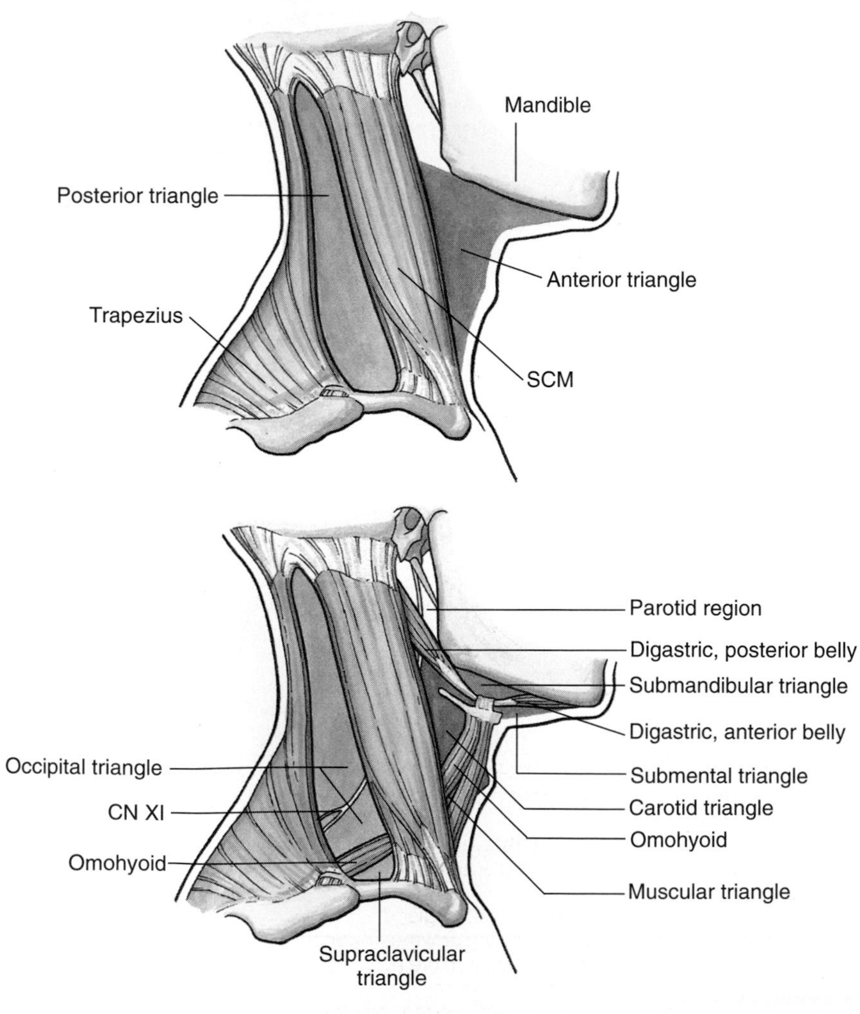

Triangles of neck

Triangles of Neck

The SCM, the main muscular landmark, bisects the neck diagonally into anterior and posterior cervical triangles.

The *posterior triangle of the neck* has

- An anterior boundary formed by the SCM
- A posterior boundary formed by the trapezius
- An inferior boundary (base) formed by the middle third of the clavicle
- Its apex where the SCM and trapezius meet on the superior nuchal line of the occipital bone
- A roof formed deeply by the investing layer of cervical fascia
- A floor formed by muscles covered by the prevertebral layer of cervical fascia

The posterior triangle is divisible into supraclavicular and occipital triangles by the inferior belly of the omohyoid muscle.

The *anterior triangle of the neck* has

- An anterior boundary formed by the median line of the neck
- A posterior boundary formed by the SCM
- A superior boundary (base) formed by the inferior border of the mandible
- Its apex at the jugular notch in the mediastinum
- A floor formed by the pharynx, larynx, and thyroid gland

The anterior triangle is subdivided into the unpaired submental triangle and three small paired triangles (submandibular, carotid, and muscular) by the digastric and omohyoid muscles.

Contents of the triangles of the neck are listed in Table 9.2.

Table 9.2
Contents of Triangles of Neck

Triangle	Main Contents
Posterior	Accessory nerve (CN XI) Cervical lymph nodes Brachial plexus (supraclavicular part) Third part of subclavian artery
Anterior	
Digastric	Submandibular gland Facial artery and vein Facial vein Portions of the parotid gland and external carotid artery Internal carotid artery, internal jugular vein, glossopharyngeal (CN IX) and vagus (CN X) nerves are situated deep in digastric triangle
Submental	Submental lymph nodes Small veins that unite to form anterior jugular vein
Carotid	External carotid artery and some of its branches Larynx and pharynx and internal and external laryngeal nerves are deep in carotid triangle
Muscular	Sternothyroid and sternohyoid muscles Thyroid gland, trachea, and esophagus
Submandibular	Submandibular gland Submandibular lymph nodes Hypoglossal nerve (CN XII) Mylohyoid nerve Parts of facial artery and vein

POSTERIOR CERVICAL TRIANGLE

Muscles in Posterior Triangle

The floor of the triangle is formed by four muscles: splenius capitis, levator scapulae, scalenus medius, and scalenus posterior (Fig. 9.2). Part of the scalenus anterior may appear in the inferomedial part of the triangle. Inferiorly, the triangle is crossed by the inferior belly of the omohyoid. The attachments, nerve supply, and actions of these muscles are given in Table 9.3.

Vessels and Nerves in Posterior Triangle

The *external jugular vein* drains most of the scalp and face on the same side. It begins near the angle of the mandible by the union of the posterior division of the retromandibular vein and the posterior auricular vein. The external jugular vein crosses the superficial surface of the SCM, deep to the platysma, and then pierces the investing layer of cervical fascia in the roof of the posterior triangle at the posterior border of the SCM. The external jugular vein descends to the inferior part of the triangle and terminates in the subclavian vein. Just superior to the clavicle, the external jugular vein receives the transverse cervical and suprascapular veins.

The *subclavian vein* (Fig. 9.2*B*), the major venous channel draining the upper limb, courses through the inferior part of the posterior triangle. It passes anterior to the scalenus anterior and unites at the medial border of the muscle with the internal jugular vein to form the *brachiocephalic vein*. The subclavian vein seldom rises superior to the level of the clavicle.

The third part of the *subclavian artery* lies posterosuperior to the subclavian vein. This is the most superficial part of the artery, and its pulsations can be felt on deep pressure. The subclavian artery is located mainly in the supraclavicular triangle, where it lies on the 1st rib.

The *transverse cervical artery* arises from the thyrocervical trunk, a branch of the subclavian artery. The transverse cervical artery runs superficially and laterally across the posterior triangle, superior to the clavicle and deep to the omohyoid muscle. The *suprascapular artery*, another branch of the thyrocervical trunk, passes inferolaterally across the inferior part of the posterior triangle, just superior to the clavicle. The suprascapular artery then runs posterior to the clavicle to supply muscles around the scapula. The *occipital artery*, a branch of the external carotid artery, enters the apex of the posterior triangle before ascending over the posterior aspect of the head to supply the posterior half of the scalp.

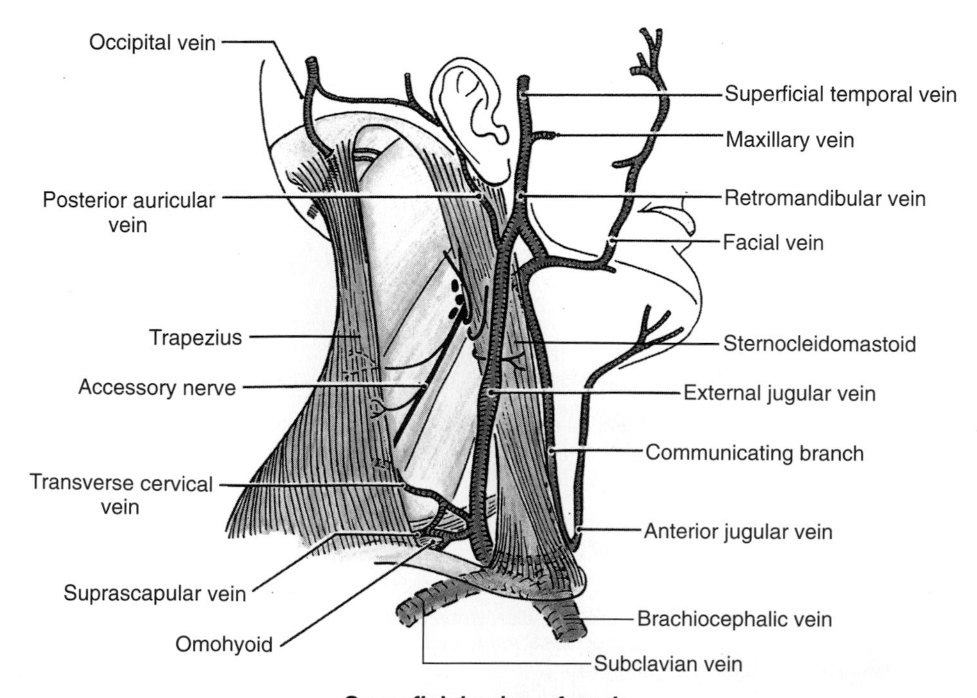

Superficial veins of neck

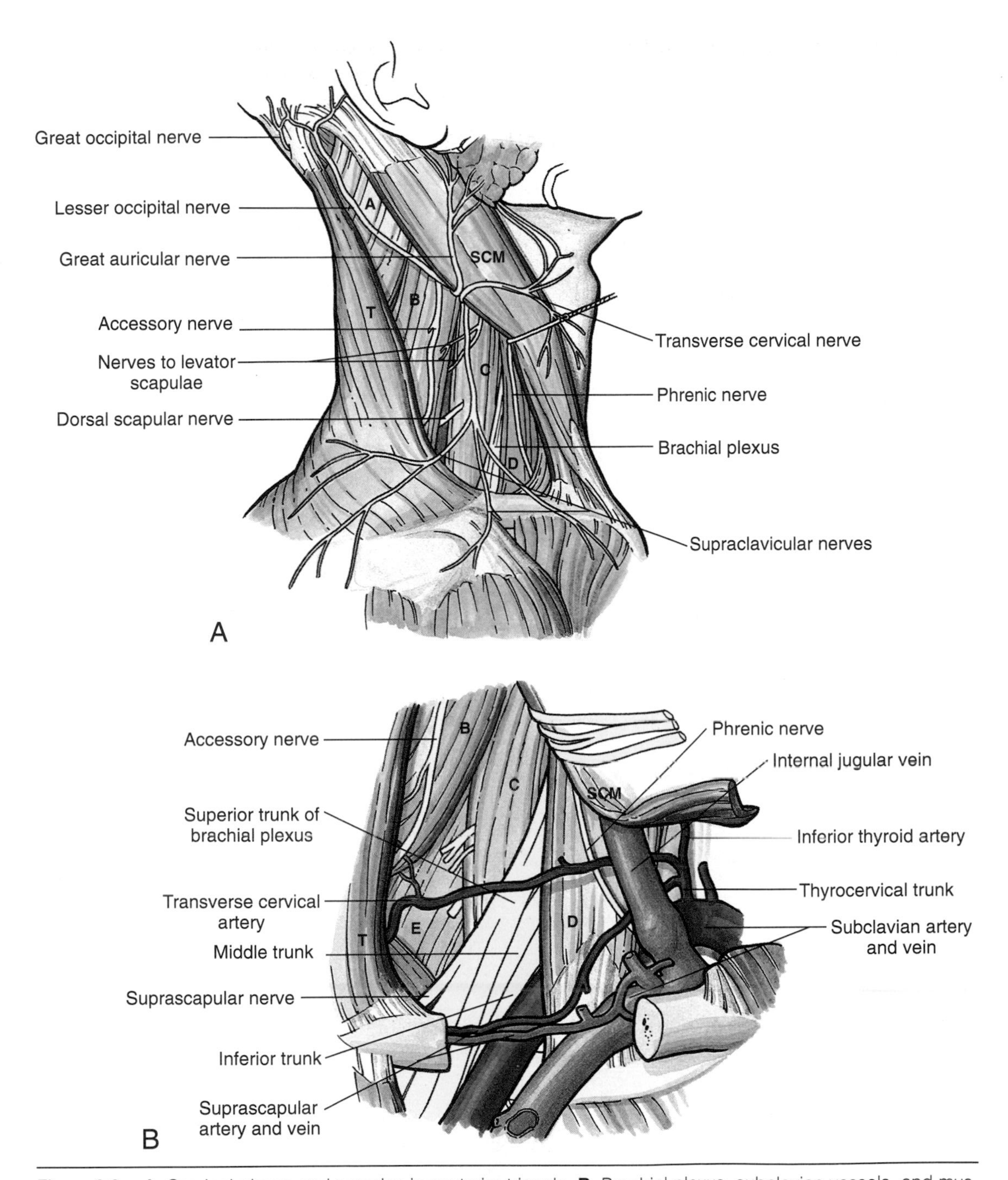

Figure 9.2. **A.** Cervical plexus and muscles in posterior triangle. **B.** Brachial plexus, subclavian vessels, and muscles. *A*, splenius; *B*, levator scapulae; *C*, scalenus medius; *D*, scalenus anterior; *E*, scalenus posterior; *T*, trapezius; *SCM*, sternocleidomastoid.

The *accessory nerve* (CN XI) runs posteroinferiorly in the neck, dividing the posterior triangle into nearly equal superior and inferior parts (Fig. 9.2*A*). It usually passes through the SCM but sometimes passes deep to it. After supplying the SCM, CN XI runs inferolaterally through the posterior triangle deep to the investing layer of cervical fascia and disappears deep to the trapezius that it supplies.

Table 9.3
Suprahyoid and Infrahyoid Muscles

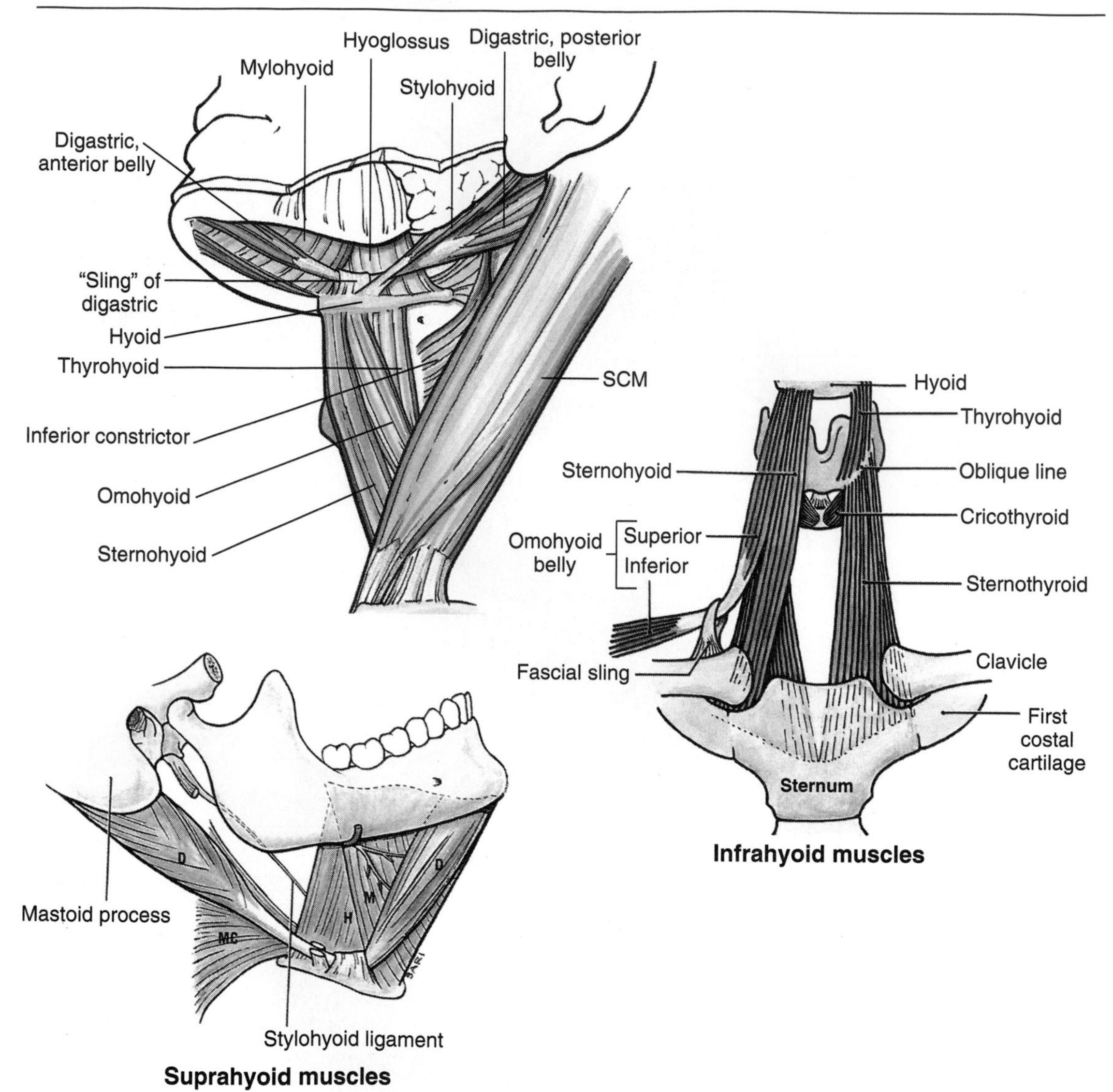

Note that suprahyoid muscles are in four layers: digastric (*D*), mylohyoid (*M*), hyoglossus (*H*), and middle constrictor of pharynx (*MC*).

Muscle	Origin	Insertion	Innervation	Actions
Mylohyoid	Mylohyoid line of mandible	Raphe and body of hyoid bone	Mylohyoid n., a branch of inferior alveolar n.	Elevates hyoid bone, floor of mouth, and tongue during swallowing and speaking
Geniohyoid	Inferior mental spine of mandible	Body of hyoid bone	C1 via the hypoglossal n. (CN XII)	Pulls hyoid bone antero-superiorly, shortens floor of mouth, and widens pharynx
Stylohyoid	Styloid process of temporal bone	Body of hyoid bone	Cervical branch of facial n. (CN VII)	Elevates and retracts hyoid bone, thereby elongating floor of mouth

Table 9.3 *Continued*

Muscle	Origin	Insertion	Innervation	Actions
Digastric	*Anterior belly:* digastric fossa of mandible *Posterior belly:* mastoid notch of temporal bone	Intermediate tendon to body and superior (greater) horn of hyoid bone	*Anterior belly*: mylohyoid n., a branch of inferior alveolar n;.*Posterior belly:* facial n. (CN VII)	Depresses mandible; raises hyoid bone and steadies it during swallowing and speaking
Sternohyoid	Manubrium of sternum and medial end of clavicle	Body of hyoid bone	C1–C3 from ansa cervicalis	Depresses hyoid bone after it has been elevated during swallowing
Sternothyroid	Posterior surface of manubrium of sternum	Oblique line of thyroid cartilage	C2 and C3 by a branch of ansa cervicalis	Depresses hyoid bone and larynx
Thyrohyoid	Oblique line of thyroid cartilage	Inferior border of body and superior (greater) horn of hyoid bone	C1 via hypoglossal n. (CN XII)	Depresses hyoid bone and elevates larynx
Omohyoid	Superior border of scapula near supra-scapular notch	Inferior border of hyoid bone	C1–C3 by a branch of ansa cervicalis	Depresses, retracts, and steadies hyoid bone

The subclavian vein is often used for insertion of a *central venous catheter* for recording venous pressure, giving hyperalimentation fluids, placing cardiac pacemaker leads (electrodes), and inserting right cardiac catheters. Consequently, the relationship of the subclavian vein to the SCM, clavicle, 1st rib, and cupula of the pleura are of great clinical importance.

When venous pressure is within the normal range, the external jugular vein is either invisible or observable for only a short distance. However, when the pressure rises (e.g., during *heart failure*), the external jugular vein becomes prominent. Should an external jugular vein be lacerated where it pierces the investing layer of cervical fascia in the roof of the posterior triangle, air may be sucked into the vein. A *venous air embolism* produced in this way fills the right heart with froth, practically stopping blood flow through it. An embolism can result in dyspnea (difficult breathing), cyanosis (blue skin), and sometimes death.

Deep pressure in the supraclavicular fossa occludes blood flow in the subclavian artery where it passes through the supraclavicular triangle and over the 1st rib. If there is severe hemorrhage in the upper limb, pressure in this triangle will control the bleeding because the subclavian artery supplies blood to all arterial vessels in the limb.

Care is essential during dissections in the posterior triangle inferior to the accessory nerve (CN XI) because of the presence of many vessels and nerves in this area. To preserve the continuity of CN XI during surgical dissections for removal of cancerous lymph nodes, the nerve is isolated at the outset and separated from the nodes.

Lesions of CN XI are uncommon, but the nerve may be damaged by traumatic injury and neck lacerations. A unilateral lesion usually does not produce an abnormal position of the head; however, weakness can occur in turning the head to one side against resistance. *Unilateral paralysis of the trapezius* is evident by the patient's inability to elevate and retract the shoulder and by difficulty in elevating the arm superior to the horizontal level. *Drooping of the shoulder* may indicate injury to the spinal root of CN XI.

The *cervical plexus* is also in the posterior triangle (Fig. 9.2*A*). This plexus, formed by union of ventral rami of the first four cervical nerves, lies deep to the internal jugular vein and the SCM. Cutaneous branches from the plexus emerge around the middle of the posterior border of the SCM and supply the skin of the neck and scalp between the auricle and external occipital protuberance. Close to their origin, the nerves of the plexus receive *rami communicantes*, most of which descend from the large *superior cervical ganglion* in the superior part of the neck. The following nerves are branches of the cervical plexus:

- Lesser occipital nerve (C2) supplies skin of the neck and scalp posterosuperior to the auricle
- Great auricular nerve (C2 and C3) ascends diagonally across the SCM onto the parotid gland, where it divides to supply skin over the gland, the posterior aspect of the auricle, and an area extending from the mandible to the mastoid process

- Transverse cervical nerve (C2 and C3) supplies skin covering the anterior triangle; the nerve curves around the middle of the posterior border of the SCM and crosses it deep to the platysma
- Supraclavicular nerves (C3 and C4) emerge as a common trunk under cover of the SCM and send small branches to the skin of the neck; they then cross the clavicle and supply the skin over the shoulder

Branches of the ventral primary rami of cervical nerves supply the rhomboids (dorsal scapular nerve), serratus anterior (long thoracic nerve), and nearby prevertebral muscles. The *suprascapular nerve* runs laterally across the posterior triangle, superior to the brachial plexus (Fig. 9.2*B*), to supply the supraspinatus and infraspinatus muscles.

The *phrenic nerve* takes origin chiefly from the fourth cervical nerve (C4) but receives contributions from the third and fifth cervical nerves (C3 and C5). The phrenic nerves contain motor, sensory, and sympathetic nerve fibers. These nerves provide the sole motor supply to the diaphragm and also sensation to its central part. In the thorax, each phrenic nerve supplies the mediastinal pleura and pericardium (Chapter 2). In the neck, each phrenic nerve receives variable communicating fibers from the cervical sympathetic ganglia or their branches. Each phrenic nerve forms at the superior part of the lateral border of the scalenus anterior at the level of the superior border of the thyroid cartilage (Fig. 9.2). It descends obliquely with the internal jugular vein across the scalenus anterior, deep to the prevertebral layer of cervical fascia and the transverse cervical and suprascapular arteries. On the left, the phrenic nerve crosses the first part of the subclavian artery, but on the right it lies on the scalenus anterior and covers the second part of this artery. The phrenic nerve crosses posterior to the subclavian vein on both sides and anterior to the internal thoracic artery as it enters the thorax.

The contribution of the fifth cervical nerve to the phrenic nerve may be derived from an *accessory phrenic nerve*. If present, it lies lateral to the main nerve and descends posterior and sometimes inferior to the subclavian vein. The accessory phrenic nerve joins the phrenic nerve either in the root of the neck or in the thorax.

The *trunks of the brachial plexus* are in the posterior triangle, just superior to the clavicle (Fig. 9.2*B*). The brachial plexus, derived from the ventral rami of C5–C8 and T1, provides innervation for most of the upper limb. The supraclavicular part of the plexus is superior to the clavicle and anterior to the scalenus medius; the infraclavicular part is in the axilla.

Severance of a phrenic nerve results in paralysis of the corresponding half of the diaphragm. To produce a short period of paralysis of the diaphragm (e.g., for a lung operation), a *phrenic nerve block* is performed. The anesthetic is injected around the nerve where it lies on the anterior surface of the middle third of the scalenus anterior. To produce a longer period of paralysis (e.g., for weeks after surgical repair of a diaphragmatic hernia) a *phrenic nerve crush* may be performed. If an accessory phrenic nerve is present, it must also be crushed to produce complete paralysis of the hemidiaphragm.

For regional anesthesia before surgery, nerve blocks are performed by injecting anesthetic around the nerves of the cervical and brachial plexuses. The anesthetic blocks nerve impulse conduction in a fashion similar to that which occurs when a dentist blocks the nerves supplying the teeth before performing dental work. To produce a *cervical plexus block*, the anesthetic is injected at several points along the posterior border of the SCM. The main injection site is at the junction of its superior and middle thirds, i.e., around the nerve point of the neck. Because the phrenic nerve supplying half the diaphragm is usually paralyzed by a cervical nerve block, this procedure is not performed on patients with pulmonary or cardiac disease.

A *supraclavicular brachial plexus block*, performed for anesthesia of the upper limb, is produced by injecting anesthetic around the supraclavicular part of the brachial plexus. The main injection site is superior to the midpoint of the clavicle.

ANTERIOR CERVICAL TRIANGLE

Muscles in Anterior Triangle

The *hyoid muscles* are concerned with steadying or moving the hyoid bone and larynx (Fig. 9.3, Table 9.3). The *suprahyoid muscles* are located superior to the hyoid bone and connect it to the skull. The suprahyoid group includes the mylohyoid, geniohyoid, stylohyoid, and digastric muscles. The *mylohyoid muscles* form the floor of the mouth and a sling inferior to the tongue. The *geniohyoid muscles* lie superior to the mylohyoids, where they reinforce the floor of the mouth. Each straplike *digastric muscle* has two bellies that descend toward the hyoid bone and are joined by an intermediate tendon. This tendon is connected to the body and superior (greater) horn of the hyoid bone by a fibrous sling that allows the

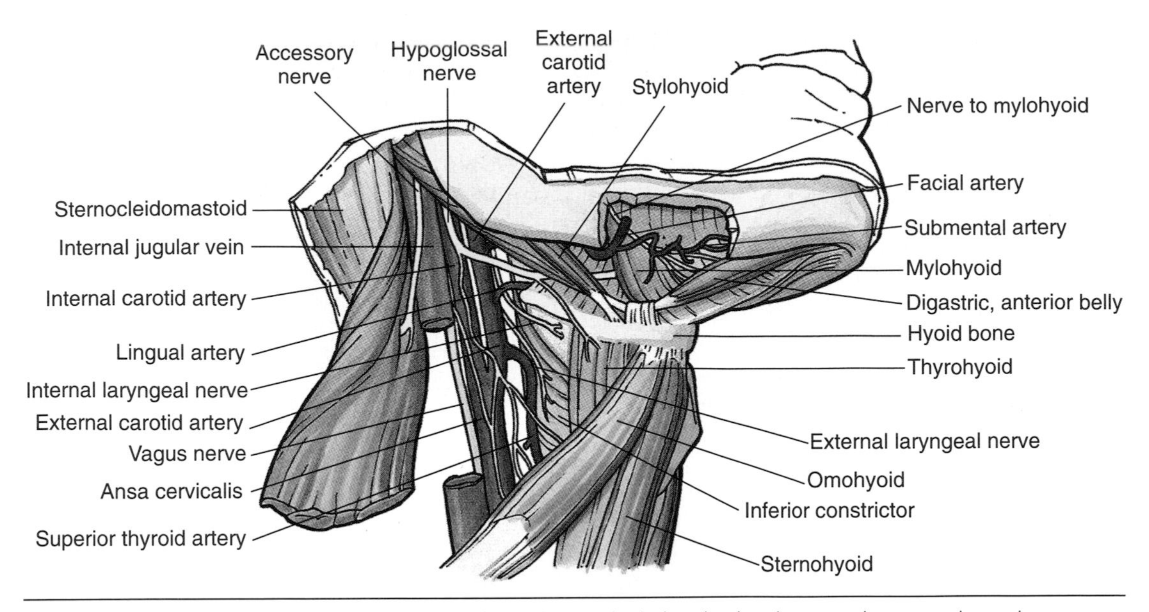

Figure 9.3. Lateral view of a deep dissection of anterior cervical triangle showing muscles, vessels, and nerves.

tendon to slide anteriorly and posteriorly. The *infrahyoid muscles* (strap muscles) are inferior to the hyoid bone. These straplike muscles anchor the hyoid bone and depress the hyoid and larynx during swallowing and speaking. The flat *sternohyoid muscles* lie superficially on each side of the anterior median line. The *omohyoid muscle* has two bellies that are united by an intermediate tendon that is connected to the clavicle by a fascial sling. The omohyoid divides the anterior and posterior triangles into smaller triangles (Fig. 9.3).

Vessels and Nerves in Anterior Triangle

The *right common carotid artery* begins at the bifurcation of the brachiocephalic trunk, posterior to the right sternoclavicular joint. The right subclavian artery is a branch of the brachiocephalic trunk, whereas the left subclavian artery arises directly from the arch of the aorta. The *left common carotid artery* also arises from the aortic arch and ascends into the neck, posterior to the left sternoclavicular joint. Within the carotid sheath (Fig. 9.1, *B* and *C*), each common carotid artery ascends with the internal jugular vein and vagus nerve to the level of the superior border of the thyroid cartilage. The common carotid artery then terminates by dividing into the internal and external carotid arteries. The internal carotid can always be distinguished from the external carotid because it has no branches in the neck, whereas the external carotid does (e.g., superior thyroid artery).

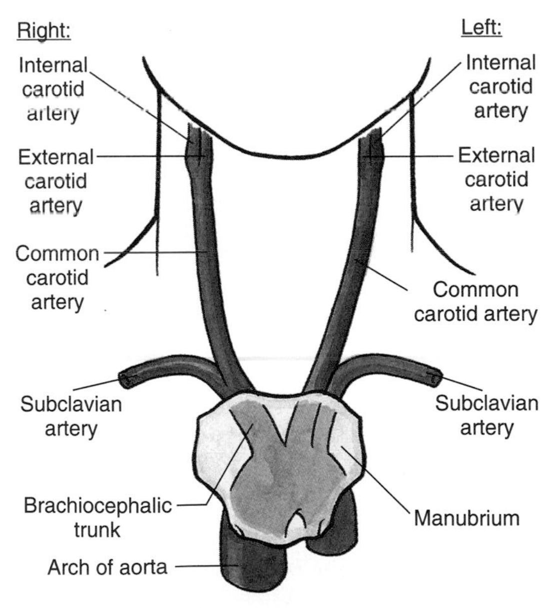

The *internal carotid arteries*, the direct continuation of the common carotid arteries, have no branches in the neck. They enter the skull and become the main arteries of the brain and structures in the orbit. Each artery arises from the common carotid artery at the level of the superior border of the thyroid cartilage. A sympathetic plexus of nerve fibers accompanies each artery. During passage through the neck, the internal carotid artery lies anterior to the longus capitis muscle and the sympathetic trunk and posterolateral to the vagus nerve (CN X).

The *carotid sinus* is a slight dilation of the proximal part of the internal carotid artery; the dilation may involve the terminal part of the common carotid artery. The wall of the sinus contains receptors that are sensitive to changes in blood pressure. The carotid sinus is innervated principally by the carotid sinus nerve, a branch of the glossopharyngeal nerve (CN IX).

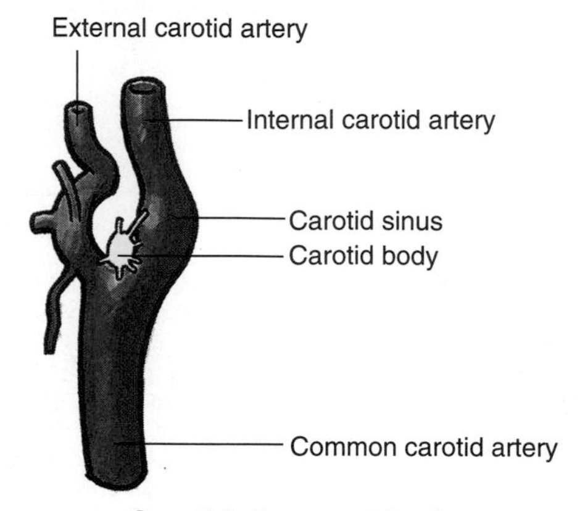

Carotid sinus and body

The *carotid body* is a small mass of tissue that is located at the bifurcation of the common carotid artery in close relation to the carotid sinus. This vascular body is a *chemoreceptor* that responds to changes in the chemical composition of the blood. The pharyngeal branch of the vagus (CN X) gives origin to the nerve to the carotid body.

The *external carotid arteries* supply structures external to the skull. Each artery runs posterosuperiorly to the region between the neck of the mandible and the lobule of the auricle and terminates by dividing into two branches, the maxillary and superficial temporal arteries. The six other branches of the external carotid artery are the ascending pharyngeal, superior thyroid, lingual, facial, occipital, and posterior auricular arteries.

Ascending Pharyngeal Artery. The ascending pharyngeal artery, the first or second branch of the external carotid artery, ascends on the pharynx deep to the internal carotid artery and sends branches to the pharynx, prevertebral muscles, middle ear, and cranial meninges.

Superior Thyroid Artery. The superior thyroid artery, the most inferior of the three anterior branches of the external carotid artery, runs anteroinferiorly deep to the infrahyoid muscles to reach the thyroid. In addition to supplying this gland, it gives off muscular branches to the SCM and infrahyoid muscles. It also gives rise to the *superior laryngeal artery*, which supplies the larynx.

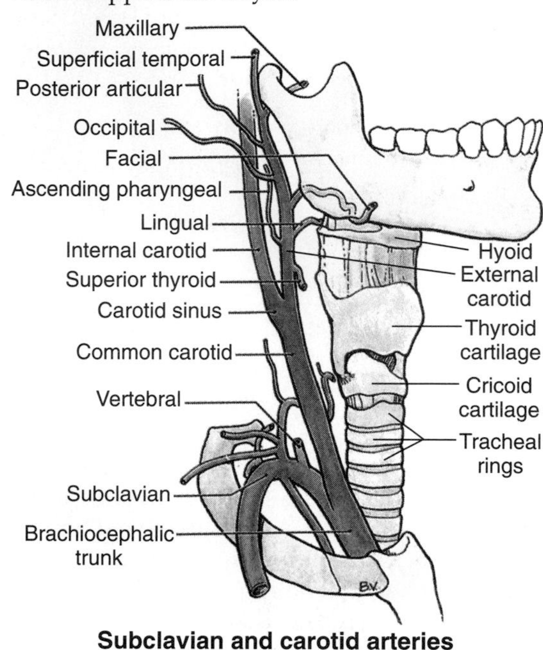

Subclavian and carotid arteries

Lingual Artery. The lingual artery arises from the external carotid artery where it lies on the middle constrictor muscle of the pharynx. It arches superoanteriorly and passes deep to the hypoglossal nerve (CN XII), stylohyoid muscle, and posterior belly of the digastric muscle. It disappears deep to the hyoglossus and turns superiorly at the anterior border of this muscle and becomes the *deep lingual artery*.

Facial Artery. The facial artery arises from the external carotid artery, either in common with the lingual artery or immediately superior to it. The facial artery gives off a *tonsillar branch* and branches to the palate and submandibular gland. It then passes superiorly under cover of the digastric and stylohyoid muscles and the angle of the mandible. The facial artery loops anteriorly and enters a deep groove in the *submandibular gland*. It then hooks around the inferior border of the mandible and enters the face.

Occipital Artery. The occipital artery arises from the posterior surface of the external carotid artery superior to the origin of the facial artery. It passes posteriorly along the inferior border of the posterior belly of the digastric muscle and ends in the posterior part of the scalp. During its course, it passes superficial to the internal carotid artery and CN IX–CN XI.

Posterior Auricular Artery. The posterior auricular artery, a small posterior branch of the external carotid artery, ascends posterior to the external acoustic meatus and supplies adjacent muscles, the parotid gland, facial nerve, and structures in the temporal bone, auricle, and scalp.

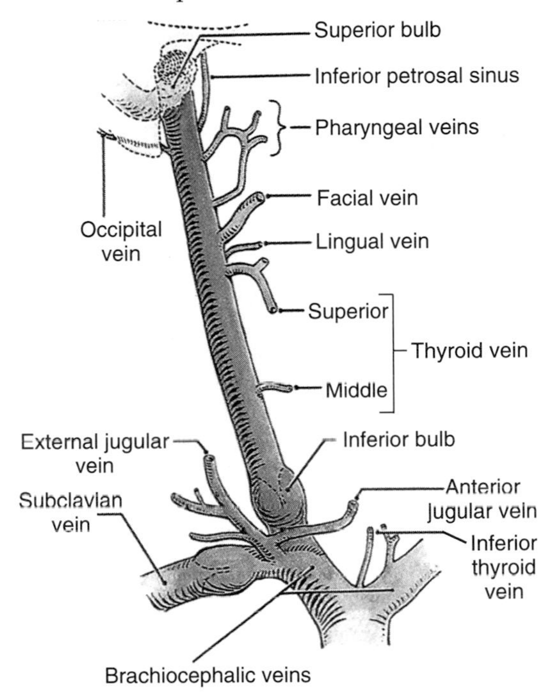

Internal jugular vein and tributaries

Veins. Most veins in the anterior triangle are tributaries of the internal jugular vein (Fig. 9.2*B*), which is usually the largest vein in the neck. The *internal jugular vein* drains blood from the brain and superficial parts of the face and neck. This large vein commences at the jugular foramen in the posterior cranial fossa as the direct continuation of the sigmoid sinus (p. 347). From the dilation at its origin, the *superior bulb of the internal jugular vein,* the vein runs inferiorly through the neck in the carotid sheath with the common carotid artery and vagus nerve (CN X). The artery is medial and the vein lateral, and the nerve lies posteriorly between these vessels (Fig. 9.1C). The internal jugular vein leaves the anterior triangle by passing deep to the SCM muscle. Posterior to the sternal end of the clavicle, the internal jugular vein unites with the subclavian vein to form the *brachiocephalic vein.* The inferior end of the internal jugular vein also dilates to form the *inferior bulb of the internal jugular vein.* The tributaries of the internal jugular vein are the inferior petrosal sinus and the facial, lingual, pharyngeal, and superior and middle thyroid veins. The *occipital vein* usually drains into the suboccipital venous plexus, but it may drain into the internal jugular vein.

Sometimes an external carotid artery has to be ligated to control bleeding from one of its relatively inaccessible branches. This decreases blood flow through the artery and its branches but does not eliminate it. Blood will flow retrogradely (pass backward) into the artery from the external carotid artery on the other side through communications between its branches (e.g., those in face and scalp). The descending branch of the occipital artery provides the main collateral circulation when the external carotid or subclavian arteries are ligated.

The *carotid triangle* provides an important surgical approach to the carotid system of arteries (Fig. 9.3). It is also important for approaches to the internal jugular vein, the vagus (CN X) and hypoglossal (CN XII) nerves, and the cervical sympathetic trunk. Damage or compression of the vagus and/or recurrent laryngeal nerves during a surgical dissection of the carotid triangle may produce an alteration in the voice because they supply the laryngeal muscles.

The *carotid pulse* (neck pulse) is easily felt by palpating the common carotid artery in the side of the neck, where it lies in a groove between the trachea and the strap muscles. It is routinely checked during cardiopulmonary resuscitation (*CPR*). Absence of a carotid pulse indicates cardiac arrest.

The carotid sinus responds to an increase in arterial pressure and to the parasympathetic outflow from the brain through the vagus nerve by slowing the heart. Pressure on the carotid sinus may cause syncope (fainting), and persons with a supersensitive carotid sinus may experience cessation of the heartbeat (temporary or permanent).

The carotid body responds either to increased carbon dioxide tension or to decreased oxygen tension in the blood. A fall in oxygen or a rise in carbon dioxide content initiates reflexes through the glossopharyngeal and vagus nerves that stimulate respiration. Reduced oxygen results in an increase in the depth and rapidity of breathing; pulse rate and blood pressure also rise.

Pulsations of the internal jugular vein caused by contraction of the right ventricle of the heart may be palpable superior to the medial end of the clavicle in the root of the neck. Because no valves are in the brachiocephalic vein or the superior vena cava, a wave of contraction passes up these vessels to the internal jugular vein. The *internal jugular pulse* increases considerably in conditions such as mitral valve disease (p. 62). This results in increased pressure in the pulmonary circulation, the right side of the heart, and the great veins.

Nerves. The transverse cervical nerve (C2 and C3) supplies skin covering the anterior triangle (Fig. 9.2). The hypoglossal nerve (CN XII), the motor nerve of the tongue, enters the submandibular triangle deep to the posterior belly of the digastric muscle to supply the tongue (Fig. 9.3). Branches of CN IX and CN X are located in the digastric and carotid triangle (Table 9.2).

Deep Structures in Neck

The deep structures in the neck include the vertebrae, muscles, vessels, and nerves. The cervical vertebrae and joints of the neck are described in Chapter 4 with the back (p. 196). The deep prevertebral muscles are covered anteriorly by the prevertebral layer of cervical fascia (Fig. 9.1C, Table 9.4). These muscles flex the neck and the head on the neck.

Root of Neck

The root of the neck is the junctional area between the thorax and neck (Fig. 9.4). It opens into the *superior thoracic aperture* through which pass all structures going from the head to the thorax and vice versa (p. 34). The root of the neck is bounded

- Laterally by the first pair of ribs and their costal cartilages
- Anteriorly by the manubrium
- Posteriorly by the body of T1 vertebra

ARTERIES

The arteries in the root of the neck originate from the arch of the aorta. They are the large brachiocephalic trunk on the right side and the common carotid and subclavian arteries on the left side (Fig. 9.4).

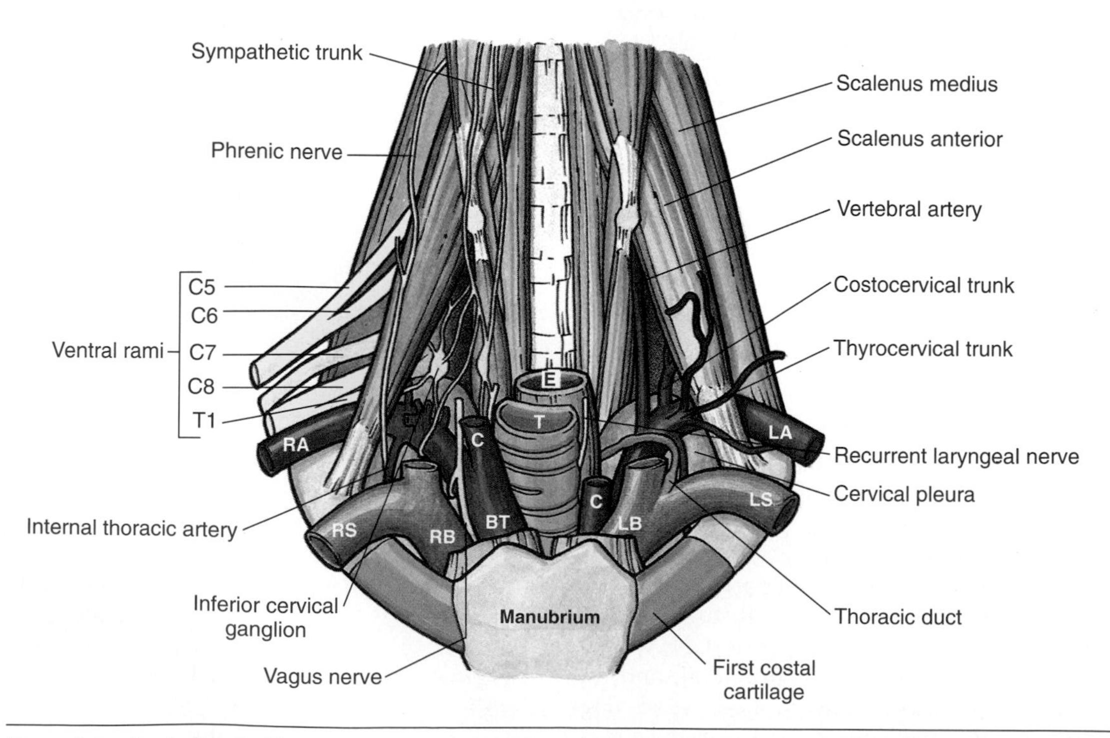

Figure 9.4. Root of neck. *E*, esophagus; *T*, trachea; *C*, left and right common carotid arteries; *RA*, right subclavian artery; *LA*, left subclavian artery; *RS*, right subclavian vein; *LS*, left subclavian vein; *RB*, right brachiocephalic vein; *LB*, left brachiocephalic vein; *BT*, brachiocephalic trunk.

Table 9.4
Prevertebral Muscles

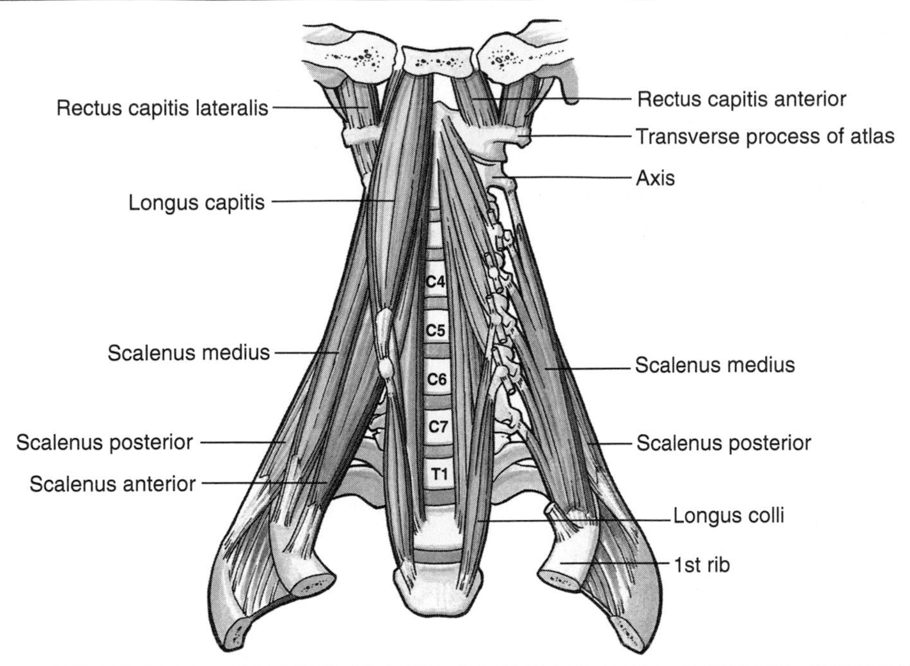

Muscle	Superior Attachment	Inferior Attachment	Innervation	Main Actions
Longus colli	Anterior tubercle of C1 vertebra (axis)	Body of T3 vertebra with attachments to bodies of C1–C3 and transverse processes of C3–C6 vertebrae	Ventral rami of C2–C6 spinal nn.	Flexes neck
Longus capitis	Basilar part of occipital bone	Anterior tubercles of C3–C6 transverse processes	Ventral rami of C2–C3 spinal nn.	Flexes head
Rectus capitis anterior	Base of skull, just anterior to occipital condyle	Anterior surface of lateral mass of C1 vertebra (atlas)	Branches from loop between C1 and C2 spinal nn.	Flexes head
Rectus capitis lateralis	Jugular process of occipital bone	Transverse process of C1 vertebra (atlas)	Branches from loop between C1 and C2 spinal nn.	Flexes head and helps to stabilize the head

The *brachiocephalic trunk* is the largest branch of the arch of the aorta. It arises posterior to the middle of the manubrium and passes superiorly and to the right and posterior to the right sternoclavicular joint, where it divides into the right common carotid and right subclavian arteries. The brachiocephalic trunk is covered anteriorly by the sternohyoid and sternothyroid muscles. The brachiocephalic trunk usually has no branches, but occasionally (10% of people) a thyroid ima (lowest thyroid) artery arises from it.

The *subclavian arteries* supply the upper limbs, but they also supply branches to the neck and brain. The right subclavian artery arises from the brachiocephalic trunk posterior to the right sternoclavicular joint. The left subclavian artery arises from the arch of the aorta and enters the root of the neck by passing superiorly, posterior to the left sternoclavicular joint.

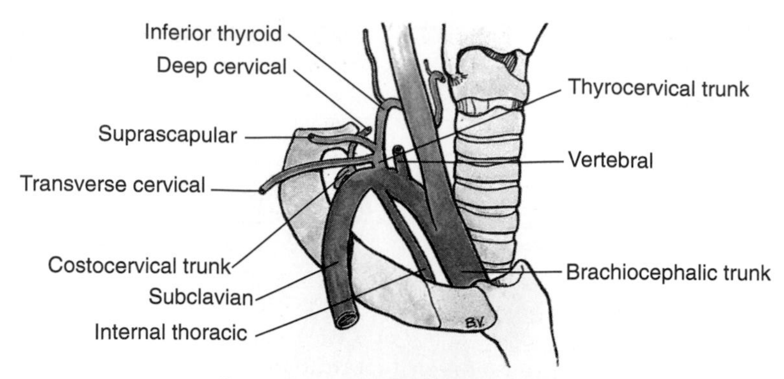

Branches of subclavian artery

Each subclavian artery arches superiorly, posteriorly, and laterally—grooving the pleura and lung—and then passes inferiorly, posterior to the midpoint of the clavicle. As these arteries ascend, they are crossed anteriorly by the scalenus anterior muscles.

For purposes of description, the scalenus anterior divides the subclavian artery into three parts: the first part is medial to the muscle; the second is posterior to it; and the third is lateral to it.

Branches of the subclavian arteries (Fig. 9.4) follow:

- Vertebral artery
- Internal thoracic artery
- Thyrocervical trunk
- Costocervical trunk
- Dorsal scapular arteries

On the left, all branches except the dorsal scapular arise from the first part of the subclavian artery. On the right, the costocervical trunk usually arises from its second part.

The *vertebral artery* arises from the first part of the subclavian artery and usually ascends through the transverse foramina of C1–C6 vertebrae (p. 196), but it may enter a foramen more superior than C6 vertebra. The artery courses in a groove on the posterior arch of the atlas before it enters the cranial cavity through the foramen magnum. At the inferior border of the pons, the vertebral arteries join to form the *basilar artery* that participates in the formation of the cerebral arterial circle.

The *internal thoracic artery* arises from the anteroinferior aspect of the subclavian artery and passes inferomedially into the thorax. The cervical portion of the internal thoracic artery has no branches; the thoracic distribution of this artery is considered on page 45.

The *thyrocervical trunk* arises from the anterosuperior aspect of the first part of the subclavian artery, just medial to the scalenus anterior muscle. It has three branches, the largest and most important of which is the *inferior thyroid artery*. Other branches of the thyrocervical trunk are the *suprascapular artery* supplying muscles around the scapula and the *transverse cervical artery* sending branches to muscles in the posterior triangle of the neck.

The *costocervical trunk* arises from the posterior aspect of the subclavian artery posterior to the scalenus anterior on the right side and usually just medial to it on the left side. The trunk passes posterosuperiorly and divides into the superior intercostal and deep cervical arteries, which supply the first two intercostal spaces and neck muscles, respectively.

The *dorsal scapular artery*, present in about 70% of people, is usually a branch of the second or third part of the subclavian artery, but it may be a deep branch of the transverse cervical artery (Fig. 9.2*B*). It passes laterally through the trunks of the brachial plexus, anterior to the scalenus medius and then deep to the levator scapulae to reach the scapula and supply the rhomboids.

VEINS

There are two large veins in the root of the neck. The external jugular vein is described on page 416. The variable *anterior jugular vein*, usually the smallest of the jugular veins, typically arises near the hyoid bone from the confluence of superficial submandibular veins. It descends in the superficial fascia between the anterior median line and the anterior border of the SCM. At the root of the neck, the vein turns laterally, posterior to the SCM, and opens into the termination of the external jugular vein or into the subclavian vein. Superior to the manubrium, the right and left anterior jugular veins may unite in the suprasternal space to form the *jugular venous arch*.

The *subclavian vein*, the continuation of the axillary vein, begins at the lateral border of the 1st rib and ends at the medial border of the scalenus anterior. Here it unites with the *internal jugular vein*, posterior to the medial end of the clavicle, to form the *brachiocephalic vein* (Figs. 9.2*B* and 9.4). The subclavian vein usually has only one named tributary, the external jugular vein. The subclavian vein passes over the 1st rib anterior to the scalene tubercle and parallel to the subclavian artery, but it is separated from it by the scalenus anterior.

The *internal jugular vein* ends posterior to the medial end of the clavicle by uniting with the subclavian vein to form the brachiocephalic vein. Throughout its course, the internal jugular vein is enclosed by the carotid sheath (Fig. 9.1, *B* and *C*).

NERVES

Important nerves pass through the root of the neck. The *vagus nerve* (CN X) passes inferiorly in the posterior part of the carotid sheath in the angle between and posterior to the internal jugular vein and carotid artery (Fig. 9.4). On the right side, the vagus crosses the origin of the subclavian artery, posterior to the brachiocephalic vein and sternoclavicular joint, to enter the thorax. The *recurrent laryngeal nerve* loops around the subclavian artery on the right side and the arch of the aorta on the left side. After looping, both recurrent nerves pass superiorly to the posteromedial aspect of the thyroid gland, where they ascend in the *tracheoesophageal groove* to supply all intrinsic muscles of the larynx except the cricothyroid (see Fig. 9.6*A* and p. 437). The cardiac branches of CN X originate in the neck as well as in the thorax and run along the arteries to the cardiac plexus of nerves.

The *phrenic nerves* are formed at the lateral border of the scalenus anterior muscles and descend anterior to them under cover of the internal jugular veins and SCMs. They pass between the subclavian arteries and veins and proceed to the thorax and supply the diaphragm.

The *sympathetic trunks* are located in the neck anterolateral to the vertebral column, beginning at the level of C1 vertebra (Fig. 9.4). These trunks receive no white rami communicantes in the neck, but they are associated with three *cervical sympathetic ganglia* (superior, middle, and inferior). These ganglia receive preganglionic fibers from the superior thoracic spinal nerves through white rami communicantes. From the sympathetic trunks, fibers pass to cervical

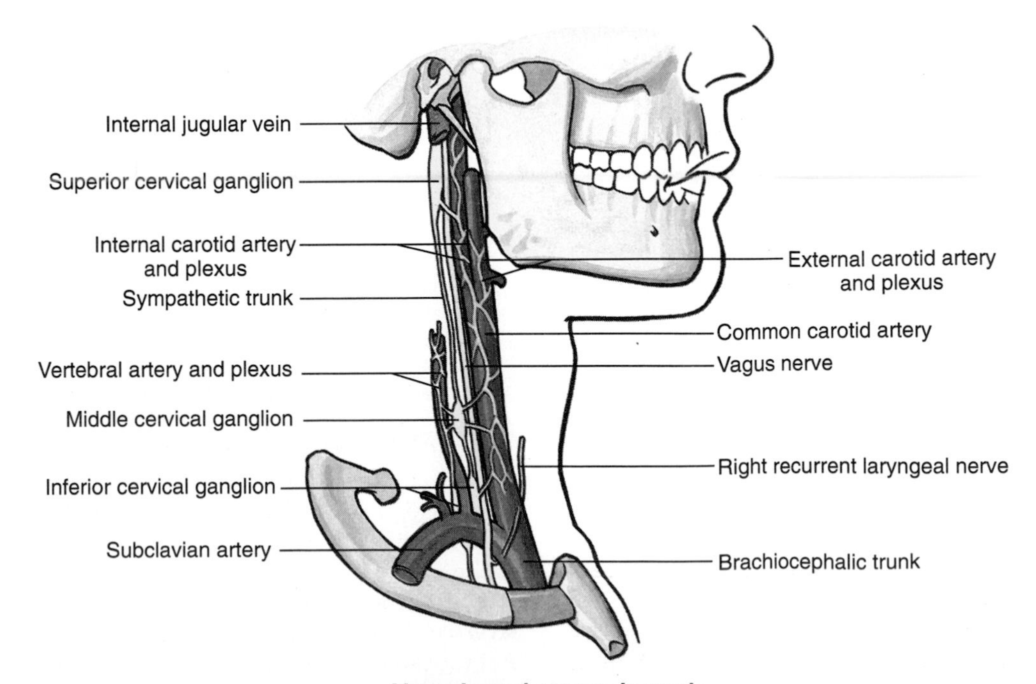

Vessels and nerves in neck

structures as postganglionic fibers in the cervical spinal nerves or leave as direct visceral branches (e.g., to thyroid gland). Branches to the head run with the arteries, especially the vertebral and internal and external carotid arteries.

The *inferior cervical ganglion* usually fuses with the first thoracic ganglion to form the large *cervicothoracic ganglion* (stellate ganglion). This ganglion lies anterior to the transverse process of C7 vertebra, just superior to the neck of the 1st rib on each side and posterior to the origin of the vertebral artery. Some postganglionic fibers from the ganglion pass into C7 and C8 spinal nerves and to the heart; other fibers contribute to the vertebral nerve plexus around the vertebral artery.

The *middle cervical ganglion* lies on the anterior aspect of the inferior thyroid artery at the level of the cricoid cartilage and the transverse process of C6 vertebra, just anterior to the vertebral artery. Postganglionic branches pass from the ganglion to C5 and C6 spinal nerves and to the heart and thyroid gland.

The *superior cervical ganglion* is located at the level of C1 and C2 vertebrae. Due to its large size, it forms a good landmark for locating the sympathetic trunk. Postganglionic branches pass from it along the internal carotid artery and enter the cranial cavity. This ganglion also sends branches to the external carotid artery and into the superior four cervical spinal nerves. Other postganglionic fibers pass from it to the cardiac plexus of nerves (pp. 53 and 67).

Anesthetic injected around the large cervicothoracic ganglion will block transmission of stimuli through the cervical and superior thoracic ganglia. A *cervicothoracic (stellate) ganglion block* may be performed to relieve vascular spasms involving the brain and upper limb.

A lesion of a sympathetic trunk in the neck results in a sympathetic disturbance known as the *Horner syndrome*, consisting of

- Pupillary constriction caused by paralysis of dilator pupillae (p. 370)
- Ptosis (drooping of upper eyelid) caused by paralysis of smooth muscle in levator palpebrae superioris (p. 369)
- Sinking in of eye, possibly from paralysis of smooth muscle (orbitalis muscle) in floor of orbit
- Vasodilation and absence of sweating on the face and neck caused by lack of a sympathetic nerve supply to the blood vessels and sweat glands

Lymphatics in Neck

All lymphatic vessels from the head and neck drain into the *deep cervical lymph nodes* (Fig. 9.5*B*). The main group forms a chain along the internal jugular vein, mostly under cover of the SCM. Other deep nodes include the prelaryngeal, pretracheal, paratracheal, and retropharyngeal nodes. The deep nodes drain into the jugular trunks and from them into the thoracic duct (left side) and right lymphatic duct (right side). The *superficial cervical lymph nodes* drain into the deep cervical lymph nodes (Fig. 9.5*A*). They are in the posterior triangle along the external jugular vein and in the anterior triangle along the anterior jugular vein.

The *thoracic duct*, a large lymphatic channel, begins in the abdomen and passes superiorly through the thorax and superior thoracic aperture at the left border of the esophagus (p. 78). It then arches laterally in the root of the neck, posterior to the carotid sheath and anterior to the sympathetic trunk and the vertebral and subclavian arteries. The thoracic duct enters the left brachiocephalic vein at the junction of the subclavian and internal jugular veins. The thoracic duct drains lymph from the entire body, except the right side of the head and neck, the right upper limb, and the right side of the thorax, all of which drain through the *right lymphatic duct*.

In *radical neck dissections* performed when cancer involves the lymphatics, the deep cervical lymph nodes and the tissues around them are removed as completely as possible. The major arteries, brachial plexus, CN X, and phrenic nerve are preserved, but most cutaneous branches of the cervical plexus are removed. The aim of the dissection is to remove all tissue that bears lymph nodes in one piece.

Cervical Viscera

The cervical viscera are disposed in three layers:

- Endocrine layer containing thyroid and parathyroid glands

Figure 9.5. Lymphatic drainage of head and neck. **A.** Superficial. **B.** Deep. *A*, pharyngeal tonsil; *P*, palatine tonsil.

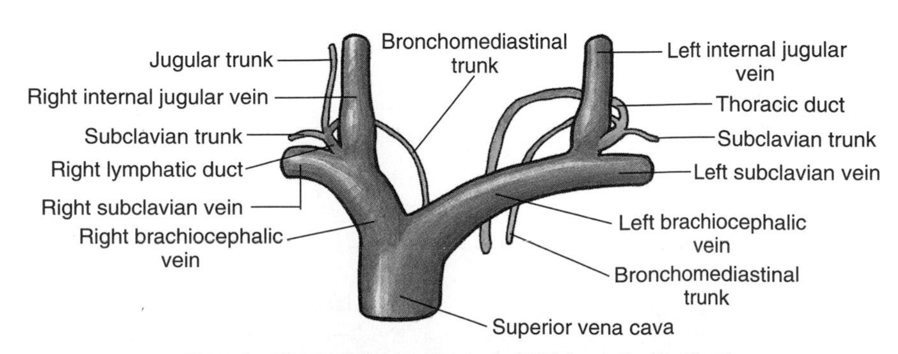

Termination of thoracic and right lymphatic ducts

- Respiratory layer containing larynx and trachea
- Alimentary layer containing pharynx and esophagus

ENDOCRINE LAYER

The endocrine layer contains the thyroid and parathyroid glands.

Thyroid Gland

The thyroid gland produces *thyroxine*, a hormone that controls the rate of metabolism. It also secretes *calcitonin*, a hormone concerned with calcium metabolism. The thyroid gland lies deep to the sternothyroid and sternohyoid muscles from the level of C5 to T1 vertebrae (Fig. 9.6). It consists of right and left lobes located anterolateral to the larynx and trachea. An *isthmus* unites the lobes over the trachea, usually anterior to the second and third tracheal rings. A *pyramidal lobe* may arise from the isthmus, usually to the left of the median plane. The thyroid gland is surrounded by a thin *fibrous capsule* that sends septa deeply into the gland. External to the capsule is a loose sheath derived from the pretracheal layer of cervical fascia (Fig. 9.1, *C* and *D*). The thyroid is attached by dense connective tissue to the cricoid cartilage and the superior tracheal rings.

The highly vascular thyroid gland is supplied by the superior and inferior *thyroid arteries* (Fig. 9.6, *A* and *B*). These vessels lie between the fibrous capsule and the pretracheal layer of cervical fascia. The first branch of the external carotid, the *superior thyroid artery*, descends to the superior pole of each lobe of the gland, pierces the pretracheal fascia, and divides into anterior and posterior branches. The *inferior thyroid artery*, a branch of the thyrocervical trunk, runs superomedially posterior to the carotid sheath to reach the posterior aspect of the gland. It divides into several branches that pierce the pretracheal layer of cervical fascia and supply the inferior pole of the gland.

Three pairs of *thyroid veins* usually drain the venous plexus on the anterior surface of the thyroid gland and trachea (Fig. 9.6C). The superior thyroid veins drain the superior poles; the middle thyroid veins drain the middle of the lobes; and the inferior thyroid veins drain the inferior poles. The superior and middle thyroid veins empty into the internal jugular veins, and the inferior thyroid veins empty into the brachiocephalic veins.

The *lymphatic vessels of the thyroid gland* run in the interlobular connective tissue, often around the arteries, and communicate with a capsular network of lymphatic vessels. From here, the vessels pass to the prelaryngeal, pretracheal, and paratracheal lymph nodes (Fig. 9.6*D*). Laterally, lymphatic vessels located along the superior thyroid veins pass to the inferior deep cervical lymph nodes. Some lymphatic vessels may drain into the brachiocephalic lymph nodes or into the thoracic duct.

Nerves of the thyroid gland are derived from the superior, middle, and inferior cervical sympathetic ganglia (Fig. 9.4). They reach the gland through the cardiac and superior and inferior laryngeal nerves and along the thyroid arteries. Some of these fibers are vasomotor.

Parathyroid Glands

The parathyroid glands secrete *parathormone*, a hormone concerned with the metabolism of calcium and phosphorus. These small, ovoid endocrine glands usually lie outside the thyroid capsule on the medial half of the posterior surface of each thyroid lobe (Fig. 9.7). Usually there are two parathyroids on each side, but the total number varies between two and six. The *superior parathyroid glands*, more constant in position than the inferior ones, are usually located at the level of the inferior border of the cricoid cartilage. The *inferior parathyroid glands* are usually near the inferior surface of the thyroid, but they may lie some distance inferior to it.

The parathyroid glands are supplied by branches of the *inferior thyroid arteries*. The parathyroid veins drain into the plexus of veins on the anterior surface of the thyroid and trachea (Fig. 9.6C). The *lymphatic vessels* drain with those from the thyroid into the deep cervical lymph nodes and paratracheal lymph nodes (Fig. 9.6*D*). *Nerves of the parathyroid glands* are derived from the thyroid branches of the cervical sympathetic ganglia.

RESPIRATORY LAYER

The respiratory layer contains the larynx and trachea.

Larynx

The larynx is in the anterior part of the neck at the level of the bodies of C3–C6 vertebrae. It connects the inferior part of the pharynx with

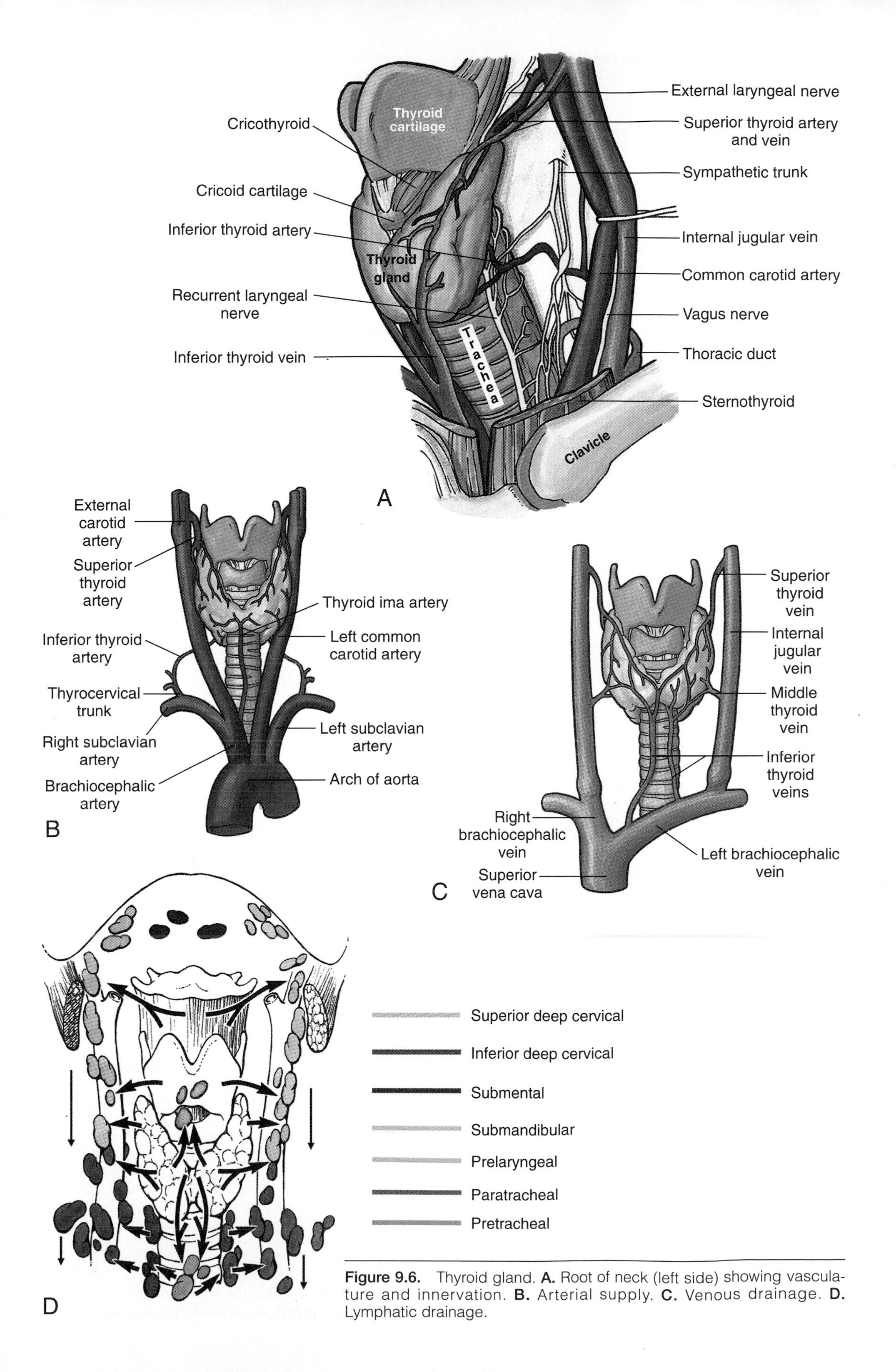

Figure 9.6. Thyroid gland. **A.** Root of neck (left side) showing vasculature and innervation. **B.** Arterial supply. **C.** Venous drainage. **D.** Lymphatic drainage.

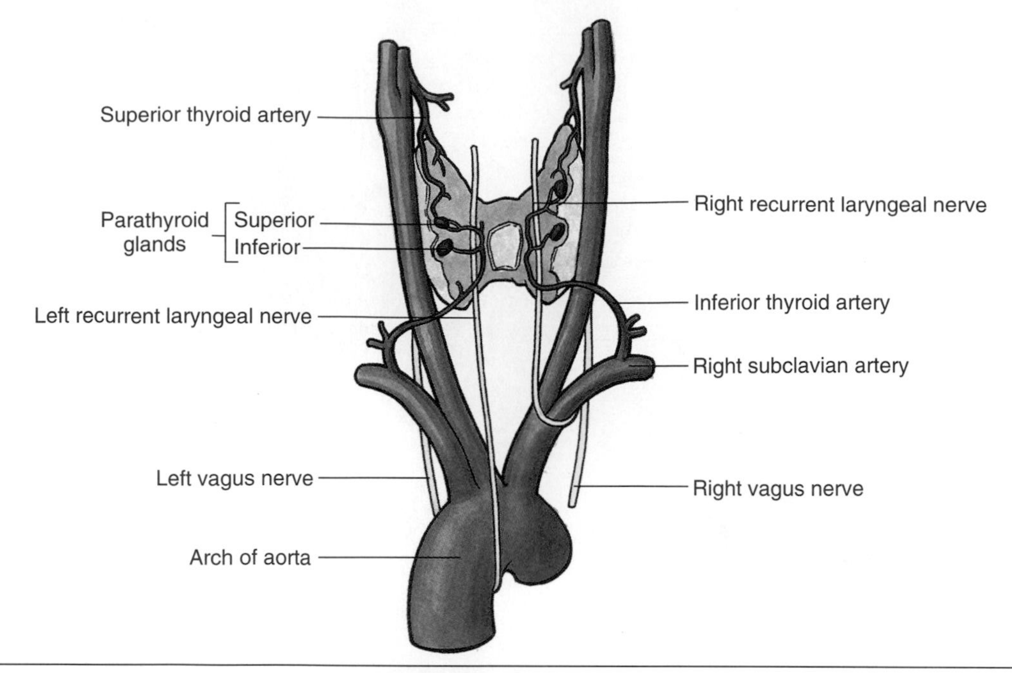

Figure 9.7. Posterior view of thyroid gland showing location of parathyroid glands and their blood supply.

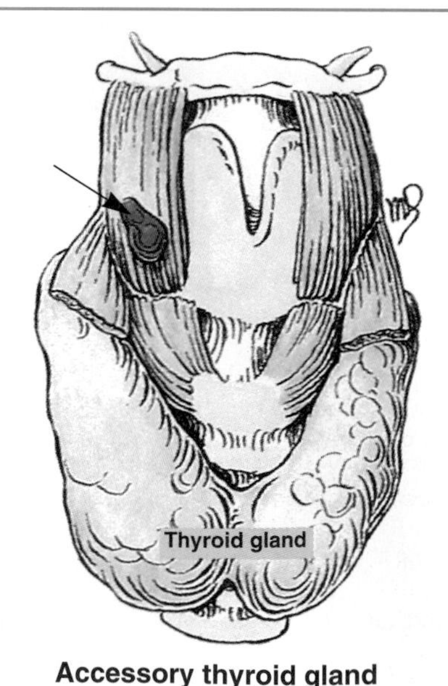

Accessory thyroid gland

During surgical operations on the thyroid, the inferior thyroid artery may have to be tied. The recurrent laryngeal nerve, which it is intimately related to it (Fig. 9.7), must be preserved because it supplies all the intrinsic muscles of the larynx, except the cricothyroid. The recurrent laryngeal nerve is also closely related posteriorly to the connective tissue attaching the thyroid gland to the cricoid cartilage and the superior tracheal rings. In addition to the named thyroid arteries, numerous small vessels pass to the thyroid gland from arteries supplying the pharynx and trachea. Hence, the thyroid oozes blood during a *thyroidectomy* (removal of all or part of the gland), even when the main arteries supplying it have been ligated.

A small detached mass of thyroid tissue may appear superior to the gland as an accessory thyroid gland (*arrow*). Abnormal enlargement of the gland (*goiter*) causes a large swelling in the neck. When the gland enlarges, the sternothyroid and sternohyoid muscles stretch and adhere to the gland. A goiter usually exerts pressure on the trachea and recurrent laryngeal nerves.

During *total thyroidectomy*, the parathyroid glands are in danger of being damaged or removed, but they are safe during *subtotal thyroidectomy* when the posterior part of the thyroid is preserved. Variability in the position of the parathyroid glands, particularly the inferior ones, may create a problem during thyroid and/or parathyroid surgery. If these glands are inadvertently removed during surgery, the patient suffers from a severe convulsive disorder known as *tetany*. The generalized convulsive muscle spasms result from a fall in blood calcium levels. In instances when it is necessary to do a total thyroidectomy, the parathyroid glands are carefully isolated with their blood vessels intact before the thyroid gland is removed.

the trachea (Fig. 9.1*A*). The larynx serves as a valve to guard the air passages and maintain a patent airway, especially during swallowing. The larynx is also the *phonating mechanism* that is designed for voice production.

The *laryngeal skeleton* consists of nine cartilages that are joined by ligaments and membranes (Fig. 9.8). Three of the cartilages are single (thyroid, cricoid, and epiglottic), and three are paired (arytenoid, corniculate, and cuneiform).

The *thyroid cartilage* is the largest of the laryngeal cartilages. The inferior two-thirds of its two platelike laminae are fused anteriorly in the median plane to form the *laryngeal prominence* (p. 412). Immediately superior to this prominence, the thyroid laminae diverge to form a V-shaped *thyroid notch*. The posterior border of each lamina projects superiorly as the superior (greater) horn and inferiorly as the inferior (lesser) horn. The superior border and superior horns (cornua) of the thyroid cartilage

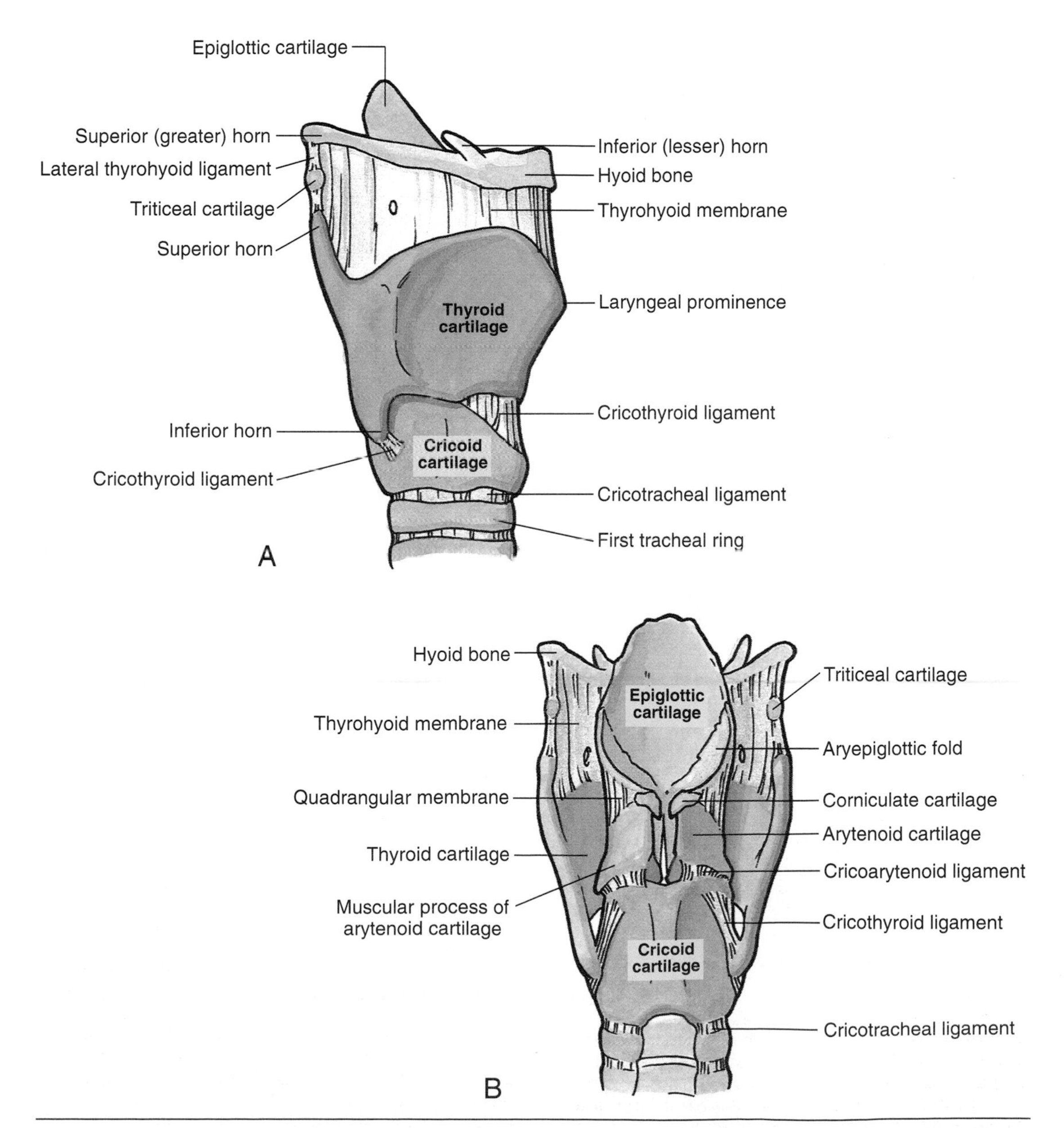

Figure 9.8. Skeleton of larynx and associated ligaments and membranes. **A.** Lateral view. **B.** Posterior view.

are attached to the hyoid bone by the *thyrohyoid membrane*. The thicker median part of this membrane is the median thyrohyoid ligament; its thickened lateral parts are the lateral thyrohyoid ligaments which may contain kernellike *triticeal cartilages* that help close the inlet of the larynx during swallowing. The inferior horns of the thyroid cartilage articulate with the lateral surfaces of the cricoid cartilage at the *cricothyroid joints*. The main movements at these joints are rotation and gliding of the thyroid cartilage, which result in changes in the length of the vocal folds.

The *cricoid cartilage* is shaped like a signet ring with its band facing anteriorly. The posterior (signet) part of the cricoid is the lamina, and the anterior (band) part is the arch. Although much smaller than the thyroid cartilage, the cricoid cartilage is thicker and stronger. The cricoid cartilage is attached to the inferior margin of the thyroid cartilage by the *cricothyroid ligament* (membrane) and to the first tracheal ring by the cricotracheal ligament. The cricothyroid ligament produces a soft spot inferior to the thyroid cartilage. The larynx is closest to the skin here and is most accessible.

The *arytenoid cartilages* are shaped like three-sided pyramids. The paired cartilages articulate with lateral parts of the superior border of the lamina of the cricoid cartilage. Each cartilage has an apex superiorly, a vocal process anteriorly, and a muscular process that projects laterally from the base. The apex is attached to the aryepiglottic fold, the vocal process to the vocal ligament, and the muscular process to the posterior and lateral cricoarytenoid muscles.

The *cricoarytenoid joints* are located between the bases of the arytenoid cartilages and the superior surfaces of the laminae of the cricoid cartilage. These joints permit the following movements of the arytenoid cartilages: sliding toward or away from one another, tilting anteriorly and posteriorly, and rotating. These movements are important in approximating, tensing, and relaxing the vocal folds (cords). The elastic *vocal ligament* extends from the junction of the laminae of the thyroid cartilage anteriorly to the vocal process of the arytenoid cartilage posteriorly. The vocal ligament forms the skeleton of the vocal fold. The triangular membrane of which the vocal ligament forms the superior border is the cricothyroid ligament (conus elasticus). This ligament blends anteriorly with the median cricothyroid ligament (Fig. 9.9C).

The *epiglottic cartilage* gives flexibility to the epiglottis (Fig. 9.8). Situated posterior to the root of the tongue and hyoid bone and anterior to the laryngeal inlet, the leaflike epiglottic cartilage forms the superior part of the anterior wall and the superior margin of the inlet. Its broad superior end is free, and its tapered inferior end is attached to the *thyroepiglottic ligament* in the angle formed by the thyroid laminae. The anterior surface of the epiglottic cartilage is attached to the hyoid bone by the *hyoepiglottic ligament*. The *quadrangular membrane* is a thin, submucosal sheet of connective tissue that extends from the arytenoid cartilage to the epiglottic cartilage. Its free inferior margin constitutes the *vestibular ligament*, which is covered loosely by the vestibular fold. This fold lies superior to the vocal fold and extends from the thyroid cartilage to the arytenoid cartilage.

The *corniculate and cuneiform cartilages* are small nodules in the posterior part of the aryepiglottic folds that are attached to the apices of the arytenoid cartilages.

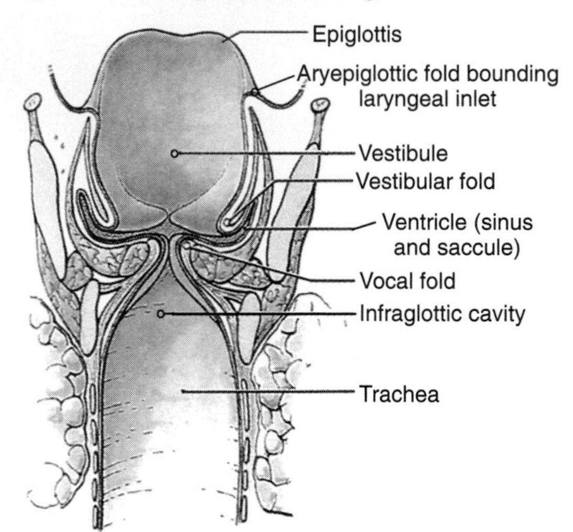

Compartments of larynx (coronal section)

Interior of Larynx. The cavity of the larynx extends from the *inlet of the larynx* (laryngeal aditus), through which it communicates with the laryngopharynx, to the level of the inferior border of the cricoid cartilage, where it is continuous with the cavity of the trachea (Fig. 9.9). The cavity of the larynx is divided into three parts:

- Vestibule of larynx, which is superior to vestibular folds
- Ventricle of larynx, which is between the

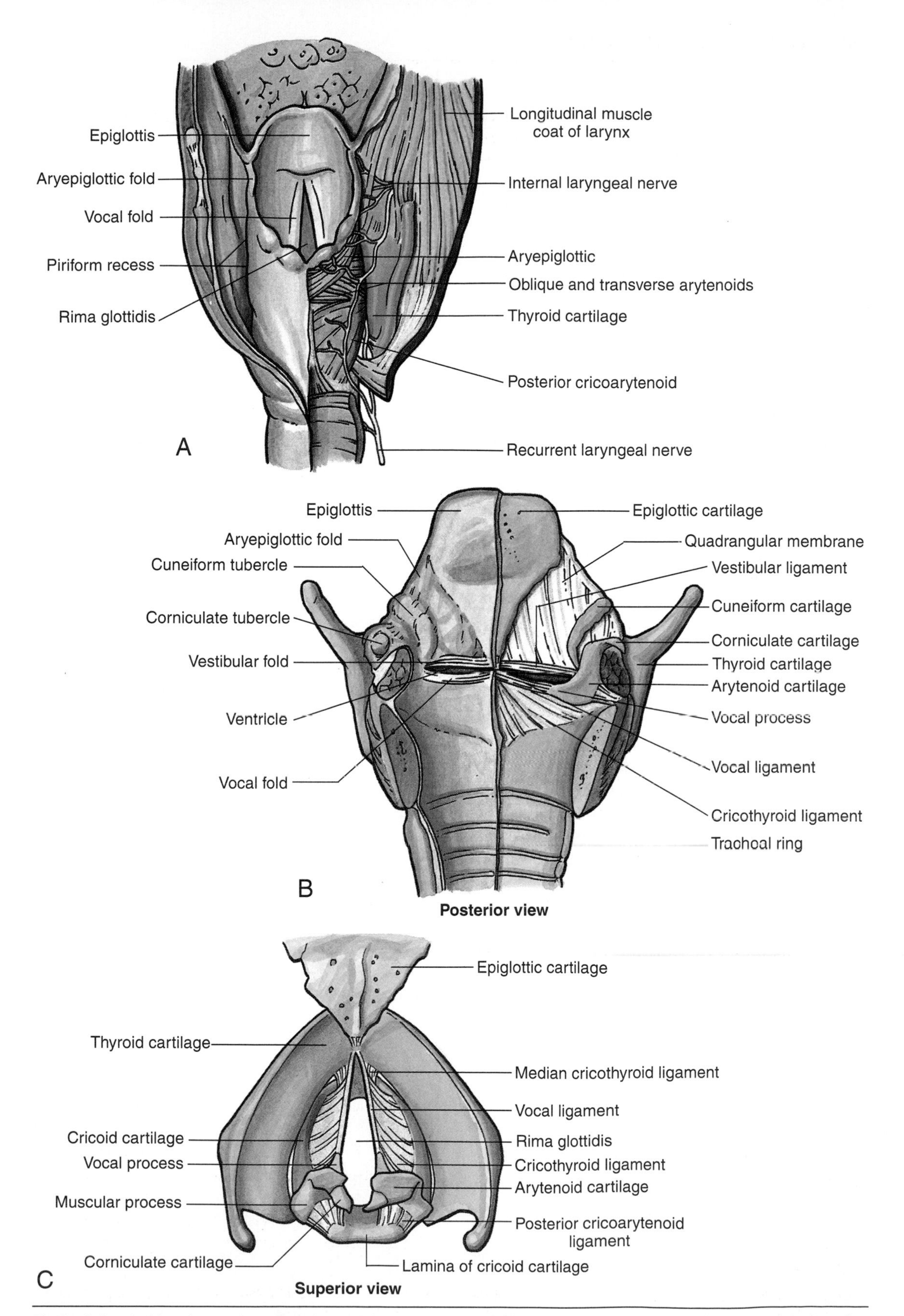

Figure 9.9. **A.** Muscles and nerves of larynx (schematic drawing). *Pink*, mucosa. **B.** Interior of larynx. **C.** Skeleton and ligaments of larynx.

vestibular folds and superior to the vocal folds (the ventricle extends laterally as the *sinus of the larynx*; from each sinus, a small blind *saccule of the larynx* passes superiorly between the vestibular fold and thyroid lamina)

- Infraglottic cavity, which is the inferior cavity of the larynx that extends from the vocal folds to the inferior border of the cricoid cartilage, where it is continuous with the cavity of the trachea

The *vocal folds* (true vocal cords) control sound production. The apex of each wedge-shaped fold projects medially into the laryngeal cavity, and its base lies against the lamina of the thyroid cartilage. Each vocal fold has

- A vocal ligament consisting of elastic tissue derived from the cricothyroid ligament
- A vocalis (vocal muscle), which is part of the thyroarytenoid muscle

The *glottis* comprises the vocal folds and processes, together with the *rima glottidis* (aperture between vocal folds). The shape of the rima glottidis varies according to the position of the vocal folds. During normal respiration it is narrow and wedge-shaped; during forced respiration it is wide. The rima glottidis is slitlike when the vocal folds are closely approximated during phonation (speaking). Variation in the tension and length of the vocal folds, in the width of the rima glottidis, and in the intensity of the expiratory effort produces changes in pitch of the voice. The lower range of pitch of the male voice results from the greater length of the vocal folds.

The *vestibular folds* (false vocal cords) extend between the thyroid and arytenoid cartilages. They play little or no part in voice production; they are protective in function. They consist of two thick folds of mucous membrane enclosing the vestibular ligaments. The space between these ligaments is the *rima vestibuli*.

Laryngeal fractures may result from blows (e.g., during boxing or karate) or from compression (e.g., by a shoulder strap during a vehicular accident). Because of the frequency of laryngeal injuries, most goalies in ice hockey have throat protectors hanging from their masks that cover their larynges. Laryngeal fractures produce submucous hemorrhage and edema, respiratory obstruction, hoarseness, and sometimes an inability to speak because of injury to the laryngeal nerves.

The larynx may be examined visually by *indirect laryngoscopy* using a laryngoscopic mirror, or it can be viewed by *direct laryngoscopy* using a tubular, endoscopic laryngoscope. The vestibular folds normally appear pink, whereas the vocal folds are usually pearly white.

If a foreign object such as food enters the larynx, the laryngeal muscles go into spasm, tensing the vocal folds. As a result, the rima glottidis closes, no air enters the trachea, and asphyxiation may occur. If the foreign object cannot be dislodged, emergency therapy must be given to open the airway. The procedure used depends on the condition of the patient, the facilities available, and the experience of the person giving first aid. A large bore needle or cannula may be inserted through the cricothyroid ligament (*needle cricothyrotomy*) to permit fast entry of air. Later a *surgical cricothyrotomy* may be performed during which the skin and cricothyroid ligament are incised and a tracheostomy tube is inserted in the trachea.

The vestibular folds are part of the protective mechanism that closes the larynx. The vestibule is very sensitive to foreign objects such as food. When objects contact the vestibular epithelium, violent coughing occurs in an attempt to expel them. If this action fails, the aspirated food or other material may lodge in the rima glottitis causing *laryngeal obstruction* (choking). Because the lungs still contain air, compression of the abdomen (*Heimlich maneuver*) causes the diaphragm to elevate and the lungs to compress, expelling air from the trachea into the larynx. This maneuver may dislodge the food or other material from the larynx.

Laryngeal Muscles. The laryngeal muscles are divided into extrinsic and intrinsic groups. The *extrinsic muscles* move the larynx as a whole (Table 9.3). The infrahyoid muscles are

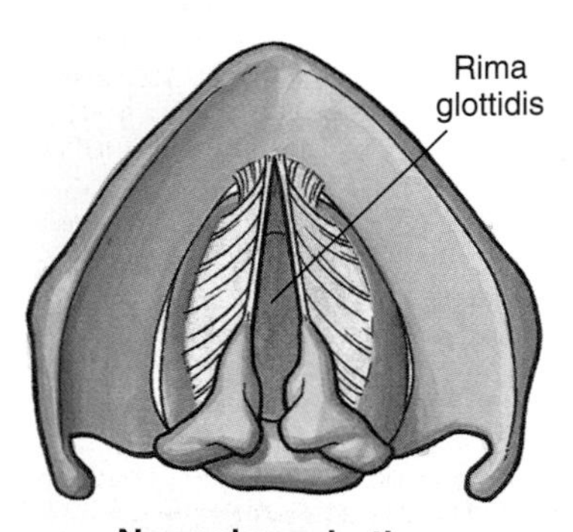

Normal respiration

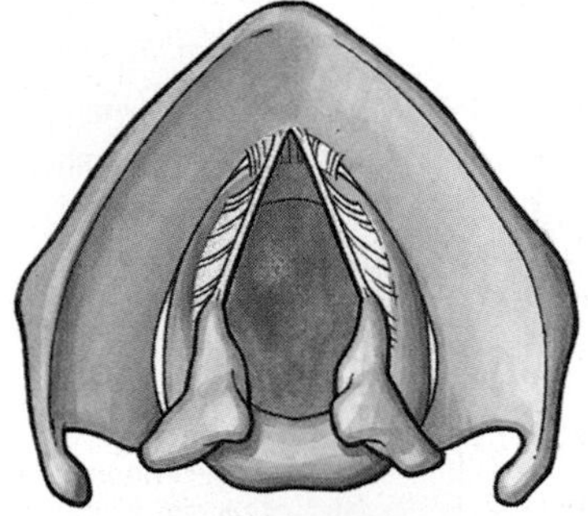

Forced inspiration

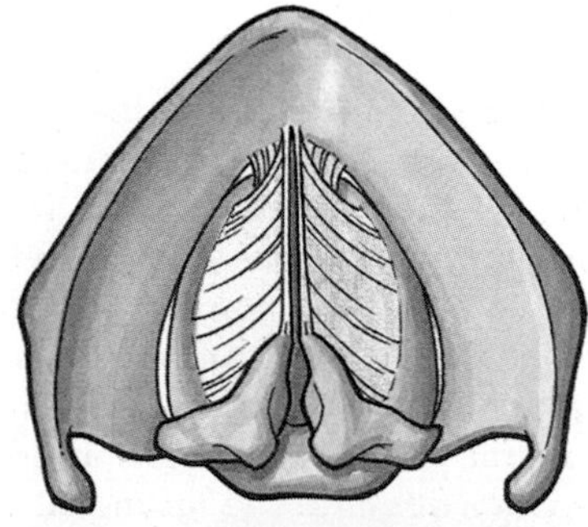

Phonation

depressors of the hyoid bone and larynx, whereas the suprahyoid muscles and the stylopharyngeus are elevators of the hyoid bone and larynx. The *intrinsic muscles* are concerned with movements of laryngeal parts, making alterations in the length and tension of the vocal folds and in the size and shape of the rima glottidis. All but one intrinsic muscle of the larynx is supplied by the *recurrent laryngeal nerve*, a branch of CN X; the cricothyroid is supplied by the *external laryngeal nerve*. Actions of the intrinsic laryngeal muscles are described in Table 9.5.

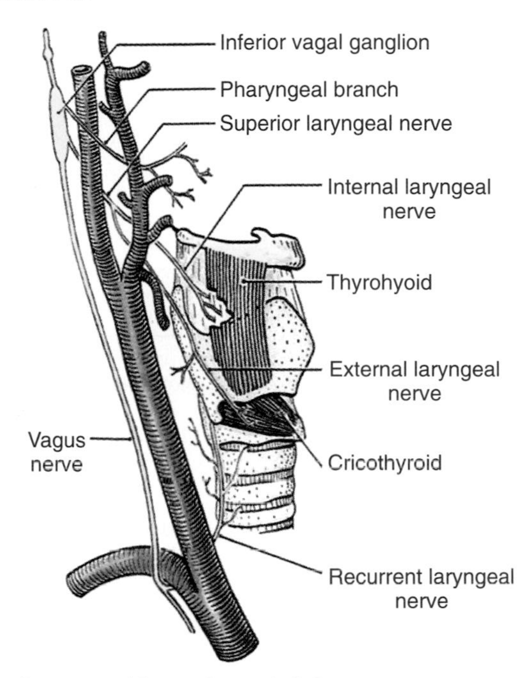

Laryngeal branches of right vagus nerve

Laryngeal Nerves. The laryngeal nerves are derived from the vagus nerve (CN X) through the internal and external branches of the superior laryngeal and recurrent laryngeal nerves. The *superior laryngeal nerve* arises from the middle of the inferior vagal ganglion located at the superior end of the carotid triangle. It divides into two terminal branches within the carotid sheath: the internal laryngeal nerve (sensory and autonomic) and the external laryngeal nerve (motor). The *internal laryngeal nerve* (Fig. 9.9*A*), the larger of the terminal branches, pierces the thyrohyoid membrane with the superior laryngeal artery and supplies sensory fibers to the laryngeal mucous membrane superior to the vocal folds, including the superior surface of these folds. The *external laryngeal nerve* descends posterior to the sternothyroid muscle in company with the superior thyroid artery (Fig. 9.6*A*). At first it lies on the inferior constrictor muscle of the pharynx, and then it pierces and supplies it and the cricothyroid muscle.

The *recurrent laryngeal nerve* supplies all intrinsic muscles of the larynx except the cricothyroid, which is supplied by the external laryngeal nerve. It also supplies sensory fibers to the laryngeal mucous membrane inferior to the vocal folds. The terminal part of the recurrent laryngeal nerve, the *inferior laryngeal nerve*, enters the larynx by passing deep to the inferior border of the inferior constrictor muscle of the pharynx. It divides into anterior and posterior branches, which accompany the inferior laryngeal artery into the larynx.

Laryngeal Vessels. The *laryngeal arteries*, branches of the superior and inferior thyroid arteries, supply the larynx (Fig. 9.6*A*). The *superior laryngeal artery* accompanies the internal branch of the superior laryngeal nerve through the thyrohyoid membrane and then branches to supply the internal surface of the larynx. The *inferior laryngeal artery* accompanies the inferior laryngeal nerve and supplies the mucous membrane and muscles in the inferior aspect of the larynx.

The *laryngeal veins* accompany the laryngeal arteries. The superior laryngeal vein usually joins the superior thyroid vein and through it drains into the internal jugular vein. The inferior laryngeal vein joins the inferior thyroid vein or the anastomosing venous plexus of thyroid veins on the anterior aspect of the trachea (Fig. 9.6*A*).

The recurrent laryngeal nerves are vulnerable to injury during thyroidectomy and other surgical operations in the anterior triangles of the neck. Because the inferior laryngeal nerve innervates the muscles moving the vocal folds, *paralysis of the vocal fold* results when the nerve is injured. The voice is poor because the paralyzed fold cannot meet the normal vocal fold. When bilateral paralysis of the vocal folds occurs, the voice is almost absent because the vocal folds cannot be adducted. Hoarseness is the most common symptom of serious disorders of the larynx (e.g., carcinoma of vocal folds). People requiring *laryngectomy* (removal of larynx), can learn esophageal speech (regurgitation of ingested air) and other rehabilitative speech techniques.

Table 9.5
Muscles of Larynx

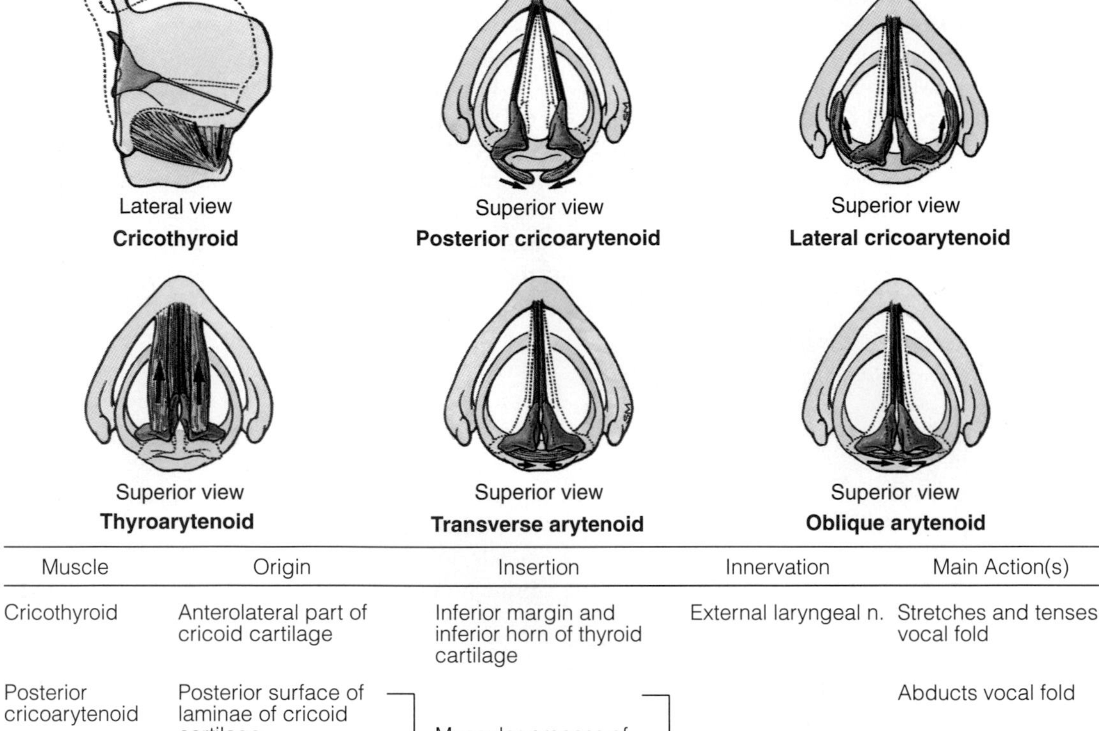

Muscle	Origin	Insertion	Innervation	Main Action(s)
Cricothyroid	Anterolateral part of cricoid cartilage	Inferior margin and inferior horn of thyroid cartilage	External laryngeal n.	Stretches and tenses vocal fold
Posterior cricoarytenoid	Posterior surface of laminae of cricoid cartilage	Muscular process of arytenoid cartilage	Recurrent laryngeal n.	Abducts vocal fold
Lateral cricoarytenoid	Arch of cricoid cartilage	Muscular process of arytenoid cartilage	Recurrent laryngeal n.	Adducts vocal fold
Thyroarytenoid[a]	Posterior surface of thyroid cartilage	Muscular process of arytenoid process	Recurrent laryngeal n.	Relaxes vocal fold
Transverse and oblique arytenoids[b]	One arytenoid cartilage	Opposite arytenoid cartilage	Recurrent laryngeal n.	Close inlet of larynx by approximating arytenoid cartilages
Vocalis[c]	Angle between laminae of thyroid cartilage	Vocal process of arytenoid cartilage	Recurrent laryngeal n.	Alters vocal fold during phonation

[a] Superior fibers of the thyroarytenoid muscle pass into the aryepiglottic fold, and some of them reach the epiglottic cartilage. These fibers consitute the thyroepiglottic muscle, which widens inlet of larynx.

[b] Some fibers of oblique arytenoid muscle continue as aryepiglottic muscle (Fig. 9.9*A*).

[c] This slender muscular slip is derived from inferior deeper fibers of the thyroarytenoid muscle.

The *laryngeal lymphatic vessels* superior to the vocal folds accompany the superior laryngeal artery through the thyrohyoid membrane and drain into the superior deep cervical lymph nodes (Fig. 9.6*D*). The lymphatic vessels inferior to the vocal folds drain into the inferior deep cervical lymph nodes.

Trachea

The trachea extends from the inferior end of the larynx at the level of the sixth cervical vertebra (Fig. 9.1*A*). It ends at the sternal angle at the level of the fifth to seventh thoracic vertebrae, where it divides into the right and left *main bronchi* (p. 50). Lateral to the trachea are the common carotid arteries and the thyroid lobes (Fig. 9.6*B*). Inferior to the isthmus of the gland are the jugular venous arch and the inferior thyroid veins. The brachiocephalic trunk is related to the right side of the trachea in the root of the neck.

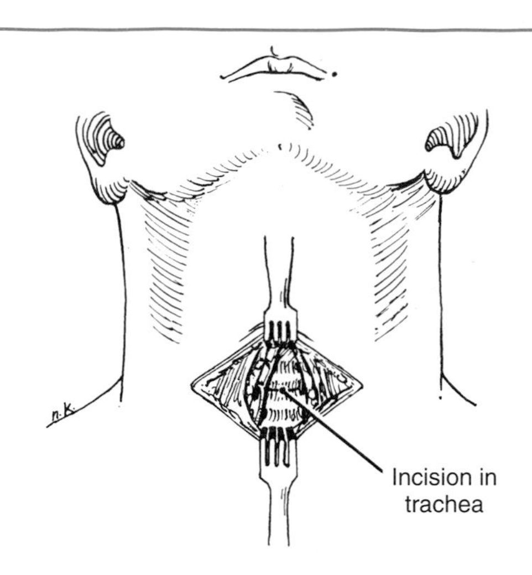

Tracheostomy

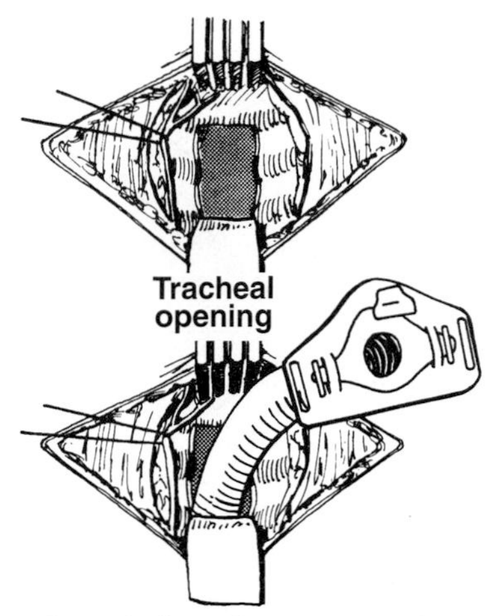

Inserted tracheostomy tube

A vertical or transverse incision through the neck and anterior wall of the trachea (*tracheostomy*) may be performed to establish an airway in patients with upper airway obstruction or respiratory failure. The infrahyoid (strap) muscles are retracted laterally, and the isthmus of the thyroid gland is either divided or retracted superiorly. An opening is made in the trachea between the first and second tracheal rings or through the second through fourth rings. The tracheostomy tube is then inserted into the trachea and secured by neck straps.

To avoid complications during a tracheostomy, the following anatomical relationships must be remembered:

- The inferior thyroid veins form a plexus anterior to the trachea
- A small thyroid ima artery is present in about 10% of people and ascends to the isthmus of the thyroid gland
- The left brachiocephalic vein, jugular venous arch, and pleurae may be encountered, particularly in infants and children
- The thymus covers the inferior part of the trachea in infants and children
- The trachea is small, mobile, and soft in infants, making it easy to cut through its posterior wall and damage the esophagus

ALIMENTARY LAYER

The alimentary layer contains the pharynx and esophagus.

Pharynx

The pharynx is the part of the digestive system located between the nasal and oral cavities and posterior to the larynx; the pharynx conducts food to the esophagus and air to the larynx, trachea, and lungs. The pharynx extends from the base of the skull to the inferior border of the cricoid cartilage anteriorly and the inferior border of C6 vertebra posteriorly (Fig. 9.10). It is widest (about 5 cm) opposite the hyoid bone and narrowest (about 1.5 cm) at its inferior end, where it is continuous with the esophagus. The posterior wall of the pharynx lies against the prevertebral layer of cervical fascia (Fig. 9.1).

The wall of the pharynx is composed mainly of two layers of pharyngeal muscle. The external circular layer of muscles consists of three constrictor muscles (Figs. 9.11 and 9.12). The internal, mainly longitudinal, layer of muscles consists of the palatopharyngeus, stylopharyngeus, and salpingopharyngeus. These muscles elevate the larynx and pharynx during swallowing and speaking. The attachments, nerve supply, and actions of the pharyngeal muscles are described in Table 9.6.

Pharyngeal Muscles. The constrictor muscles of the pharynx contract involuntarily so that contraction takes place sequentially from the superior to the inferior end of the pharynx. This action propels food into the esophagus. All three constrictors are supplied by the pharyngeal plexus of nerves that lies on the lateral wall of the pharynx, mainly on the middle constrictor. The overlapping of the constrictor muscles leaves the following four gaps in the musculature for structures to enter the pharynx.

(*a*) Superior to the superior constrictor, the levator veli palatini, auditory tube, and ascending palatine artery pass through the gap

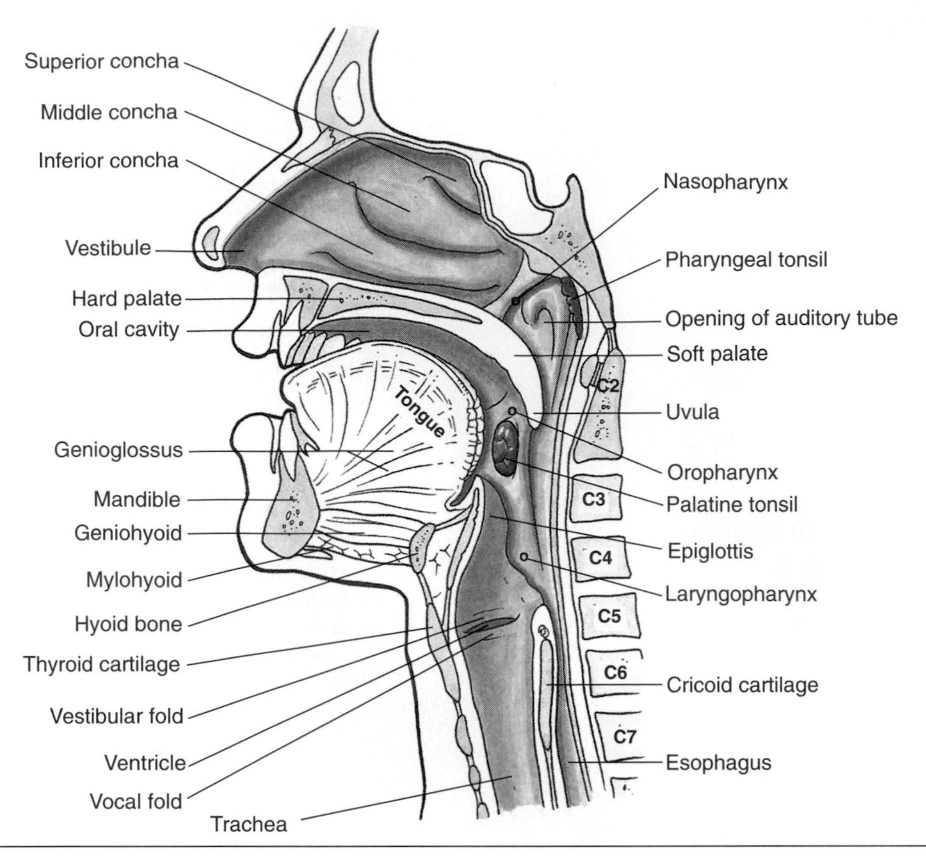

Figure 9.10. Sagittal section of head and neck, showing nose, mouth, pharynx, larynx, and esophagus.

between the superior constrictor and the skull. Superior to the superior constrictor, the pharyngobasilar fascia blends with the buccopharyngeal fascia to form, with the mucous membrane, the thin wall of the *pharyngeal recess* (Fig. 9.12*B*).

(*b*) Between the superior and middle constrictors is a gap that forms the gateway to the mouth, through which pass the stylopharyngeus, glossopharyngeal nerve (CN IX), and stylohyoid ligament.

(*c*) Between the middle and inferior constrictors is a gap for the internal laryngeal nerve and superior laryngeal artery and vein to pass to the larynx.

(*d*) Inferior to the inferior constrictor is a gap for the recurrent laryngeal nerve and inferior laryngeal artery to pass superiorly into the larynx.

Pharyngeal Nerves. Nerve supply to the pharynx (motor and most of sensory) is derived from the *pharyngeal plexus of nerves*. This plexus is formed by pharyngeal branches of the vagus (CN X) and glossopharyngeal (CN IX) nerves and by sympathetic branches from the superior cervical ganglion. *Motor fibers in the pharyngeal plexus* are derived from the cranial root of CN XI (accessory nerve) and are carried by the vagus nerve (CN X) to all muscles of the pharynx and soft palate, except the stylopharyngeus (supplied by CN IX) and the tensor veli palatini (supplied by CN V^3). *Sensory fibers in the pharyngeal plexus* are derived from the glossopharyngeal nerve (CN IX). They supply most of the mucosa of all three parts of the pharynx. The sensory nerve supply of the mucous membrane of the nasopharynx is mainly from the maxillary nerve (CN V^2), a purely sensory nerve.

Interior of Pharynx. The pharynx is divided into three parts:

- Nasopharynx, posterior to the nose and superior to the soft palate
- Oropharynx, posterior to the mouth
- Laryngopharynx, posterior to the larynx

The *nasopharynx* has a respiratory function. It lies superior to the soft palate and is the posterior extension of the nasal cavity (Fig. 9.10). The

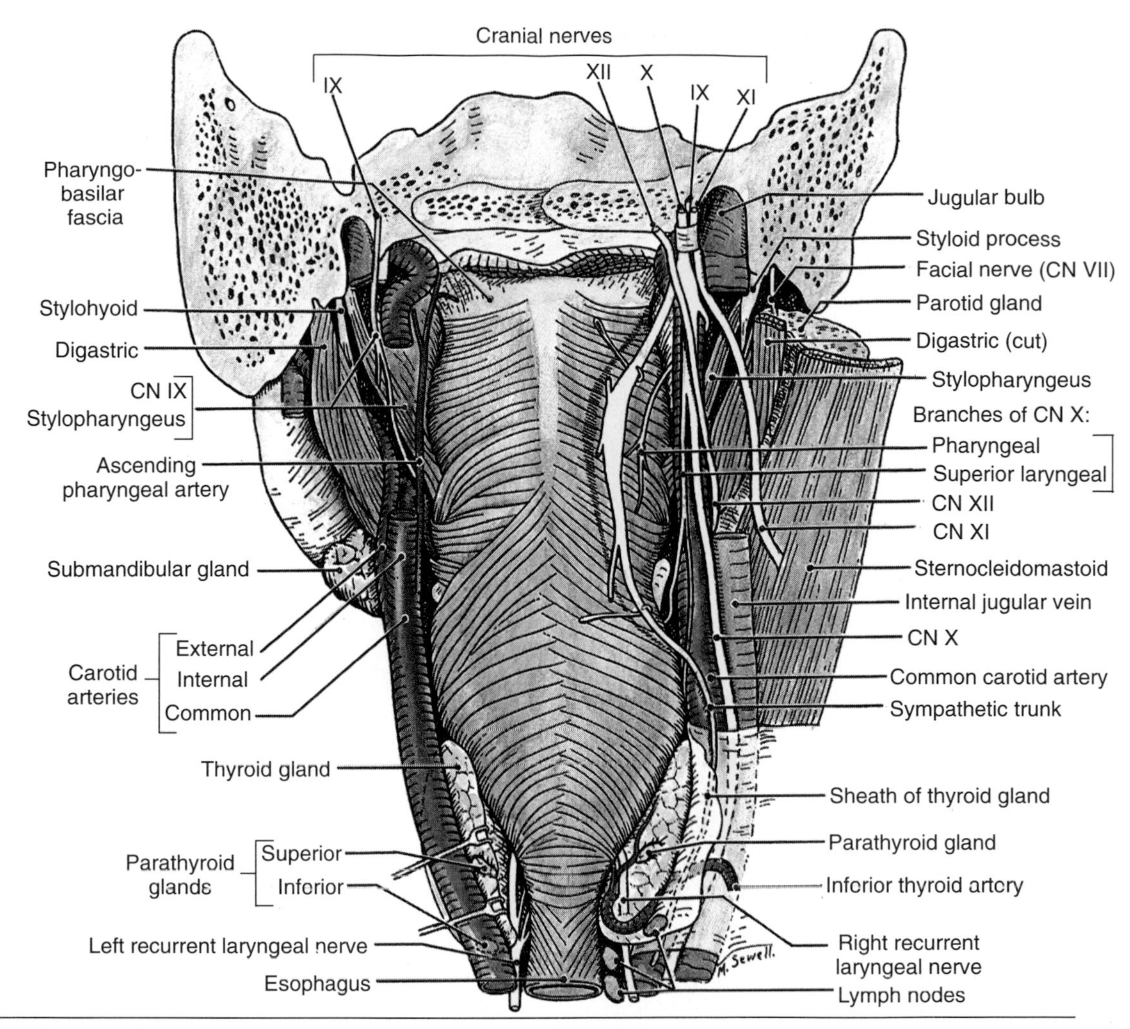

Figure 9.11. Posterior view of pharynx and related vasculature and nerves.

nose opens into the nasopharynx via *choanae* (paired openings between the nasal cavity and nasopharynx). The roof and posterior wall of the nasopharynx form a continuous surface that lies inferior to the body of the sphenoid bone and the basilar part of the occipital bone. A collection of lymphoid tissue, the *pharyngeal tonsil,* is in the mucous membrane of the roof and posterior wall of the nasopharynx (Fig. 9.12*B*). Extending inferiorly from the medial end of the auditory tube is a vertical fold of mucous membrane, the *salpingopharyngeal fold.* It covers the salpingopharyngeus muscle that opens the pharyngeal orifice of the tube during swallowing. The collection of lymphoid tissue in the submucosa of the pharynx near the pharyngeal orifice of the auditory tube is the *tubal tonsil.* Posterior to the torus of the auditory tube and the salpingopharyngeal fold is a slitlike lateral projection of the pharynx, the *pharyngeal recess,* that extends laterally and posteriorly.

The *oropharynx* has a digestive function. It is continuous with the oral cavity through the oropharyngeal isthmus. The oropharynx is bounded by the soft palate superiorly, the base of the tongue inferiorly, and the palatoglossal and palatopharyngeal arches laterally (Fig. 9.13*A*). It extends from the soft palate to the superior border of the epiglottis.

Deglutition (swallowing) is the process whereby food (often referred to as a bolus) is transferred from the mouth through the pharynx and esophagus into the stomach. Solid food is masticated and mixed with saliva to form a soft bolus during chewing. Deglutition occurs in three stages: (*a*) first stage is volun-

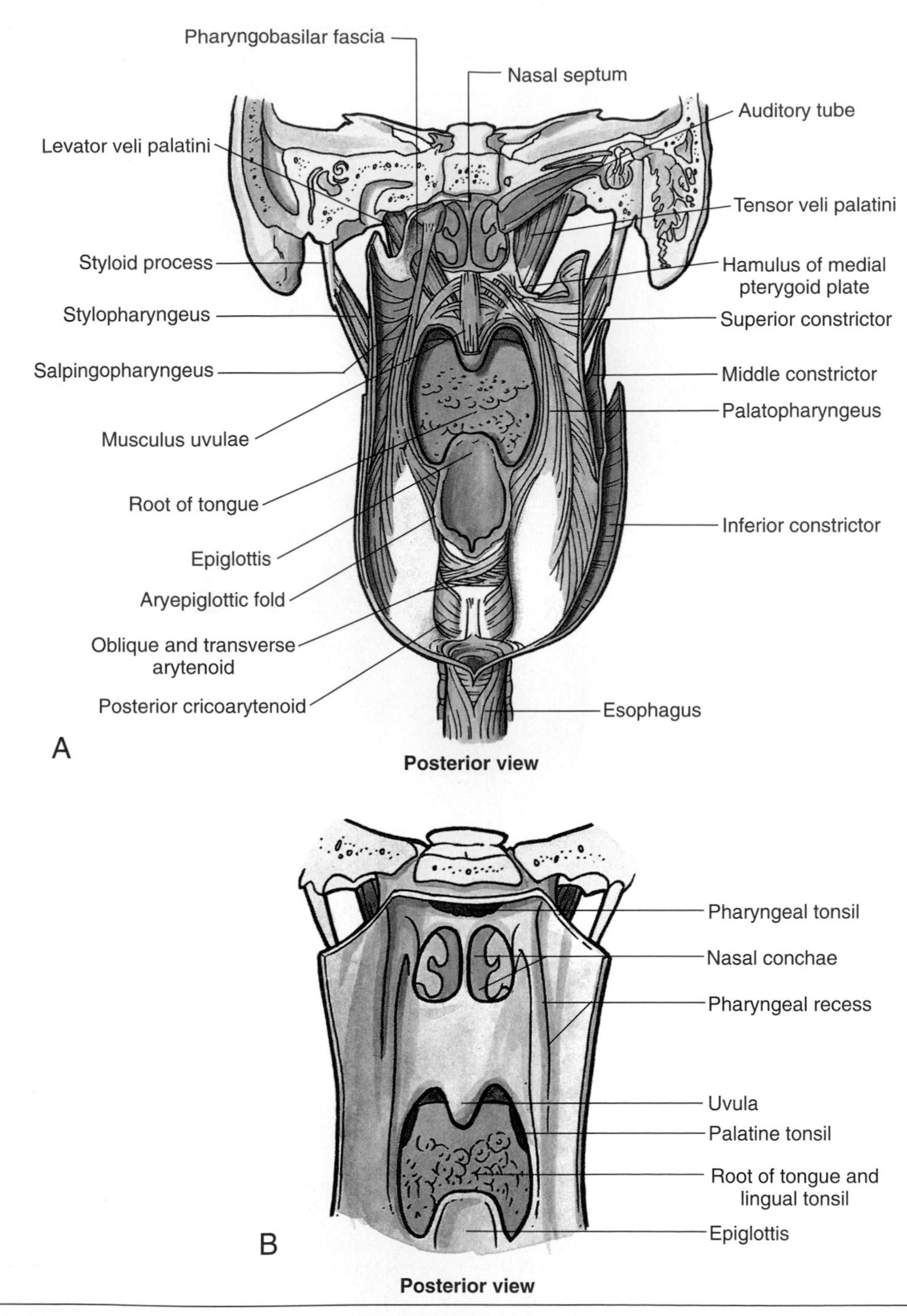

Figure 9.12. **A.** Muscles of soft palate and interior of pharynx. Posterior wall of pharynx has been cut in midline and reflected laterally. **B.** Interior of pharynx showing pharyngeal recess and palatine, pharyngeal, and lingual tonsils.

tary; the bolus is pushed from the mouth into the oropharynx, mainly by movements of the tongue; (*b*) second stage is involuntary and usually rapid; walls of the pharynx are contracted; (*c*) third stage is involuntary and squeezes the bolus from the laryngopharynx into the esophagus; this is produced by the inferior constrictor muscle of the pharynx.

The *palatine tonsils* ("tonsils") are collections of lymphoid tissue on each side of the orophar-

Table 9.6
Muscles of Pharynx

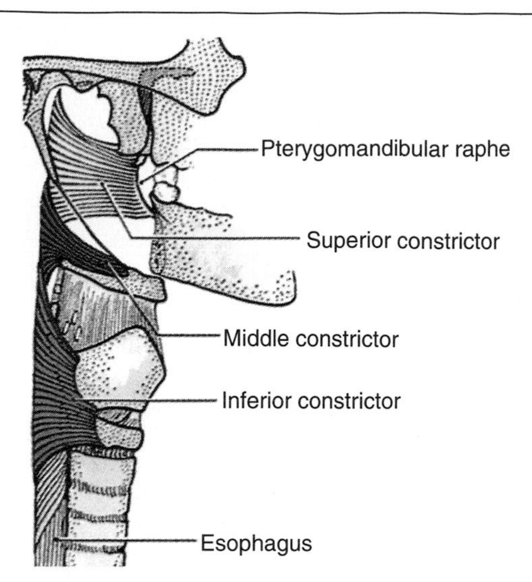

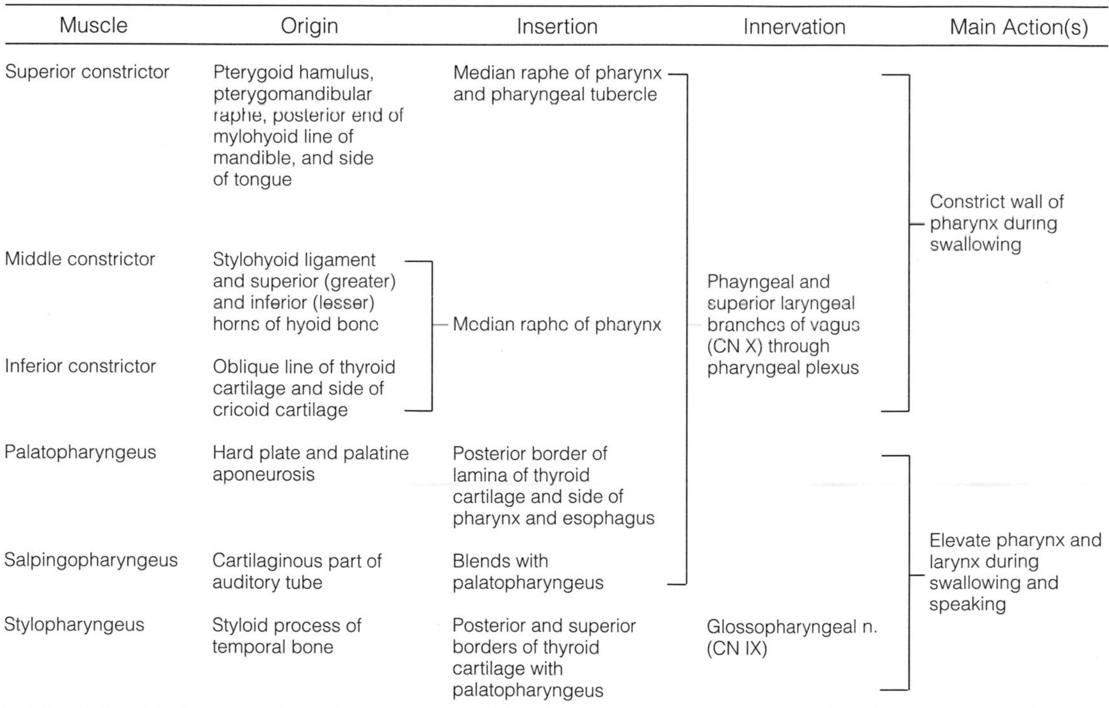

Muscle	Origin	Insertion	Innervation	Main Action(s)
Superior constrictor	Pterygoid hamulus, pterygomandibular raphe, posterior end of mylohyoid line of mandible, and side of tongue	Median raphe of pharynx and pharyngeal tubercle		
Middle constrictor	Stylohyoid ligament and superior (greater) and inferior (lesser) horns of hyoid bone	Median raphe of pharynx	Phayngeal and superior laryngeal branches of vagus (CN X) through pharyngeal plexus	Constrict wall of pharynx during swallowing
Inferior constrictor	Oblique line of thyroid cartilage and side of cricoid cartilage			
Palatopharyngeus	Hard plate and palatine aponeurosis	Posterior border of lamina of thyroid cartilage and side of pharynx and esophagus		
Salpingopharyngeus	Cartilaginous part of auditory tube	Blends with palatopharyngeus		Elevate pharynx and larynx during swallowing and speaking
Stylopharyngeus	Styloid process of temporal bone	Posterior and superior borders of thyroid cartilage with palatopharyngeus	Glossopharyngeal n. (CN IX)	

ynx in the interval between the palatine arches (Fig. 9.13*A*). The tonsil does not fill the *tonsillar cleft* between these arches. The *tonsillar bed*, in which the palatine tonsil lies, is between the palatoglossal and palatopharyngeal arches. The tonsillar bed contains two muscles, the palatopharyngeus and superior constrictor muscle of the pharynx (Fig. 9.13*B*). The thin, fibrous sheet covering the tonsillar bed is part of the *pharyngobasilar fascia*. This fascia blends with the periosteum of the base of the skull and defines the limits of the pharyngeal wall in its superior part.

The *laryngopharynx* lies posterior to the larynx, extending from the superior border of the epiglottis to the inferior border of the cricoid

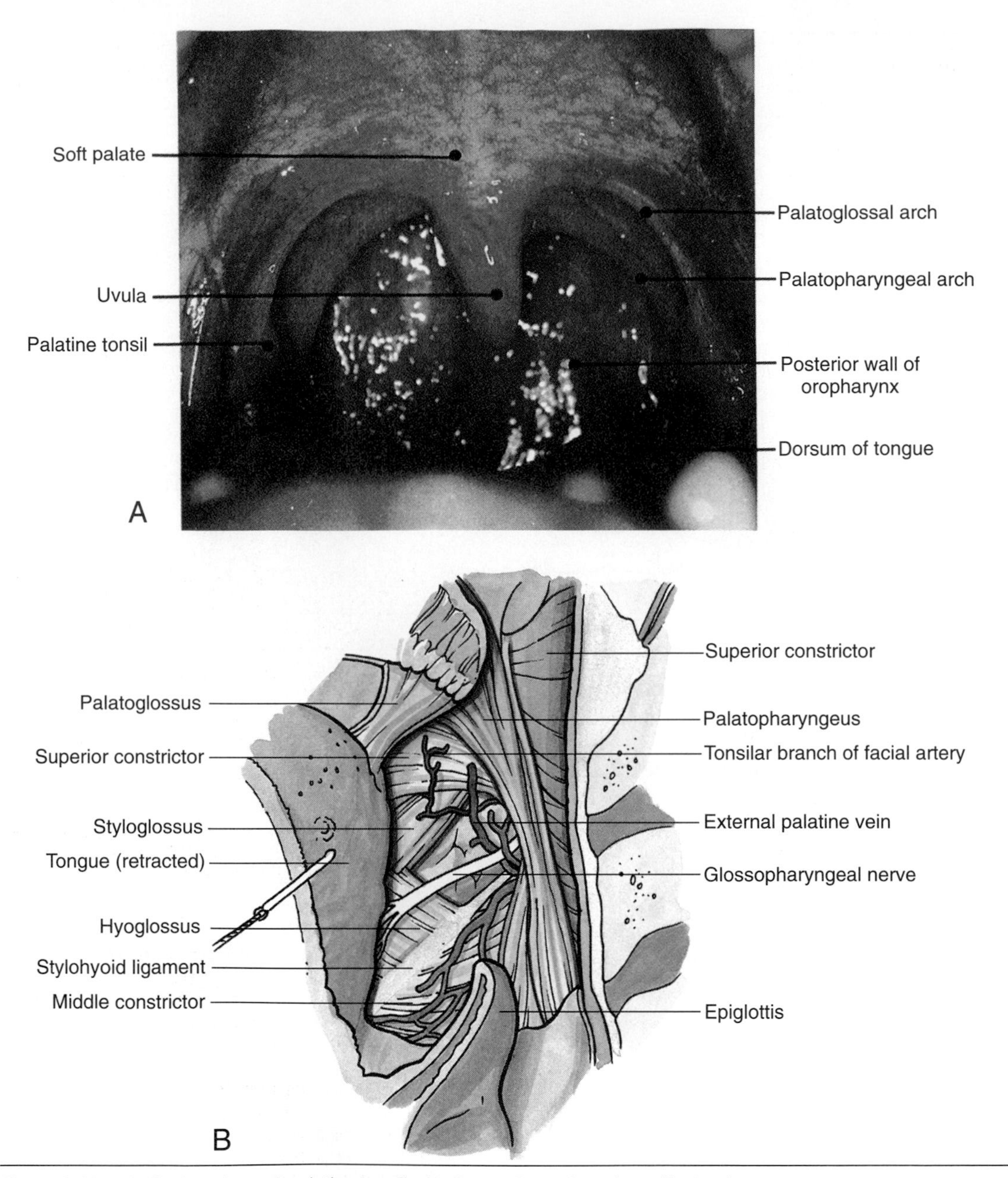

Figure 9.13. **A.** Oral cavity and palatine tonsils. **B.** Deep dissection of tonsillar bed.

cartilage, where it narrows and is continuous with the esophagus. Posteriorly the laryngopharynx is related to the bodies of C4–C6 vertebrae. Its posterior and lateral walls are formed by the middle and inferior constrictor muscles, and internally the wall is formed by the palatopharyngeus and stylopharyngeus muscles. The laryngopharynx communicates with the larynx through the inlet of the larynx. The *piriform recess* is a small depression of the laryngopharyngeal cavity on each side of the inlet (Fig. 9.9*A*). This mucosa-lined fossa is separated from the inlet by the *aryepiglottic fold*. Laterally the piriform recess is bounded by the medial surfaces of the thyroid cartilage and the *thyrohyoid membrane*. Branches of the internal laryngeal and recurrent laryngeal nerves lie deep to the mucous membrane of the piriform recess.

Pharyngeal Vessels. The *tonsillar artery*, a branch of the facial artery (Fig. 9.13*B*), passes

Foreign bodies (e.g., a chicken bone) entering the pharynx may become lodged in the piriform recess. If sharp, they may pierce the mucous membrane and injure the internal laryngeal nerve. This may result in anesthesia of the laryngeal mucous membrane as far inferiorly as the vocal folds. Similarly, the nerve may be injured if the instrument used to remove the foreign body accidentally pierces the mucous membrane.

Young children swallow various objects, most of which reach the stomach and pass through the gastrointestinal tract without difficulty. In some cases, the foreign body stops at the inferior end of the laryngopharynx, its narrowest part. A radiographic examination and/or a magnetic resonance image will reveal the presence of a radiopaque foreign body (e.g., a bone in the piriform recess). Foreign bodies in the pharynx are often removed under direct vision through a *pharyngoscope.*

through the superior constrictor muscle and enters the inferior pole of the tonsil. The tonsil also receives arterial twigs from the ascending palatine, lingual, descending palatine, and ascending pharyngeal arteries.

. The large *external palatine vein* (paratonsillar vein) descends from the soft palate and passes close to the lateral surface of the tonsil before it enters the pharyngeal venous plexus.

Tonsillar nerves are derived from the tonsillar plexus of nerves formed by branches of the glossopharyngeal and vagus nerves. Other branches are derived from the pharyngeal plexus of nerves.

The *tonsillar lymphatic vessels* pass laterally and inferiorly to the lymph nodes near the angle of the mandible and the jugulodigastric node (Fig. 9.5*B*). The *jugulodigastric node* is often referred to as the *tonsillar node* because of its frequent enlargement when the tonsil is inflamed (*tonsillitis*). The palatine, lingual, and pharyngeal tonsils form the *tonsillar ring,* an incomplete circular band of lymphoid tissue around the superior part of the pharynx. The anteroinferior part of the ring is formed by the *lingual tonsil,* a collection of lymphoid tissue in the posterior part of the tongue (Fig. 9.12*B*). Lateral parts of the ring are formed by the palatine and tubal tonsils, and posterior and superior parts are formed by the pharyngeal tonsil.

Esophagus

The esophagus extends from the pharynx to the stomach. It begins in the median plane at the inferior border of the cricoid cartilage, passes inferiorly, and enters the stomach at the cardiac orifice (p. 100). It lies between the trachea and the cervical vertebral bodies (Fig. 9.10). On the right side, the esophagus is in contact with the cervical pleura at the root of the neck, whereas on the left side, posterior to the subclavian artery, the thoracic duct lies between the pleura and the esophagus (p. 77).

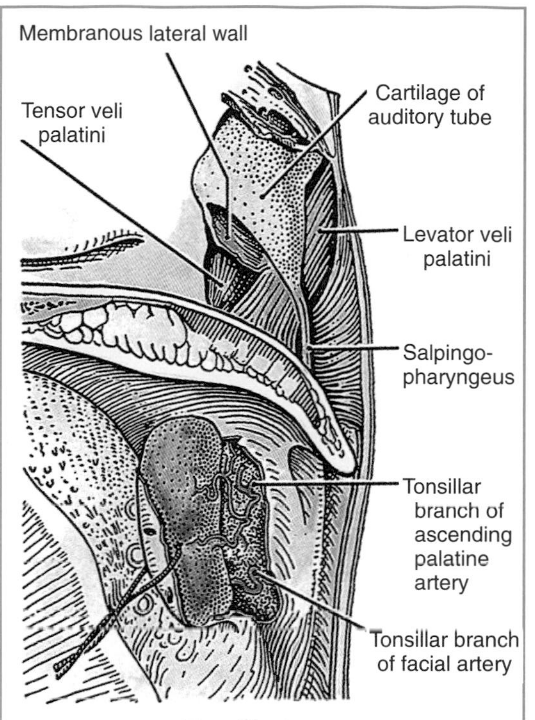

Tonsillectomy

Tonsillectomy is performed by dissecting the tonsil from its bed or by a *guillotine operation.* In each case the tonsil and the fascial sheet covering the tonsillar bed are removed. Because of the rich blood supply of the tonsil, bleeding may arise from the tonsillar artery or other arterial twigs, but bleeding commonly arises from the large external palatine vein (Fig. 9.13*B*). The glossopharyngeal nerve (CN IX) accompanies the tonsillar artery on the lateral wall of the pharynx. Because this wall is thin, CN IX is vulnerable to injury. The *internal carotid artery* is especially vulnerable when it is tortuous and lies directly lateral to the tonsil.

Enlarged pharyngeal tonsils, *adenoids*, can obstruct the passage of air from the nasal cavity through the choanae into the nasopharynx. The adenoids obstruct nasal respiration, making mouth breathing necessary. Infection from the enlarged pharyngeal tonsils may spread to the tubal tonsils, causing swelling and closure of the auditory tubes. Impairment of hearing may result from nasal obstruction and blockage of the auditory tubes. Infection spreading from the nasopharynx to the middle ear causes *otitis media* (middle ear infection), which may produce temporary or permanent hearing loss.

10 / SUMMARY OF CRANIAL NERVES

Although the cranial nerves have been described in preceding chapters, it is helpful to summarize them in schematic and tabular forms. In addition, selected cranial nerve lesions are included to illustrate important clinical features of the nerves.

Cranial nerves carry one or more of the following functional components.

- Somatic motor (general somatic efferent): supplies striated muscles in orbit and tongue
- Branchial motor (special visceral efferent): supplies muscles derived from embryonic branchial (pharyngeal) arches (face, larynx, and pharynx)
- Visceral motor (general visceral efferent): gives rise to cranial parasympathetic system

Table 10.1
Summary of Cranial Nerves

Nerve	Components	Cells of Origin	Cranial Exit	Distribution and Functions
Olfactory (CN I)	Special sensory	Olfactory epithelium (olfactory cells)	Foramina in cribriform plate of ethmoid bone	Smell from nasal mucosa of roof of each nasal cavity and superior sides of nasal septum and superior concha
Optic (CN II)	Special sensory	Retina (ganglion cells)	Optic canal	Vision from retina
Oculomotor (CN III)	Somatic motor	Midbrain	Superior orbital fissure	Motor to superior, inferior, and medial rectus, inferior oblique, and levator palpebrae superioris muscles; raises upper eyelid; turns eyeball superiorly, inferiorly, and medially
	Visceral motor	Preganglionic: midbrain; postganglionic: ciliary ganglion	Superior orbital fissure	Parasympathetic innervation to sphincter pupillae and ciliary muscle; constricts pupil and accommodates lens of eye
Trochlear (CN IV)	Somatic motor	Midbrain	Superior orbital fissure	Motor to superior oblique that assists in turning eye inferolaterally
Trigeminal (CN V)				
Ophthalmic division (CN V^1)	General sensory	Trigeminal ganglion	Superior orbital fissure	Sensation from cornea, skin of forehead, scalp, eyelids, nose, and mucosa of nasal cavity and paranasal sinuses
Maxillary division (CN V^2)	General sensory	Trigeminal ganglion	Foramen rotundum	Sensation from skin of face over maxilla including upper lip, maxillary teeth, mucosa of nose, maxillary sinuses, and palate
Mandibular division (CN V^3)	Branchial motor	Pons	Foramen ovale	Motor to muscles of mastication, mylohyoid, anterior belly of digastric, tensor veli palatini, and tensor tympani
	General sensory	Trigeminal ganglion	Foramen ovale	Sensation from the skin over mandible, including lower lip and side of head, mandibular teeth, temporomandibular joint, and mucosa of mouth and anterior two-thirds of tongue

Table 10.1. *Continued*

Nerve	Components	Cells of Origin	Cranial Exit	Distribution and Functions
Abducent (CN VI)	Somatic motor	Pons	Superior orbital fissure	Motor to lateral rectus that turns eye laterally
Facial (CN VII)	Branchial motor	Pons	Internal acoustic meatus, facial canal, and stylomastoid foramen	Motor to muscles of facial expression and scalp; also supplies stapedius of middle ear, stylohyoid, and posterior belly of digastric
	Special sensory	Geniculate ganglion		Taste from anterior two-thirds of tongue, floor of mouth, and palate
	General sensory	Geniculate ganglion		Sensation from skin of external acoustic meatus
	Visceral motor	Preganglionic: pons; postganglionic: pterygopalatine ganglion and submandibular ganglion		Parasympathetic innervation to submandibular and sublingual salivary glands, lacrimal gland, and glands of nose and palate
Vestibulocochlear (CN VIII)				
Vestibular	Special sensory	Vestibular ganglion	Internal acoustic meatus	Vestibular sensation from semicircular ducts, utricle, and saccule related to position and movement of head
Cochlear	Special sensory	Spiral ganglion		Hearing from spiral organ
Glossopharyngeal (CN IX)	Branchial motor	Medulla	Jugular foramen	Motor to stylopharyngeus that assists with swallowing
	Visceral motor	Preganglionic: medulla; postganglionic: otic ganglion		Parasympathetic innervation to parotid gland
	Visceral sensory	Superior ganglion		Visceral sensation from parotid gland, carotid body and sinus, pharynx, and middle ear
	Special sensory	Inferior ganglion		Taste from posterior third of tongue
	General sensory	Inferior ganglion		Cutaneous sensation from external ear
Vagus (CN X)	Branchial motor	Medulla	Jugular foramen	Motor to constrictor muscles of pharynx, intrinsic muscles of larynx, and muscles of palate, except tensor veli palatini, and striated muscle in superior two-thirds of esophagus
	Visceral motor	Preganglionic: medulla; postganglionic: neurons in, on, or near viscera		Parasympathetic innervation to smooth muscle of trachea, bronchi, digestive tract, and cardiac muscle of heart
	Visceral sensory	Superior ganglion		Visceral sensation from base of tongue, pharynx, larynx, trachea, bronchi, heart, esophagus, stomach, and intestine
	Special sensory	Inferior ganglion		Taste from epiglottis and palate
	General sensory	Superior ganglion		Sensation from auricle, external acoustic meatus, and dura mater of posterior cranial fossa
Accessory (CN XI)				
Cranial root	Somatic motor	Medulla	Jugular foramen	Motor to striated muscles of soft palate, pharynx, and larynx via fibers that join CN X
Spinal root	Branchial motor	Spinal cord		Motor to sternocleidomastoid and trapezius
Hypoglossal (CN XII)	Somatic motor	Medulla	Hypoglossal canal	Motor to muscles of tongue (except palatoglossus)

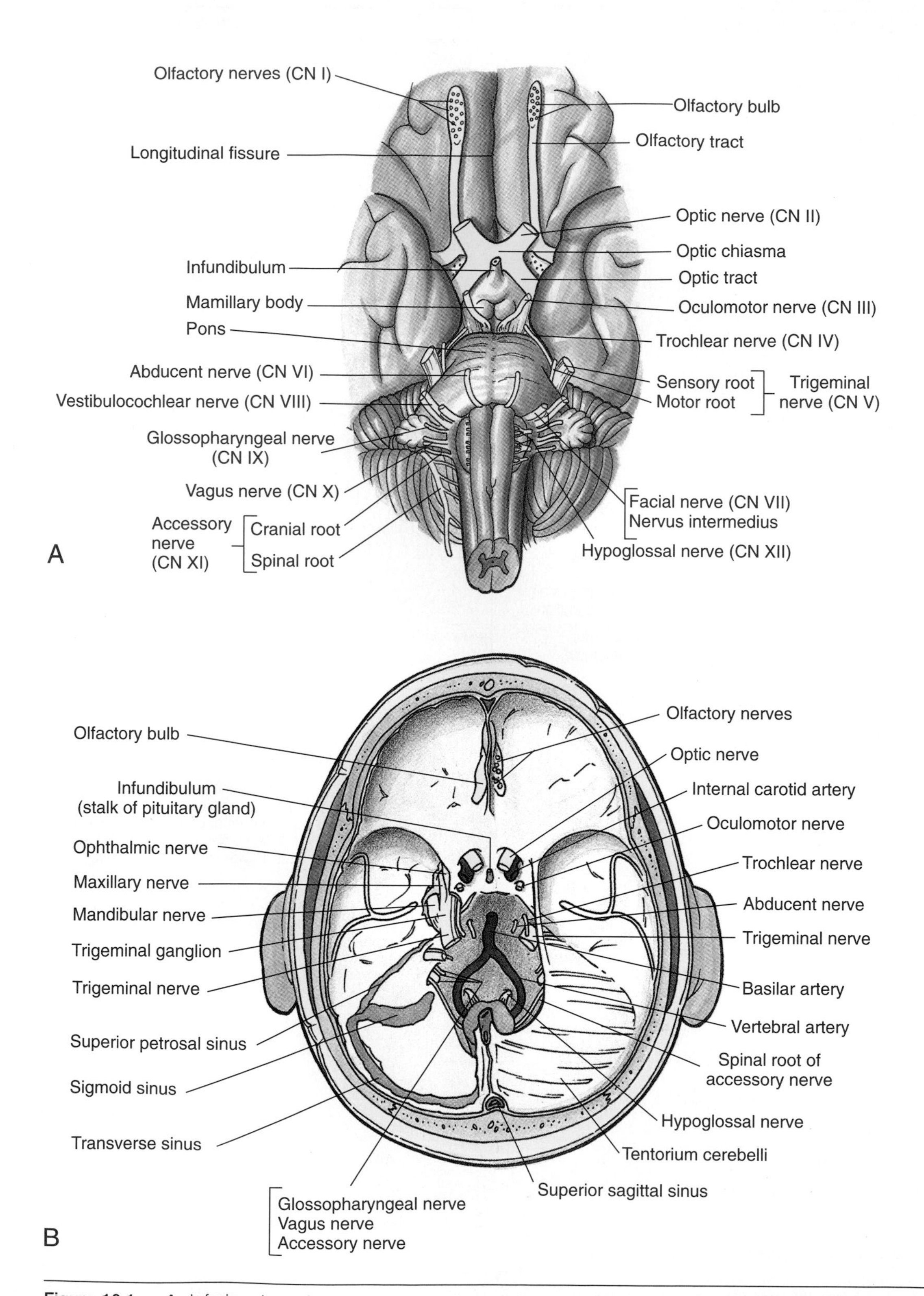

Figure 10.1. **A.** Inferior view of brain showing the cranial nerves. **B.** Interior of base of skull showing cranial nerves, dura mater, and blood vessels.

- Visceral sensory (general visceral afferent): receives visceral sensation from parotid gland, carotid body and sinus, middle ear, pharynx, larynx, trachea, bronchi, lungs, heart, esophagus, stomach, and intestines as far as the left colic flexure
- General sensory (general somatic afferent): receives general sensation from skin and mucous membranes, mainly via the trigeminal nerve (CN V) but also via CN VII, CN IX, and CN X
- Special sensory [special visceral afferent (taste and smell) and special somatic afferent (vision, hearing, and balance)]

Figure 10.1 and Table 10.1 summarize the cranial nerves. Refer to them as each nerve is considered.

Olfactory Nerve (CN I)

The olfactory nerve is the nerve of smell. The olfactory neurosensory cells are in the *olfactory epithelium* in the superior part of the lateral and septal walls of the nasal cavity (Fig. 10.2). The central processes of the bipolar *olfactory neurosensory cells* form about 20 bundles on each side that collectively form the olfactory nerves. They pass through foramina in the *cribriform plate* of the ethmoid bone, pierce the dura and arachnoid mater, and enter the *olfactory bulbs* in the anterior cranial fossa.

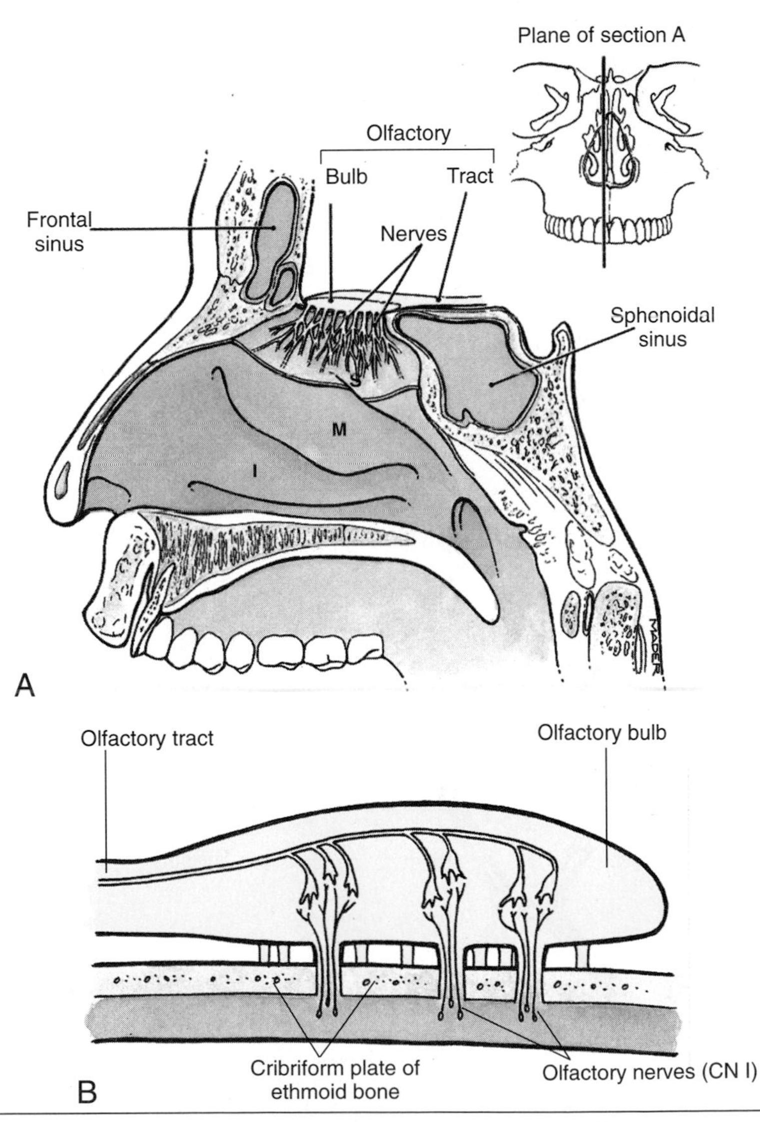

Figure 10.2. A. Olfactory area and passage of olfactory nerves through cribriform plate to end in olfactory bulb. *S*, superior concha; *M*, middle concha; *I*, inferior concha. **B.** Enlargement of olfactory area showing bundles of olfactory nerve fibers arising from olfactory epithelium, passing through cribriform plate, and synapsing with neurons in olfactory bulb.

An extension of the cranial meninges surrounds the olfactory nerve as it leaves the cribriform plate. Hence, there is a communication with the *subarachnoid space* that provides a route for loss of cerebrospinal fluid (cerebrospinal fluid rhinorrhea) when there is a fracture of the cribriform plate of the ethmoid that tears the meninges. This communication also provides a potential route for microorganisms from the nose to enter the cerebrospinal fluid. Fractures of the cribriform plate often pull the axons of the olfactory neurosensory cells from the olfactory bulb, producing *anosmia* (loss of smell). Unilateral anosmia may also result from a tumor of the meninges (*meningioma*) that compresses the olfactory bulb or tract.

Optic Nerve (CN II)

The optic nerve is the nerve of sight (Fig. 10.3). CN II is surrounded by the cranial meninges containing an extension of the subarachnoid space. CN II begins where the axons of the retinal ganglion cells pierce the sclera. The nerve passes posteromedially through the orbit and runs through the *optic canal* to the middle cranial fossa to join the *optic chiasma*.

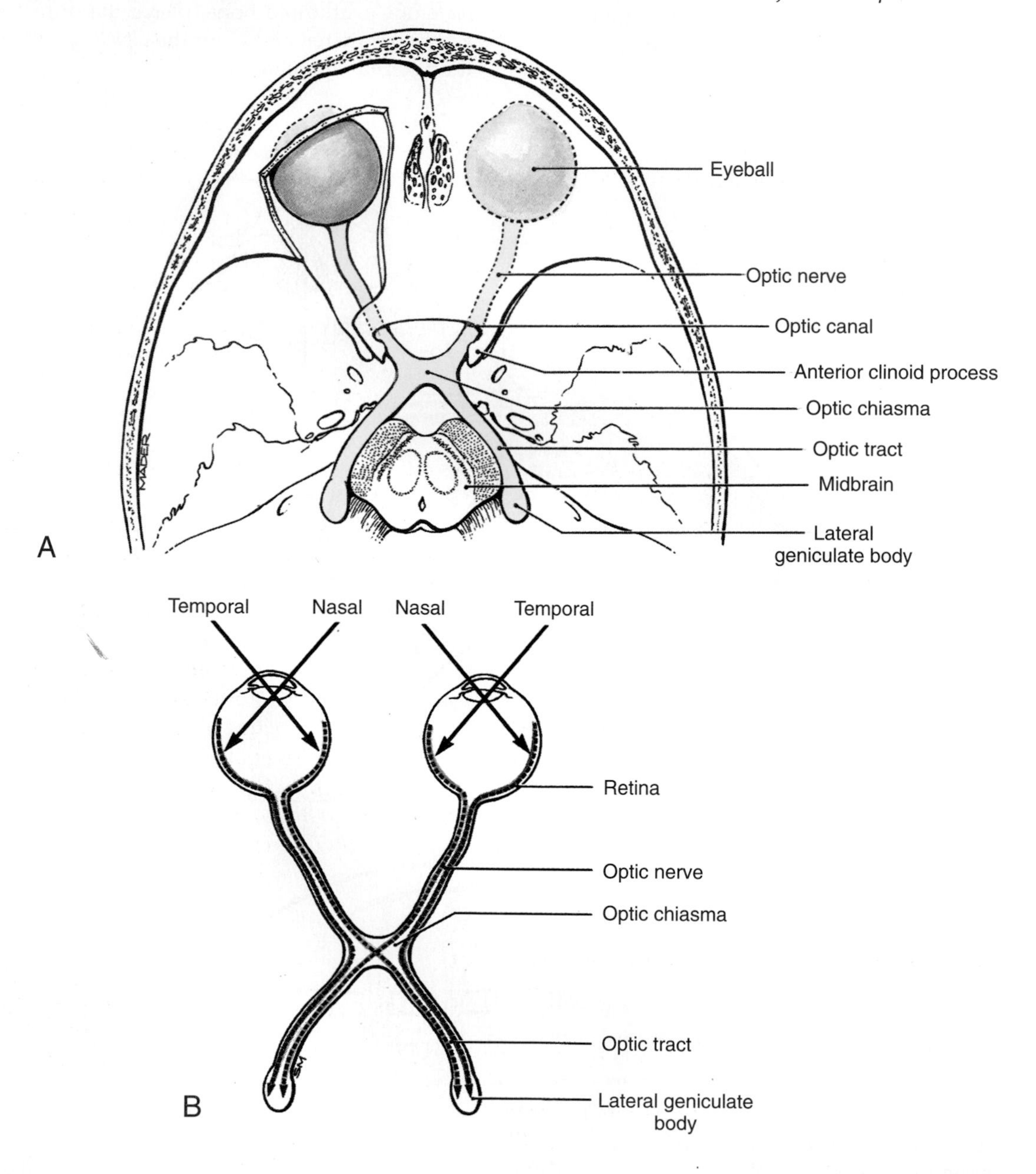

Figure 10.3. A. Anterior half of internal surface of skull showing optic nerves, optic chiasma, and optic tracts. **B.** Horizontal section through visual apparatus. *Arrows*, rays of light from nasal and temporal halves of person's field of vision.

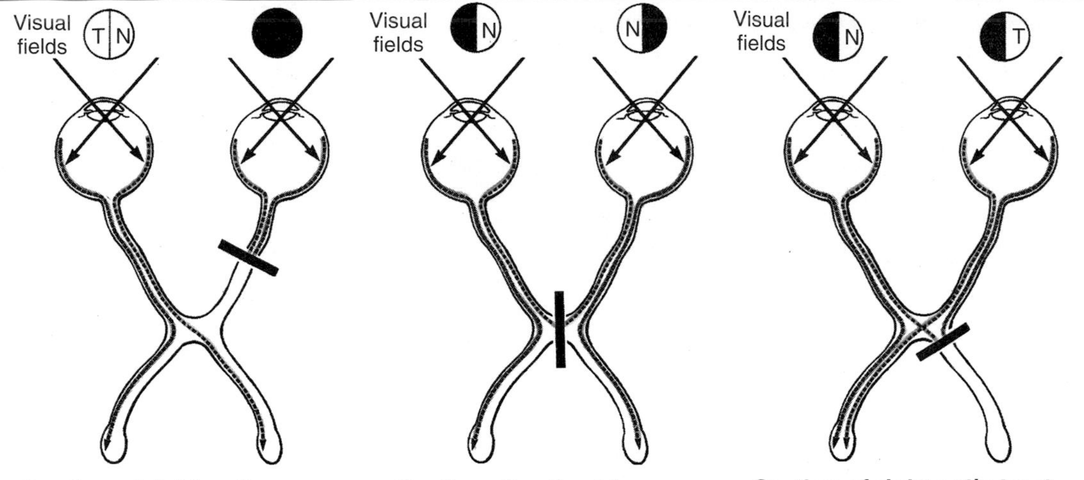

Because CN II is surrounded by an extension of the cranial meninges, an increase in cerebrospinal fluid pressure slows the return of venous blood, causing *edema of the retina*. This condition is apparent on ophthalmoscope examination as a swelling of the optic disc or papilla (*papilledema*). This is an indication of increased intracranial pressure.

Section of the right optic nerve results in blindness in the temporal (*T*) and nasal (*N*) visual fields of the right eye. Section of the right optic tract eliminates vision from the left temporal and right nasal visual fields, and section of the optic chiasma reduces peripheral vision that results in bitemporal hemianopsia.

Defects of vision caused by compression of the optic chiasma may result from tumors of the pituitary gland and berry aneurysms of the internal carotid or anterior cerebral arteries.

Posterior to the optic chiasma, the optic nerves continue as the *optic tracts*. Within the optic chiasma there is partial decussation of optic nerve fibers. Fibers from the nasal half of each retina cross to the opposite side, whereas those from the temporal half of each retina are uncrossed (Fig. 10.3*B*). Thus fibers from the right halves of both retinas form the right optic tract, and those from the left halves form the left optic tract. The decussation of nerve fibers in the chiasma results in the right optic tract conveying impulses from the left *visual field* and vice versa. Most fibers in the optic tracts terminate in the lateral *geniculate bodies* of the thalamus. From these nuclei axons are relayed to the *visual cortices* of the occipital lobes of the brain.

Oculomotor Nerve (CN III)

The oculomotor nerve is the chief motor nerve to the ocular muscles (Fig. 10.4). CN III emerges from the midbrain, pierces the dura, and runs in the lateral wall of the *cavernous sinus*. CN III leaves the skull through the *superior orbital fissure* and enters the orbit; within this fissure it divides into (*a*) a superior division that supplies the superior rectus and levator palpebrae superioris and (*b*) an inferior division that supplies the inferior and medial rectus and inferior oblique. The inferior division also carries autonomic fibers to the *ciliary ganglion* where the parasympathetic fibers synapse. The postganglionic fibers emerge from this ganglion and pass to the eyeball in the *short ciliary nerves* to supply the ciliary muscle (accommodation of lens) and sphincter pupillae (constriction of pupil).

Signs of complete division of CN III are

- Ptosis (drooping) of upper eyelid caused by paralysis of levator palpebrae superioris
- No pupillary reflexes
- Dilation of pupil caused by interruption of parasympathetic fibers to iris
- Eyeball abducted and directed slightly inferiorly because of unopposed actions of lateral rectus and superior oblique
- No accommodation of lens because of paralysis of ciliary muscle

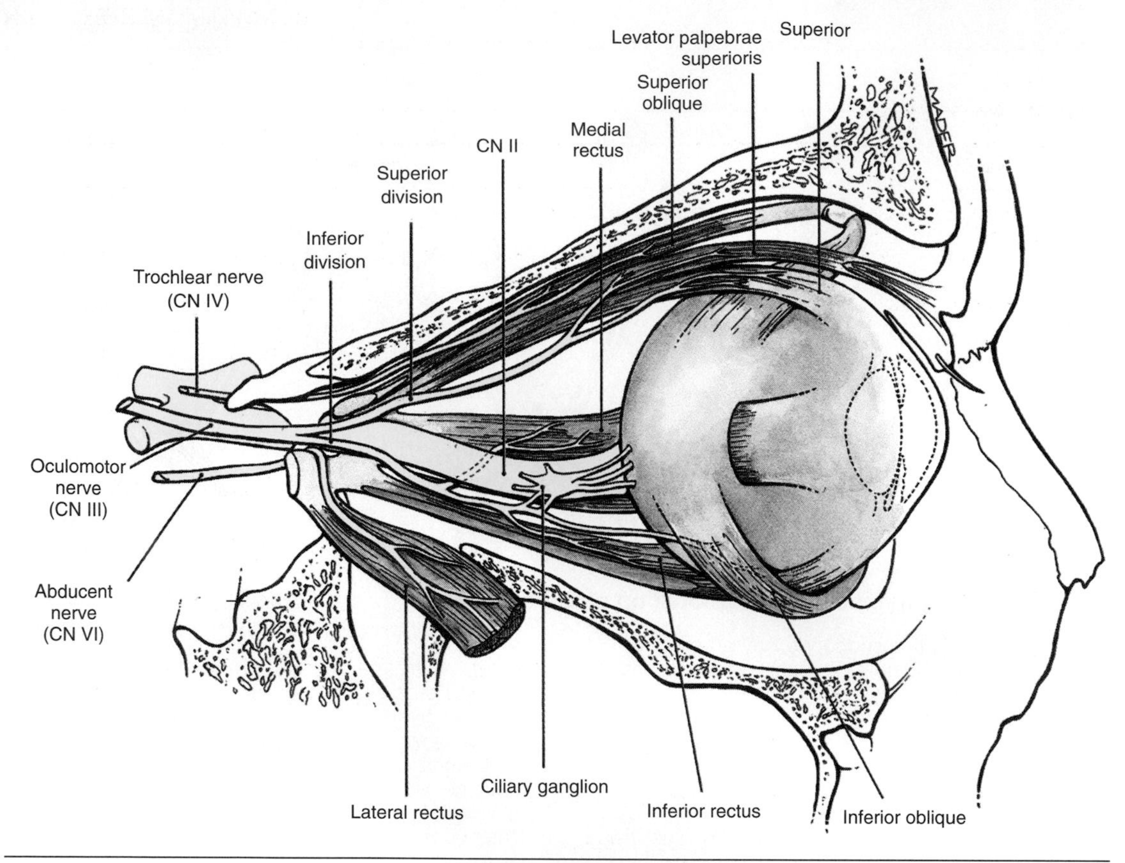

Figure 10.4. Overview of oculomotor (CN III), trochlear (CN IV), and abducent nerves (CN VI).

Trochlear Nerve (CN IV)

The trochlear nerve emerges from the dorsal surface of the midbrain, winds around the brainstem, pierces the dura, and passes anteriorly in the lateral wall of the cavernous sinus. CN IV leaves the cavernous sinus and passes through the superior orbital fissure into the orbit to supply the superior oblique (Fig. 10.4).

CN IV may be torn when there are severe head injuries because of its long intracranial course; however, it is rarely paralyzed alone. Damage to CN IV prevents the patient from looking inferolaterally with the affected eye. The characteristic sign of trochlear nerve injury is *diplopia* (double vision) when looking down, e.g., when going downstairs.

Trigeminal Nerve (CN V)

The trigeminal nerve is both motor and sensory. It emerges from the pons by a small motor root and a large sensory root (Fig. 10.5). CN V is motor to the muscles of mastication and is the principal general sensory nerve for the head (face, teeth, mouth, nasal cavity, and dura mater). Fibers in the sensory root are mainly axons of neurons in the *trigeminal ganglion*. The peripheral processes of these cells form the ophthalmic (CN

Trigeminal neuralgia (tic douloureux), the principal condition affecting the sensory root (part) of CN V, is usually characterized by attacks of excruciating pain in the area of distribution of the maxillary and/or mandibular divisions. The maxillary nerve (CN V^2) is the most frequently involved; the sudden attack is often set off by touching an especially sensitive facial area. Usually the cause of the neuralgia is undetected, but often there is inflammation of the petrous part of the temporal bone (osteitis) or an aberrant artery that lies close to the sensory root of CN V and compresses it. Another painful condition that may affect CN V is *herpes zoster* (p. 44).

SUMMARY OF CN V

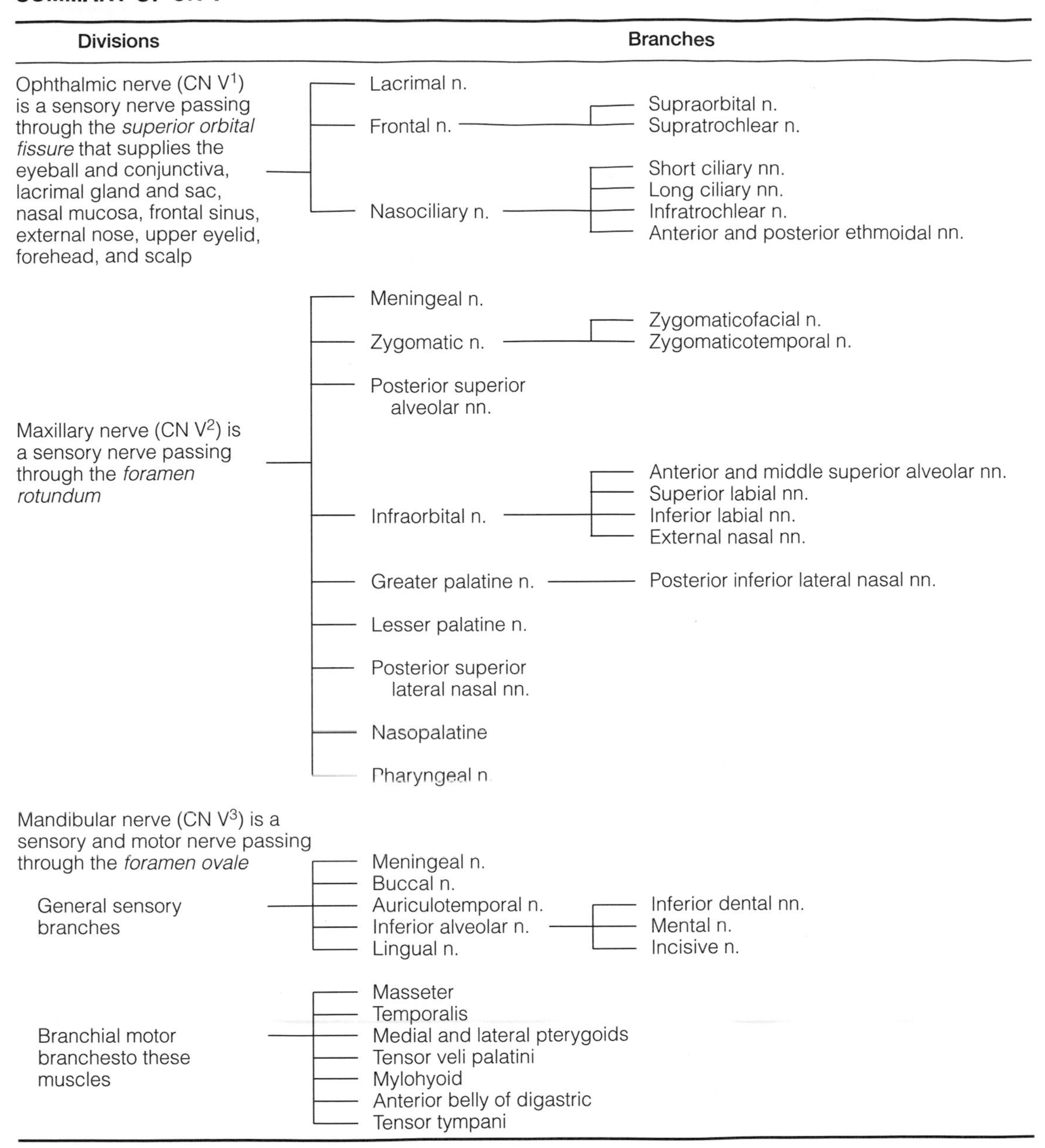

Divisions	Branches	
Ophthalmic nerve (CN V^1) is a sensory nerve passing through the *superior orbital fissure* that supplies the eyeball and conjunctiva, lacrimal gland and sac, nasal mucosa, frontal sinus, external nose, upper eyelid, forehead, and scalp	Lacrimal n.	
	Frontal n.	Supraorbital n. Supratrochlear n.
	Nasociliary n.	Short ciliary nn. Long ciliary nn. Infratrochlear n. Anterior and posterior ethmoidal nn.
Maxillary nerve (CN V^2) is a sensory nerve passing through the *foramen rotundum*	Meningeal n.	
	Zygomatic n.	Zygomaticofacial n. Zygomaticotemporal n.
	Posterior superior alveolar nn.	
	Infraorbital n.	Anterior and middle superior alveolar nn. Superior labial nn. Inferior labial nn. External nasal nn.
	Greater palatine n.	Posterior inferior lateral nasal nn.
	Lesser palatine n.	
	Posterior superior lateral nasal nn.	
	Nasopalatine	
	Pharyngeal n.	
Mandibular nerve (CN V^3) is a sensory and motor nerve passing through the *foramen ovale*		
General sensory branches	Meningeal n. Buccal n. Auriculotemporal n. Inferior alveolar n. Lingual n.	Inferior dental nn. Mental n. Incisive n. (branches of inferior alveolar n.)
Branchial motor branchesto these muscles	Masseter Temporalis Medial and lateral pterygoids Tensor veli palatini Mylohyoid Anterior belly of digastric Tensor tympani	

V^1) and maxillary (CN V^2) nerves and the sensory component of the mandibular nerve (CN V^3).

Abducent Nerve (CN VI)

The abducent (abducens) nerve emerges from the brainstem between the pons and medulla, pierces the dura, and passes through the cavernous sinus. CN VI enters the orbit through the superior orbital fissure and runs anteriorly to supply the lateral rectus that abducts the eye (Fig. 10.4).

CN VI has a long intracranial course and may be stretched when intracranial pressure rises, partly because of the sharp bend it makes over the crest of the petrous part of the temporal bone. A space-occupying lesion (e.g., brain tumor) may compress CN VI, causing paralysis. Complete paralysis causes medial deviation of the affected eye (i.e., it is fully adducted), and the patient cannot abduct the eye.

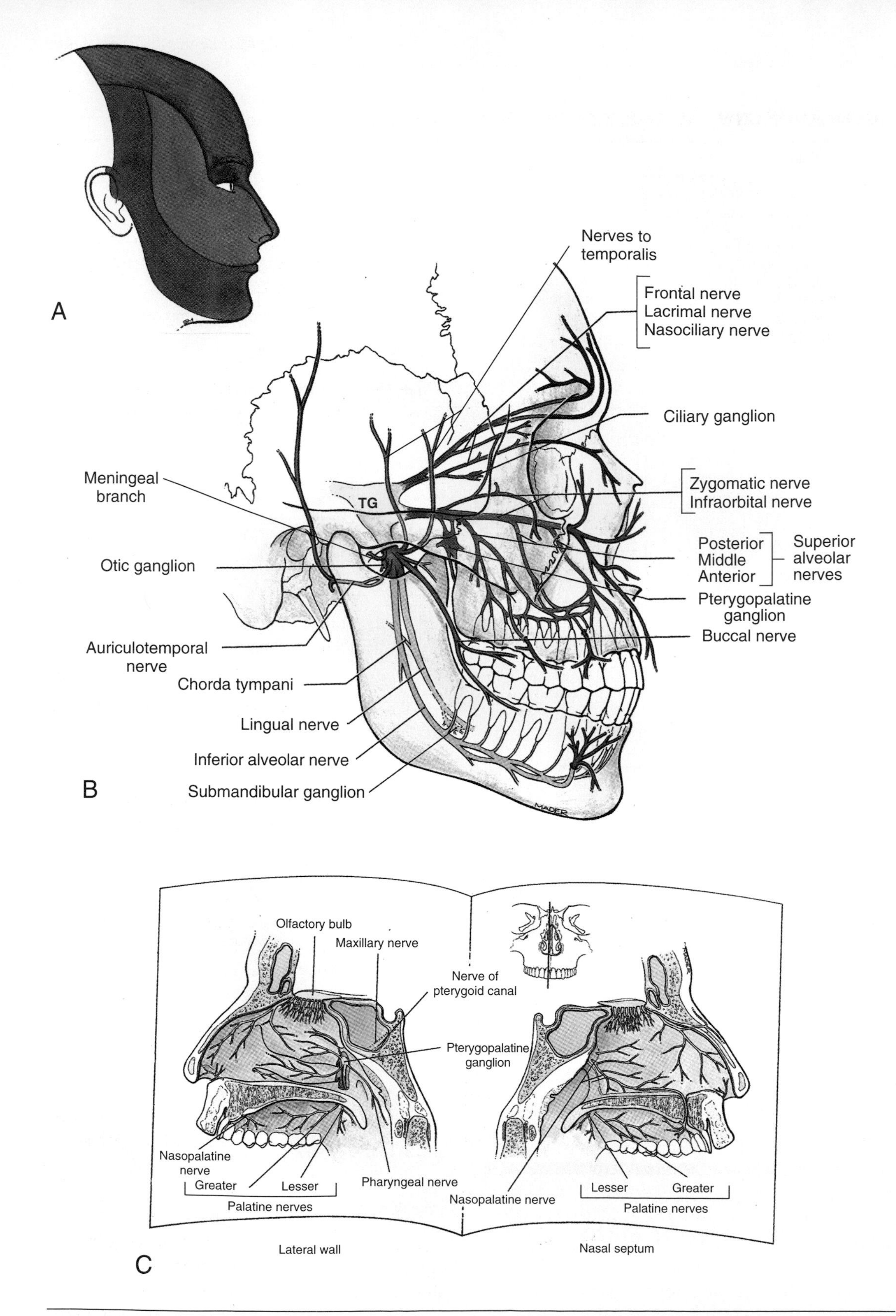

Figure 10.5. **A** and **B.** Distribution of trigeminal nerve (CN V). *Red*, ophthalmic nerve (CN V^1); *blue*, maxillary nerve (CN V^2); *green*, mandibular nerve (CN V^3); *TG*, trigeminal ganglion. **C.** Distribution of maxillary nerve (CN V^2) to walls of nasal cavity.

Facial Nerve (CN VII)

The facial nerve emerges from the junction of the pons and medulla; the larger motor root innervates the muscles of facial expression; the smaller root, *nervus intermedius*, carries taste, parasympathetic, and somatic sensory fibers (Figs. 10.1 and 10.6). During its course, CN VII traverses the posterior cranial fossa, the internal acoustic meatus, the facial canal in the temporal bone, and the parotid gland. At the medial wall of the tympanic cavity, the facial canal bends posteroinferiorly where the *geniculate ganglion* (sensory ganglion of CN VII) is located. Within the facial canal, CN VII gives rise to the *greater petrosal nerve*, the *nerve to the stapedius*, and the *chorda tympani nerve*. CN VII then emerges from the skull via the *stylomastoid foramen* and enters the parotid gland. Here it divides into the following terminal branches: posterior auricular, temporal, zygomatic, buccal, mandibular, and cervical.

SUMMARY OF CN VII

Branchial Motor

The terminal branches innervate the muscles of facial expression, occipitalis and auricular muscles, and posterior belly of digastric, stylohyoid, and stapedius muscles.

General Sensory

Some fibers from the geniculate ganglion supply a small area of skin around the external acoustic meatus.

Parasympathetic (Visceral Motor) to Lacrimal Gland

The *greater petrosal nerve* arises from CN VII at the geniculate ganglion and emerges from the superior surface of the petrous part of the temporal bone to enter the middle cranial fossa;

↓

The greater petrosal nerve joins the *deep petrosal nerve* (sympathetic) at the foramen lacerum to form the nerve of the pterygoid canal;

↓

The *nerve of the pterygoid canal* travels through the pterygoid canal and enters the pterygopalatine fossa;

↓

Parasympathetic fibers from this nerve in the pterygopalatine fossa synapse in the *pterygopalatine ganglion;*

↓

Postganglionic parasympathetic fibers from the pterygopalatine ganglion innervate the lacrimal gland via the zygomatic branch of CN V^2.

Parasympathetic (Visceral Motor) to Submandibular and Sublingual Glands

The *chorda tympani* branch arises from CN VII just superior to the stylomastoid foramen;

↓

The chorda tympani crosses the tympanic membrane medial to the handle of the malleus;

↓

The chorda tympani passes through the petrotympanic fissure of the temporal bone to join the lingual nerve; parasympathetic fibers of the chorda tympani synapse in the submandibular ganglion;

↓

Postsynaptic fibers from this ganglion innervate the sublingual and submandibular salivary glands.

Taste (Special Sensory)

Fibers of the chorda tympani nerve join the lingual nerve to supply taste sensation from the anterior two-thirds of the tongue.

The main features of parasympathetic ganglia associated with the facial nerve and other cranial nerves are summarized in Table 10.2.

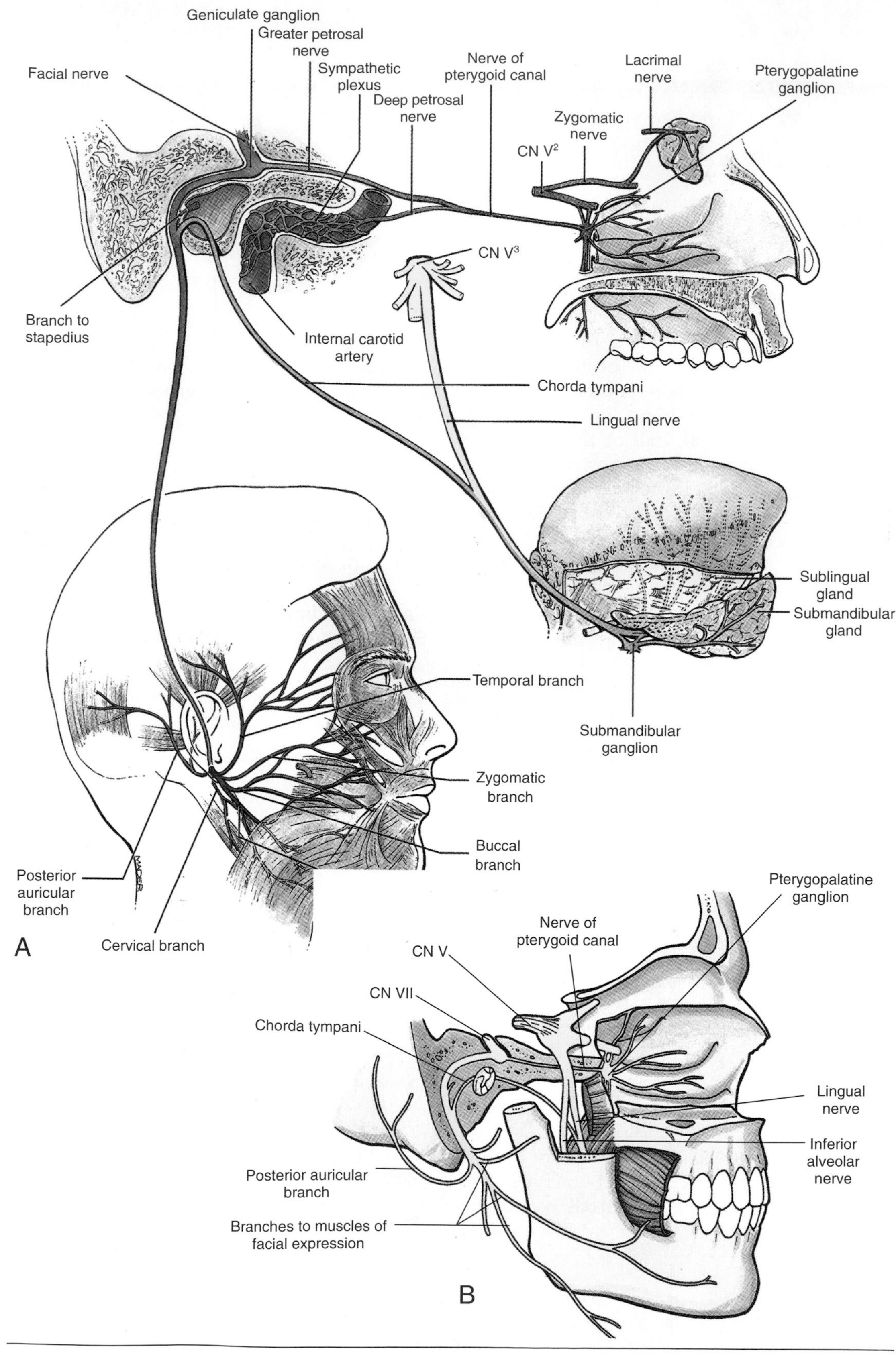

Figure 10.6. **A.** Distribution of facial nerve (CN VII). **B.** Relationships of chorda tympani and nerve of pterygoid canal.

Parasympathetic fibers synapse in these ganglia, whereas sympathetic and other fibers pass through them.

CN VII is the most frequently paralyzed nerve of all cranial nerves. Its passage through the facial canal makes it vulnerable to compression when viral infection produces inflammation and swelling of the nerve and its surrounding connective tissue just before it emerges from the stylomastoid foramen. *Bell's palsy* is the common disorder resulting from a facial nerve lesion. The paralysis usually disappears in a few weeks; however, in severe cases recovery (often incomplete) takes about 3 months.

A lesion of CN VII proximal to the origin of the chorda tympani will result in loss of taste in the anterior part of the tongue in addition to facial paralysis.

Vestibulocochlear Nerve (CN VIII)

The vestibulocochlear nerve emerges from the junction of the pons and medulla and enters the *internal acoustic meatus* with the facial nerve. Here CN VIII separates into the vestibular and cochlear nerves (Fig. 10.7). The vestibular fibers, concerned with equilibrium, are axons of neurons in the *vestibular ganglion*; the peripheral processes enter the maculae of the utricle and saccule and the ampullae of the semicircular ducts. The cochlear fibers, concerned with hearing, are axons of neurons in the *spiral ganglion*; the peripheral processes enter the spiral organ (of Corti).

Although the vestibular and cochlear nerves are essentially independent, peripheral lesions often produce concurrent clinical effects because of their close relationship. Hence, lesions of CN VIII may cause tinnitus (ringing or buzzing in the ears), vertigo (loss of balance), and impairment or loss of hearing. Central lesions of CN VIII may involve either the cochlear or vestibular divisions.

There are two kinds of deafness:

- Conductive deafness involving the external or middle ear [e.g., caused by otitis media (inflammation in middle ear)]
- Sensorineural deafness caused by disease in the cochlea or in the pathway from the cochlea to the brain

An *acoustic neuroma*, a slow-growing benign tumor of Schwann cells, begins around the vestibular nerve while it is in the internal acoustic meatus; an early symptom of this lesion is loss of hearing.

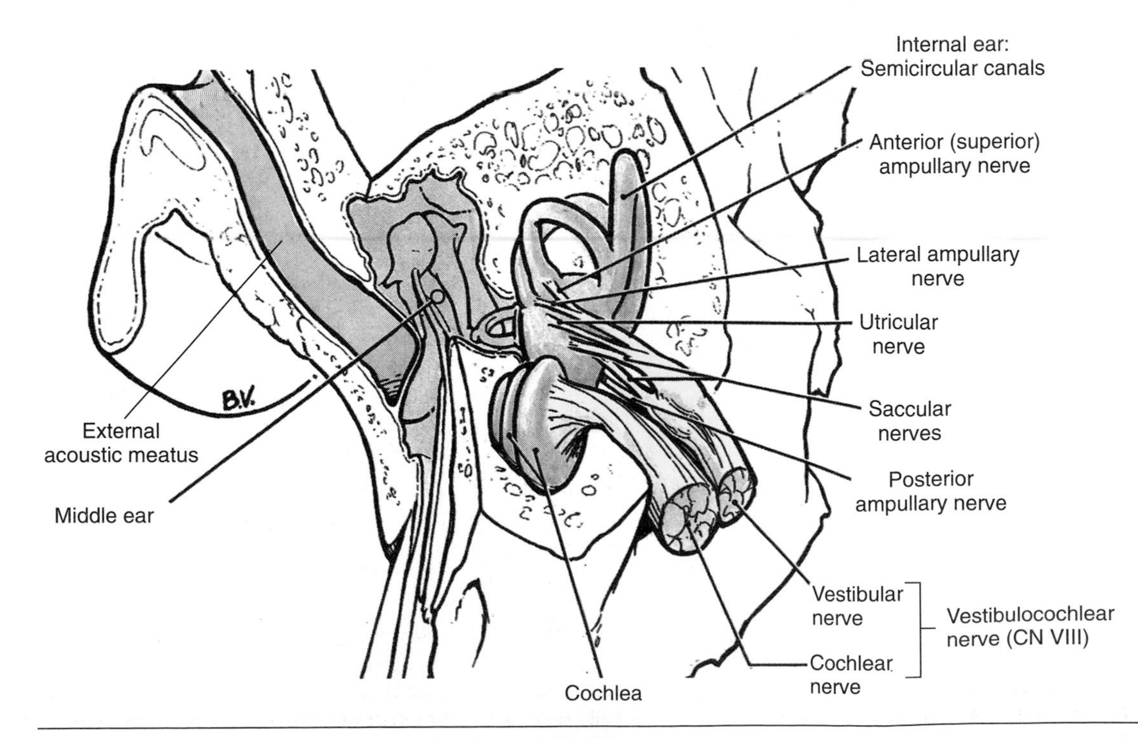

Figure 10.7. Distribution of vestibulocochlear nerve (CN VIII).

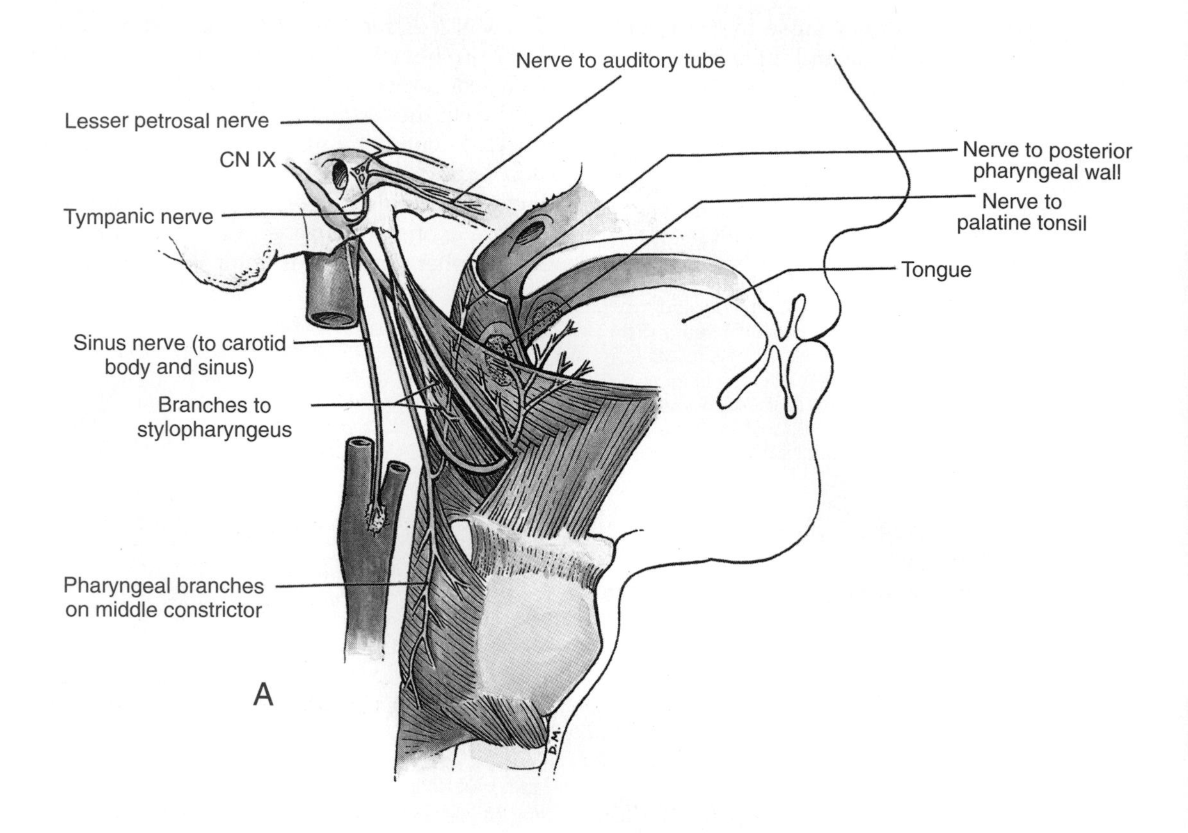

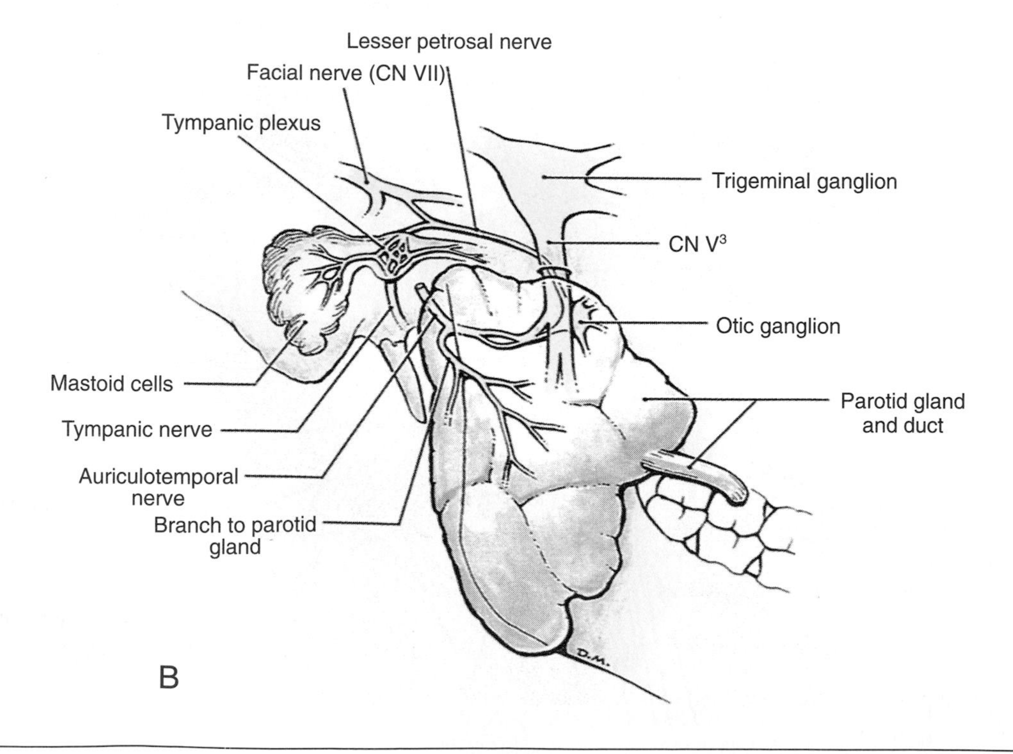

Figure 10.8. **A.** Distribution of glossopharyngeal nerve (CN IX). **B.** Parasympathetic component of CN IX supplies secretory fibers through otic ganglion to parotid gland.

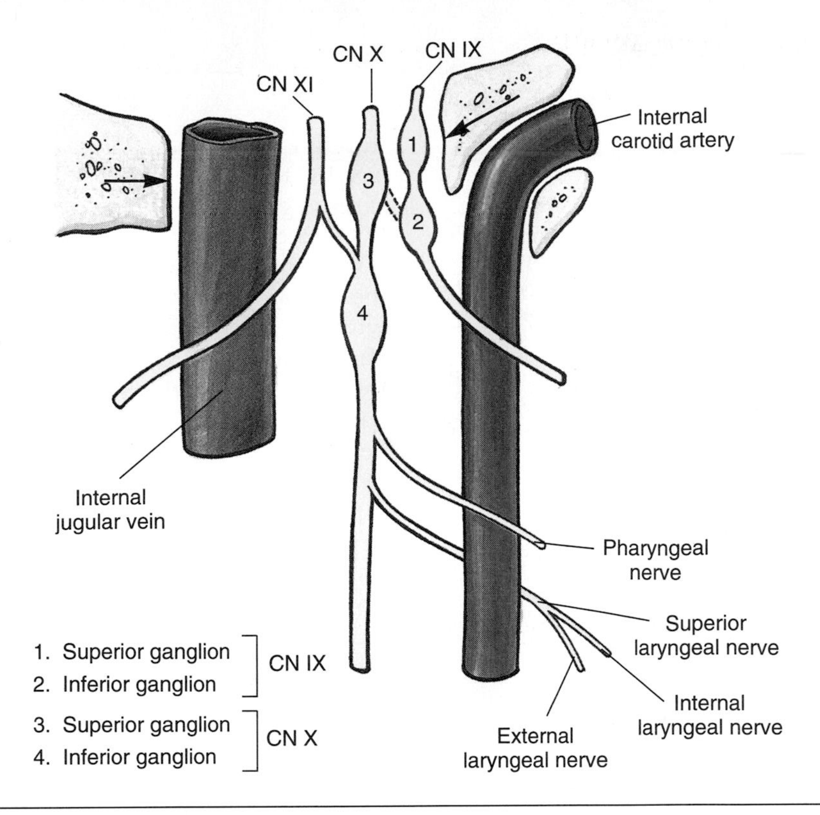

Figure 10.9. Relationship of internal jugular vein and CN IX, X, and XI at jugular foramen. *Arrows*, margins of foramen. Note relationship of these nerves to internal carotid artery.

Glossopharyngeal Nerve (CN IX)

The glossopharyngeal nerve emerges from the medulla and passes anterolaterally to leave the skull through the *jugular foramen* (Figs. 10.8*A* and 10.9). At this foramen it has superior and inferior ganglia that contain cell bodies for the afferent components of the nerve. CN IX follows the stylopharyngeus muscle and passes between the superior and middle constrictor muscles of the pharynx to reach the oropharynx and tongue. It contributes to the *pharyngeal plexus* of nerves.

SUMMARY OF CN IX

The glossopharyngeal nerve is afferent from the tongue and pharynx and efferent to the stylopharyngeus and parotid gland.

Sensory (General Visceral)

The branches are

- Tympanic nerve
- Carotid sinus nerve to carotid sinus and carotid body
- Nerves to mucosa of tongue and oropharynx, palatine tonsil, soft palate, and posterior third of tongue

Taste (Special Sensory)

Branches to posterior third of tongue.

Branchial Motor

Nerve to stylopharyngeus

Parasympathetic (Visceral Motor)

The *tympanic nerve* arises from CN IX before it emerges from the jugular foramen;

↓

The tympanic nerve enters the middle ear via a foramen in the petrous temporal bone;

↓

The tympanic nerve joins the *tympanic plexus* on the promontory of the middle ear;

↓

From the tympanic plexus the tympanic nerve continues as the lesser petrosal nerve;

↓

The *lesser petrosal nerve* is joined by a branch of the facial nerve from the geniculate ganglion and emerges from the superior surface of the petrous part of the temporal bone to enter the middle cranial fossa;

↓

The lesser petrosal nerve exits the skull through the foramen ovale;

↓

Parasympathetic fibers synapse in *otic ganglion;*

↓

Postsynaptic fibers from otic ganglion innervate parotid gland via auriculotemporal nerve (CN V^3).

Vagus Nerve (CN X)

The vagus nerve arises by a series of rootlets from the medulla and leaves the skull through the *jugular foramen* in company with CN IX and CN XI (Fig. 10.9). CN X has a *superior ganglion* in this foramen that is mainly concerned with the general sensory component of the nerve. Inferior to the foramen there is an *inferior ganglion* that is concerned with the visceral sensory component of the nerve. In the region of the superior ganglion there are connections with CN IX, CN XI, and the superior cervical ganglion. It continues inferiorly in the *carotid sheath* to the root of the neck. The course of CN X in the thorax differs on the two sides (p. 70). The vagus supplies branches to the heart, bronchi, and lungs (Fig. 10.10). In the abdomen, the anterior and posterior vagal trunks break up into branches to the esophagus, stomach, and intestinal tract as far as the left colic flexure.

Isolated lesions of CN X are uncommon. Paralysis of the recurrent laryngeal nerves may result from cancer of the larynx and thyroid gland and from injury during surgery on the thyroid gland, neck, esophagus, heart, or lungs. Because of its longer course, lesions of the left recurrent laryngeal nerve are more common than those of the right. Proximal lesions of the vagus nerve affect the pharyngeal and superior laryngeal nerves, as well as the recurrent laryngeal nerves, causing difficulty in swallowing and speaking.

SUMMARY OF CN X

Divisions	Branches
Arises by a series of rootlets from medulla	
Leaves skull through jugular foramen	Meningeal branch to dura mater Auricular n
Enters carotid sheath and continues to root of neck	Pharyngeal nn. Superior laryngeal nn. Right recurrent laryngeal n. Cardiac nn.
Passes through superior thoracic aperture into thorax	Left recurrent laryngeal n. Cardiac nn. Pulmonary branches to bronchi and lungs Esophageal nn.
Passes through esophageal hiatus in diaphragm and enters abdomen	Esophageal branches Gastric branches Pancreatic branches Branches to gallbladder Branches to intestine as far as left colic flexure

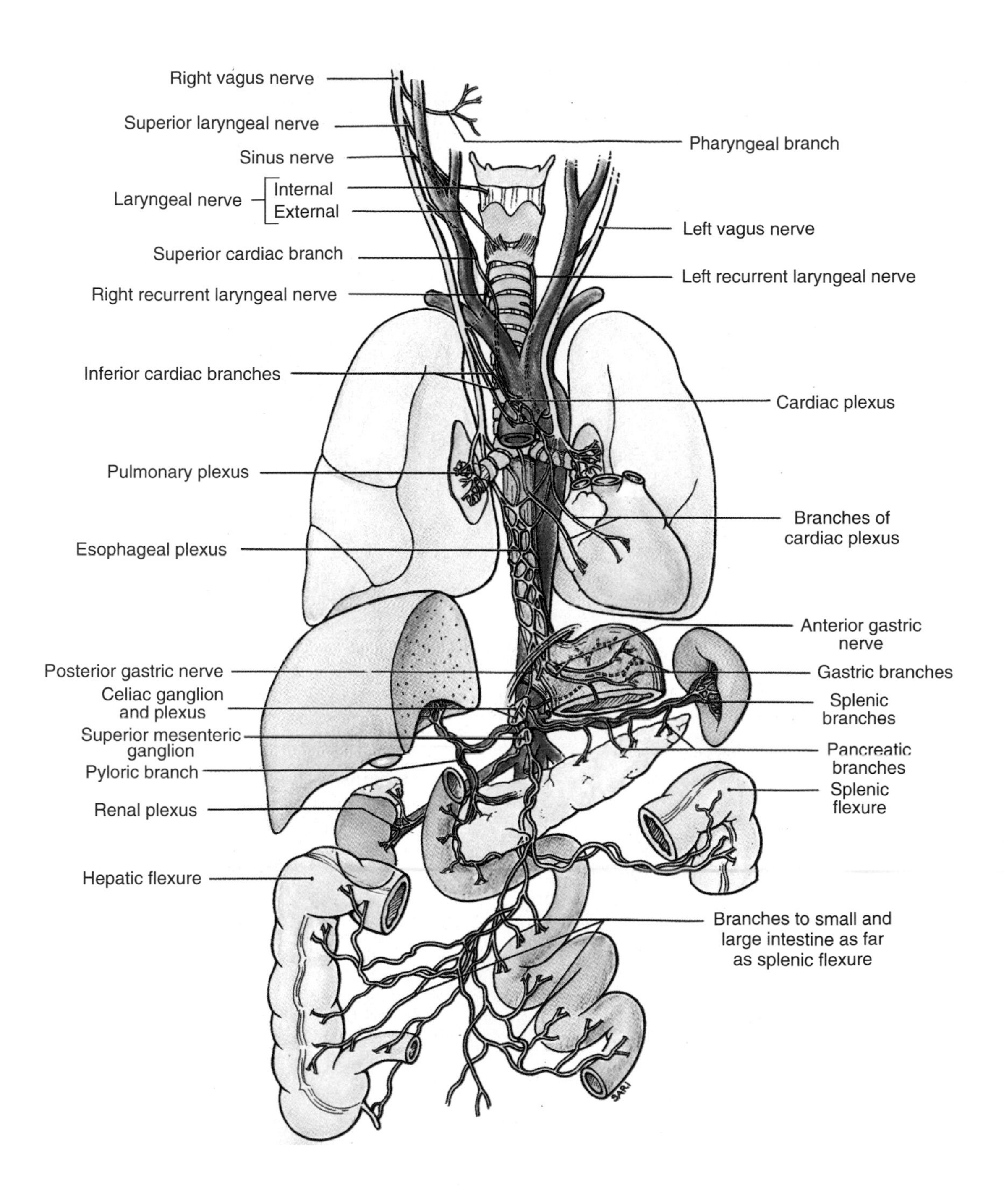

Figure 10.10. Distribution of vagus nerve (CN X).

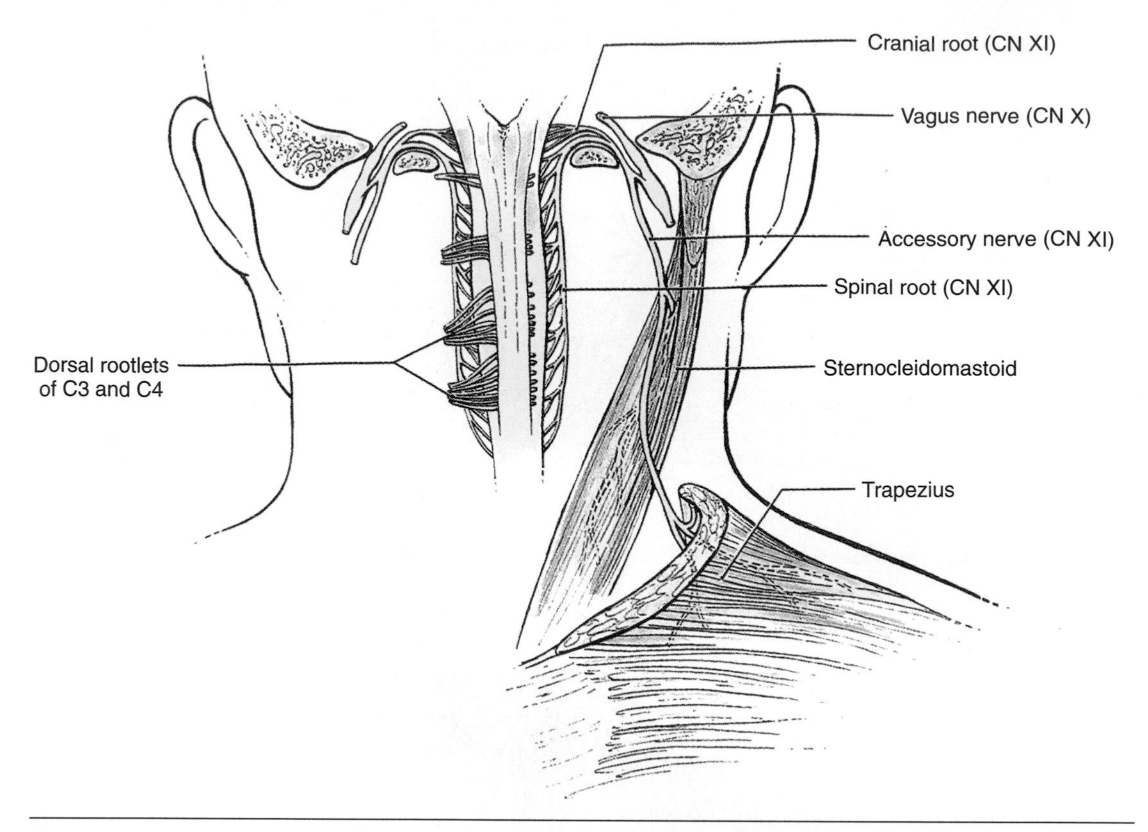

Figure 10.11. Distribution of accessory nerve (CN XI).

Accessory Nerve (CN XI)

The accessory nerve has cranial and spinal roots (Fig. 10.11). The *cranial root* arises from the medulla by a series of rootlets, and the *spinal root* emerges as a series of rootlets from the first five cervical segments of the spinal cord. The cranial and spinal roots join as they pass through the *jugular foramen* and then separate. The cranial root joins the vagus, and its fibers are distributed by vagal branches to striated muscle of the soft palate, pharynx, larynx, and esophagus. The spinal root descends to supply the sternocleidomastoid and trapezius.

Hypoglossal Nerve (CN XII)

The hypoglossal nerve arises by several rootlets from the medulla and leaves the skull through the *hypoglossal canal*. It passes inferolaterally to the angle of the mandible and then curves anteriorly to enter the tongue (Fig. 10.12). CN XII ends in many branches that supply all extrinsic and intrinsic muscles of the tongue, except the palatoglossus. The hypoglossal nerve is joined in the hypoglossal canal by the superior division of C1 nerve. CN XII has the following branches:

- Meningeal branch returns to skull through hypoglossal canal and innervates dura on floor and posterior wall of posterior cranial fossa
- Descending branch joins ansa cervicalis to supply infrahyoid muscles
- Terminal branches to the styloglossus, hyoglossus, genioglossus muscles, and intrinsic muscles of tongue

Injury to CN XII paralyses the ipsilateral half of the tongue. After some time the tongue atrophies, making it appear shrunken and wrinkled. When the tongue is protruded, its tip deviates toward the paralyzed side because of the unopposed action of the genioglossus in the normal side of the tongue.

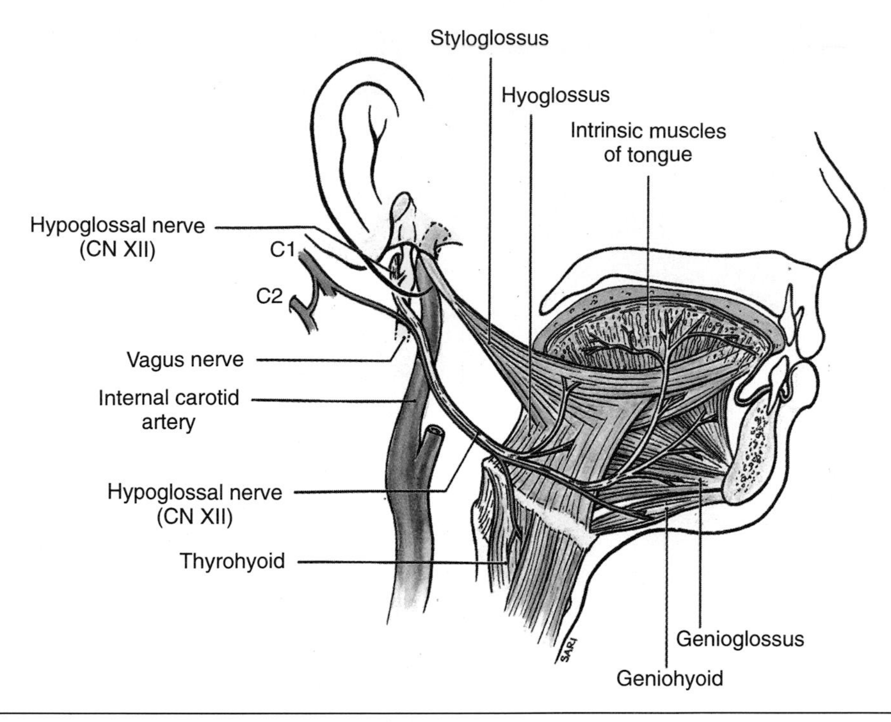

Figure 10.12. Distribution of hypoglossal nerve (CN XII).

Table 10.2
Parasympathetic Ganglia Associated with CN III, CN V, CN VII, and CN IX

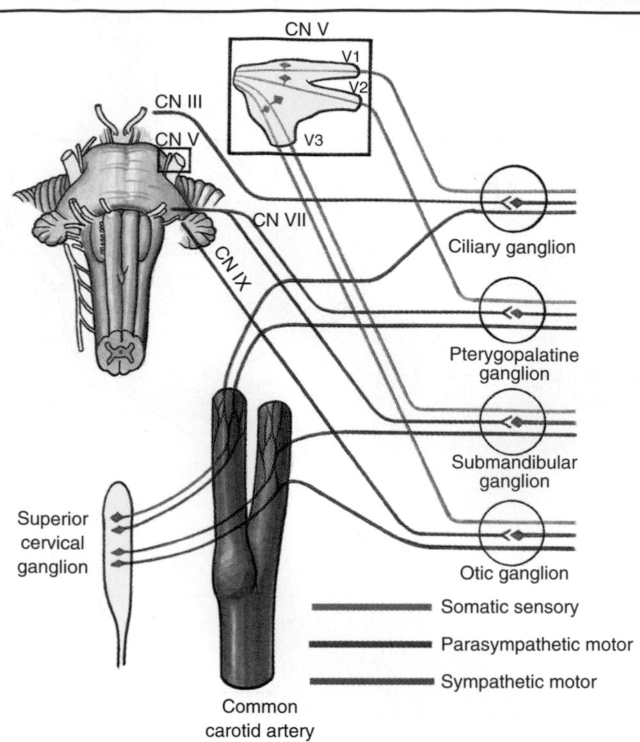

Ganglion	Location	Parasympathetic Root	Sympathetic Root	Main Distribution
Ciliary	Located between optic nerve and lateral rectus, close to apex of orbit	Inferior branch of oculomotor nerve (CN III)	Branch from internal carotid plexus in cavernous sinus	Parasympathetic postganglionic fibers from ciliary ganglion pass to ciliary muscle and sphincter pupillae of iris; sympathetic postganglionic fibers from superior cervical ganglion pass to dilator pupillae and blood vessels of eye
Pterygopalatine	Located in pterygopalatine fossa where it is attached by pterygopalatine branches of maxillary nerve; located just anterior to opening of pterygoid canal and inferior to CN V^2	Greater petrosal nerve from facial nerve (CN VII)	Deep petrosal nerve, a branch of internal carotid plexus that is continuation of postsynaptic fibers of cervical sympathetic trunk; fibers from superior cervical ganglion pass through pterygopalatine ganglion and enter branches of CN V^2	Parasympathetic postganglionic fibers from pterygopalatine ganglion innervate lacrimal gland via zygomatic branch of CN V^2; sympathetic postganglionic fibers from superior cervical ganglion accompany those branches of pterygopalatine nerve that are distributed to the nasal cavity, palate, and superior part of the pharynx
Otic	Located between tensor veli palatini and mandibular nerve (CN V^3); lies inferior to foramen ovale sphenoid bone	Tympanic nerve from glossopharyngeal nerve (CN IX); from tympanic plexus tympanic nerve continues as lesser petrosal nerve	Fibers from superior cervical ganglion come from plexus on middle meningeal artery	Parasympathetic postganglionic fibers from otic ganglion are distributed to parotid gland via auriculotemporal nerve (branch of CN V^3); sympathetic postganglionic fibers from superior cervical ganglion pass to parotid gland and supply its blood vessels

Table 10.2. *Continued*

Ganglion	Location	Parasympathetic Root	Sympathetic Root	Main Distribution
Submandibular	Suspended from lingual nerve by two short roots; lies on surface of hyoglossus muscle inferior to submandibular duct	Parasympathetic fibers join facial nerve (CN VII) and leave it in its chorda tympani branch, which unites with lingual nerve	Sympathetic fibers from superior cervical ganglion come from the plexus on facial artery	Postganglionic parasympathetic fibers from submandibular ganglion are distributed to the sublingual and submandibular glands; sympathetic fibers supply sublingual and submandibular glands and appear to be secretomotor

SUGGESTED READINGS

Agur AMR. Grant's atlas of anatomy, 9th ed. Baltimore: Williams & Wilkins, 1991.

Barr ML, Kiernan JA. The human nervous system: an anatomical viewpoint, 6th ed. Philadelphia: Lippincott, 1993.

Bergman RA, Thompson SA, Afifi AK, Saadeh FA. Compendium of human anatomic variation: text, atlas, and world literature. Baltimore: Urban & Schwarzenberg, 1988.

Bertram EG, Moore KL. An atlas of the human brain and spinal cord. Baltimore: Williams & Wilkins, 1982.

Cormack DH. Essential histology. Philadelphia: Lippincott, 1993.

Davies MF, Anderson RH, Becker AE. The conduction system of the heart. London: Butterworths, 1983.

Devinsky O, Feldman E. Examination of the cranial and peripheral nerves. New York: Churchill Livingstone, 1988.

Ellis H. Clinical anatomy, 8th ed. Oxford: Blackwell, 1993.

Fitzgerald MJT. Neuroanatomy: basic and clinical, 2nd ed. London: Baillière-Tindall, 1992.

Fujimura I, De Souza RR, Ferraz De Carvalho CA, Rodriguez AJ Jr. A method for locating the marginal branch of the facial nerve in the neck. Clin Anat 1990;3:143–147.

Ger R. Surgical anatomy of hepatic venous system. Clin Anat 1988;1:15–22.

Ger R, Evans JT. Tracheostomy: an anatomico-clinical review. Clin Anat 1993;6:337.

Griffith HW. Complete guide to sports injuries. Los Angeles: Price/Stern/Sloan, 1986.

Haagensen CD. Diseases of the breast, 3rd ed. Philadelphia: Saunders, 1986.

Haines DE. Neuroanatomy: an atlas of structures, sections, and systems, 4nd ed. Baltimore: Williams & Wilkins, 1995.

Haines DE. On the question of a subdural space. Anat Rec 1990;230:3–21.

Hew E. Anesthesia. In: Gross A, Gross P, Langer B, eds. Surgery. A complete guide for patients and their families. Toronto: Harper & Collins, 1989.

Holinger LD. Tracheotomy. In: Raffensperger JG, ed. Swenson's pediatric surgery, 5th ed. Norwalk CT: Appleton & Lange, 1990.

Jenkins DB. Functional anatomy of the limbs and back, 6th ed. Philadelphia: Saunders, 1990.

Kaplan LJ, Bellows CF, Whitman GJR, Barnes AU. Coexistent diaphragmatic herniation and eventration: embryoloic rationale for therapeutic interventions. Clin Anat 1994;7:143–151.

Mace SE. Cricothyrotomy. In: Roberts JR, Hedges JR. Clinical procedures in emergency medicine, 2nd ed. Philadelphia: Saunders, 1991.

McKee NH, Fish JS, Manktelow RT, McAvoy GV, Young S, Zuker RM. Gracilis muscle anatomy as related to function of a free functioning muscle transplant. Clin Anat 1990:3:87–92.

Moore KL. Clinically oriented anatomy, 3rd ed. Baltimore: Williams & Wilkins, 1992.

Moore KL, Persaud TVN. The developing human: clinical oriented embryology, 5th ed. Philadelphia: Saunders, 1993.

Moore KL, Persaud TVN, Shiota K. Color atlas of clinical embryology. Philadelphia: Saunders, 1994.

Raffensperger JG. Congenital cysts and sinuses of the neck. In: Raffensperger RG, ed. Swenson's pediatric surgery, 5th ed. Norwalk CT: Appleton & Lange, 1990.

Rowland LP, ed. Merritt's textbook of neurology, 9th ed. Baltimore: Williams & Wilkins, 1995.

Sabiston DC Jr, Lyerly HK. Essentials of surgery, 2nd ed. Philadelphia: Saunders, 1994.

Salter RB. Textbook of disorders and injuries of the musculoskeletal system, 2nd ed. Baltimore: Williams & Wilkins, 1983.

Scott-Connor C, Dawson DL. Operative anatomy. Philadelphia: Lippincott, 1993.

Skandalakis JE, Gray SW, Rowe JS. Anatomical complications in general surgery. New York: McGraw-Hill, 1983.

Skandalakis JE, Skandalakis PN, Skandalakis LJ. Surgical anatomy and technique: a pocket manual. New York: Springer-Verlag, 1995.

Solomon J, Rangecroft L. Thyroglossal duct lesions in children. J Pediatr Surg 1984;19:555.

Swartz MH. Textbook of physical diagnosis, 2nd ed. Philadelphia: Saunders, 1994.

Williams PL, Warwick R, Dyson M, Bannister LH. Gray's anatomy, 37th ed. New York: Churchill Livingstone, 1989.

Willms JL, Schneiderman H, Algranati PS. Physical diagnosis: bedside evaluation of diagnosis and function. Baltimore: Williams & Wilkins, 1994.

Wilson-Pauwels L, Akesson EJ, Stewart PA. Cranial nerves: gross anatomy and clinical comments. St Louis: Mosby, 1988:3

Woodburne RT, Burkel WE. Essentials of human anatomy, 9th ed. New York: Oxford University Press, 1994.

FIGURE AND TABLE CREDITS

Figures

Liberal use has been made of illustrations from *Grant's Atlas of Anatomy* and *Clinically Oriented Anatomy*:

Agur AMR. Clinically oriented anatomy, 3rd ed. Baltimore: Williams & Wilkins, 1992.

Moore KL. Grant's atlas of anatomy, 9th ed. Baltimore: Williams & Wilkins, 1991.

Figure 2.9. Courtesy of Drs. DE Sanders, S Herman, and EL Lansdown, Department of Radiology, University of Toronto, Toronto, Ontario, Canada.

Figure 2.13. Courtesy of Dr. EL Lansdown, Professor of Radiology, University of Toronto, Toronto, Ontario, Canada.

Figure 2.17. Courtesy of Dr. W Kucharczyk, Clinical Director of Tri-Hospital Magnetic Resonance Centre, Toronto, Ontario, Canada.

Figure 2.19. Courtesy of Dr. W Kucharczyk, Clinical Director of Tri-Hospital Resonance Centre, Toronto, Ontario, Canada.

Figure 3.7. Courtesy of Dr. EL Lansdown, Professor of Radiology, University of Toronto, Toronto, Ontario, Canada.

Figure 3.11. Courtesy of Dr. J Heslin, Assistant Professor of Anatomy, University of Toronto, Toronto, Ontario, Canada.

Figure 3.14. Courtesy of Dr. GB Habler, Assistant Professor of Medicine, University of Toronto, Toronto, Ontario, Canada.

Figure 3.24. Courtesy of Dr. W Kucharczyk, Clinical Director of Tri-Hospital Resonance Centre, Toronto, Ontario, Canada.

Figure 3.25. Courtesy of Dr. AM Arenson, Assistant Professor of Radiology, University of Toronto, Toronto, Ontario, Canada.

Figure 3.26. Courtesy of Dr. Tom White, Department of Radiology, The Health Sciences Center, The University of Tennessee, Memphis, TN.

Figure 4.11. Courtesy of Dr. AM Arenson, Assistant Professor of Radiology, University of Toronto, Toronto, Canada.

Figure 4.14. B. Courtesy of Dr. AM Arenson, Assistant Professor of Radiology, University of Toronto, Toronto, Canada. D. Courtesy of Dr. W Kucharczyk, Clinical Director of Tri-Hospital Resonance Centre, Toronto, Ontario, Canada.

Unnumbered figures in box on page 167 showing palpation of prostate and cervical examination. From: Laurenson RD. An introduction to clinical anatomy by dissection of the human body. Philadelphia: Saunders, 1968.

Unnumbered figure in box on page 193 showing spina bifida cystica. Moore KL, Persaud TVN. The developing human: clinical oriented embryology, 5th ed. Philadelphia: Saunders, 1993 (courtesy of D Parkinson, Children's Centre, Winnipeg, Ontario).

Unnumbered figure on page 372. Courtesy of R Bunzic, Professor of Ophthalmology, University of Toronto.

Figure 8.23. Courtesy of Dr. D Armstrong, Associate Professor of Radiology, University of Toronto, Toronto, Ontario, Canada.

Figure 9.13. A. From Liebgott B. The anatomical basis of dentistry. Philadelphia: Saunders, 1982.

Tables

Table 1.3, upper right. Courtesy of WW Kucharzyck, Clinical Director of Tri-Hospital Resource Center, Toronto, Ontario.

Table 1.3, lower right. Courtesy of AM Arenson, Assistant Professor of Radiology, University of Toronto, Toronto, Ontario.

INDEX

Page numbers in *italics* denote figures; those in **boldface** refer to major descriptions; those followed by "t" denote tables.

B

C

D

O

Q

R

S